INTELLECTUAL PROPERTY IN THE NEW TECHNOLOGICAL AGE

2013 Case and Statutory Supplement

INTELLECTUAL PROPERTY IN THE NEW TECHNOLOGICAL AGE

2013 Case and Statutory Supplement

ROBERT P. MERGES
Wilson Sonsini Goodrich & Rosati Professor of Law and Technology
Director, Berkeley Center for Law & Technology
University of California at Berkeley

PETER S. MENELL
Koret Professor of Law
Director, Berkeley Center for Law & Technology
University of California at Berkeley

MARK A. LEMLEY
William H. Neukom Professor of Law
Director, Stanford Program in Law, Science, and Technology
Stanford University

Wolters Kluwer
Law & Business

Published by Wolters Kluwer Law & Business in New York.

Wolters Kluwer Law & Business serves customers worldwide with CCH, Aspen Publishers, and Kluwer Law International products. (www.wolterskluwerlb.com)

To contact Customer Service, e-mail customer.service@wolterskluwer.com, call 1-800-234-1660, fax 1-800-901-9075, or mail correspondence to:

Wolters Kluwer Law & Business
Attn: Order Department
PO Box 990
Frederick, MD 21705

Printed in the United States of America.

1 2 3 4 5 6 7 8 9 0

ISBN 978-1-4548-2790-0

ISSN 1540-2932

About Wolters Kluwer Law & Business

Wolters Kluwer Law & Business is a leading global provider of intelligent information and digital solutions for legal and business professionals in key specialty areas, and respected educational resources for professors and law students. Wolters Kluwer Law & Business connects legal and business professionals as well as those in the education market with timely, specialized authoritative content and information-enabled solutions to support success through productivity, accuracy and mobility.

Serving customers worldwide, Wolters Kluwer Law & Business products include those under the Aspen Publishers, CCH, Kluwer Law International, Loislaw, ftwilliam.com and MediRegs family of products.

CCH products have been a trusted resource since 1913, and are highly regarded resources for legal, securities, antitrust and trade regulation, government contracting, banking, pension, payroll, employment and labor, and healthcare reimbursement and compliance professionals.

Aspen Publishers products provide essential information to attorneys, business professionals and law students. Written by preeminent authorities, the product line offers analytical and practical information in a range of specialty practice areas from securities law and intellectual property to mergers and acquisitions and pension/benefits. Aspen's trusted legal education resources provide professors and students with high-quality, up-to-date and effective resources for successful instruction and study in all areas of the law.

Kluwer Law International products provide the global business community with reliable international legal information in English. Legal practitioners, corporate counsel and business executives around the world rely on Kluwer Law journals, looseleafs, books, and electronic products for comprehensive information in many areas of international legal practice.

Loislaw is a comprehensive online legal research product providing legal content to law firm practitioners of various specializations. Loislaw provides attorneys with the ability to quickly and efficiently find the necessary legal information they need, when and where they need it, by facilitating access to primary law as well as state-specific law, records, forms and treatises.

ftwilliam.com offers employee benefits professionals the highest quality plan documents (retirement, welfare and non-qualified) and government forms (5500/PBGC, 1099 and IRS) software at highly competitive prices.

MediRegs products provide integrated health care compliance content and software solutions for professionals in healthcare, higher education and life sciences, including professionals in accounting, law and consulting.

Wolters Kluwer Law & Business, a division of Wolters Kluwer, is headquartered in New York. Wolters Kluwer is a market-leading global information services company focused on professionals.

CONTENTS

PREFACE

The impetus for writing Intellectual Property in the New Technological Age was our desire to have a book that would encompass and keep pace with the rapid technological advances driving the U.S. economy and changes in intellectual property law. The rapid diffusion of new technology has continued since our book was released in 1997, and the goal of the revised sixth edition, released in 2012, was to fully integrate these developments.

This supplement includes all of the changes that were made to the intellectual property statutes up to June 1, 2013. That includes the America Invents Act, the most important change to patent law in the last sixty years. Because the AIA has a complex array of provisions, and because many of them apply only to applications filed in 2013 or later, we have chosen to reprint three different versions of the Patent Act: (1) the old Act as it existed on September 1, 2011; (2) a clean version of the patent law as amended by the AIA; and (3) a blackline version showing the changes made by the AIA.

This supplement also includes excerpts of three important 2013 Supreme Court decisions in copyright and patent law.

As an additional way for professors and students to keep abreast of the myriad developments in the field of law and technology, we recommend the Annual Review of Law & Technology. Co-sponsored by the Berkeley Center for Law & Technology and the Berkeley Technology Law Journal, the Annual Review provides a compendium of case comments on important cases, new legislation, international treatises, and developments in foreign law over the prior year. The most recent Annual Review will be available in July 2013.

Robert P. Merges
Peter S. Menell
Mark A. Lemley

August 2013

INTELLECTUAL PROPERTY
IN THE NEW TECHNOLOGICAL AGE
2013 Case and Statutory Supplement

Part I
Cases and Notes

3 *Patent Law*

Insert at page 144 just before ***Mayo Collaborative Services v. Prometheus Laboratories, Inc:***

Association for Molecular Pathology, Inc. v. Myriad Genetics, Inc.

U.S. Supreme Court

2013 WL 2631062 (No. 12–398, June 13, 2013)

Thomas, J.

Respondent Myriad Genetics, Inc. (Myriad), discovered the precise location and sequence of two human genes, mutations of which can substantially increase the risks of breast and ovarian cancer. Myriad obtained a number of patents based upon its discovery. This case involves claims from three of them and requires us to resolve whether a naturally occurring segment of deoxyribonucleic acid (DNA) is patent eligible under 35 U.S.C. § 101 by virtue of its isolation from the rest of the human genome. We also address the patent eligibility of synthetically created DNA known as complementary DNA (cDNA), which contains the same protein-coding information found in a segment of natural DNA but omits portions within the DNA segment that do not code for proteins. For the reasons that follow, we hold that a naturally occurring DNA segment is a product of nature and not patent eligible merely because it has been isolated, but that cDNA is patent eligible because it is not naturally occurring. We, therefore, affirm in part and reverse in part the decision of the United States Court of Appeals for the Federal Circuit.

I

A

Genes form the basis for hereditary traits in living organisms. The human genome consists of approximately 22,000 genes packed into 23 pairs of chromosomes.

Each gene is encoded as DNA, which takes the shape of the familiar "double helix" that Doctors James Watson and Francis Crick first described in 1953. Each "cross-bar" in the DNA helix consists of two chemically joined nucleotides. The possible nucleotides are adenine (A), thymine (T), cytosine (C), and guanine (G), each of which binds naturally with another nucleotide: A pairs with T; C pairs with G. The nucleotide cross-bars are chemically connected to a sugar-phosphate backbone that forms the outside framework of the DNA helix. Sequences of DNA nucleotides contain the information necessary to create strings of amino acids, which in turn are used in the body to build proteins. Only some DNA nucleotides, however, code for amino acids; these nucleotides are known as "exons." Nucleotides that do not code for amino acids, in contrast, are known as "introns."

Creation of proteins from DNA involves two principal steps, known as transcription and translation. In transcription, the bonds between DNA nucleotides separate, and the DNA helix unwinds into two single strands. A single strand is used as a template to create a complementary ribonucleic acid (RNA) strand. The nucleotides on the DNA strand pair naturally with their counterparts, with the exception that RNA uses the nucleotide base uracil (U) instead of thymine (T). Transcription results in a single strand RNA molecule, known as pre-RNA, whose nucleotides form an inverse image of the DNA strand from which it was created. Pre-RNA still contains nucleotides corresponding to both the exons and introns in the DNA molecule. The pre-RNA is then naturally "spliced" by the physical removal of the introns. The resulting product is a strand of RNA that contains nucleotides corresponding only to the exons from the original DNA strand. The exons-only strand is known as messenger RNA (mRNA), which creates amino acids through translation. In translation, cellular structures known as ribosomes read each set of three nucleotides, known as codons, in the mRNA. Each codon either tells the ribosomes which of the 20 possible amino acids to synthesize or provides a stop signal that ends amino acid production.

DNA's informational sequences and the processes that create mRNA, amino acids, and proteins occur naturally within cells. Scientists can, however, extract DNA from cells using well known laboratory methods. These methods allow scientists to isolate specific segments of DNA—for instance, a particular gene or part of a gene—which can then be further studied, manipulated, or used. It is also possible to create DNA synthetically through processes similarly well known in the field of genetics. One such method begins with an mRNA molecule and uses the natural bonding properties of nucleotides to create a new, synthetic DNA molecule. The result is the inverse of the mRNA's inverse image of the original DNA, with one important distinction: Because the natural creation of mRNA involves splicing that removes introns, the synthetic DNA created from mRNA also contains only the exon sequences. This synthetic DNA created in the laboratory from mRNA is known as complementary DNA (cDNA).

Changes in the genetic sequence are called mutations. Mutations can be as small as the alteration of a single nucleotide—a change affecting only one letter in the genetic code. Such small-scale changes can produce an entirely different amino acid or can end protein production altogether. Large changes, involving the deletion, rearrangement, or duplication of hundreds or even millions of nucleotides, can result in the elimination, misplacement, or duplication of entire genes. Some mutations are harmless, but others can cause disease or increase the risk of disease. As a result, the study of genetics can lead to valuable medical breakthroughs.

B

This case involves patents filed by Myriad after it made one such medical breakthrough. Myriad discovered the precise location and sequence of what are now known as the BRCA1 and BRCA2 genes. Mutations in these genes can dramatically increase an individual's risk of developing breast and ovarian cancer. The average American woman has a 12- to 13-percent risk of developing breast cancer, but for women with certain genetic mutations, the risk can range between 50 and 80 percent for breast cancer and between 20 and 50 percent for ovarian cancer. Before Myriad's discovery of the BRCA1 and BRCA2 genes, scientists knew that heredity played a role in establishing a woman's risk of developing breast and ovarian cancer, but they did not know which genes were associated with those cancers.

Myriad identified the exact location of the BRCA1 and BRCA2 genes on chromosomes 17 and 13. Chromosome 17 has approximately 80 million nucleotides, and chromosome 13 has approximately 114 million. Association for Molecular Pathology v. United States Patent and Trademark Office, 689 F.3d 1303, 1328 (C.A. Fed. 2012). Within those chromosomes, the BRCA1 and BRCA2 genes are each about 80,000 nucleotides long. If just exons are counted, the BRCA1 gene is only about 5,500 nucleotides long; for the BRCA2 gene, that number is about 10,200. Ibid. Knowledge of the location of the BRCA1 and BRCA2 genes allowed Myriad to determine their typical nucleotide sequence. That information, in turn, enabled Myriad to develop medical tests that are useful for detecting mutations in a patient's BRCA1 and BRCA2 genes and thereby assessing whether the patient has an increased risk of cancer.

Once it found the location and sequence of the BRCA1 and BRCA2 genes, Myriad sought and obtained a number of patents. Nine composition claims from three of those patents are at issue in this case. [From Fn. 2: At issue are claims 1, 2, 5, 6, and 7 of U.S. Patent 5,747,282 (the '282 patent), claim 1 of U.S. Patent 5,693,473 (the '473 patent), and claims 1, 6, and 7 of U.S. Patent 5,837,492 (the '492 patent).] See id., at 1309, and n. 1 (noting composition claims). Claims 1, 2, 5, and 6 from the '282 patent are representative. The first claim asserts a patent on "[a]n isolated DNA coding for a BRCA1 polypeptide," which has "the amino acid sequence set forth in SEQ ID NO:2." App. 822. SEQ ID NO:2 sets forth a list of 1,863 amino acids that the typical BRCA1 gene encodes. See id., at 785–790. Put differently, claim 1 asserts a patent claim on the DNA code that tells a cell to produce the string of BRCA1 amino acids listed in SEQ ID NO:2.

Claim 2 of the '282 patent operates similarly. It claims "[t]he isolated DNA of claim 1, wherein said DNA has the nucleotide sequence set forth in SEQ ID NO:1." Id., at 822. Like SEQ ID NO:2, SEQ ID NO:1 sets forth a long list of data, in this instance the sequence of cDNA that codes for the BRCA1 amino acids listed in claim

1. Importantly, SEQ ID NO:1 lists only the cDNA exons in the BRCA1 gene, rather than a full DNA sequence containing both exons and introns. See id., at 779 (stating that SEQ ID NO:1's "MOLECULE TYPE:" is "cDNA"). As a result, the Federal Circuit recognized that claim 2 asserts a patent on the cDNA nucleotide sequence listed in SEQ ID NO:1, which codes for the typical BRCA1 gene.

Claim 5 of the '282 patent claims a subset of the data in claim 1. In particular, it claims "[a]n isolated DNA having at least 15 nucleotides of the DNA of claim 1." App. 822. The practical effect of claim 5 is to assert a patent on any series of 15 nucleotides that exist in the typical BRCA1 gene. Because the BRCA1 gene is thousands of nucleotides long, even BRCA1 genes with substantial mutations are likely to contain at least one segment of 15 nucleotides that correspond to the typical BRCA1 gene. Similarly, claim 6 of the '282 patent claims "[a]n isolated DNA having at least 15 nucleotides of the DNA of claim 2." Ibid. This claim operates similarly to claim 5, except that it references the cDNA-based claim 2. The remaining claims at issue are similar, though several list common mutations rather than typical BRCA1 and BRCA2 sequences. See ibid. (claim 7 of the '282 patent); id., at 930 (claim 1 of the '473 patent); id., at 1028 (claims 1, 6, and 7 of the '492 patent).

C

Myriad's patents would, if valid, give it the exclusive right to isolate an individual's BRCA1 and BRCA2 genes (or any strand of 15 or more nucleotides within the genes) by breaking the covalent bonds that connect the DNA to the rest of the individual's genome. The patents would also give Myriad the exclusive right to synthetically create BRCA cDNA.

But isolation is necessary to conduct genetic testing, and Myriad was not the only entity to offer BRCA testing after it discovered the genes. The University of Pennsylvania's Genetic Diagnostic Laboratory (GDL) and others provided genetic testing services to women. Petitioner Dr. Harry Ostrer, then a researcher at New York University School of Medicine, routinely sent his patients' DNA samples to GDL for testing. After learning of GDL's testing and Ostrer's activities, Myriad sent letters to them asserting that the genetic testing infringed Myriad's patents. App. 94–95 (Ostrer letter). In response, GDL agreed to stop testing and informed Ostrer that it would no longer accept patient samples. Myriad also filed patent infringement suits against other entities that performed BRCA testing, resulting in settlements in which the defendants agreed to cease all allegedly infringing activity. 689 F.3d, at 1315. Myriad, thus, solidified its position as the only entity providing BRCA testing.

Some years later, petitioner Ostrer, along with medical patients, advocacy groups, and other doctors, filed this lawsuit seeking a declaration that Myriad's patents are invalid under 35 U.S.C. § 101. The District Court . . . granted summary judgment to petitioners on the composition claims at issue in this case based on its conclusion that Myriad's claims, including claims related to cDNA, were invalid because they covered products of nature. The Federal Circuit reversed, and this Court granted the petition for certiorari, vacated the judgment, and remanded the case in light of Mayo Collaborative Services v. Prometheus Laboratories, Inc., 566 U.S. —— (2012).

On remand, the Federal Circuit affirmed the District Court in part and reversed in part, with each member of the panel writing separately.

[In three separate opinions,] the court held that both isolated DNA and cDNA were patent eligible under § 101. The central dispute among the panel members was whether the act of *isolating* DNA—separating a specific gene or sequence of nucleotides from the rest of the chromosome—is an inventive act that entitles the individual who first isolates it to a patent. Each of the judges on the panel had a different view on that question. Judges Lourie and Moore agreed that Myriad's claims were patent eligible under § 101 but disagreed on the rationale. Judge Lourie relied on the fact that the entire DNA molecule is held together by chemical bonds and that the covalent bonds at both ends of the segment must be severed in order to isolate segments of DNA. This process technically creates new molecules with unique chemical compositions. Judge Lourie found this chemical alteration to be dispositive, because isolating a particular strand of DNA creates a nonnaturally occurring molecule, even though the chemical alteration does not change the information-transmitting quality of the DNA.

Judge Moore concurred in part but did not rely exclusively on Judge Lourie's conclusion that chemically breaking covalent bonds was sufficient to render isolated DNA patent eligible. Instead, Judge Moore also relied on the United States Patent and Trademark Office's (PTO) practice of granting such patents and on the reliance interests of patent holders. Id., at 1343. However, she acknowledged that her vote might have come out differently if she "were deciding this case on a blank canvas." Ibid.

Finally, Judge Bryson concurred in part and dissented in part, concluding that isolated DNA is not patent eligible. As an initial matter, he emphasized that the breaking of chemical bonds was not dispositive: "[T]here is no magic to a chemical bond that requires us to recognize a new product when a chemical bond is created or broken." Id., at 1351. Instead, he relied on the fact that "[t]he nucleotide sequences of the claimed molecules are the same as the nucleotide sequences found in naturally occurring human genes." Id., at 1355. Judge Bryson then concluded that genetic "structural similarity dwarfs the significance of the structural differences between isolated DNA and naturally occurring DNA, especially where the structural differences are merely ancillary to the breaking of covalent bonds, a process that is itself not inventive." Ibid.

Although the judges expressed different views concerning the patentability of isolated DNA, all three agreed that patent claims relating to cDNA met the patent eligibility requirements of § 101.

II

A

Section 101 of the Patent Act provides:

> "Whoever invents or discovers any new and useful . . . composition of matter, or any new and useful improvement thereof, may obtain a patent therefor, subject to the conditions and requirements of this title."

35 U.S.C. § 101.

We have "long held that this provision contains an important implicit exception[:] Laws of nature, natural phenomena, and abstract ideas are not patentable." Mayo, 566 U.S., at —— (slip op., at 1). Rather, " 'they are the basic tools of scientific and technological work' " that lie beyond the domain of patent protection. Id., at —— (slip op., at 2). As the Court has explained, without this exception, there would be considerable danger that the grant of patents would "tie up" the use of such tools and thereby "inhibit future innovation premised upon them." Id., at —— (slip op., at 17). This would be at odds with the very point of patents, which exist to promote creation.

The rule against patents on naturally occurring things is not without limits, however, for "all inventions at some level embody, use, reflect, rest upon, or apply laws of nature, natural phenomena, or abstract ideas," and "too broad an interpretation of this exclusionary principle could eviscerate patent law." 566 U.S., at —— (slip op., at 2). As we have recognized before, patent protection strikes a delicate balance between creating "incentives that lead to creation, invention, and discovery" and "imped[ing] the flow of information that might permit, indeed spur, invention." Id., at —— (slip op., at 23). We must apply this well-established standard to determine whether Myriad's patents claim any "new and useful . . . composition of matter," § 101, or instead claim naturally occurring phenomena.

B

It is undisputed that Myriad did not create or alter any of the genetic information encoded in the BRCA1 and BRCA2 genes. The location and order of the nucleotides existed in nature before Myriad found them. Nor did Myriad create or alter the genetic structure of DNA. Instead, Myriad's principal contribution was uncovering the precise location and genetic sequence of the BRCA1 and BRCA2 genes within chromosomes 17 and 13. The question is whether this renders the genes patentable.

Myriad recognizes that our decision in [*Diamond v.*] *Chakrabarty*, 447 U.S. 303 (1980)] is central to this inquiry. In *Chakrabarty*, scientists added four plasmids to a bacterium, which enabled it to break down various components of crude oil. 447 U.S., at 305, and n. 1. The Court held that the modified bacterium was patentable. The Chakrabarty bacterium was new "with markedly different characteristics from any found in nature," 447 U.S., at 310, due to the additional plasmids and resultant "capacity for degrading oil." Id., at 305, n. 1. In this case, by contrast, Myriad did not create anything. To be sure, it found an important and useful gene, but separating that gene from its surrounding genetic material is not an act of invention.

Groundbreaking, innovative, or even brilliant discovery does not by itself satisfy the § 101 inquiry. In Funk Brothers Seed Co. v. Kalo Inoculant Co., 333 U.S. 127 (1948), this Court considered a composition patent that claimed a mixture of naturally occurring strains of bacteria that helped leguminous plants take nitrogen from the air and fix it in the soil. Id., at 128-129. The Court held that the composition was not patent eligible because the patent holder did not alter the bacteria in any way. Id., at 132. His patent claim thus fell squarely within the law of nature exception. So do Myriad's. Myriad found the location of the BRCA1 and BRCA2 genes, but that discovery, by itself, does not render the BRCA genes "new . . . composition[s] of matter," § 101, that are patent eligible.

Indeed, Myriad's patent descriptions highlight the problem with its claims. For example, a section of the '282 patent's Detailed Description of the Invention indicates that Myriad found the location of a gene associated with increased risk of breast cancer and identified mutations of that gene that increase the risk. In subsequent language Myriad explains that the location of the gene was unknown until Myriad found it among the approximately eight million nucleotide pairs contained in a subpart of chromosome 17. Many of Myriad's patent descriptions simply detail the "iterative process" of discovery by which Myriad narrowed the possible locations for the gene sequences that it sought. See, e.g., id., at 750. Myriad seeks to import these extensive research efforts into the § 101 patent-eligibility inquiry. But extensive effort alone is insufficient to satisfy the demands of § 101.

Nor are Myriad's claims saved by the fact that isolating DNA from the human genome severs chemical bonds and thereby creates a nonnaturally occurring molecule. Myriad's claims are simply not expressed in terms of chemical composition, nor do they rely in any way on the chemical changes that result from the isolation of a particular section of DNA. Instead, the claims understandably focus on the genetic information encoded in the BRCA1 and BRCA2 genes. If the patents depended upon the creation of a unique molecule, then a would-be infringer could arguably avoid at least Myriad's patent claims on entire genes (such as claims 1 and 2 of the '282 patent) by isolating a DNA sequence that included both the BRCA1 or BRCA2 gene and one additional nucleotide pair. Such a molecule would not be chemically identical to the molecule "invented" by Myriad. But Myriad obviously would resist that outcome because its claim is concerned primarily with the information contained in the genetic sequence, not with the specific chemical composition of a particular molecule.

Finally, Myriad argues that the PTO's past practice of awarding gene patents is entitled to deference, citing J.E.M. Ag Supply, Inc. v. Pioneer Hi–Bred Int'l, Inc., 534 U.S. 124 (2001). We disagree. Congress has not endorsed the views of the PTO in subsequent legislation.

Further undercutting the PTO's practice, the United States argued in the Federal Circuit and in this Court that isolated DNA was not patent eligible under § 101, Brief for United States as Amicus Curiae 20-33, and that the PTO's practice was not "a sufficient reason to hold that isolated DNA is patent-eligible." Id., at 26. See also id., at 28–29. These concessions weigh against deferring to the PTO's determination.

C

cDNA does not present the same obstacles to patentability as naturally occurring, isolated DNA segments. As already explained, creation of a cDNA sequence from mRNA results in an exons-only molecule that is not naturally occurring.FN8 Petitioners concede that cDNA differs from natural DNA in that "the non-coding regions have been removed." They nevertheless argue that cDNA is not patent eligible because "[t]he nucleotide sequence of cDNA is dictated by nature, not by the lab technician." That may be so, but the lab technician unquestionably creates something new when cDNA is made. cDNA retains the naturally occurring exons of DNA, but it is distinct from the DNA from which it was derived. As a result, cDNA is not a "product of nature" and is patent eligible under § 101, except insofar as very short series of DNA may have no intervening introns to remove when creating cDNA. In that situation, a short strand of cDNA may be indistinguishable from natural DNA.

III

It is important to note what is not implicated by this decision. First, there are no method claims before this Court. Had Myriad created an innovative method of manipulating genes while searching for the BRCA1 and BRCA2 genes, it could possibly have sought a method patent. But the processes used by Myriad to isolate DNA were well understood by geneticists at the time of Myriad's patents [and so method claims were not available].

Similarly, this case does not involve patents on new applications of knowledge about the BRCA1 and BRCA2 genes. Judge Bryson aptly noted that, "[a]s the first party with knowledge of the [BRCA1 and BRCA2] sequences, Myriad was in an excellent position to claim applications of that knowledge. Many of its unchallenged claims are limited to such applications." 689 F.3d, at 1349.

Nor do we consider the patentability of DNA in which the order of the naturally occurring nucleotides has been altered. Scientific alteration of the genetic code presents a different inquiry, and we express no opinion about the application of § 101 to such endeavors. We merely hold that genes and the information they encode are not patent eligible under § 101 simply because they have been isolated from the surrounding genetic material.

NOTES AND QUESTIONS

1. Standing Issues. There were questions concerning standing in this case, but the Court resolved them under the 2007 *MedImmune* decision. From note 3:

> Myriad continues to challenge Dr. Ostrer's Declaratory Judgment Act standing in this Court. Brief for Respondents 17–22. But we find that, under the Court's decision in MedImmune, Inc. v. Genentech, Inc., Dr. Ostrer has alleged sufficient facts "under all the circumstances, [to] show that there is a substantial controversy, between parties having

adverse legal interests, of sufficient immediacy and reality to warrant the issuance of a declaratory judgment." 549 U.S. 118, 127 (2007) (internal quotation marks omitted).

2. Justice Scalia concurred in part, and concurred in the judgment. Here is the entirety of his opinion:

> I join the judgment of the Court, and all of its opinion except Part I–A and some portions of the rest of the opinion going into fine details of molecular biology. I am unable to affirm those details on my own knowledge or even my own belief. It suffices for me to affirm, having studied the opinions below and the expert briefs presented here, that the portion of DNA isolated from its natural state sought to be patented is identical to that portion of the DNA in its natural state; and that complementary DNA (cDNA) is a synthetic creation not normally present in nature.

What do you make of this? Is Justice Scalia concerned that his colleagues on the Court have entwined patent law with technical details that he, and presumably other judges, may find difficult to understand? Is he expressing concern that the Supreme Court is verging on territory beyond its expertise in deciding this case?

3. An omitted footnote in the case reads as follows:

> Myriad also argues that we should uphold its patents so as not to disturb the reliance interests of patent holders like itself. Brief for Respondents 38–39. Concerns about reliance interests arising from PTO determinations, insofar as they are relevant, are better directed to Congress.

Does this mean the Supreme Court should never take into account reliance interests of various parties? Note that the Court says the reliance interests here "aris[e] from PTO determinations" – and not, presumably, earlier opinions from the Supreme Court itself. Is it accurate to imply that the Supreme Court had never even indirectly condoned DNA patent claims? What about *Chakrabarty*? What about the many times the Court denied cert in earlier cases centering on § 101 challenges to gene patents?

4. The Court makes much of the distinction between a claim to DNA as a chemical entity, a molecule, and a claim to the informational content of the DNA. The former, it implies, is patentable, partly because it is presumably a more limited claim. The latter, meanwhile, is not, because it is already present in nature and (the Court believes) it is also inherently broader. For example, the Court says "[Myriad's] claim is concerned primarily with the information contained in the genetic sequence, not with the specific chemical composition of a particular molecule." Can you come up with some drafting strategies that take advantage of this distinction? Is it always true that a "specific chemical composition" claim is narrower than a claim that encompasses "information contained" in a genetic sequence? Does the Court imply that a very broad claim drafted in molecular terms – e.g., a "Markush group" for DNA molecules – would also be invalid under § 101? Is it the nature of the claim that is the problem from the Court's perspective – chemical versus informational – or instead the breadth of the claim? If breadth is the problem, what is the difference between the "metascope" focus of § 101 and the traditional scope-limiting doctrines of § 112?

Insert at page 384, just after the end of Problem 3-14:

Bowman v. Monsanto Co.

133 S.Ct. 1761 (2013)

Kagan, J.

Under the doctrine of patent exhaustion, the authorized sale of a patented article gives the purchaser, or any subsequent owner, a right to use or resell that article. Such a sale, however, does not allow the purchaser to make new copies of the patented invention. The question in this case is whether a farmer who buys patented seeds may reproduce them through planting and harvesting without the patent holder's permission. We hold that he may not.

I

Respondent Monsanto invented a genetic modification that enables soybean plants to survive exposure to glyphosate, the active ingredient in many herbicides (including Monsanto's own Roundup). Monsanto markets soybean seed containing this altered genetic material as Roundup Ready seed. Farmers planting that seed can use a glyphosate-based herbicide to kill weeds without damaging their crops. Two patents issued to Monsanto cover various aspects of its Roundup Ready technology, including a seed incorporating the genetic alteration.

Monsanto sells, and allows other companies to sell, Roundup Ready soybean seeds to growers who assent to a special licensing agreement. See App. 27a. That agreement permits a grower to plant the purchased seeds in one (and only one) season. He can then consume the resulting crop or sell it as a commodity, usually to a grain elevator or agricultural processor. See 657 F.3d, at 1344–1345. But under the agreement, the farmer may not save any of the harvested soybeans for replanting, nor may he supply them to anyone else for that purpose. These restrictions reflect the ease of producing new generations of Roundup Ready seed. Because glyphosate resistance comes from the seed's genetic material, that trait is passed on from the planted seed to the *1765 harvested soybeans: Indeed, a single Roundup Ready seed can grow a plant containing dozens of genetically identical beans, each of which, if replanted, can grow another such plant—and so on and so on. See App. 100a. The agreement's terms prevent the farmer from co-opting that process to produce his own Roundup Ready seeds, forcing him instead to buy from Monsanto each season.

Petitioner Vernon Bowman is a farmer in Indiana who, it is fair to say, appreciates Roundup Ready soybean seed. He purchased Roundup Ready each year, from a company affiliated with Monsanto, for his first crop of the season. In accord with the agreement just described, he used all of that seed for planting, and sold his entire crop to a grain elevator (which typically would resell it to an agricultural processor for human or animal consumption).

Bowman, however, devised a less orthodox approach for his second crop of each season. Because he thought such late-season planting "risky," he did not want to pay the premium price that Monsanto charges for Roundup Ready seed. He therefore went to a grain elevator; purchased "commodity soybeans" intended for human or animal consumption; and planted them in his fields. Those soybeans came from prior harvests of other local farmers. And because most of those farmers also used Roundup Ready seed, Bowman could anticipate that many of the purchased soybeans would contain Monsanto's patented technology. When he applied a glyphosate-based herbicide to his fields, he confirmed that this was so; a significant proportion of the new plants survived the treatment, and produced in their turn a new crop of soybeans with the Roundup Ready trait. Bowman saved seed from that crop to use in his late-season planting the next year—and then the next, and the next, until he had harvested eight crops in that way. Each year, that is, he planted saved seed from the year before (sometimes adding more soybeans bought from the grain elevator), sprayed his fields with glyphosate to kill weeds (and any non-resistant plants), and produced a new crop of glyphosate-resistant—i.e., Roundup Ready—soybeans.

After discovering this practice, Monsanto sued Bowman for infringing its patents on Roundup Ready seed. Bowman raised patent exhaustion as a defense, arguing that Monsanto could not control his use of the soybeans because they were the subject of a prior authorized sale (from local farmers to the grain elevator). The District Court rejected that argument, and awarded damages to Monsanto of $84,456. The Federal Circuit affirmed. It reasoned that patent exhaustion did not protect Bowman because he had "created a newly infringing article." 657 F.3d, at 1348. The "right to use" a patented article following an authorized sale, the court explained, "does not include the right to construct an essentially new article on the template of the original, for the right to make the article remains with the patentee." Ibid. (brackets and internal quotation marks omitted). Accordingly, Bowman could not "'replicate' Monsanto's patented technology by planting it in the ground to create newly infringing genetic material, seeds, and plants." Ibid.

The doctrine of patent exhaustion limits a patentee's right to control what others can do with an article embodying or containing an invention. Under the doctrine, "the initial authorized sale of a patented item terminates all patent rights to that item." Quanta Computer, Inc. v. LG Electronics, Inc., 553 U.S. 617, 625 (2008). And by "exhaust[ing] the [patentee's] monopoly" in that item, the sale confers on the purchaser, or any subsequent owner, "the right to use [or] sell" the thing as he sees fit. United States v. Univis Lens Co., 316 U.S. 241, 249–250 (1942). We have explained the basis for the doctrine as follows: "[T]he purpose of the patent law is fulfilled with respect to any particular article when the patentee has received his reward . . . by the sale of the article"; once that "purpose is realized the patent law affords no basis for restraining the use and enjoyment of the thing sold." Id., at 251.

Consistent with that rationale, the doctrine restricts a patentee's rights only as to the "particular article" sold, ibid.; it leaves untouched the patentee's ability to prevent a buyer from making new copies of the patented item. "[T]he purchaser of the [patented] machine . . . does not acquire any right to construct another machine either for his own use or to be vended to another." Mitchell v. Hawley, 16 Wall. 544, 548, 21 L.Ed. 322 (1873). Rather, "a second creation" of the patented item "call[s] the monopoly, conferred by the patent grant, into play for a second time." Aro Mfg. Co. v. Convertible Top Replacement Co., 365 U.S. 336, 346 (1961). That is because the patent holder has "received his reward" only for the actual article sold, and not for subsequent recreations of it. Univis, 316 U.S., at 251. If the purchaser of that article could make and sell endless copies, the patent would effectively protect the invention for just a single sale. Bowman himself disputes none of this analysis as a general matter: He forthrightly acknowledges the "well settled" principle "that the exhaustion doctrine does not extend to the right to 'make' a new product." Brief for Petitioner 37.

Unfortunately for Bowman, that principle decides this case against him. Under the patent exhaustion doctrine, Bowman could resell the patented soybeans he purchased from the grain elevator; so too he could consume the beans himself or feed them to his animals. Monsanto, although the patent holder, would have no business interfering in those uses of Roundup Ready beans. But the exhaustion doctrine does not enable Bowman to make *additional* patented soybeans without Monsanto's permission (either express or implied). And that is precisely what Bowman did. He took the soybeans he purchased home; planted them in his fields at the time he thought best; applied glyphosate to kill weeds (as well as any soy plants lacking the Roundup Ready trait); and finally harvested more (many more) beans than he started with. That is how "to 'make' a new product," to use Bowman's words, when the original product is a seed. Brief for Petitioner 37; see Webster's Third New International Dictionary 1363 (1961) ("make" means "cause to exist, occur, or appear," or more specifically, "plant and raise (a crop)"). Because Bowman thus reproduced Monsanto's patented invention, the exhaustion doctrine does not protect him.

Were the matter otherwise, Monsanto's patent would provide scant benefit. After inventing the Roundup Ready trait, Monsanto would, to be sure, "receiv[e] [its] reward" for the first seeds it sells. Univis, 316 U.S., at 251. But in short order, other seed companies could reproduce the product and market it to growers, thus depriving Monsanto of its monopoly. And farmers themselves need only buy the seed once, whether from Monsanto, a competitor, or (as here) a grain elevator. The grower could multiply his initial purchase, and then multiply that new creation, *ad infinitum*—each time profiting from the patented seed without compensating its inventor. Bowman's late-season plantings offer a prime illustration. After buying beans for a single harvest, Bowman saved enough seed each year to reduce or eliminate the need for additional purchases. Monsanto still held its patent, but received no gain from Bowman's annual production and sale of Roundup Ready soybeans. The exhaustion doctrine is limited to the "particular item" sold to avoid just such a mismatch between invention and reward.

Bowman principally argues that exhaustion should apply here because seeds are meant to be planted. The exhaustion doctrine, he reminds us, typically prevents a patentee from controlling the use of a patented product following an authorized sale. And in planting Roundup Ready seeds, Bowman continues, he is merely using them in the normal way farmers do. Bowman thus concludes that allowing Monsanto to interfere with that use would "creat[e] an impermissible exception to the exhaustion doctrine" for patented seeds and other "self-replicating technologies." Brief for Petitioner 16.

But it is really Bowman who is asking for an unprecedented exception. Reproducing a patented article no doubt "uses" it after a fashion. But as already explained, we have always drawn the boundaries of the exhaustion doctrine to exclude that activity, so that the patentee retains an undiminished right to prohibit others from making the thing his patent protects. See, e.g., Cotton–Tie Co. v. Simmons, 106 U.S. 89, 93–94 (1882) (holding that a purchaser could not "use" the buckle from a patented cotton-bale tie to "make" a new tie). That is because, once again, if simple copying were a protected use, a patent would plummet in value after the first sale of the first item containing the invention. The undiluted patent monopoly, it might be said, would extend not for 20 years (as the Patent Act promises), but for only one transaction. And that would result in less incentive for innovation than Congress wanted. Hence our repeated insistence that exhaustion applies only to the particular item sold, and not to reproductions.

Nor do we think that rule will prevent farmers from making appropriate use of the Roundup Ready seed they buy. Bowman himself stands in a peculiarly poor position to assert such a claim. As noted earlier, the commodity soybeans he purchased were intended not for planting, but for consumption. [I]n the more ordinary case, when a farmer purchases Roundup Ready seed qua seed—that is, seed intended to grow a crop—he will be able to plant it. Monsanto, to be sure, conditions the farmer's ability to reproduce Roundup Ready; but it does not—could not realistically—preclude all planting. No sane farmer, after all, would buy the product without some ability to grow soybeans from it. And so Monsanto, predictably enough, sells Roundup Ready seed to farmers with a license to use it to make a crop. Applying our usual rule in this context therefore will allow farmers to benefit from Roundup Ready, even as it rewards Monsanto for its innovation.

Still, Bowman has another seeds-are-special argument: that soybeans naturally "self-replicate or 'sprout' unless stored in a controlled manner," and thus "it was the planted soybean, not Bowman" himself. Brief for Petitioner 42. But we think that blame-the-bean defense tough to credit. Bowman was not a passive observer of his soybeans' multiplication; or put another way, the seeds he purchased (miraculous though they might be in other respects) did not spontaneously create eight successive soybean crops. As we have explained, Bowman devised and executed a novel way to harvest crops from Roundup Ready seeds without paying the usual premium. In all this, the bean surely figured. But it was Bowman, and not the bean, who controlled the reproduction (unto the eighth generation) of Monsanto's patented invention.

Our holding today is limited—addressing the situation before us, rather than every one involving a self-replicating product. We recognize that such inventions are becoming ever more prevalent, complex, and diverse. In another case, the article's self-replication might occur outside the purchaser's control. Or it might be a necessary but incidental step in using the item for another purpose. We need not address here whether or how the doctrine of patent exhaustion would apply in such circumstances. In the case at hand, Bowman planted Monsanto's patented soybeans solely to make and market replicas of them. Patent exhaustion provides no haven for that conduct. We accordingly affirm the judgment of the Court of Appeals for the Federal Circuit.

NOTES

1. Footnote 3 from the Court's opinion, excluded from the excerpt above, reads as follows:

> This conclusion applies however Bowman acquired Roundup Ready seed: The doctrine of patent exhaustion no more protected Bowman's reproduction of the seed he purchased for his first crop (from a Monsanto-affiliated seed company) than the beans he bought for his second (from a grain elevator). The difference between the two purchases was that the first—but not the second—came with a license from Monsanto to plant the seed and then harvest and market one crop of beans. We do not here confront a case in which Monsanto (or an affiliated seed company) sold Roundup Ready to a farmer without an express license agreement. For reasons we explain below, we think that case unlikely to arise. And in the event it did, the farmer might reasonably claim that the sale came with an implied license to plant and harvest one soybean crop.

The Court later explains that a farmer is unlikely to purchase patented seeds from Monsanto without an express license to plant them and grow a crop. It also states that the doctrine of implied license would probably protect the farmer in this situation anyway. The point is important, because it touches on the difference between behavior that is explicitly licensed by a patent owner and acts that are permissible even in the absence of a license. It is interesting that the Court does not say that an express contract may undercut rights otherwise protected under patent exhaustion – a theory that some cases, and more than a few business models, have relied on in recent years.

2. Another omitted part of the opinion emphasizes the differences between a patent and protection of plant varieties under the Plant Variety Protection Act (PVPA). The latter statute includes an explicit "saved seed exemption" from infringement liability, and the Court points out that a ruling in favor of Bowman in this case would effectively undermine the different regimes for plant variety protection and patents. The idea is that, compared to plant variety protection, patent law applies higher standards to an invention seeking protection, but, once a patent issues, confers broader rights.

4 *Copyright Law*

Insert on page 587 before note 7.

Kirtsaeng v. John Wiley & Sons, Inc.
Supreme Court of the United States
133 S.Ct. 1351 (2013)

Justice BREYER delivered the opinion of the Court.

Section 106 of the Copyright Act grants "the owner of copyright under this title" certain "exclusive rights," including the right "to distribute copies . . . of the copyrighted work to the public by sale or other transfer of ownership." 17 U.S.C. § 106(3). These rights are qualified, however, by the application of various limitations set forth in the next several sections of the Act, §§ 107 through 122. Those sections, typically entitled "Limitations on exclusive rights," include, for example, the principle of "fair use" (§ 107), permission for limited library archival reproduction, (§ 108), and the doctrine at issue here, the "first sale" doctrine (§ 109).

Section 109(a) sets forth the "first sale" doctrine as follows:

> "Notwithstanding the provisions of section 106(3) [the section that grants the owner exclusive distribution rights], the owner of a particular copy or phonorecord *lawfully made under this title* . . . is entitled, without the authority of the copyright owner, to sell or otherwise dispose of the possession of that copy or phonorecord." (Emphasis added.)

Thus, even though § 106(3) forbids distribution of a copy of, say, the copyrighted novel Herzog without the copyright owner's permission, § 109(a) adds that, once a copy of Herzog has been lawfully sold (or its ownership otherwise lawfully transferred), the buyer of *that copy* and subsequent owners are free to dispose of it as they wish. In copyright jargon, the "first sale" has "exhausted" the copyright owner's § 106(3) exclusive distribution right.

What, however, if the copy of Herzog was printed abroad and then initially sold with the copyright owner's permission? Does the "first sale" doctrine still apply? Is the buyer, like the buyer of a domestically manufactured copy, free to bring the copy into the United States and dispose of it as he or she wishes?

To put the matter technically, an "importation" provision, § 602(a)(1), says that

> "[i]mportation into the United States, without the authority of the owner of copyright under this title, of copies . . . of a work that have been acquired outside the United States is an infringement of the exclusive right to distribute copies . . . *under section 106* . . . ." (emphasis added).

Thus § 602(a)(1) makes clear that importing a copy without permission violates the owner's exclusive distribution right. But in doing so, § 602(a)(1) refers explicitly to the *§ 106(3)* exclusive distribution right. As we have just said, § 106 is by its terms "[s]ubject to" the various doctrines and principles contained in §§ 107 through 122, including § 109(a)'s "first sale" limitation. Do those same modifications apply—in particular, does the "first sale" modification apply—when considering whether § 602(a)(1) prohibits importing a copy?

In *Quality King Distributors, Inc. v. L'anza Research Int'l, Inc.*, 523 U.S. 135, 145 (1998), we held that § 602(a)(1)'s reference to § 106(3)'s exclusive distribution right incorporates the later subsections' limitations, including, in particular, the "first sale" doctrine of § 109. Thus, it might seem that, § 602(a)(1) notwithstanding, one who buys a copy abroad can freely import that copy into the United States and dispose of it, just as he could had he bought the copy in the United States.

But *Quality King* considered an instance in which the copy, though purchased abroad, was initially manufactured in the United States (and then sent abroad and sold). This case is like *Quality King* but for one important fact. The copies at issue here were manufactured abroad. That fact is important because § 109(a) says that the "first sale" doctrine applies to "a particular copy or phonorecord *lawfully made under this title.*" And we must decide here whether the five words, "lawfully made under this title," make a critical legal difference.

Putting section numbers to the side, we ask whether the "first sale" doctrine applies to protect a buyer or other lawful owner of a copy (of a copyrighted work) lawfully manufactured abroad. Can that buyer bring that copy into the United States (and sell it or give it away) without obtaining permission to do so from the copyright owner? Can, for example, someone who purchases, say at a used bookstore, a book printed abroad subsequently resell it without the copyright owner's permission?

In our view, the answers to these questions are, yes. We hold that the "first sale" doctrine applies to copies of a copyrighted work lawfully made abroad.

I

A

Respondent, John Wiley & Sons, Inc., publishes academic textbooks. Wiley obtains from its authors various foreign and domestic copyright assignments, licenses and permissions—to the point that we can, for present purposes, refer to

Wiley as the relevant American copyright owner. Wiley often assigns to its wholly owned foreign subsidiary, John Wiley & Sons (Asia) Pte Ltd., rights to publish, print, and sell Wiley's English language textbooks abroad. Each copy of a Wiley Asia foreign edition will likely contain language making clear that the copy is to be sold only in a particular country or geographical region outside the United States. . . .

Petitioner, Supap Kirtsaeng, a citizen of Thailand, moved to the United States in 1997 . . . . While he was studying in the United States, Kirtsaeng asked his friends and family in Thailand to buy copies of foreign edition English-language textbooks at Thai book shops, where they sold at low prices, and mail them to him in the United States. Kirtsaeng would then sell them, reimburse his family and friends, and keep the profit. . . .

II

We must decide whether the words "lawfully made under this title" restrict the scope of § 109(a)'s "first sale" doctrine geographically. The Second Circuit, the Ninth Circuit, Wiley, and the Solicitor General (as *amicus*) all read those words as imposing a form of *geographical* limitation. The Second Circuit held that they limit the "first sale" doctrine to particular copies "made in territories *in which the Copyright Act is law,*" which (the Circuit says) are copies "manufactured domestically," not "outside of the United States." (emphasis added). . . .

Under [this] geographical interpretation[], § 109(a)'s "first sale" doctrine would not apply to the Wiley Asia books at issue here. And, despite an American copyright owner's permission to *make* copies abroad, one who *buys* a copy of any such book or other copyrighted work—whether at a retail store, over the Internet, or at a library sale—could not resell (or otherwise dispose of) that particular copy without further permission.

Kirtsaeng, however, reads the words "lawfully made under this title" as imposing a *non*-geographical limitation. He says that they mean made "in accordance with" or "in compliance with" the Copyright Act. In that case, § 109(a)'s "first sale" doctrine would apply to copyrighted works as long as their manufacture met the requirements of American copyright law. In particular, the doctrine would apply where, as here, copies are manufactured abroad with the permission of the copyright owner. See § 106 (referring to the owner's right to authorize).

In our view, § 109(a)'s language, its context, and the common-law history of the "first sale" doctrine, taken together, favor a *non*-geographical interpretation. We also doubt that Congress would have intended to create the practical copyright-related harms with which a geographical interpretation would threaten ordinary scholarly, artistic, commercial, and consumer activities. We consequently conclude that Kirtsaeng's nongeographical reading is the better reading of the Act.

A

The language of § 109(a) read literally favors Kirtsaeng's nongeographical interpretation, namely, that "lawfully made under this title" means made "in accordance with" or "in compliance with" the Copyright Act. The language of § 109(a) says nothing about geography. The word "under" can mean "[i]n accordance with." 18 Oxford English Dictionary 950 (2d ed. 1989). See also Black's Law Dictionary 1525 (6th ed. 1990) ("according to"). And a nongeographical interpretation provides each word of the five-word phrase with a distinct purpose. The first two words of the phrase, "lawfully made," suggest an effort to distinguish those copies that were made lawfully from those that were not, and the last three words, "under this title," set forth the standard of "lawful[ness]." Thus, the nongeographical reading is simple, it promotes a traditional copyright objective (combatting piracy), and it makes word-by-word linguistic sense.

The geographical interpretation, however, bristles with linguistic difficulties. It gives the word "lawfully" little, if any, linguistic work to do. (How could a book be *un*lawfully "made under this title"?) It imports geography into a statutory provision that says nothing explicitly about it. And it is far more complex than may at first appear.

To read the clause geographically, Wiley . . . must first emphasize the word "under." Indeed, Wiley reads "under this title" to mean "in conformance with the Copyright Act *where the Copyright Act is applicable*." Wiley must then take a second step, arguing that the Act "is applicable" only in the United States. . . .

One difficulty is that neither "under" nor any other word in the phrase means "where." It might mean "subject to," but as this Court has repeatedly acknowledged, the word evades a uniform, consistent meaning. See *Kucana v. Holder*, 558 U.S. 233, 245 (2010) ("'under' is chameleon"); *Ardestani v. INS*, 502 U.S. 129, 135 (1991) ("under" has "many dictionary definitions" and "must draw its meaning from its context").

A far more serious difficulty arises out of the uncertainty and complexity surrounding the second step's effort to read the necessary geographical limitation into the word "applicable" (or the equivalent). Where, precisely, is the Copyright Act "applicable"? The Act does not instantly *protect* an American copyright holder from unauthorized piracy taking place abroad. But that fact does not mean the Act is *inapplicable* to copies made abroad. As a matter of ordinary English, one can say that a statute imposing, say, a tariff upon "any rhododendron grown in Nepal" applies to *all* Nepalese rhododendrons. And, similarly, one can say that the American Copyright Act is *applicable* to *all* pirated copies, including those printed overseas. Indeed, the Act itself makes clear that (in the Solicitor General's language) foreign-printed pirated copies are "subject to" the Act. § 602(a)(2) (referring to importation of copies "the making of which either constituted an infringement of copyright, or which would have constituted an infringement of copyright if this title had been applicable").

The appropriateness of this linguistic usage is underscored by the fact that § 104 of the Act itself says that works "*subject to protection under this title*" include unpublished works "without regard to the nationality or domicile of the author," and works "first published" in any one of the nearly 180 nations that have signed a copyright treaty with the United States. §§ 104(a), (b) (2006 ed.) (emphasis added); § 101 (defining "treaty party"). Thus, ordinary English permits us to say that the Act "applies" to an Irish manuscript lying in its author's Dublin desk drawer as well as to an original recording of a ballet performance first made in Japan and now on display in a Kyoto art gallery. Cf. 4 M. Nimmer & D. Nimmer, Copyright § 17.02, pp. 17–18, 17–19 (2012) (noting that the principle that "copyright laws do not have any extraterritorial operation" "requires some qualification"). . . .

In sum, we believe that geographical interpretations create more linguistic problems than they resolve. And considerations of simplicity and coherence tip the purely linguistic balance in Kirtsaeng's, nongeographical, favor.

B

Both historical and contemporary statutory context indicate that Congress, when writing the present version of § 109(a), did not have geography in mind. In respect to history, we compare § 109(a)'s present language with the language of its immediate predecessor. That predecessor said:

> "[N]othing in this Act shall be deemed to forbid, prevent, or restrict the transfer of any copy of a copyrighted work *the possession of which has been lawfully obtained.*" Copyright Act of 1909, § 41, 35 Stat. 1084 (emphasis added).

The predecessor says nothing about geography (and Wiley does not argue that it does). So we ask whether Congress, in changing its language implicitly *introduced* a geographical limitation that previously was lacking.

A comparison of language indicates that it did not. The predecessor says that the "first sale" doctrine protects "the transfer of any copy *the possession of which has been lawfully obtained.*" The present version says that "*the owner* of a particular copy or phonorecord lawfully made under this title is entitled to sell or otherwise dispose of the possession of that copy or phonorecord." What does this change in language accomplish?

The language of the former version referred to those *who are not owners* of a copy, but mere possessors who "lawfully obtained" a copy. The present version covers only those who are *owners* of a "lawfully made" copy. Whom does the change leave out? Who might have lawfully *obtained* a copy of a copyrighted work but not *owned* that copy? One answer is owners of movie theaters, who during the 1970's (and before) often *leased* films from movie distributors or filmmakers. Because the theater owners had "lawfully obtained" their copies, the earlier version could be read as allowing them to sell that copy, *i.e.*, it might have given them "first sale" protection. Because the theater owners were lessees, not owners, of their copies, the change in language makes clear that they (like bailees and other lessees) cannot take advantage of the "first sale" doctrine.

This objective perfectly well explains the new language of the present version, including the five words here at issue. Section 109(a) now makes clear that a lessee of a copy will *not* receive "first sale" protection but one who *owns* a copy *will* receive "first sale" protection, *provided,* of course, that the copy was "*lawfully made*" and not pirated. . . .

Finally, we normally presume that the words "lawfully made under this title" carry the same meaning when they appear in different but related sections. But doing so here produces surprising consequences. Consider:

(1) Section 109(c) says that, despite the copyright owner's exclusive right "to display" a copyrighted work (provided in § 106(5)), the owner of a particular copy "lawfully made under this title" may publicly display it without further authorization. To interpret these words geographically would mean that one who buys a copyrighted work of art, a poster, or even a bumper sticker, in Canada, in Europe, in Asia, could not display it in America without the copyright owner's further authorization.

(2) Section 109(e) specifically provides that the owner of a particular copy of a copyrighted video arcade game "lawfully made under this title" may "publicly perform or display that game in coin-operated equipment" without the authorization of the copyright owner. To interpret these words geographically means that an arcade owner could not ("without the authority of the copyright owner") perform or display arcade games (whether new or used) originally made in Japan.

(3) Section 110(1) says that a teacher, without the copyright owner's authorization, is allowed to perform or display a copyrighted work (say, an audiovisual work) "in the course of face-to-face teaching activities"—unless the teacher knowingly used "a copy that was not lawfully made under this title." To interpret these words geographically would mean that the teacher could not (without further authorization) use a copy of a film during class if the copy was lawfully made in Canada, Mexico, Europe, Africa, or Asia.

(4) In its introductory sentence, § 106 provides the Act's basic exclusive rights to an "owner of a copyright under this title." The last three words cannot support a geographic interpretation.

Wiley basically accepts the first three readings, but argues that Congress intended the restrictive consequences. And it argues that context simply requires that the words of the fourth example receive a different interpretation. Leaving the fourth example to the side, we shall explain in Part II–D why we find it unlikely that Congress would have intended these, and other related consequences.

C

A relevant canon of statutory interpretation favors a nongeographical reading. "[W]hen a statute covers an issue previously governed by the common law," we must presume that "Congress intended to retain the substance of the common law." *Samantar v. Yousuf*, 560 U.S. ——, ——, n. 13 (2010). See also *Isbrandtsen Co. v. Johnson*, 343 U. S. 779, 783 (1952) ("Statutes which invade the common law . . . are to be read with a presumption favoring the retention of long-established and familiar principles, except when a statutory purpose to the contrary is evident").

The "first sale" doctrine is a common-law doctrine with an impeccable historic pedigree. In the early 17th century Lord Coke explained the common law's refusal to permit restraints on the alienation of chattels. . . .

. . . Coke emphasizes the importance of leaving buyers of goods free to compete with each other when reselling or otherwise disposing of those goods. American law too has generally thought that competition, including freedom to resell, can work to the advantage of the consumer. See, *e.g.*, *Leegin Creative Leather Products, Inc. v. PSKS, Inc., 551 U.S. 877, 886 (2007)* (restraints with "manifestly anticompetitive effects" are *per se* illegal; others are subject to the rule of reason (internal quotation marks omitted)); 1 P. Areeda & H. Hovenkamp, Antitrust Law ¶ 100, p. 4 (3d ed. 2006) ("[T]he principal objective of antitrust policy is to maximize consumer welfare by encouraging firms to behave competitively").

The "first sale" doctrine also frees courts from the administrative burden of trying to enforce restrictions upon difficult-to-trace, readily movable goods. And it avoids the selective enforcement inherent in any such effort. Thus, it is not surprising that for at least a century the "first sale" doctrine has played an important role in American copyright law. See *Bobbs–Merrill Co. v. Straus*, 210 U.S. 339 (1908); Copyright Act of 1909, § 41, 35 Stat. 1084. See also Copyright Law Revision, Further Discussions and Comments on Preliminary Draft for Revised U.S. Copyright Law, 88th Cong., 2d Sess., pt. 4, p. 212 (Comm. Print 1964) (Irwin Karp of Authors' League of America expressing concern for "the very basic concept of copyright law that, once you've sold a copy legally, you can't restrict its resale").

The common-law doctrine makes no geographical distinctions; nor can we find any in *Bobbs–Merrill* (where this Court first applied the "first sale" doctrine) or in § 109(a)'s predecessor provision, which Congress enacted a year later. . . .

D

Associations of libraries, used-book dealers, technology companies, consumer-goods retailers, and museums point to various ways in which a geographical interpretation would fail to further basic constitutional copyright objectives, in particular "promot[ing] the Progress of Science and useful Arts." U.S. Const., Art. I, § 8, cl. 8.

The American Library Association tells us that library collections contain at least 200 million books published abroad (presumably, many were first published in one of the nearly 180 copyright-treaty nations and enjoy American copyright protection under 17 U.S.C. § 104); that many others were first published in the United States but printed abroad because of lower costs; and that a geographical interpretation will likely require the libraries to obtain permission (or at least create significant uncertainty) before circulating or otherwise distributing these books.

How, the American Library Association asks, are the libraries to obtain permission to distribute these millions of books? How can they find, say, the copyright owner of a foreign book, perhaps written decades ago? . . . Are the libraries to stop circulating or distributing or displaying the millions of books in their collections that were printed abroad?

Used-book dealers tell us that, from the time when Benjamin Franklin and Thomas Jefferson built commercial and personal libraries of foreign books, American readers have bought used books published and printed abroad. The dealers say that they have "operat[ed] . . . for centuries" under the assumption that the "first sale" doctrine applies. But under a geographical interpretation a contemporary tourist who buys, say, at Shakespeare and Co. (in Paris), a dozen copies of a foreign book for American friends might find that she had violated the copyright law. . . .

Technology companies tell us that "automobiles, microwaves, calculators, mobile phones, tablets, and personal computers" contain copyrightable software programs or packaging. Many of these items are made abroad with the American copyright holder's permission and then sold and imported (with that permission) to the United States. A geographical interpretation would prevent the resale of, say, a car, without the permission of the holder of each copyright on each piece of copyrighted automobile software. . . .

Art museum directors ask us to consider their efforts to display foreign-produced works by, say, Cy Twombly, René Magritte, Henri Matisse, Pablo Picasso, and others. A geographical interpretation, they say, would require the museums to obtain permission from the copyright owners before they could display the work—even if the copyright owner has already sold or donated the work to a foreign museum. What are the museums to do, they ask, if the artist retained the copyright, if the artist cannot be found, or if a group of heirs is arguing about who owns which copyright?

These examples, and others previously mentioned, help explain *why* Lord Coke considered the "first sale" doctrine necessary to protect "Trade and Traffi[c], and bargaining and contracting," and they help explain *why* American copyright law has long applied that doctrine.

Neither Wiley nor any of its many *amici* deny that a geographical interpretation could bring about these "horribles"—at least in principle. Rather, Wiley essentially says that the list is artificially invented. It points out that a federal court first adopted a geographical interpretation more than 30 years ago. Yet, it adds, these problems have not occurred. Why not? Because, says Wiley, the problems and threats are purely theoretical; they are unlikely to reflect reality.

We are less sanguine. For one thing, the law has not been settled for long in Wiley's favor. . . .

For another thing, reliance upon the "first sale" doctrine is deeply embedded in the practices of those, such as booksellers, libraries, museums, and retailers, who have long relied upon its protection. . . . That inertia means a dramatic change is likely necessary before these institutions, instructed by their counsel, would begin to engage in the complex permission-verifying process that a geographical interpretation would demand. And this Court's adoption of the geographical interpretation could provide that dramatic change. . . .

Thus, we believe that the practical problems that petitioner and his *amici* have described are too serious, too extensive, and too likely to come about for us to dismiss them as insignificant—particularly in light of the ever-growing importance of foreign trade to America. . . . The upshot is that copyright-related consequences along with language, context, and interpretive canons argue strongly against a geographical interpretation of § 109(a). . . .

For these reasons we conclude that the considerations supporting Kirtsaeng's nongeographical interpretation of the words "lawfully made under this title" are the more persuasive. The judgment of the Court of Appeals is reversed, and the case is remanded for further proceedings consistent with this opinion.

It is so ordered.

Justice GINSBURG, with whom Justice KENNEDY joins, and with whom Justice SCALIA joins except as to Parts III and V–B–1, dissenting.

"In the interpretation of statutes, the function of the courts is easily stated. It is to construe the language so as to give effect to the intent of Congress." Instead of adhering to the Legislature's design, the Court today adopts an interpretation of the Copyright Act at odds with Congress' aim to protect copyright owners against the unauthorized importation of low-priced, foreign-made copies of their copyrighted works. The Court's bold departure from Congress' design is all the more stunning, for it places the United States at the vanguard of the movement for "international exhaustion" of copyrights—a movement the United States has steadfastly resisted on the world stage.

To justify a holding that shrinks to insignificance copyright protection against the unauthorized importation of foreign-made copies, the Court identifies several "practical problems." The Court's parade of horribles, however, is largely imaginary. Congress' objective in enacting 17 U.S.C. § 602(a)(1)'s importation prohibition can be honored without generating the absurd consequences hypothesized in the Court's opinion. I dissent from the Court's embrace of "international exhaustion" . . . .

Because economic conditions and demand for particular goods vary across the globe, copyright owners have a financial incentive to charge different prices for copies of their works in different geographic regions. Their ability to engage in such price discrimination, however, is undermined if arbitrageurs are permitted to import copies from low-price regions and sell them in high-price regions. The question in this case is whether the unauthorized importation of foreign-made copies constitutes copyright infringement under U.S. law. . . .

The text of the Copyright Act demonstrates that Congress intended to provide copyright owners with a potent remedy against the importation of foreign-made copies of their copyrighted works. . . .

The Court's point of departure is similar to mine. According to the Court, the phrase "'lawfully made under this title' means made 'in accordance with' or 'in compliance with' the Copyright Act." But the Court overlooks that, according to the very dictionaries it cites, the word "under" commonly signals a relationship of subjection, where one thing is governed or regulated by another. See Black's Law Dictionary 1525 (6th ed. 1990) ("under" "frequently" means "inferior" or "subordinate"; 18 Oxford English Dictionary 950 (2d ed. 1989) ("under" means, among other things, "[i]n accordance with (*some regulative power or principle*)" (emphasis added)). Only by disregarding this established meaning of "under" can the Court arrive at the conclusion that Wiley's foreign-manufactured textbooks were "lawfully made under" U.S. copyright law, even though that law did not govern their creation. It is anomalous, however, to speak of particular conduct as "lawful" under an inapplicable law. For example, one might say that driving on the right side of the road in England is "lawful" under U.S. law, but that would be so only because U.S. law has nothing to say about the subject. The governing law is English law, and English law demands that driving be done on the left side of the road.[1]

The logical implication of the Court's definition of the word "under" is that any copy manufactured abroad—even a piratical one made without the copyright owner's authorization and in violation of the law of the country where it was created—would fall within the scope of § 109(a). Any such copy would have been made "in accordance with" or "in compliance with" the U.S. Copyright Act, in the sense that manufacturing the copy did not violate the Act (because the Act does not apply extraterritorially).

The Court rightly refuses to accept such an absurd conclusion. Instead, it interprets § 109(a) as applying only to copies whose making actually complied with Title 17, or would have complied with Title 17 had Title 17 been applicable (*i.e.*, had the copies been made in the United States). Congress, however, used express language when it called for such a counterfactual inquiry in 17 U.S.C. §§ 602(a)(2) and (b). See § 602(a)(2) ("Importation into the United States or exportation from the United States, without the authority of the owner of copyright under this title, of copies or phonorecords, the making of which either constituted an infringement of copyright, or *which would have constituted an infringement of copyright if this title had been applicable,* is an infringement of the exclusive right to distribute copies or phonorecords under section 106." (emphasis added)); § 602(b) ("In a case where the making of the copies or phonorecords *would have constituted an infringement of copyright if this title had been applicable,* their importation is prohibited." (emphasis added)). Had Congress intended courts to engage in a similarly hypothetical inquiry under § 109(a), Congress would presumably have included similar language in that section. See *Russello v. United States,* 464 U.S. 16, 23 (1983) ("'[W]here Congress includes particular language in one section of a statute but omits it in another section of the same Act, it is generally presumed that Congress acts intentionally and purposely in the disparate inclusion or exclusion.'").

Not only does the Court adopt an unnatural construction of the § 109(a) phrase "lawfully made under this title." Concomitantly, the Court reduces § 602(a)(1) to insignificance. As the Court appears to acknowledge, the only independent effect § 602(a)(1) has under today's decision is to prohibit unauthorized importations carried out by persons who merely have possession of, but do not own, the imported copies. See 17 U.S.C. § 109(a) (§ 109(a) applies to any "*owner* of a particular copy or phonorecord lawfully made under this title" (emphasis added)). If this is enough to avoid rendering § 602(a)(1) entirely "superfluous," it hardly suffices to give the

1. The Court asserts that my position gives the word "lawfully" in § 109(a) "little, if any, linguistic work to do." That is not so. My reading gives meaning to each word in the phrase "lawfully made under this title." The word "made" signifies that the conduct at issue is the creation or manufacture of a copy. See Webster's Third New International Dictionary 1356 (1961) (defining "made" as "artificially produced by a manufacturing process"). The word "lawfully" indicates that for § 109(a) to apply, the copy's creation must have complied with some body of law. Finally, the prepositional phrase "under this title" clarifies what that body of law is—namely, the copyright prescriptions contained in Title 17 of the U.S. Code.

owner's importation right the scope Congress intended it to have. Congress used broad language in § 602(a)(1); it did so to achieve a broad objective. Had Congress intended simply to provide a copyright remedy against larcenous lessees, licensees, consignees, and bailees of films and other copyright-protected goods, it likely would have used language tailored to that narrow purpose.[2]

The Court's decision also overwhelms 17 U.S.C. § 602(a)(3)'s exceptions to § 602(a)(1)'s importation prohibition. Those exceptions permit the importation of copies without the copyright owner's authorization for certain governmental, personal, scholarly, educational, and religious purposes. 17 U.S.C. § 602(a)(3). Copies imported under these exceptions "will often be lawfully made gray market goods purchased through normal market channels abroad." 2 Goldstein § 7.6.1.2(a), at 7:141.[3] But if, as the Court holds, such copies can in any event be imported by virtue of § 109(a), § 602(a)(3)'s work has already been done. For example, had Congress conceived of § 109(a)'s sweep as the Court does, what earthly reason would there be to provide, as Congress did in § 602(a)(3)(C), that a library may import "no more than five copies" of a non-audiovisual work for its "lending or archival purposes"?

The far more plausible reading of §§ 109(a) and 602(a), then, is that Congress intended § 109(a) to apply to copies made in the United States, not to copies manufactured and sold abroad. That reading of the first sale and importation provisions leaves § 602(a)(3)'s exceptions with real, meaningful work to do. See *TRW Inc. v. Andrews*, 534 U.S. 19, 31 (2001) ("It is a cardinal principle of statutory construction that a statute ought, upon the whole, to be so construed that, if it can be prevented, no clause, sentence, or word shall be superfluous, void, or insignificant." (internal quotation marks omitted)). . . .

III

The history of § 602(a)(1) reinforces the conclusion I draw from the text of the relevant provisions: § 109(a) does not apply to copies manufactured abroad. Section 602(a)(1) was enacted as part of the Copyright Act of 1976. . . .

. . . The House and Senate Committee Reports on the 1976 Act demonstrate that Congress understood, as did the Copyright Office, just what that text meant. Both Reports state:

2. Notably, the Court ignores the history of § 602(a)(1), which reveals that the primary purpose of the prescription was not to provide a remedy against rogue licensees, consignees, and bailees, against whom copyright owners could frequently assert breach-of-contract claims even in the absence of § 602(a)(1). Instead, the primary purpose of § 602(a)(1) was to reach third-party importers, enterprising actors like Kirtsaeng, against whom copyright owners could not assert contract claims due to lack of privity.

3. The term "gray market good" refers to a good that is "imported outside the distribution channels that have been contractually negotiated by the intellectual property owner." Such goods are also commonly called "parallel imports."

> "Section 602 [deals] with two separate situations: importation of 'piratical' articles (that is, copies or phonorecords made without any authorization of the copyright owner), and unauthorized importation of copies or phonorecords that were lawfully made. *The general approach of section 602 is to make unauthorized importation an act of infringement in both cases,* but to permit the Bureau of Customs to prohibit importation only of 'piratical' articles." S. Rep. No. 94–473, p. 151 (1975) (emphasis added). See also H.R. Rep. No. 94–1476, p. 169 (1976) (same).

In sum, the legislative history of the Copyright Act of 1976 is hardly "inconclusive." To the contrary, it confirms what the plain text of the Act conveys: Congress intended § 602(a)(1) to provide copyright owners with a remedy against the unauthorized importation of foreign-made copies of their works, even if those copies were made and sold abroad with the copyright owner's authorization.

IV

Unlike the Court's holding, my position is consistent with the stance the United States has taken in international-trade negotiations. . . .

V

I turn now to the Court's justifications for a decision difficult to reconcile with the Copyright Act's text and history. . . .

B

The Court sees many "horribles" following from a holding that the § 109(a) phrase "lawfully made under this title" does not encompass foreign-made copies. If § 109(a) excluded foreign-made copies, the Court fears, then copyright owners could exercise perpetual control over the downstream distribution or public display of such copies. A ruling in Wiley's favor, the Court asserts, would shutter libraries, put used-book dealers out of business, cripple art museums, and prevent the resale of a wide range of consumer goods, from cars to calculators. Copyright law and precedent, however, erect barriers to the anticipated horribles.

1

Recognizing that foreign-made copies fall outside the ambit of § 109(a) would not mean they are forever free of the first sale doctrine. . . .

Under the logic of *Bobbs–Merrill,* the sale of a foreign-manufactured copy in the United States carried out with the copyright owner's authorization would exhaust the copyright owner's right to "vend" that copy. The copy could thenceforth be resold, lent out, or otherwise redistributed without further authorization from the copyright owner. . . .

2

Other statutory prescriptions provide further protection against the absurd consequences imagined by the Court. For example, § 602(a)(3)(C) permits "an organization operated for scholarly, educational, or religious purposes" to import, without the copyright owner's authorization, up to five foreign-made copies of a non-audiovisual work—notably, a book—for "library lending or archival purposes."

The Court also notes that *amici* representing art museums fear that a ruling in Wiley's favor would prevent museums from displaying works of art created abroad. . . .

Limiting § 109(c) to U.S.-made works, however, does not bar art museums from lawfully displaying works made in other countries. Museums can, of course, seek the copyright owner's permission to display a work. Furthermore, the sale of a work of art to a U.S. museum may carry with it an implied license to publicly display the work. . . .

The Court worries about the resale of foreign-made consumer goods "contain [ing] copyrightable software programs or packaging." For example, the Court observes that a car might be programmed with diverse forms of software, the copyrights to which might be owned by individuals or entities other than the manufacturer of the car. Must a car owner, the Court asks, obtain permission from all of these various copyright owners before reselling her car? Although this question strays far from the one presented in this case and briefed by the parties, principles of fair use and implied license (to the extent that express licenses do not exist) would likely permit the car to be resold without the copyright owners' authorization. . . .

Part II
Rules and Statutes

Restatement of Torts

§757 Liability for Disclosure or Use of Another's Trade Secret—General Principle

One who discloses or uses another's trade secret without privilege to do so, is liable to the other if

(a) he discovered the secret by improper means, or

(b) his disclosure or use constitutes a breach of confidence reposed in him by the other in disclosing the secret to him, or

(c) he learned the secret from a third person with notice of the facts that it was a secret and that the third person discovered it by improper means or that the third person's disclosure of it was a breach of his duty to the other, or

(d) he learned the secret with notice of the facts that it was a secret and that disclosure was made to him by mistake.

§758 Innocent Discovery of Secret—Effect of Subsequent Notice or Change of Position

One who learns another's trade secret from a third person without notice that it is a secret and that the third person's disclosure is a breach of his duty to the other, or who learns the secret through a mistake without notice of the secrecy and the mistake,

(a) is not liable to the other for a disclosure or use of the secret prior to receipt of such notice, and

(b) is liable to the other for a disclosure or use of the secret after the receipt of such notice, unless prior thereto he has in good faith paid value for the secret or has so changed his position that to subject him to liability would be inequitable.

§759 Procuring Information by Improper Means

One who, for the purpose of advancing a rival business interest, procures by improper means information about another's business is liable to the other for the harm caused by his possession, disclosure, or use of the information.

Uniform Trade Secrets Act

(with California Amendments)

California Civil Code

§§3426-3426.11

§3426

This title may be cited as the Uniform Trade Secrets Act.

§3426.1

As used in this title, unless the context requires otherwise:

(a) "Improper means" includes theft, bribery, misrepresentation, breach or inducement of a breach of a duty to maintain secrecy, or espionage through electronic or other means. Reverse engineering or independent derivation alone shall not be considered improper means.

(b) "Misappropriation" means:

(1) Acquisition of a trade secret of another by a person who knows or has reason to know that the trade secret was acquired by improper means; or

(2) Disclosure or use of a trade secret of another without express or implied consent by a person who:

(A) Used improper means to acquire knowledge of the trade secret; or

(B) At the time of disclosure or use, knew or had reason to know that his or her knowledge of the trade secret was:

(i) Derived from or through a person who had utilized improper means to acquire it;

(ii) Acquired under circumstances giving rise to a duty to maintain its secrecy or limit its use; or

(iii) Derived from or through a person who owed a duty to the person seeking relief to maintain its secrecy or limit its use; or

(C) Before a material change of his or her position, knew or had reason to know that it was a trade secret and that knowledge of it had been acquired by accident or mistake.

(c) "Person" means a natural person, corporation, business trust, estate, trust, partnership, limited liability company, association, joint venture, government, governmental subdivision or agency, or any other legal or commercial entity.

(d) "Trade secret" means information, including a formula, pattern, compilation, program, device, method, technique, or process, that:

(1) Derives independent economic value, actual or potential, from not being generally known to the public or to other persons who can obtain economic value from its disclosure or use; and*

(2) Is the subject of efforts that are reasonable under the circumstances to maintain its secrecy.

§3426.2

(a) Actual or threatened misappropriation may be enjoined. Upon application to the court, an injunction shall be terminated when the trade secret has ceased to exist, but the injunction may be continued for an additional period of time in order to eliminate commercial advantage that otherwise would be derived from the misappropriation.

(b) If the court determines that it would be unreasonable to prohibit future use, an injunction may condition future use upon payment of a reasonable royalty for no longer than the period of time the use could have been prohibited.

(c) In appropriate circumstances, affirmative acts to protect a trade secret may be compelled by court order.

§3426.3

(a) A complainant may recover damages for the actual loss caused by misappropriation. A complainant also may recover for the unjust enrichment caused by misappropriation that is not taken into account in computing damages for actual loss.

(b) If neither damages nor unjust enrichment caused by misappropriation are provable, the court may order payment of a reasonable royalty for no longer than the period of time the use could have been prohibited.

(c) If willful and malicious misappropriation exists, the court may award exemplary damages in an amount not exceeding twice any award made under subdivision (a) or (b).

---

* The original version of the Uniform Act reads "not being generally known to or readily ascertainable by proper means by the public. . . ." — Eds.

§3426.4

If a claim of misappropriation is made in bad faith, a motion to terminate an injunction is made or resisted in bad faith, or willful and malicious misappropriation exists, the court may award reasonable attorney's fees to the prevailing party.

§3426.5

In an action under this title, a court shall preserve the secrecy of an alleged trade secret by reasonable means, which may include granting protective orders in connection with discovery proceedings, holding in-camera hearings, sealing the records of the action, and ordering any person involved in the litigation not to disclose an alleged trade secret without prior court approval.

§3426.6

An action for misappropriation must be brought within three years after the misappropriation is discovered or by the exercise of reasonable diligence should have been discovered. For the purposes of this section, a continuing misappropriation constitutes a single claim.

§3426.7

(a) Except as otherwise expressly provided, this title does not supersede any statute relating to misappropriation of a trade secret, or any statute otherwise regulating trade secrets.

(b) This title does not affect (1) contractual remedies, whether or not based upon misappropriation of a trade secret, (2) other civil remedies that are not based upon misappropriation of a trade secret, or (3) criminal remedies, whether or not based upon misappropriation of a trade secret.

(c) This title does not affect the disclosure of a record by a state or local agency under the California Public Records Act (Chapter 3.5 (commencing with Section 6250) of Division 7 of Title 1 of the Government Code). Any determination as to whether the disclosure of a record under the California Public Records Act constitutes a misappropriation of a trade secret and the rights and remedies with respect thereto shall be made pursuant to the law in effect before the operative date of this title.

§3426.8

This title shall be applied and construed to effectuate its general purpose to make uniform the law with respect to the subject of this title among states enacting it.

§3426.9

If any provision of this title or its application to any person or circumstances is held invalid, the invalidity does not affect other provisions or applications of the title which can be given effect without the invalid provision or application, and to this end the provisions of this title are severable.

§3426.10

This title does not apply to misappropriation occurring prior to January 1, 1985. If a continuing misappropriation otherwise covered by this title began before January 1, 1985, this title does not apply to the part of the misappropriation occurring before that date. This title does apply to the part of the misappropriation occurring on or after that date unless the appropriation was not a misappropriation under the law in effect before the operative date of this title.

§3426.11

Notwithstanding subdivision (b) of Section 47, in any legislative or judicial proceeding, or in any other official proceeding authorized by law, or in the initiation or course of any other proceeding authorized by law and reviewable pursuant to Chapter 2 (commencing with Section 1084) of Title 1 of Part 3 of the Code of Civil Procedure, the voluntary, intentional disclosure of trade secret information, unauthorized by its owner, to a competitor or potential competitor of the owner of the trade secret information or the agent or representative of such a competitor or potential competitor is not privileged and is not a privileged communication for purposes of Part 2 (commencing with Section 43) of Division 1. This section does not in any manner limit, restrict, impair, or otherwise modify either the application of the other subdivisions of Section 47 to the conduct to which this section applies or the court's authority to control, order, or permit access to evidence in any case before it. Nothing in this section shall be construed to limit, restrict, or otherwise impair, the capacity of persons employed by public entities to report improper government activity, as defined in Section 10542 of the Government Code, or the capacity of private persons to report improper activities of a private business.

Economic Espionage Act of 1996

18 U.S.C. §1831 et seq.

Sec. 1831. Economic Espionage

(a) In General.—Whoever, intending or knowing that the offense will benefit any foreign government, foreign instrumentality, or foreign agent, knowingly—

(1) steals, or without authorization appropriates, takes, carries away, or conceals, or by fraud, artifice, or deception obtains a trade secret;

(2) without authorization copies, duplicates, sketches, draws, photographs, downloads, uploads, alters, destroys, photocopies, replicates, transmits, delivers, sends, mails, communicates, or conveys a trade secret;

(3) receives, buys, or possesses a trade secret, knowing the same to have been stolen or appropriated, obtained, or converted without authorization;

(4) attempts to commit any offense described in any of paragraphs (1) through (3); or

(5) conspires with one or more other persons to commit any offense described in any of paragraphs (1) through (3), and one or more of such persons do any act to effect the object of the conspiracy, shall, except as provided in subsection (b), be fined not more than $500,000 or imprisoned not more than 15 years, or both.

(b) Organizations.—Any organization that commits any offense described in subsection (a) shall be fined not more than the greater of $10,000,000 or 3 times the value of the stolen trade secret to the organization, including expenses for research and design and other costs of reproducing the trade secret that the organization has thereby avoided.

Sec. 1832. Theft of Trade Secrets

(a) Whoever, with intent to convert a trade secret, that is related to a product or service used in or intended for use in interstate or foreign commerce, to the economic benefit of anyone other than the owner thereof, and intending or knowing that the offense will, injure any owner of that trade secret, knowingly—

(1) steals, or without authorization appropriates, takes, carries away, or conceals, or by fraud, artifice, or deception obtains such information;

(2) without authorization copies, duplicates, sketches, draws, photographs, downloads, uploads, alters, destroys, photocopies, replicates, transmits, delivers, sends, mails, communicates, or conveys such information;

(3) receives, buys, or possesses such information, knowing the same to have been stolen or appropriated, obtained, or converted without authorization;

(4) attempts to commit any offense described in paragraphs (1) through (3); or

(5) conspires with one or more other persons to commit any offense described in paragraphs (1) through (3), and one or more of such persons do any act to effect the object of the conspiracy, shall, except as provided in subsection (b), be fined under this title or imprisoned not more than 10 years, or both.

(b) Any organization that commits any offense described in subsection (a) shall be fined not more than $5,000,000.

Sec. 1833. Exceptions to Prohibitions

This chapter does not prohibit—

(1) any otherwise lawful activity conducted by a governmental entity of the United States, a State, or a political subdivision of a State; or

(2) the reporting of a suspected violation of law to any governmental entity of the United States, a State, or a political subdivision of a State, if such entity has lawful authority with respect to that violation.

Sec. 1834. Criminal Forfeiture

Forfeiture, destruction, and restitution relating to this chapter shall be subject to section 2323, to the extent provided in that section, in addition to any other similar remedies provided by law.

Sec. 1835. Orders to Preserve Confidentiality

In any prosecution or other proceeding under this chapter, the court shall enter such orders and take such other action as may be necessary and appropriate to preserve the confidentiality of trade secrets, consistent with the requirements of the Federal Rules of Criminal and Civil Procedure, the Federal Rules of Evidence, and all other applicable laws. An interlocutory appeal by the United States shall lie from a decision or order of a district court authorizing or directing the disclosure of any trade secret.

Sec. 1836. Civil Proceedings to Enjoin Violations

(a) The Attorney General may, in a civil action, obtain appropriate injunctive relief against any violation of this section.

(b) The district courts of the United States shall have exclusive original jurisdiction of civil actions under this subsection.

Sec. 1837. Applicability to Conduct Outside the United States

This chapter also applies to conduct occurring outside the United States if—

(1) the offender is a natural person who is a citizen or permanent resident alien of the United States, or an organization organized under the laws of the United States or a State or political subdivision thereof; or

(2) an act in furtherance of the offense was committed in the United States.

Sec. 1838. Construction with Other Laws

This chapter shall not be construed to preempt or displace any other remedies, whether civil or criminal, provided by United States Federal, State, commonwealth, possession, or territory law for the misappropriation of a trade secret, or to affect the otherwise lawful disclosure of information by any Government employee under section 552 of title 5 (commonly known as the Freedom of Information Act).

Sec. 1839. Definitions

As used in this chapter—

(1) the term "foreign instrumentality" means any agency, bureau, ministry, component, institution, association, or any legal, commercial, or business organization, corporation, firm, or entity that is substantially owned, controlled, sponsored, commanded, managed, or dominated by a foreign government;

(2) the term "foreign agent" means any officer, employee, proxy, servant, delegate, or representative of a foreign government;

(3) the term "trade secret" means all forms and types of financial, business, scientific, technical, economic, or engineering information, including patterns, plans, compilations, program devices, formulas, designs, prototypes, methods, techniques, processes, procedures, programs, or codes, whether tangible or intangible, and whether or how stored, compiled, or memorialized physically, electronically, graphically, photographically, or in writing if—

(A) the owner thereof has taken reasonable measures to keep such information secret; and

(B) the information derives independent economic value, actual or potential, from not being generally known to, and not being readily ascertainable through proper means by, the public; and

(4) the term "owner", with respect to a trade secret, means the person or entity in whom or in which rightful legal or equitable title to, or license in, the trade secret is reposed.

Sec. 2323.

(a) Civil Forfeiture.—

(1) Property subject to forfeiture.— The following property is subject to forfeiture to the United States Government:

(A) Any article, the making or trafficking of which is, prohibited under section 506 of title 17, or section 2318, 2319, 2319A, 2319B, or 2320, or chapter 90section 2318, 2319, 2319A, 2319B, or 2320, or chapter 90, of this title.

(B) Any property used, or intended to be used, in any manner or part to commit or facilitate the commission of an offense referred to in subparagraph (A).

(C) Any property constituting or derived from any proceeds obtained directly or indirectly as a result of the commission of an offense referred to in subparagraph (A).

(2) Procedures.— The provisions of chapter 46 relating to civil forfeitures shall extend to any seizure or civil forfeiture under this section. For seizures made under this section, the court shall enter an appropriate protective order with respect to discovery and use of any records or information that has been seized. The protective order shall provide for appropriate procedures to ensure that confidential, private, proprietary, or privileged information contained in such records is not improperly disclosed or used. At the conclusion of the forfeiture proceedings, unless otherwise requested by an agency of the United States, the court shall order that any property forfeited under paragraph (1) be destroyed, or otherwise disposed of according to law.

(b) Criminal Forfeiture.—

(1) Property subject to forfeiture.— The court, in imposing sentence on a person convicted of an offense under section 506 of title 17, or section 2318, 2319, 2319A, 2319B, or 2320, or chapter 90section 2318, 2319, 2319A, 2319B, or 2320, or chapter 90, of this title, shall order, in addition to any other sentence imposed, that the person forfeit to the United States Government any property subject to forfeiture under subsection (a) for that offense.

(2) Procedures.—

(A) In general.— The forfeiture of property under paragraph (1), including any seizure and disposition of the property and any related judicial or administrative proceeding, shall be governed by the procedures set forth in section 413 of the Comprehensive Drug Abuse Prevention and Control Act of 1970 (21 U.S.C. 853), other than subsection (d) of that section.

(B) Destruction.— At the conclusion of the forfeiture proceedings, the court, unless otherwise requested by an agency of the United States shall order that any—

(i) forfeited article or component of an article bearing or consisting of a counterfeit mark be destroyed or otherwise disposed of according to law; and

(ii) infringing items or other property described in subsection (a)(1)(A) and forfeited under paragraph (1) of this subsection be destroyed or otherwise disposed of according to law.

(c) Restitution.— When a person is convicted of an offense under section 506 of title 17 or section 2318, 2319, 2319A, 2319B, or 2320, or chapter 90section 2318, 2319, 2319A, 2319B, or 2320, or chapter 90, of this title, the court, pursuant to sections 3556, 3663A, and 3664 of this title, shall order the person to pay restitution to any victim of the offense as an offense against property referred to in section 3663A (c)(1)(A)(ii) of this title.

Patent Act

TITLE 35, U.S.C.

Part I. United States Patent and Trademark Office

Chapter 1. Establishment, Officers and Employees, Functions

Sec. 1. Establishment

(a) Establishment.—The United States Patent and Trademark Office is established as an agency of the United States, within the Department of Commerce. In carrying out its functions, the United States Patent and Trademark Office shall be subject to the policy direction of the Secretary of Commerce, but otherwise shall retain responsibility for decisions regarding the management and administration of its operations and shall exercise independent control of its budget allocations and expenditures, personnel decisions and processes, procurements, and other administrative and management functions in accordance with this title and applicable provisions of law. Those operations designed to grant and issue patents and those operations which are designed to facilitate the registration of trademarks shall be treated as separate operating units within the Office.

(b) Offices.—The United States Patent and Trademark Office shall maintain its principal office in the metropolitan Washington, D.C., area, for the service of process and papers and for the purpose of carrying out its functions. The United States Patent and Trademark Office shall be deemed, for purposes of venue in civil actions, to be a resident of the district in which its principal office is located, except where jurisdiction is otherwise provided by law. The United States Patent and Trademark Office may establish satellite offices in such other places in the United States as it considers necessary and appropriate in the conduct of its business.

(c) Reference.—For purposes of this title, the United States Patent and Trademark Office shall also be referred to as the "Office" and the "Patent and Trademark Office".

Sec. 2. Powers and Duties

(a) In general.—The United States Patent and Trademark Office, subject to the policy direction of the Secretary of Commerce—

(1) shall be responsible for the granting and issuing of patents and the registration of trademarks; and

(2) shall be responsible for disseminating to the public information with respect to patents and trademarks.

(b) Specific powers.—The Office—

(1) shall adopt and use a seal of the Office, which shall be judicially noticed and with which letters patent, certificates of trademark registrations, and papers issued by the Office shall be authenticated;

(2) may establish regulations, not inconsistent with law, which—

(A) shall govern the conduct of proceedings in the Office;

(B) shall be made in accordance with section 553 of title 5;

(C) shall facilitate and expedite the processing of patent applications, particularly those which can be filed, stored, processed, searched, and retrieved electronically, subject to the provisions of section 122 relating to the confidential status of applications;

(D) may govern the recognition and conduct of agents, attorneys, or other persons representing applicants or other parties before the Office, and may require them, before being recognized as representatives of applicants or other persons, to show that they are of good moral character and reputation and are possessed of the necessary qualifications to render to applicants or other persons valuable service, advice, and assistance in the presentation or prosecution of their applications or other business before the Office;

(E) shall recognize the public interest in continuing to safeguard broad access to the United States patent system through the reduced fee structure for small entities under section 41(h)(1);

(F) provide for the development of a performance-based process that includes quantitative and qualitative measures and standards for evaluating cost-effectiveness and is consistent with the principles of impartiality and competitiveness; and

(G) may, subject to any conditions prescribed by the Director and at the request of the patent applicant, provide for prioritization of examination of applications for products, processes, or technologies that are important to the national economy or national competitiveness without recovering the aggregate extra cost of providing such prioritization, notwithstanding section 41 or any other provision of law;

(3) may acquire, construct, purchase, lease, hold, manage, operate, improve, alter, and renovate any real, personal, or mixed property, or any interest therein, as it considers necessary to carry out its functions;

(4)(A) may make such purchases, contracts for the construction, maintenance, or management and operation of facilities, and contracts for supplies or services, without regard to the provisions of subtitle I and chapter 33 of title 40, division C (except sections 3302, 3501(b), 3509, 3906, 4710, and 4711) of subtitle I of title 41, and the McKinney-Vento Homeless Assistance Act (42 U.S.C. 11301 et seq.); and

(B) may enter into and perform such purchases and contracts for printing services, including the process of composition, platemaking, presswork, silk screen processes, binding, microform, and the products of such processes, as it considers necessary to carry out the functions of the Office, without regard to sections 501 through 517 and 1101 through 1123 of title 44;

(5) may use, with their consent, services, equipment, personnel, and facilities of other departments, agencies, and instrumentalities of the Federal Government,

on a reimbursable basis, and cooperate with such other departments, agencies, and instrumentalities in the establishment and use of services, equipment, and facilities of the Office;

(6) may, when the Director determines that it is practicable, efficient, and cost-effective to do so, use, with the consent of the United States and the agency, instrumentality, Patent and Trademark Office, or international organization concerned, the services, records, facilities, or personnel of any State or local government agency or instrumentality or foreign patent and trademark office or international organization to perform functions on its behalf;

(7) may retain and use all of its revenues and receipts, including revenues from the sale, lease, or disposal of any real, personal, or mixed property, or any interest therein, of the Office;

(8) shall advise the President, through the Secretary of Commerce, on national and certain international intellectual property policy issues;

(9) shall advise Federal departments and agencies on matters of intellectual property policy in the United States and intellectual property protection in other countries;

(10) shall provide guidance, as appropriate, with respect to proposals by agencies to assist foreign governments and international intergovernmental organizations on matters of intellectual property protection;

(11) may conduct programs, studies, or exchanges of items or services regarding domestic and international intellectual property law and the effectiveness of intellectual property protection domestically and throughout the world, and the Office is authorized to expend funds to cover the subsistence expenses and travel-related expenses, including per diem, lodging costs, and transportation costs, of persons attending such programs who are not Federal employees;

(12)(A) shall advise the Secretary of Commerce on programs and studies relating to intellectual property policy that are conducted, or authorized to be conducted, cooperatively with foreign intellectual property offices and international intergovernmental organizations; and

(B) may conduct programs and studies described in subparagraph (A); and

(13)(A) in coordination with the Department of State, may conduct programs and studies cooperatively with foreign intellectual property offices and international intergovernmental organizations; and

(B) with the concurrence of the Secretary of State, may authorize the transfer of not to exceed $100,000 in any year to the Department of State for the purpose of making special payments to international intergovernmental organizations for studies and programs for advancing international cooperation concerning patents, trademarks, and other matters.

(c) Clarification of specific powers.—(1) The special payments under subsection (b)(13)(B) shall be in addition to any other payments or contributions to international organizations described in subsection (b)(13)(B) and shall not be subject to any limitations imposed by law on the amounts of such other payments or contributions by the United States Government.

(2) Nothing in subsection (b) shall derogate from the duties of the Secretary of State or from the duties of the United States Trade Representative as set forth in section 141 of the Trade Act of 1974 (19 U.S.C. 2171).

(3) Nothing in subsection (b) shall derogate from the duties and functions of the Register of Copyrights or otherwise alter current authorities relating to copyright matters.

(4) In exercising the Director's powers under paragraphs (3) and (4)(A) of subsection (b), the Director shall consult with the Administrator of General Services.

(5) In exercising the Director's powers and duties under this section, the Director shall consult with the Register of Copyrights on all copyright and related matters.

(d) Construction.—Nothing in this section shall be construed to nullify, void, cancel, or interrupt any pending request-for-proposal let or contract issued by the General Services Administration for the specific purpose of relocating or leasing space to the United States Patent and Trademark Office.

Sec. 3. Officers and Employees

(a) Under Secretary and Director.—

(1) In general.—The powers and duties of the United States Patent and Trademark Office shall be vested in an Under Secretary of Commerce for Intellectual Property and Director of the United States Patent and Trademark Office (in this title referred to as the "Director"), who shall be a citizen of the United States and who shall be appointed by the President, by and with the advice and consent of the Senate. The Director shall be a person who has a professional background and experience in patent or trademark law.

(2) Duties.—

(A) In general.—The Director shall be responsible for providing policy direction and management supervision for the Office and for the issuance of patents and the registration of trademarks. The Director shall perform these duties in a fair, impartial, and equitable manner.

(B) Consulting with the Public Advisory Committees.—The Director shall consult with the Patent Public Advisory Committee established in section 5 on a regular basis on matters relating to the patent operations of the Office, shall consult with the Trademark Public Advisory Committee established in section 5 on a regular basis on matters relating to the trademark operations of the Office, and shall consult with the respective Public Advisory Committee before submitting budgetary proposals to the Office of Management and Budget or changing or proposing to change patent or trademark user fees or patent or trademark regulations which are subject to the requirement to provide notice and opportunity for public comment under section 553 of title 5, as the case may be.

(3) Oath.—The Director shall, before taking office, take an oath to discharge faithfully the duties of the Office.

(4) Removal.—The Director may be removed from office by the President. The President shall provide notification of any such removal to both Houses of Congress.

(b) Officers and employees of the Office.—

(1) Deputy Under Secretary and Deputy Director.—The Secretary of Commerce, upon nomination by the Director, shall appoint a Deputy Under Secretary of Commerce for Intellectual Property and Deputy Director of the United States Patent and Trademark Office who shall be vested with the authority to act in the capacity of the Director in the event of the absence or incapacity of the Director. The Deputy Director shall be a citizen of the United States who has a professional background and experience in patent or trademark law.

(2) Commissioners.—

(A) Appointment and duties.—The Secretary of Commerce shall appoint a Commissioner for Patents and a Commissioner for Trademarks, without regard to chapter 33, 51, or 53 of title 5. The Commissioner for Patents shall be a citizen of the United States with demonstrated management ability and professional background and experience in patent law and serve for a term of 5 years. The Commissioner for Trademarks shall be a citizen of the United States with demonstrated management ability and professional background and experience in trademark law and serve for a term of 5 years. The Commissioner for Patents and the Commissioner for Trademarks shall serve as the chief operating officers for the operations of the Office relating to patents and trademarks, respectively, and shall be responsible for the management and direction of all aspects of the activities of the Office that affect the administration of patent and trademark operations, respectively. The Secretary may reappoint a Commissioner to subsequent terms of 5 years as long as the performance of the Commissioner as set forth in the performance agreement in subparagraph (B) is satisfactory.

(B) Salary and performance agreement.—The Commissioners shall be paid an annual rate of basic pay not to exceed the maximum rate of basic pay for the Senior Executive Service established under section 5382 of title 5, including any applicable locality-based comparability payment that may be authorized under section 5304(h)(2)(C) of title 5. The compensation of the Commissioners shall be considered, for purposes of section 207(c)(2)(A) of title 18, to be the equivalent of that described under clause (ii) of section 207(c)(2)(A) of title 18. In addition, the Commissioners may receive a bonus in an amount of up to, but not in excess of, 50 percent of the Commissioners' annual rate of basic pay, based upon an evaluation by the Secretary of Commerce, acting through the Director, of the Commissioners' performance as defined in an annual performance agreement between the Commissioners and the Secretary. The annual performance agreements shall incorporate measurable organization and individual goals in key operational areas as delineated in an annual performance plan agreed to by the Commissioners and the Secretary. Payment of a bonus under this subparagraph may be made to the Commissioners only to the extent that such payment does not cause the Commissioners' total aggregate compensation in a calendar year to equal or exceed the amount of the salary of the Vice President under section 104 of title 3.

(C) Removal.—The Commissioners may be removed from office by the Secretary for misconduct or nonsatisfactory performance under the performance agreement described in subparagraph (B), without regard to the provisions of title 5. The Secretary shall provide notification of any such removal to both Houses of Congress.

(3) Other officers and employees.—The Director shall—

(A) appoint such officers, employees (including attorneys), and agents of the Office as the Director considers necessary to carry out the functions of the Office; and

(B) define the title, authority, and duties of such officers and employees and delegate to them such of the powers vested in the Office as the Director may determine.

The Office shall not be subject to any administratively or statutorily imposed limitation on positions or personnel, and no positions or personnel of the Office shall be taken into account for purposes of applying any such limitation.

(4) Training of examiners.—The Office shall submit to the Congress a proposal to provide an incentive program to retain as employees patent and trademark examiners of the primary examiner grade or higher who are eligible for retirement, for the sole purpose of training patent and trademark examiners.

(5) National security positions.—The Director, in consultation with the Director of the Office of Personnel Management, shall maintain a program for identifying national security positions and providing for appropriate security clearances, in order to maintain the secrecy of certain inventions, as described in section 181, and to prevent disclosure of sensitive and strategic information in the interest of national security.

(6) Administrative patent judges and administrative trademark judges.—The Director may fix the rate of basic pay for the administrative patent judges appointed pursuant to section 6 and the administrative trademark judges appointed pursuant to section 17 of the Trademark Act of 1946 (15 U.S.C. 1067) at not greater than the rate of basic pay payable for level III of the Executive Schedule under section 5314 of title 5. The payment of a rate of basic pay under this paragraph shall not be subject to the pay limitation under section 5306(e) or 5373 of title 5.

(c) Continued applicability of title 5.—Officers and employees of the Office shall be subject to the provisions of title 5, relating to Federal employees.

(d) Adoption of existing labor agreements.—The Office shall adopt all labor agreements which are in effect, as of the day before the effective date of the Patent and Trademark Office Efficiency Act, with respect to such Office (as then in effect).

(e) Carryover of personnel.—

(1) From PTO.—Effective as of the effective date of the Patent and Trademark Office Efficiency Act, all officers and employees of the Patent and Trademark Office on the day before such effective date shall become officers and employees of the Office, without a break in service.

(2) Other personnel.—Any individual who, on the day before the effective date of the Patent and Trademark Office Efficiency Act, is an officer or employee

of the Department of Commerce (other than an officer or employee under paragraph (1)) shall be transferred to the Office, as necessary to carry out the purposes of that Act, if—

(A) such individual serves in a position for which a major function is the performance of work reimbursed by the Patent and Trademark Office, as determined by the Secretary of Commerce;

(B) such individual serves in a position that performed work in support of the Patent and Trademark Office during at least half of the incumbent's work time, as determined by the Secretary of Commerce; or

(C) such transfer would be in the interest of the Office, as determined by the Secretary of Commerce in consultation with the Director.

Any transfer under this paragraph shall be effective as of the same effective date as referred to in paragraph (1), and shall be made without a break in service

(f) Transition provisions.—

(1) Interim appointment of Director.—On or after the effective date of the Patent and Trademark Office Efficiency Act, the President shall appoint an individual to serve as the Director until the date on which a Director qualifies under subsection (a). The President shall not make more than one such appointment under this subsection.

(2) Continuation in office of certain officers.—(A) The individual serving as the Assistant Commissioner for Patents on the day before the effective date of the Patent and Trademark Office Efficiency Act may serve as the Commissioner for Patents until the date on which a Commissioner for Patents is appointed under subsection (b).

(B) The individual serving as the Assistant Commissioner for Trademarks on the day before the effective date of the Patent and Trademark Office Efficiency Act may serve as the Commissioner for Trademarks until the date on which a Commissioner for Trademarks is appointed under subsection (b).

Sec. 4. Restrictions on Officers and Employees as to Interest in Patents

Officers and employees of the Patent and Trademark Office shall be incapable, during the period of their appointments and for one year thereafter, of applying for a patent and of acquiring, directly or indirectly, except by inheritance or bequest, any patent or any right or interest in any patent, issued or to be issued by the Office. In patents applied for thereafter they shall not be entitled to any priority date earlier than one year after the termination of their appointment.

Sec. 5. Patent and Trademark Office Public Advisory Committees

(a) Establishment of Public Advisory Committees.—

(1) Appointment.—The United States Patent and Trademark Office shall have a Patent Public Advisory Committee and a Trademark Public Advisory

Committee, each of which shall have nine voting members who shall be appointed by the Secretary of Commerce and serve at the pleasure of the Secretary of Commerce. In each year, 3 members shall be appointed to each Advisory Committee for 3-year terms that shall begin on December 1 of that year. Any vacancy on an Advisory Committee shall be filled within 90 days after it occurs. A new member who is appointed to fill a vacancy shall be appointed to serve for the remainder of the predecessor's term.

(2) Chair.—The Secretary of Commerce, in consultation with the Director, shall designate a Chair and Vice Chair of each Advisory Committee from among the members appointed under paragraph (1). If the Chair resigns before the completion of his or her term, or is otherwise unable to exercise the functions of the Chair, the Vice Chair shall exercise the functions of the Chair. (3) Repealed.

(b) Basis for appointments.—Members of each Advisory Committee—

(1) shall be citizens of the United States who shall be chosen so as to represent the interests of diverse users of the United States Patent and Trademark Office with respect to patents, in the case of the Patent Public Advisory Committee, and with respect to trademarks, in the case of the Trademark Public Advisory Committee;

(2) shall include members who represent small and large entity applicants located in the United States in proportion to the number of applications filed by such applicants, but in no case shall members who represent small entity patent applicants, including small business concerns, independent inventors, and nonprofit organizations, constitute less than 25 percent of the members of the Patent Public Advisory Committee, and such members shall include at least one independent inventor; and

(3) shall include individuals with substantial background and achievement in finance, management, labor relations, science, technology, and office automation.

In addition to the voting members, each Advisory Committee shall include a representative of each labor organization recognized by the United States Patent and Trademark Office. Such representatives shall be nonvoting members of the Advisory Committee to which they are appointed.

(c) Meetings.—Each Advisory Committee shall meet at the call of the chair to consider an agenda set by the chair.

(d) Duties.—Each Advisory Committee shall—

(1) review the policies, goals, performance, budget, and user fees of the United States Patent and Trademark Office with respect to patents, in the case of the Patent Public Advisory Committee, and with respect to Trademarks, in the case of the Trademark Public Advisory Committee, and advise the Director on these matters;

(2) within 60 days after the end of each fiscal year—

(A) prepare an annual report on the matters referred to in paragraph (1);

(B) transmit the report to the Secretary of Commerce, the President, and the Committees on the Judiciary of the Senate and the House of Representatives; and

(C) publish the report in the Official Gazette of the United States Patent and Trademark Office.

(e) Compensation.—Each member of each Advisory Committee shall be compensated for each day (including travel time) during which such member is attending meetings or conferences of that Advisory Committee or otherwise engaged in the business of that Advisory Committee, at the rate which is the daily equivalent of the annual rate of basic pay in effect for level III of the Executive Schedule under section 5314 of title 5. While away from such member's home or regular place of business such member shall be allowed travel expenses, including per diem in lieu of subsistence, as authorized by section 5703 of title 5.

(f) Access to information.—Members of each Advisory Committee shall be provided access to records and information in the United States Patent and Trademark Office, except for personnel or other privileged information and information concerning patent applications required to be kept in confidence by section 122.

(g) Applicability of certain ethics laws.—Members of each Advisory Committee shall be special Government employees within the meaning of section 202 of title 18.

(h) Inapplicability of Federal Advisory Committee Act.—The Federal Advisory Committee Act (5 U.S.C. App.) shall not apply to each Advisory Committee.

(i) Open meetings.—The meetings of each Advisory Committee shall be open to the public, except that each Advisory Committee may by majority vote meet in executive session when considering personnel, privileged, or other confidential information.

(j) Inapplicability of patent prohibition.—Section 4 shall not apply to voting members of the Advisory Committees.

Sec. 6. Patent Trial and Appeal Board

(a) In general.—There shall be in the Office a Patent Trial and Appeal Board. The Director, the Deputy Director, the Commissioner for Patents, the Commissioner for Trademarks, and the administrative patent judges shall constitute the Patent Trial and Appeal Board. The administrative patent judges shall be persons of competent legal knowledge and scientific ability who are appointed by the Secretary, in consultation with the Director. Any reference in any Federal law, Executive order, rule, regulation, or delegation of authority, or any document of or pertaining to the Board of Patent Appeals and Interferences is deemed to refer to the Patent Trial and Appeal Board.

(b) Duties.—The Patent Trial and Appeal Board shall—

(1) on written appeal of an applicant, review adverse decisions of examiners upon applications for patents pursuant to section 134(a);

(2) review appeals of reexaminations pursuant to section 134(b);

(3) conduct derivation proceedings pursuant to section 135; and

(4) conduct inter partes reviews and post-grant reviews pursuant to chapters 31 and 32.

(c) 3-member panels.—Each appeal, derivation proceeding, post-grant review, and inter partes review shall be heard by at least 3 members of the Patent Trial and Appeal Board, who shall be designated by the Director. Only the Patent Trial and Appeal Board may grant rehearings.

(d) Treatment of prior appointments.—The Secretary of Commerce may, in the Secretary's discretion, deem the appointment of an administrative patent judge who, before the date of the enactment of this subsection, held office pursuant to an appointment by the Director to take effect on the date on which the Director initially appointed the administrative patent judge. It shall be a defense to a challenge to the appointment of an administrative patent judge on the basis of the judge's having been originally appointed by the Director that the administrative patent judge so appointed was acting as a de facto officer.

Sec. 7. Library

The Director shall maintain a library of scientific and other works and periodicals, both foreign and domestic, in the Patent and Trademark Office to aid the officers in the discharge of their duties.

Sec. 8. Classification of Patents

The Director may revise and maintain the classification by subject matter of United States letters patent, and such other patents and printed publications as may be necessary or practicable, for the purpose of determining with readiness and accuracy the novelty of inventions for which applications for patent are filed.

Sec. 9. Certified Copies of Records

The Director may furnish certified copies of specifications and drawings of patents issued by the Patent and Trademark Office, and of other records available either to the public or to the person applying therefor.

Sec. 10. Publications

(a) The Director may publish in printed, typewritten, or electronic form, the following:

1. Patents and published applications for patents, including specifications and drawings, together with copies of the same. The Patent and Trademark Office may print the headings of the drawings for patents for the purpose of photolithography.

2. Certificates of trade-mark registrations, including statements and drawings, together with copies of the same.

3. The Official Gazette of the United States Patent and Trademark Office.

4. Annual indexes of patents and patentees, and of trade-marks and registrants.

5. Annual volumes of decisions in patent and trade-mark cases.

6. Pamphlet copies of the patent laws and rules of practice, laws and rules relating to trade-marks, and circulars or other publications relating to the business of the Office.

(b) The Director may exchange any of the publications specified in items 3, 4, 5, and 6 of subsection (a) of this section for publications desirable for the use of the Patent and Trademark Office.

Sec. 11. Exchange of Copies of Patents and Applications with Foreign Countries

The Director may exchange copies of specifications and drawings of United States patents and published applications for patents for those of foreign countries. The Director shall not enter into an agreement to provide such copies of specifications and drawings of United States patents and applications to a foreign country, other than a NAFTA country or a WTO member country, without the express authorization of the Secretary of Commerce. For purposes of this section, the terms "NAFTA country" and "WTO member country" have the meanings given those terms in section 104(b).

Sec. 12. Copies of Patents and Applications for Public Libraries

The Director may supply copies of specifications and drawings of patents and published applications for patents in printed or electronic form to public libraries in the United States which shall maintain such copies for the use of the public, at the rate for each year's issue established for this purpose in section 41(d) of this title.

Sec. 13. Annual Report to Congress

The Director shall report to the Congress, not later than 180 days after the end of each fiscal year, the moneys received and expended by the Office, the purposes for which the moneys were spent, the quality and quantity of the work of the Office, the nature of training provided to examiners, the evaluation of the Commissioner of Patents and the Commissioner of Trademarks by the Secretary of Commerce, the compensation of the Commissioners, and other information relating to the Office.

Chapter 2. Proceedings in the Patent and Trademark Office

Sec. 21. Filing Date and Day for Taking Action

(a) The Director may by rule prescribe that any paper or fee required to be filed in the Patent and Trademark Office will be considered filed in the Office on the date on which it was deposited with the United States Postal Service or would have been deposited with the United States Postal Service but for postal service interruptions or emergencies designated by the Director.

(b) When the day, or the last day, for taking any action or paying any fee in the United States Patent and Trademark Office falls on Saturday, Sunday, or a federal holiday within the District of Columbia, the action may be taken, or the fee paid, on the next succeeding secular or business day.

Sec. 22. Printing of Papers Filed

The Director may require papers filed in the Patent and Trademark Office to be printed, typewritten, or on an electronic medium.

Sec. 23. Testimony in Patent and Trademark Office Cases

The Director may establish rules for taking affidavits and depositions required in cases in the Patent and Trademark Office. Any officer authorized by law to take depositions to be used in the courts of the United States, or of the State where he resides, may take such affidavits and depositions.

Sec. 24. Subpoenas, Witnesses

The clerk of any United States court for the district wherein testimony is to be taken for use in any contested case in the Patent and Trademark Office, shall, upon the application of any party thereto, issue a subpoena for any witness residing or being within such district, commanding him to appear and testify before an officer in such district authorized to take depositions and affidavits, at the time and place stated in the subpoena. The provisions of the Federal Rules of Civil Procedure relating to the attendance of witnesses and to the production of documents and things shall apply to contested cases in the Patent and Trademark Office.

Every witness subpoenaed and in attendance shall be allowed the fees and traveling expenses allowed to witnesses attending the United States district courts.

A judge of a court whose clerk issued a subpoena may enforce obedience to the process or punish disobedience as in other like cases, on proof that a witness, served with such subpoena, neglected or refused to appear or to testify. No witness shall be deemed guilty of contempt for disobeying such subpoena unless his fees and traveling expenses in going to, and returning from, and one day's attendance at the place of examination, are paid or tendered him at the time of the service of the subpoena; nor for refusing to disclose any secret matter except upon appropriate order of the court which issued the subpoena.

Sec. 25. Declaration in Lieu of Oath

(a) The Director may by rule prescribe that any document to be filed in the Patent and Trademark Office and which is required by any law, rule, or other regulation to be under oath may be subscribed to by a written declaration in such form as the Director may prescribe, such declaration to be in lieu of the oath otherwise required.

(b) Whenever such written declaration is used, the document must warn the declarant that willful false statements and the like are punishable by fine or imprisonment, or both (18 U.S.C. 1001).

Sec. 26. Effect of Defective Execution

Any document to be filed in the Patent and Trademark Office and which is required by any law, rule, or other regulation to be executed in a specified manner may be provisionally accepted by the Director despite a defective execution, provided a properly executed document is submitted within such time as may be prescribed.

Chapter 3. Practice Before Patent and Trademark Office

Sec. 32. Suspension or Exclusion from Practice

The Director may, after notice and opportunity for a hearing, suspend or exclude, either generally or in any particular case, from further practice before the Patent and Trademark Office, any person, agent, or attorney shown to be incompetent or disreputable, or guilty of gross misconduct, or who does not comply with the regulations established under section 2(b)(2)(D), or who shall, by word, circular, letter, or advertising, with intent to defraud in any manner, deceive, mislead, or threaten any applicant or prospective applicant, or other person having immediate or prospective business before the Office. The reasons for any such suspension or exclusion shall be duly recorded. The Director shall have the discretion to designate any attorney who is an officer or employee of the United States Patent and Trademark Office to conduct the hearing required by this section. A proceeding under this section shall be commenced not later than the earlier of either the date that is 10 years after the date on which the misconduct forming the basis for the proceeding occurred, or 1 year after the date on which the misconduct forming the basis for the proceeding is made known to an officer or employee of the Office as prescribed in the regulations established under section 2(b)(2)(D). The United States District Court for the Eastern District of Virginia, under such conditions and upon such proceedings as it by its rules determines, may review the action of the Director upon the petition of the person so refused recognition or so suspended or excluded.

Sec. 33. Unauthorized Representation as Practitioner

Whoever, not being recognized to practice before the Patent and Trademark Office, holds himself out or permits himself to be held out as so recognized, or as being qualified to prepare or prosecute applications for patent, shall be fined not more than $1,000 for each offense.

Chapter 4. Patent Fees; Funding; Search Systems

Sec. 41. Patent Fees; Patent and Trademark Search Systems

(a) General fees.—The Director shall charge the following fees:

(1) Filing and basic national fees.—

(A) On filing each application for an original patent, except for design, plant, or provisional applications, $330.

(B) On filing each application for an original design patent, $220.

(C) On filing each application for an original plant patent, $220.

(D) On filing each provisional application for an original patent, $ 220.

(E) On filing each application for the reissue of a patent, $330.

(F) The basic national fee for each international application filed under the treaty defined in section 351(a) entering the national stage under section 371, $330.

(G) In addition, excluding any sequence listing or computer program listing filed in an electronic medium as prescribed by the Director, for any application the specification and drawings of which exceed 100 sheets of paper (or equivalent as prescribed by the Director if filed in an electronic medium), $270 for each additional 50 sheets of paper (or equivalent as prescribed by the Director if filed in an electronic medium) or fraction thereof.

(2) Excess claims fees.—

(A) In general.—In addition to the fee specified in paragraph (1)—

(i) on filing or on presentation at any other time, $220 for each claim in independent form in excess of 3;

(ii) on filing or on presentation at any other time, $52 for each claim (whether dependent or independent) in excess of 20; and

(iii) for each application containing a multiple dependent claim, $390.

(B) Multiple dependent claims.—For the purpose of computing fees under subparagraph (A), a multiple dependent claim referred to in section 112 or any claim depending therefrom shall be considered as separate dependent claims in accordance with the number of claims to which reference is made.

(C) Refunds; errors in payment.—The Director may by regulation provide for a refund of any part of the fee specified in subparagraph (A) for any claim that is canceled before an examination on the merits, as prescribed by the Director, has been made of the application under section 131. Errors in payment of the additional fees under this paragraph may be rectified in accordance with regulations prescribed by the Director.

(3) Examination fees.—

(A) In general.—

(i) For examination of each application for an original patent, except for design, plant, provisional, or international applications, $220.

(ii) For examination of each application for an original design patent, $140.

(iii) For examination of each application for an original plant patent, $170.

(iv) For examination of the national stage of each international application, $220.

(v) For examination of each application for the reissue of a patent, $650.

(B) Applicability of other fee provisions.—The provisions of paragraphs (3) and (4) of section 111(a) relating to the payment of the fee for filing the application shall apply to the payment of the fee specified in subparagraph (A) with respect to an application filed under section 111(a). The provisions of section 371(d) relating to the payment of the national fee shall apply to the payment of the fee specified in subparagraph (A) with respect to an international application.

(4) Issue fees.—

(A) For issuing each original patent, except for design or plant patents, $1,510.

(B) For issuing each original design patent, $860.

(C) For issuing each original plant patent, $1,190.

(D) For issuing each reissue patent, $1,510.

(5) Disclaimer fee.—On filing each disclaimer, $140.

(6) Appeal fees.—

(A) On filing an appeal from the examiner to the Patent Trial and Appeal Board, $540.

(B) In addition, on filing a brief in support of the appeal, $540, and on requesting an oral hearing in the appeal before the Patent Trial and Appeal Board, $1,080.

(7) Revival fees.—On filing each petition for the revival of an unintentionally abandoned application for a patent, for the unintentionally delayed payment of the fee for issuing each patent, or for an unintentionally delayed response by the patent owner in any reexamination proceeding, $1,620, unless the petition is filed under section 133 or 151, in which case the fee shall be $540.

(8) Extension fees.—For petitions for 1-month extensions of time to take actions required by the Director in an application—

(A) on filing a first petition, $130;

(B) on filing a second petition, $360; and

(C) on filing a third or subsequent petition, $620.

(b) Maintenance fees.—

(1) In general.—The Director shall charge the following fees for maintaining in force all patents based on applications filed on or after December 12, 1980:

(A) Three years and 6 months after grant, $980.

(B) Seven years and 6 months after grant, $2,480.

(C) Eleven years and 6 months after grant, $4,110.

(2) Grace period; surcharge.—Unless payment of the applicable maintenance fee under paragraph (1) is received in the Office on or before the date the fee is due or within a grace period of 6 months thereafter, the patent shall expire as of the end of such grace period. The Director may require the payment of a surcharge as a condition of accepting within such 6-month grace period the payment of an applicable maintenance fee.

(3) No maintenance fee for design or plant patent.—No fee may be established for maintaining a design or plant patent in force.

(c) Delays in payment of maintenance fees.—

(1) Acceptance.—The Director may accept the payment of any maintenance fee required by subsection (b) of this section which is made within twenty-four months after the six-month grace period if the delay is shown to the satisfaction of the Director to have been unintentional, or at any time after the six-month grace period if the delay is shown to the satisfaction of the Director to have been unavoidable. The Director may require the payment of a surcharge as a condition of accepting payment of any maintenance fee after the six-month grace period. If the Director accepts payment of a maintenance fee after the six-month grace period, the patent shall be considered as not having expired at the end of the grace period.

(2) Effect on rights of others.—A patent, the term of which has been maintained as a result of the acceptance of a payment of a maintenance fee under this subsection, shall not abridge or affect the right of any person or that person's successors in business who made, purchased, offered to sell, or used anything protected by the patent within the United States, or imported anything protected by the patent into the United States after the 6-month grace period but prior to the acceptance of a maintenance fee under this subsection, to continue the use of, to offer for sale, or to sell to others to be used, offered for sale, or sold, the specific thing so made, purchased, offered for sale, used, or imported. The court before which such matter is in question may provide for the continued manufacture, use, offer for sale, or sale of the thing made, purchased, offered for sale, or used within the United States, or imported into the United States, as specified, or for the manufacture, use, offer for sale, or sale in the United States of which substantial preparation was made after the 6-month grace period but before the acceptance of a maintenance fee under this subsection, and the court may also provide for the continued practice of any process that is practiced, or for the practice of which substantial preparation was made, after the 6-month grace period but before the acceptance of a maintenance fee under this subsection, to the extent and under such terms as the court deems equitable for the protection of investments made or business commenced after the 6-month grace period but before the acceptance of a maintenance fee under this subsection.

(d) Patent search and other fees.—

(1) Patent search fees.—

(A) In general.—The Director shall charge the fees specified under subparagraph (B) for the search of each application for a patent, except for provisional applications. The Director shall adjust the fees charged under this paragraph to ensure that the fees recover an amount not to exceed the estimated average cost to the Office of searching applications for patent by Office personnel.

(B) Specific fees.—The fees referred to in subparagraph (A) are—

(i) $540 for each application for an original patent, except for design, plant, provisional, or international applications;

(ii) $100 for each application for an original design patent;

(iii) $330 for each application for an original plant patent;

(iv) $540 for the national stage of each international application; and

(v) $540 for each application for the reissue of a patent.

(C) Applicability of other provisions.—The provisions of paragraphs (3) and (4) of section 111(a) relating to the payment of the fee for filing the application shall apply to the payment of the fee specified in this paragraph with respect to an application filed under section 111(a). The provisions of section 371(d) relating to the payment of the national fee shall apply to the payment of the fee specified in this paragraph with respect to an international application.

(D) Refunds.—The Director may by regulation provide for a refund of any part of the fee specified in this paragraph for any applicant who files a written declaration of express abandonment as prescribed by the Director before an examination has been made of the application under section 131.

(2) Other fees.—

(A) In general.—The Director shall establish fees for all other processing, services, or materials relating to patents not specified in this section to recover the estimated average cost to the Office of such processing, services, or materials, except that the Director shall charge the following fees for the following services:

(i) For recording a document affecting title, $40 per property.

(ii) For each photocopy, $.25 per page.

(iii) For each black and white copy of a patent, $3.

(B) Copies for libraries.—The yearly fee for providing a library specified in section 12 with uncertified printed copies of the specifications and drawings for all patents in that year shall be $50.

(e) Waiver of fees; copies regarding notice.—The Director may waive the payment of any fee for any service or material related to patents in connection with an occasional or incidental request made by a department or agency of the Government, or any officer thereof. The Director may provide any applicant issued a notice under section 132 with a copy of the specifications and drawings for all patents referred to in that notice without charge.

(f) Adjustment of fees.—The fees established in subsections (a) and (b) of this section may be adjusted by the Director on October 1, 1992, and every year thereafter, to reflect any fluctuations occurring during the previous 12 months in the Consumer Price Index, as determined by the Secretary of Labor. Changes of less than 1 per centum may be ignored.

(g) Repealed. Pub.L. 112-29, § 11(e)(3), Sept. 16, 2011, 125 Stat. 323

(h) Fees for small entities.—

(1) Reductions in fees.—Subject to paragraph (3), fees charged under subsections (a), (b), and (d)(1) shall be reduced by 50 percent with respect to their application to any small business concern as defined under section 3 of the Small Business Act, and to any independent inventor or nonprofit organization as defined in regulations issued by the Director.

(2) Surcharges and other fees.—With respect to its application to any entity described in paragraph (1), any surcharge or fee charged under subsection (c) or (d) shall not be higher than the surcharge or fee required of any other entity under the same or substantially similar circumstances.

(3) Reduction for electronic filing.—The fee charged under subsection (a)(1)(A) shall be reduced by 75 percent with respect to its application to any entity to which paragraph (1) applies, if the application is filed by electronic means as prescribed by the Director.

(i) Electronic patent and trademark data.—

(1) Maintenance of collections.—The Director shall maintain, for use by the public, paper, microform, or electronic collections of United States patents, foreign patent documents, and United States trademark registrations arranged to permit search for and retrieval of information. The Director may not impose fees directly for the use of such collections, or for the use of the public patent or trademark search rooms or libraries.

(2) Availability of automated search systems.—The Director shall provide for the full deployment of the automated search systems of the Patent and Trademark Office so that such systems are available for use by the public, and shall assure full access by the public to, and dissemination of, patent and trademark information, using a variety of automated methods, including electronic bulletin boards and remote access by users to mass storage and retrieval systems.

(3) Access fees.—The Director may establish reasonable fees for access by the public to the automated search systems of the Patent and Trademark Office. If such fees are established, a limited amount of free access shall be made available to users of the systems for purposes of education and training. The Director may waive the payment by an individual of fees authorized by this subsection upon a showing of need or hardship, and if such a waiver is in the public interest.

(4) Annual report to Congress.—The Director shall submit to the Congress an annual report on the automated search systems of the Patent and Trademark Office and the access by the public to such systems. The Director shall also publish such report in the Federal Register. The Director shall provide an opportunity for the submission of comments by interested persons on each such report.

Sec. 42. Patent and Trademark Office Funding

(a) All fees for services performed by or materials furnished by the Patent and Trademark Office will be payable to the Director.

(b) All fees paid to the Director and all appropriations for defraying the costs of the activities of the Patent and Trademark Office will be credited to the Patent and Trademark Office Appropriation Account in the Treasury of the United States.

(c)(1) To the extent and in the amounts provided in advance in appropriations Acts, fees authorized in this title or any other Act to be charged or established by the Director shall be collected by and shall, subject to paragraph (3), be available to the Director to carry out the activities of the Patent and Trademark Office.

(2) There is established in the Treasury a Patent and Trademark Fee Reserve Fund. If fee collections by the Patent and Trademark Office for a fiscal year exceed the amount appropriated to the Office for that fiscal year, fees collected in excess of the appropriated amount shall be deposited in the Patent and Trademark Fee Reserve Fund. To the extent and in the amounts provided in appropriations Acts, amounts in the Fund shall be made available until expended only for obligation and expenditure by the Office in accordance with paragraph (3).

(3)(A) Any fees that are collected under this title, and any surcharges on such fees, may only be used for expenses of the Office relating to the processing of patent applications and for other activities, services, and materials relating to patents and to cover a proportionate share of the administrative costs of the Office.

(B) Any fees that are collected under section 31 of the Trademark Act of 1946, and any surcharges on such fees, may only be used for expenses of the Office relating to the processing of trademark registrations and for other activities, services, and materials relating to trademarks and to cover a proportionate share of the administrative costs of the Office.

(d) The Director may refund any fee paid by mistake or any amount paid in excess of that required.

(e) The Secretary of Commerce shall, on the day each year on which the President submits the annual budget to the Congress, provide to the Committees on the Judiciary of the Senate and the House of Representatives—

(1) a list of patent and trademark fee collections by the Patent and Trademark Office during the preceding fiscal year;

(2) a list of activities of the Patent and Trademark Office during the preceding fiscal year which were supported by patent fee expenditures, trademark fee expenditures, and appropriations;

(3) budget plans for significant programs, projects, and activities of the Office, including out-year funding estimates;

(4) any proposed disposition of surplus fees by the Office; and

(5) such other information as the committees consider necessary.

Part II. Patentability of Inventions and Grant of Patents

Chapter 10. Patentability of Inventions (Refs & Annos)

Sec. 100. Definitions

When used in this title unless the context otherwise indicates—

(a) The term "invention" means invention or discovery.

(b) The term "process" means process, art or method, and includes a new use of a known process, machine, manufacture, composition of matter, or material.

(c) The terms "United States" and "this country" mean the United States of America, its territories and possessions.

(d) The word "patentee" includes not only the patentee to whom the patent was issued but also the successors in title to the patentee.

(e) The term "third-party requester" means a person requesting ex parte reexamination under section 302 or inter partes reexamination under section 311 who is not the patent owner.

Sec. 101. Inventions Patentable

Whoever invents or discovers any new and useful process, machine, manufacture, or composition of matter, or any new and useful improvement thereof, may obtain a patent therefor, subject to the conditions and requirements of this title.

Sec. 102. Conditions for patentability; novelty

(a) Novelty; prior art.—A person shall be entitled to a patent unless—

(1) the claimed invention was patented, described in a printed publication, or in public use, on sale, or otherwise available to the public before the effective filing date of the claimed invention; or

(2) the claimed invention was described in a patent issued under section 151, or in an application for patent published or deemed published under section 122(b), in which the patent or application, as the case may be, names another inventor and was effectively filed before the effective filing date of the claimed invention.

(b) Exceptions.—

(1) Disclosures made 1 year or less before the effective filing date of the claimed invention.—A disclosure made 1 year or less before the effective filing date of a claimed invention shall not be prior art to the claimed invention under subsection (a)(1) if—

(A) the disclosure was made by the inventor or joint inventor or by another who obtained the subject matter disclosed directly or indirectly from the inventor or a joint inventor; or

(B) the subject matter disclosed had, before such disclosure, been publicly disclosed by the inventor or a joint inventor or another who obtained the subject matter disclosed directly or indirectly from the inventor or a joint inventor.

(2) Disclosures appearing in applications and patents.—A disclosure shall not be prior art to a claimed invention under subsection (a)(2) if—

(A) the subject matter disclosed was obtained directly or indirectly from the inventor or a joint inventor;

(B) the subject matter disclosed had, before such subject matter was effectively filed under subsection (a)(2), been publicly disclosed by the inventor or a joint inventor or another who obtained the subject matter disclosed directly or indirectly from the inventor or a joint inventor; or

(C) the subject matter disclosed and the claimed invention, not later than the effective filing date of the claimed invention, were owned by the same person or subject to an obligation of assignment to the same person.

(c) Common ownership under joint research agreements.—Subject matter disclosed and a claimed invention shall be deemed to have been owned by the same person or subject to an obligation of assignment to the same person in applying the provisions of subsection (b)(2)(C) if—

(1) the subject matter disclosed was developed and the claimed invention was made by, or on behalf of, 1 or more parties to a joint research agreement that was in effect on or before the effective filing date of the claimed invention;

(2) the claimed invention was made as a result of activities undertaken within the scope of the joint research agreement; and

(3) the application for patent for the claimed invention discloses or is amended to disclose the names of the parties to the joint research agreement.

(d) Patents and published applications effective as prior art.—For purposes of determining whether a patent or application for patent is prior art to a claimed invention under subsection (a)(2), such patent or application shall be considered to have been effectively filed, with respect to any subject matter described in the patent or application—

(1) if paragraph (2) does not apply, as of the actual filing date of the patent or the application for patent; or

(2) if the patent or application for patent is entitled to claim a right of priority under section 119, 365(a), or 365(b), or to claim the benefit of an earlier filing date under section 120, 121, or 365(c), based upon 1 or more prior filed applications for patent, as of the filing date of the earliest such application that describes the subject matter.

Sec. 103. Conditions for Patentability; Non-Obvious Subject Matter

A patent for a claimed invention may not be obtained, notwithstanding that the claimed invention is not identically disclosed as set forth in section 102, if the differences between the claimed invention and the prior art are such that the claimed invention as a whole would have been obvious before the effective filing date of the claimed invention to a person having ordinary skill in the art to which the claimed invention pertains. Patentability shall not be negated by the manner in which the invention was made.

Sec. 104. Repealed

Sec. 105. Inventions in Outer Space

(a) Any invention made, used or sold in outer space on a space object or component thereof under the jurisdiction or control of the United States shall be considered to be made, used or sold within the United States for the purposes of this title, except with respect to any space object or component thereof that is specifically identified and otherwise provided for by an international agreement to which the United States is a party, or with respect to any space object or component thereof that is carried on the registry of a foreign state in accordance with the Convention on Registration of Objects Launched into Outer Space.

(b) Any invention made, used or sold in outer space on a space object or component thereof that is carried on the registry of a foreign state in accordance with the Convention on Registration of Objects Launched into Outer Space, shall be considered to be made, used or sold within the United States for the purposes of this title if specifically so agreed in an international agreement between the United States and the state of registry.

Chapter 11. Application for Patent

Sec. 111. Application

(a) In general.—

(1) Written application.—An application for patent shall be made, or authorized to be made, by the inventor, except as otherwise provided in this title, in writing to the Director.

(2) Contents.—Such application shall include—

(A) a specification as prescribed by section 112;

(B) a drawing as prescribed by section 113; and

(C) an oath or declaration as prescribed by section 115.

(3) Fee and oath or declaration.—The application must be accompanied by the fee required by law. The fee and oath or declaration may be submitted after the specification and any required drawing are submitted, within such period and under such conditions, including the payment of a surcharge, as may be prescribed by the Director.

(4) Failure to submit.—Upon failure to submit the fee and oath or declaration within such prescribed period, the application shall be regarded as abandoned, unless it is shown to the satisfaction of the Director that the delay in submitting the fee and oath or declaration was unavoidable or unintentional. The filing date of an application shall be the date on which the specification and any required drawing are received in the Patent and Trademark Office.

(b) Provisional application.—

(1) Authorization.—A provisional application for patent shall be made or authorized to be made by the inventor, except as otherwise provided in this title, in writing to the Director. Such application shall include—

(A) a specification as prescribed by section 112(a); and

(B) a drawing as prescribed by section 113.

(2) Claim.—A claim, as required by subsections (b) through (e) of section 112, shall not be required in a provisional application.

(3) Fee.—(A) The application must be accompanied by the fee required by law.

(B) The fee may be submitted after the specification and any required drawing are submitted, within such period and under such conditions, including the payment of a surcharge, as may be prescribed by the Director.

(C) Upon failure to submit the fee within such prescribed period, the application shall be regarded as abandoned, unless it is shown to the satisfaction of the Director that the delay in submitting the fee was unavoidable or unintentional.

(4) Filing date.—The filing date of a provisional application shall be the date on which the specification and any required drawing are received in the Patent and Trademark Office.

(5) Abandonment.—Notwithstanding the absence of a claim, upon timely request and as prescribed by the Director, a provisional application may be treated as an application filed under subsection (a). Subject to section 119(e)(3), if no such request is made, the provisional application shall be regarded as abandoned 12 months after the filing date of such application and shall not be subject to revival after such 12-month period.

(6) Other basis for provisional application.—Subject to all the conditions in this subsection and section 119(e), and as prescribed by the Director, an application for patent filed under subsection (a) may be treated as a provisional application for patent.

(7) No right of priority or benefit of earliest filing date.—A provisional application shall not be entitled to the right of priority of any other application under section 119 or 365(a) or to the benefit of an earlier filing date in the United States under section 120, 121, or 365(c).

(8) Applicable provisions.—The provisions of this title relating to applications for patent shall apply to provisional applications for patent, except as otherwise provided, and except that provisional applications for patent shall not be subject to sections 131 and 135.

Sec. 112. Specification

(a) In general.—The specification shall contain a written description of the invention, and of the manner and process of making and using it, in such full, clear, concise, and exact terms as to enable any person skilled in the art to which it pertains, or with which it is most nearly connected, to make and use the same, and shall set forth the best mode contemplated by the inventor or joint inventor of carrying out the invention.

(b) Conclusion.—The specification shall conclude with one or more claims particularly pointing out and distinctly claiming the subject matter which the inventor or a joint inventor regards as the invention.

(c) Form.—A claim may be written in independent or, if the nature of the case admits, in dependent or multiple dependent form.

(d) Reference in dependent forms.—Subject to subsection (e), a claim in dependent form shall contain a reference to a claim previously set forth and then specify a further limitation of the subject matter claimed. A claim in dependent form shall be construed to incorporate by reference all the limitations of the claim to which it refers.

(e) Reference in multiple dependent form.—A claim in multiple dependent form shall contain a reference, in the alternative only, to more than one claim previously set forth and then specify a further limitation of the subject matter claimed. A multiple dependent claim shall not serve as a basis for any other multiple dependent claim. A multiple dependent claim shall be construed to incorporate by reference all the limitations of the particular claim in relation to which it is being considered.

(f) Element in claim for a combination.—An element in a claim for a combination may be expressed as a means or step for performing a specified function without the recital of structure, material, or acts in support thereof, and such claim shall be construed to cover the corresponding structure, material, or acts described in the specification and equivalents thereof.

Sec. 113. Drawings

The applicant shall furnish a drawing where necessary for the understanding of the subject matter sought to be patented. When the nature of such subject matter admits of illustration by a drawing and the applicant has not furnished such a drawing, the Director may require its submission within a time period of not less than two months from the sending of a notice thereof. Drawings submitted after the filing date of the application may not be used (i) to overcome any insufficiency of the specification due to lack of an enabling disclosure or otherwise inadequate disclosure therein, or (ii) to supplement the original disclosure thereof for the purpose of interpretation of the scope of any claim.

Sec. 114. Models, Specimens

The Director may require the applicant to furnish a model of convenient size to exhibit advantageously the several parts of his invention.

When the invention relates to a composition of matter, the Director may require the applicant to furnish specimens or ingredients for the purpose of inspection or experiment.

Sec. 115. Inventor's Oath or Declaration

(a) Naming the inventor; inventor's oath or declaration.—An application for patent that is filed under section 111(a) or commences the national stage under section 371 shall include, or be amended to include, the name of the inventor for any invention claimed in the application. Except as otherwise provided in this section, each individual who is the inventor or a joint inventor of a claimed invention in an application for patent shall execute an oath or declaration in connection with the application.

(b) Required statements.—An oath or declaration under subsection (a) shall contain statements that—

(1) the application was made or was authorized to be made by the affiant or declarant; and

(2) such individual believes himself or herself to be the original inventor or an original joint inventor of a claimed invention in the application.

(c) Additional requirements.—The Director may specify additional information relating to the inventor and the invention that is required to be included in an oath or declaration under subsection (a).

(d) Substitute statement.—

(1) In general.—In lieu of executing an oath or declaration under subsection (a), the applicant for patent may provide a substitute statement under the circumstances described in paragraph (2) and such additional circumstances that the Director may specify by regulation.

(2) Permitted circumstances.—A substitute statement under paragraph (1) is permitted with respect to any individual who—

(A) is unable to file the oath or declaration under subsection (a) because the individual—

(i) is deceased;

(ii) is under legal incapacity; or

(iii) cannot be found or reached after diligent effort; or

(B) is under an obligation to assign the invention but has refused to make the oath or declaration required under subsection (a).

(3) Contents.—A substitute statement under this subsection shall—

(A) identify the individual with respect to whom the statement applies;

(B) set forth the circumstances representing the permitted basis for the filing of the substitute statement in lieu of the oath or declaration under subsection (a); and

(C) contain any additional information, including any showing, required by the Director.

(e) Making required statements in assignment of record.—An individual who is under an obligation of assignment of an application for patent may include the required statements under subsections (b) and (c) in the assignment executed by the individual, in lieu of filing such statements separately.

(f) Time for filing.—The applicant for patent shall provide each required oath or declaration under subsection (a), substitute statement under subsection (d), or recorded assignment meeting the requirements of subsection (e) no later than the date on which the issue fee for the patent is paid.

(g) Earlier-filed application containing required statements or substitute statement.—

(1) Exception.—The requirements under this section shall not apply to an individual with respect to an application for patent in which the individual is

named as the inventor or a joint inventor and that claims the benefit under section 120, 121, or 365(c) of the filing of an earlier-filed application, if—

(A) an oath or declaration meeting the requirements of subsection (a) was executed by the individual and was filed in connection with the earlier-filed application;

(B) a substitute statement meeting the requirements of subsection (d) was filed in connection with the earlier filed application with respect to the individual; or

(C) an assignment meeting the requirements of subsection (e) was executed with respect to the earlier-filed application by the individual and was recorded in connection with the earlier-filed application.

(2) Copies of oaths, declarations, statements, or assignments.—Notwithstanding paragraph (1), the Director may require that a copy of the executed oath or declaration, the substitute statement, or the assignment filed in connection with the earlier-filed application be included in the later-filed application.

(h) Supplemental and corrected statements; filing additional statements.—

(1) In general.—Any person making a statement required under this section may withdraw, replace, or otherwise correct the statement at any time. If a change is made in the naming of the inventor requiring the filing of 1 or more additional statements under this section, the Director shall establish regulations under which such additional statements may be filed.

(2) Supplemental statements not required.—If an individual has executed an oath or declaration meeting the requirements of subsection (a) or an assignment meeting the requirements of subsection (e) with respect to an application for patent, the Director may not thereafter require that individual to make any additional oath, declaration, or other statement equivalent to those required by this section in connection with the application for patent or any patent issuing thereon.

(3) Savings clause.—A patent shall not be invalid or unenforceable based upon the failure to comply with a requirement under this section if the failure is remedied as provided under paragraph (1).

(i) Acknowledgment of penalties.—Any declaration or statement filed pursuant to this section shall contain an acknowledgment that any willful false statement made in such declaration or statement is punishable under section 1001 of title 18 by fine or imprisonment of not more than 5 years, or both.

Sec. 116. Inventors

(a) Joint inventions.—When an invention is made by two or more persons jointly, they shall apply for patent jointly and each make the required oath, except as otherwise provided in this title. Inventors may apply for a patent jointly even though (1) they did not physically work together or at the same time, (2) each did not make the same type or amount of contribution, or (3) each did not make a contribution to the subject matter of every claim of the patent.

(b) Omitted inventor.—If a joint inventor refuses to join in an application for patent or cannot be found or reached after diligent effort, the application may be made by the other inventor on behalf of himself and the omitted inventor. The Director, on proof of the pertinent facts and after such notice to the omitted inventor as he prescribes, may grant a patent to the inventor making the application, subject to the same rights which the omitted inventor would have had if he had been joined. The omitted inventor may subsequently join in the application.

(c) Correction of errors in application—Whenever through error a person is named in an application for patent as the inventor, or through error an inventor is not named in an application, the Director may permit the application to be amended accordingly, under such terms as he prescribes.

Sec. 117. Death or Incapacity of Inventor

Legal representatives of deceased inventors and of those under legal incapacity may make application for patent upon compliance with the requirements and on the same terms and conditions applicable to the inventor.

Sec. 118. Filing by Other than Inventor

A person to whom the inventor has assigned or is under an obligation to assign the invention may make an application for patent. A person who otherwise shows sufficient proprietary interest in the matter may make an application for patent on behalf of and as agent for the inventor on proof of the pertinent facts and a showing that such action is appropriate to preserve the rights of the parties. If the Director grants a patent on an application filed under this section by a person other than the inventor, the patent shall be granted to the real party in interest and upon such notice to the inventor as the Director considers to be sufficient.

Sec. 119. Benefit of Earlier Filing Date; Right of Priority

(a) An application for patent for an invention filed in this country by any person who has, or whose legal representatives or assigns have, previously regularly filed an application for a patent for the same invention in a foreign country which affords similar privileges in the case of applications filed in the United States or to citizens of the United States, or in a WTO member country, shall have the same effect as the same application would have if filed in this country on the date on which the application for patent for the same invention was first filed in such foreign country, if the application in this country is filed within twelve months from the earliest date on which such foreign application was filed.

(b)(1) No application for patent shall be entitled to this right of priority unless a claim is filed in the Patent and Trademark Office, identifying the foreign application by specifying the application number on that foreign application, the intellectual property authority or country in or for which the application was filed, and the date of filing the application, at such time during the pendency of the application as required by the Director.

(2) The Director may consider the failure of the applicant to file a timely claim for priority as a waiver of any such claim. The Director may establish procedures, including the payment of a surcharge, to accept an unintentionally delayed claim under this section.

(3) The Director may require a certified copy of the original foreign application, specification, and drawings upon which it is based, a translation if not in the English language, and such other information as the Director considers necessary. Any such certification shall be made by the foreign intellectual property authority in which the foreign application was filed and show the date of the application and of the filing of the specification and other papers.

(c) In like manner and subject to the same conditions and requirements, the right provided in this section may be based upon a subsequent regularly filed application in the same foreign country instead of the first filed foreign application, provided that any foreign application filed prior to such subsequent application has been withdrawn, abandoned, or otherwise disposed of, without having been laid open to public inspection and without leaving any rights outstanding, and has not served, nor thereafter shall serve, as a basis for claiming a right of priority.

(d) Applications for inventors' certificates filed in a foreign country in which applicants have a right to apply, at their discretion, either for a patent or for an inventor's certificate shall be treated in this country in the same manner and have the same effect for purpose of the right of priority under this section as applications for patents, subject to the same conditions and requirements of this section as apply to applications for patents, provided such applicants are entitled to the benefits of the Stockholm Revision of the Paris Convention at the time of such filing.

(e)(1) An application for patent filed under section 111(a) or section 363 for an invention disclosed in the manner provided by section 112(a) (other than the requirement to disclose the best mode) in a provisional application filed under section 111(b), by an inventor or inventors named in the provisional application, shall have the same effect, as to such invention, as though filed on the date of the provisional application filed under section 111(b), if the application for patent filed under section 111(a) or section 363 is filed not later than 12 months after the date on which the provisional application was filed and if it contains or is amended to contain a specific reference to the provisional application. No application shall be entitled to the benefit of an earlier filed provisional application under this subsection unless an amendment containing the specific reference to the earlier filed provisional application is submitted at such time during the pendency of the application as required by the Director. The Director may consider the failure to submit such an amendment within that time period as a waiver of any benefit under this subsection. The Director may establish procedures, including the payment of a surcharge, to accept an unintentionally delayed submission of an amendment under this subsection during the pendency of the application.

(2) A provisional application filed under section 111(b) of this title may not be relied upon in any proceeding in the Patent and Trademark Office unless the fee set forth in subparagraph (A) or (C) of section 41(a)(1) of this title has been paid.

(3) If the day that is 12 months after the filing date of a provisional application falls on a Saturday, Sunday, or Federal holiday within the District of Columbia, the period of pendency of the provisional application shall be extended to the next succeeding secular or business day.

(f) Applications for plant breeder's rights filed in a WTO member country (or in a foreign UPOV Contracting Party) shall have the same effect for the purpose of the right of priority under subsections (a) through (c) of this section as applications for patents, subject to the same conditions and requirements of this section as apply to applications for patents.

(g) As used in this section—

(1) the term "WTO member country" has the same meaning as the term is defined in section 104(b)(2) of this title; and

(2) the term "UPOV Contracting Party" means a member of the International Convention for the Protection of New Varieties of Plants.

Sec. 120. Benefit of Earlier Filing Date in the United States

An application for patent for an invention disclosed in the manner provided by section 112(a) (other than the requirement to disclose the best mode) in an application previously filed in the United States, or as provided by section 363 of this title, which names an inventor or joint inventor in the previously filed application shall have the same effect, as to such invention, as though filed on the date of the prior application, if filed before the patenting or abandonment of or termination of proceedings on the first application or on an application similarly entitled to the benefit of the filing date of the first application and if it contains or is amended to contain a specific reference to the earlier filed application. No application shall be entitled to the benefit of an earlier filed application under this section unless an amendment containing the specific reference to the earlier filed application is submitted at such time during the pendency of the application as required by the Director. The Director may consider the failure to submit such an amendment within that time period as a waiver of any benefit under this section. The Director may establish procedures, including the payment of a surcharge, to accept an unintentionally delayed submission of an amendment under this section.

Sec. 121. Divisional Applications

If two or more independent and distinct inventions are claimed in one application, the Director may require the application to be restricted to one of the inventions. If the other invention is made the subject of a divisional application which complies with the requirements of section 120 it shall be entitled to the benefit of the filing date of the original application. A patent issuing on an application with respect to which a requirement for restriction under this section has been made, or on an application filed as a result of such a requirement, shall not be used as a reference either in the Patent and Trademark Office or in the courts against a divisional application or against the original application or any patent issued on either of them, if the divisional application is filed before the issuance of the patent on the other application. The validity of a patent shall not be questioned for failure of the Director to require the application to be restricted to one invention.

Sec. 122. Confidential Status of Applications; Publication of Patent Applications

(a) Confidentiality.—Except as provided in subsection (b), applications for patents shall be kept in confidence by the Patent and Trademark Office and no information concerning the same given without authority of the applicant or owner unless necessary to carry out the provisions of an Act of Congress or in such special circumstances as may be determined by the Director.

(b) Publication—

(1) In general.—(A) Subject to paragraph (2), each application for a patent shall be published, in accordance with procedures determined by the Director, promptly after the expiration of a period of 18 months from the earliest filing date for which a benefit is sought under this title. At the request of the applicant, an application may be published earlier than the end of such 18-month period.

(B) No information concerning published patent applications shall be made available to the public except as the Director determines.

(C) Notwithstanding any other provision of law, a determination by the Director to release or not to release information concerning a published patent application shall be final and nonreviewable.

(2) Exceptions.—(A) An application shall not be published if that application is—

(i) no longer pending;

(ii) subject to a secrecy order under section 181;

(iii) a provisional application filed under section 111(b); or

(iv) an application for a design patent filed under chapter 16.

(B)(i) If an applicant makes a request upon filing, certifying that the invention disclosed in the application has not and will not be the subject of an application filed in another country, or under a multilateral international agreement, that requires publication of applications 18 months after filing, the application shall not be published as provided in paragraph (1).

(ii) An applicant may rescind a request made under clause (i) at any time.

(iii) An applicant who has made a request under clause (i) but who subsequently files, in a foreign country or under a multilateral international agreement specified in clause (i), an application directed to the invention disclosed in the application filed in the Patent and Trademark

Office, shall notify the Director of such filing not later than 45 days after the date of the filing of such foreign or international application. A failure of the applicant to provide such notice within the prescribed period shall result in the application being regarded as abandoned, unless it is shown to the satisfaction of the Director that the delay in submitting the notice was unintentional.

(iv) If an applicant rescinds a request made under clause (i) or notifies the Director that an application was filed in a foreign country or under a multilateral international agreement specified in clause (i), the application shall be published in accordance with the provisions of paragraph (1) on or as soon as is practical after the date that is specified in clause (i).

(v) If an applicant has filed applications in one or more foreign countries, directly or through a multilateral international agreement, and such foreign filed applications corresponding to an application filed in the Patent and Trademark Office or the description of the invention in such foreign filed applications is less extensive than the application or description of the invention in the application filed in the Patent and Trademark Office, the applicant may submit a redacted copy of the application filed in the Patent and Trademark Office eliminating any part or description of the invention in such application that is not also contained in any of the corresponding applications filed in a foreign country. The Director may only publish the redacted copy of the application unless the redacted copy of the application is not received within 16 months after the earliest effective filing date for which a benefit is sought under this title. The provisions of section 154(d) shall not apply to a claim if the description of the invention published in the redacted application filed under this clause with respect to the claim does not enable a person skilled in the art to make and use the subject matter of the claim.

(c) Protest and pre-issuance opposition.—The Director shall establish appropriate procedures to ensure that no protest or other form of pre-issuance opposition to the grant of a patent on an application may be initiated after publication of the application without the express written consent of the applicant.

(d) National security.—No application for patent shall be published under subsection (b)(1) if the publication or disclosure of such invention would be detrimental to the national security. The Director shall establish appropriate procedures to ensure that such applications are promptly identified and the secrecy of such inventions is maintained in accordance with chapter 17.

(e) Preissuance submissions by third parties.—

(1) In general.—Any third party may submit for consideration and inclusion in the record of a patent application, any patent, published patent application, or other printed publication of potential relevance to the examination of the application, if such submission is made in writing before the earlier of—

(A) the date a notice of allowance under section 151 is given or mailed in the application for patent; or

(B) the later of—

(i) 6 months after the date on which the application for patent is first published under section 122 by the Office, or

(ii) the date of the first rejection under section 132 of any claim by the examiner during the examination of the application for patent.

(2) Other requirements.—Any submission under paragraph (1) shall—

(A) set forth a concise description of the asserted relevance of each submitted document;

(B) be accompanied by such fee as the Director may prescribe; and

(C) include a statement by the person making such submission affirming that the submission was made in compliance with this section.

Sec. 123. Micro Entity Defined

(a) In general.—For purposes of this title, the term "micro entity" means an applicant who makes a certification that the applicant—

(1) qualifies as a small entity, as defined in regulations issued by the Director;

(2) has not been named as an inventor on more than 4 previously filed patent applications, other than applications filed in another country, provisional applications under section 111(b), or international applications filed under the treaty defined in section 351(a) for which the basic national fee under section 41(a) was not paid;

(3) did not, in the calendar year preceding the calendar year in which the applicable fee is being paid, have a gross income, as defined in section 61(a) of the Internal Revenue Code of 1986, exceeding 3 times the median household income for that preceding calendar year, as most recently reported by the Bureau of the Census; and

(4) has not assigned, granted, or conveyed, and is not under an obligation by contract or law to assign, grant, or convey, a license or other ownership interest in the application concerned to an entity that, in the calendar year preceding the calendar year in which the applicable fee is being paid, had a gross income, as defined in section 61(a) of the Internal Revenue Code of 1986, exceeding 3 times the median household income for that preceding calendar year, as most recently reported by the Bureau of the Census.

(b) Applications resulting from prior employment.—An applicant is not considered to be named on a previously filed application for purposes of subsection (a)(2) if the applicant has assigned, or is under an obligation by contract or law to assign, all ownership rights in the application as the result of the applicant's previous employment.

(c) Foreign currency exchange rate.—If an applicant's or entity's gross income in the preceding calendar year is not in United States dollars, the average currency exchange rate, as reported by the Internal Revenue Service, during that calendar year shall be used to determine whether the applicant's or entity's gross income exceeds the threshold specified in paragraphs (3) or (4) of subsection (a).

(d) Institutions of higher education.—For purposes of this section, a micro entity shall include an applicant who certifies that—

(1) the applicant's employer, from which the applicant obtains the majority of the applicant's income, is an institution of higher education as defined in section 101(a) of the Higher Education Act of 1965 (20 U.S.C. 1001(a)); or

(2) the applicant has assigned, granted, conveyed, or is under an obligation by contract or law, to assign, grant, or convey, a license or other ownership interest in the particular applications to such an institution of higher education.

(e) Director's authority.—In addition to the limits imposed by this section, the Director may, in the Director's discretion, impose income limits, annual filing limits, or other limits on who may qualify as a micro entity pursuant to this section if the Director determines that such additional limits are reasonably necessary to avoid an undue impact on other patent applicants or owners or are otherwise reasonably necessary and appropriate. At least 3 months before any limits proposed to be imposed pursuant to this subsection take effect, the Director shall inform the Committee on the Judiciary of the House of Representatives and the Committee on the Judiciary of the Senate of any such proposed limits.

Chapter 12. Examination of Application

Sec. 132. Notice of Rejection; Reexamination

(a) Whenever, on examination, any claim for a patent is rejected, or any objection or requirement made, the Director shall notify the applicant thereof, stating the reasons for such rejection, or objection or requirement, together with such information and references as may be useful in judging of the propriety of continuing the prosecution of his application; and if after receiving such notice, the applicant persists in his claim for a patent, with or without amendment, the application shall be reexamined. No amendment shall introduce new matter into the disclosure of the invention.

(b) The Director shall prescribe regulations to provide for the continued examination of applications for patent at the request of the applicant. The Director may establish appropriate fees for such continued examination and shall provide a 50 percent reduction in such fees for small entities that qualify for reduced fees under section 41(h)(1) of this title.

Sec. 133. Time for Prosecuting Application

Upon failure of the applicant to prosecute the application within six months after any action therein, of which notice has been given or mailed to the applicant, or within such shorter time, not less than thirty days, as fixed by the Director in such action, the application shall be regarded as abandoned by the parties thereto, unless it be shown to the satisfaction of the Director that such delay was unavoidable.

Sec. 134. Appeal to the Patent Trial and Appeal Board

(a) Patent applicant.—An applicant for a patent, any of whose claims has been twice rejected, may appeal from the decision of the primary examiner to the Patent Trial and Appeal Board, having once paid the fee for such appeal.

(b) Patent owner.—A patent owner in a reexamination may appeal from the final rejection of any claim by the primary examiner to the Patent Trial and Appeal Board, having once paid the fee for such appeal.

Sec. 135. Derivation proceedings

(a) Institution of proceeding.—

(1) In general.—An applicant for patent may file a petition with respect to an invention to institute a derivation proceeding in the Office. The petition shall set forth with particularity the basis for finding that an individual named in an earlier application as the inventor or a joint inventor derived such invention from an individual named in the petitioner's application as the inventor or a joint inventor and, without authorization, the earlier application claiming such invention was filed. Whenever the Director determines that a petition filed under this subsection demonstrates that the standards for instituting a derivation proceeding are met, the Director may institute a derivation proceeding.

(2) Time for filing.—A petition under this section with respect to an invention that is the same or substantially the same invention as a claim contained in a patent issued on an earlier application, or contained in an earlier application when published or deemed published under section 122(b), may not be filed unless such petition is filed during the 1-year period following the date on which the patent containing such claim was granted or the earlier application containing such claim was published, whichever is earlier.

(3) Earlier application.—For purposes of this section, an application shall not be deemed to be an earlier application with respect to an invention, relative to another application, unless a claim to the invention was or could have been made in such application having an effective filing date that is earlier than the effective filing date of any claim to the invention that was or could have been made in such other application.

(4) No appeal.—A determination by the Director whether to institute a derivation proceeding under paragraph (1) shall be final and not appealable.

(b)(1) A claim which is the same as, or for the same or substantially the same subject matter as, a claim of an issued patent may not be made in any application unless such a claim is made prior to one year from the date on which the patent was granted.

(2) A claim which is the same as, or for the same or substantially the same subject matter as, a claim of an application published under section 122(b) of this title may be made in an application filed after the application is published only if the claim is made before 1 year after the date on which the application is published.

(c) Any agreement or understanding between parties to an interference, including any collateral agreements referred to therein, made in connection with or in contemplation of the termination of the interference, shall be in writing and a true copy thereof filed in the Patent and Trademark Office before the termination of the interference as between the said parties to the agreement or understanding. If any party filing the same so requests, the copy shall be kept separate from the file of the interference, and made available only to Government agencies on written request, or to any person on a showing of good cause. Failure to file the copy of such agreement or understanding shall render permanently unenforceable such agreement or understanding and any patent of such parties involved in the interference or any patent subsequently issued on any application of such parties so involved. The Director may, however, on a showing of good cause for failure to file within the time prescribed, permit the filing of the agreement or understanding during the six-month period subsequent to the termination of the interference as between the parties to the agreement or understanding.

The Director shall give notice to the parties or their attorneys of record, a reasonable time prior to said termination, of the filing requirement of this section. If the Director gives such notice at a later time, irrespective of the right to file such agreement or understanding within the six-month period on a showing of good cause, the parties may file such agreement or understanding within sixty days of the receipt of such notice.

Any discretionary action of the Director under this subsection shall be reviewable under section 10 of the Administrative Procedure Act.

(d) Parties to a patent interference, within such time as may be specified by the Director by regulation, may determine such contest or any aspect thereof by arbitration. Such arbitration shall be governed by the provisions of title 9 to the extent such title is not inconsistent with this section. The parties shall give notice of any arbitration award to the Director, and such award shall, as between the parties to the arbitration, be dispositive of the issues to which it relates. The arbitration award shall be unenforceable until such notice is given. Nothing in this subsection shall preclude the Director from determining patentability of the invention involved in the interference.

(e) Settlement.—Parties to a proceeding instituted under subsection (a) may terminate the proceeding by filing a written statement reflecting the agreement of the parties as to the correct inventor of the claimed invention in dispute. Unless the Patent Trial and Appeal Board finds the agreement to be inconsistent with the evidence of record, if any, it shall take action consistent with the agreement. Any written settlement or understanding of the parties shall be filed with the Director. At the request of a party to the proceeding, the agreement or understanding shall be treated as business confidential information, shall be kept separate from the file of the involved patents or applications, and shall be made available only to Government agencies on written request, or to any person on a showing of good cause.

(f) Arbitration.—Parties to a proceeding instituted under subsection (a) may, within such time as may be specified by the Director by regulation, determine such contest or any aspect thereof by arbitration. Such arbitration shall be governed by the provisions of title 9, to the extent such title is not inconsistent with this section. The parties shall give notice of any arbitration award to the Director, and such award shall, as between the parties to the arbitration, be dispositive of the issues to which it relates. The arbitration award shall be unenforceable until such notice is given. Nothing in this subsection shall preclude the Director from determining the patentability of the claimed inventions involved in the proceeding.

Chapter 13. Review of Patent and Trademark Office Decisions

Sec. 141. Appeal to Court of Appeals for the Federal Circuit

(a) Examinations.—An applicant who is dissatisfied with the final decision in an appeal to the Patent Trial and Appeal Board undersection 134(a) may appeal the Board's decision to the United States Court of Appeals for the Federal Circuit. By filing such an appeal, the applicant waives his or her right to proceed under section 145.

(b) Reexaminations.—A patent owner who is dissatisfied with the final decision in an appeal of a reexamination to the Patent Trial and Appeal Board under section 134(b) may appeal the Board's decision only to the United States Court of Appeals for the Federal Circuit.

(c) Post-grant and inter partes reviews.—A party to an inter partes review or a post-grant review who is dissatisfied with the final written decision of the Patent Trial and Appeal Board under section 318(a) or 328(a) (as the case may be) may appeal the Board's decision only to the United States Court of Appeals for the Federal Circuit.

(d) Derivation proceedings.—A party to a derivation proceeding who is dissatisfied with the final decision of the Patent Trial and Appeal Board in the proceeding may appeal the decision to the United States Court of Appeals for the Federal Circuit, but such appeal shall be dismissed if any adverse party to such derivation proceeding, within 20 days after the appellant has filed notice of appeal in accordance with section 142, files notice with the Director that the party elects to have all further proceedings conducted as provided in section 146. If the appellant does not, within 30 days after the filing of such notice by the adverse party, file a civil action under section 146, the Board's decision shall govern the further proceedings in the case.

Sec. 142. Notice of Appeal

When an appeal is taken to the United States Court of Appeals for the Federal Circuit, the appellant shall file in the Patent and Trademark Office a written notice of appeal directed to the Director, within such time after the date of the decision from which the appeal is taken as the Director prescribes, but in no case less than 60 days after that date.

Sec. 143. Proceedings on Appeal

With respect to an appeal described in section 142, the Director shall transmit to the United States Court of Appeals for the Federal Circuit a certified list of the documents comprising the record in the Patent and Trademark Office. The court may request that the Director forward the original or certified copies of such documents during pendency of the appeal. In an ex parte case, the Director shall submit to the court in writing the grounds for the decision of the Patent and Trademark Office, addressing all of the issues raised in the appeal. The Director shall have the right to intervene in an appeal from a decision entered by the Patent Trial and Appeal Board in a derivation proceeding under section 135 or in an inter partes or post-grant review under chapter 31 or 32. The court shall, before hearing an appeal, give notice of the time and place of the hearing to the Director and the parties in the appeal.

Sec. 144. Decision on Appeal

The United States Court of Appeals for the Federal Circuit shall review the decision from which an appeal is taken on the record before the Patent and Trademark Office. Upon its determination the court shall issue to the Director its mandate and opinion, which shall be entered of record in the Patent and Trademark Office and shall govern the further proceedings in the case.

Sec. 145. Civil Action to Obtain Patent

An applicant dissatisfied with the decision of the Board of Patent Appeals and Interferences in an appeal under section 134(a) of this title may, unless appeal has been taken to the United States Court of Appeals for the Federal Circuit, have remedy by civil action against the Director in the United States District Court for the Eastern District of Virginia if commenced within such time after such decision, not less than sixty days, as the Director appoints. The court may adjudge that such applicant is entitled to receive a patent for his invention, as specified in any of his claims involved in the decision of the Patent Trial and Appeal Board, as the facts in the case may appear and such adjudication shall authorize the Director to issue such patent on compliance with the requirements of law. All the expenses of the proceedings shall be paid by the applicant.

Sec. 146. Civil action in case of derivation proceeding

Any party to a derivation proceeding dissatisfied with the decision of the Patent Trial and Appeal Board on the derivation proceeding, may have remedy by civil action, if commenced within such time after such decision, not less than sixty days, as the Director appoints or as provided in section 141, unless he has appealed to the United States Court of Appeals for the Federal Circuit, and such appeal is pending or has been decided. In such suits the record in the Patent and Trademark Office shall be admitted on motion of either party upon the terms and conditions as to costs, expenses, and the further cross-examination of the witnesses as the court imposes, without prejudice to the right of the parties to take further testimony. The testimony and exhibits of the record in the Patent and Trademark Office when admitted shall have the same effect as if originally taken and produced in the suit.

Such suit may be instituted against the party in interest as shown by the records of the Patent and Trademark Office at the time of the decision complained of, but any party in interest may become a party to the action. If there be adverse parties residing in a plurality of districts not embraced within the same state, or an adverse party residing in a foreign country, the United States District Court for the Eastern District of Virginia shall have jurisdiction and may issue summons against the adverse parties directed to the marshal of any district in which any adverse party resides. Summons against adverse parties residing in foreign countries may be served by publication or otherwise as the court directs. The Director shall not be a necessary party but he shall be notified of the filing of the suit by the clerk of the court in which it is filed and shall have the right to intervene. Judgment of the court in favor of the right of an applicant to a patent shall authorize the Director to issue such patent on the filing in the Patent and Trademark Office of a certified copy of the judgment and on compliance with the requirements of law.

Chapter 14. Issue of Patent

Sec. 151. Issue of Patent

If it appears that applicant is entitled to a patent under the law, a written notice of allowance of the application shall be given or mailed to the applicant. The notice shall specify a sum, constituting the issue fee or a portion thereof, which shall be paid within three months thereafter.

Upon payment of this sum the patent shall issue, but if payment is not timely made, the application shall be regarded as abandoned.

Any remaining balance of the issue fee shall be paid within three months from the sending of a notice thereof and, if not paid, the patent shall lapse at the termination of this three-month period. In calculating the amount of a remaining balance, charges for a page or less may be disregarded.

If any payment required by this section is not timely made, but is submitted with the fee for delayed payment and the delay in payment is shown to have been unavoidable, it may be accepted by the Director as though no abandonment or lapse had ever occurred.

Sec. 152. Issue of Patent to Assignee

Patents may be granted to the assignee of the inventor of record in the Patent and Trademark Office, upon the application made and the specification sworn to by the inventor, except as otherwise provided in this title.

Sec. 153. How Issued

Patents shall be issued in the name of the United States of America, under the seal of the Patent and Trademark Office, and shall be signed by the Director or have his signature placed thereon and shall be recorded in the Patent and Trademark Office.

Sec. 154. Contents and Term of Patent; Provisional Rights

(a) In general.—

(1) Contents.— Every patent shall contain a short title of the invention and a grant to the patentee, his heirs or assigns, of the right to exclude others from making, using, offering for sale, or selling the invention throughout the United States or importing the invention into the United States, and, if the invention is a process, of the right to exclude others from using, offering for sale or selling throughout the United States, or importing into the United States, products made by that process, referring to the specification for the particulars thereof.

(2) Term.— Subject to the payment of fees under this title, such grant shall be for a term beginning on the date on which the patent issues and ending 20 years from the date on which the application for the patent was filed in the United States or, if the application contains a specific reference to an earlier filed application or applications under section 120, 121, or 365(c) of this title, from the date on which the earliest such application was filed.

(3) Priority.— Priority under section 119, 365(a), or 365(b) of this title shall not be taken into account in determining the term of a patent.

(4) Specification and drawing.— A copy of the specification and drawing shall be annexed to the patent and be a part of such patent.

(b) Adjustment of patent term.—

(1) Patent term guarantees.—

(A) Guarantee of prompt Patent and Trademark Office responses.—Subject to the limitations under paragraph (2), if the issue of an original patent is delayed due to the failure of the Patent and Trademark Office to—

(i) provide at least one of the notifications under section 132 of this title or a notice of allowance under section 151 of this title not later than 14 months after—

(I) the date on which an application was filed under section 111(a) of this title; or

(II) the date of commencement of the national stage under section 371 in an international application;

(ii) respond to a reply under section 132, or to an appeal taken under section 134, within 4 months after the date on which the reply was filed or the appeal was taken;

(iii) act on an application within 4 months after the date of a decision by the Patent Trial and Appeal Board under section 134 or 135 or a decision by a Federal court under section 141, 145, or 146 in a case in which allowable claims remain in the application; or

(iv) issue a patent within 4 months after the date on which the issue fee was paid under section 151 and all outstanding requirements were satisfied,

the term of the patent shall be extended 1 day for each day after the end of the period specified in clause (i), (ii), (iii), or (iv), as the case may be, until the action described in such clause is taken.

(B) Guarantee of no more than 3-year application pendency.—Subject to the limitations under paragraph (2), if the issue of an original patent is delayed due to the failure of the United States Patent and Trademark Office to issue a patent within 3 years after the actual filing date of the application under section 111(a) in the United States or, in the case of an international application the date of commencement of the national stage under section 371 in the international application, not including—

(i) any time consumed by continued examination of the application requested by the applicant under section 132(b);

(ii) any time consumed by a proceeding under section 135(a), any time consumed by the imposition of an order under section 181, or any time consumed by appellate review by the Patent Trial and Appeal Board or by a Federal court; or

(iii) any delay in the processing of the application by the United States Patent and Trademark Office requested by the applicant except as permitted by paragraph (3)(C), the term of the patent shall be extended 1 day for each day after the end of that 3-year period until the patent is issued.

(C) Guarantee of adjustments for delays due to derivation proceedings, secrecy orders, and appeals.—Subject to the limitations under paragraph (2), if the issue of an original patent is delayed due to—

(i) a proceeding under section 135(a);

(ii) the imposition of an order under section 181; or

(iii) appellate review by the Patent Trial and Appeal Board or by a Federal court in a case in which the patent was issued under a decision in the review reversing an adverse determination of patentability,

the term of the patent shall be extended 1 day for each day of the pendency of the proceeding, order, or review, as the case may be.

(2) Limitations.—

(A) In general.—To the extent that periods of delay attributable to grounds specified in paragraph (1) overlap, the period of any adjustment granted under this subsection shall not exceed the actual number of days the issuance of the patent was delayed.

(B) Disclaimed term.— No patent the term of which has been disclaimed beyond a specified date may be adjusted under this section beyond the expiration date specified in the disclaimer.

(C) Reduction of period of adjustment.—

(i) The period of adjustment of the term of a patent under paragraph (1) shall be reduced by a period equal to the period of time during which the applicant failed to engage in reasonable efforts to conclude prosecution of the application.

(ii) With respect to adjustments to patent term made under the authority of paragraph (1)(B), an applicant shall be deemed to have failed to engage in reasonable efforts to conclude processing or examination of an application for the cumulative total of any periods of time in excess of 3 months that are taken to respond to a notice from the Office making any rejection, objection, argument, or other request, measuring such 3-month period from the date the notice was given or mailed to the applicant.

(iii) The Director shall prescribe regulations establishing the circumstances that constitute a failure of an applicant to engage in reasonable efforts to conclude processing or examination of an application.

(3) Procedures for patent term adjustment determination.—

(A) The Director shall prescribe regulations establishing procedures for the application for and determination of patent term adjustments under this subsection.

(B) Under the procedures established under subparagraph (A), the Director shall—

(i) make a determination of the period of any patent term adjustment under this subsection, and shall transmit a notice of that determination no later than the date of issuance of the patent; and

(ii) provide the applicant one opportunity to request reconsideration of any patent term adjustment determination made by the Director.

(C) The Director shall reinstate all or part of the cumulative period of time of an adjustment under paragraph (2)(C) if the applicant, prior to the issuance of the patent, makes a showing that, in spite of all due care, the applicant was unable to respond within the 3-month period, but in no case shall more than three additional months for each such response beyond the original 3-month period be reinstated.

(D) The Director shall proceed to grant the patent after completion of the Director's determination of a patent term adjustment under the procedures established under this subsection, notwithstanding any appeal taken by the applicant of such determination.

(4) Appeal of patent term adjustment determination.—

(A) An applicant dissatisfied with the Director's decision on the applicant's request for reconsideration under paragraph (3)(B)(ii) shall have exclusive remedy by a civil action against the Director filed in the United States District Court for the Eastern District of Virginia within 180 days after the date of the Director's decision on the applicant's request for reconsideration. Chapter 7 of title 5 shall apply to such action. Any final judgment resulting in a change to the period of adjustment of the patent term shall be served on the Director, and the Director shall thereafter alter the term of the patent to reflect such change.

(B) The determination of a patent term adjustment under this subsection shall not be subject to appeal or challenge by a third party prior to the grant of the patent.

(c) Continuation.—

(1) Determination.— The term of a patent that is in force on or that results from an application filed before the date that is 6 months after the date of the enactment of the Uruguay Round Agreements Act shall be the greater of the 20-year term as provided in subsection (a), or 17 years from grant, subject to any terminal disclaimers.

(2) Remedies.— The remedies of sections 283, 284, and 285 of this title shall not apply to acts which—

(A) were commenced or for which substantial investment was made before the date that is 6 months after the date of the enactment of the Uruguay Round Agreements Act; and

(B) became infringing by reason of paragraph (1).

(3) Remuneration.— The acts referred to in paragraph (2) may be continued only upon the payment of an equitable remuneration to the patentee that is determined in an action brought under chapter 28 and chapter 29 (other than those provisions excluded by paragraph (2)) of this title.

(d) Provisional rights.—

(1) In general.— In addition to other rights provided by this section, a patent shall include the right to obtain a reasonable royalty from any person who, during the period beginning on the date of publication of the application for such patent under section 122(b), or in the case of an international application filed under the treaty defined in section 351(a) designating the United States under Article 21(2)(a) of such treaty, the date of publication of the application, and ending on the date the patent is issued—

(A)(i) makes, uses, offers for sale, or sells in the United States the invention as claimed in the published patent application or imports such an invention into the United States; or

(ii) if the invention as claimed in the published patent application is a process, uses, offers for sale, or sells in the United States or imports into the

United States products made by that process as claimed in the published patent application; and

(B) had actual notice of the published patent application and, in a case in which the right arising under this paragraph is based upon an international application designating the United States that is published in a language other than English, had a translation of the international application into the English language.

(2) Right based on substantially identical inventions.— The right under paragraph (1) to obtain a reasonable royalty shall not be available under this subsection unless the invention as claimed in the patent is substantially identical to the invention as claimed in the published patent application.

(3) Time limitation on obtaining a reasonable royalty.— The right under paragraph (1) to obtain a reasonable royalty shall be available only in an action brought not later than 6 years after the patent is issued. The right under paragraph (1) to obtain a reasonable royalty shall not be affected by the duration of the period described in paragraph (1).

(4) Requirements for international applications.—

(A) Effective date.— The right under paragraph (1) to obtain a reasonable royalty based upon the publication under the treaty defined in section 351(a) of an international application designating the United States shall commence on the date of publication under the treaty of the international application, or, if the publication under the treaty of the international application is in a language other than English, on the date on which the Patent and Trademark Office receives a translation of the publication in the English language.

(B) Copies.— The Director may require the applicant to provide a copy of the international application and a translation thereof.

Sec. 155. Repealed

Sec. 155A. Repealed

Sec. 156. Extension of Patent Term

(a) The term of a patent which claims a product, a method of using a product, or a method of manufacturing a product shall be extended in accordance with this section from the original expiration date of the patent, which shall include any patent term adjustment granted under section 154(b), if—

(1) the term of the patent has not expired before an application is submitted under subsection (d)(1) for its extension;

(2) the term of the patent has never been extended under subsection (e)(1) of this section;

(3) an application for extension is submitted by the owner of record of the patent or its agent and in accordance with the requirements of paragraphs (1) through (4) of subsection (d);

(4) the product has been subject to a regulatory review period before its commercial marketing or use;

(5)(A) except as provided in subparagraph (B) or (C), the permission for the commercial marketing or use of the product after such regulatory review period is the first permitted commercial marketing or use of the product under the provision of law under which such regulatory review period occurred;

(B) in the case of a patent which claims a method of manufacturing the product which primarily uses recombinant DNA technology in the manufacture of the product, the permission for the commercial marketing or use of the product after such regulatory review period is the first permitted commercial marketing or use of a product manufactured under the process claimed in the patent; or

(C) for purposes of subparagraph (A), in the case of a patent which—

(i) claims a new animal drug or a veterinary biological product which (I) is not covered by the claims in any other patent which has been extended, and (II) has received permission for the commercial marketing or use in non-food-producing animals and in food-producing animals, and

(ii) was not extended on the basis of the regulatory review period for use in non-food-producing animals,

the permission for the commercial marketing or use of the drug or product after the regulatory review period for use in food-producing animals is the first permitted commercial marketing or use of the drug or product for administration to a food-producing animal.

The product referred to in paragraphs (4) and (5) is hereinafter in this section referred to as the "approved product".

(b) Except as provided in subsection (d)(5)(F), the rights derived from any patent the term of which is extended under this section shall during the period during which the term of the patent is extended—

(1) in the case of a patent which claims a product, be limited to any use approved for the product—

(A) before the expiration of the term of the patent—

(i) under the provision of law under which the applicable regulatory review occurred, or

(ii) under the provision of law under which any regulatory review described in paragraph (1), (4), or (5) of subsection (g) occurred, and

(B) on or after the expiration of the regulatory review period upon which the extension of the patent was based;

(2) in the case of a patent which claims a method of using a product, be limited to any use claimed by the patent and approved for the product—

(A) before the expiration of the term of the patent—

(i) under any provision of law under which an applicable regulatory review occurred, and

(ii) under the provision of law under which any regulatory review described in paragraph (1), (4), or (5) of subsection (g) occurred, and

(B) on or after the expiration of the regulatory review period upon which the extension of the patent was based; and

(3) in the case of a patent which claims a method of manufacturing a product, be limited to the method of manufacturing as used to make—

(A) the approved product, or

(B) the product if it has been subject to a regulatory review period described in paragraph (1), (4), or (5) of subsection (g).

As used in this subsection, the term "product" includes an approved product.

(c) The term of a patent eligible for extension under subsection (a) shall be extended by the time equal to the regulatory review period for the approved product which period occurs after the date the patent is issued, except that—

(1) each period of the regulatory review period shall be reduced by any period determined under subsection (d)(2)(B) during which the applicant for the patent extension did not act with due diligence during such period of the regulatory review period;

(2) after any reduction required by paragraph (1), the period of extension shall include only one-half of the time remaining in the periods described in paragraphs (1)(B)(i), (2)(B)(i), (3)(B)(i), (4)(B)(i), and (5)(B)(i) of subsection (g);

(3) if the period remaining in the term of a patent after the date of the approval of the approved product under the provision of law under which such regulatory review occurred when added to the regulatory review period as revised under paragraphs (1) and (2) exceeds fourteen years, the period of extension shall be reduced so that the total of both such periods does not exceed fourteen years; and

(4) in no event shall more than one patent be extended under subsection (e)(1) for the same regulatory review period for any product.

(d)(1) To obtain an extension of the term of a patent under this section, the owner of record of the patent or its agent shall submit an application to the Director. Except as provided in paragraph (5), such an application may only be submitted within the sixty-day period beginning on the date the product received permission under the provision of law under which the applicable regulatory review period occurred for commercial marketing or use. The application shall contain—

(A) the identity of the approved product and the Federal statute under which regulatory review occurred;

(B) the identity of the patent for which an extension is being sought and the identity of each claim of such patent which claims the approved product or a method of using or manufacturing the approved product;

(C) information to enable the Director to determine under subsections (a) and (b) the eligibility of a patent for extension and the rights that will be derived from the extension and information to enable the Director and the Secretary of Health and Human Services or the Secretary of Agriculture to determine the period of the extension under subsection (g);

(D) a brief description of the activities undertaken by the applicant during the applicable regulatory review period with respect to the approved product and the significant dates applicable to such activities; and

(E) such patent or other information as the Director may require.

For purposes of determining the date on which a product receives permission under the second sentence of this paragraph, if such permission is transmitted after 4:30 P.M., Eastern Time, on a business day, or is transmitted on a day that is not a business day, the product shall be deemed to receive such permission on the next business day. For purposes of the preceding sentence, the term "business day" means any Monday, Tuesday, Wednesday, Thursday, or Friday, excluding any legal holiday under section 6103 of title 5.

(2) (A) Within 60 days of the submittal of an application for extension of the term of a patent under paragraph (1), the Director shall notify—

(i) the Secretary of Agriculture if the patent claims a drug product or a method of using or manufacturing a drug product and the drug product is subject to the Virus-Serum-Toxin Act, and

(ii) the Secretary of Health and Human Services if the patent claims any other drug product, a medical device, or a food additive or color additive or a method of using or manufacturing such a product, device, or additive and if the product, device, and additive are subject to the Federal Food, Drug, and Cosmetic Act, of the extension application and shall submit to the Secretary who is so notified a copy of the application. Not later than 30 days after the receipt of an application from the Director, the Secretary receiving the application shall review the dates contained in the application pursuant to paragraph (1)(C) and determine the applicable regulatory review period, shall notify the Director of the determination, and shall publish in the Federal Register a notice of such determination.

(B) (i) If a petition is submitted to the Secretary making the determination under subparagraph (A), not later than 180 days after the publication of the determination under subparagraph (A), upon which it may reasonably be determined that the applicant did not act with due diligence during the applicable regulatory review period, the Secretary making the determination shall, in accordance with regulations promulgated by such Secretary, determine if the applicant acted with due diligence during the applicable regulatory review period. The Secretary making the determination shall make such determination not later than 90 days after the receipt of such a petition. For a drug product, device, or additive subject to the Federal Food, Drug, and Cosmetic Act or the Public Health Service Act, the Secretary may not delegate the authority to make the determination prescribed by this clause to an office below the Office of the Director of Food and Drugs. For a product subject to the Virus-Serum-Toxin Act, the Secretary of Agriculture may not delegate the authority to make the determination prescribed by this clause to an office below the Office of the Assistant Secretary for Marketing and Inspection Services.

(ii) The Secretary making a determination under clause (i) shall notify the Director of the determination and shall publish in the Federal Register a notice of such determination together with the factual and legal basis for such determination. Any interested person may request, within the 60-day period beginning on the publication of a determination, the Secretary making the determination to hold an informal hearing on the determination. If such a request is made within such period, such Secretary shall hold such hearing not later than 30 days after the date of the request, or at the request of the person making the request, not later than 60 days after such date. The Secretary who is holding the hearing shall provide notice of the hearing to the owner of the patent involved and to any interested person and provide the owner and any interested person an opportunity to participate in the hearing. Within 30 days after the completion of the hearing, such Secretary shall affirm or revise the determination which was the subject of the hearing and shall notify the Director of any revision of the determination and shall publish any such revision in the Federal Register.

(3) For the purposes of paragraph (2)(B), the term "due diligence" means that degree of attention, continuous directed effort, and timeliness as may reasonably be expected from, and are ordinarily exercised by, a person during a regulatory review period.

(4) An application for the extension of the term of a patent is subject to the disclosure requirements prescribed by the Director.

(5)(A) If the owner of record of the patent or its agent reasonably expects that the applicable regulatory review period described in paragraph (1)(B)(ii), (2)(B)(ii), (3)(B)(ii), (4)(B)(ii), or (5)(B)(ii) of subsection (g) that began for a product that is the subject of such patent may extend beyond the expiration of the patent term in effect, the owner or its agent may submit an application to the Director for an interim extension during the period beginning 6 months, and ending 15 days, before such term is due to expire. The application shall contain—

(i) the identity of the product subject to regulatory review and the Federal statute under which such review is occurring;

(ii) the identity of the patent for which interim extension is being sought and the identity of each claim of such patent which claims the product under regulatory review or a method of using or manufacturing the product;

(iii) information to enable the Director to determine under subsection (a)(1), (2), and (3) the eligibility of a patent for extension;

(iv) a brief description of the activities undertaken by the applicant during the applicable regulatory review period to date with respect to the product under review and the significant dates applicable to such activities; and

(v) such patent or other information as the Director may require.

(B) If the Director determines that, except for permission to market or use the product commercially, the patent would be eligible for an extension of the patent term under this section, the Director shall publish in the Federal Register a notice of such determination, including the identity of the product under regulatory review, and shall issue to the applicant a certificate of interim extension for a period of not more than 1 year.

(C) The owner of record of a patent, or its agent, for which an interim extension has been granted under subparagraph (B), may apply for not more than 4 subsequent interim extensions under this paragraph, except that, in the case of a patent subject to subsection (g)(6)(C), the owner of record of the patent, or its agent, may apply for only 1 subsequent interim extension under this paragraph. Each such subsequent application shall be made during the period beginning 60 days before, and ending 30 days before, the expiration of the preceding interim extension.

(D) Each certificate of interim extension under this paragraph shall be recorded in the official file of the patent and shall be considered part of the original patent.

(E) Any interim extension granted under this paragraph shall terminate at the end of the 60-day period beginning on the date on which the product involved receives permission for commercial marketing or use, except that, if within that 60-day period the applicant notifies the Director of such permission and submits any additional information under paragraph (1) of this subsection not previously contained in the application for interim extension, the patent shall be further extended, in accordance with the provisions of this section—

(i) for not to exceed 5 years from the date of expiration of the original patent term; or

(ii) if the patent is subject to subsection (g)(6)(C), from the date on which the product involved receives approval for commercial marketing or use.

(F) The rights derived from any patent the term of which is extended under this paragraph shall, during the period of interim extension—

(i) in the case of a patent which claims a product, be limited to any use then under regulatory review;

(ii) in the case of a patent which claims a method of using a product, be limited to any use claimed by the patent then under regulatory review; and

(iii) in the case of a patent which claims a method of manufacturing a product, be limited to the method of manufacturing as used to make the product then under regulatory review.

(e)(1) a determination that a patent is eligible for extension may be made by the Director solely on the basis of the representations contained in the application for the extension. If the Director determines that a patent is eligible for extension under subsection (a) and that the requirements of paragraphs (1) through (4) of subsection (d) have been complied with, the Director shall issue to the applicant for the extension of the term of the patent a certificate of extension, under seal, for the period prescribed by subsection (c). Such certificate shall be recorded in the official file of the patent and shall be considered as part of the original patent.

(2) If the term of a patent for which an application has been submitted under subsection (d)(1) would expire before a certificate of extension is issued or denied under paragraph (1) respecting the application, the Director shall extend, until such determination is made, the term of the patent for periods of up to one year if he determines that the patent is eligible for extension.

(f) For purposes of this section:

(1) The term "product" means:

(A) A drug product.

(B) Any medical device, food additive, or color additive subject to regulation under the Federal Food, Drug, and Cosmetic Act.

(2) The term "drug product" means the active ingredient of—

(A) a new drug, antibiotic drug, or human biological product (as those terms are used in the Federal Food, Drug, and Cosmetic Act and the Public Health Service Act), or

(B) a new animal drug or veterinary biological product (as those terms are used in the Federal Food, Drug, and Cosmetic Act and the Virus-Serum-Toxin Act) which is not primarily manufactured using recombinant DNA, recombinant RNA, hybridoma technology, or other processes involving site specific genetic manipulation techniques, including any salt or ester of the active ingredient, as a single entity or in combination with another active ingredient.

(3) The term "major health or environmental effects test" means a test which is reasonably related to the evaluation of the health or environmental effects of a product, which requires at least six months to conduct, and the data from which is submitted to receive permission for commercial marketing or use. Periods of analysis or evaluation of test results are not to be included in determining if the conduct of a test required at least six months.

(4) (A) Any reference to section 351 is a reference to section 351 of the Public Health Service Act.

(B) Any reference to section 503, 505, 512, or 515 is a reference to section 503, 505, 512, or 515 of the Federal Food, Drug, and Cosmetic Act.

(C) Any reference to the Virus-Serum-Toxin Act is a reference to the Act of March 4, 1913 (21 U.S.C. 151-158).

(5) The term "informal hearing" has the meaning prescribed for such term by section 201(y) of the Federal Food, Drug, and Cosmetic Act.

(6) The term "patent" means a patent issued by the United States Patent and Trademark Office.

(7) The term "date of enactment" as used in this section means September 24, 1984, for a human drug product, a medical device, food additive, or color additive.

(8) The term "date of enactment" as used in this section means the date of enactment of the Generic Animal Drug and Patent Term Restoration Act for an animal drug or a veterinary biological product.

(g) For purposes of this section, the term "regulatory review period" has the following meanings:

(1) (A) In the case of a product which is a new drug, antibiotic drug, or human biological product, the term means the period described in subparagraph (B) to which the limitation described in paragraph (6) applies.

(B) The regulatory review period for a new drug, antibiotic drug, or human biological product is the sum of—

(i) the period beginning on the date an exemption under subsection (i) of section 505 or subsection (d) of section 507 became effective for the approved product and ending on the date an application was initially submitted for such drug product under section 351, 505, or 507, and

(ii) the period beginning on the date the application was initially submitted for the approved product under section 351, subsection (b) of section 505, or section 507 and ending on the date such application was approved under such section.

(2)(A) In the case of a product which is a food additive or color additive, the term means the period described in subparagraph (B) to which the limitation described in paragraph (6) applies.

(B) The regulatory review period for a food or color additive is the sum of—

(i) the period beginning on the date a major health or environmental effects test on the additive was initiated and ending on the date a petition was initially submitted with respect to the product under the Federal Food, Drug, and Cosmetic Act requesting the issuance of a regulation for use of the product, and

(ii) the period beginning on the date a petition was initially submitted with respect to the product under the Federal Food, Drug, and Cosmetic Act requesting the issuance of a regulation for use of the product, and ending on the date such regulation became effective or, if objections were filed to such regulation, ending on the date such objections were resolved and commercial marketing was permitted or, if commercial marketing was permitted and later revoked pending further proceedings as a result of such objections, ending on the date such proceedings were finally resolved and commercial marketing was permitted.

(3)(A) In the case of a product which is a medical device, the term means the period described in subparagraph (B) to which the limitation described in paragraph (6) applies.

(B) The regulatory review period for a medical device is the sum of—

(i) the period beginning on the date a clinical investigation on humans involving the device was begun and ending on the date an application was initially submitted with respect to the device under section 515, and

(ii) the period beginning on the date an application was initially submitted with respect to the device under section 515 and ending on the date such application was approved under such Act or the period beginning on the date a notice of completion of a product development protocol was initially submitted under section 515(f)(5) and ending on the date the protocol was declared completed under section 515(f)(6).

(4)(A) In the case of a product which is a new animal drug, the term means the period described in subparagraph (B) to which the limitation described in paragraph (6) applies.

(B) The regulatory review period for a new animal drug product is the sum of—

(i) the period beginning on the earlier of the date a major health or environmental effects test on the drug was initiated or the date an exemption under subsection (j) of section 512 became effective for the approved new animal drug product and ending on the date an application was initially submitted for such animal drug product under section 512, and

(ii) the period beginning on the date the application was initially submitted for the approved animal drug product under subsection (b) of section 512 and ending on the date such application was approved under such section.

(5) (A) In the case of a product which is a veterinary biological product, the term means the period described in subparagraph (B) to which the limitation described in paragraph (6) applies.

(B) The regulatory period for a veterinary biological product is the sum of—

(i) the period beginning on the date the authority to prepare an experimental biological product under the Virus-Serum-Toxin Act became effective and ending on the date an application for a license was submitted under the Virus-Serum-Toxin Act, and

(ii) the period beginning on the date an application for a license was initially submitted for approval under the Virus-Serum-Toxin Act and ending on the date such license was issued.

(6) A period determined under any of the preceding paragraphs is subject to the following limitations:

(A) If the patent involved was issued after the date of the enactment of this section, the period of extension determined on the basis of the regulatory review period determined under any such paragraph may not exceed five years.

(B) If the patent involved was issued before the date of the enactment of this section and—

(i) no request for an exemption described in paragraph (1) (B) or (4) (B) was submitted and no request for the authority described in paragraph (5) (B) was submitted,

(ii) no major health or environmental effects test described in paragraph (2) (B) or (4) (B) was initiated and no petition for a regulation or application for registration described in such paragraph was submitted, or

(iii) no clinical investigation described in paragraph (3) was begun or product development, protocol described in such paragraph was submitted,

before such date for the approved product the period of extension determined on the basis of the regulatory review period determined under any such paragraph may not exceed five years.

(C) If the patent involved was issued before the date of the enactment of this section and if an action described in subparagraph (B) was taken before the date of the enactment of this section with respect to the approved product and the commercial marketing or use of the product has not been approved before such date, the period of extension determined on the basis of the regulatory review period determined under such paragraph may not exceed two years or in the case of an approved product which is a new animal drug or veterinary biological product (as those terms are used in the Federal Food, Drug, and Cosmetic Act or the Virus-Serum-Toxin Act), three years.

(h) The Director may establish such fees as the Director determines appropriate to cover the costs to the Office of receiving and acting upon applications under this section.

Sec. 157. Repealed

Chapter 15. Plant Patents

Sec. 161. Patents for Plants

Whoever invents or discovers and asexually reproduces any distinct and new variety of plant, including cultivated sports, mutants, hybrids, and newly found seedlings, other than a tuber propagated plant or a plant found in an uncultivated state, may obtain a patent therefor, subject to the conditions and requirements of this title.

The provisions of this title relating to patents for inventions shall apply to patents for plants, except as otherwise provided.

Sec. 162. Description, Claim

No plant patent shall be declared invalid for noncompliance with section 112 of this title if the description is as complete as is reasonably possible.

The claim in the specification shall be in formal terms to the plant shown and described.

Sec. 163. Grant

In the case of a plant patent, the grant shall include the right to exclude others from asexually reproducing the plant, and from using, offering for sale, or selling the plant so reproduced, or any of its parts, throughout the United States, or from importing the plant so reproduced, or any parts thereof, into the United States.

Sec. 164. Assistance of Department of Agriculture

The President may by Executive order direct the Secretary of Agriculture, in accordance with the requests of the Director, for the purpose of carrying into effect the provisions of this title with respect to plants (1) to furnish available information of the Department of Agriculture, (2) to conduct through the appropriate bureau or division of the Department research upon special problems, or (3) to detail to the Director officers and employees of the Department.

Chapter 16. Designs

Sec. 171. Patents for Designs

Whoever invents any new, original and ornamental design for an article of manufacture may obtain a patent therefor, subject to the conditions and requirements of this title.

The provisions of this title relating to patents for inventions shall apply to patents for designs, except as otherwise provided.

Sec. 172. Right of Priority

The right of priority provided for by subsections (a) through (d) of section 119 of this title shall be six months in the case of designs. The right of priority provided for by section 119(e) of this title shall not apply to designs.

Sec. 173. Term of Design Patent

Patents for designs shall be granted for the term of fourteen years from the date of grant.

Chapter 17. Secrecy of Certain Inventions and Filing Applications in Foreign Country

Sec. 181. Secrecy of Certain Inventions and Withholding of Patent

Whenever publication or disclosure by the publication of an application or by the grant of a patent on an invention in which the Government has a property interest might, in the opinion of the head of the interested Government agency, be detrimental to the national security, the Commissioner of Patents upon being so notified shall order that the invention be kept secret and shall withhold the publication of the application or the grant of a patent therefor under the conditions set forth hereinafter.

Whenever the publication or disclosure of an invention by the publication of an application or by the granting of a patent, in which the Government does not have a property interest, might, in the opinion of the Commissioner of Patents, be detrimental to the national security, he shall make the application for patent in which such invention is disclosed available for inspection to the Atomic Energy Commission, the Secretary of Defense, and the chief officer of any other department or agency of the Government designated by the President as a defense agency of the United States.

Each individual to whom the application is disclosed shall sign a dated acknowledgment thereof, which acknowledgment shall be entered in the file of the application. If, in the opinion of the Atomic Energy Commission, the Secretary of a Defense Department, or the chief officer of another department or agency so designated, the publication or disclosure of the invention by the publication of an application or by the granting of a patent therefor would be detrimental to the national security, the Atomic Energy Commission, the Secretary of a Defense Department, or such other chief officer shall notify the Commissioner of Patents and the Commissioner of Patents shall order that the invention be kept secret and shall withhold the publication of the application or the grant of a patent for such period as the national interest requires, and notify the applicant thereof. Upon proper showing by the head of the department or agency who caused the secrecy order to be issued that the examination of the application might jeopardize the national interest, the Commissioner of Patents shall thereupon maintain the application in a sealed condition and notify the applicant thereof. The owner of an application which has been placed under a secrecy order shall have a right to appeal from the order to the Secretary of Commerce under rules prescribed by him.

An invention shall not be ordered kept secret and the publication of the application or the grant of a patent withheld for a period of more than one year. The Commissioner of Patents shall renew the order at the end thereof, or at the end of any renewal period, for additional periods of one year upon notification by the head of the department or the chief officer of the agency who caused the order to be issued that an affirmative determination has been made that the national interest continues so to require. An order in effect, or issued, during a time when the United States is at war, shall remain in effect for the duration of hostilities and one year following cessation of hostilities. An order in effect, or issued, during a national emergency declared by the President shall remain in effect for the duration of the national emergency and six months thereafter. The Commissioner of Patents may rescind any order upon notification by the heads of the departments and the chief officers of the agencies who caused the order to be issued that the publication or disclosure of the invention is no longer deemed detrimental to the national security.

Sec. 182. Abandonment of Invention for Unauthorized Disclosure

The invention disclosed in an application for patent subject to an order made pursuant to section 181 of this title may be held abandoned upon its being established by the Commissioner of Patents that in violation of said order the invention has been published or disclosed or that an application for a patent therefor has been filed in a foreign country by the inventor, his successors, assigns, or legal representatives, or anyone in privity with him or them, without the consent of the Commissioner of Patents. The abandonment shall be held to have occurred as of the time of violation. The consent of the Commissioner of Patents shall not be given without the concurrence of the heads of the departments and the chief officers of the agencies who caused the order to be issued. A holding of abandonment shall constitute forfeiture by the applicant, his successors, assigns, or legal representatives, or anyone in privity with him or them, of all claims against the United States based upon such invention.

Sec. 183. Right to Compensation

An applicant, his successors, assigns, or legal representatives, whose patent is withheld as herein provided, shall have the right, beginning at the date the applicant is notified that, except for such order, his application is otherwise in condition for allowance, or February 1, 1952, whichever is later, and ending six years after a patent is issued thereon, to apply to the head of any department or agency who caused the order to be issued for compensation for the damage caused by the order of secrecy and/or for the use of the invention by the Government, resulting from his disclosure. The right to compensation for use shall begin on the date of the first use of the invention by the Government. The head of the department or agency is authorized, upon the presentation of a claim, to enter into an agreement with the applicant, his successors, assigns, or legal representatives, in full settlement for the damage and/or use. This settlement agreement shall be conclusive for all purposes notwithstanding any other provision of law to the contrary. If full settlement of the claim cannot be effected, the head of the department or agency may award and pay to such applicant, his successors, assigns, or legal representatives, a sum not exceeding 75 per centum of the sum which the head of the department or agency considers just compensation for the damage and/or use. A claimant may bring suit against the United States in the United States Court of Federal Claims or in the District Court of the United States for the district in which such claimant is a resident for an amount which when added to the award shall constitute just compensation for the damage and/or use of the invention by the Government. The owner of any patent issued upon an application that was subject to a secrecy order issued pursuant to section 181 of this title, who did not apply for compensation as above provided, shall have the right, after the date of issuance of such patent, to bring suit in the United States Court of Federal Claims for just compensation for the damage caused by reason of the order of secrecy and/or use by the Government of the invention resulting from his disclosure. The right to compensation for use shall begin on the date of the first use of the invention by the Government. In a suit under the provisions of this section the United States may avail itself of all defenses it may plead in an action under section 1498 of title 28. This section shall not confer a right of action on anyone or his successors, assigns, or legal representatives who, while in the full-time employment or service of the United States, discovered, invented, or developed the invention on which the claim is based.

Sec. 184. Filing of Application in Foreign Country

(a) Filing in a Foreign Country.—Except when authorized by a license obtained from the Commissioner of Patents a person shall not file or cause or authorize to be filed in any foreign country prior to six months after filing in the United States an application for patent or for the registration of a utility model, industrial design, or model in respect of an invention made in this country. A license shall not be granted with respect to an invention subject to an order issued by the Commissioner of Patents pursuant to section 181 of this title without the concurrence of the head of the departments and the chief officers of the agencies who caused the order to be issued. The license may be granted retroactively where an application has been filed abroad through error and the application does not disclose an invention within the scope of section 181 of this title.

(b) Application.—The term "application" when used in this chapter includes applications and any modifications, amendments, or supplements thereto, or divisions thereof.

(c) Subsequent Modification, Amendments, and Supplements.—The scope of a license shall permit subsequent modifications, amendments, and supplements containing additional subject matter if the application upon which the request for the license is based is not, or was not, required to be made available for inspection under section 181 of this title and if such modifications, amendments, and supplements do not change the general nature of the invention in a manner which would require such application to be made available for inspection under such section 181. In any case in which a license is not, or was not, required in order to file an application in any foreign country, such subsequent modifications, amendments, and supplements may be made, without a license, to the application filed in the foreign country if the United States application was not required to be made available for inspection under section 181 and if such modifications, amendments, and supplements do not, or did not, change the general nature of the invention in a manner which would require the United States application to have been made available for inspection under such section 181.

Sec. 185. Patent Barred for Filing Without License

Notwithstanding any other provisions of law any person, and his successors, assigns, or legal representatives, shall not receive a United States patent for an invention if that person, or his successors, assigns, or legal representatives shall, without procuring the license prescribed in section 184 of this title, have made, or consented to or assisted another's making, application in a foreign country for a patent or for the registration of a utility model, industrial design, or model in respect of the invention. A United States patent issued to such person, his successors, assigns, or legal representatives shall be invalid, unless the failure to procure such license was through error, and the patent does not disclose subject matter within the scope of section 181 of this title.

Sec. 186. Penalty

Whoever, during the period or periods of time an invention has been ordered to be kept secret and the grant of a patent thereon withheld pursuant to section 181 of this title, shall, with knowledge of such order and without due authorization, willfully publish or disclose or authorize or cause to be published or disclosed the invention, or material information with respect thereto, or whoever willfully, in violation of the provisions of section 184 of this title, shall file or cause or authorize to be filed in any foreign country an application for patent or for the registration of a utility model, industrial design, or model in respect of any invention made in the United States, shall, upon conviction, be fined not more than $10,000 or imprisoned for not more than two years, or both.

Sec. 187. Nonapplicability to Certain Persons

The prohibitions and penalties of this chapter shall not apply to any officer or agent of the United States acting within the scope of his authority, nor to any person acting upon his written instructions or permission.

Sec. 188. Rules and Regulations, Delegation of Power

The Atomic Energy Commission, the Secretary of a defense department, the chief officer of any other department or agency of the Government designated by the President as a defense agency of the United States, and the Secretary of Commerce, may separately issue rules and regulations to enable the respective department or agency to carry out the provisions of this chapter, and may delegate any power conferred by this chapter.

Chapter 18. Patent Rights in Inventions Made with Federal Assistance

Sec. 200. Policy and Objective

It is the policy and objective of the Congress to use the patent system to promote the utilization of inventions arising from federally supported research or development; to encourage maximum participation of small business firms in federally supported research and development efforts; to promote collaboration between commercial concerns and nonprofit organizations, including universities; to ensure that inventions made by nonprofit organizations and small business firms are used in a manner to promote free competition and enterprise without unduly encumbering future research and discovery; to promote the commercialization and public availability of inventions made in the United States by United States industry and labor; to ensure that the Government obtains sufficient rights in federally supported inventions to meet the needs of the Government and protect the public against nonuse or unreasonable use of inventions; and to minimize the costs of administering policies in this area.

Sec. 201. Definitions

As used in this chapter—

(a) The term "Federal agency" means any executive agency as defined in section 105 of title 5, and the military departments as defined by section 102 of title 5.

(b) The term "funding agreement" means any contract, grant, or cooperative agreement entered into between any Federal agency, other than the Tennessee Valley Authority, and any contractor for the performance of experimental, developmental, or research work funded in whole or in part by the Federal Government. Such term includes any assignment, substitution of parties, or subcontract of any type entered into for the performance of experimental, developmental, or research work under a funding agreement as herein defined.

(c) The term "contractor" means any person, small business firm, or nonprofit organization that is a party to a funding agreement.

(d) The term "invention" means any invention or discovery which is or may be patentable or otherwise protectable under this title or any novel variety of plant which is or may be protectable under the Plant Variety Protection Act (7 U.S.C. 2321 et seq.).

(e) The term "subject invention" means any invention of the contractor conceived or first actually reduced to practice in the performance of work under a funding agreement: *Provided,* That in the case of a variety of plant, the date of determination (as defined in section 41(d) of the Plant Variety Protection Act (7 U.S.C. 2401(d))) must also occur during the period of contract performance.

(f) The term "practical application" means to manufacture in the case of a composition or product, to practice in the case of a process or method, or to operate in the case of a machine or system; and, in each case, under such conditions as to establish that the invention is being utilized and that its benefits are to the extent permitted by law or Government regulations available to the public on reasonable terms.

(g) The term "made" when used in relation to any invention means the conception or first actual reduction to practice of such invention.

(h) The term "small business firm" means a small business concern as defined at section 2 of Public Law 85-536 (15 U.S.C. 632) and implementing regulations of the Administrator of the Small Business Administration.

(i) The term "nonprofit organization" means universities and other institutions of higher education or an organization of the type described in section 501(c)(3) of the Internal Revenue Code of 1986 (26 U.S.C. 501(c)) and exempt from taxation under section 501(a) of the Internal Revenue Code (26 U.S.C. 501(a)) or any nonprofit scientific or educational organization qualified under a State nonprofit organization statute.

Sec. 202. Disposition of Rights

(a) Each nonprofit organization or small business firm may, within a reasonable time after disclosure as required by paragraph (c)(1) of this section, elect to retain title to any subject invention: *Provided, however,* That a funding agreement may provide otherwise (i) when the contractor is not located in the United States or does not have a place of business located in the United States or is subject to the control of a foreign government, (ii) in exceptional circumstances when it is determined by the agency that restriction or elimination of the right to retain title to any subject invention will better promote the policy and objectives of this chapter (iii) when it is determined by a Government authority which is authorized by statute or Executive order to conduct foreign intelligence or counter-intelligence activities that the restriction or elimination of the right to retain title to any subject invention is necessary to protect the security of such activities or, (iv) when the funding agreement includes the operation of a Government-owned, contractor-operated facility of the Department of Energy primarily dedicated to that Department's naval nuclear propulsion or weapons related programs and all funding agreement limitations under this subparagraph on the contractor's right to elect title to a subject invention are limited to inventions occurring under the above two programs of the Department of Energy. The rights of the nonprofit organization or small business firm shall be subject to the provisions of paragraph (c) of this section and the other provisions of this chapter.

(b)(1) The rights of the Government under subsection (a) shall not be exercised by a Federal agency unless it first determines that at least one of the conditions identified in clauses (i) through (iv) of subsection (a) exists. Except in the case of subsection (a)(iii), the agency shall file with the Secretary of Commerce, within thirty days after the award of the applicable funding agreement, a copy of such determination. In the case of a determination under subsection (a)(ii), the statement shall include an analysis justifying the determination. In the case of determinations applicable to funding agreements with small business firms, copies shall also be sent to the Chief Counsel for Advocacy of the Small Business Administration. If the Secretary of Commerce believes that any individual determination or pattern of determinations is contrary to the policies and objectives of this chapter or otherwise not in conformance with this chapter, the Secretary shall so advise the head of the agency concerned and the Administrator of the Office of Federal Procurement Policy, and recommend corrective actions.

(2) Whenever the Administrator of the Office of Federal Procurement Policy has determined that one or more Federal agencies are utilizing the authority of clause (i) or (ii) of subsection (a) of this section in a manner that is contrary to the policies and objectives of this chapter, the Administrator is authorized to issue regulations describing classes of situations in which agencies may not exercise the authorities of those clauses.

(3) If the contractor believes that a determination is contrary to the policies and objectives of this chapter or constitutes an abuse of discretion by the agency, the determination shall be subject to section 203(b).

[(4) Redesignated (3)]

(c) Each funding agreement with a small business firm or nonprofit organization shall contain appropriate provisions to effectuate the following:

(1) That the contractor disclose each subject invention to the Federal agency within a reasonable time after it becomes known to contractor personnel responsible for the administration of patent matters, and that the Federal Government may receive title to any subject invention not disclosed to it within such time.

(2) That the contractor make a written election within two years after disclosure to the Federal agency (or such additional time as may be approved by the Federal agency) whether the contractor will retain title to a subject invention:

Provided, That in any case where the 1-year period referred to in section 102(b) would end before the end of that 2-year period, the period for election may be shortened by the Federal agency to a date that is not more than sixty days before the end of that 1-year period: *And provided further,* That the Federal Government may receive title to any subject invention in which the contractor does not elect to retain rights or fails to elect rights within such times.

(3) That a contractor electing rights in a subject invention agrees to file a patent application prior to the expiration of the 1-year period referred to in section 102(b), and shall thereafter file corresponding patent applications in other countries in which it wishes to retain title within reasonable times, and that the Federal Government may receive title to any subject inventions in the United States or other countries in which the contractor has not filed patent applications on the subject invention within such times.

(4) With respect to any invention in which the contractor elects rights, the Federal agency shall have a nonexclusive, nontransferable, irrevocable, paid-up license to practice or have practiced for or on behalf of the United States any subject invention throughout the world: *Provided,* That the funding agreement may provide for such additional rights, including the right to assign or have assigned foreign patent rights in the subject invention, as are determined by the agency as necessary for meeting the obligations of the United States under any treaty, international agreement, arrangement of cooperation, memorandum of understanding, or similar arrangement, including military agreement relating to weapons development and production.

(5) The right of the Federal agency to require periodic reporting on the utilization or efforts at obtaining utilization that are being made by the contractor or his licensees or assignees: *Provided,* That any such information as well as any information on utilization or efforts at obtaining utilization obtained as part of a proceeding under section 203 of this chapter shall be treated by the Federal agency as commercial and financial information obtained from a person and privileged and confidential and not subject to disclosure under section 552 of title 5.

(6) An obligation on the part of the contractor, in the event a United States patent application is filed by or on its behalf or by any assignee of the contractor, to include within the specification of such application and any patent issuing thereon, a statement specifying that the invention was made with Government support and that the Government has certain rights in the invention.

(7) In the case of a nonprofit organization, (A) a prohibition upon the assignment of rights to a subject invention in the United States without the approval of the Federal agency, except where such assignment is made to an organization which has as one of its primary functions the management of inventions (provided that such assignee shall be subject to the same provisions as the contractor); (B) a requirement that the contractor share royalties with the inventor; (C) except with respect to a funding agreement for the operation of a Government-owned-contractor-operated facility, a requirement that the balance of any royalties or income earned by the contractor with respect to subject inventions, after payment of expenses (including payments to inventors) incidental to the administration of subject inventions, be utilized for the support of scientific research or

education; (D) a requirement that, except where it is determined to be infeasible following a reasonable inquiry, a preference in the licensing of subject inventions shall be given to small business firms; and (E) with respect to a funding agreement for the operation of a Government-owned-contractor-operated facility, requirements (i) that after payment of patenting costs, licensing costs, payments to inventors, and other expenses incidental to the administration of subject inventions, 100 percent of the balance of any royalties or income earned and retained by the contractor during any fiscal year up to an amount equal to 5 percent of the annual budget of the facility, shall be used by the contractor for scientific research, development, and education consistent with the research and development mission and objectives of the facility, including activities that increase the licensing potential of other inventions of the facility; provided that if said balance exceeds 5 percent of the annual budget of the facility, that 15 percent of such excess shall be paid to the Treasury of the United States and the remaining 85 percent shall be used for the same purposes described above in this clause; and (ii) that, to the extent it provides the most effective technology transfer, the licensing of subject inventions shall be administered by contractor employees on location at the facility.

(8) The requirements of sections 203 and 204 of this chapter.

(d) If a contractor does not elect to retain title to a subject invention in cases subject to this section, the Federal agency may consider and after consultation with the contractor grant requests for retention of rights by the inventor subject to the provisions of this Act and regulations promulgated hereunder.

(e) In any case when a Federal employee is a coinventor of any invention made with a nonprofit organization, a small business firm, or a non-Federal inventor, the Federal agency employing such coinventor may, for the purpose of consolidating rights in the invention and if it finds that it would expedite the development of the invention—

(1) license or assign whatever rights it may acquire in the subject invention to the nonprofit organization, small business firm, or non-Federal inventor in accordance with the provisions of this chapter; or

(2) acquire any rights in the subject invention from the nonprofit organization, small business firm, or non-Federal inventor, but only to the extent the party from whom the rights are acquired voluntarily enters into the transaction and no other transaction under this chapter is conditioned on such acquisition.

(f) (1) No funding agreement with a small business firm or nonprofit organization shall contain a provision allowing a Federal agency to require the licensing to third parties of inventions owned by the contractor that are not subject inventions unless such provision has been approved by the head of the agency and a written justification has been signed by the head of the agency. Any such provision shall clearly state whether the licensing may be required in connection with the practice of a subject invention, a specifically identified work object, or both. The head of the agency may not delegate the authority to approve provisions or sign justifications required by this paragraph.

(2) A Federal agency shall not require the licensing of third parties under any such provision unless the head of the agency determines that the use of the invention by others is necessary for the practice of a subject invention or for the use of a work object of the funding agreement and that such action is necessary to achieve the practical application of the subject invention or work object. Any such determination shall be on the record after an opportunity for an agency hearing. Any action commenced for judicial review of such determination shall be brought within sixty days after notification of such determination.

Sec. 203. March-in Rights

(a) With respect to any subject invention in which a small business firm or nonprofit organization has acquired title under this chapter, the Federal agency under whose funding agreement the subject invention was made shall have the right, in accordance with such procedures as are provided in regulations promulgated hereunder to require the contractor, an assignee or exclusive licensee of a subject invention to grant a nonexclusive, partially exclusive, or exclusive license in any field of use to a responsible applicant or applicants, upon terms that are reasonable under the circumstances, and if the contractor, assignee, or exclusive licensee refuses such request, to grant such a license itself, if the Federal agency determines that such—

(1) action is necessary because the contractor or assignee has not taken, or is not expected to take within a reasonable time, effective steps to achieve practical application of the subject invention in such field of use;

(2) action is necessary to alleviate health or safety needs which are not reasonably satisfied by the contractor, assignee, or their licensees;

(3) action is necessary to meet requirements for public use specified by Federal regulations and such requirements are not reasonably satisfied by the contractor, assignee, or licensees; or

(4) action is necessary because the agreement required by section 204 has not been obtained or waived or because a licensee of the exclusive right to use or sell any subject invention in the United States is in breach of its agreement obtained pursuant to section 204.

(b) A determination pursuant to this section or section 202(b)(4) shall not be subject to chapter 71 of title 41. An administrative appeals procedure shall be established by regulations promulgated in accordance with section 206. Additionally, any contractor, inventor, assignee, or exclusive licensee adversely affected by a determination under this section may, at any time within sixty days after the determination is issued, file a petition in the United States Court of Federal Claims, which shall have jurisdiction to determine the appeal on the record and to affirm, reverse, remand or modify, as appropriate, the determination of the Federal agency. In cases described in paragraphs (1) and (3) of subsection (a), the agency's determination shall be held in abeyance pending the exhaustion of appeals or petitions filed under the preceding sentence.

Sec. 204. Preference for United States Industry

Notwithstanding any other provision of this chapter, no small business firm or nonprofit organization which receives title to any subject invention and no assignee of any such small business firm or nonprofit organization shall grant to any person the exclusive right to use or sell any subject invention in the United States unless such person agrees that any products embodying the subject invention or produced through the use of the subject invention will be manufactured substantially in the United States. However, in individual cases, the requirement for such an agreement may be waived by the Federal agency under whose funding agreement the invention was made upon a showing by the small business firm, nonprofit organization, or assignee that reasonable but unsuccessful efforts have been made to grant licenses on similar terms to potential licensees that would be likely to manufacture substantially in the United States or that under the circumstances domestic manufacture is not commercially feasible.

Sec. 205. Confidentiality

Federal agencies are authorized to withhold from disclosure to the public information disclosing any invention in which the Federal Government owns or may own a right, title, or interest (including a nonexclusive license) for a reasonable time in order for a patent application to be filed. Furthermore, Federal agencies shall not be required to release copies of any document which is part of an application for patent filed with the United States Patent and Trademark Office or with any foreign patent office.

Sec. 206. Uniform Clauses and Regulations

The Secretary of Commerce may issue regulations which may be made applicable to Federal agencies implementing the provisions of sections 202 through 204 of this chapter and shall establish standard funding agreement provisions required under this chapter. The regulations and the standard funding agreement shall be subject to public comment before their issuance.

Sec. 207. Domestic and Foreign Protection of Federally Owned Inventions

(a) Each Federal agency is authorized to—

(1) apply for, obtain, and maintain patents or other forms of protection in the United States and in foreign countries on inventions in which the Federal Government owns a right, title, or interest;

(2) grant nonexclusive, exclusive, or partially exclusive licenses under federally owned inventions, royalty-free or for royalties or other consideration, and

on such terms and conditions, including the grant to the licensee of the right of enforcement pursuant to the provisions of chapter 29 of this title as determined appropriate in the public interest;

(3) undertake all other suitable and necessary steps to protect and administer rights to federally owned inventions on behalf of the Federal Government either directly or through contract, including acquiring rights for and administering royalties to the Federal Government in any invention, but only to the extent the party from whom the rights are acquired voluntarily enters into the transaction, to facilitate the licensing of a federally owned invention; and

(4) transfer custody and administration, in whole or in part, to another Federal agency, of the right, title, or interest in any federally owned invention.

(b) For the purpose of assuring the effective management of Government-owned inventions, the Secretary of Commerce is authorized to—

(1) assist Federal agency efforts to promote the licensing and utilization of Government-owned inventions;

(2) assist Federal agencies in seeking protection and maintaining inventions in foreign countries, including the payment of fees and costs connected therewith; and

(3) consult with and advise Federal agencies as to areas of science and technology research and development with potential for commercial utilization.

Sec. 208. Regulations Governing Federal Licensing

The Secretary of Commerce is authorized to promulgate regulations specifying the terms and conditions upon which any federally owned invention, other than inventions owned by the Tennessee Valley Authority, may be licensed on a nonexclusive, partially exclusive, or exclusive basis.

Sec. 209. Licensing Federally Owned Inventions

(a) Authority.— A Federal agency may grant an exclusive or partially exclusive license on a federally owned invention under section 207(a)(2) only if—

(1) Granting the license is a reasonable and necessary incentive to—

(A) call forth the investment capital and expenditures needed to bring the invention to practical application; or

(B) otherwise promote the invention's utilization by the public;

(2) the Federal agency finds that the public will be served by the granting of the license, as indicated by the applicant's intentions, plans, and ability to bring the invention to practical application or otherwise promote the invention's utilization by the public, and that the proposed scope of exclusivity is not greater than reasonably necessary to provide the incentive for bringing the invention to practical application, as proposed by the applicant, or otherwise to promote the invention's utilization by the public;

(3) the applicant makes a commitment to achieve practical application of the invention within a reasonable time, which time may be extended by the agency upon the applicant's request and the applicant's demonstration that the refusal of such extension would be unreasonable;

(4) granting the license will not tend to substantially lessen competition or create or maintain a violation of the Federal antitrust laws; and

(5) in the case of an invention covered by a foreign patent application or patent, the interests of the Federal Government or United States industry in foreign commerce will be enhanced.

(b) Manufacture in United States.— A Federal agency shall normally grant a license under section 207(a)(2) to use or sell any federally owned invention in the United States only to a licensee who agrees that any products embodying the invention or produced through the use of the invention will be manufactured substantially in the United States.

(c) Small business.— First preference for the granting of any exclusive or partially exclusive licenses under section 207(a)(2) shall be given to small business firms having equal or greater likelihood as other applicants to bring the invention to practical application within a reasonable time.

(d) Terms and conditions.— Any licenses granted under section 207(a)(2) shall contain such terms and conditions as the granting agency considers appropriate, and shall include provisions—

(1) retaining a nontransferable, irrevocable, paid-up license for any Federal agency to practice the invention or have the invention practiced throughout the world by or on behalf of the Government of the United States;

(2) requiring periodic reporting on utilization of the invention, and utilization efforts, by the licensee, but only to the extent necessary to enable the Federal agency to determine whether the terms of the license are being complied with, except that any such report shall be treated by the Federal agency as commercial and financial information obtained from a person and privileged and confidential and not subject to disclosure under section 552 of title 5; and

(3) Empowering the Federal agency to terminate the license in whole or in part if the agency determines that—

(A) the licensee is not executing its commitment to achieve practical application of the invention, including commitments contained in any plan submitted in support of its request for a license, and the licensee cannot otherwise demonstrate to the satisfaction of the Federal agency that it has taken, or can be expected to take within a reasonable time, effective steps to achieve practical application of the invention;

(B) the licensee is in breach of an agreement described in subsection (b);

(C) termination is necessary to meet requirements for public use specified by Federal regulations issued after the date of the license, and such requirements are not reasonably satisfied by the licensee; or

(D) the licensee has been found by a court of competent jurisdiction to have violated the Federal antitrust laws in connection with its performance under the license agreement.

(e) Public notice.— No exclusive or partially exclusive license may be granted under section 207(a)(2) unless public notice of the intention to grant an exclusive or partially exclusive license on a federally owned invention has been provided in an appropriate manner at least 15 days before the license is granted, and the Federal agency has considered all comments received before the end of the comment period in response to that public notice. This subsection shall not apply to the licensing of inventions made under a cooperative research and development agreement entered into under section 12 of the Stevenson-Wydler Technology Innovation Act of 1980 (15 U.S.C. 3710a).

(f) Plan.— No Federal agency shall grant any license under a patent or patent application on a federally owned invention unless the person requesting the license has supplied the agency with a plan for development or marketing of the invention, except that any such plan shall be treated by the Federal agency as commercial and financial information obtained from a person and privileged and confidential and not subject to disclosure under section 552 of title 5.

Sec. 210. Precedence of Chapter

(a) This chapter shall take precedence over any other Act which would require a disposition of rights in subject inventions of small business firms or nonprofit organizations contractors in a manner that is inconsistent with this chapter, including but not necessarily limited to the following:

(1) section 10(a) of the Act of June 29, 1935, as added by title I of the Act of August 14, 1946 (7 U.S.C. 427i(a); 60 Stat. 1085);

(2) section 205(a) of the Act of August 14, 1946 (7 U.S.C. 1624(a); 60 Stat. 1090);

(3) section 501(c) of the Federal Mine Safety and Health Act of 1977 (30 U.S.C. 951(c); 83 Stat. 742);

(4) section 30168(e) of title 49;

(5) section 12 of the National Science Foundation Act of 1950 (42 U.S.C. 1871(a); 82 Stat. 360);

(6) section 152 of the Atomic Energy Act of 1954 (42 U.S.C. 2182; 68 Stat. 943);

(7) section 20135 of title 51;

(8) section 6 of the Coal Research and Development Act of 1960 (30 U.S.C. 666; 74 Stat. 337);

(9) section 4 of the Helium Act Amendments of 1960 (50 U.S.C. 167b; 74 Stat. 920);

(10) section 32 of the Arms Control and Disarmament Act of 1961 (22 U.S.C. 2572; 75 Stat. 634);

(11) section 9 of the Federal Nonnuclear Energy Research and Development Act of 1974 (42 U.S.C. 5908; 88 Stat. 1878);

(12) section 5(d) of the Consumer Product Safety Act (15 U.S.C. 2054(d); 86 Stat. 1211);

(13) section 3 of the Act of April 5, 1944 (30 U.S.C. 323; 58 Stat. 191);

(14) section 8001(c)(3) of the Solid Waste Disposal Act (42 U.S.C. 6981(c); 90 Stat. 2829);

(15) section 219 of the Foreign Assistance Act of 1961 (22 U.S.C. 2179; 83 Stat. 806);

(16) section 427(b) of the Federal Mine Health and Safety Act of 1977 (30 U.S.C. 937(b); 86 Stat. 155);

(17) section 306(d) of the Surface Mining and Reclamation Act of 1977 (30 U.S.C. 1226(d); 91 Stat. 455);

(18) section 21(d) of the Federal Fire Prevention and Control Act of 1974 (15 U.S.C. 2218(d); 88 Stat. 1548);

(19) section 6(b) of the Solar Photovoltaic Energy Research Development and Demonstration Act of 1978 (42 U.S.C. 5585(b); 92 Stat. 2516);

(20) section 12 of the Native Latex Commercialization and Economic Development Act of 1978 (7 U.S.C. 178j, 92 Stat. 2533); and

(21) section 408 of the Water Resources and Development Act of 1978 (42 U.S.C. 7879; 92 Stat. 1360).

The Act creating this chapter shall be construed to take precedence over any future Act unless that Act specifically cites this Act and provides that it shall take precedence over this Act.

(b) Nothing in this chapter is intended to alter the effect of the laws cited in paragraph (a) of this section or any other laws with respect to the disposition of rights in inventions made in the performance of funding agreements with persons other than nonprofit organizations or small business firms.

(c) Nothing in this chapter is intended to limit the authority of agencies to agree to the disposition of rights in inventions made in the performance of work under funding agreements with persons other than nonprofit organizations or small business firms in accordance with the Statement of Government Patent Policy issued on February 18, 1983, agency regulations, or other applicable regulations or to otherwise limit the authority of agencies to allow such persons to retain ownership of inventions except that all funding agreements, including those with other than small business firms and nonprofit organizations, shall include the requirements established in section 202(c)(4) and section 203 of this title. Any disposition of rights in inventions made in accordance with the Statement or implementing regulations, including any disposition occurring before enactment of this section, are hereby authorized.

(d) Nothing in this chapter shall be construed to require the disclosure of intelligence sources or methods or to otherwise affect the authority granted to the Director of Central Intelligence by statute or Executive order for the protection of intelligence sources or methods.

(e) The provisions of the Stevenson-Wydler Technology Innovation Act of 1980 shall take precedence over the provisions of this chapter to the extent that they permit or require a disposition of rights in subject inventions which is inconsistent with this chapter.

Sec. 211. Relationship to Antitrust Laws

Nothing in this chapter shall be deemed to convey to any person immunity from civil or criminal liability, or to create any defenses to actions, under any antitrust law.

Sec. 212. Disposition of Rights in Educational Awards

No scholarship, fellowship, training grant, or other funding agreement made by a Federal agency primarily to an awardee for educational purposes will contain any provision giving the Federal agency any rights to inventions made by the awardee.

Part III. Patents and Protection of Patent Rights

Chapter 25. Amendment and Correction of Patents (Refs & Annos)

Sec. 251. Reissue of Defective Patents

(a) In general.—Whenever any patent is, through error, deemed wholly or partly inoperative or invalid, by reason of a defective specification or drawing, or by reason of the patentee claiming more or less than he had a right to claim in the patent, the Director shall, on the surrender of such patent and the payment of the fee required by law, reissue the patent for the invention disclosed in the original patent, and in accordance with a new and amended application, for the unexpired part of the term of the original patent. No new matter shall be introduced into the application for reissue.

(b) Multiple reissued patents.—The Director may issue several reissued patents for distinct and separate parts of the thing patented, upon demand of the applicant, and upon payment of the required fee for a reissue for each of such reissued patents.

(c) Applicability of this title.—The provisions of this title relating to applications for patent shall be applicable to applications for reissue of a patent, except that application for reissue may be made and sworn to by the assignee of the entire interest if the application does not seek to enlarge the scope of the claims of the original patent or the application for the original patent was filed by the assignee of the entire interest.

(d) Reissue patent enlarging scope of claims.—No reissued patent shall be granted enlarging the scope of the claims of the original patent unless applied for within two years from the grant of the original patent.

Sec. 252. Effect of Reissue

The surrender of the original patent shall take effect upon the issue of the reissued patent, and every reissued patent shall have the same effect and operation in law, on the trial of actions for causes thereafter arising, as if the same had been originally granted in such amended form, but in so far as the claims of the original and reissued patents are substantially identical, such surrender shall not affect any action then pending nor abate any cause of action then existing, and the reissued patent, to the extent that its claims are substantially identical with the original patent, shall constitute a continuation thereof and have effect continuously from the date of the original patent.

A reissued patent shall not abridge or affect the right of any person or that person's successors in business who, prior to the grant of a reissue, made, purchased, offered to sell, or used within the United States, or imported into the United States, anything patented by the reissued patent, to continue the use of, to offer to sell, or to sell to others to be used, offered for sale, or sold, the specific thing so made, purchased, offered for sale, used, or imported unless the making, using, offering for sale, or selling of such thing infringes a valid claim of the reissued patent which was in the original patent. The court before which such matter is in question may provide for the continued manufacture, use, offer for sale, or sale of the thing made, purchased, offered for sale, used, or imported as specified, or for the manufacture, use, offer for sale, or sale in the United States of which substantial preparation was made before the grant of the reissue, and the court may also provide for the continued practice of any process patented by the reissue that is practiced, or for the practice of which substantial preparation was made, before the grant of the reissue, to the extent and under such terms as the court deems equitable for the protection of investments made or business commenced before the grant of the reissue.

Sec. 253. Disclaimer

(a) In general.—Whenever a claim of a patent is invalid the remaining claims shall not thereby be rendered invalid. A patentee, whether of the whole or any sectional interest therein, may, on payment of the fee required by law, make disclaimer of any complete claim, stating therein the extent of his interest in such patent. Such disclaimer shall be in writing, and recorded in the Patent and Trademark Office; and it shall thereafter be considered as part of the original patent to the extent of the interest possessed by the disclaimant and by those claiming under him.

(b) Additional disclaimer or dedication.—In the manner set forth in subsection (a), any patentee or applicant may disclaim or dedicate to the public the entire term, or any terminal part of the term, of the patent granted or to be granted.

Sec. 254. Certificate of Correction of Patent and Trademark Office Mistake

Whenever a mistake in a patent, incurred through the fault of the Patent and Trademark Office, is clearly disclosed by the records of the Office, the Director may issue a certificate of correction stating the fact and nature of such mistake, under seal, without charge, to be recorded in the records of patents. A printed copy thereof shall be attached to each printed copy of the patent, and such certificate shall be considered as part of the original patent. Every such patent, together with such certificate, shall have the same effect and operation in law on the trial of actions for causes thereafter arising as if the same had been originally issued in such corrected form. The Director may issue a corrected patent without charge in lieu of and with like effect as a certificate of correction.

Sec. 255. Certificate of Correction of Applicant's Mistake

Whenever a mistake of a clerical or typographical nature, or of minor character, which was not the fault of the Patent and Trademark Office, appears in a patent and a showing has been made that such mistake occurred in good faith, the Director may, upon payment of the required fee, issue a certificate of correction, if the correction does not involve such changes in the patent as would constitute new matter or would require re-examination. Such patent, together with the certificate, shall have the same effect and operation in law on the trial of actions for causes thereafter arising as if the same had been originally issued in such corrected form.

Sec. 256. Correction of Named Inventor

(a) Correction.—Whenever through error a person is named in an issued patent as the inventor, or through error an inventor is not named in an issued patent, the Director may, on application of all the parties and assignees, with proof of the facts and such other requirements as may be imposed, issue a certificate correcting such error.

(b) Patent valid if error corrected.—The error of omitting inventors or naming persons who are not inventors shall not invalidate the patent in which such error occurred if it can be corrected as provided in this section. The court before which such matter is called in question may order correction of the patent on notice and hearing of all parties concerned and the Director shall issue a certificate accordingly.

Sec. 257. Supplemental examinations to consider, reconsider, or correct information

(a) Request for supplemental examination.—A patent owner may request supplemental examination of a patent in the Office to consider, reconsider, or correct information believed to be relevant to the patent, in accordance with such requirements as the Director may establish. Within 3 months after the date a request for supplemental examination meeting the requirements of this section is received, the Director shall conduct the supplemental examination and shall conclude such examination by issuing a certificate indicating whether the information presented in the request raises a substantial new question of patentability.

(b) Reexamination ordered.—If the certificate issued under subsection (a) indicates that a substantial new question of patentability is raised by 1 or more items of information in the request, the Director shall order reexamination of the patent. The reexamination shall be conducted according to procedures established by chapter 30, except that the patent owner shall not have the right to file a statement pursuant to section 304. During the reexamination, the Director shall address each substantial new question of patentability identified during the supplemental examination, notwithstanding the limitations in chapter 30 relating to patents and printed publication or any other provision of such chapter.

(c) Effect.—

(1) In general.—A patent shall not be held unenforceable on the basis of conduct relating to information that had not been considered, was inadequately considered, or was incorrect in a prior examination of the patent if the information was considered, reconsidered, or corrected during a supplemental examination of the patent. The making of a request under subsection (a), or the absence thereof, shall not be relevant to enforceability of the patent under section 282.

(2) Exceptions.—

(A) Prior allegations.—Paragraph (1) shall not apply to an allegation pled with particularity in a civil action, or set forth with particularity in a notice received by the patent owner under section 505(j)(2)(B)(iv)(II) of the Federal Food, Drug, and Cosmetic Act (21 U.S.C. 355(j)(2)(B)(iv)(II)), before the date of a supplemental examination request under subsection (a) to consider, reconsider, or correct information forming the basis for the allegation.

(B) Patent enforcement actions.—In an action brought under section 337(a) of the Tariff Act of 1930 (19 U.S.C. 1337(a)), or section 281, paragraph (1) shall not apply to any defense raised in the action that is based upon information that was considered, reconsidered, or corrected pursuant to a supplemental examination request under subsection (a), unless the supplemental examination, and any reexamination ordered pursuant to the request, are concluded before the date on which the action is brought.

(d) Fees and regulations.—

(1) Fees.—The Director shall, by regulation, establish fees for the submission of a request for supplemental examination of a patent, and to consider each item of information submitted in the request. If reexamination is ordered under subsection (b), fees established and applicable to ex parte reexamination proceedings under chapter 30 shall be paid, in addition to fees applicable to supplemental examination.

(2) Regulations.—The Director shall issue regulations governing the form, content, and other requirements of requests for supplemental examination, and establishing procedures for reviewing information submitted in such requests.

(e) Fraud.—If the Director becomes aware, during the course of a supplemental examination or reexamination proceeding ordered under this section, that a material fraud on the Office may have been committed in connection with the patent that is the subject of the supplemental examination, then in addition to any other actions the Director is authorized to take, including the cancellation of any claims found to be invalid under section 307 as a result of a reexamination ordered under this section, the Director shall also refer the matter to the Attorney General for such further action as the Attorney General may deem appropriate. Any such referral shall be treated as confidential, shall not be included in the file of the patent, and shall not be disclosed to the public unless the United States charges a person with a criminal offense in connection with such referral.

(f) Rule of construction.—Nothing in this section shall be construed—

(1) to preclude the imposition of sanctions based upon criminal or antitrust laws (including section 1001(a) of title 18, the first section of the Clayton Act, and section 5 of the Federal Trade Commission Act to the extent that section relates to unfair methods of competition);

(2) to limit the authority of the Director to investigate issues of possible misconduct and impose sanctions for misconduct in connection with matters or proceedings before the Office; or

(3) to limit the authority of the Director to issue regulations under chapter 3 relating to sanctions for misconduct by representatives practicing before the Office.

Chapter 26. Ownership and Assignment

Sec. 261. Ownership; Assignment

Subject to the provisions of this title, patents shall have the attributes of personal property.

Applications for patent, patents, or any interest therein, shall be assignable in law by an instrument in writing. The applicant, patentee, or his assigns or legal representatives may in like manner grant and convey an exclusive right under his application for patent, or patents, to the whole or any specified part of the United States.

A certificate of acknowledgment under the hand and official seal of a person authorized to administer oaths within the United States, or, in a foreign country, of a diplomatic or consular officer of the United States or an officer authorized to administer oaths whose authority is proved by a certificate of a diplomatic or consular officer of the United States, or apostille of an official designated by a foreign country which, by treaty or convention, accords like effect to apostilles of designated officials in the United States, shall be prima facie evidence of the execution of an assignment, grant or conveyance of a patent or application for patent.

An assignment, grant or conveyance shall be void as against any subsequent purchaser or mortgagee for a valuable consideration, without notice, unless it is recorded in the Patent and Trademark Office within three months from its date or prior to the date of such subsequent purchase or mortgage.

Sec. 262. Joint Owners

In the absence of any agreement to the contrary, each of the joint owners of a patent may make, use, offer to sell, or sell the patented invention within the United States, or import the patented invention into the United States, without the consent of and without accounting to the other owners.

Chapter 27. Government Interests in Patents

Sec. 267. Time for Taking action in Government Applications

Notwithstanding the provisions of sections 133 and 151 of this title, the Director may extend the time for taking any action to three years, when an application has become the property of the United States and the head of the appropriate department or agency of the Government has certified to the Director that the invention disclosed therein is important to the armament or defense of the United States.

Chapter 28. Infringement of Patents

Sec. 271. Infringement of Patent

(a) Except as otherwise provided in this title, whoever without authority makes, uses, offers to sell, or sells any patented invention, within the United States or imports into the United States any patented invention during the term of the patent therefor, infringes the patent.

(b) Whoever actively induces infringement of a patent shall be liable as an infringer.

(c) Whoever offers to sell or sells within the United States or imports into the United States a component of a patented machine, manufacture, combination or composition, or a material or apparatus for use in practicing a patented process, constituting a material part of the invention, knowing the same to be especially made or especially adapted for use in an infringement of such patent, and not a staple article or commodity of commerce suitable for substantial noninfringing use, shall be liable as a contributory infringer.

(d) No patent owner otherwise entitled to relief for infringement or contributory infringement of a patent shall be denied relief or deemed guilty of misuse or illegal extension of the patent right by reason of his having done one or more of the following: (1) derived revenue from acts which if performed by another without his consent would constitute contributory infringement of the patent; (2) licensed or authorized another to perform acts which if performed without his consent would constitute contributory infringement of the patent; (3) sought to enforce his patent rights against infringement or contributory infringement; (4) refused to license or use any rights to the patent; or (5) conditioned the license of any rights to the patent or the sale of the patented product on the acquisition of a license to rights in another patent or purchase of a separate product, unless, in view of the circumstances, the patent owner has market power in the relevant market for the patent or patented product on which the license or sale is conditioned.

(e)(1) It shall not be an act of infringement to make, use, offer to sell, or sell within the United States or import into the United States a patented invention (other than a new animal drug or veterinary biological product (as those terms are used in the Federal Food, Drug, and Cosmetic Act and the Act of March 4, 1913) which is primarily manufactured using recombinant DNA, recombinant RNA, hybridoma technology, or other processes involving site specific genetic manipulation techniques) solely for uses reasonably related to the development and submission of information under a Federal law which regulates the manufacture, use, or sale of drugs or veterinary biological products.

(2) It shall be an act of infringement to submit—

(A) an application under section 505(j) of the Federal Food, Drug, and Cosmetic Act or described in section 505(b)(2) of such Act for a drug claimed in a patent or the use of which is claimed in a patent,

(B) an application under section 512 of such Act or under the Act of March 4, 1913 (21 U.S.C. 151-158) for a drug or veterinary biological product which is not primarily manufactured using recombinant DNA, recombinant RNA, hybridoma technology, or other processes involving site specific genetic manipulation techniques and which is claimed in a patent or the use of which is claimed in a patent, or

(C)(i) with respect to a patent that is identified in the list of patents described in section 351(l)(3) of the Public Health Service Act (including as provided under section 351(l)(7) of such Act), an application seeking approval of a biological product, or

(ii) if the applicant for the application fails to provide the application and information required under section 351(l)(2)(A) of such Act, an application seeking approval of a biological product for a patent that could be identified pursuant to section 351(l)(3)(A)(i) of such Act,

if the purpose of such submission is to obtain approval under such Act to engage in the commercial manufacture, use, or sale of a drug, veterinary biological product, or biological product claimed in a patent or the use of which is claimed in a patent before the expiration of such patent.

(3) In any action for patent infringement brought under this section, no injunctive or other relief may be granted which would prohibit the making, using, offering to sell, or selling within the United States or importing into the United States of a patented invention under paragraph (1).

(4) For an act of infringement described in paragraph (2)—

(A) the court shall order the effective date of any approval of the drug or veterinary biological product involved in the infringement to be a date which is not earlier than the date of the expiration of the patent which has been infringed,

(B) injunctive relief may be granted against an infringer to prevent the commercial manufacture, use, offer to sell, or sale within the United States or importation into the United States of an approved drug, veterinary biological product, or biological product,

(C) damages or other monetary relief may be awarded against an infringer only if there has been commercial manufacture, use, offer to sell, or sale within the United States or importation into the United States of an approved drug, veterinary biological product, or biological product, and

(D) the court shall order a permanent injunction prohibiting any infringement of the patent by the biological product involved in the infringement until a date which is not earlier than the date of the expiration of the patent that has been infringed under paragraph (2)(C), provided the patent is the subject of a final court decision, as defined in section 351(k)(6) of the Public Health Service Act, in an action for infringement of the patent under section 351(l)(6) of such Act, and the biological product has not yet been approved because of section 351(k)(7) of such Act.

The remedies prescribed by subparagraphs (A), (B), (C), and (D) are the only remedies which may be granted by a court for an act of infringement described in paragraph (2), except that a court may award attorney fees under section 285.

(5) Where a person has filed an application described in paragraph (2) that includes a certification under subsection (b)(2)(A)(iv) or (j)(2)(A)(vii)(IV) of section 505 of the Federal Food, Drug, and Cosmetic Act (21 U.S.C. 355), and neither the owner of the patent that is the subject of the certification nor the holder of the approved application under subsection (b) of such section for the drug that is claimed by the patent or a use of which is claimed by the patent brought an action for infringement of such patent before the expiration of 45 days after the date on which the notice given under subsection (b)(3) or (j)(2)(B) of such section was received, the courts of the United States shall, to the extent consistent with the Constitution, have subject matter jurisdiction in any action brought by such person under section 2201 of title 28 for a declaratory judgment that such patent is invalid or not infringed.

(6)(A) Subparagraph (B) Applies, in lieu of paragraph (4), in the case of a patent—

(i) that is identified, as applicable, in the list of patents described in section 351(l)(4) of the Public Health Service Act or the lists of patents described in section 351(l)(5)(B) of such Act with respect to a biological product; and

(ii) for which an action for infringement of the patent with respect to the biological product—

(I) was brought after the expiration of the 30-day period described in subparagraph (A) or (B), as applicable, of section 351(l)(6) of such Act; or

(II) was brought before the expiration of the 30-day period described in subclause (I), but which was dismissed without prejudice or was not prosecuted to judgment in good faith.

(B) In an action for infringement of a patent described in subparagraph (A), the sole and exclusive remedy that may be granted by a court, upon a finding that the making, using, offering to sell, selling, or importation into the United States of the biological product that is the subject of the action infringed the patent, shall be a reasonable royalty.

(C) The owner of a patent that should have been included in the list described in section 351(l)(3)(A) of the Public Health Service Act, including as provided under section 351(l)(7) of such Act for a biological product, but was not timely included in such list, may not bring an action under this section for infringement of the patent with respect to the biological product.

(f)(1) Whoever without authority supplies or causes to be supplied in or from the United States all or a substantial portion of the components of a patented invention, where such components are uncombined in whole or in part, in such manner as to actively induce the combination of such components outside of the United States in a manner that would infringe the patent if such combination occurred within the United States, shall be liable as an infringer.

(2) Whoever without authority supplies or causes to be supplied in or from the United States any component of a patented invention that is especially made or especially adapted for use in the invention and not a staple article or commodity of commerce suitable for substantial noninfringing use, where such component is uncombined in whole or in part, knowing that such component is so made or adapted and intending that such component will be combined outside of the United States in a manner that would infringe the patent if such combination occurred within the United States, shall be liable as an infringer.

(g) Whoever without authority imports into the United States or offers to sell, sells, or uses within the United States a product which is made by a process patented in the United States shall be liable as an infringer, if the importation, offer to sell, sale, or use of the product occurs during the term of such process patent. In an action for infringement of a process patent, no remedy may be granted for infringement on account of the noncommercial use or retail sale of a product unless there is no adequate remedy under this title for infringement on account of the importation or other use, offer to sell, or sale of that product. A product which is made by a patented process will, for purposes of this title, not be considered to be so made after—

(1) it is materially changed by subsequent processes; or

(2) it becomes a trivial and nonessential component of another product.

(h) As used in this section, the term "whoever" includes any State, any instrumentality of a State, and any officer or employee of a State or instrumentality of a State acting in his official capacity. Any State, and any such instrumentality, officer, or employee, shall be subject to the provisions of this title in the same manner and to the same extent as any nongovernmental entity.

(i) As used in this section, an "offer for sale" or an "offer to sell" by a person other than the patentee, or any designee of the patentee, is that in which the sale will occur before the expiration of the term of the patent.

Sec. 272. Temporary Presence in the United States

The use of any invention in any vessel, aircraft or vehicle of any country which affords similar privileges to vessels, aircraft or vehicles of the United States, entering the United States temporarily or accidentally, shall not constitute infringement of any patent, if the invention is used exclusively for the needs of the vessel, aircraft or vehicle and is not offered for sale or sold in or used for the manufacture of anything to be sold in or exported from the United States.

Sec. 273. Defense to Infringement Based on Prior Commercial Use

(a) In general.—A person shall be entitled to a defense under section 282(b) with respect to subject matter consisting of a process, or consisting of a machine, manufacture, or composition of matter used in a manufacturing or other commercial process, that would otherwise infringe a claimed invention being asserted against the person if—

(1) such person, acting in good faith, commercially used the subject matter in the United States, either in connection with an internal commercial use or an actual arm's length sale or other arm's length commercial transfer of a useful end result of such commercial use; and

(2) such commercial use occurred at least 1 year before the earlier of either—

(A) the effective filing date of the claimed invention; or

(B) the date on which the claimed invention was disclosed to the public in a manner that qualified for the exception from prior art under section 102(b).

(b) Burden of proof.—A person asserting a defense under this section shall have the burden of establishing the defense by clear and convincing evidence.

(c) Additional commercial uses.—

(1) Premarketing regulatory review.—Subject matter for which commercial marketing or use is subject to a premarketing regulatory review period during which the safety or efficacy of the subject matter is established, including any period specified in section 156(g), shall be deemed to be commercially used for purposes of subsection (a)(1) during such regulatory review period.

(2) Nonprofit laboratory use.—A use of subject matter by a nonprofit research laboratory or other nonprofit entity, such as a university or hospital, for which the public is the intended beneficiary, shall be deemed to be a commercial use for purposes of subsection (a)(1), except that a defense under this section may be asserted pursuant to this paragraph only for continued and noncommercial use by and in the laboratory or other nonprofit entity.

(d) Exhaustion of rights.—Notwithstanding subsection (e)(1), the sale or other disposition of a useful end result by a person entitled to assert a defense under this section in connection with a patent with respect to that useful end result shall exhaust the patent owner's rights under the patent to the extent that such rights would have been exhausted had such sale or other disposition been made by the patent owner.

(e) Limitations and exceptions.—

(1) Personal defense.—

(A) In general.—A defense under this section may be asserted only by the person who performed or directed the performance of the commercial use described in subsection (a), or by an entity that controls, is controlled by, or is under common control with such person.

(B) Transfer of right.—Except for any transfer to the patent owner, the right to assert a defense under this section shall not be licensed or assigned or transferred to another person except as an ancillary and subordinate part of a good-faith assignment or transfer for other reasons of the entire enterprise or line of business to which the defense relates.

(C) Restriction on sites.—A defense under this section, when acquired by a person as part of an assignment or transfer described in subparagraph (B), may only be asserted for uses at sites where the subject matter that would otherwise infringe a claimed invention is in use before the later of the effective filing date of the claimed invention or the date of the assignment or transfer of such enterprise or line of business.

(2) Derivation.—A person may not assert a defense under this section if the subject matter on which the defense is based was derived from the patentee or persons in privity with the patentee.

(3) Not a general license.—The defense asserted by a person under this section is not a general license under all claims of the patent at issue, but extends only to the specific subject matter for which it has been established that a commercial use that qualifies under this section occurred, except that the defense shall also extend to variations in the quantity or volume of use of the claimed subject matter, and to improvements in the claimed subject matter that do not infringe additional specifically claimed subject matter of the patent.

(4) Abandonment of use.—A person who has abandoned commercial use (that qualifies under this section) of subject matter may not rely on activities performed before the date of such abandonment in establishing a defense under this section with respect to actions taken on or after the date of such abandonment.

(5) University exception.—

(A) In general.—A person commercially using subject matter to which subsection (a) applies may not assert a defense under this section if the claimed invention with respect to which the defense is asserted was, at the time the invention was made, owned or subject to an obligation of assignment to either an institution of higher education (as defined in section 101(a) of the Higher Education Act of 1965 (20 U.S.C. 1001(a)), or a technology transfer organization whose primary purpose is to facilitate the commercialization of technologies developed by one or more such institutions of higher education.

(B) Exception.—Subparagraph (A) shall not apply if any of the activities required to reduce to practice the subject matter of the claimed invention could not have been undertaken using funds provided by the Federal Government.

(f) Unreasonable assertion of defense.—If the defense under this section is pleaded by a person who is found to infringe the patent and who subsequently fails to demonstrate a reasonable basis for asserting the defense, the court shall find the case exceptional for the purpose of awarding attorney fees under section 285.

(g) Invalidity.—A patent shall not be deemed to be invalid under section 102 or 103 solely because a defense is raised or established under this section.

Chapter 29. Remedies for Infringement of Patent, and Other Actions

Sec. 281. Remedy for Infringement of Patent

A patentee shall have remedy by civil action for infringement of his patent.

Sec. 282. Presumption of Validity; Defenses

(a) In general.—A patent shall be presumed valid. Each claim of a patent (whether in independent, dependent, or multiple dependent form) shall be presumed valid independently of the validity of other claims; dependent or multiple dependent claims shall be presumed valid even though dependent upon an invalid claim. The burden of establishing invalidity of a patent or any claim thereof shall rest on the party asserting such invalidity.

(b) Defenses.—The following shall be defenses in any action involving the validity or infringement of a patent and shall be pleaded:

(1) Noninfringement, absence of liability for infringement or unenforceability.

(2) Invalidity of the patent or any claim in suit on any ground specified in part II as a condition for patentability.

(3) Invalidity of the patent or any claim in suit for failure to comply with—

(A) any requirement of section 112, except that the failure to disclose the best mode shall not be a basis on which any claim of a patent may be canceled or held invalid or otherwise unenforceable; or

(B) any requirement of section 251.

(4) Any other fact or act made a defense by this title.

(c) Notice of actions; actions during extension of patent term.—In an action involving the validity or infringement of a patent the party asserting invalidity or noninfringement shall give notice in the pleadings or otherwise in writing to the adverse party at least thirty days before the trial, of the country, number, date, and name of the patentee of any patent, the title, date, and page numbers of any publication to be relied upon as anticipation of the patent in suit or, except in actions in the United States Court of Federal Claims, as showing the state of the art, and the name and address of any person who may be relied upon as the prior inventor or as having prior knowledge of or as having previously used or offered for sale the invention of the patent in suit. In the absence of such notice proof of the said matters may not be made at the trial except on such terms as the court requires. Invalidity of the extension of a patent term or any portion thereof under section 154(b) or 156 because of the material failure—

(1) by the applicant for the extension, or

(2) by the Director,

to comply with the requirements of such section shall be a defense in any action involving the infringement of a patent during the period of the extension of its term and shall be pleaded. A due diligence determination under section 156(d)(2) is not subject to review in such an action.

Sec. 283. Injunction

The several courts having jurisdiction of cases under this title may grant injunctions in accordance with the principles of equity to prevent the violation of any right secured by patent, on such terms as the court deems reasonable.

Sec. 284. Damages

Upon finding for the claimant the court shall award the claimant damages adequate to compensate for the infringement, but in no event less than a reasonable royalty for the use made of the invention by the infringer, together with interest and costs as fixed by the court.

When the damages are not found by a jury, the court shall assess them. In either event the court may increase the damages up to three times the amount found or assessed. Increased damages under this paragraph shall not apply to provisional rights under section 154(d) of this title.

The court may receive expert testimony as an aid to the determination of damages or of what royalty would be reasonable under the circumstances.

Sec. 285. Attorney Fees

The court in exceptional cases may award reasonable attorney fees to the prevailing party.

Sec. 286. Time Limitation on Damages

Except as otherwise provided by law, no recovery shall be had for any infringement committed more than six years prior to the filing of the complaint or counterclaim for infringement in the action.

In the case of claims against the United States Government for use of a patented invention, the period before bringing suit, up to six years, between the date of receipt of a written claim for compensation by the department or agency of the Government having authority to settle such claim, and the date of mailing by the Government of a notice to the claimant that his claim has been denied shall not be counted as part of the period referred to in the preceding paragraph.

Sec. 287. Limitation on Damages and other Remedies; Marking and Notice

(a) Patentees, and persons making, offering for sale, or selling within the United States any patented article for or under them, or importing any patented article into the United States, may give notice to the public that the same is patented, either by fixing thereon the word "patent" or the abbreviation "pat.", together with the number of the patent, or by fixing thereon the word "patent" or the abbreviation "pat." together with an address of a posting on the Internet, accessible to the public without charge for accessing the address, that associates the patented article with the number of the patent, or when, from the character of the article, this can not be done, by fixing to it, or to the package wherein one or more of them is contained, a label containing a like notice. In the event of failure so to mark, no damages shall be recovered by the patentee in any action for infringement, except on proof that the infringer was notified of the infringement and continued to infringe thereafter, in which event damages may be recovered only for infringement occurring after such notice. Filing of an action for infringement shall constitute such notice.

(b)(1) An infringer under section 271(g) shall be subject to all the provisions of this title relating to damages and injunctions except to the extent those remedies are modified by this subsection or section 9006 of the Process Patent Amendments Act of 1988. The modifications of remedies provided in this subsection shall not be available to any person who—

(A) practiced the patented process;

(B) owns or controls, or is owned or controlled by, the person who practiced the patented process; or

(C) had knowledge before the infringement that a patented process was used to make the product the importation, use, offer for sale, or sale of which constitutes the infringement.

(2) No remedies for infringement under section 271(g) of this title shall be available with respect to any product in the possession of, or in transit to, the person subject to liability under such section before that person had notice of infringement with respect to that product. The person subject to liability shall bear the burden of proving any such possession or transit.

(3)(A) In making a determination with respect to the remedy in an action brought for infringement under section 271(g), the court shall consider—

(i) the good faith demonstrated by the defendant with respect to a request for disclosure,

(ii) the good faith demonstrated by the plaintiff with respect to a request for disclosure, and

(iii) the need to restore the exclusive rights secured by the patent.

(B) For purposes of subparagraph (A), the following are evidence of good faith:

(i) a request for disclosure made by the defendant;

(ii) a response within a reasonable time by the person receiving the request for disclosure; and

(iii) the submission of the response by the defendant to the manufacturer, or if the manufacturer is not known, to the supplier, of the product to be purchased by the defendant, together with a request for a written statement that the process claimed in any patent disclosed in the response is not used to produce such product.

The failure to perform any acts described in the preceding sentence is evidence of absence of good faith unless there are mitigating circumstances. Mitigating circumstances include the case in which, due to the nature of the product, the number of sources for the product, or like commercial circumstances, a request for disclosure is not necessary or practicable to avoid infringement.

(4)(A) For purposes of this subsection, a "request for disclosure" means a written request made to a person then engaged in the manufacture of a product to identify all process patents owned by or licensed to that person, as of the time of the request, that the person then reasonably believes could be asserted to be infringed under section 271(g) if that product were imported into, or sold, offered for sale, or used in, the United States by an unauthorized person. A request for disclosure is further limited to a request—

(i) which is made by a person regularly engaged in the United States in the sale of the same type of products as those manufactured by the person to whom the request is directed, or which includes facts showing that the person making the request plans to engage in the sale of such products in the United States;

(ii) which is made by such person before the person's first importation, use, offer for sale, or sale of units of the product produced by an infringing process and before the person had notice of infringement with respect to the product; and

(iii) which includes a representation by the person making the request that such person will promptly submit the patents identified pursuant to the request to the manufacturer, or if the manufacturer is not known, to the supplier, of the product to be purchased by the person making the request, and will request from that manufacturer or supplier a written statement that none of the processes claimed in those patents is used in the manufacture of the product.

(B) In the case of a request for disclosure received by a person to whom a patent is licensed, that person shall either identify the patent or promptly notify the licensor of the request for disclosure.

(C) A person who has marked, in the manner prescribed by subsection (a), the number of the process patent on all products made by the patented process which have been offered for sale or sold by that person in the United States, or imported by the person into the United States, before a request for disclosure is received is not required to respond to the request for disclosure. For purposes of the preceding sentence, the term "all products" does not include products made before the effective date of the Process Patent Amendments Act of 1988.

(5)(A) For purposes of this subsection, notice of infringement means actual knowledge, or receipt by a person of a written notification, or a combination thereof, of information sufficient to persuade a reasonable person that it is likely that a product was made by a process patented in the United States.

(B) A written notification from the patent holder charging a person with infringement shall specify the patented process alleged to have been used and the reasons for a good faith belief that such process was used. The patent holder shall include in the notification such information as is reasonably necessary to explain fairly the patent holder's belief, except that the patent holder is not required to disclose any trade secret information.

(C) A person who receives a written notification described in subparagraph (B) or a written response to a request for disclosure described in paragraph (4) shall be deemed to have notice of infringement with respect to any patent referred to in such written notification or response unless that person, absent mitigating circumstances—

(i) promptly transmits the written notification or response to the manufacturer or, if the manufacturer is not known, to the supplier, of the product purchased or to be purchased by that person; and

(ii) receives a written statement from the manufacturer or supplier which on its face sets forth a well grounded factual basis for a belief that the identified patents are not infringed.

(D) For purposes of this subsection, a person who obtains a product made by a process patented in the United States in a quantity which is abnormally large in relation to the volume of business of such person or an efficient inventory level shall be rebuttably presumed to have actual knowledge that the product was made by such patented process.

(6) A person who receives a response to a request for disclosure under this subsection shall pay to the person to whom the request was made a reasonable fee to cover actual costs incurred in complying with the request, which may not exceed the cost of a commercially available automated patent search of the matter involved, but in no case more than $500.

(c)(1) With respect to a medical practitioner's performance of a medical activity that constitutes an infringement under section 271(a) or (b) of this title, the provisions of sections 281, 283, 284, and 285 of this title shall not apply against the medical practitioner or against a related health care entity with respect to such medical activity.

(2) For the purposes of this subsection:

(A) the term "medical activity" means the performance of a medical or surgical procedure on a body, but shall not include (i) the use of a patented machine, manufacture, or composition of matter in violation of such patent, (ii) the practice of a patented use of a composition of matter in violation of such patent, or (iii) the practice of a process in violation of a biotechnology patent.

(B) the term "medical practitioner" means any natural person who is licensed by a State to provide the medical activity described in subsection (c)(1) or who is acting under the direction of such person in the performance of the medical activity.

(C) the term "related health care entity" shall mean an entity with which a medical practitioner has a professional affiliation under which the medical practitioner performs the medical activity, including but not limited to a nursing home, hospital, university, medical school, health maintenance organization, group medical practice, or a medical clinic.

(D) the term "professional affiliation" shall mean staff privileges, medical staff membership, employment or contractual relationship, partnership or ownership interest, academic appointment, or other affiliation under which a medical practitioner provides the medical activity on behalf of, or in association with, the health care entity.

(E) the term "body" shall mean a human body, organ or cadaver, or a nonhuman animal used in medical research or instruction directly relating to the treatment of humans.

(F) the term "patented use of a composition of matter" does not include a claim for a method of performing a medical or surgical procedure on a body that recites the use of a composition of matter where the use of that composition of matter does not directly contribute to achievement of the objective of the claimed method.

(G) the term "State" shall mean any State or territory of the United States, the District of Columbia, and the Commonwealth of Puerto Rico.

(3) This subsection does not apply to the activities of any person, or employee or agent of such person (regardless of whether such person is a tax exempt organization under section 501(c) of the Internal Revenue Code), who is engaged in the commercial development, manufacture, sale, importation, or distribution of a machine, manufacture, or composition of matter or the provision of pharmacy or clinical laboratory services (other than clinical laboratory services provided in a physician's office), where such activities are:

(A) directly related to the commercial development, manufacture, sale, importation, or distribution of a machine, manufacture, or composition of matter or the provision of pharmacy or clinical laboratory services (other than clinical laboratory services provided in a physician's office), and

(B) regulated under the Federal Food, Drug, and Cosmetic Act, the Public Health Service Act, or the Clinical Laboratories Improvement Act.

(4) This subsection shall not apply to any patent issued based on an application which has an effective filing date before September 30, 1996.

Sec. 288. Action for Infringement of a Patent Containing an Invalid Claim

Whenever, a claim of a patent is invalid, an action may be maintained for the infringement of a claim of the patent which may be valid. The patentee shall recover no costs unless a disclaimer of the invalid claim has been entered at the Patent and Trademark Office before the commencement of the suit.

Sec. 289. Additional Remedy for Infringement of Design Patent

Whoever during the term of a patent for a design, without license of the owner, (1) applies the patented design, or any colorable imitation thereof, to any article of manufacture for the purpose of sale, or (2) sells or exposes for sale any article of manufacture to which such design or colorable imitation has been applied shall be liable to the owner to the extent of his total profit, but not less than $250, recoverable in any United States district court having jurisdiction of the parties.

Nothing in this section shall prevent, lessen, or impeach any other remedy which an owner of an infringed patent has under the provisions of this title, but he shall not twice recover the profit made from the infringement.

Sec. 290. Notice of Patent Suits

The clerks of the courts of the United States, within one month after the filing of an action under this title shall give notice thereof in writing to the Director, setting forth so far as known the names and addresses of the parties, name of the inventor, and the designating number of the patent upon which the action has been brought. If any other patent is subsequently included in the action he shall give like notice thereof. Within one month after the decision is rendered or a judgment issued the clerk of the court shall give notice thereof to the Director. The Director shall, on receipt of such notices, enter the same in the file of such patent.

Sec. 291. Derived patents

(a) In general.—The owner of a patent may have relief by civil action against the owner of another patent that claims the same invention and has an earlier effective filing date, if the invention claimed in such other patent was derived from the inventor of the invention claimed in the patent owned by the person seeking relief under this section.

(b) Filing limitation.—An action under this section may be filed only before the end of the 1-year period beginning on the date of the issuance of the first patent containing a claim to the allegedly derived invention and naming an individual alleged to have derived such invention as the inventor or joint inventor.

Sec. 292. False Marking

(a) Whoever, without the consent of the patentee, marks upon, or affixes to, or uses in advertising in connection with anything made, used, offered for sale, or sold by such person within the United States, or imported by the person into the United States, the name or any imitation of the name of the patentee, the patent number, or the words "patent," "patentee," or the like, with the intent of counterfeiting or imitating the mark of the patentee, or of deceiving the public and inducing them to believe that the thing was made, offered for sale, sold, or imported into the United States by or with the consent of the patentee; or

Whoever marks upon, or affixes to, or uses in advertising in connection with any unpatented article, the word "patent" or any word or number importing that the same is patented, for the purpose of deceiving the public; or

Whoever marks upon, or affixes to, or uses in advertising in connection with any article, the words "patent applied for," "patent pending," or any word importing that an application for patent has been made, when no application for patent has been made, or if made, is not pending, for the purpose of deceiving the public—

Shall be fined not more than $500 for every such offense.

Only the United States may sue for the penalty authorized by this subsection.

(b) A person who has suffered a competitive injury as a result of a violation of this section may file a civil action in a district court of the United States for recovery of damages adequate to compensate for the injury.

(c) The marking of a product, in a manner described in subsection (a), with matter relating to a patent that covered that product but has expired is not a violation of this section.

Sec. 293. Nonresident patentee; service and notice

Every patentee not residing in the United States may file in the Patent and Trademark Office a written designation stating the name and address of a person residing within the United States on whom may be served process or notice of proceedings affecting the patent or rights thereunder. If the person designated cannot be found at the address given in the last designation, or if no person has been designated, the United States District Court for the Eastern District of Virginia shall have jurisdiction and summons shall be served by publication or otherwise as the court directs. The court shall have the same jurisdiction to take any action respecting the patent or rights thereunder that it would have if the patentee were personally within the jurisdiction of the court.

Sec. 294. Voluntary Arbitration

(a) A contract involving a patent or any right under a patent may contain a provision requiring arbitration of any dispute relating to patent validity or infringement arising under the contract. In the absence of such a provision, the parties to an existing patent validity or infringement dispute may agree in writing to settle such dispute by arbitration. Any such provision or agreement shall be valid, irrevocable, and enforceable, except for any grounds that exist at law or in equity for revocation of a contract.

(b) Arbitration of such disputes, awards by arbitrators and confirmation of awards shall be governed by title 9, to the extent such title is not inconsistent with this section. In any such arbitration proceeding, the defenses provided for under section 282 of this title shall be considered by the arbitrator if raised by any party to the proceeding.

(c) An award by an arbitrator shall be final and binding between the parties to the arbitration but shall have no force or effect on any other person. The parties to an arbitration may agree that in the event a patent which is the subject matter of an award is subsequently determined to be invalid or unenforceable in a judgment rendered by a court of competent jurisdiction from which no appeal can or has been taken, such award may be modified by any court of competent jurisdiction upon application by any party to the arbitration. Any such modification shall govern the rights and obligations between such parties from the date of such modification.

(d) When an award is made by an arbitrator, the patentee, his assignee or licensee shall give notice thereof in writing to the Director. There shall be a separate notice prepared for each patent involved in such proceeding. Such notice shall set forth the names and addresses of the parties, the name of the inventor, and the name of the patent owner, shall designate the number of the patent, and shall contain a copy of the award. If an award is modified by a court, the party requesting such modification shall give notice of such modification to the Director. The Director shall, upon receipt of either notice, enter the same in the record of the prosecution of such patent. If the required notice is not filed with the Director, any party to the proceeding may provide such notice to the Director.

(e) The award shall be unenforceable until the notice required by subsection (d) is received by the Director.

Sec. 295. Presumption: Product Made by Patented Process

In actions alleging infringement of a process patent based on the importation, sale, offer for sale, or use of a product which is made from a process patented in the United States, if the court finds—

(1) that a substantial likelihood exists that the product was made by the patented process, and

(2) that the plaintiff has made a reasonable effort to determine the process actually used in the production of the product and was unable to so determine,

the product shall be presumed to have been so made, and the burden of establishing that the product was not made by the process shall be on the party asserting that it was not so made.

Sec. 296. Liability of States, Instrumentalities of States, and State Officials for Infringement of Patents

State acting in his official capacity, shall not be immune, under the eleventh amendment of the Constitution of the United States or under any other doctrine of sovereign immunity, from suit in Federal court by any person, including any governmental or nongovernmental entity, for infringement of a patent under section 271, or for any other violation under this title.

(b) Remedies.—In a suit described in subsection (a) for a violation described in that subsection, remedies (including remedies both at law and in equity) are available for the violation to the same extent as such remedies are available for such a violation in a suit against any private entity. Such remedies include damages, interest, costs, and treble damages under section 284, attorney fees under section 285, and the additional remedy for infringement of design patents under section 289.

Sec. 297. Improper and Deceptive Invention Promotion

(a) In general.— An invention promoter shall have a duty to disclose the following information to a customer in writing, prior to entering into a contract for invention promotion services:

(1) the total number of inventions evaluated by the invention promoter for commercial potential in the past 5 years, as well as the number of those inventions that received positive evaluations, and the number of those inventions that received negative evaluations;

(2) the total number of customers who have contracted with the invention promoter in the past 5 years, not including customers who have purchased trade show services, research, advertising, or other nonmarketing services from the invention promoter, or who have defaulted in their payment to the invention promoter;

(3) the total number of customers known by the invention promoter to have received a net financial profit as a direct result of the invention promotion services provided by such invention promoter;

(4) the total number of customers known by the invention promoter to have received license agreements for their inventions as a direct result of the invention promotion services provided by such invention promoter; and

(5) the names and addresses of all previous invention promotion companies with which the invention promoter or its officers have collectively or individually been affiliated in the previous 10 years.

(b) Civil action.— (1) Any customer who enters into a contract with an invention promoter and who is found by a court to have been injured by any material false or fraudulent statement or representation, or any omission of material fact, by that invention promoter (or any agent, employee, director, officer, partner, or independent contractor of such invention promoter), or by the failure of that invention promoter to disclose such information as required under subsection (a), may recover in a civil action against the invention promoter (or the officers, directors, or partners of such invention promoter), in addition to reasonable costs and attorneys' fees—

(A) the amount of actual damages incurred by the customer; or

(B) at the election of the customer at any time before final judgment is rendered, statutory damages in a sum of not more than $5,000, as the court considers just.

(2) Notwithstanding paragraph (1), in a case where the customer sustains the burden of proof, and the court finds, that the invention promoter intentionally misrepresented or omitted a material fact to such customer, or willfully failed to disclose such information as required under subsection (a), with the purpose of deceiving that customer, the court may increase damages to not more than three times the amount awarded, taking into account past complaints made against the invention promoter that resulted in regulatory sanctions or other corrective actions based on those records compiled by the Commissioner of Patents under subsection (d).

(c) Definitions.— For purposes of this section—

(1) a "contract for invention promotion services" means a contract by which an invention promoter undertakes invention promotion services for a customer;

(2) a "customer" is any individual who enters into a contract with an invention promoter for invention promotion services;

(3) the term "invention promoter" means any person, firm, partnership, corporation, or other entity who offers to perform or performs invention promotion services for, or on behalf of, a customer, and who holds itself out through advertising in any mass media as providing such services, but does not include—

(A) any department or agency of the Federal Government or of a State or local government;

(B) any nonprofit, charitable, scientific, or educational organization, qualified under applicable State law or described under section 170(b)(1)(A) of the Internal Revenue Code of 1986;

(C) any person or entity involved in the evaluation to determine commercial potential of, or offering to license or sell, a utility patent or a previously filed nonprovisional utility patent application;

(D) any party participating in a transaction involving the sale of the stock or assets of a business; or

(E) any party who directly engages in the business of retail sales of products or the distribution of products; and

(4) the term "invention promotion services" means the procurement or attempted procurement for a customer of a firm, corporation, or other entity to develop and market products or services that include the invention of the customer.

(d) Records of complaints .—

(1) Release of complaints.— The Commissioner of Patents shall make all complaints received by the Patent and Trademark Office involving invention promoters publicly available, together with any response of the invention promoters. The Commissioner of Patents shall notify the invention promoter of a complaint

and provide a reasonable opportunity to reply prior to making such complaint publicly available.

(2) Request for complaints.— The Commissioner of Patents may request complaints relating to invention promotion services from any Federal or State agency and include such complaints in the records maintained under paragraph (1), together with any response of the invention promoters.

Sec. 298. Advice of Counsel

The failure of an infringer to obtain the advice of counsel with respect to any allegedly infringed patent, or the failure of the infringer to present such advice to the court or jury, may not be used to prove that the accused infringer willfully infringed the patent or that the infringer intended to induce infringement of the patent.

Sec. 299. Joinder of Parties

(a) Joinder of accused infringers.— With respect to any civil action arising under any Act of Congress relating to patents, other than an action or trial in which an act of infringement under section 271(e)(2) has been pled, parties that are accused infringers may be joined in one action as defendants only if—

(1) any right to relief is asserted against the parties jointly, severally, or in the alternative with respect to or arising out of the same transaction, occurrence, or series of transactions or occurrences relating to the making, using, importing into the United States, offering for sale, or selling of the same accused product or process; and

(2) questions of fact common to all defendants or counterclaim defendants will arise in the action.

(b) Allegations insufficient for joinder.— For purposes of this subsection, accused infringers may not be joined in one action as defendants or counterclaim defendants, or have their actions consolidated for trial, based solely on allegations that they each have infringed the patent or patents in suit.

(c) Waiver.— A party that is an accused infringer may waive the limitations set forth in this section with respect to that party.

Chapter 30. Prior Art Citations to Office and Ex Parte Reexamination of Patents

Sec. 301. Citation of prior art and written statements

(a) In general.—Any person at any time may cite to the Office in writing—

(1) prior art consisting of patents or printed publications which that person believes to have a bearing on the patentability of any claim of a particular patent; or

(2) statements of the patent owner filed in a proceeding before a Federal court or the Office in which the patent owner took a position on the scope of any claim of a particular patent.

(b) Official file.—If the person citing prior art or written statements pursuant to subsection (a) explains in writing the pertinence and manner of applying the prior art or written statements to at least 1 claim of the patent, the citation of the prior art or written statements and the explanation thereof shall become a part of the official file of the patent.

(c) Additional information.—A party that submits a written statement pursuant to subsection (a)(2) shall include any other documents, pleadings, or evidence from the proceeding in which the statement was filed that addresses the written statement.

(d) Limitations.—A written statement submitted pursuant to subsection (a)(2), and additional information submitted pursuant to subsection (c), shall not be considered by the Office for any purpose other than to determine the proper meaning of a patent claim in a proceeding that is ordered or instituted pursuant to section 304, 314, or 324. If any such written statement or additional information is subject to an applicable protective order, such statement or information shall be redacted to exclude information that is subject to that order.

(e) Confidentiality.—Upon the written request of the person citing prior art or written statements pursuant to subsection (a), that person's identity shall be excluded from the patent file and kept confidential.

Sec. 302. Request for Reexamination

Any person at any time may file a request for reexamination by the Office of any claim of a patent on the basis of any prior art cited under the provisions of section 301 of this title. The request must be in writing and must be accompanied by payment of a reexamination fee established by the Director pursuant to the provisions of section 41 of this title. The request must set forth the pertinency and manner of applying cited prior art to every claim for which reexamination is requested. Unless the requesting person is the owner of the patent, the Director promptly will send a copy of the request to the owner of record of the patent.

Sec. 303. Determination of Issue by Director

(a) Within three months following the filing of a request for reexamination under the provisions of section 302, the Director will determine whether a substantial new question of patentability affecting any claim of the patent concerned is raised by the request, with or without consideration of other patents or printed publications. On his own initiative, and any time, the Director may determine whether a substantial new question of patentability is raised by patents and publications discovered by him or cited under the provisions of section 301 or 302. The existence of a substantial new question of patentability is not precluded by the fact that a patent or printed publication was previously cited by or to the Office or considered by the Office.

(b) A record of the Director's determination under subsection (a) of this section will be placed in the official file of the patent, and a copy promptly will be given or mailed to the owner of record of the patent and to the person requesting reexamination, if any.

(c) A determination by the Director pursuant to subsection (a) of this section that no substantial new question of patentability has been raised will be final and nonappealable. Upon such a determination, the Director may refund a portion of the reexamination fee required under section 302.

Sec. 304. Reexamination Order by Director

If, in a determination made under the provisions of subsection 303(a) of this title, the Director finds that a substantial new question of patentability affecting any claim of a patent is raised, the determination will include an order for reexamination of the patent for resolution of the question. The patent owner will be given a reasonable period, not less than two months from the date a copy of the determination is given or mailed to him, within which he may file a statement on such question, including any amendment to his patent and new claim or claims he may wish to propose, for consideration in the reexamination. If the patent owner files such a statement, he promptly will serve a copy of it on the person who has requested reexamination under the provisions of section 302 of this title. Within a period of two months from the date of service, that person may file and have considered in the reexamination a reply to any statement filed by the patent owner. That person promptly will serve on the patent owner a copy of any reply filed.

Sec. 305. Conduct of Reexamination Proceedings

After the times for filing the statement and reply provided for by section 304 of this title have expired, reexamination will be conducted according to the procedures established for initial examination under the provisions of sections 132 and 133 of this title. In any reexamination proceeding under this chapter, the patent owner will be permitted to propose any amendment to his patent and a new claim or claims thereto, in order to distinguish the invention as claimed from the prior art cited under the provisions of section 301 of this title, or in response to a decision adverse to the patentability of a claim of a patent. No proposed amended or new claim enlarging the scope of a claim of the patent will be permitted in a reexamination proceeding under this chapter. All reexamination proceedings under this section, including any appeal to the Patent Trial and Appeal Board, will be conducted with special dispatch within the Office.

Sec. 306. Appeal

The patent owner involved in a reexamination proceeding under this chapter may appeal under the provisions of section 134 of this title, and may seek court review under the provisions of sections 141 to 144 of this title, with respect to any decision adverse to the patentability of any original or proposed amended or new claim of the patent.

Sec. 307. Certificate of Patentability, Unpatentability, and Claim Cancellation

(a) In a reexamination proceeding under this chapter, when the time for appeal has expired or any appeal proceeding has terminated, the Director will issue and publish a certificate canceling any claim of the patent finally determined to be unpatentable, confirming any claim of the patent determined to be patentable, and incorporating in the patent any proposed amended or new claim determined to be patentable.

(b) Any proposed amended or new claim determined to be patentable and incorporated into a patent following a reexamination proceeding will have the same effect as that specified in section 252 of this title for reissued patents on the right of any person who made, purchased, or used within the United States, or imported into the United States, anything patented by such proposed amended or new claim, or who made substantial preparation for the same, prior to issuance of a certificate under the provisions of subsection (a) of this section.

Chapter 31. Inter Partes Review

§ 311. Inter partes review

(a) In general.—Subject to the provisions of this chapter, a person who is not the owner of a patent may file with the Office a petition to institute an inter partes review of the patent. The Director shall establish, by regulation, fees to be paid by the person requesting the review, in such amounts as the Director determines to be reasonable, considering the aggregate costs of the review.

(b) Scope.—A petitioner in an inter partes review may request to cancel as unpatentable 1 or more claims of a patent only on a ground that could be raised under section 102 or 103 and only on the basis of prior art consisting of patents or printed publications.

(c) Filing deadline.—A petition for inter partes review shall be filed after the later of either—

(1) the date that is 9 months after the grant of a patent; or

(2) if a post-grant review is instituted under chapter 32, the date of the termination of such post-grant review. [Added by AIA Technical amendments, 2013: Section 311(c) of title 35, United States Code, shall not apply to a petition to institute an inter partes review of a patent that is not a patent described in section 3(n)(1) of the Leahy-Smith America Invents Act (35 U.S.C. 100 note).]

§ 312. Petitions

(a) Requirements of petition.—A petition filed under section 311 may be considered only if—

(1) the petition is accompanied by payment of the fee established by the Director under section 311;

(2) the petition identifies all real parties in interest;

(3) the petition identifies, in writing and with particularity, each claim challenged, the grounds on which the challenge to each claim is based, and the evidence that supports the grounds for the challenge to each claim, including—

(A) copies of patents and printed publications that the petitioner relies upon in support of the petition; and

(B) affidavits or declarations of supporting evidence and opinions, if the petitioner relies on expert opinions;

(4) the petition provides such other information as the Director may require by regulation; and

(5) the petitioner provides copies of any of the documents required under paragraphs (2), (3), and (4) to the patent owner or, if applicable, the designated representative of the patent owner.

(b) Public availability.—As soon as practicable after the receipt of a petition under section 311, the Director shall make the petition available to the public.

§ 313. Preliminary response to petition

If an inter partes review petition is filed under section 311, the patent owner shall have the right to file a preliminary response to the petition, within a time period set by the Director, that sets forth reasons why no inter partes review should be instituted based upon the failure of the petition to meet any requirement of this chapter.

§ 314. Institution of inter partes review

(a) Threshold.—The Director may not authorize an inter partes review to be instituted unless the Director determines that the information presented in the petition filed under section 311 and any response filed under section 313 shows that there is a reasonable likelihood that the petitioner would prevail with respect to at least 1 of the claims challenged in the petition.

(b) Timing.—The Director shall determine whether to institute an inter partes review under this chapter pursuant to a petition filed under section 311 within 3 months after—

(1) receiving a preliminary response to the petition under section 313; or

(2) if no such preliminary response is filed, the last date on which such response may be filed.

(c) Notice.—The Director shall notify the petitioner and patent owner, in writing, of the Director's determination under subsection (a), and shall make such notice available to the public as soon as is practicable. Such notice shall include the date on which the review shall commence.

(d) No appeal.—The determination by the Director whether to institute an inter partes review under this section shall be final and nonappealable.

§ 315. Relation to other proceedings or actions

(a) Infringer's civil action.—

(1) Inter partes review barred by civil action.—An inter partes review may not be instituted if, before the date on which the petition for such a review is filed, the petitioner or real party in interest filed a civil action challenging the validity of a claim of the patent.

(2) Stay of civil action.—If the petitioner or real party in interest files a civil action challenging the validity of a claim of the patent on or after the date on which the petitioner files a petition for inter partes review of the patent, that civil action shall be automatically stayed until either—

(A) the patent owner moves the court to lift the stay;

(B) the patent owner files a civil action or counterclaim alleging that the petitioner or real party in interest has infringed the patent; or

(C) the petitioner or real party in interest moves the court to dismiss the civil action.

(3) Treatment of counterclaim.—A counterclaim challenging the validity of a claim of a patent does not constitute a civil action challenging the validity of a claim of a patent for purposes of this subsection.

(b) Patent owner's action.—An inter partes review may not be instituted if the petition requesting the proceeding is filed more than 1 year after the date on which the petitioner, real party in interest, or privy of the petitioner is served with a complaint alleging infringement of the patent. The time limitation set forth in the preceding sentence shall not apply to a request for joinder under subsection (c).

(c) Joinder.—If the Director institutes an inter partes review, the Director, in his or her discretion, may join as a party to that inter partes review any person who properly files a petition under section 311 that the Director, after receiving a preliminary response under section 313 or the expiration of the time for filing such a response, determines warrants the institution of an inter partes review under section 314.

(d) Multiple proceedings.—Notwithstanding sections 135(a), 251, and 252, and chapter 30, during the pendency of an inter partes review, if another proceeding or matter involving the patent is before the Office, the Director may determine the manner in which the inter partes review or other proceeding or matter may proceed, including providing for stay, transfer, consolidation, or termination of any such matter or proceeding.

§ 316. Conduct of inter partes review

(a) Regulations.—The Director shall prescribe regulations—

(1) providing that the file of any proceeding under this chapter shall be made available to the public, except that any petition or document filed with the intent that it be sealed shall, if accompanied by a motion to seal, be treated as sealed pending the outcome of the ruling on the motion;

(2) setting forth the standards for the showing of sufficient grounds to institute a review under section 314(a);

(3) establishing procedures for the submission of supplemental information after the petition is filed;

(4) establishing and governing inter partes review under this chapter and the relationship of such review to other proceedings under this title;

(5) setting forth standards and procedures for discovery of relevant evidence, including that such discovery shall be limited to—

(A) the deposition of witnesses submitting affidavits or declarations; and

(B) what is otherwise necessary in the interest of justice;

(6) prescribing sanctions for abuse of discovery, abuse of process, or any other improper use of the proceeding, such as to harass or to cause unnecessary delay or an unnecessary increase in the cost of the proceeding;

(7) providing for protective orders governing the exchange and submission of confidential information;

(8) providing for the filing by the patent owner of a response to the petition under section 313 after an inter partes review has been instituted, and requiring that the patent owner file with such response, through affidavits or declarations, any additional factual evidence and expert opinions on which the patent owner relies in support of the response;

(9) setting forth standards and procedures for allowing the patent owner to move to amend the patent under subsection (d) to cancel a challenged claim or propose a reasonable number of substitute claims, and ensuring that any information submitted by the patent owner in support of any amendment entered under subsection (d) is made available to the public as part of the prosecution history of the patent;

(10) providing either party with the right to an oral hearing as part of the proceeding;

(11) requiring that the final determination in an inter partes review be issued not later than 1 year after the date on which the Director notices the institution of a review under this chapter, except that the Director may, for good cause shown, extend the 1-year period by not more than 6 months, and may adjust the time periods in this paragraph in the case of joinder under section 315(c);

(12) setting a time period for requesting joinder under section 315(c); and

(13) providing the petitioner with at least 1 opportunity to file written comments within a time period established by the Director.

(b) Considerations.—In prescribing regulations under this section, the Director shall consider the effect of any such regulation on the economy, the integrity of the patent system, the efficient administration of the Office, and the ability of the Office to timely complete proceedings instituted under this chapter.

(c) Patent Trial and Appeal Board.—The Patent Trial and Appeal Board shall, in accordance with section 6, conduct each inter partes review instituted under this chapter.

(d) Amendment of the patent.—

(1) In general.—During an inter partes review instituted under this chapter, the patent owner may file 1 motion to amend the patent in 1 or more of the following ways:

(A) Cancel any challenged patent claim.

(B) For each challenged claim, propose a reasonable number of substitute claims.

(2) Additional motions.—Additional motions to amend may be permitted upon the joint request of the petitioner and the patent owner to materially advance the settlement of a proceeding under section 317, or as permitted by regulations prescribed by the Director.

(3) Scope of claims.—An amendment under this subsection may not enlarge the scope of the claims of the patent or introduce new matter.

(e) Evidentiary standards.—In an inter partes review instituted under this chapter, the petitioner shall have the burden of proving a proposition of unpatentability by a preponderance of the evidence.

§ 317. Settlement

(a) In general.—An inter partes review instituted under this chapter shall be terminated with respect to any petitioner upon the joint request of the petitioner and the patent owner, unless the Office has decided the merits of the proceeding before the request for termination is filed. If the inter partes review is terminated with respect to a petitioner under this section, no estoppel under section 315(e) shall attach to the petitioner, or to the real party in interest or privy of the petitioner, on the basis of that petitioner's institution of that inter partes review. If no petitioner remains in the inter partes review, the Office may terminate the review or proceed to a final written decision under section 318(a).

(b) Agreements in writing.—Any agreement or understanding between the patent owner and a petitioner, including any collateral agreements referred to in such agreement or understanding, made in connection with, or in contemplation of, the termination of an inter partes review under this section shall be in writing and a true copy of such agreement or understanding shall be filed in the Office before the termination of the inter partes review as between the parties. At the request of a party to the proceeding, the agreement or understanding shall be treated as business confidential information, shall be kept separate from the file of the involved patents, and shall be made available only to Federal Government agencies on written request, or to any person on a showing of good cause.

§ 318. Decision of the Board

(a) Final written decision.—If an inter partes review is instituted and not dismissed under this chapter, the Patent Trial and Appeal Board shall issue a final written decision with respect to the patentability of any patent claim challenged by the petitioner and any new claim added under section 316(d).

(b) Certificate.—If the Patent Trial and Appeal Board issues a final written decision under subsection (a) and the time for appeal has expired or any appeal has terminated, the Director shall issue and publish a certificate canceling any claim of the patent finally determined to be unpatentable, confirming any claim of the patent determined to be patentable, and incorporating in the patent by operation of the certificate any new or amended claim determined to be patentable.

(c) Intervening rights.—Any proposed amended or new claim determined to be patentable and incorporated into a patent following an inter partes review under this chapter shall have the same effect as that specified in section 252 for reissued patents on the right of any person who made, purchased, or used within the United States, or imported into the United States, anything patented by such proposed amended or new claim, or who made substantial preparation therefor, before the issuance of a certificate under subsection (b).

(d) Data on length of review.—The Office shall make available to the public data describing the length of time between the institution of, and the issuance of a final written decision under subsection (a) for, each inter partes review.

§ 319. Appeal

A party dissatisfied with the final written decision of the Patent Trial and Appeal Board under section 318(a) may appeal the decision pursuant to sections 141 through 144. Any party to the inter partes review shall have the right to be a party to the appeal.

Chapter 32. Post-Grant Review

§ 321. Post-grant review

(a) In general.—Subject to the provisions of this chapter, a person who is not the owner of a patent may file with the Office a petition to institute a post-grant review of the patent. The Director shall establish, by regulation, fees to be paid by the person requesting the review, in such amounts as the Director determines to be reasonable, considering the aggregate costs of the post-grant review.

(b) Scope.—A petitioner in a post-grant review may request to cancel as unpatentable 1 or more claims of a patent on any ground that could be raised under paragraph (2) or (3) of section 282(b) (relating to invalidity of the patent or any claim).

(c) Filing deadline.—A petition for a post-grant review may only be filed not later than the date that is 9 months after the date of the grant of the patent or of the issuance of a reissue patent (as the case may be).

§ 322. Petitions

(a) Requirements of petition.—A petition filed under section 321 may be considered only if—

(1) the petition is accompanied by payment of the fee established by the Director under section 321;

(2) the petition identifies all real parties in interest;

(3) the petition identifies, in writing and with particularity, each claim challenged, the grounds on which the challenge to each claim is based, and the evidence that supports the grounds for the challenge to each claim, including—

(A) copies of patents and printed publications that the petitioner relies upon in support of the petition; and

(B) affidavits or declarations of supporting evidence and opinions, if the petitioner relies on other factual evidence or on expert opinions;

(4) the petition provides such other information as the Director may require by regulation; and

(5) the petitioner provides copies of any of the documents required under paragraphs (2), (3), and (4) to the patent owner or, if applicable, the designated representative of the patent owner.

(b) Public availability.—As soon as practicable after the receipt of a petition under section 321, the Director shall make the petition available to the public.

§ 323. Preliminary response to petition

If a post-grant review petition is filed under section 321, the patent owner shall have the right to file a preliminary response to the petition, within a time period set by the Director, that sets forth reasons why no post-grant review should be instituted based upon the failure of the petition to meet any requirement of this chapter.

§ 324. Institution of post-grant review

(a) Threshold.—The Director may not authorize a post-grant review to be instituted unless the Director determines that the information presented in the petition filed under section 321, if such information is not rebutted, would demonstrate that it is more likely than not that at least 1 of the claims challenged in the petition is unpatentable.

(b) Additional grounds.—The determination required under subsection (a) may also be satisfied by a showing that the petition raises a novel or unsettled legal question that is important to other patents or patent applications.

(c) Timing.—The Director shall determine whether to institute a post-grant review under this chapter pursuant to a petition filed under section 321 within 3 months after—

(1) receiving a preliminary response to the petition under section 323; or

(2) if no such preliminary response is filed, the last date on which such response may be filed.

(d) Notice.—The Director shall notify the petitioner and patent owner, in writing, of the Director's determination under subsection (a) or (b), and shall make such notice available to the public as soon as is practicable. Such notice shall include the date on which the review shall commence.

(e) No appeal.—The determination by the Director whether to institute a post-grant review under this section shall be final and nonappealable.

§ 325. Relation to other proceedings or actions

(a) Infringer's civil action.—

(1) Post-grant review barred by civil action.—A post-grant review may not be instituted under this chapter if, before the date on which the petition for such a review is filed, the petitioner or real party in interest filed a civil action challenging the validity of a claim of the patent.

(2) Stay of civil action.—If the petitioner or real party in interest files a civil action challenging the validity of a claim of the patent on or after the date on which the petitioner files a petition for post-grant review of the patent, that civil action shall be automatically stayed until either—

(A) the patent owner moves the court to lift the stay;

(B) the patent owner files a civil action or counterclaim alleging that the petitioner or real party in interest has infringed the patent; or

(C) the petitioner or real party in interest moves the court to dismiss the civil action.

(3) Treatment of counterclaim.—A counterclaim challenging the validity of a claim of a patent does not constitute a civil action challenging the validity of a claim of a patent for purposes of this subsection.

(b) Preliminary injunctions.—If a civil action alleging infringement of a patent is filed within 3 months after the date on which the patent is granted, the court may not stay its consideration of the patent owner's motion for a preliminary injunction against infringement of the patent on the basis that a petition for post-grant review has been filed under this chapter or that such a post-grant review has been instituted under this chapter.

(c) Joinder.—If more than 1 petition for a post-grant review under this chapter is properly filed against the same patent and the Director determines that more than 1 of these petitions warrants the institution of a post-grant review under section 324, the Director may consolidate such reviews into a single post-grant review.

(d) Multiple proceedings.—Notwithstanding sections 135(a), 251, and 252, and chapter 30, during the pendency of any post-grant review under this chapter, if another proceeding or matter involving the patent is before the Office, the Director may determine the manner in which the post- grant review or other proceeding or matter may proceed, including providing for the stay, transfer, consolidation, or termination of any such matter or proceeding. In determining whether to institute or order a proceeding under this chapter, chapter 30, or chapter 31, the Director may take into account whether, and reject the petition or request because, the same or substantially the same prior art or arguments previously were presented to the Office.

(e) Estoppel.—

(1) Proceedings before the Office.—The petitioner in a post-grant review of a claim in a patent under this chapter that results in a final written decision under section 328(a), or the real party in interest or privy of the petitioner, may not request or maintain a proceeding before the Office with respect to that claim on any ground that the petitioner raised or reasonably could have raised during that post-grant review.

(2) Civil actions and other proceedings.—The petitioner in a post- grant review of a claim in a patent under this chapter that results in a final written decision under section 328(a), or the real party in interest or privy of the petitioner, may not assert either in a civil action arising in whole or in part under section 1338 of title 28 or in a proceeding before the International Trade Commission under section 337 of the Tariff Act of 1930 that the claim is invalid on any ground that the petitioner raised or reasonably could have raised during that post-grant review.

(f) Reissue patents.—A post-grant review may not be instituted under this chapter if the petition requests cancellation of a claim in a reissue patent that is identical to or narrower than a claim in the original patent from which the reissue patent was issued, and the time limitations in section 321(c) would bar filing a petition for a post-grant review for such original patent.

Sec.326. Conduct of post-grant review

(a) Regulations.—The Director shall prescribe regulations—

(1) providing that the file of any proceeding under this chapter shall be made available to the public, except that any petition or document filed with the intent that it be sealed shall, if accompanied by a motion to seal, be treated as sealed pending the outcome of the ruling on the motion;

(2) setting forth the standards for the showing of sufficient grounds to institute a review under subsections (a) and (b) of section 324;

(3) establishing procedures for the submission of supplemental information after the petition is filed;

(4) establishing and governing a post-grant review under this chapter and the relationship of such review to other proceedings under this title;

(5) setting forth standards and procedures for discovery of relevant evidence, including that such discovery shall be limited to evidence directly related to factual assertions advanced by either party in the proceeding;

(6) prescribing sanctions for abuse of discovery, abuse of process, or any other improper use of the proceeding, such as to harass or to cause unnecessary delay or an unnecessary increase in the cost of the proceeding;

(7) providing for protective orders governing the exchange and submission of confidential information;

(8) providing for the filing by the patent owner of a response to the petition under section 323 after a post-grant review has been instituted, and requiring that the patent owner file with such response, through affidavits or declarations, any additional factual evidence and expert opinions on which the patent owner relies in support of the response;

(9) setting forth standards and procedures for allowing the patent owner to move to amend the patent under subsection (d) to cancel a challenged claim or propose a reasonable number of substitute claims, and ensuring that any information submitted by the patent owner in support of any amendment entered under subsection (d) is made available to the public as part of the prosecution history of the patent;

(10) providing either party with the right to an oral hearing as part of the proceeding;

(11) requiring that the final determination in any post-grant review be issued not later than 1 year after the date on which the Director notices the institution of a proceeding under this chapter, except that the Director may, for good cause shown, extend the 1-year period by not more than 6 months, and may adjust the time periods in this paragraph in the case of joinder under section 325(c); and

(12) providing the petitioner with at least 1 opportunity to file written comments within a time period established by the Director.

(b) Considerations.—In prescribing regulations under this section, the Director shall consider the effect of any such regulation on the economy, the integrity of the patent system, the efficient administration of the Office, and the ability of the Office to timely complete proceedings instituted under this chapter.

(c) Patent Trial and Appeal Board.—The Patent Trial and Appeal Board shall, in accordance with section 6, conduct each post-grant review instituted under this chapter.

(d) Amendment of the patent.—

(1) In general.—During a post-grant review instituted under this chapter, the patent owner may file 1 motion to amend the patent in 1 or more of the following ways:

(A) Cancel any challenged patent claim.

(B) For each challenged claim, propose a reasonable number of substitute claims.

(2) Additional motions.—Additional motions to amend may be permitted upon the joint request of the petitioner and the patent owner to materially advance the settlement of a proceeding under section 327, or upon the request of the patent owner for good cause shown.

(3) Scope of claims.—An amendment under this subsection may not enlarge the scope of the claims of the patent or introduce new matter.

(e) Evidentiary standards.—In a post-grant review instituted under this chapter, the petitioner shall have the burden of proving a proposition of unpatentability by a preponderance of the evidence.

Sec. 327. Settlement

(a) In general.—A post-grant review instituted under this chapter shall be terminated with respect to any petitioner upon the joint request of the petitioner and the patent owner, unless the Office has decided the merits of the proceeding before the request for termination is filed. If the post-grant review is terminated with respect to a petitioner under this section, no estoppel under section 325(e) shall attach to the petitioner, or to the real party in interest or privy of the petitioner, on the basis of that petitioner's institution of that post-grant review. If no petitioner remains in the post-grant review, the Office may terminate the post-grant review or proceed to a final written decision under section 328(a).

(b) Agreements in writing.—Any agreement or understanding between the patent owner and a petitioner, including any collateral agreements referred to in such agreement or understanding, made in connection with, or in contemplation of, the termination of a post-grant review under this section shall be in writing, and a true copy of such agreement or understanding shall be filed in the Office before the termination of the post-grant review as between the parties. At the request of a party to the proceeding, the agreement or understanding shall be treated as business confidential information, shall be kept separate from the file of the involved patents, and shall be made available only to Federal Government agencies on written request, or to any person on a showing of good cause.

Sec. 328. Decision of the Board

(a) Final written decision.—If a post-grant review is instituted and not dismissed under this chapter, the Patent Trial and Appeal Board shall issue a final written decision with respect to the patentability of any patent claim challenged by the petitioner and any new claim added under section 326(d).

(b) Certificate.—If the Patent Trial and Appeal Board issues a final written decision under subsection (a) and the time for appeal has expired or any appeal has terminated, the Director shall issue and publish a certificate canceling any claim of the patent finally determined to be unpatentable, confirming any claim of the patent determined to be patentable, and incorporating in the patent by operation of the certificate any new or amended claim determined to be patentable.

(c) Intervening rights.—Any proposed amended or new claim determined to be patentable and incorporated into a patent following a post-grant review under this chapter shall have the same effect as that specified in section 252 for reissued patents on the right of any person who made, purchased, or used within the United States, or imported into the United States, anything patented by such proposed amended or new claim, or who made substantial preparation therefor, before the issuance of a certificate under subsection (b).

(d) Data on length of review.—The Office shall make available to the public data describing the length of time between the institution of, and the issuance of a final written decision under subsection (a) for, each post-grant review.

Sec. 329. Appeal

A party dissatisfied with the final written decision of the Patent Trial and Appeal Board under section 328(a) may appeal the decision pursuant to sections 141 through 144. Any party to the post-grant review shall have the right to be a party to the appeal.

Part IV. Patent Cooperation Treaty

Chapter 35. Definitions

Sec. 351. Definitions

When used in this part unless the context otherwise indicates—

(a) The term "treaty" means the Patent Co-operation Treaty done at Washington, on June 19, 1970.

(b) The term "Regulations", when capitalized, means the Regulations under the treaty, done at Washington on the same date as the treaty. The term "regulations", when not capitalized, means the regulations established by the Director under this title.

(c) The term "international application" means an application filed under the treaty.

(d) The term "international application originating in the United States" means an international application filed in the Patent and Trademark Office when it is acting as a Receiving Office under the treaty, irrespective of whether or not the United States has been designated in that international application.

(e) The term "international application designating the United States" means an international application specifying the United States as a country in which a patent is sought, regardless where such international application is filed.

(f) The term "Receiving Office" means a national patent office or intergovernmental organization which receives and processes international applications as prescribed by the treaty and the Regulations.

(g) The terms "International Searching Authority" and "International Preliminary Examining Authority" mean a national patent office or intergovernmental organization as appointed under the treaty which processes international applications as prescribed by the treaty and the Regulations.

(h) The term "International Bureau" means the international intergovernmental organization which is recognized as the coordinating body under the treaty and the Regulations.

(i) Terms and expressions not defined in this part are to be taken in the sense indicated by the treaty and the Regulations.

Chapter 36. International Stage

Sec. 361. Receiving Office

(a) The Patent and Trademark Office shall act as a Receiving Office for international applications filed by nationals or residents of the United States. In accordance with any agreement made between the United States and another country, the Patent and Trademark Office may also act as a Receiving Office for international applications filed by residents or nationals of such country who are entitled to file international applications.

(b) The Patent and Trademark Office shall perform all acts connected with the discharge of duties required of a Receiving Office, including the collection of international fees and their transmittal to the International Bureau.

(c) International applications filed in the Patent and Trademark Office shall be in the English language.

(d) The international fee, and the transmittal and search fees prescribed under section 376(a) of this part, shall either be paid on filing of an international application or within such later time as may be fixed by the Director.

Sec. 362. International Searching Authority and International Preliminary Examining Authority

(a) The Patent and Trademark Office may act as an International Searching Authority and International Preliminary Examining Authority with respect to international applications in accordance with the terms and conditions of an agreement which may be concluded with the International Bureau, and may discharge all duties required of such Authorities, including the collection of handling fees and their transmittal to the International Bureau.

(b) The handling fee, preliminary examination fee, and any additional fees due for international preliminary examination shall be paid within such time as may be fixed by the Director.

Sec. 363. International Application Designating the United States: Effect

An international application designating the United States shall have the effect, from its international filing date under article 11 of the treaty, of a national application for patent regularly filed in the Patent and Trademark Office.

Sec. 364. International Stage: Procedure

(a) International applications shall be processed by the Patent and Trademark Office when acting as a Receiving Office, International Searching Authority, or

International Preliminary Examining Authority, in accordance with the applicable provisions of the treaty, the Regulations, and this title.

(b) An applicant's failure to act within prescribed time limits in connection with requirements pertaining to a pending international application may be excused upon a showing satisfactory to the Director of unavoidable delay, to the extent not precluded by the treaty and the Regulations, and provided the conditions imposed by the treaty and the Regulations regarding the excuse of such failure to act are complied with.

Sec. 365. Right of Priority; Benefit of the Filing Date of a Prior Application

(a) In accordance with the conditions and requirements of subsections (a) through (d) of section 119 of this title, a national application shall be entitled to the right of priority based on a prior filed international application which designated at least one country other than the United States.

(b) In accordance with the conditions and requirement of section 119(a) of this title and the treaty and the Regulations, an international application designating the United States shall be entitled to the right of priority based on a prior foreign application, or a prior international application designating at least one country other than the United States.

(c) In accordance with the conditions and requirements of section 120 of this title, an international application designating the United States shall be entitled to the benefit of the filing date of a prior national application or a prior international application designating the United States, and a national application shall be entitled to the benefit of the filing date of a prior international application designating the United States. If any claim for the benefit of an earlier filing date is based on a prior international application which designated but did not originate in the United States, the Director may require the filing in the Patent and Trademark Office of a certified copy of such application together with a translation thereof into the English language, if it was filed in another language.

Sec. 366. Withdrawn International Application

Subject to section 367 of this part, if an international application designating the United States is withdrawn or considered withdrawn, either generally or as to the United States, under the conditions of the treaty and the Regulations, before the applicant has complied with the applicable requirements prescribed by section 371(c) of this part, the designation of the United States shall have no effect after the date of withdrawal, and shall be considered as not having been made, unless a claim for the benefit of a prior filing date under section 365(c) of this part was made in a national application, or an international application designating the United States, filed before the date of such withdrawal. However, such withdrawn international application may serve as the basis for a claim of priority under section 365(a) and (b) of this part, if it designated a country other than the United States.

Sec. 367. Actions of Other Authorities: Review

(a) Where a Receiving Office other than the Patent and Trademark Office has refused to accord an international filing date to an international application designating the United States or where it has held such application to be withdrawn either generally or as to the United States, the applicant may request review of the matter by the Director, on compliance with the requirements of and within the time limits specified by the treaty and the Regulations. Such review may result in a determination that such application be considered as pending in the national stage.

(b) The review under subsection (a) of this section, subject to the same requirements and conditions, may also be requested in those instances where an international application designating the United States is considered withdrawn due to a finding by the International Bureau under article 12(3) of the treaty.

Sec. 368. Secrecy of Certain Inventions; Filing International Applications in Foreign Countries

(a) International applications filed in the Patent and Trademark Office shall be subject to the provisions of chapter 17 of this title.

(b) In accordance with article 27(8) of the treaty, the filing of an international application in a country other than the United States on the invention made in this country shall be considered to constitute the filing of an application in a foreign country within the meaning of chapter 17 of this title, whether or not the United States is designated in that international application.

(c) If a license to file in a foreign country is refused or if an international application is ordered to be kept secret and a permit refused, the Patent and Trademark Office when acting as a Receiving Office, International Searching Authority, or International Preliminary Examining Authority, may not disclose the contents of such application to anyone not authorized to receive such disclosure.

Chapter 37. National Stage

Sec. 371. National Stage: Commencement

(a) Receipt from the International Bureau of copies of international applications with any amendments to the claims, international search reports, and international preliminary examination reports including any annexes thereto may be required in the case of international applications designating or electing the United States.

(b) Subject to subsection (f) of this section, the national stage shall commence with the expiration of the applicable time limit under article 22(1) or (2), or under article 39(1)(a) of the treaty.

(c) The applicant shall file in the Patent and Trademark Office—

(1) the national fee provided in section 41(a) of this title;

(2) a copy of the international application, unless not required under subsection (a) of this section or already communicated by the International Bureau, and a translation into the English language of the international application, if it was filed in another language;

(3) amendments, if any, to the claims in the international application, made under article 19 of the treaty, unless such amendments have been communicated to the Patent and Trademark Office by the International Bureau, and a translation into the English language if such amendments were made in another language;

(4) an oath or declaration of the inventor (or other person authorized under chapter 11 of this title) complying with the requirements of section 115 of this title and with regulations prescribed for oaths or declarations of applicants;

(5) a translation into the English language of any annexes to the international preliminary examination report, if such annexes were made in another language.

(d) The requirements with respect to the national fee referred to in subsection (c)(1), the translation referred to in subsection (c)(2), and the oath or declaration referred to in subsection (c)(4) of this section shall be complied with by the date of the commencement of the national stage or by such later time as may be fixed by the Director. The copy of the international application referred to in subsection (c)(2) shall be submitted by the date of the commencement of the national stage. Failure to comply with these requirements shall be regarded as abandonment of the application by the parties thereof, unless it be shown to the satisfaction of the Director that such failure to comply was unavoidable. The payment of a surcharge may be required as a condition of accepting the national fee referred to in subsection (c)(1) or the oath or declaration referred to in subsection (c)(4) of this section if these requirements are not met by the date of the commencement of the national stage. The requirements of subsection (c)(3) of this section shall be complied with by the date of the commencement of the national stage, and failure to do so shall be regarded as a cancellation of the amendments to the claims in the international application made under article 19 of the treaty. The requirement of subsection (c)(5) shall be complied with at such time as may be fixed by the Director and failure to do so shall be regarded as cancellation of the amendments made under article 34(2)(b) of the treaty.

(e) After an international application has entered the national stage, no patent may be granted or refused thereon before the expiration of the applicable time limit under article 28 or article 41 of the treaty, except with the express consent of the applicant. The applicant may present amendments to the specification, claims and drawings of the application after the national stage has commenced.

(f) At the express request of the applicant, the national stage of processing may be commenced at any time at which the application is in order for such purpose and the applicable requirements of subsection (c) of this section have been complied with.

Sec. 372. National Stage: Requirements and Procedure

(a) All questions of substance and, within the scope of the requirements of the treaty and Regulations, procedure in an international application designating the United States shall be determined as in the case of national applications regularly filed in the Patent and Trademark Office.

(b) In case of international applications designating but not originating in, the United States—

(1) the Director may cause to be reexamined questions relating to form and contents of the application in accordance with the requirements of the treaty and the Regulations;

(2) the Director may cause the question of unity of invention to be reexamined under section 121 of this title, within the scope of the requirements of the treaty and the Regulations; and

(3) the Director may require a verification of the translation of the international application or any other document pertaining to the application if the application or other document was filed in a language other than English.

Sec. 373. Repealed

Sec. 374. Publication of International Application

The publication under the treaty defined in section 351(a) of this title, of an international application designating the United States shall be deemed a publication under section 122(b), except as provided in section 154(d) of this title.

Sec. 375. Patent issued on International Application: Effect

(a) A patent may be issued by the Director based on an international application designating the United States, in accordance with the provisions of this title. Such patent shall have the force and effect of a patent issued on a national application filed under the provisions of chapter 11 of this title.

(b) Where due to an incorrect translation the scope of a patent granted on an international application designating the United States, which was not originally filed in the English language, exceeds the scope of the international application in its original language, a court of competent jurisdiction may retroactively limit the scope of the patent, by declaring it unenforceable to the extent that it exceeds the scope of the international application in its original language.

Sec. 376. Fees

(a) The required payment of the international fee and the handling fee, which amounts are specified in the Regulations, shall be paid in United States currency. The Patent and Trademark Office shall charge a national fee as provided in section 41(a), and may also charge the following fees:

(1) A transmittal fee (see section 361(d)).

(2) A search fee (see section 361(d)).

(3) A supplemental search fee (to be paid when required).

(4) A preliminary examination fee and any additional fees (see section 362(b)).

(5) Such other fees as established by the Director.

(b) The amounts of fees specified in subsection (a) of this section, except the international fee and the handling fee, shall be prescribed by the Director. He may refund any sum paid by mistake or in excess of the fees so specified, or if required under the treaty and the Regulations. The Director may also refund any part of the search fee, the national fee, the preliminary examination fee, and any additional fees, where he determines such refund to be warranted.

Leahy-Smith America Invents Act (as passed 2011)

Section 1. Short Title; Table of Contents.

(a) Short Title—This Act may be cited as the "Leahy-Smith America Invents Act".
(b) Table of Contents—The table of contents for this Act is as follows:

Sec. 2. Definitions.

In this Act:

(1) DIRECTOR—The term "Director" means the Under Secretary of Commerce for Intellectual Property and Director of the United States Patent and Trademark Office.

(2) OFFICE—The term "Office" means the United States Patent and Trademark Office.

(3) PATENT PUBLIC ADVISORY COMMITTEE—The term "Patent Public Advisory Committee" means the Patent Public Advisory Committee established under section 5(a) of title 35, United States Code.

(4) TRADEMARK ACT OF 1946—The term "Trademark Act of 1946" means the Act entitled "An Act to provide for the registration and protection of trademarks used in commerce, to carry out the provisions of certain international conventions, and for other purposes", approved July 5, 1946 (15 U.S.C. 1051 et seq.) (commonly referred to as the "Trademark Act of 1946" or the "Lanham Act").

(5) TRADEMARK PUBLIC ADVISORY COMMITTEE—The term "Trademark Public Advisory Committee" means the Trademark Public Advisory Committee established under section 5(a) of title 35, United States Code.

Sec. 3. First Inventor to File.

(a) Definitions—Section 100 of title 35, United States Code, is amended—

(1) in subsection (e), by striking "or inter partes reexamination under section 311"; and

(2) by adding at the end the following:

"(f) The term 'inventor' means the individual or, if a joint invention, the individuals collectively who invented or discovered the subject matter of the invention.

(g) The terms 'joint inventor' and 'coinventor' mean any 1 of the individuals who invented or discovered the subject matter of a joint invention.

(h) The term 'joint research agreement' means a written contract, grant, or cooperative agreement entered into by 2 or more persons or entities for the performance of experimental, developmental, or research work in the field of the claimed invention.

(i)(1) The term 'effective filing date' for a claimed invention in a patent or application for patent means—

(A) if subparagraph (B) does not apply, the actual filing date of the patent or the application for the patent containing a claim to the invention; or

(B) the filing date of the earliest application for which the patent or application is entitled, as to such invention, to a right of priority under section 119, 365(a), or 365(b) or to the benefit of an earlier filing date under section 120, 121, or 365(c).

(2) The effective filing date for a claimed invention in an application for reissue or reissued patent shall be determined by deeming the claim to the invention to have been contained in the patent for which reissue was sought.

(j) The term 'claimed invention' means the subject matter defined by a claim in a patent or an application for a patent.".

(b) Conditions for Patentability—

(1) IN GENERAL—Section 102 of title 35, United States Code, is amended to read as follows:

"Sec. 102. Conditions for Patentability; Novelty

(a) Novelty; Prior Art—A person shall be entitled to a patent unless—

(1) the claimed invention was patented, described in a printed publication, or in public use, on sale, or otherwise available to the public before the effective filing date of the claimed invention; or

(2) the claimed invention was described in a patent issued under section 151, or in an application for patent published or deemed published under section 122(b), in which the patent or application, as the case may be, names another inventor and was effectively filed before the effective filing date of the claimed invention.

(b) Exceptions—

(1) DISCLOSURES MADE 1 YEAR OR LESS BEFORE THE EFFECTIVE FILING DATE OF THE CLAIMED INVENTION—A disclosure made 1 year or less before the effective filing date of a claimed invention shall not be prior art to the claimed invention under subsection (a)(1) if—

(A) the disclosure was made by the inventor or joint inventor or by another who obtained the subject matter disclosed directly or indirectly from the inventor or a joint inventor; or

(B) the subject matter disclosed had, before such disclosure, been publicly disclosed by the inventor or a joint inventor or another who obtained the subject matter disclosed directly or indirectly from the inventor or a joint inventor.

(2) DISCLOSURES APPEARING IN APPLICATIONS AND PATENTS—A disclosure shall not be prior art to a claimed invention under subsection (a)(2) if—

(A) the subject matter disclosed was obtained directly or indirectly from the inventor or a joint inventor;

(B) the subject matter disclosed had, before such subject matter was effectively filed under subsection (a)(2), been publicly disclosed by the inventor or a joint inventor or another who obtained the subject matter disclosed directly or indirectly from the inventor or a joint inventor; or

(C) the subject matter disclosed and the claimed invention, not later than the effective filing date of the claimed invention, were owned by the same person or subject to an obligation of assignment to the same person.

(c) Common Ownership Under Joint Research Agreements—Subject matter disclosed and a claimed invention shall be deemed to have been owned by the same person or subject to an obligation of assignment to the same person in applying the provisions of subsection (b)(2)(C) if—

(1) the subject matter disclosed was developed and the claimed invention was made by, or on behalf of, 1 or more parties to a joint research agreement that was in effect on or before the effective filing date of the claimed invention;

(2) the claimed invention was made as a result of activities undertaken within the scope of the joint research agreement; and

(3) the application for patent for the claimed invention discloses or is amended to disclose the names of the parties to the joint research agreement.

(d) Patents and Published Applications Effective as Prior Art—For purposes of determining whether a patent or application for patent is prior art to a claimed invention under subsection (a)(2), such patent or application shall be considered to have been effectively filed, with respect to any subject matter described in the patent or application—

(1) if paragraph (2) does not apply, as of the actual filing date of the patent or the application for patent; or

(2) if the patent or application for patent is entitled to claim a right of priority under section 119, 365(a), or 365(b), or to claim the benefit of an earlier filing date under section 120, 121, or 365(c), based upon 1 or more prior filed applications for patent, as of the filing date of the earliest such application that describes the subject matter.".

(2) CONTINUITY OF INTENT UNDER THE CREATE ACT—The enactment of section 102(c) of title 35, United States Code, under paragraph (1) of this subsection is done with the same intent to promote joint research activities that was expressed, including in the legislative history, through the enactment of the Cooperative Research and Technology Enhancement Act of 2004 (Public Law 108-453; the "CREATE Act"), the amendments of which are stricken by subsection (c) of this section. The United States Patent and Trademark Office shall administer section 102(c) of title 35, United States Code, in a manner consistent with the legislative history of the CREATE Act that was relevant to its administration by the United States Patent and Trademark Office.

(3) CONFORMING AMENDMENT—The item relating to section 102 in the table of sections for chapter 10 of title 35, United States Code, is amended to read as follows:

"102. Conditions for patentability; novelty.".

(c) Conditions for Patentability; Nonobvious Subject Matter—Section 103 of title 35, United States Code, is amended to read as follows:

"Sec. 103. Conditions for Patentability; Non-Obvious Subject Matter

A patent for a claimed invention may not be obtained, notwithstanding that the claimed invention is not identically disclosed as set forth in section 102, if the differences between the claimed invention and the prior art are such that the claimed invention as a whole would have been obvious before the effective filing date of the claimed invention to a person having ordinary skill in the art to which the claimed invention pertains. Patentability shall not be negated by the manner in which the invention was made.".

(d) Repeal of Requirements for Inventions Made Abroad—Section 104 of title 35, United States Code, and the item relating to that section in the table of sections for chapter 10 of title 35, United States Code, are repealed.

(e) Repeal of Statutory Invention Registration—

(1) IN GENERAL—Section 157 of title 35, United States Code, and the item relating to that section in the table of sections for chapter 14 of title 35, United States Code, are repealed.

(2) REMOVAL OF CROSS REFERENCES—Section 111(b)(8) of title 35, United States Code, is amended by striking "sections 115, 131, 135, and 157" and inserting "sections 131 and 135".

(3) EFFECTIVE DATE—The amendments made by this subsection shall take effect upon the expiration of the 18-month period beginning on the date of the enactment of this Act, and shall apply to any request for a statutory invention registration filed on or after that effective date.

(f) Earlier Filing Date for Inventor and Joint Inventor—Section 120 of title 35, United States Code, is amended by striking "which is filed by an inventor or inventors named" and inserting "which names an inventor or joint inventor".

(g) Conforming Amendments—

(1) RIGHT OF PRIORITY—Section 172 of title 35, United States Code, is amended by striking "and the time specified in section 102(d)".

(2) LIMITATION ON REMEDIES—Section 287(c)(4) of title 35, United States Code, is amended by striking "the earliest effective filing date of which is prior to" and inserting "which has an effective filing date before".

(3) INTERNATIONAL APPLICATION DESIGNATING THE UNITED STATES: EFFECT—Section 363 of title 35, United States Code, is amended by striking "except as otherwise provided in section 102(e) of this title".

(4) PUBLICATION OF INTERNATIONAL APPLICATION: EFFECT—Section 374 of title 35, United States Code, is amended by striking "sections 102(e) and 154(d)" and inserting "section 154(d)".

(5) PATENT ISSUED ON INTERNATIONAL APPLICATION: EFFECT—The second sentence of section 375(a) of title 35, United States Code, is amended by striking "Subject to section 102(e) of this title, such" and inserting "Such".

(6) LIMIT ON RIGHT OF PRIORITY—Section 119(a) of title 35, United States Code, is amended by striking "; but no patent shall be granted" and all that follows through "one year prior to such filing".

(7) INVENTIONS MADE WITH FEDERAL ASSISTANCE—Section 202(c) of title 35, United States Code, is amended—

(A) in paragraph (2)—

(i) by striking "publication, on sale, or public use," and all that follows through "obtained in the United States" and inserting "the 1-year period referred to in section 102(b) would end before the end of that 2-year period"; and

(ii) by striking "prior to the end of the statutory" and inserting "before the end of that 1-year"; and

(B) in paragraph (3), by striking "any statutory bar date that may occur under this title due to publication, on sale, or public use" and inserting "the expiration of the 1-year period referred to in section 102(b)".

(h) Derived Patents—

(1) IN GENERAL—Section 291 of title 35, United States Code, is amended to read as follows:

"Sec. 291. Derived Patents

(a) In General—The owner of a patent may have relief by civil action against the owner of another patent that claims the same invention and has an earlier effective filing date, if the invention claimed in such other patent was derived from the inventor of the invention claimed in the patent owned by the person seeking relief under this section.

(b) Filing Limitation—An action under this section may be filed only before the end of the 1-year period beginning on the date of the issuance of the first patent containing a claim to the allegedly derived invention and naming an individual alleged to have derived such invention as the inventor or joint inventor.".

(2) CONFORMING AMENDMENT—The item relating to section 291 in the table of sections for chapter 29 of title 35, United States Code, is amended to read as follows:

"291. Derived patents.".

(i) Derivation Proceedings—Section 135 of title 35, United States Code, is amended to read as follows:

"Sec. 135. Derivation Proceedings

(a) Institution of Proceeding—An applicant for patent may file a petition to institute a derivation proceeding in the Office. The petition shall set forth with particularity the basis for finding that an inventor named in an earlier application derived the claimed invention from an inventor named in the petitioner's application and, without authorization, the earlier application claiming such invention was filed. Any such petition may be filed only within the 1-year period beginning on the date of the first publication of a claim to an invention that is the same or substantially the same as the earlier application's claim to the invention, shall be made under oath, and shall be supported by substantial evidence. Whenever the Director determines that a petition filed under this subsection demonstrates that the standards for instituting a derivation proceeding are met, the Director may institute a derivation proceeding. The determination by the Director whether to institute a derivation proceeding shall be final and nonappealable.

(b) Determination by Patent Trial and Appeal Board—In a derivation proceeding instituted under subsection (a), the Patent Trial and Appeal Board shall determine whether an inventor named in the earlier application derived the claimed invention from an inventor named in the petitioner's application and, without authorization, the earlier application claiming such invention was filed. In appropriate circumstances, the Patent Trial and Appeal Board may correct the naming of the inventor in any application or patent at issue. The Director shall prescribe regulations setting forth standards for the conduct of derivation proceedings, including requiring parties to provide sufficient evidence to prove and rebut a claim of derivation.

(c) Deferral of Decision—The Patent Trial and Appeal Board may defer action on a petition for a derivation proceeding until the expiration of the 3-month period beginning on the date on which the Director issues a patent that includes the claimed invention that is the subject of the petition. The Patent Trial and Appeal Board also may defer action on a petition for a derivation proceeding, or stay the proceeding after it has been instituted, until the termination of a proceeding under chapter 30, 31, or 32 involving the patent of the earlier applicant.

(d) Effect of Final Decision—The final decision of the Patent Trial and Appeal Board, if adverse to claims in an application for patent, shall constitute the final refusal by the Office on those claims. The final decision of the Patent Trial and Appeal Board, if adverse to claims in a patent, shall, if no appeal or other review of the decision has been or can be taken or had, constitute cancellation of those claims, and notice of such cancellation shall be endorsed on copies of the patent distributed after such cancellation.

(e) Settlement—Parties to a proceeding instituted under subsection (a) may terminate the proceeding by filing a written statement reflecting the agreement of the parties as to the correct inventors of the claimed invention in dispute. Unless the Patent Trial and Appeal Board finds the agreement to be inconsistent with the evidence of record, if any, it shall take action consistent with the agreement. Any written settlement or understanding of the parties shall be filed with the Director. At the request of a party to the proceeding, the agreement or understanding shall be treated as business confidential information, shall be kept separate from the file of the involved patents or applications, and shall be made available only to Government agencies on written request, or to any person on a showing of good cause.

(f) Arbitration—Parties to a proceeding instituted under subsection (a) may, within such time as may be specified by the Director by regulation, determine such contest or any aspect thereof by arbitration. Such arbitration shall be governed by the provisions of title 9, to the extent such title is not inconsistent with this section. The parties shall give notice of any arbitration award to the Director, and such award shall, as between the parties to the arbitration, be dispositive of the issues to which it relates. The arbitration award shall be unenforceable until such notice is given. Nothing in this subsection shall preclude the Director from determining the patentability of the claimed inventions involved in the proceeding.".

(j) Elimination of References to Interferences—(1) Sections 134, 145, 146, 154, and 305 of title 35, United States Code, are each amended by striking 'Board of Patent Appeals and Interferences' each place it appears and inserting 'Patent Trial and Appeal Board'.

(2) (A) Section 146 of title 35, United States Code, is amended—

(i) by striking "an interference" and inserting "a derivation proceeding"; and

(ii) by striking "the interference" and inserting "the derivation proceeding".

(B) The subparagraph heading for section 154(b)(1)(C) of title 35, United States Code, is amended to read as follows:

"(C) GUARANTEE OF ADJUSTMENTS FOR DELAYS DUE TO DERIVATION PROCEEDINGS, SECRECY ORDERS, AND APPEALS—".

(3) The section heading for section 134 of title 35, United States Code, is amended to read as follows:

"Sec. 134. Appeal to the Patent Trial and Appeal Board".

(4) The section heading for section 146 of title 35, United States Code, is amended to read as follows:

"Sec. 146. Civil Action In Case of Derivation Proceeding".

(5) The items relating to sections 134 and 135 in the table of sections for chapter 12 of title 35, United States Code, are amended to read as follows:

"134. Appeal to the Patent Trial and Appeal Board.

135. Derivation proceedings.".

(6) The item relating to section 146 in the table of sections for chapter 13 of title 35, United States Code, is amended to read as follows:

"146. Civil action in case of derivation proceeding.".

(k) Statute of Limitations—

(1) IN GENERAL—Section 32 of title 35, United States Code, is amended by inserting between the third and fourth sentences the following: "A proceeding under this section shall be commenced not later than the earlier of either the date that is 10 years after the date on which the misconduct forming the basis for the proceeding occurred, or 1 year after the date on which the misconduct forming

the basis for the proceeding is made known to an officer or employee of the Office as prescribed in the regulations established under section 2(b)(2)(D).".

(2) REPORT TO CONGRESS—The Director shall provide on a biennial basis to the Judiciary Committees of the Senate and House of Representatives a report providing a short description of incidents made known to an officer or employee of the Office as prescribed in the regulations established under section 2(b)(2)(D) of title 35, United States Code, that reflect substantial evidence of misconduct before the Office but for which the Office was barred from commencing a proceeding under section 32 of title 35, United States Code, by the time limitation established by the fourth sentence of that section.

(3) EFFECTIVE DATE—The amendment made by paragraph (1) shall apply in any case in which the time period for instituting a proceeding under section 32 of title 35, United States Code, had not lapsed before the date of the enactment of this Act.

(l) Small Business Study—

(1) DEFINITIONS- In this subsection—

(A) the term "Chief Counsel" means the Chief Counsel for Advocacy of the Small Business Administration;

(B) the term "General Counsel" means the General Counsel of the United States Patent and Trademark Office; and

(C) the term "small business concern" has the meaning given that term under section 3 of the Small Business Act (15 U.S.C. 632).

(2) STUDY—

(A) IN GENERAL—The Chief Counsel, in consultation with the General Counsel, shall conduct a study of the effects of eliminating the use of dates of invention in determining whether an applicant is entitled to a patent under title 35, United States Code.

(B) AREAS OF STUDY—The study conducted under subparagraph (A) shall include examination of the effects of eliminating the use of invention dates, including examining—

(i) how the change would affect the ability of small business concerns to obtain patents and their costs of obtaining patents;

(ii) whether the change would create, mitigate, or exacerbate any disadvantages for applicants for patents that are small business concerns relative to applicants for patents that are not small business concerns, and whether the change would create any advantages for applicants for patents that are small business concerns relative to applicants for patents that are not small business concerns;

(iii) the cost savings and other potential benefits to small business concerns of the change; and

(iv) the feasibility and costs and benefits to small business concerns of alternative means of determining whether an applicant is entitled to a patent under title 35, United States Code.

(3) REPORT—Not later than the date that is 1 year after the date of the enactment of this Act, the Chief Counsel shall submit to the Committee on Small

Business and Entrepreneurship and the Committee on the Judiciary of the Senate and the Committee on Small Business and the Committee on the Judiciary of the House of Representatives a report on the results of the study under paragraph (2).

(m) Report on Prior User Rights—

(1) IN GENERAL—Not later than the end of the 4-month period beginning on the date of the enactment of this Act, the Director shall report, to the Committee on the Judiciary of the Senate and the Committee on the Judiciary of the House of Representatives, the findings and recommendations of the Director on the operation of prior user rights in selected countries in the industrialized world. The report shall include the following:

(A) A comparison between patent laws of the United States and the laws of other industrialized countries, including members of the European Union and Japan, Canada, and Australia.

(B) An analysis of the effect of prior user rights on innovation rates in the selected countries.

(C) An analysis of the correlation, if any, between prior user rights and start-up enterprises and the ability to attract venture capital to start new companies.

(D) An analysis of the effect of prior user rights, if any, on small businesses, universities, and individual inventors.

(E) An analysis of legal and constitutional issues, if any, that arise from placing trade secret law in patent law.

(F) An analysis of whether the change to a first-to-file patent system creates a particular need for prior user rights.

(2) CONSULTATION WITH OTHER AGENCIES—In preparing the report required under paragraph (1), the Director shall consult with the United States Trade Representative, the Secretary of State, and the Attorney General.

(n) Effective Date—

(1) IN GENERAL—Except as otherwise provided in this section, the amendments made by this section shall take effect upon the expiration of the 18-month period beginning on the date of the enactment of this Act, and shall apply to any application for patent, and to any patent issuing thereon, that contains or contained at any time—

(A) a claim to a claimed invention that has an effective filing date as defined in section 100(i) of title 35, United States Code, that is on or after the effective date described in this paragraph; or

(B) a specific reference under section 120, 121, or 365(c) of title 35, United States Code, to any patent or application that contains or contained at any time such a claim.

(2) INTERFERING PATENTS—The provisions of sections 102(g), 135, and 291 of title 35, United States Code, as in effect on the day before the effective date set forth in paragraph (1) of this subsection, shall apply to each claim of an application for patent, and any patent issued thereon, for which the amendments made by this section also apply, if such application or patent contains or contained at any time—

(A) a claim to an invention having an effective filing date as defined in section 100(i) of title 35, United States Code, that occurs before the effective date set forth in paragraph (1) of this subsection; or

(B) a specific reference under section 120, 121, or 365(c) of title 35, United States Code, to any patent or application that contains or contained at any time such a claim.

(o) Sense of Congress—It is the sense of the Congress that converting the United States patent system from 'first to invent' to a system of 'first inventor to file' will promote the progress of science and the useful arts by securing for limited times to inventors the exclusive rights to their discoveries and provide inventors with greater certainty regarding the scope of protection provided by the grant of exclusive rights to their discoveries.

(p) Sense of Congress—It is the sense of the Congress that converting the United States patent system from "first to invent" to a system of "first inventor to file" will improve the United States patent system and promote harmonization of the United States patent system with the patent systems commonly used in nearly all other countries throughout the world with whom the United States conducts trade and thereby promote greater international uniformity and certainty in the procedures used for securing the exclusive rights of inventors to their discoveries.

Sec. 4. Inventor's Oath or Declaration.

(a) Inventor's Oath or Declaration—

(1) IN GENERAL—Section 115 of title 35, United States Code, is amended to read as follows:

"Sec. 115. Inventor's Oath or Declaration

(a) Naming the Inventor; Inventor's Oath or Declaration—An application for patent that is filed under section 111(a) or commences the national stage under section 371 shall include, or be amended to include, the name of the inventor for any invention claimed in the application. Except as otherwise provided in this section, each individual who is the inventor or a joint inventor of a claimed invention in an application for patent shall execute an oath or declaration in connection with the application.

(b) Required Statements—An oath or declaration under subsection (a) shall contain statements that—

(1) the application was made or was authorized to be made by the affiant or declarant; and

(2) such individual believes himself or herself to be the original inventor or an original joint inventor of a claimed invention in the application.

(c) Additional Requirements—The Director may specify additional information relating to the inventor and the invention that is required to be included in an oath or declaration under subsection (a).

(d) Substitute Statement—

(1) IN GENERAL—In lieu of executing an oath or declaration under subsection (a), the applicant for patent may provide a substitute statement under the circumstances described in paragraph (2) and such additional circumstances that the Director may specify by regulation.

(2) PERMITTED CIRCUMSTANCES—A substitute statement under paragraph (1) is permitted with respect to any individual who—

(A) is unable to file the oath or declaration under subsection (a) because the individual—

(i) is deceased;

(ii) is under legal incapacity; or

(iii) cannot be found or reached after diligent effort; or

(B) is under an obligation to assign the invention but has refused to make the oath or declaration required under subsection (a).

(3) CONTENTS—A substitute statement under this subsection shall—

(A) identify the individual with respect to whom the statement applies;

(B) set forth the circumstances representing the permitted basis for the filing of the substitute statement in lieu of the oath or declaration under subsection (a); and

(C) contain any additional information, including any showing, required by the Director.

(e) Making Required Statements in Assignment of Record—An individual who is under an obligation of assignment of an application for patent may include the required statements under subsections (b) and (c) in the assignment executed by the individual, in lieu of filing such statements separately.

(f) Time for Filing—A notice of allowance under section 151 may be provided to an applicant for patent only if the applicant for patent has filed each required oath or declaration under subsection (a) or has filed a substitute statement under subsection (d) or recorded an assignment meeting the requirements of subsection (e).

(g) Earlier-Filed Application Containing Required Statements or Substitute Statement—

(1) EXCEPTION—The requirements under this section shall not apply to an individual with respect to an application for patent in which the individual is named as the inventor or a joint inventor and who claims the benefit under section 120, 121, or 365(c) of the filing of an earlier-filed application, if—

(A) an oath or declaration meeting the requirements of subsection (a) was executed by the individual and was filed in connection with the earlier-filed application;

(B) a substitute statement meeting the requirements of subsection (d) was filed in connection with the earlier filed application with respect to the individual; or

(C) an assignment meeting the requirements of subsection (e) was executed with respect to the earlier-filed application by the individual and was recorded in connection with the earlier-filed application.

(2) COPIES OF OATHS, DECLARATIONS, STATEMENTS, OR ASSIGNMENTS—Notwithstanding paragraph (1), the Director may require that a copy

of the executed oath or declaration, the substitute statement, or the assignment filed in connection with the earlier-filed application be included in the later-filed application.

(h) Supplemental and Corrected Statements; Filing Additional Statements—

(1) IN GENERAL—Any person making a statement required under this section may withdraw, replace, or otherwise correct the statement at any time. If a change is made in the naming of the inventor requiring the filing of 1 or more additional statements under this section, the Director shall establish regulations under which such additional statements may be filed.

(2) SUPPLEMENTAL STATEMENTS NOT REQUIRED—If an individual has executed an oath or declaration meeting the requirements of subsection (a) or an assignment meeting the requirements of subsection (e) with respect to an application for patent, the Director may not thereafter require that individual to make any additional oath, declaration, or other statement equivalent to those required by this section in connection with the application for patent or any patent issuing thereon.

(3) SAVINGS CLAUSE—A patent shall not be invalid or unenforceable based upon the failure to comply with a requirement under this section if the failure is remedied as provided under paragraph (1).

(i) Acknowledgment of Penalties—Any declaration or statement filed pursuant to this section shall contain an acknowledgment that any willful false statement made in such declaration or statement is punishable under section 1001 of title 18 by fine or imprisonment of not more than 5 years, or both.".

(2) RELATIONSHIP TO DIVISIONAL APPLICATIONS—Section 121 of title 35, United States Code, is amended by striking "If a divisional application" and all that follows through "inventor.".

(3) REQUIREMENTS FOR NONPROVISIONAL APPLICATIONS—Section 111(a) of title 35, United States Code, is amended—

(A) in paragraph (2)(C), by striking "by the applicant" and inserting "or declaration";

(B) in the heading for paragraph (3), by inserting "OR DECLARATION" after "AND OATH"; and

(C) by inserting "or declaration" after "and oath" each place it appears.

(4) CONFORMING AMENDMENT—The item relating to section 115 in the table of sections for chapter 11 of title 35, United States Code, is amended to read as follows:

"115. Inventor's oath or declaration.".

(b) Filing by Other Than Inventor—

(1) IN GENERAL—Section 118 of title 35, United States Code, is amended to read as follows:

"Sec. 118. Filing by Other Than Inventor

A person to whom the inventor has assigned or is under an obligation to assign the invention may make an application for patent. A person who otherwise shows sufficient proprietary interest in the matter may make an application for patent on behalf of and as agent for the inventor on proof of the pertinent facts and a showing that such action is appropriate to preserve the rights of the parties. If the Director grants a patent on an application filed under this section by a person other than the inventor, the patent shall be granted to the real party in interest and upon such notice to the inventor as the Director considers to be sufficient.".

(2) CONFORMING AMENDMENT—Section 251 of title 35, United States Code, is amended in the third undesignated paragraph by inserting "or the application for the original patent was filed by the assignee of the entire interest" after "claims of the original patent".

(c) Specification—Section 112 of title 35, United States Code, is amended—

(1) in the first undesignated paragraph—

(A) by striking "The specification" and inserting "(a) In General—The specification"; and

(B) by striking "of carrying out his invention" and inserting "or joint inventor of carrying out the invention";

(2) in the second undesignated paragraph—

(A) by striking "The specification" and inserting "(b) Conclusion- The specification"; and

(B) by striking "applicant regards as his invention" and inserting "inventor or a joint inventor regards as the invention";

(3) in the third undesignated paragraph, by striking "A claim" and inserting "(c) Form- A claim";

(4) in the fourth undesignated paragraph, by striking "Subject to the following paragraph," and inserting "(d) Reference in Dependent Forms—Subject to subsection (e),";

(5) in the fifth undesignated paragraph, by striking "A claim" and inserting "(e) Reference in Multiple Dependent Form—A claim"; and

(6) in the last undesignated paragraph, by striking "An element" and inserting "(f) Element in Claim for a Combination—An element".

(d) Conforming Amendments—

(1) Sections 111(b)(1)(A) of title 35, United States Code, is amended by striking "the first paragraph of section 112 of this title" and inserting "section 112(a)".

(2) Section 111(b)(2) of title 35, United States Code, is amended by striking "the second through fifth paragraphs of section 112," and inserting "subsections (b) through (e) of section 112,".

(e) Effective Date—The amendments made by this section shall take effect upon the expiration of the 1-year period beginning on the date of the enactment of this Act and shall apply to any patent application that is filed on or after that effective date.

Sec. 5. Defense to Infringement Based on Prior Commercial Use.

(a) In General—Section 273 of title 35, United States Code, is amended to read as follows:

"Sec. 273. Defense to Infringement Based on Prior Commercial Use

(a) In General—A person shall be entitled to a defense under section 282(b) with respect to subject matter consisting of a process, or consisting of a machine, manufacture, or composition of matter used in a manufacturing or other commercial process, that would otherwise infringe a claimed invention being asserted against the person if—

(1) such person, acting in good faith, commercially used the subject matter in the United States, either in connection with an internal commercial use or an actual arm's length sale or other arm's length commercial transfer of a useful end result of such commercial use; and

(2) such commercial use occurred at least 1 year before the earlier of either—

(A) the effective filing date of the claimed invention; or

(B) the date on which the claimed invention was disclosed to the public in a manner that qualified for the exception from prior art under section 102(b).

(b) Burden of Proof—A person asserting a defense under this section shall have the burden of establishing the defense by clear and convincing evidence.

(c) Additional Commercial Uses—

(1) PREMARKETING REGULATORY REVIEW—Subject matter for which commercial marketing or use is subject to a premarketing regulatory review period during which the safety or efficacy of the subject matter is established, including any period specified in section 156(g), shall be deemed to be commercially used for purposes of subsection (a)(1) during such regulatory review period.

(2) NONPROFIT LABORATORY USE—A use of subject matter by a nonprofit research laboratory or other nonprofit entity, such as a university or hospital, for which the public is the intended beneficiary, shall be deemed to be a commercial use for purposes of subsection (a)(1), except that a defense under this section may be asserted pursuant to this paragraph only for continued and noncommercial use by and in the laboratory or other nonprofit entity.

(d) Exhaustion of Rights—Notwithstanding subsection (e)(1), the sale or other disposition of a useful end result by a person entitled to assert a defense under this section in connection with a patent with respect to that useful end result shall exhaust the patent owner's rights under the patent to the extent that such rights would have been exhausted had such sale or other disposition been made by the patent owner.

(e) Limitations and Exceptions—

(1) PERSONAL DEFENSE—

(A) IN GENERAL—A defense under this section may be asserted only by the person who performed or directed the performance of the commercial use described in subsection (a), or by an entity that controls, is controlled by, or is under common control with such person.

(B) TRANSFER OF RIGHT—Except for any transfer to the patent owner, the right to assert a defense under this section shall not be licensed or assigned or transferred to another person except as an ancillary and subordinate part of a good-faith assignment or transfer for other reasons of the entire enterprise or line of business to which the defense relates.

(C) RESTRICTION ON SITES—A defense under this section, when acquired by a person as part of an assignment or transfer described in subparagraph (B), may only be asserted for uses at sites where the subject matter that would otherwise infringe a claimed invention is in use before the later of the effective filing date of the claimed invention or the date of the assignment or transfer of such enterprise or line of business.

(2) DERIVATION—A person may not assert a defense under this section if the subject matter on which the defense is based was derived from the patentee or persons in privity with the patentee.

(3) NOT A GENERAL LICENSE—The defense asserted by a person under this section is not a general license under all claims of the patent at issue, but extends only to the specific subject matter for which it has been established that a commercial use that qualifies under this section occurred, except that the defense shall also extend to variations in the quantity or volume of use of the claimed subject matter, and to improvements in the claimed subject matter that do not infringe additional specifically claimed subject matter of the patent.

(4) ABANDONMENT OF USE—A person who has abandoned commercial use (that qualifies under this section) of subject matter may not rely on activities performed before the date of such abandonment in establishing a defense under this section with respect to actions taken on or after the date of such abandonment.

(5) UNIVERSITY EXCEPTION—

(A) IN GENERAL—A person commercially using subject matter to which subsection (a) applies may not assert a defense under this section if the claimed invention with respect to which the defense is asserted was, at the time the invention was made, owned or subject to an obligation of assignment to either an institution of higher education (as defined in section 101(a) of the Higher Education Act of 1965 (20 U.S.C. 1001(a)), or a technology transfer organization whose primary purpose is to facilitate the commercialization of technologies developed by one or more such institutions of higher education.

(B) EXCEPTION—Subparagraph (A) shall not apply if any of the activities required to reduce to practice the subject matter of the claimed invention could not have been undertaken using funds provided by the Federal Government.

(f) Unreasonable Assertion of Defense—If the defense under this section is pleaded by a person who is found to infringe the patent and who subsequently fails to demonstrate a reasonable basis for asserting the defense, the court shall find the case exceptional for the purpose of awarding attorney fees under section 285.

(g) Invalidity—A patent shall not be deemed to be invalid under section 102 or 103 solely because a defense is raised or established under this section.".

(b) Conforming Amendment—The item relating to section 273 in the table of sections for chapter 28 of title 35, United States Code, is amended to read as follows:

"273. Defense to Infringement Based on Prior Commercial Use.".

(c) Effective Date—The amendments made by this section shall apply to any patent issued on or after the date of the enactment of this Act.

Sec. 6. Post-Grant Review Proceedings.

(a) Inter Partes Review—Chapter 31 of title 35, United States Code, is amended to read as follows:

"Chapter 31—Inter Partes Review

Sec.

311.	Inter partes review.
312.	Petitions.
313.	Preliminary response to petition.
314.	Institution of inter partes review.
315.	Relation to other proceedings or actions.
316.	Conduct of inter partes review.
317.	Settlement.
318.	Decision of the Board.
319.	Appeal.

Sec. 311. Inter Partes Review

(a) In General—Subject to the provisions of this chapter, a person who is not the owner of a patent may file with the Office a petition to institute an inter partes review of the patent. The Director shall establish, by regulation, fees to be paid by the person requesting the review, in such amounts as the Director determines to be reasonable, considering the aggregate costs of the review.

(b) Scope—A petitioner in an inter partes review may request to cancel as unpatentable 1 or more claims of a patent only on a ground that could be raised under section 102 or 103 and only on the basis of prior art consisting of patents or printed publications.

(c) Filing Deadline—A petition for inter partes review shall be filed after the later of either—

(1) the date that is 9 months after the grant of a patent or issuance of a reissue of a patent; or

(2) if a post-grant review is instituted under chapter 32, the date of the termination of such post-grant review.

Sec. 312. Petitions

(a) Requirements of Petition—A petition filed under section 311 may be considered only if—

(1) the petition is accompanied by payment of the fee established by the Director under section 311;

(2) the petition identifies all real parties in interest;

(3) the petition identifies, in writing and with particularity, each claim challenged, the grounds on which the challenge to each claim is based, and the evidence that supports the grounds for the challenge to each claim, including—

(A) copies of patents and printed publications that the petitioner relies upon in support of the petition; and

(B) affidavits or declarations of supporting evidence and opinions, if the petitioner relies on expert opinions;

(4) the petition provides such other information as the Director may require by regulation; and

(5) the petitioner provides copies of any of the documents required under paragraphs (2), (3), and (4) to the patent owner or, if applicable, the designated representative of the patent owner.

(b) Public Availability—As soon as practicable after the receipt of a petition under section 311, the Director shall make the petition available to the public.

Sec. 313. Preliminary Response to Petition

If an inter partes review petition is filed under section 311, the patent owner shall have the right to file a preliminary response to the petition, within a time period set by the Director, that sets forth reasons why no inter partes review should be instituted based upon the failure of the petition to meet any requirement of this chapter.

Sec. 314. Institution of Inter Partes Review

(a) Threshold—The Director may not authorize an inter partes review to be instituted unless the Director determines that the information presented in the petition filed under section 311 and any response filed under section 313 shows that there is a reasonable likelihood that the petitioner would prevail with respect to at least 1 of the claims challenged in the petition.

(b) Timing—The Director shall determine whether to institute an inter partes review under this chapter pursuant to a petition filed under section 311 within 3 months after—

(1) receiving a preliminary response to the petition under section 313; or

(2) if no such preliminary response is filed, the last date on which such response may be filed.

(c) Notice—The Director shall notify the petitioner and patent owner, in writing, of the Director's determination under subsection (a), and shall make such notice available to the public as soon as is practicable. Such notice shall include the date on which the review shall commence.

(d) No Appeal—The determination by the Director whether to institute an inter partes review under this section shall be final and nonappealable.

Sec. 315. Relation to Other Proceedings or Actions

(a) Infringer's Civil Action—

(1) INTER PARTES REVIEW BARRED BY CIVIL ACTION—An inter partes review may not be instituted if, before the date on which the petition for such a review is filed, the petitioner or real party in interest filed a civil action challenging the validity of a claim of the patent.

(2) STAY OF CIVIL ACTION—If the petitioner or real party in interest files a civil action challenging the validity of a claim of the patent on or after the date on which the petitioner files a petition for inter partes review of the patent, that civil action shall be automatically stayed until either—

(A) the patent owner moves the court to lift the stay;

(B) the patent owner files a civil action or counterclaim alleging that the petitioner or real party in interest has infringed the patent; or

(C) the petitioner or real party in interest moves the court to dismiss the civil action.

(3) TREATMENT OF COUNTERCLAIM—A counterclaim challenging the validity of a claim of a patent does not constitute a civil action challenging the validity of a claim of a patent for purposes of this subsection.

(b) Patent Owner's Action—An inter partes review may not be instituted if the petition requesting the proceeding is filed more than 1 year after the date on which the petitioner, real party in interest, or privy of the petitioner is served with a complaint alleging infringement of the patent. The time limitation set forth in the preceding sentence shall not apply to a request for joinder under subsection (c).

(c) Joinder—If the Director institutes an inter partes review, the Director, in his or her discretion, may join as a party to that inter partes review any person who properly files a petition under section 311 that the Director, after receiving a preliminary response under section 313 or the expiration of the time for filing such a response, determines warrants the institution of an inter partes review under section 314.

(d) Multiple Proceedings—Notwithstanding sections 135(a), 251, and 252, and chapter 30, during the pendency of an inter partes review, if another proceeding or matter involving the patent is before the Office, the Director may determine the manner in which the inter partes review or other proceeding or matter may proceed, including providing for stay, transfer, consolidation, or termination of any such matter or proceeding.

(e) Estoppel—

(1) PROCEEDINGS BEFORE THE OFFICE—The petitioner in an inter partes review of a claim in a patent under this chapter that results in a final written decision under section 318(a), or the real party in interest or privy of the petitioner, may not request or maintain a proceeding before the Office with respect

to that claim on any ground that the petitioner raised or reasonably could have raised during that inter partes review.

(2) CIVIL ACTIONS AND OTHER PROCEEDINGS—The petitioner in an inter partes review of a claim in a patent under this chapter that results in a final written decision under section 318(a), or the real party in interest or privy of the petitioner, may not assert either in a civil action arising in whole or in part under section 1338 of title 28 or in a proceeding before the International Trade Commission under section 337 of the Tariff Act of 1930 that the claim is invalid on any ground that the petitioner raised or reasonably could have raised during that inter partes review.

Sec. 316. Conduct of Inter Partes Review

(a) Regulations—The Director shall prescribe regulations—

(1) providing that the file of any proceeding under this chapter shall be made available to the public, except that any petition or document filed with the intent that it be sealed shall, if accompanied by a motion to seal, be treated as sealed pending the outcome of the ruling on the motion;

(2) setting forth the standards for the showing of sufficient grounds to institute a review under section 314(a);

(3) establishing procedures for the submission of supplemental information after the petition is filed;

(4) establishing and governing inter partes review under this chapter and the relationship of such review to other proceedings under this title;

(5) setting forth standards and procedures for discovery of relevant evidence, including that such discovery shall be limited to—

(A) the deposition of witnesses submitting affidavits or declarations; and

(B) what is otherwise necessary in the interest of justice;

(6) prescribing sanctions for abuse of discovery, abuse of process, or any other improper use of the proceeding, such as to harass or to cause unnecessary delay or an unnecessary increase in the cost of the proceeding;

(7) providing for protective orders governing the exchange and submission of confidential information;

(8) providing for the filing by the patent owner of a response to the petition under section 313 after an inter partes review has been instituted, and requiring that the patent owner file with such response, through affidavits or declarations, any additional factual evidence and expert opinions on which the patent owner relies in support of the response;

(9) setting forth standards and procedures for allowing the patent owner to move to amend the patent under subsection (d) to cancel a challenged claim or propose a reasonable number of substitute claims, and ensuring that any information submitted by the patent owner in support of any amendment entered under subsection (d) is made available to the public as part of the prosecution history of the patent;

(10) providing either party with the right to an oral hearing as part of the proceeding;

(11) requiring that the final determination in an inter partes review be issued not later than 1 year after the date on which the Director notices the institution of a review under this chapter, except that the Director may, for good cause shown, extend the 1-year period by not more than 6 months, and may adjust the time periods in this paragraph in the case of joinder under section 315(c);

(12) setting a time period for requesting joinder under section 315(c); and

(13) providing the petitioner with at least 1 opportunity to file written comments within a time period established by the Director.

(b) Considerations—In prescribing regulations under this section, the Director shall consider the effect of any such regulation on the economy, the integrity of the patent system, the efficient administration of the Office, and the ability of the Office to timely complete proceedings instituted under this chapter.

(c) Patent Trial and Appeal Board—The Patent Trial and Appeal Board shall, in accordance with section 6, conduct each inter partes review instituted under this chapter.

(d) Amendment of the Patent—

(1) IN GENERAL—During an inter partes review instituted under this chapter, the patent owner may file 1 motion to amend the patent in 1 or more of the following ways:

(A) Cancel any challenged patent claim.

(B) For each challenged claim, propose a reasonable number of substitute claims.

(2) ADDITIONAL MOTIONS—Additional motions to amend may be permitted upon the joint request of the petitioner and the patent owner to materially advance the settlement of a proceeding under section 317, or as permitted by regulations prescribed by the Director.

(3) SCOPE OF CLAIMS—An amendment under this subsection may not enlarge the scope of the claims of the patent or introduce new matter.

(e) Evidentiary Standards—In an inter partes review instituted under this chapter, the petitioner shall have the burden of proving a proposition of unpatentability by a preponderance of the evidence.

Sec. 317. Settlement

(a) In General—An inter partes review instituted under this chapter shall be terminated with respect to any petitioner upon the joint request of the petitioner and the patent owner, unless the Office has decided the merits of the proceeding before the request for termination is filed. If the inter partes review is terminated with respect to a petitioner under this section, no estoppel under section 315(e) shall attach to the petitioner, or to the real party in interest or privy of the petitioner, on the basis of that petitioner's institution of that inter partes review. If no petitioner remains in the inter partes review, the Office may terminate the review or proceed to a final written decision under section 318(a).

(b) Agreements in Writing—Any agreement or understanding between the patent owner and a petitioner, including any collateral agreements referred to in such agreement or understanding, made in connection with, or in contemplation of, the termination of an inter partes review under this section shall be in writing and a true copy of such agreement or understanding shall be filed in the Office before the termination of the inter partes review as between the parties. At the request of a party to the proceeding, the agreement or understanding shall be treated as business confidential information, shall be kept separate from the file of the involved patents, and shall be made available only to Federal Government agencies on written request, or to any person on a showing of good cause.

Sec. 318. Decision of the Board

(a) Final Written Decision—If an inter partes review is instituted and not dismissed under this chapter, the Patent Trial and Appeal Board shall issue a final written decision with respect to the patentability of any patent claim challenged by the petitioner and any new claim added under section 316(d).

(b) Certificate—If the Patent Trial and Appeal Board issues a final written decision under subsection (a) and the time for appeal has expired or any appeal has terminated, the Director shall issue and publish a certificate canceling any claim of the patent finally determined to be unpatentable, confirming any claim of the patent determined to be patentable, and incorporating in the patent by operation of the certificate any new or amended claim determined to be patentable.

(c) Intervening Rights—Any proposed amended or new claim determined to be patentable and incorporated into a patent following an inter partes review under this chapter shall have the same effect as that specified in section 252 for reissued patents on the right of any person who made, purchased, or used within the United States, or imported into the United States, anything patented by such proposed amended or new claim, or who made substantial preparation therefor, before the issuance of a certificate under subsection (b).

(d) Data on Length of Review—The Office shall make available to the public data describing the length of time between the institution of, and the issuance of a final written decision under subsection (a) for, each inter partes review.

Sec. 319. Appeal

A party dissatisfied with the final written decision of the Patent Trial and Appeal Board under section 318(a) may appeal the decision pursuant to sections 141 through 144. Any party to the inter partes review shall have the right to be a party to the appeal.".

(b) Conforming Amendment—The table of chapters for part III of title 35, United States Code, is amended by striking the item relating to chapter 31 and inserting the following:

"31. Inter Partes Review 311".

(c) Regulations and Effective Date—

(1) REGULATIONS—The Director shall, not later than the date that is 1 year after the date of the enactment of this Act, issue regulations to carry out chapter 31 of title 35, United States Code, as amended by subsection (a) of this section.

(2) APPLICABILITY—

(A) IN GENERAL—The amendments made by subsection (a) shall take effect upon the expiration of the 1-year period beginning on the date of the enactment of this Act and shall apply to any patent issued before, on, or after that effective date.

(B) GRADUATED IMPLEMENTATION—The Director may impose a limit on the number of inter partes reviews that may be instituted under chapter 31 of title 35, United States Code, during each of the first 4 1-year periods in which the amendments made by subsection (a) are in effect, if such number in each year equals or exceeds the number of inter partes reexaminations that are ordered under chapter 31 of title 35, United States Code, in the last fiscal year ending before the effective date of the amendments made by subsection (a).

(3) TRANSITION—

(A) IN GENERAL—Chapter 31 of title 35, United States Code, is amended—

(i) in section 312—

(I) in subsection (a)—

(aa) in the first sentence, by striking "a substantial new question of patentability affecting any claim of the patent concerned is raised by the request," and inserting "the information presented in the request shows that there is a reasonable likelihood that the requester would prevail with respect to at least 1 of the claims challenged in the request,"; and

(bb) in the second sentence, by striking "The existence of a substantial new question of patentability" and inserting "A showing that there is a reasonable likelihood that the requester would prevail with respect to at least 1 of the claims challenged in the request"; and

(II) in subsection (c), in the second sentence, by striking "no substantial new question of patentability has been raised," and inserting "the showing required by subsection (a) has not been made,"; and

(ii) in section 313, by striking "a substantial new question of patentability affecting a claim of the patent is raised" and inserting "it has been shown that there is a reasonable likelihood that the requester would prevail with respect to at least 1 of the claims challenged in the request".

(B) APPLICATION—The amendments made by this paragraph–

(i) shall take effect on the date of the enactment of this Act; and

(ii) shall apply to requests for inter partes reexamination that are filed on or after such date of enactment, but before the effective date set forth in paragraph (2)(A) of this subsection.

(C) CONTINUED APPLICABILITY OF PRIOR PROVISIONS—The provisions of chapter 31 of title 35, United States Code, as amended by this paragraph, shall continue to apply to requests for inter partes reexamination that are filed before the effective date set forth in paragraph (2)(A) as if subsection (a) had not been enacted.

(d) Post-Grant Review—Part III of title 35, United States Code, is amended by adding at the end the following:

Chapter 32—Post-Grant Review

Sec.

Sec. 321. Post-Grant Review

(a) In General—Subject to the provisions of this chapter, a person who is not the owner of a patent may file with the Office a petition to institute a post-grant review of the patent. The Director shall establish, by regulation, fees to be paid by the person requesting the review, in such amounts as the Director determines to be reasonable, considering the aggregate costs of the post-grant review.

(b) Scope—A petitioner in a post-grant review may request to cancel as unpatentable 1 or more claims of a patent on any ground that could be raised under paragraph (2) or (3) of section 282(b) (relating to invalidity of the patent or any claim).

(c) Filing Deadline—A petition for a post-grant review may only be filed not later than the date that is 9 months after the date of the grant of the patent or of the issuance of a reissue patent (as the case may be).

Sec. 322. Petitions

(a) Requirements of Petition—A petition filed under section 321 may be considered only if—

(1) the petition is accompanied by payment of the fee established by the Director under section 321;

(2) the petition identifies all real parties in interest;

(3) the petition identifies, in writing and with particularity, each claim challenged, the grounds on which the challenge to each claim is based, and the evidence that supports the grounds for the challenge to each claim, including—

(A) copies of patents and printed publications that the petitioner relies upon in support of the petition; and

(B) affidavits or declarations of supporting evidence and opinions, if the petitioner relies on other factual evidence or on expert opinions;

(4) the petition provides such other information as the Director may require by regulation; and

(5) the petitioner provides copies of any of the documents required under paragraphs (2), (3), and (4) to the patent owner or, if applicable, the designated representative of the patent owner.

(b) Public Availability—As soon as practicable after the receipt of a petition under section 321, the Director shall make the petition available to the public.

Sec. 323. Preliminary Response to Petition

If a post-grant review petition is filed under section 321, the patent owner shall have the right to file a preliminary response to the petition, within a time period set by the Director, that sets forth reasons why no post-grant review should be instituted based upon the failure of the petition to meet any requirement of this chapter.

Sec. 324. Institution of Post-Grant Review

(a) Threshold—The Director may not authorize a post-grant review to be instituted unless the Director determines that the information presented in the petition filed under section 321, if such information is not rebutted, would demonstrate that it is more likely than not that at least 1 of the claims challenged in the petition is unpatentable.

(b) Additional Grounds—The determination required under subsection (a) may also be satisfied by a showing that the petition raises a novel or unsettled legal question that is important to other patents or patent applications.

(c) Timing—The Director shall determine whether to institute a post-grant review under this chapter pursuant to a petition filed under section 321 within 3 months after—

(1) receiving a preliminary response to the petition under section 323; or

(2) if no such preliminary response is filed, the last date on which such response may be filed.

(d) Notice—The Director shall notify the petitioner and patent owner, in writing, of the Director's determination under subsection (a) or (b), and shall make such notice available to the public as soon as is practicable. Such notice shall include the date on which the review shall commence.

(e) No Appeal—The determination by the Director whether to institute a post-grant review under this section shall be final and nonappealable.

Sec. 325. Relation to other Proceedings or Actions

(a) Infringer's Civil Action—

(1) POST-GRANT REVIEW BARRED BY CIVIL ACTION—A post-grant review may not be instituted under this chapter if, before the date on which the petition for such a review is filed, the petitioner or real party in interest filed a civil action challenging the validity of a claim of the patent.

(2) STAY OF CIVIL ACTION—If the petitioner or real party in interest files a civil action challenging the validity of a claim of the patent on or after the date on which the petitioner files a petition for post-grant review of the patent, that civil action shall be automatically stayed until either—

(A) the patent owner moves the court to lift the stay;

(B) the patent owner files a civil action or counterclaim alleging that the petitioner or real party in interest has infringed the patent; or

(C) the petitioner or real party in interest moves the court to dismiss the civil action.

(3) TREATMENT OF COUNTERCLAIM—A counterclaim challenging the validity of a claim of a patent does not constitute a civil action challenging the validity of a claim of a patent for purposes of this subsection.

(b) Preliminary Injunctions—If a civil action alleging infringement of a patent is filed within 3 months after the date on which the patent is granted, the court may not stay its consideration of the patent owner's motion for a preliminary injunction against infringement of the patent on the basis that a petition for post-grant review has been filed under this chapter or that such a post-grant review has been instituted under this chapter.

(c) Joinder—If more than 1 petition for a post-grant review under this chapter is properly filed against the same patent and the Director determines that more than 1 of these petitions warrants the institution of a post-grant review under section 324, the Director may consolidate such reviews into a single post-grant review.

(d) Multiple Proceedings—Notwithstanding sections 135(a), 251, and 252, and chapter 30, during the pendency of any post-grant review under this chapter, if another proceeding or matter involving the patent is before the Office, the Director may determine the manner in which the post-grant review or other proceeding or matter may proceed, including providing for the stay, transfer, consolidation, or termination of any such matter or proceeding. In determining whether to institute or order a proceeding under this chapter, chapter 30, or chapter 31, the Director may take into account whether, and reject the petition or request because, the same or substantially the same prior art or arguments previously were presented to the Office.

(e) Estoppel—

(1) PROCEEDINGS BEFORE THE OFFICE—The petitioner in a post-grant review of a claim in a patent under this chapter that results in a final written

decision under section 328(a), or the real party in interest or privy of the petitioner, may not request or maintain a proceeding before the Office with respect to that claim on any ground that the petitioner raised or reasonably could have raised during that post-grant review.

(2) CIVIL ACTIONS AND OTHER PROCEEDINGS—The petitioner in a post-grant review of a claim in a patent under this chapter that results in a final written decision under section 328(a), or the real party in interest or privy of the petitioner, may not assert either in a civil action arising in whole or in part under section 1338 of title 28 or in a proceeding before the International Trade Commission under section 337 of the Tariff Act of 1930 that the claim is invalid on any ground that the petitioner raised or reasonably could have raised during that post-grant review.

(f) Reissue Patents—A post-grant review may not be instituted under this chapter if the petition requests cancellation of a claim in a reissue patent that is identical to or narrower than a claim in the original patent from which the reissue patent was issued, and the time limitations in section 321(c) would bar filing a petition for a post-grant review for such original patent.

Sec. 326. Conduct of Post-Grant Review

(a) Regulations—The Director shall prescribe regulations—

(1) providing that the file of any proceeding under this chapter shall be made available to the public, except that any petition or document filed with the intent that it be sealed shall, if accompanied by a motion to seal, be treated as sealed pending the outcome of the ruling on the motion;

(2) setting forth the standards for the showing of sufficient grounds to institute a review under subsections (a) and (b) of section 324;

(3) establishing procedures for the submission of supplemental information after the petition is filed;

(4) establishing and governing a post-grant review under this chapter and the relationship of such review to other proceedings under this title;

(5) setting forth standards and procedures for discovery of relevant evidence, including that such discovery shall be limited to evidence directly related to factual assertions advanced by either party in the proceeding;

(6) prescribing sanctions for abuse of discovery, abuse of process, or any other improper use of the proceeding, such as to harass or to cause unnecessary delay or an unnecessary increase in the cost of the proceeding;

(7) providing for protective orders governing the exchange and submission of confidential information;

(8) providing for the filing by the patent owner of a response to the petition under section 323 after a post-grant review has been instituted, and requiring that the patent owner file with such response, through affidavits or declarations, any additional factual evidence and expert opinions on which the patent owner relies in support of the response;

(9) setting forth standards and procedures for allowing the patent owner to move to amend the patent under subsection (d) to cancel a challenged claim or propose a reasonable number of substitute claims, and ensuring that any information submitted by the patent owner in support of any amendment entered under subsection (d) is made available to the public as part of the prosecution history of the patent;

(10) providing either party with the right to an oral hearing as part of the proceeding;

(11) requiring that the final determination in any post-grant review be issued not later than 1 year after the date on which the Director notices the institution of a proceeding under this chapter, except that the Director may, for good cause shown, extend the 1-year period by not more than 6 months, and may adjust the time periods in this paragraph in the case of joinder under section 325(c); and

(12) providing the petitioner with at least 1 opportunity to file written comments within a time period established by the Director.

(b) Considerations- In prescribing regulations under this section, the Director shall consider the effect of any such regulation on the economy, the integrity of the patent system, the efficient administration of the Office, and the ability of the Office to timely complete proceedings instituted under this chapter.

(c) Patent Trial and Appeal Board—The Patent Trial and Appeal Board shall, in accordance with section 6, conduct each post-grant review instituted under this chapter.

(d) Amendment of the Patent—

(1) IN GENERAL—During a post-grant review instituted under this chapter, the patent owner may file 1 motion to amend the patent in 1 or more of the following ways:

(A) Cancel any challenged patent claim.

(B) For each challenged claim, propose a reasonable number of substitute claims.

(2) ADDITIONAL MOTIONS—Additional motions to amend may be permitted upon the joint request of the petitioner and the patent owner to materially advance the settlement of a proceeding under section 327, or upon the request of the patent owner for good cause shown.

(3) SCOPE OF CLAIMS—An amendment under this subsection may not enlarge the scope of the claims of the patent or introduce new matter.

(e) Evidentiary Standards—In a post-grant review instituted under this chapter, the petitioner shall have the burden of proving a proposition of unpatentability by a preponderance of the evidence.

Sec. 327. Settlement

(a) In General—A post-grant review instituted under this chapter shall be terminated with respect to any petitioner upon the joint request of the petitioner and the patent owner, unless the Office has decided the merits of the proceeding before the request for termination is filed. If the post-grant review is terminated with respect to a petitioner under this section, no estoppel under section 325(e) shall attach to the petitioner, or to the real party in interest or privy of the petitioner, on the basis of that petitioner's institution of that post-grant review. If no petitioner remains in the post-grant review, the Office may terminate the post-grant review or proceed to a final written decision under section 328(a).

(b) Agreements in Writing—Any agreement or understanding between the patent owner and a petitioner, including any collateral agreements referred to in such agreement or understanding, made in connection with, or in contemplation of, the termination of a post-grant review under this section shall be in writing, and a true copy of such agreement or understanding shall be filed in the Office before the termination of the post-grant review as between the parties. At the request of a party to the proceeding, the agreement or understanding shall be treated as business confidential information, shall be kept separate from the file of the involved patents, and shall be made available only to Federal Government agencies on written request, or to any person on a showing of good cause.

Sec. 328. Decision of the Board

(a) Final Written Decision—If a post-grant review is instituted and not dismissed under this chapter, the Patent Trial and appeal Board shall issue a final written decision with respect to the patentability of any patent claim challenged by the petitioner and any new claim added under section 326(d).

(b) Certificate—If the Patent Trial and Appeal Board issues a final written decision under subsection (a) and the time for appeal has expired or any appeal has terminated, the Director shall issue and publish a certificate canceling any claim of the patent finally determined to be unpatentable, confirming any claim of the patent determined to be patentable, and incorporating in the patent by operation of the certificate any new or amended claim determined to be patentable.

(c) Intervening Rights—Any proposed amended or new claim determined to be patentable and incorporated into a patent following a post-grant review under this chapter shall have the same effect as that specified in section 252 of this title for reissued patents on the right of any person who made, purchased, or used within the United States, or imported into the United States, anything patented by such proposed amended or new claim, or who made substantial preparation therefor, before the issuance of a certificate under subsection (b).

(d) Data on Length of Review—The Office shall make available to the public data describing the length of time between the institution of, and the issuance of a final written decision under subsection (a) for, each post-grant review.

Sec. 329. Appeal

A party dissatisfied with the final written decision of the Patent Trial and Appeal Board under section 328(a) may appeal the decision pursuant to sections 141 through 144. Any party to the post-grant review shall have the right to be a party to the appeal.".

(e) Conforming Amendment—The table of chapters for part III of title 35, United States Code, is amended by adding at the end the following:
"32. Post-Grant Review 321".

(f) Regulations and Effective Date—

(1) REGULATIONS—The Director shall, not later than the date that is 1 year after the date of the enactment of this Act, issue regulations to carry out chapter 32 of title 35, United States Code, as added by subsection (d) of this section.

(2) APPLICABILITY—

(A) IN GENERAL—The amendments made by subsection (d) shall take effect upon the expiration of the 1-year period beginning on the date of the enactment of this Act and, except as provided in section 18 and in paragraph (3), shall apply only to patents described in section 3(n)(1).

(B) LIMITATION—The Director may impose a limit on the number of post-grant reviews that may be instituted under chapter 32 of title 35, United States Code, during each of the first 4 1-year periods in which the amendments made by subsection (d) are in effect.

(3) PENDING INTERFERENCES—

(A) PROCEDURES IN GENERAL—The Director shall determine, and include in the regulations issued under paragraph (1), the procedures under which an interference commenced before the effective date set forth in paragraph (2)(A) is to proceed, including whether such interference—

(i) is to be dismissed without prejudice to the filing of a petition for a post-grant review under chapter 32 of title 35, United States Code; or

(ii) is to proceed as if this Act had not been enacted.

(B) PROCEEDINGS BY PATENT TRIAL AND APPEAL BOARD—For purposes of an interference that is commenced before the effective date set forth in paragraph (2)(A), the Director may deem the Patent Trial and Appeal Board to be the Board of Patent Appeals and Interferences, and may allow the Patent Trial and Appeal Board to conduct any further proceedings in that interference.

(C) APPEALS—The authorization to appeal or have remedy from derivation proceedings in sections 141(d) and 146 of title 35, United States Code, as amended by this Act, and the jurisdiction to entertain appeals from derivation proceedings in section 1295(a)(4)(A) of title 28, United States Code, as amended by this Act, shall be deemed to extend to any final decision in an interference that is commenced before the effective date set forth in paragraph (2)(A) of this subsection and that is not dismissed pursuant to this paragraph.

(g) Citation of Prior Art and Written Statements—

(1) IN GENERAL—Section 301 of title 35, United States Code, is amended to read as follows:

"Sec. 301. Citation of Prior Art and Written Statements

(a) In General—Any person at any time may cite to the Office in writing—

(1) prior art consisting of patents or printed publications which that person believes to have a bearing on the patentability of any claim of a particular patent; or

(2) statements of the patent owner filed in a proceeding before a Federal court or the Office in which the patent owner took a position on the scope of any claim of a particular patent.

(b) Official File—If the person citing prior art or written statements pursuant to subsection (a) explains in writing the pertinence and manner of applying the prior art or written statements to at least 1 claim of the patent, the citation of the prior art or written statements and the explanation thereof shall become a part of the official file of the patent.

(c) Additional Information—A party that submits a written statement pursuant to subsection (a)(2) shall include any other documents, pleadings, or evidence from the proceeding in which the statement was filed that addresses the written statement.

(d) Limitations—A written statement submitted pursuant to subsection (a)(2), and additional information submitted pursuant to subsection (c), shall not be considered by the Office for any purpose other than to determine the proper meaning of a patent claim in a proceeding that is ordered or instituted pursuant to section 304, 314, or 324. If any such written statement or additional information is subject to an applicable protective order, such statement or information shall be redacted to exclude information that is subject to that order.

(e) Confidentiality—Upon the written request of the person citing prior art or written statements pursuant to subsection (a), that person's identity shall be excluded from the patent file and kept confidential.".

(2) CONFORMING AMENDMENT—The item relating to section 301 in the table of sections for chapter 30 of title 35, United States Code, is amended to read as follows:

"301. Citation of prior art and written statements.".

(3) EFFECTIVE DATE—The amendments made by this subsection shall take effect upon the expiration of the 1-year period beginning on the date of the enactment of this Act and shall apply to any patent issued before, on, or after that effective date.

(h) Reexamination—

(1) DETERMINATION BY DIRECTOR—

(A) IN GENERAL—Section 303(a) of title 35, United States Code, is amended by striking "section 301 of this title" and inserting "section 301 or 302".

(B) EFFECTIVE DATE—The amendment made by this paragraph shall take effect upon the expiration of the 1-year period beginning on the date of the enactment of this Act and shall apply to any patent issued before, on, or after that effective date.

(2) APPEAL—

(A) IN GENERAL—Section 306 of title 35, United States Code, is amended by striking "145" and inserting "144".

(B) EFFECTIVE DATE—The amendment made by this paragraph shall take effect on the date of the enactment of this Act and shall apply to any appeal of a reexamination before the Board of Patent Appeals and Interferences or the Patent Trial and Appeal Board that is pending on, or brought on or after, the date of the enactment of this Act.

Sec. 7. Patent Trial and Appeal Board.

(a) Composition and Duties—

(1) IN GENERAL—Section 6 of title 35, United States Code, is amended to read as follows:

"Sec. 6. Patent Trial and Appeal Board

(a) In General—There shall be in the Office a Patent Trial and Appeal Board. The Director, the Deputy Director, the Commissioner for Patents, the Commissioner for Trademarks, and the administrative patent judges shall constitute the Patent Trial and Appeal Board. The administrative patent judges shall be persons of competent legal knowledge and scientific ability who are appointed by the Secretary, in consultation with the Director. Any reference in any Federal law, Executive order, rule, regulation, or delegation of authority, or any document of or pertaining to the Board of Patent Appeals and Interferences is deemed to refer to the Patent Trial and Appeal Board.

(b) Duties—The Patent Trial and Appeal Board shall—

(1) on written appeal of an applicant, review adverse decisions of examiners upon applications for patents pursuant to section 134(a);

(2) review appeals of reexaminations pursuant to section 134(b);

(3) conduct derivation proceedings pursuant to section 135; and

(4) conduct inter partes reviews and post-grant reviews pursuant to chapters 31 and 32.

(c) 3-Member Panels—Each appeal, derivation proceeding, post-grant review, and inter partes review shall be heard by at least 3 members of the Patent Trial and Appeal Board, who shall be designated by the Director. Only the Patent Trial and Appeal Board may grant rehearings.

(d) Treatment of Prior Appointments—The Secretary of Commerce may, in the Secretary's discretion, deem the appointment of an administrative patent judge who, before the date of the enactment of this subsection, held office pursuant to an appointment by the Director to take effect on the date on which the Director initially appointed the administrative patent judge. It shall be a defense to a challenge to the appointment of an administrative patent judge on the basis of the judge's having been originally appointed by the Director that the administrative patent judge so appointed was acting as a de facto officer.".

(2) CONFORMING AMENDMENT—The item relating to section 6 in the table of sections for chapter 1 of title 35, United States Code, is amended to read as follows:

"6. Patent Trial and Appeal Board.".

(b) Administrative Appeals—Section 134 of title 35, United States Code, is amended—

(1) in subsection (b), by striking "any reexamination proceeding" and inserting "a reexamination"; and

(2) by striking subsection (c).

(c) Circuit Appeals—

(1) IN GENERAL—Section 141 of title 35, United States Code, is amended to read as follows:

"Sec. 141. Appeal to Court of Appeals for the Federal Circuit

(a) Examinations—An applicant who is dissatisfied with the final decision in an appeal to the Patent Trial and Appeal Board under section 134(a) may appeal the Board's decision to the United States Court of Appeals for the Federal Circuit. By filing such an appeal, the applicant waives his or her right to proceed under section 145.

(b) Reexaminations—A patent owner who is dissatisfied with the final decision in an appeal of a reexamination to the Patent Trial and Appeal Board under section 134(b) may appeal the Board's decision only to the United States Court of Appeals for the Federal Circuit.

(c) Post-Grant and Inter Partes Reviews—A party to an inter partes review or a post-grant review who is dissatisfied with the final written decision of the Patent Trial and Appeal Board under section 318(a) or 328(a) (as the case may be) may appeal the Board's decision only to the United States Court of Appeals for the Federal Circuit.

(d) Derivation Proceedings—A party to a derivation proceeding who is dissatisfied with the final decision of the Patent Trial and Appeal Board in the proceeding may appeal the decision to the United States Court of Appeals for the Federal Circuit, but such appeal shall be dismissed if any adverse party to such derivation proceeding, within 20 days after the appellant has filed notice of appeal in accordance with section 142, files notice with the Director that the party elects to have all further proceedings conducted as provided in section 146. If the appellant does not, within 30 days after the filing of such notice by the adverse party, file a civil action under section 146, the Board's decision shall govern the further proceedings in the case.".

(2) JURISDICTION—Section 1295(a)(4)(A) of title 28, United States Code, is amended to read as follows:

"(A) the Patent Trial and Appeal Board of the United States Patent and Trademark Office with respect to a patent application, derivation proceeding, reexamination, post-grant review, or inter partes review under title 35, at the instance of a party who exercised that party's right to participate in the applicable proceeding before or appeal to the Board, except that an applicant or a party to a derivation proceeding may also have remedy by civil action pursuant to section 145 or 146 of title 35; an appeal under this subparagraph of a decision of the Board with respect to an application or derivation proceeding shall waive the right of such applicant or party to proceed under section 145 or 146 of title 35;".

(3) PROCEEDINGS ON APPEAL—Section 143 of title 35, United States Code, is amended—

(A) by striking the third sentence and inserting the following: "In an ex parte case, the Director shall submit to the court in writing the grounds for the decision of the Patent and Trademark Office, addressing all of the issues raised in the appeal. The Director shall have the right to intervene in an appeal from a decision entered by the Patent Trial and Appeal Board in a derivation proceeding under section 135 or in an inter partes or post-grant review under chapter 31 or 32."; and

(B) by striking the last sentence.

(d) Conforming Amendments—

(1) ATOMIC ENERGY ACT OF 1954—Section 152 of the Atomic Energy Act of 1954 (42 U.S.C. 2182) is amended in the third undesignated paragraph—

(A) by striking "Board of Patent Appeals and Interferences" each place it appears and inserting "Patent Trial and Appeal Board"; and

(B) by inserting "and derivation" after "established for interference".

(2) TITLE 51—Section 20135 of title 51, United States Code, is amended—

(A) in subsections (e) and (f), by striking "Board of Patent Appeals and Interferences" each place it appears and inserting "Patent Trial and Appeal Board"; and

(B) in subsection (e), by inserting "and derivation" after "established for interference".

(e) Effective Date—The amendments made by this section shall take effect upon the expiration of the 1-year period beginning on the date of the enactment of this Act and shall apply to proceedings commenced on or after that effective date, except that—

(1) the extension of jurisdiction to the United States Court of Appeals for the Federal Circuit to entertain appeals of decisions of the Patent Trial and Appeal Board in reexaminations under the amendment made by subsection (c)(2) shall be deemed to take effect on the date of the enactment of this Act and shall extend to any decision of the Board of Patent Appeals and Interferences with respect to a reexamination that is entered before, on, or after the date of the enactment of this Act;

(2) the provisions of sections 6, 134, and 141 of title 35, United States Code, as in effect on the day before the effective date of the amendments made by this section shall continue to apply to inter partes reexaminations that are requested under section 311 of such title before such effective date;

(3) the Patent Trial and Appeal Board may be deemed to be the Board of Patent Appeals and Interferences for purposes of appeals of inter partes reexaminations that are requested under section 311 of title 35, United States Code, before the effective date of the amendments made by this section; and

(4) the Director's right under the fourth sentence of section 143 of title 35, United States Code, as amended by subsection (c)(3) of this section, to intervene in an appeal from a decision entered by the Patent Trial and Appeal Board shall be deemed to extend to inter partes reexaminations that are requested under section 311 of such title before the effective date of the amendments made by this section.

Sec. 8. Preissuance Submissions by Third Parties.

(a) In General—Section 122 of title 35, United States Code, is amended by adding at the end the following:

"(e) Preissuance Submissions by Third Parties—

(1) IN GENERAL—Any third party may submit for consideration and inclusion in the record of a patent application, any patent, published patent application, or other printed publication of potential relevance to the examination of the application, if such submission is made in writing before the earlier of—

(A) the date a notice of allowance under section 151 is given or mailed in the application for patent; or

(B) the later of—

(i) 6 months after the date on which the application for patent is first published under section 122 by the Office, or

(ii) the date of the first rejection under section 132 of any claim by the examiner during the examination of the application for patent.

(2) OTHER REQUIREMENTS—Any submission under paragraph (1) shall—

(A) set forth a concise description of the asserted relevance of each submitted document;

(B) be accompanied by such fee as the Director may prescribe; and

(C) include a statement by the person making such submission affirming that the submission was made in compliance with this section.".

(b) Effective Date—The amendments made by this section shall take effect upon the expiration of the 1-year period beginning on the date of the enactment of this Act and shall apply to any patent application filed before, on, or after that effective date.

Sec. 9. Venue.

(a) Technical Amendments Relating to Venue—Sections 32, 145, 146, 154(b)(4)(A), and 293 of title 35, United States Code, and section 21(b)(4) of the Trademark Act of 1946 (15 U.S.C. 1071(b)(4)), are each amended by striking "United States District Court for the District of Columbia" each place that term appears and inserting "United States District Court for the Eastern District of Virginia".

(b) Effective Date—The amendments made by this section shall take effect on the date of the enactment of this Act and shall apply to any civil action commenced on or after that date.

Sec. 10. Fee Setting Authority.

(a) Fee Setting—

(1) IN GENERAL—The Director may set or adjust by rule any fee established, authorized, or charged under title 35, United States Code, or the Trademark Act of 1946 (15 U.S.C. 1051 et seq.), for any services performed by or materials furnished by, the Office, subject to paragraph (2).

(2) FEES TO RECOVER COSTS—Fees may be set or adjusted under paragraph (1) only to recover the aggregate estimated costs to the Office for processing, activities, services, and materials relating to patents (in the case of patent fees) and trademarks (in the case of trademark fees), including administrative costs of the Office with respect to such patent or trademark fees (as the case may be).

(b) Small and Micro Entities—The fees set or adjusted under subsection (a) for filing, searching, examining, issuing, appealing, and maintaining patent applications and patents shall be reduced by 50 percent with respect to the application of such fees to any small entity that qualifies for reduced fees under section 41(h)(1) of title 35, United States Code, and shall be reduced by 75 percent with respect to the application of such fees to any micro entity as defined in section 123 of that title (as added by subsection (g) of this section).

(c) Reduction of Fees in Certain Fiscal Years—In each fiscal year, the Director—

(1) shall consult with the Patent Public Advisory Committee and the Trademark Public Advisory Committee on the advisability of reducing any fees described in subsection (a); and

(2) after the consultation required under paragraph (1), may reduce such fees.

(d) Role of the Public Advisory Committee—The Director shall—

(1) not less than 45 days before publishing any proposed fee under subsection (a) in the Federal Register, submit the proposed fee to the Patent Public Advisory Committee or the Trademark Public Advisory Committee, or both, as appropriate;

(2)(A) provide the relevant advisory committee described in paragraph (1) a 30-day period following the submission of any proposed fee, in which to deliberate, consider, and comment on such proposal;

(B) require that, during that 30-day period, the relevant advisory committee hold a public hearing relating to such proposal; and

(C) assist the relevant advisory committee in carrying out that public hearing, including by offering the use of the resources of the Office to notify and promote the hearing to the public and interested stakeholders;

(3) require the relevant advisory committee to make available to the public a written report setting forth in detail the comments, advice, and recommendations of the committee regarding the proposed fee; and

(4) consider and analyze any comments, advice, or recommendations received from the relevant advisory committee before setting or adjusting (as the case may be) the fee.

(e) Publication in the Federal Register—

(1) PUBLICATION AND RATIONALE—The Director shall—

(A) publish any proposed fee change under this section in the Federal Register;

(B) include, in such publication, the specific rationale and purpose for the proposal, including the possible expectations or benefits resulting from the proposed change; and

(C) notify, through the Chair and Ranking Member of the Committees on the Judiciary of the Senate and the House of Representatives, the Congress of the proposed change not later than the date on which the proposed change is published under subparagraph (A).

(2) PUBLIC COMMENT PERIOD—The Director shall, in the publication under paragraph (1), provide the public a period of not less than 45 days in which to submit comments on the proposed change in fees.

(3) PUBLICATION OF FINAL RULE—The final rule setting or adjusting a fee under this section shall be published in the Federal Register and in the Official Gazette of the Patent and Trademark Office.

(4) CONGRESSIONAL COMMENT PERIOD—A fee set or adjusted under subsection (a) may not become effective—

(A) before the end of the 45-day period beginning on the day after the date on which the Director publishes the final rule adjusting or setting the fee under paragraph (3); or

(B) if a law is enacted disapproving such fee.

(5) RULE OF CONSTRUCTION—Rules prescribed under this section shall not diminish—

(A) the rights of an applicant for a patent under title 35, United States Code, or for a mark under the Trademark Act of 1946; or

(B) any rights under a ratified treaty.

(f) Retention of Authority—The Director retains the authority under subsection (a) to set or adjust fees only during such period as the Patent and Trademark Office remains an agency within the Department of Commerce.

(g) Micro Entity Defined—

(1) IN GENERAL—Chapter 11 of title 35, United States Code, is amended by adding at the end the following new section:

"*Sec. 123. Micro Entity Defined*

(a) In General—For purposes of this title, the term 'micro entity' means an applicant who makes a certification that the applicant—

(1) qualifies as a small entity, as defined in regulations issued by the Director;

(2) has not been named as an inventor on more than 4 previously filed patent applications, other than applications filed in another country, provisional applications under section 111(b), or international applications filed under the treaty defined in section 351(a) for which the basic national fee under section 41(a) was not paid;

(3) did not, in the calendar year preceding the calendar year in which the applicable fee is being paid, have a gross income, as defined in section 61(a) of the Internal Revenue Code of 1986, exceeding 3 times the median household income for that preceding calendar year, as most recently reported by the Bureau of the Census; and

(4) has not assigned, granted, or conveyed, and is not under an obligation by contract or law to assign, grant, or convey, a license or other ownership interest in the application concerned to an entity that, in the calendar year preceding the calendar year in which the applicable fee is being paid, had a gross income, as defined in section 61(a) of the Internal Revenue Code of 1986, exceeding 3 times the median household income for that preceding calendar year, as most recently reported by the Bureau of the Census.

(b) Applications Resulting From Prior Employment—An applicant is not considered to be named on a previously filed application for purposes of subsection (a)(2) if the applicant has assigned, or is under an obligation by contract or law to assign, all ownership rights in the application as the result of the applicant's previous employment.

(c) Foreign Currency Exchange Rate—If an applicant's or entity's gross income in the preceding calendar year is not in United States dollars, the average currency exchange rate, as reported by the Internal Revenue Service, during that calendar year shall be used to determine whether the applicant's or entity's gross income exceeds the threshold specified in paragraphs (3) or (4) of subsection (a).

(d) Institutions of Higher Education—For purposes of this section, a micro entity shall include an applicant who certifies that—

(1) the applicant's employer, from which the applicant obtains the majority of the applicant's income, is an institution of higher education as defined in section 101(a) of the Higher Education Act of 1965 (20 U.S.C. 1001(a)); or

(2) the applicant has assigned, granted, conveyed, or is under an obligation by contract or law, to assign, grant, or convey, a license or other ownership interest in the particular applications to such an institution of higher education.

(e) Director's Authority—In addition to the limits imposed by this section, the Director may, in the Director's discretion, impose income limits, annual filing limits, or other limits on who may qualify as a micro entity pursuant to this section if the Director determines that such additional limits are reasonably necessary to avoid an undue impact on other patent applicants or owners or are otherwise reasonably necessary and appropriate. At least 3 months before any limits proposed to be imposed pursuant to this subsection take effect, the Director shall inform the Committee on the Judiciary of the House of Representatives and the Committee on the Judiciary of the Senate of any such proposed limits.".

(2) CONFORMING AMENDMENT—Chapter 11 of title 35, United States Code, is amended by adding at the end the following new item:

"123. Micro entity defined.".

(h) Electronic Filing Incentive—

(1) IN GENERAL—Notwithstanding any other provision of this section, an additional fee of $400 shall be established for each application for an original patent, except for a design, plant, or provisional application, that is not filed by electronic means as prescribed by the Director. The fee established by this subsection shall be reduced by 50 percent for small entities that qualify for reduced fees under section 41(h)(1) of title 35, United States Code. All fees paid under this subsection shall be deposited in the Treasury as an offsetting receipt that shall not be available for obligation or expenditure.

(2) EFFECTIVE DATE—This subsection shall take effect upon the expiration of the 60-day period beginning on the date of the enactment of this Act.

(i) Effective Date; Sunset—

(1) EFFECTIVE DATE—Except as provided in subsection (h), this section and the amendments made by this section shall take effect on the date of the enactment of this Act.

(2) SUNSET—The authority of the Director to set or adjust any fee under subsection (a) shall terminate upon the expiration of the 7-year period beginning on the date of the enactment of this Act.

(3) PRIOR REGULATIONS NOT AFFECTED—The termination of authority under this subsection shall not affect any regulations issued under this section before the effective date of such termination or any rulemaking proceeding for the issuance of regulations under this section that is pending on such date.

Sec. 11. Fees for Patent Services.

(a) General Patent Services—Subsections (a) and (b) of section 41 of title 35, United States Code, are amended to read as follows:

"(a) General Fees—The Director shall charge the following fees:

(1) FILING AND BASIC NATIONAL FEES—

(A) On filing each application for an original patent, except for design, plant, or provisional applications, $330.

(B) On filing each application for an original design patent, $220.

(C) On filing each application for an original plant patent, $220.

(D) On filing each provisional application for an original patent, $220.

(E) On filing each application for the reissue of a patent, $330.

(F) The basic national fee for each international application filed under the treaty defined in section 351(a) entering the national stage under section 371, $330.

(G) In addition, excluding any sequence listing or computer program listing filed in an electronic medium as prescribed by the Director, for any application the specification and drawings of which exceed 100 sheets of paper (or equivalent as prescribed by the Director if filed in an electronic medium), $270 for each additional 50 sheets of paper (or equivalent as prescribed by the Director if filed in an electronic medium) or fraction thereof.

(2) EXCESS CLAIMS FEES—

(A) IN GENERAL—In addition to the fee specified in paragraph (1)—

(i) on filing or on presentation at any other time, $220 for each claim in independent form in excess of 3;

(ii) on filing or on presentation at any other time, $52 for each claim (whether dependent or independent) in excess of 20; and

(iii) for each application containing a multiple dependent claim, $390.

(B) MULTIPLE DEPENDENT CLAIMS—For the purpose of computing fees under subparagraph (A), a multiple dependent claim referred to in section 112 or any claim depending therefrom shall be considered as separate dependent claims in accordance with the number of claims to which reference is made.

(C) REFUNDS; ERRORS IN PAYMENT—The Director may by regulation provide for a refund of any part of the fee specified in subparagraph (A) for any claim that is canceled before an examination on the merits, as prescribed by the Director, has been made of the application under section 131. Errors in payment of the additional fees under this paragraph may be rectified in accordance with regulations prescribed by the Director.

(3) EXAMINATION FEES—

(A) IN GENERAL—

(i) For examination of each application for an original patent, except for design, plant, provisional, or international applications, $220.

(ii) For examination of each application for an original design patent, $140.

(iii) For examination of each application for an original plant patent, $170.

(iv) For examination of the national stage of each international application, $220.

(v) For examination of each application for the reissue of a patent, $650.

(B) APPLICABILITY OF OTHER FEE PROVISIONS—The provisions of paragraphs (3) and (4) of section 111(a) relating to the payment of the fee for filing the application shall apply to the payment of the fee specified in subparagraph (A) with respect to an application filed under section 111(a). The provisions of section 371(d) relating to the payment of the national fee shall apply to the payment of the fee specified in subparagraph (A) with respect to an international application.

(4) ISSUE FEES—

(A) For issuing each original patent, except for design or plant patents, $1,510.

(B) For issuing each original design patent, $860.

(C) For issuing each original plant patent, $1,190.

(D) For issuing each reissue patent, $1,510.

(5) DISCLAIMER FEE—On filing each disclaimer, $140.

(6) APPEAL FEES—

(A) On filing an appeal from the examiner to the Patent Trial and Appeal Board, $540.

(B) In addition, on filing a brief in support of the appeal, $540, and on requesting an oral hearing in the appeal before the Patent Trial and Appeal Board, $1,080.

(7) REVIVAL FEES—On filing each petition for the revival of an unintentionally abandoned application for a patent, for the unintentionally delayed payment of the fee for issuing each patent, or for an unintentionally delayed response by the patent owner in any reexamination proceeding, $1,620, unless the petition is filed under section 133 or 151, in which case the fee shall be $540.

(8) EXTENSION FEES—For petitions for 1-month extensions of time to take actions required by the Director in an application—

(A) on filing a first petition, $130;

(B) on filing a second petition, $360; and

(C) on filing a third or subsequent petition, $620.

(b) Maintenance Fees—

(1) IN GENERAL—The Director shall charge the following fees for maintaining in force all patents based on applications filed on or after December 12, 1980:

(A) Three years and 6 months after grant, $980.

(B) Seven years and 6 months after grant, $2,480.

(C) Eleven years and 6 months after grant, $4,110.

(2) GRACE PERIOD; SURCHARGE—Unless payment of the applicable maintenance fee under paragraph (1) is received in the Office on or before the date the fee is due or within a grace period of 6 months thereafter, the patent shall expire as of the end of such grace period. The Director may require the payment of a surcharge as a condition of accepting within such 6-month grace period the payment of an applicable maintenance fee.

(3) NO MAINTENANCE FEE FOR DESIGN OR PLANT PATENT—No fee may be established for maintaining a design or plant patent in force.".

(b) Delays in Payment—Subsection (c) of section 41 of title 35, United States Code, is amended—

(1) by striking "(c)(1) The Director" and inserting:

"(c) Delays in Payment of Maintenance Fees—

(1) ACCEPTANCE—The Director"; and

(2) by striking "(2) A patent" and inserting:

"(2) EFFECT ON RIGHTS OF OTHERS—A patent".

(c) Patent Search Fees—Subsection (d) of section 41 of title 35, United States Code, is amended to read as follows:

"(d) Patent Search and Other Fees—

(1) PATENT SEARCH FEES—

(A) IN GENERAL—The Director shall charge the fees specified under subparagraph (B) for the search of each application for a patent, except for provisional applications. The Director shall adjust the fees charged under this paragraph to ensure that the fees recover an amount not to exceed the estimated average cost to the Office of searching applications for patent by Office personnel.

(B) SPECIFIC FEES—The fees referred to in subparagraph (A) are—

(i) $540 for each application for an original patent, except for design, plant, provisional, or international applications;

(ii) $100 for each application for an original design patent;

(iii) $330 for each application for an original plant patent;

(iv) $540 for the national stage of each international application; and

(v) $540 for each application for the reissue of a patent.

(C) APPLICABILITY OF OTHER PROVISIONS—The provisions of paragraphs (3) and (4) of section 111(a) relating to the payment of the fee for filing the application shall apply to the payment of the fee specified in this paragraph with respect to an application filed under section 111(a). The provisions of section 371(d) relating to the payment of the national fee shall apply to the payment of the fee specified in this paragraph with respect to an international application.

(D) REFUNDS—The Director may by regulation provide for a refund of any part of the fee specified in this paragraph for any applicant who files a written declaration of express abandonment as prescribed by the Director before an examination has been made of the application under section 131.

(2) OTHER FEES—

(A) IN GENERAL—The Director shall establish fees for all other processing, services, or materials relating to patents not specified in this section to recover the estimated average cost to the Office of such processing, services, or materials, except that the Director shall charge the following fees for the following services:

(i) For recording a document affecting title, $40 per property.

(ii) For each photocopy, $.25 per page.

(iii) For each black and white copy of a patent, $3.

(B) COPIES FOR LIBRARIES—The yearly fee for providing a library specified in section 12 with uncertified printed copies of the specifications and drawings for all patents in that year shall be $50.".

(d) Fees for Small Entities—Subsection (h) of section 41 of title 35, United States Code, is amended to read as follows:

"(h) Fees for Small Entities—

(1) REDUCTIONS IN FEES—Subject to paragraph (3), fees charged under subsections (a), (b), and (d)(1) shall be reduced by 50 percent with respect to

their application to any small business concern as defined under section 3 of the Small Business Act, and to any independent inventor or nonprofit organization as defined in regulations issued by the Director.

(2) SURCHARGES AND OTHER FEES—With respect to its application to any entity described in paragraph (1), any surcharge or fee charged under subsection (c) or (d) shall not be higher than the surcharge or fee required of any other entity under the same or substantially similar circumstances.

(3) REDUCTION FOR ELECTRONIC FILING—The fee charged under subsection (a)(1)(A) shall be reduced by 75 percent with respect to its application to any entity to which paragraph (1) applies, if the application is filed by electronic means as prescribed by the Director.".

(e) Technical Amendments—Section 41 of title 35, United States Code, is amended—

(1) in subsection (e), in the first sentence, by striking "The Director" and inserting "Waiver of Fees; Copies Regarding Notice—The Director";

(2) in subsection (f), by striking "The fees" and inserting "Adjustment of Fees—The fees";

(3) by repealing subsection (g); and

(4) in subsection (i)—

(A) by striking "(i)(1) The Director" and inserting the following:

"(i) Electronic Patent and Trademark Data—

(1) MAINTENANCE OF COLLECTIONS—The Director";

(B) by striking "(2) The Director" and inserting the following:

"(2) AVAILABILITY OF AUTOMATED SEARCH SYSTEMS—The Director";

(C) by striking "(3) The Director" and inserting the following:

"(3) ACCESS FEES—The Director"; and

(D) by striking "(4) The Director" and inserting the following:

"(4) ANNUAL REPORT TO CONGRESS—The Director".

(f) Adjustment of Trademark Fees—Section 802(a) of division B of the Consolidated Appropriations Act, 2005 (Public Law 108-447) is amended—

(1) in the first sentence, by striking "During fiscal years 2005, 2006, and 2007,", and inserting "Until such time as the Director sets or adjusts the fees otherwise,"; and

(2) in the second sentence, by striking "During fiscal years 2005, 2006, and 2007, the" and inserting "The".

(g) Effective Date, Applicability, and Transition Provisions—Section 803(a) of division B of the Consolidated Appropriations Act, 2005 (Public Law 108-447) is amended by striking "and shall apply only with respect to the remaining portion of fiscal year 2005 and fiscal year 2006".

(h) Prioritized Examination Fee—

(1) IN GENERAL—

(A) FEE—

(i) PRIORITIZED EXAMINATION FEE—A fee of $4,800 shall be established for filing a request, pursuant to section 2(b)(2)(G) of title 35, United States Code, for prioritized examination of a nonprovisional application for an original utility or plant patent.

(ii) ADDITIONAL FEES—In addition to the prioritized examination fee under clause (i), the fees due on an application for which prioritized examination is being sought are the filing, search, and examination fees (including any applicable excess claims and application size fees), processing fee, and publication fee for that application.

(B) REGULATIONS; LIMITATIONS—

(i) REGULATIONS—The Director may by regulation prescribe conditions for acceptance of a request under subparagraph (A) and a limit on the number of filings for prioritized examination that may be accepted.

(ii) LIMITATION ON CLAIMS—Until regulations are prescribed under clause (i), no application for which prioritized examination is requested may contain or be amended to contain more than 4 independent claims or more than 30 total claims.

(iii) LIMITATION ON TOTAL NUMBER OF REQUESTS—The Director may not accept in any fiscal year more than 10,000 requests for prioritization until regulations are prescribed under this subparagraph setting another limit.

(2) REDUCTION IN FEES FOR SMALL ENTITIES—The Director shall reduce fees for providing prioritized examination of nonprovisional applications for original utility and plant patents by 50 percent for small entities that qualify for reduced fees under section 41(h)(1) of title 35, United States Code.

(3) DEPOSIT OF FEES—All fees paid under this subsection shall be credited to the United States Patent and Trademark Office Appropriation Account, shall remain available until expended, and may be used only for the purposes specified in section 42(c)(3)(A) of title 35, United States Code.

(4) EFFECTIVE DATE AND TERMINATION—

(A) EFFECTIVE DATE—This subsection shall take effect on the date that is 10 days after the date of the enactment of this Act.

(B) TERMINATION—The fee imposed under paragraph (1)(A)(i), and the reduced fee under paragraph (2), shall terminate on the effective date of the setting or adjustment of the fee under paragraph (1)(A)(i) pursuant to the exercise of the authority under section 10 for the first time with respect to that fee.

(i) Appropriation Account Transition Fees—

(1) SURCHARGE—

(A) IN GENERAL—There shall be a surcharge of 15 percent, rounded by standard arithmetic rules, on all fees charged or authorized by subsections (a), (b), and (d)(1) of section 41, and section 132(b), of title 35, United States Code. Any surcharge imposed under this subsection is, and shall be construed to be, separate from and in addition to any other surcharge imposed under this Act or any other provision of law.

(B) DEPOSIT OF AMOUNTS—Amounts collected pursuant to the surcharge imposed under subparagraph (A) shall be credited to the United States Patent and Trademark Appropriation Account, shall remain available until expended, and may be used only for the purposes specified in section 42(c)(3)(A) of title 35, United States Code.

(2) EFFECTIVE DATE AND TERMINATION OF SURCHARGE—The surcharge provided for in paragraph (1)—

(A) shall take effect on the date that is 10 days after the date of the enactment of this Act; and

(B) shall terminate, with respect to a fee to which paragraph (1)(A) applies, on the effective date of the setting or adjustment of that fee pursuant to the exercise of the authority under section 10 for the first time with respect to that fee.

(j) Effective Date—Except as otherwise provided in this section, this section and the amendments made by this section shall take effect on the date of the enactment of this Act.

Sec. 12. Supplemental Examination.

(a) In General—Chapter 25 of title 35, United States Code, is amended by adding at the end the following:

"Sec. 257. Supplemental Examinations to Consider, Reconsider, or Correct Information

(a) Request for Supplemental Examination—A patent owner may request supplemental examination of a patent in the Office to consider, reconsider, or correct information believed to be relevant to the patent, in accordance with such requirements as the Director may establish. Within 3 months after the date a request for supplemental examination meeting the requirements of this section is received, the Director shall conduct the supplemental examination and shall conclude such examination by issuing a certificate indicating whether the information presented in the request raises a substantial new question of patentability.

(b) Reexamination Ordered—If the certificate issued under subsection (a) indicates that a substantial new question of patentability is raised by 1 or more items of information in the request, the Director shall order reexamination of the patent. The reexamination shall be conducted according to procedures established by chapter 30, except that the patent owner shall not have the right to file a statement pursuant to section 304. During the reexamination, the Director shall address each substantial new question of patentability identified during the supplemental examination, notwithstanding the limitations in chapter 30 relating to patents and printed publication or any other provision of such chapter.

(c) Effect—

(1) IN GENERAL—A patent shall not be held unenforceable on the basis of conduct relating to information that had not been considered, was inadequately considered, or was incorrect in a prior examination of the patent if the information was considered, reconsidered, or corrected during a supplemental examination of the patent. The making of a request under subsection (a), or the absence thereof, shall not be relevant to enforceability of the patent under section 282.

(2) EXCEPTIONS—

(A) PRIOR ALLEGATIONS—Paragraph (1) shall not apply to an allegation pled with particularity in a civil action, or set forth with particularity in a notice received by the patent owner under section 505(j)(2)(B)(iv)(II) of the Federal Food, Drug, and Cosmetic Act (21 U.S.C. 355(j)(2)(B)(iv)(II)), before the date of a supplemental examination request under subsection (a) to consider, reconsider, or correct information forming the basis for the allegation.

(B) PATENT ENFORCEMENT ACTIONS—In an action brought under section 337(a) of the Tariff Act of 1930 (19 U.S.C. 1337(a)), or section 281 of this title, paragraph (1) shall not apply to any defense raised in the action that is based upon information that was considered, reconsidered, or corrected pursuant to a supplemental examination request under subsection (a), unless the supplemental examination, and any reexamination ordered pursuant to the request, are concluded before the date on which the action is brought.

(d) Fees and Regulations—

(1) FEES—The Director shall, by regulation, establish fees for the submission of a request for supplemental examination of a patent, and to consider each item of information submitted in the request. If reexamination is ordered under subsection (b), fees established and applicable to ex parte reexamination proceedings under chapter 30 shall be paid, in addition to fees applicable to supplemental examination.

(2) REGULATIONS—The Director shall issue regulations governing the form, content, and other requirements of requests for supplemental examination, and establishing procedures for reviewing information submitted in such requests.

(e) Fraud—If the Director becomes aware, during the course of a supplemental examination or reexamination proceeding ordered under this section, that a material fraud on the Office may have been committed in connection with the patent that is the subject of the supplemental examination, then in addition to any other actions the Director is authorized to take, including the cancellation of any claims found to be invalid under section 307 as a result of a reexamination ordered under this section, the Director shall also refer the matter to the Attorney General for such further action as the Attorney General may deem appropriate. Any such referral shall be treated as confidential, shall not be included in the file of the patent, and shall not be disclosed to the public unless the United States charges a person with a criminal offense in connection with such referral.

(f) Rule of Construction—Nothing in this section shall be construed—

(1) to preclude the imposition of sanctions based upon criminal or antitrust laws (including section 1001(a) of title 18, the first section of the Clayton Act, and section 5 of the Federal Trade Commission Act to the extent that section relates to unfair methods of competition);

(2) to limit the authority of the Director to investigate issues of possible misconduct and impose sanctions for misconduct in connection with matters or proceedings before the Office; or

(3) to limit the authority of the Director to issue regulations under chapter 3 relating to sanctions for misconduct by representatives practicing before the Office.".

(b) Conforming Amendment—The table of sections for chapter 25 of title 35, United States Code, is amended by adding at the end the following new item:

"257. Supplemental examinations to consider, reconsider, or correct information.".

(c) Effective Date—The amendments made by this section shall take effect upon the expiration of the 1-year period beginning on the date of the enactment of this Act and shall apply to any patent issued before, on, or after that effective date.

Sec. 13. Funding Agreements.

(a) In General—Section 202(c)(7)(E)(i) of title 35, United States Code, is amended—

(1) by striking "75 percent" and inserting "15 percent";

(2) by striking "25 percent" and inserting "85 percent"; and

(3) by striking "as described above in this clause (D);" and inserting "described above in this clause;".

(b) Effective Date—The amendments made by this section shall take effect on the date of the enactment of this Act and shall apply to any patent issued before, on, or after that date.

Sec. 14. Tax Strategies Deemed Within the Prior Art.

(a) In General—For purposes of evaluating an invention under section 102 or 103 of title 35, United States Code, any strategy for reducing, avoiding, or deferring tax liability, whether known or unknown at the time of the invention or application for patent, shall be deemed insufficient to differentiate a claimed invention from the prior art.

(b) Definition—For purposes of this section, the term "tax liability" refers to any liability for a tax under any Federal, State, or local law, or the law of any foreign jurisdiction, including any statute, rule, regulation, or ordinance that levies, imposes, or assesses such tax liability.

(c) Exclusions—This section does not apply to that part of an invention that–

(1) is a method, apparatus, technology, computer program product, or system, that is used solely for preparing a tax or information return or other tax filing, including one that records, transmits, transfers, or organizes data related to such filing; or

(2) is a method, apparatus, technology, computer program product, or system used solely for financial management, to the extent that it is severable from

any tax strategy or does not limit the use of any tax strategy by any taxpayer or tax advisor.

(d) Rule of Construction—Nothing in this section shall be construed to imply that other business methods are patentable or that other business method patents are valid.

(e) Effective Date; Applicability—This section shall take effect on the date of the enactment of this Act and shall apply to any patent application that is pending on, or filed on or after, that date, and to any patent that is issued on or after that date.

Sec. 15. Best Mode Requirement.

(a) In General—Section 282 of title 35, United States Code, is amended in the second undesignated paragraph by striking paragraph (3) and inserting the following:

"(3) Invalidity of the patent or any claim in suit for failure to comply with—

(A) any requirement of section 112, except that the failure to disclose the best mode shall not be a basis on which any claim of a patent may be canceled or held invalid or otherwise unenforceable; or

(B) any requirement of section 251.".

(b) Conforming Amendment—Sections 119(e)(1) and 120 of title 35, United States Code, are each amended by striking "the first paragraph of section 112 of this title" and inserting "section 112(a) (other than the requirement to disclose the best mode)".

(c) Effective Date—The amendments made by this section shall take effect upon the date of the enactment of this Act and shall apply to proceedings commenced on or after that date.

Sec. 16. Marking.

(a) Virtual Marking—

(1) IN GENERAL—Section 287(a) of title 35, United States Code, is amended by striking "or when," and inserting "or by fixing thereon the word 'patent' or the abbreviation 'pat.' together with an address of a posting on the Internet, accessible to the public without charge for accessing the address, that associates the patented article with the number of the patent, or when,".

(2) EFFECTIVE DATE—The amendment made by this subsection shall apply to any case that is pending on, or commenced on or after, the date of the enactment of this Act.

(3) REPORT—Not later than the date that is 3 years after the date of the enactment of this Act, the Director shall submit a report to Congress that provides—

(A) an analysis of the effectiveness of "virtual marking", as provided in the amendment made by paragraph (1) of this subsection, as an alternative to the physical marking of articles;

(B) an analysis of whether such virtual marking has limited or improved the ability of the general public to access information about patents;

(C) an analysis of the legal issues, if any, that arise from such virtual marking; and

(D) an analysis of the deficiencies, if any, of such virtual marking.

(b) False Marking—

(1) CIVIL PENALTY—Section 292(a) of title 35, United States, Code, is amended by adding at the end the following: "Only the United States may sue for the penalty authorized by this subsection.".

(2) CIVIL ACTION FOR DAMAGES—Subsection (b) of section 292 of title 35, United States Code, is amended to read as follows:

"(b) A person who has suffered a competitive injury as a result of a violation of this section may file a civil action in a district court of the United States for recovery of damages adequate to compensate for the injury.".

(3) EXPIRED PATENTS—Section 292 of title 35, United States Code, is amended by adding at the end the following:

"(c) The marking of a product, in a manner described in subsection (a), with matter relating to a patent that covered that product but has expired is not a violation of this section.".

(4) EFFECTIVE DATE—The amendments made by this subsection shall apply to all cases, without exception, that are pending on, or commenced on or after, the date of the enactment of this Act.

Sec. 17. Advice of Counsel.

(a) In General—Chapter 29 of title 35, United States Code, is amended by adding at the end the following:

"Sec. 298. Advice of Counsel

The failure of an infringer to obtain the advice of counsel with respect to any allegedly infringed patent, or the failure of the infringer to present such advice to the court or jury, may not be used to prove that the accused infringer willfully infringed the patent or that the infringer intended to induce infringement of the patent.".

(b) Conforming Amendment—The table of sections for chapter 29 of title 35, United States Code, is amended by adding at the end the following:
"298. Advice of counsel.".

Sec. 18. Transitional Program for Covered Business Method Patents.

(a) Transitional Program—

(1) ESTABLISHMENT—Not later than the date that is 1 year after the date of the enactment of this Act, the Director shall issue regulations establishing and implementing a transitional post-grant review proceeding for review of the validity of covered business method patents. The transitional proceeding implemented pursuant to this subsection shall be regarded as, and shall employ the standards and procedures of, a post-grant review under chapter 32 of title 35, United States Code, subject to the following:

(A) Section 321(c) of title 35, United States Code, and subsections (b), (e)(2), and (f) of section 325 of such title shall not apply to a transitional proceeding.

(B) A person may not file a petition for a transitional proceeding with respect to a covered business method patent unless the person or the person's real party in interest or privy has been sued for infringement of the patent or has been charged with infringement under that patent.

(C) A petitioner in a transitional proceeding who challenges the validity of 1 or more claims in a covered business method patent on a ground raised under section 102 or 103 of title 35, United States Code, as in effect on the day before the effective date set forth in section 3(n)(1), may support such ground only on the basis of—

(i) prior art that is described by section 102(a) of such title of such title (as in effect on the day before such effective date); or

(ii) prior art that—

(I) discloses the invention more than 1 year before the date of the application for patent in the United States; and

(II) would be described by section 102(a) of such title (as in effect on the day before the effective date set forth in section 3(n)(1)) if the disclosure had been made by another before the invention thereof by the applicant for patent.

(D) The petitioner in a transitional proceeding that results in a final written decision under section 328(a) of title 35, United States Code, with respect to a claim in a covered business method patent, or the petitioner's real party in interest, may not assert, either in a civil action arising in whole or in part under section 1338 of title 28, United States Code, or in a proceeding before the International Trade Commission under section 337 of the Tariff Act of 1930 (19 U.S.C. 1337), that the claim is invalid on any ground that the petitioner raised during that transitional proceeding.

(E) The Director may institute a transitional proceeding only for a patent that is a covered business method patent.

(2) EFFECTIVE DATE—The regulations issued under paragraph (1) shall take effect upon the expiration of the 1-year period beginning on the date of the enactment of this Act and shall apply to any covered business method patent

issued before, on, or after that effective date, except that the regulations shall not apply to a patent described in section 6(f)(2)(A) of this Act during the period in which a petition for post-grant review of that patent would satisfy the requirements of section 321(c) of title 35, United States Code.

(3) SUNSET—

(A) IN GENERAL—This subsection, and the regulations issued under this subsection, are repealed effective upon the expiration of the 8-year period beginning on the date that the regulations issued under to paragraph (1) take effect.

(B) APPLICABILITY—Notwithstanding subparagraph (A), this subsection and the regulations issued under this subsection shall continue to apply, after the date of the repeal under subparagraph (A), to any petition for a transitional proceeding that is filed before the date of such repeal.

(b) Request for Stay—

(1) IN GENERAL—If a party seeks a stay of a civil action alleging infringement of a patent under section 281 of title 35, United States Code, relating to a transitional proceeding for that patent, the court shall decide whether to enter a stay based on—

(A) whether a stay, or the denial thereof, will simplify the issues in question and streamline the trial;

(B) whether discovery is complete and whether a trial date has been set;

(C) whether a stay, or the denial thereof, would unduly prejudice the nonmoving party or present a clear tactical advantage for the moving party; and

(D) whether a stay, or the denial thereof, will reduce the burden of litigation on the parties and on the court.

(2) REVIEW—A party may take an immediate interlocutory appeal from a district court's decision under paragraph (1). The United States Court of Appeals for the Federal Circuit shall review the district court's decision to ensure consistent application of established precedent, and such review may be de novo.

(c) ATM Exemption for Venue Purposes—In an action for infringement under section 281 of title 35, United States Code, of a covered business method patent, an automated teller machine shall not be deemed to be a regular and established place of business for purposes of section 1400(b) of title 28, United States Code.

(d) Definition—

(1) IN GENERAL—For purposes of this section, the term "covered business method patent" means a patent that claims a method or corresponding apparatus for performing data processing or other operations used in the practice, administration, or management of a financial product or service, except that the term does not include patents for technological inventions.

(2) REGULATIONS—To assist in implementing the transitional proceeding authorized by this subsection, the Director shall issue regulations for determining whether a patent is for a technological invention.

(e) Rule of Construction—Nothing in this section shall be construed as amending or interpreting categories of patent-eligible subject matter set forth under section 101 of title 35, United States Code.

Sec. 19. Jurisdiction and Procedural Matters.

(a) State Court Jurisdiction—Section 1338(a) of title 28, United States Code, is amended by striking the second sentence and inserting the following: "No State court shall have jurisdiction over any claim for relief arising under any Act of Congress relating to patents, plant variety protection, or copyrights. For purposes of this subsection, the term 'State' includes any State of the United States, the District of Columbia, the Commonwealth of Puerto Rico, the United States Virgin Islands, American Samoa, Guam, and the Northern Mariana Islands.".

(b) Court of Appeals for the Federal Circuit—Section 1295(a)(1) of title 28, United States Code, is amended to read as follows:

(1) of an appeal from a final decision of a district court of the United States, the District Court of Guam, the District Court of the Virgin Islands, or the District Court of the Northern Mariana Islands, in any civil action arising under, or in any civil action in which a party has asserted a compulsory counterclaim arising under, any Act of Congress relating to patents or plant variety protection;".

(c) Removal—

(1) IN GENERAL—Chapter 89 of title 28, United States Code, is amended by adding at the end the following new section:

"Sec. 1454. Patent, Plant Variety Protection, and Copyright Cases

(a) In General—A civil action in which any party asserts a claim for relief arising under any Act of Congress relating to patents, plant variety protection, or copyrights may be removed to the district court of the United States for the district and division embracing the place where the action is pending.

(b) Special Rules—The removal of an action under this section shall be made in accordance with section 1446, except that if the removal is based solely on this section—

(1) the action may be removed by any party; and

(2) the time limitations contained in section 1446(b) may be extended at any time for cause shown.

(c) Clarification of Jurisdiction in Certain Cases—The court to which a civil action is removed under this section is not precluded from hearing and determining any claim in the civil action because the State court from which the civil action is removed did not have jurisdiction over that claim.

(d) Remand—If a civil action is removed solely under this section, the district court–

(1) shall remand all claims that are neither a basis for removal under subsection (a) nor within the original or supplemental jurisdiction of the district court under any Act of Congress; and

(2) may, under the circumstances specified in section 1367(c), remand any claims within the supplemental jurisdiction of the district court under section 1367.".

(2) CONFORMING AMENDMENT—The table of sections for chapter 89 of title 28, United States Code, is amended by adding at the end the following new item:

"1454. Patent, plant variety protection, and copyright cases.".

(d) Procedural Matters in Patent Cases—

(1) JOINDER OF PARTIES AND STAY OF ACTIONS—Chapter 29 of title 35, United States Code, as amended by this Act, is further amended by adding at the end the following new section:

"Sec. 299. Joinder of Parties

(a) Joinder of Accused Infringers—With respect to any civil action arising under any Act of Congress relating to patents, other than an action or trial in which an act of infringement under section 271(e)(2) has been pled, parties that are accused infringers may be joined in one action as defendants or counterclaim defendants, or have their actions consolidated for trial, or counterclaim defendants only if—

(1) any right to relief is asserted against the parties jointly, severally, or in the alternative with respect to or arising out of the same transaction, occurrence, or series of transactions or occurrences relating to the making, using, importing into the United States, offering for sale, or selling of the same accused product or process; and

(2) questions of fact common to all defendants or counterclaim defendants will arise in the action.

(b) Allegations Insufficient for Joinder—For purposes of this subsection, accused infringers may not be joined in one action as defendants or counterclaim defendants, or have their actions consolidated for trial, based solely on allegations that they each have infringed the patent or patents in suit.

(c) Waiver—A party that is an accused infringer may waive the limitations set forth in this section with respect to that party.".

(2) CONFORMING AMENDMENT—The table of sections for chapter 29 of title 35, United States Code, as amended by this Act, is further amended by adding at the end the following new item:

"299. Joinder of parties.".

(e) Effective Date—The amendments made by this section shall apply to any civil action commenced on or after the date of the enactment of this Act.

Sec. 20. Technical Amendments.

(a) Joint Inventions—Section 116 of title 35, United States Code, is amended—

(1) in the first—undesignated paragraph, by striking "When" and inserting "(a) Joint Inventions—When";

(2) in the second undesignated paragraph, by striking "If a joint inventor" and inserting "(b) Omitted Inventor—If a joint inventor"; and

(3) in the third undesignated paragraph—

(A) by striking "Whenever" and inserting "(c) Correction of Errors in Application—Whenever"; and

(B) by striking "and such error arose without any deceptive intention on his part,".

(b) Filing of Application in Foreign Country—Section 184 of title 35, United States Code, is amended—

(1) in the first undesignated paragraph—

(A) by striking "Except when" and inserting "(a) Filing in Foreign Country- Except when"; and

(B) by striking "and without deceptive intent";

(2) in the second undesignated paragraph, by striking "The term" and inserting "(b) Application—The term"; and

(3) in the third undesignated paragraph, by striking "The scope" and inserting "(c) Subsequent Modifications, Amendments, and Supplements—The scope".

(c) Filing Without a License—Section 185 of title 35, United States Code, is amended by striking "and without deceptive intent".

(d) Reissue of Defective Patents—Section 251 of title 35, United States Code, is amended—

(1) in the first undesignated paragraph—

(A) by striking "Whenever" and inserting "(a) In General- Whenever"; and

(B) by striking "without any deceptive intention";

(2) in the second undesignated paragraph, by striking "The Director" and inserting "(b) Multiple Reissued Patents—The Director";

(3) in the third undesignated paragraph, by striking "The provisions" and inserting "(c) Applicability of This Title—The provisions"; and

(4) in the last undesignated paragraph, by striking "No reissued patent" and inserting "(d) Reissue Patent Enlarging Scope of Claims—No reissued patent".

(e) Effect of Reissue—Section 253 of title 35, United States Code, is amended—

(1) in the first undesignated paragraph, by striking "Whenever, without any deceptive intention," and inserting "(a) In General—Whenever"; and

(2) in the second undesignated paragraph, by striking "In like manner" and inserting "(b) Additional Disclaimer or Dedication—In the manner set forth in subsection (a),".

(f) Correction of Named Inventor—Section 256 of title 35, United States Code, is amended—

(1) in the first undesignated paragraph—

(A) by striking "Whenever" and inserting "(a) Correction- Whenever"; and

(B) by striking "and such error arose without any deceptive intention on his part"; and

(2) in the second undesignated paragraph, by striking "The error" and inserting "(b) Patent Valid if Error Corrected—The error".

(g) Presumption of Validity—Section 282 of title 35, United States Code, is amended—

(1) in the first undesignated paragraph—

(A) by striking "A patent" and inserting "(a) In General—A patent"; and

(B) by striking the third sentence;

(2) in the second undesignated paragraph—

(A) by striking "The following" and inserting "(b) Defenses—The following";

(B) in paragraph (1), by striking "uneforceability," and inserting "unenforceability."; and

(C) in paragraph (2), by striking "patentability," and inserting "patentability." ; and

(3) in the third undesignated paragraph—

(A) by striking "In actions involving the validity or infringement of a patent" and inserting "(c) Notice of Actions; Actions During Extension of Patent Term- In an action involving the validity or infringement of a patent"; and

(B) by striking "Claims Court" and inserting "Court of Federal Claims".

(h) Action for Infringement—Section 288 of title 35, United States Code, is amended by striking ", without deceptive intention,".

(i) Reviser's Notes—

(1) Section 3(e)(2) of title 35, United States Code, is amended by striking "this Act," and inserting "that Act,".

(2) Section 202 of title 35, United States Code, is amended—

(A) in subsection (b)(3), by striking "the section 203(b)" and inserting "section 203(b)"; and

(B) in subsection (c)(7)(D), by striking "except where it proves" and all that follows through "small business firms; and" and inserting: "except where it is determined to be infeasible following a reasonable inquiry, a preference in the licensing of subject inventions shall be given to small business firms; and".

(3) Section 209(d)(1) of title 35, United States Code, is amended by striking "nontransferrable" and inserting "nontransferable".

(4) Section 287(c)(2)(G) of title 35, United States Code, is amended by striking "any state" and inserting "any State".

(5) Section 371(b) of title 35, United States Code, is amended by striking "of the treaty" and inserting "of the treaty.".

(j) Unnecessary References—

(1) IN GENERAL—Title 35, United States Code, is amended by striking "of this title" each place that term appears.

(2) EXCEPTION—The amendment made by paragraph (1) shall not apply to the use of such term in the following sections of title 35, United States Code:

(A) Section 1(c).

(B) Section 101.

(C) Subsections (a) and (b) of section 105.

(D) The first instance of the use of such term in section 111(b)(8).

(E) Section 161.

(F) Section 164.

(G) Section 171.

(H) Section 251(c), as so designated by this section.

(I) Section 261.

(J) Subsections (g) and (h) of section 271.

(K) Section 287(b)(1).

(L) Section 289.

(M) The first instance of the use of such term in section 375(a).

(k) Additional Technical Amendments—Sections 155 and 155A of title 35, United States Code, and the items relating to those sections in the table of sections for chapter 14 of such title, are repealed.

(l) Effective Date—The amendments made by this section shall take effect upon the expiration of the 1-year period beginning on the date of the enactment of this Act and shall apply to proceedings commenced on or after that effective date.

Sec. 21. Travel Expenses and Payment of Administrative Judges.

(a) Authority To Cover Certain Travel Related Expenses—Section 2(b)(11) of title 35, United States Code, is amended by inserting ", and the Office is authorized to expend funds to cover the subsistence expenses and travel-related expenses, including per diem, lodging costs, and transportation costs, of persons attending such programs who are not Federal employees" after "world".

(b) Payment of Administrative Judges—Section 3(b) of title 35, United States Code, is amended by adding at the end the following:

"(6) ADMINISTRATIVE PATENT JUDGES AND ADMINISTRATIVE TRADEMARK JUDGES—The Director may fix the rate of basic pay for the administrative patent judges appointed pursuant to section 6 and the administrative trademark judges appointed pursuant to section 17 of the Trademark Act of 1946 (15 U.S.C. 1067) at not greater than the rate of basic pay payable for level III of the Executive Schedule under section 5314 of title 5. The payment of a rate of basic pay under this paragraph shall not be subject to the pay limitation under section 5306(e) or 5373 of title 5.".

Sec. 22. Patent and Trademark Office Funding.

(a) In General—Section 42(c) of title 35, United States Code, is amended—

(1) by striking '(c)' and inserting '(c)(1)';

(2) in the first sentence, by striking 'shall be available' and inserting 'shall, subject to paragraph (3), be available';

(3) by striking the second sentence; and

(4) by adding at the end the following:

"(2) There is established in the Treasury a Patent and Trademark Fee Reserve Fund. If fee collections by the Patent and Trademark Office for a fiscal year exceed the amount appropriated to the Office for that fiscal year, fees collected in excess of the appropriated amount shall be deposited in the Patent and Trademark Fee Reserve Fund. To the extent and in the amounts provided in appropriations Acts, amounts in the Fund shall be made available until expended only for obligation and expenditure by the Office in accordance with paragraph (3).

(3)(A) Any fees that are collected under sections 41, 42, and 376, and any surcharges on such fees, may only be used for expenses of the Office relating to the processing of patent applications and for other activities, services, and materials relating to patents and to cover a share of the administrative costs of the Office relating to patents.

(B) Any fees that are collected under section 31 of the Trademark Act of 1946, and any surcharges on such fees, may only be used for expenses of the Office relating to the processing of trademark registrations and for other activities, services, and materials relating to trademarks and to cover a share of the administrative costs of the Office relating to trademarks.".

(b) Effective Date—The amendments made by this section shall take effect on October 1, 2011.

Sec. 23. Satellite Offices.

(a) Establishment—Subject to available resources, the Director shall, by not later than the date that is 3 years after the date of the enactment of this Act, establish 3 or more satellite offices in the United States to carry out the responsibilities of the Office.

(b) Purposes—The purposes of the satellite offices established under subsection (a) are to—

(1) increase outreach activities to better connect patent filers and innovators with the Office;

(2) enhance patent examiner retention;

(3) improve recruitment of patent examiners;

(4) decrease the number of patent applications waiting for examination; and

(5) improve the quality of patent examination.

(c) Required Considerations-

(1) IN GENERAL—In selecting the location of each satellite office to be established under subsection (a), the Director—

(A) shall ensure geographic diversity among the offices, including by ensuring that such offices are established in different States and regions throughout the Nation;

(B) may rely upon any previous evaluations by the Office of potential locales for satellite offices, including any evaluations prepared as part of the Office's Nationwide Workforce Program that resulted in the 2010 selection of Detroit, Michigan, as the first satellite office of the Office;

(C) shall evaluate and consider the extent to which the purposes of satellite offices listed under subsection (b) will be achieved;

(D) shall consider the availability of scientific and technically knowledgeable personnel in the region from which to draw new patent examiners at minimal recruitment cost; and

(E) shall consider the economic impact to the region.

(2) OPEN SELECTION PROCESS—Nothing in paragraph (1) shall constrain the Office to only consider its evaluations in selecting the Detroit, Michigan, satellite office.

(d) Report to Congress—Not later than the end of the third fiscal year that begins after the date of the enactment of this Act, the Director shall submit a report to Congress on—

(1) the rationale of the Director in selecting the location of any satellite office required under subsection (a), including an explanation of how the selected location will achieve the purposes of satellite offices listed under subsection (b) and how the required considerations listed under subsection (c) were met;

(2) the progress of the Director in establishing all such satellite offices; and

(3) whether the operation of existing satellite offices is achieving the purposes under subsection (b).

Sec. 24. Designation of Detroit Satellite Office.

(a) Designation—The satellite office of the United States Patent and Trademark Office to be located in Detroit, Michigan, shall be known and designated as the "Elijah J. McCoy United States Patent and Trademark Office".

(b) References—Any reference in a law, map, regulation, document, paper, or other record of the United States to the satellite office of the United States Patent and Trademark Office to be located in Detroit, Michigan, referred to in subsection (a) shall be deemed to be a reference to the "Elijah J. McCoy United States Patent and Trademark Office".

Sec. 25. Priority Examination for Important Technologies.

Section 2(b)(2) of title 35, United States Code, is amended—

(1) in subparagraph (E), by striking "and" after the semicolon;

(2) in subparagraph (F), by inserting "and" after the semicolon; and

(3) by adding at the end the following:

"(G) may, subject to any conditions prescribed by the Director and at the request of the patent applicant, provide for prioritization of examination of applications for products, processes, or technologies that are important to the national economy or national competitiveness without recovering the aggregate extra cost of providing such prioritization, notwithstanding section 41 or any other provision of law;".

Sec. 26. Study on Implementation.

(a) PTO Study—The Director shall conduct a study on the manner in which this Act and the amendments made by this Act are being implemented by the Office, and on such other aspects of the patent policies and practices of the Federal Government with respect to patent rights, innovation in the United States, competitiveness of United States markets, access by small businesses to capital for investment, and such other issues, as the Director considers appropriate.

(b) Report to Congress—The Director shall, not later than the date that is 4 years after the date of the enactment of this Act, submit to the Committees on the Judiciary of the House of Representatives and the Senate a report on the results of the study conducted under subsection (a), including recommendations for any changes to laws and regulations that the Director considers appropriate.

Sec. 27. Study on Genetic Testing.

(a) In General—The Director shall conduct a study on effective ways to provide independent, confirming genetic diagnostic test activity where gene patents and exclusive licensing for primary genetic diagnostic tests exist.

(b) Items Included in Study—The study shall include an examination of at least the following:

(1) The impact that the current lack of independent second opinion testing has had on the ability to provide the highest level of medical care to patients and recipients of genetic diagnostic testing, and on inhibiting innovation to existing testing and diagnoses.

(2) The effect that providing independent second opinion genetic diagnostic testing would have on the existing patent and license holders of an exclusive genetic test.

(3) The impact that current exclusive licensing and patents on genetic testing activity has on the practice of medicine, including but not limited to: the interpretation of testing results and performance of testing procedures.

(4) The role that cost and insurance coverage have on access to and provision of genetic diagnostic tests.

(c) Confirming Genetic Diagnostic Test Activity Defined—For purposes of this section, the term "confirming genetic diagnostic test activity" means the performance of a genetic diagnostic test, by a genetic diagnostic test provider, on an individual solely for the purpose of providing the individual with an independent confirmation of results obtained from another test provider's prior performance of the test on the individual.

(d) Report—Not later than 9 months after the date of enactment of this Act, the Director shall report to the Committee on the Judiciary of the Senate and the Committee on the Judiciary of the House of Representatives on the findings of the study and provide recommendations for establishing the availability of such independent confirming genetic diagnostic test activity.

Sec. 28. Patent Ombudsman Program for Small Business Concerns.

Using available resources, the Director shall establish and maintain in the Office a Patent Ombudsman Program. The duties of the Program's staff shall include providing support and services relating to patent filings to small business concerns and independent inventors.

Sec. 29. Establishment of Methods for Studying the Diversity of Applicants.

The Director shall, not later than the end of the 6-month period beginning on the date of the enactment of this Act, establish methods for studying the diversity of patent applicants, including those applicants who are minorities, women, or veterans. The Director shall not use the results of such study to provide any preferential treatment to patent applicants.

Sec. 30. Sense of Congress.

It is the sense of Congress that the patent system should promote industries to continue to develop new technologies that spur growth and create jobs across the country which includes protecting the rights of small businesses and inventors from predatory behavior that could result in the cutting off of innovation.

Sec. 31. Uspto Study on International Patent Protections for Small Businesses.

(a) Study Required—The Director, in consultation with the Secretary of Commerce and the Administrator of the Small Business Administration, shall, using the existing resources of the Office, carry out a study—

(1) to determine how the Office, in coordination with other Federal departments and agencies, can best help small businesses with international patent protection; and

(2) whether, in order to help small businesses pay for the costs of filing, maintaining, and enforcing international patent applications, there should be established either—

(A) a revolving fund loan program to make loans to small businesses to defray the costs of such applications, maintenance, and enforcement and related technical assistance; or

(B) a grant program to defray the costs of such applications, maintenance, and enforcement and related technical assistance.

(b) Report—Not later than 120 days after the date of the enactment of this Act, the Director shall issue a report to the Congress containing—

(1) all findings and determinations made in carrying out the study required under subsection (a);

(2) a statement of whether the determination was made that—

(A) a revolving fund loan program described under subsection (a)(2)(A) should be established;

(B) a grant program described under subsection (a)(2)(B) should be established; or

(C) neither such program should be established; and

(3) any legislative recommendations the Director may have developed in carrying out such study.

Sec. 32. Pro Bono Program.

(a) In General—The Director shall work with and support intellectual property law associations across the country in the establishment of pro bono programs designed to assist financially under-resourced independent inventors and small businesses.

(b) Effective Date—This section shall take effect on the date of the enactment of this Act.

Sec. 33. Limitation on Issuance of Patents.

(a) Limitation—Notwithstanding any other provision of law, no patent may issue on a claim directed to or encompassing a human organism.

(b) Effective Date—

(1) IN GENERAL—Subsection (a) shall apply to any application for patent that is pending on, or filed on or after, the date of the enactment of this Act.

(2) PRIOR APPLICATIONS—Subsection (a) shall not affect the validity of any patent issued on an application to which paragraph (1) does not apply.

Sec. 34. Study of Patent Litigation.

(a) GAO Study—The Comptroller General of the United States shall conduct a study of the consequences of litigation by non-practicing entities, or by patent assertion entities, related to patent claims made under title 35, United States Code, and regulations authorized by that title.

(b) Contents of Study—The study conducted under this section shall include the following:

(1) The annual volume of litigation described in subsection (a) over the 20-year period ending on the date of the enactment of this Act.

(2) The volume of cases comprising such litigation that are found to be without merit after judicial review.

(3) The impacts of such litigation on the time required to resolve patent claims.

(4) The estimated costs, including the estimated cost of defense, associated with such litigation for patent holders, patent licensors, patent licensees, and inventors, and for users of alternate or competing innovations.

(5) The economic impact of such litigation on the economy of the United States, including the impact on inventors, job creation, employers, employees, and consumers.

(6) The benefit to commerce, if any, supplied by non-practicing entities or patent assertion entities that prosecute such litigation.

(c) Report to Congress—The Comptroller General shall, not later than the date that is 1 year after the date of the enactment of this Act, submit to the Committee on the Judiciary of the House of Representatives and the Committee on the Judiciary of the Senate a report on the results of the study required under this section, including recommendations for any changes to laws and regulations that will minimize any negative impact of patent litigation that was the subject of such study.

Sec. 35. Effective Date.

Except as otherwise provided in this Act, the provisions of this Act shall take effect upon the expiration of the 1-year period beginning on the date of the enactment of this Act and shall apply to any patent issued on or after that effective date.

Sec. 36. Budgetary Effects.

The budgetary effects of this Act, for the purpose of complying with the Statutory Pay-As-You-Go Act of 2010, shall be determined by reference to the latest statement titled "Budgetary Effects of PAYGO Legislation" for this Act, submitted for printing in the Congressional Record by the Chairman of the House Budget Committee, provided that such statement has been submitted prior to the vote on passage.

Sec. 37. Calculation of 60-Day Period for Application of Patent Term Extension.

(a) In General—Section 156(d)(1) of title 35, United States Code, is amended by adding at the end the following flush sentence:

"For purposes of determining the date on which a product receives permission under the second sentence of this paragraph, if such permission is transmitted after 4:30 P.M., Eastern Time, on a business day, or is transmitted on a day that is not a business day, the product shall be deemed to receive such permission on the next business day. For purposes of the preceding sentence, the term 'business day' means any Monday, Tuesday, Wednesday, Thursday, or Friday, excluding any legal holiday under section 6103 of title 5.".

(b) Applicability—The amendment made by subsection (a) shall apply to any application for extension of a patent term under section 156 of title 35, United States Code, that is pending on, that is filed after, or as to which a decision regarding the application is subject to judicial review on, the date of the enactment of this Act.

"Blackline" Version of Selected Sections of 1952 Act Showing Changes Under AIA

Sec. 100. Definitions

When used in this title unless the context otherwise indicates—

(a) The term "invention" means invention or discovery.

(b) The term "process" means process, art or method, and includes a new use of a known process, machine, manufacture, composition of matter, or material.

(c) The terms "United States" and "this country" mean the United States of America, its territories and possessions.

(d) The word "patentee" includes not only the patentee to whom the patent was issued but also the successors in title to the patentee.

(e) The term "third-party requester" means a person requesting ex parte reexamination under ~~section 302 or inter partes reexamination under section 311~~ who is not the patent owner.

(f) The term 'inventor' means the individual or, if a joint invention, the individuals collectively who invented or discovered the subject matter of the invention.

(g) The terms 'joint inventor' and 'coinventor' mean any 1 of the individuals who invented or discovered the subject matter of a joint invention.

(h) The term 'joint research agreement' means a written contract, grant, or cooperative agreement entered into by 2 or more persons or entities for the performance of experimental, developmental, or research work in the field of the claimed invention.

(i)(1) The term 'effective filing date' for a claimed invention in a patent or application for patent means-

(A) if subparagraph (B) does not apply, the actual filing date of the patent or the application for the patent containing a claim to the invention; or

(B) the filing date of the earliest application for which the patent or application is entitled, as to such invention, to a right of priority under section

119, 365(a), or 365(b) or to the benefit of an earlier filing date under section 120, 121, or 365(c).

(2) The effective filing date for a claimed invention in an application for reissue or reissued patent shall be determined by deeming the claim to the invention to have been contained in the patent for which reissue was sought.

(j) The term 'claimed invention' means the subject matter defined by a claim in a patent or an application for a patent.

Sec. 102. Conditions for Patentability; Novelty and Loss of Right to Patent

~~A person shall be entitled to a patent unless—~~

~~(a) the invention was known or used by others in this country, or patented or described in a printed publication in this or a foreign country, before the invention thereof by the applicant for patent, or~~

~~(b) the invention was patented or described in a printed publication in this or a foreign country or in public use or on sale in this country, more than one year prior to the date of the application for patent in the United States, or~~

~~(c) he has abandoned the invention, or~~

~~(d) the invention was first patented or caused to be patented, or was the subject of an inventor's certificate, by the applicant or his legal representatives or assigns in a foreign country prior to the date of the application for patent in this country on an application for patent or inventor's certificate filed more than twelve months before the filing of the application in the United States, or~~

~~(e) the invention was described in (1) an application for patent, published under section 122(b), by another filed in the United States before the invention by the applicant for patent or (2) a patent granted on an application for patent by another filed in the United States before the invention by the applicant for patent, except that an international application filed under the treaty defined in section 351(a) shall have the effects for the purposes of this subsection of an application filed in the United States only if the international application designated the United States and was published under Article 21(2) of such treaty in the English language;[1] or~~

~~(f) he did not himself invent the subject matter sought to be patented, or~~

~~(g)(1) during the course of an interference conducted under section 135 or section 291, another inventor involved therein establishes, to the extent permitted in section 104, that before such person's invention thereof the invention was made by such other inventor and not abandoned, suppressed, or concealed, or (2) before such person's invention thereof, the invention was made in this country by another inventor who had not abandoned, suppressed, or concealed it. In determining priority of invention under this subsection, there shall be considered not only the respective dates of conception and reduction to practice of the invention, but also the reasonable diligence of one who was first to conceive and last to reduce to practice, from a time prior to conception by the other.~~

Sec. 102~~. Conditions for Patentability; Novelty~~

(a) Novelty; Prior Art. A person shall be entitled to a patent unless—

'(1) the claimed invention was patented, described in a printed publication, or in public use, on sale, or otherwise available to the public before the effective filing date of the claimed invention; or

'(2) the claimed invention was described in a patent issued under section 151, or in an application for patent published or deemed published under section 122(b), in which the patent or application, as the case may be, names another inventor and was effectively filed before the effective filing date of the claimed invention.

(b) Exceptions.—

(1) Disclosures made 1 year or less before the effective filing date of the claimed invention. A disclosure made 1 year or less before the effective filing date of a claimed invention shall not be prior art to the claimed invention under subsection (a)(1) if—

(A) the disclosure was made by the inventor or joint inventor or by another who obtained the subject matter disclosed directly or indirectly from the inventor or a joint inventor; or

(B) the subject matter disclosed had, before such disclosure, been publicly disclosed by the inventor or a joint inventor or another who obtained the subject matter disclosed directly or indirectly from the inventor or a joint inventor.

(2) Disclosures appearing in applications and patents. A disclosure shall not be prior art to a claimed invention under subsection (a)(2) if—

(A) the subject matter disclosed was obtained directly or indirectly from the inventor or a joint inventor;

(B) the subject matter disclosed had, before such subject matter was effectively filed under subsection (a)(2), been publicly disclosed by the inventor or a joint inventor or another who obtained the subject matter disclosed directly or indirectly from the inventor or a joint inventor; or

(C) the subject matter disclosed and the claimed invention, not later than the effective filing date of the claimed invention, were owned by the same person or subject to an obligation of assignment to the same person.

(c) Common Ownership Under Joint Research Agreements. Subject matter disclosed and a claimed invention shall be deemed to have been owned by the same person or subject to an obligation of assignment to the same person in applying the provisions of subsection (b)(2)(C) if—

(1) the subject matter disclosed was developed and the claimed invention was made by, or on behalf of, 1 or more parties to a joint research agreement that was in effect on or before the effective filing date of the claimed invention;

(2) the claimed invention was made as a result of activities undertaken within the scope of the joint research agreement; and

(3) the application for patent for the claimed invention discloses or is amended to disclose the names of the parties to the joint research agreement.

(d) Patents and Published Applications Effective as Prior Art. For purposes of determining whether a patent or application for patent is prior art to a claimed invention under subsection (a)(2), such patent or application shall be considered to have been effectively filed, with respect to any subject matter described in the patent or application—

(1) if paragraph (2) does not apply, as of the actual filing date of the patent or the application for patent; or

(2) if the patent or application for patent is entitled to claim a right of priority under section 119, 365(a), or 365(b), or to claim the benefit of an earlier filing date under section 120, 121, or 365(c), based upon 1 or more prior filed applications for patent, as of the filing date of the earliest such application that describes the subject matter.'.

Sec. 103. Conditions for Patentability; Non-Obvious Subject Matter

~~(a) A patent may not be obtained though the invention is not identically disclosed or described as set forth in section 102 of this title, if the differences between the subject matter sought to be patented and the prior art are such that the subject matter as a whole would have been obvious at the time the invention was made to a person having ordinary skill in the art to which said subject matter pertains. Patentability shall not be negatived by the manner in which the invention was made.~~

~~(b)(1) Notwithstanding subsection (a), and upon timely election by the applicant for patent to proceed under this subsection, a biotechnological process using or resulting in a composition of matter that is novel under section 102 and nonobvious under subsection (a) of this section shall be considered nonobvious if-~~

~~(A) claims to the process and the composition of matter are contained in either the same application for patent or in separate applications having the same effective filing date; and~~

~~(B) the composition of matter, and the process at the time it was invented, were owned by the same person or subject to an obligation of assignment to the same person.~~

~~(2) A patent issued on a process under paragraph (1)-~~

~~(A) shall also contain the claims to the composition of matter used in or made by that process, or~~

~~(B) shall, if such composition of matter is claimed in another patent, be set to expire on the same date as such other patent, notwithstanding section 154.~~

~~(3) For purposes of paragraph (1), the term "biotechnological process" means-~~

~~(A) a process of genetically altering or otherwise inducing a single- or multi-celled organism to-~~

~~(i) express an exogenous nucleotide sequence,~~

~~(ii) inhibit, eliminate, augment, or alter expression of an endogenous nucleotide sequence, or~~

~~(iii) express a specific physiological characteristic not naturally associated with said organism;~~

~~(B) cell fusion procedures yielding a cell line that expresses a specific protein, such as a monoclonal antibody; and~~

~~(C) a method of using a product produced by a process defined by subparagraph (A) or (B), or a combination of subparagraphs (A) and (B).~~

~~(c)(1) Subject matter developed by another person, which qualifies as prior art only under one or more of subsections (e), (f), and (g) of section 102 of this title, shall not preclude patentability under this section where the subject matter and the claimed invention were, at the time the claimed invention was made, owned by the same person or subject to an obligation of assignment to the same person.~~

~~(2) For purposes of this subsection, subject matter developed by another person and a claimed invention shall be deemed to have been owned by the same person or subject to an obligation of assignment to the same person if—~~

~~(A) the claimed invention was made by or on behalf of parties to a joint research agreement that was in effect on or before the date the claimed invention was made;~~

~~(B) the claimed invention was made as a result of activities undertaken within the scope of the joint research agreement; and~~

~~(C) the application for patent for the claimed invention discloses or is amended to disclose the names of the parties to the joint research agreement.~~

~~(3) For purposes of paragraph (2), the term "joint research agreement" means a written contract, grant, or cooperative agreement entered into by two or more persons or entities for the performance of experimental, developmental, or research work in the field of the claimed invention.~~

~~Sec. 103. Conditions for Patentability; Non-Obvious Subject Matter~~

A patent for a claimed invention may not be obtained, notwithstanding that the claimed invention is not identically disclosed as set forth in section 102, if the differences between the claimed invention and the prior art are such that the claimed invention as a whole would have been obvious before the effective filing date of the claimed invention to a person having ordinary skill in the art to which the claimed invention pertains. Patentability shall not be negated by the manner in which the invention was made.'.

Sec. 104. Invention Made Abroad

~~(a) In general.—~~

~~(1) Proceedings.—In proceedings in the Patent and Trademark Office, in the courts, and before any other competent authority, an applicant for a patent, or a patentee, may not establish a date of invention by reference to knowledge or use thereof, or other activity with respect thereto, in a foreign country other than a NAFTA country or a WTO member country, except as provided in sections 119 and 365 of this title.~~

~~(2) Rights.—If an invention was made by a person, civil or military—~~

~~(A) while domiciled in the United States, and serving in any other country in connection with operations by or on behalf of the United States,~~

~~(B) while domiciled in a NAFTA country and serving in another country in connection with operations by or on behalf of that NAFTA country, or~~

~~(C) while domiciled in a WTO member country and serving in another country in connection with operations by or on behalf of that WTO member country,~~

~~that person shall be entitled to the same rights of priority in the United States with respect to such invention as if such invention had been made in the United States, that NAFTA country, or that WTO member country, as the case may be.~~

~~(3) Use of information.–To the extent that any information in a NAFTA country or a WTO member country concerning knowledge, use, or other activity relevant to proving or disproving a date of invention has not been made available for use in a proceeding in the Patent and Trademark Office, a court, or any other competent authority to the same extent as such information could be made available in the United States, the Director, court, or such other authority shall draw appropriate inferences, or take other action permitted by statute, rule,~~ or regula~~tion, in favor of the party that requested the information in the proceeding.~~

~~(b) Definitions.–As used in this section–~~

~~(1) the term "NAFTA country" has the meaning given that term in section 2(4) of the North American Free Trade Agreement Implementation Act; and~~

~~(2) the term "WTO member country" has the meaning given that term in section 2(10) of the Uruguay Round Agreements Act.~~

~~Sec. 118. Filing by other Than Inventor~~

~~Whenever an inventor refuses to execute an application for patent, or cannot be found or reached after diligent effort, a person to whom the inventor has assigned or agreed in writing to assign the invention or who otherwise shows sufficient proprietary interest in the matter justifying such action, may make application for patent on behalf of and as agent for the inventor on proof of the pertinent facts and a showing that such action is necessary to preserve the rights of the parties or to prevent irreparable damage; and the Director may grant a patent to such inventor upon such notice to him as the Director deems sufficient, and on compliance with such regulations as he prescribes.~~

A person to whom the inventor has assigned or is under an obligation to assign the invention may make an application for patent. A person who otherwise shows sufficient proprietary interest in the matter may make an application for patent on behalf of and as agent for the inventor on proof of the pertinent facts and a showing that such action is appropriate to preserve the rights of the parties. If the Director grants a patent on an application filed under this section by a person other than the inventor, the patent shall be granted to the real party in interest and upon such notice to the inventor as the Director considers to be sufficient.

~~Sec. 135. Interferences~~

~~(a) Whenever an application is made for a patent which, in the opinion of the Director, would interfere with any pending application, or with any unexpired~~

~~patent, an interference may be declared and the Director shall give notice of such declaration to the applicants, or applicant and patentee, as the case may be. The Board of Patent Appeals and Interferences shall determine questions of priority of the inventions and may determine questions of patentability. Any final decision, if adverse to the claim of an applicant, shall constitute the final refusal by the Patent and Trademark Office of the claims involved, and the Director may issue a patent to the applicant who is adjudged the prior inventor. A final judgment adverse to a patentee from which no appeal or other review has been or can be taken or had shall constitute cancellation of the claims involved in the patent, and notice of such cancellation shall be endorsed on copies of the patent distributed after such cancellation by the Patent and Trademark Office.~~

~~(b)(1) A claim which is the same as, or for the same or substantially the same subject matter as, a claim of an issued patent may not be made in any application unless such a claim is made prior to one year from the date on which the patent was granted.~~

~~(2) A claim which is the same as, or for the same or substantially the same subject matter as, a claim of an application published under section 122(b) of this title may be made in an application filed after the application is published only if the claim is made before 1 year after the date on which the application is published.~~

~~(c) Any agreement or understanding between parties to an interference, including any collateral agreements referred to therein, made in connection with or in contemplation of the termination of the interference, shall be in writing and a true copy thereof filed in the Patent and Trademark Office before the termination of the interference as between the said parties to the agreement or understanding. If any party filing the same so requests, the copy shall be kept separate from the file of the interference, and made available only to Government agencies on written request, or to any person on a showing of good cause. Failure to file the copy of such agreement or understanding shall render permanently unenforceable such agreement or understanding and any patent of such parties involved in the interference or any patent subsequently issued on any application of such parties so involved. The Director may, however, on a showing of good cause for failure to file within the time prescribed, permit the filing of the agreement or understanding during the six-month period subsequent to the termination of the interference as between the parties to the agreement or understanding.~~

~~The Director shall give notice to the parties or their attorneys of record, a reasonable time prior to said termination, of the filing requirement of this section. If the Director gives such notice at a later time, irrespective of the right to file such agreement or understanding within the six-month period on a showing of good cause, the parties may file such agreement or understanding within sixty days of the receipt of such notice.~~

~~Any discretionary action of the Director under this subsection shall be reviewable under section 10 of the Administrative Procedure Act.~~

~~(d) Parties to a patent interference, within such time as may be specified by the Director by regulation, may determine such contest or any aspect thereof by arbitration. Such arbitration shall be governed by the provisions of title 9 to the extent~~

~~such title is not inconsistent with this section. The parties shall give notice of any arbitration award to the Director, and such award shall, as between the parties to the arbitration, be dispositive of the issues to which it relates. The arbitration award shall be unenforceable until such notice is given. Nothing in this subsection shall preclude the Director from determining patentability of the invention involved in the interference.~~

Sec. 135. Derivation Proceedings

(a) Institution of Proceeding. An applicant for patent may file a petition to institute a derivation proceeding in the Office. The petition shall set forth with particularity the basis for finding that an inventor named in an earlier application derived the claimed invention from an inventor named in the petitioner's application and, without authorization, the earlier application claiming such invention was filed. Any such petition may be filed only within the 1-year period beginning on the date of the first publication of a claim to an invention that is the same or substantially the same as the earlier application's claim to the invention, shall be made under oath, and shall be supported by substantial evidence. Whenever the Director determines that a petition filed under this subsection demonstrates that the standards for instituting a derivation proceeding are met, the Director may institute a derivation proceeding. The determination by the Director whether to institute a derivation proceeding shall be final and nonappealable.

(b) Determination by Patent Trial and Appeal Board. In a derivation proceeding instituted under subsection (a), the Patent Trial and Appeal Board shall determine whether an inventor named in the earlier application derived the claimed invention from an inventor named in the petitioner's application and, without authorization, the earlier application claiming such invention was filed. In appropriate circumstances, the Patent Trial and Appeal Board may correct the naming of the inventor in any application or patent at issue. The Director shall prescribe regulations setting forth standards for the conduct of derivation proceedings, including requiring parties to provide sufficient evidence to prove and rebut a claim of derivation.

(c) Deferral of Decision. The Patent Trial and Appeal Board may defer action on a petition for a derivation proceeding until the expiration of the 3-month period beginning on the date on which the Director issues a patent that includes the claimed invention that is the subject of the petition. The Patent Trial and Appeal Board also may defer action on a petition for a derivation proceeding, or stay the proceeding after it has been instituted, until the termination of a proceeding under chapter 30, 31, or 32 involving the patent of the earlier applicant.

(d) Effect of Final Decision. The final decision of the Patent Trial and Appeal Board, if adverse to claims in an application for patent, shall constitute the final refusal by the Office on those claims. The final decision of the Patent Trial and Appeal Board, if adverse to claims in a patent, shall, if no appeal or other review of the decision has been or can be taken or had, constitute cancellation of those claims, and notice of such cancellation shall be endorsed on copies of the patent distributed after such cancellation.

(e) Settlement. Parties to a proceeding instituted under subsection (a) may terminate the proceeding by filing a written statement reflecting the agreement of the parties as to the correct inventors of the claimed invention in dispute. Unless the Patent Trial and Appeal Board finds the agreement to be inconsistent with the evidence of record, if any, it shall take action consistent with the agreement. Any written settlement or understanding of the parties shall be filed with the Director. At the request of a party to the proceeding, the agreement or understanding shall be treated as business confidential information, shall be kept separate from the file of the involved patents or applications, and shall be made available only to Government agencies on written request, or to any person on a showing of good cause.

(f) Arbitration. Parties to a proceeding instituted under subsection (a) may, within such time as may be specified by the Director by regulation, determine such contest or any aspect thereof by arbitration. Such arbitration shall be governed by the provisions of title 9, to the extent such title is not inconsistent with this section. The parties shall give notice of any arbitration award to the Director, and such award shall, as between the parties to the arbitration, be dispositive of the issues to which it relates. The arbitration award shall be unenforceable until such notice is given. Nothing in this subsection shall preclude the Director from determining the patentability of the claimed inventions involved in the proceeding.'

Sec. 301. Citation of Prior Art

~~Any person at any time may cite to the Office in writing prior art consisting of patents or printed publications which that person believes to have a bearing on the patentability of any claim of a particular patent. If the person explains in writing the pertinency and manner of applying such prior art to at least one claim of the patent, the citation of such prior art and the explanation thereof will become a part of the official file of the patent. At the written request of the person citing the prior art, his or her identity will be excluded from the patent file and kept confidential.~~

~~Sec. 301. Citation of Prior Art and Written Statements~~

(a) In General. Any person at any time may cite to the Office in writing—

(1) prior art consisting of patents or printed publications which that person believes to have a bearing on the patentability of any claim of a particular patent; or

(2) statements of the patent owner filed in a proceeding before a Federal court or the Office in which the patent owner took a position on the scope of any claim of a particular patent.

(b) Official File. If the person citing prior art or written statements pursuant to subsection (a) explains in writing the pertinence and manner of applying the prior art or written statements to at least 1 claim of the patent, the citation of the prior art or written statements and the explanation thereof shall become a part of the official file of the patent.

(c) Additional Information. A party that submits a written statement pursuant to subsection (a)(2) shall include any other documents, pleadings, or evidence from the proceeding in which the statement was filed that addresses the written statement.

(d) Limitations. A written statement submitted pursuant to subsection (a)(2), and additional information submitted pursuant to subsection (c), shall not be considered by the Office for any purpose other than to determine the proper meaning of a patent claim in a proceeding that is ordered or instituted pursuant to section 304, 314, or 324. If any such written statement or additional information is subject to an applicable protective order, such statement or information shall be redacted to exclude information that is subject to that order.

(e) Confidentiality. Upon the written request of the person citing prior art or written statements pursuant to subsection (a), that person's identity shall be excluded from the patent file and kept confidential.

Sec. 311. Request for Inter Partes Reexamination

~~(a) In general.–Any third-party requester at any time may file a request for inter partes reexamination by the Office of a patent on the basis of any prior art cited under the provisions of section 301.~~

~~(b) Requirements.–The request shall–~~

~~(1) be in writing, include the identity of the real party in interest, and be accompanied by payment of an inter partes reexamination fee established by the Director under section 41; and~~

~~(2) set forth the pertinency and manner of applying cited prior art to every claim for which reexamination is requested.~~

~~(c) Copy.–The Director promptly shall send a copy of the request to the owner of record of the patent.~~

~~'Sec. 311. Inter Partes Review~~

(a) In General. Subject to the provisions of this chapter, a person who is not the owner of a patent may file with the Office a petition to institute an inter partes review of the patent. The Director shall establish, by regulation, fees to be paid by the person requesting the review, in such amounts as the Director determines to be reasonable, considering the aggregate costs of the review.

(b) Scope. A petitioner in an inter partes review may request to cancel as unpatentable 1 or more claims of a patent only on a ground that could be raised under section 102 or 103 and only on the basis of prior art consisting of patents or printed publications.

(c) Filing Deadline. A petition for inter partes review shall be filed after the later of either—

(1) the date that is 9 months after the grant of a patent or issuance of a reissue of a patent; or

(2) if a post-grant review is instituted under chapter 32, the date of the termination of such post-grant review.

~~Sec. 312. Determination of Issue by Director~~

~~(a) Reexamination. Not later than 3 months after the filing of a request for inter partes reexamination under section 311, the Director shall determine whether a substantial new question of patentability affecting any claim of the patent concerned is raised by the request, with or without consideration of other patents or printed publications. The existence of a substantial new question of patentability is not precluded by the fact that a patent or printed publication was previously cited by or to the Office or considered by the Office.~~

~~(b) Record. A record of the Director's determination under subsection (a) shall be placed in the official file of the patent, and a copy shall be promptly given or mailed to the owner of record of the patent and to the third-party requester.~~

~~(c) Final Decision. A determination by the Director under subsection (a) shall be final and non-appealable. Upon a determination that no substantial new question of patentability has been raised, the Director may refund a portion of the inter partes reexamination fee required under section 311.~~

Sec. 312. Petitions

(a) Requirements of Petition. A petition filed under section 311 may be considered only if—

(1) the petition is accompanied by payment of the fee established by the Director under section 311;

(2) the petition identifies all real parties in interest;

(3) the petition identifies, in writing and with particularity, each claim challenged, the grounds on which the challenge to each claim is based, and the evidence that supports the grounds for the challenge to each claim, including—

(A) copies of patents and printed publications that the petitioner relies upon in support of the petition; and

(B) affidavits or declarations of supporting evidence and opinions, if the petitioner relies on expert opinions;

(4) the petition provides such other information as the Director may require by regulation; and

(5) the petitioner provides copies of any of the documents required under paragraphs (2), (3), and (4) to the patent owner or, if applicable, the designated representative of the patent owner.

(b) Public Availability. As soon as practicable after the receipt of a petition under section 311, the Director shall make the petition available to the public.

Sec. 312. Determination of issue by Director [Transitional Provision]

(a) Reexamination. Not later than 3 months after the filing of a request for inter partes reexamination under section 311, the Director shall determine whether ~~a substantial new question of patentability affecting any claim of the patent concerned is raised by the request,~~ <u>the information presented in the request shows that there is a reasonable likelihood that the requester would prevail with respect to at least 1 of the claims challenged in the request</u>, with or without consideration of other patents or printed publications. ~~The existence of a substantial new question of patentability~~ <u>A showing that there is a reasonable likelihood that the requester would prevail with respect to at least 1 of the claims challenged in the request</u> is not precluded by the fact that a patent or printed publication was previously cited by or to the Office or considered by the Office.

(b) Record. A record of the Director's determination under subsection (a) shall be placed in the official file of the patent, and a copy shall be promptly given or mailed to the owner of record of the patent and to the third-party requester.

(c) Final Decision. A determination by the Director under subsection (a) shall be final and non-appealable. Upon a determination that ~~no substantial new question of patentability has been raised,~~ <u>the showing required by subsection (a) has not been made,</u> the Director may refund a portion of the inter partes reexamination fee required under section 311.

Sec. 313. Inter Partes Reexamination Order by Director

~~If, in a determination made under section 312(a), the Director finds that a substantial new question of patentability affecting a claim of a patent is raised, the determination shall include an order for inter partes reexamination of the patent for resolution of the question. The order may be accompanied by the initial action of the Patent and Trademark Office on the merits of the inter partes reexamination conducted in accordance with section 314.~~

~~Sec. 313. Preliminary Response to Petition~~

<u>If an inter partes review petition is filed under section 311, the patent owner shall have the right to file a preliminary response to the petition, within a time period set by the Director, that sets forth reasons why no inter partes review should be instituted based upon the failure of the petition to meet any requirement of this chapter.</u>

<u>Sec. 313. Inter Partes Reexamination Order by Director [Transitional Provision]</u>

If, in a determination made under section 312(a), the Director finds that <u>it has been shown that there is a reasonable likelihood that the requester would prevail</u>

with respect to at least 1 of the claims challenged in the request ~~a substantial new question of patentability affecting a claim of a patent is raised~~, the determination shall include an order for inter partes reexamination of the patent for resolution of the question. The order may be accompanied by the initial action of the Patent and Trademark Office on the merits of the inter partes reexamination conducted in accordance with section 314.

~~Sec. 314. Conduct of Inter Partes Reexamination Proceedings~~

~~(a) In general.–Except as otherwise provided in this section, reexamination shall be conducted according to the procedures established for initial examination under the provisions of sections 132 and 133. In any inter partes reexamination proceeding under this chapter, the patent owner shall be permitted to propose any amendment to the patent and a new claim or claims, except that no proposed amended or new claim enlarging the scope of the claims of the patent shall be permitted.~~

~~(b) Response.–(1) With the exception of the inter partes reexamination request, any document filed by either the patent owner or the third-party requester shall be served on the other party. In addition, the Office shall send to the third-party requester a copy of any communication sent by the Office to the patent owner concerning the patent subject to the inter partes reexamination proceeding.~~

~~(2) Each time that the patent owner files a response to an action on the merits from the Patent and Trademark Office, the third-party requester shall have one opportunity to file written comments addressing issues raised by the action of the Office or the patent owner's response thereto, if those written comments are received by the Office within 30 days after the date of service of the patent owner's response.~~

~~[(3) Redesignated (2)]~~

~~(c) Special dispatch.–Unless otherwise provided by the Director for good cause, all inter partes reexamination proceedings under this section, including any appeal to the Board of Patent Appeals and Interferences, shall be conducted with special dispatch within the Office.~~

Sec. 314. Institution of Inter Partes Review

(a) Threshold. The Director may not authorize an inter partes review to be instituted unless the Director determines that the information presented in the petition filed under section 311 and any response filed under section 313 shows that there is a reasonable likelihood that the petitioner would prevail with respect to at least 1 of the claims challenged in the petition.

(b) Timing. The Director shall determine whether to institute an inter partes review under this chapter pursuant to a petition filed under section 311 within 3 months after—

(1) receiving a preliminary response to the petition under section 313; or

(2) if no such preliminary response is filed, the last date on which such response may be filed.

(c) Notice. The Director shall notify the petitioner and patent owner, in writing, of the Director's determination under subsection (a), and shall make such notice available to the public as soon as is practicable. Such notice shall include the date on which the review shall commence.

(d) No Appeal. The determination by the Director whether to institute an inter partes review under this section shall be final and nonappealable.

Sec. 315. Appeal

~~(a) Patent owner.–The patent owner involved in an inter partes reexamination proceeding under this chapter–~~

~~(1) may appeal under the provisions of section 134 and may appeal under the provisions of sections 141 through 144, with respect to any decision adverse to the patentability of any original or proposed amended or new claim of the patent; and~~

~~(2) may be a party to any appeal taken by a third-party requester under subsection (b).~~

~~(b) Third-party requester.–A third-party requester–~~

~~(1) may appeal under the provisions of section 134, and may appeal under the provisions of sections 141 through 144, with respect to any final decision favorable to the patentability of any original or proposed amended or new claim of the patent; and~~

~~(2) may, subject to subsection (c), be a party to any appeal taken by the patent owner under the provisions of section 134 or sections 141 through 144.~~

~~(c) Civil action.–A third-party requester whose request for an inter partes reexamination results in an order under section 313 is estopped from asserting at a later time, in any civil action arising in whole or in part under section 1338 of title 28, the invalidity of any claim finally determined to be valid and patentable on any ground which the third-party requester raised or could have raised during the inter partes reexamination proceedings. This subsection does not prevent the assertion of invalidity based on newly discovered prior art unavailable to the third-party requester and the Patent and Trademark Office at the time of the inter partes reexamination proceedings.~~

~~Sec. 315. Relation to Other Proceedings or Actions~~

(a) Infringer's Civil Action.—

(1) Inter partes review barred by civil action. An inter partes review may not be instituted if, before the date on which the petition for such a review is filed, the petitioner or real party in interest filed a civil action challenging the validity of a claim of the patent.

(2) Stay of civil action. If the petitioner or real party in interest files a civil action challenging the validity of a claim of the patent on or after the date on which the petitioner files a petition for inter partes review of the patent, that civil action shall be automatically stayed until either-

(A) the patent owner moves the court to lift the stay;

(B) the patent owner files a civil action or counterclaim alleging that the petitioner or real party in interest has infringed the patent; or

(C) the petitioner or real party in interest moves the court to dismiss the civil action.

(3) Treatment of counterclaim. A counterclaim challenging the validity of a claim of a patent does not constitute a civil action challenging the validity of a claim of a patent for purposes of this subsection.

(b) Patent Owner's Action. An inter partes review may not be instituted if the petition requesting the proceeding is filed more than 1 year after the date on which the petitioner, real party in interest, or privy of the petitioner is served with a complaint alleging infringement of the patent. The time limitation set forth in the preceding sentence shall not apply to a request for joinder under subsection (c).

(c) Joinder. If the Director institutes an inter partes review, the Director, in his or her discretion, may join as a party to that inter partes review any person who properly files a petition under section 311 that the Director, after receiving a preliminary response under section 313 or the expiration of the time for filing such a response, determines warrants the institution of an inter partes review under section 314.

(d) Multiple Proceedings. Notwithstanding sections 135(a), 251, and 252, and chapter 30, during the pendency of an inter partes review, if another proceeding or matter involving the patent is before the Office, the Director may determine the manner in which the inter partes review or other proceeding or matter may proceed, including providing for stay, transfer, consolidation, or termination of any such matter or proceeding.

(e) Estoppel.—

(1) Proceedings before the office. The petitioner in an inter partes review of a claim in a patent under this chapter that results in a final written decision under section 318(a), or the real party in interest or privy of the petitioner, may not request or maintain a proceeding before the Office with respect to that claim on any ground that the petitioner raised or reasonably could have raised during that inter partes review.

(2) Civil actions and other proceedings. The petitioner in an inter partes review of a claim in a patent under this chapter that results in a final written decision under section 318(a), or the real party in interest or privy of the petitioner, may not assert either in a civil action arising in whole or in part under section 1338 of title 28 or in a proceeding before the International Trade Commission under section 337 of the Tariff Act of 1930 that the claim is invalid on any ground that the petitioner raised or reasonably could have raised during that inter partes review.

Sec. 316. Certificate of Patentability, Unpatentability, and Claim Cancellation

~~(a) In general.–In an inter partes reexamination proceeding under this chapter, when the time for appeal has expired or any appeal proceeding has terminated, the Director shall issue and publish a certificate canceling any claim of the patent finally determined to be unpatentable, confirming any claim of the patent determined to be patentable, and incorporating in the patent any proposed amended or new claim determined to be patentable.~~

~~(b) Amended or new claim.–Any proposed amended or new claim determined to be patentable and incorporated into a patent following an inter partes reexamination proceeding shall have the same effect as that specified in section 252 of this title for reissued patents on the right of any person who made, purchased, or used within the United States, or imported into the United States, anything patented by such proposed amended or new claim, or who made substantial preparation therefor, prior to issuance of a certificate under the provisions of subsection (a) of this section.~~

~~Sec. 316. Conduct of Inter Partes Review~~

(a) Regulations. The Director shall prescribe regulations—

(1) providing that the file of any proceeding under this chapter shall be made available to the public, except that any petition or document filed with the intent that it be sealed shall, if accompanied by a motion to seal, be treated as sealed pending the outcome of the ruling on the motion;

(2) setting forth the standards for the showing of sufficient grounds to institute a review under section 314(a);

(3) establishing procedures for the submission of supplemental information after the petition is filed;

(4) establishing and governing inter partes review under this chapter and the relationship of such review to other proceedings under this title;

(5) setting forth standards and procedures for discovery of relevant evidence, including that such discovery shall be limited to—

(A) the deposition of witnesses submitting affidavits or declarations; and

(B) what is otherwise necessary in the interest of justice;

(6) prescribing sanctions for abuse of discovery, abuse of process, or any other improper use of the proceeding, such as to harass or to cause unnecessary delay or an unnecessary increase in the cost of the proceeding;

(7) providing for protective orders governing the exchange and submission of confidential information;

(8) providing for the filing by the patent owner of a response to the petition under section 313 after an inter partes review has been instituted, and requiring that the patent owner file with such response, through affidavits or declarations, any additional factual evidence and expert opinions on which the patent owner relies in support of the response;

(9) setting forth standards and procedures for allowing the patent owner to move to amend the patent under subsection (d) to cancel a challenged claim or propose a reasonable number of substitute claims, and ensuring that any information submitted by the patent owner in support of any amendment entered under subsection (d) is made available to the public as part of the prosecution history of the patent;

(10) providing either party with the right to an oral hearing as part of the proceeding;

(11) requiring that the final determination in an inter partes review be issued not later than 1 year after the date on which the Director notices the institution of a review under this chapter, except that the Director may, for good cause shown, extend the 1-year period by not more than 6 months, and may adjust the time periods in this paragraph in the case of joinder under section 315(c);

(12) setting a time period for requesting joinder under section 315(c); and

(13) providing the petitioner with at least 1 opportunity to file written comments within a time period established by the Director.

(b) Considerations. In prescribing regulations under this section, the Director shall consider the effect of any such regulation on the economy, the integrity of the patent system, the efficient administration of the Office, and the ability of the Office to timely complete proceedings instituted under this chapter.

(c) Patent Trial and Appeal Board. The Patent Trial and Appeal Board shall, in accordance with section 6, conduct each inter partes review instituted under this chapter.

(d) Amendment of the Patent.—

(1) In general. During an inter partes review instituted under this chapter, the patent owner may file 1 motion to amend the patent in 1 or more of the following ways:

(A) Cancel any challenged patent claim.

(B) For each challenged claim, propose a reasonable number of substitute claims.

(2) Additional motions. Additional motions to amend may be permitted upon the joint request of the petitioner and the patent owner to materially advance the settlement of a proceeding under section 317, or as permitted by regulations prescribed by the Director.

(3) Scope of claims. An amendment under this subsection may not enlarge the scope of the claims of the patent or introduce new matter.

(e) Evidentiary Standards. In an inter partes review instituted under this chapter, the petitioner shall have the burden of proving a proposition of unpatentability by a preponderance of the evidence.

Sec. 317. Inter Partes Reexamination Prohibited

~~(a) Order for reexamination.–Notwithstanding any provision of this chapter, once an order for inter partes reexamination of a patent has been issued under section 313, neither the third-party requester nor its privies, may file a subsequent~~

~~request for inter partes reexamination of the patent until an inter partes reexamination certificate is issued and published under section 316, unless authorized by the Director.~~

~~(b) Final decision. Once a final decision has been entered against a party in a civil action arising in whole or in part under section 1338 of title 28, that the party has not sustained its burden of proving the invalidity of any patent claim in suit or if a final decision in an inter partes reexamination proceeding instituted by a third-party requester is favorable to the patentability of any original or proposed amended or new claim of the patent, then neither that party nor its privies may thereafter request an inter partes reexamination of any such patent claim on the basis of issues which that party or its privies raised or could have raised in such civil action or inter partes reexamination proceeding, and an inter partes reexamination requested by that party or its privies on the basis of such issues may not thereafter be maintained by the Office, notwithstanding any other provision of this chapter. This subsection does not prevent the assertion of invalidity based on newly discovered prior art unavailable to the third-party requester and the Patent and Trademark Office at the time of the inter partes reexamination proceedings.~~

~~Sec. 317.~~ Settlement

(a) In General. An inter partes review instituted under this chapter shall be terminated with respect to any petitioner upon the joint request of the petitioner and the patent owner, unless the Office has decided the merits of the proceeding before the request for termination is filed. If the inter partes review is terminated with respect to a petitioner under this section, no estoppel under section 315(e) shall attach to the petitioner, or to the real party in interest or privy of the petitioner, on the basis of that petitioner's institution of that inter partes review. If no petitioner remains in the inter partes review, the Office may terminate the review or proceed to a final written decision under section 318(a).

(b) Agreements in Writing. Any agreement or understanding between the patent owner and a petitioner, including any collateral agreements referred to in such agreement or understanding, made in connection with, or in contemplation of, the termination of an inter partes review under this section shall be in writing and a true copy of such agreement or understanding shall be filed in the Office before the termination of the inter partes review as between the parties. At the request of a party to the proceeding, the agreement or understanding shall be treated as business confidential information, shall be kept separate from the file of the involved patents, and shall be made available only to Federal Government agencies on written request, or to any person on a showing of good cause.

Sec. 318. Stay of Litigation

~~Once an order for inter partes reexamination of a patent has been issued under section 313, the patent owner may obtain a stay of any pending litigation which~~

~~involves an issue of patentability of any claims of the patent which are the subject of the inter partes reexamination order, unless the court before which such litigation is pending determines that a stay would not serve the interests of justice.~~

~~Sec. 318. Decision of the Board~~

(a) Final Written Decision. If an inter partes review is instituted and not dismissed under this chapter, the Patent Trial and Appeal Board shall issue a final written decision with respect to the patentability of any patent claim challenged by the petitioner and any new claim added under section 316(d).

(b) Certificate. If the Patent Trial and Appeal Board issues a final written decision under subsection (a) and the time for appeal has expired or any appeal has terminated, the Director shall issue and publish a certificate canceling any claim of the patent finally determined to be unpatentable, confirming any claim of the patent determined to be patentable, and incorporating in the patent by operation of the certificate any new or amended claim determined to be patentable.

(c) Intervening Rights. Any proposed amended or new claim determined to be patentable and incorporated into a patent following an inter partes review under this chapter shall have the same effect as that specified in section 252 for reissued patents on the right of any person who made, purchased, or used within the United States, or imported into the United States, anything patented by such proposed amended or new claim, or who made substantial preparation therefor, before the issuance of a certificate under subsection (b).

(d) Data on Length of Review. The Office shall make available to the public data describing the length of time between the institution of, and the issuance of a final written decision under subsection (a) for, each inter partes review.

Sec. 319. Appeal

A party dissatisfied with the final written decision of the Patent Trial and Appeal Board under section 318(a) may appeal the decision pursuant to sections 141 through 144. Any party to the inter partes review shall have the right to be a party to the appeal.

Chapter 32. Post-Grant Review

Sec. 321. Post-Grant Review

(a) In General. Subject to the provisions of this chapter, a person who is not the owner of a patent may file with the Office a petition to institute a post-grant review of the patent. The Director shall establish, by regulation, fees to be paid by the person requesting the review, in such amounts as the Director determines to be reasonable, considering the aggregate costs of the post-grant review.

(b) Scope. A petitioner in a post-grant review may request to cancel as unpatentable 1 or more claims of a patent on any ground that could be raised under paragraph (2) or (3) of section 282(b) (relating to invalidity of the patent or any claim).

(c) Filing Deadline. A petition for a post-grant review may only be filed not later than the date that is 9 months after the date of the grant of the patent or of the issuance of a reissue patent (as the case may be).

Sec. 322. Petitions

(a) Requirements of Petition. A petition filed under section 321 may be considered only if—

(1) the petition is accompanied by payment of the fee established by the Director under section 321;

(2) the petition identifies all real parties in interest;

(3) the petition identifies, in writing and with particularity, each claim challenged, the grounds on which the challenge to each claim is based, and the evidence that supports the grounds for the challenge to each claim, including-

(A) copies of patents and printed publications that the petitioner relies upon in support of the petition; and

(B) affidavits or declarations of supporting evidence and opinions, if the petitioner relies on other factual evidence or on expert opinions;

(4) the petition provides such other information as the Director may require by regulation; and

(5) the petitioner provides copies of any of the documents required under paragraphs (2), (3), and (4) to the patent owner or, if applicable, the designated representative of the patent owner.

(b) Public Availability. As soon as practicable after the receipt of a petition under section 321, the Director shall make the petition available to the public.

Sec. 323. Preliminary Response to Petition

If a post-grant review petition is filed under section 321, the patent owner shall have the right to file a preliminary response to the petition, within a time period set by the Director, that sets forth reasons why no post-grant review should be instituted based upon the failure of the petition to meet any requirement of this chapter.

Sec. 324. Institution of Post-Grant Review

(a) Threshold. The Director may not authorize a post-grant review to be instituted unless the Director determines that the information presented in the petition filed under section 321, if such information is not rebutted, would demonstrate that it is more likely than not that at least 1 of the claims challenged in the petition is unpatentable.

(b) Additional Grounds. The determination required under subsection (a) may also be satisfied by a showing that the petition raises a novel or unsettled legal question that is important to other patents or patent applications.

(c) Timing. The Director shall determine whether to institute a post-grant review under this chapter pursuant to a petition filed under section 321 within 3 months after—

(1) receiving a preliminary response to the petition under section 323; or

(2) if no such preliminary response is filed, the last date on which such response may be filed.

(d) Notice. The Director shall notify the petitioner and patent owner, in writing, of the Director's determination under subsection (a) or (b), and shall make such notice available to the public as soon as is practicable. Such notice shall include the date on which the review shall commence.

(e) No Appeal. The determination by the Director whether to institute a post-grant review under this section shall be final and nonappealable.

Sec. 325. Relation to Other Proceedings or Actions

(a) Infringer's Civil Action.—

(1) Post-grant review barred by civil action. A post-grant review may not be instituted under this chapter if, before the date on which the petition for such a review is filed, the petitioner or real party in interest filed a civil action challenging the validity of a claim of the patent.

(2) Stay of civil action. If the petitioner or real party in interest files a civil action challenging the validity of a claim of the patent on or after the date on which the petitioner files a petition for post-grant review of the patent, that civil action shall be automatically stayed until either—

(A) the patent owner moves the court to lift the stay;

(B) the patent owner files a civil action or counterclaim alleging that the petitioner or real party in interest has infringed the patent; or

(C) the petitioner or real party in interest moves the court to dismiss the civil action.

(3) Treatment of counterclaim. A counterclaim challenging the validity of a claim of a patent does not constitute a civil action challenging the validity of a claim of a patent for purposes of this subsection.

(b) Preliminary Injunctions. If a civil action alleging infringement of a patent is filed within 3 months after the date on which the patent is granted, the court may not stay its consideration of the patent owner's motion for a preliminary injunction against infringement of the patent on the basis that a petition for post-grant review has been filed under this chapter or that such a post-grant review has been instituted under this chapter.

(c) Joinder. If more than 1 petition for a post-grant review under this chapter is properly filed against the same patent and the Director determines that more than 1 of these petitions warrants the institution of a post-grant review under section 324, the Director may consolidate such reviews into a single post-grant review.

(d) Multiple Proceedings. Notwithstanding sections 135(a), 251, and 252, and chapter 30, during the pendency of any post-grant review under this chapter, if another proceeding or matter involving the patent is before the Office, the Director may determine the manner in which the post-grant review or other proceeding or matter may proceed, including providing for the stay, transfer, consolidation, or termination of any such matter or proceeding. In determining whether to institute or order a proceeding under this chapter, chapter 30, or chapter 31, the Director may take into account whether, and reject the petition or request because, the same or substantially the same prior art or arguments previously were presented to the Office.

(e) Estoppel.—

(1) Proceedings before the office. The petitioner in a post-grant review of a claim in a patent under this chapter that results in a final written decision under section 328(a), or the real party in interest or privy of the petitioner, may not request or maintain a proceeding before the Office with respect to that claim on any ground that the petitioner raised or reasonably could have raised during that post-grant review.

(2) Civil actions and other proceedings. The petitioner in a post-grant review of a claim in a patent under this chapter that results in a final written decision under section 328(a), or the real party in interest or privy of the petitioner, may not assert either in a civil action arising in whole or in part under section 1338 of title 28 or in a proceeding before the International Trade Commission under section 337 of the Tariff Act of 1930 that the claim is invalid on any ground that the petitioner raised or reasonably could have raised during that post-grant review.

(f) Reissue Patents. A post-grant review may not be instituted under this chapter if the petition requests cancellation of a claim in a reissue patent that is identical to or narrower than a claim in the original patent from which the reissue patent was issued, and the time limitations in section 321(c) would bar filing a petition for a post-grant review for such original patent.

Sec. 326. Conduct of Post-Grant Review

(a) Regulations. The Director shall prescribe regulations-

(1) providing that the file of any proceeding under this chapter shall be made available to the public, except that any petition or document filed with the intent that it be sealed shall, if accompanied by a motion to seal, be treated as sealed pending the outcome of the ruling on the motion;

(2) setting forth the standards for the showing of sufficient grounds to institute a review under subsections (a) and (b) of section 324;

(3) establishing procedures for the submission of supplemental information after the petition is filed;

(4) establishing and governing a post-grant review under this chapter and the relationship of such review to other proceedings under this title;

(5) setting forth standards and procedures for discovery of relevant evidence, including that such discovery shall be limited to evidence directly related to factual assertions advanced by either party in the proceeding;

(6) prescribing sanctions for abuse of discovery, abuse of process, or any other improper use of the proceeding, such as to harass or to cause unnecessary delay or an unnecessary increase in the cost of the proceeding;

(7) providing for protective orders governing the exchange and submission of confidential information;

(8) providing for the filing by the patent owner of a response to the petition under section 323 after a post-grant review has been instituted, and requiring that the patent owner file with such response, through affidavits or declarations, any additional factual evidence and expert opinions on which the patent owner relies in support of the response;

(9) setting forth standards and procedures for allowing the patent owner to move to amend the patent under subsection (d) to cancel a challenged claim or propose a reasonable number of substitute claims, and ensuring that any information submitted by the patent owner in support of any amendment entered under subsection (d) is made available to the public as part of the prosecution history of the patent;

(10) providing either party with the right to an oral hearing as part of the proceeding;

(11) requiring that the final determination in any post-grant review be issued not later than 1 year after the date on which the Director notices the institution of a proceeding under this chapter, except that the Director may, for good cause shown, extend the 1-year period by not more than 6 months, and may adjust the time periods in this paragraph in the case of joinder under section 325(c); and

(12) providing the petitioner with at least 1 opportunity to file written comments within a time period established by the Director.

(b) Considerations. In prescribing regulations under this section, the Director shall consider the effect of any such regulation on the economy, the integrity of the patent system, the efficient administration of the Office, and the ability of the Office to timely complete proceedings instituted under this chapter.

(c) Patent Trial and Appeal Board. The Patent Trial and Appeal Board shall, in accordance with section 6, conduct each post-grant review instituted under this chapter.

(d) Amendment of the Patent.-

(1) In general. During a post-grant review instituted under this chapter, the patent owner may file 1 motion to amend the patent in 1 or more of the following ways:

(A) Cancel any challenged patent claim.

(B) For each challenged claim, propose a reasonable number of substitute claims.

(2) Additional motions. Additional motions to amend may be permitted upon the joint request of the petitioner and the patent owner to materially advance the settlement of a proceeding under section 327, or upon the request of the patent owner for good cause shown.

(3) Scope of claims. An amendment under this subsection may not enlarge the scope of the claims of the patent or introduce new matter.

(e) Evidentiary Standards. In a post-grant review instituted under this chapter, the petitioner shall have the burden of proving a proposition of unpatentability by a preponderance of the evidence.

Sec. 327. Settlement

(a) In General. A post-grant review instituted under this chapter shall be terminated with respect to any petitioner upon the joint request of the petitioner and the patent owner, unless the Office has decided the merits of the proceeding before the request for termination is filed. If the post-grant review is terminated with respect to a petitioner under this section, no estoppel under section 325(e) shall attach to the petitioner, or to the real party in interest or privy of the petitioner, on the basis of that petitioner's institution of that post-grant review. If no petitioner remains in the post-grant review, the Office may terminate the post-grant review or proceed to a final written decision under section 328(a).

(b) Agreements in Writing. Any agreement or understanding between the patent owner and a petitioner, including any collateral agreements referred to in such agreement or understanding, made in connection with, or in contemplation of, the termination of a post-grant review under this section shall be in writing, and a true copy of such agreement or understanding shall be filed in the Office before the termination of the post-grant review as between the parties. At the request of a party to the proceeding, the agreement or understanding shall be treated as business confidential information, shall be kept separate from the file of the involved patents, and shall be made available only to Federal Government agencies on written request, or to any person on a showing of good cause.

Sec. 328. Decision of the Board

(a) Final Written Decision. If a post-grant review is instituted and not dismissed under this chapter, the Patent Trial and Appeal Board shall issue a final written decision with respect to the patentability of any patent claim challenged by the petitioner and any new claim added under section 326(d).

(b) Certificate. If the Patent Trial and Appeal Board issues a final written decision under subsection (a) and the time for appeal has expired or any appeal has terminated, the Director shall issue and publish a certificate canceling any claim of the patent finally determined to be unpatentable, confirming any claim of the patent determined to be patentable, and incorporating in the patent by operation of the certificate any new or amended claim determined to be patentable.

(c) Intervening Rights. Any proposed amended or new claim determined to be patentable and incorporated into a patent following a post-grant review under this chapter shall have the same effect as that specified in section 252 of this title for reissued patents on the right of any person who made, purchased, or used within the United States, or imported into the United States, anything patented by such proposed amended or new claim, or who made substantial preparation therefor, before the issuance of a certificate under subsection (b).

(d) Data on Length of Review. The Office shall make available to the public data describing the length of time between the institution of, and the issuance of a final written decision under subsection (a) for, each post-grant review.

Sec. 329. Appeal

A party dissatisfied with the final written decision of the Patent Trial and Appeal Board under section 328(a) may appeal the decision pursuant to sections 141 through 144. Any party to the post-grant review shall have the right to be a party to the appeal.

Copyright Act

Title 17 of the United States Code

Chapter 1 Subject Matter and Scope of Copyright

§ 101 Definitions

Except as otherwise provided in this title, as used in this title, the following terms and their variant forms mean the following:

An "anonymous work" is a work on the copies or phonorecords of which no natural person is identified as author.

An "architectural work" is the design of a building as embodied in any tangible medium of expression, including a building, architectural plans, or drawings. The work includes the overall form as well as the arrangement and composition of spaces and elements in the design, but does not include individual standard features.

"Audiovisual works" are works that consist of a series of related images which are intrinsically intended to be shown by the use of machines or devices such as projectors, viewers, or electronic equipment, together with accompanying sounds, if any, regardless of the nature of the material objects, such as films or tapes, in which the works are embodied.

The "Berne Convention" is the Convention for the Protection of Literary and Artistic Works, signed at Berne, Switzerland, on September 9, 1886, and all acts, protocols, and revisions thereto.

The "best edition" of a work is the edition, published in the United States at any time before the date of deposit, that the Library of Congress determines to be most suitable for its purposes.

A person's "children" are that person's immediate offspring, whether legitimate or not, and any children legally adopted by that person.

A "collective work" is a work, such as a periodical issue, anthology, or encyclopedia, in which a number of contributions, constituting separate and independent works in themselves, are assembled into a collective whole.

A "compilation" is a work formed by the collection and assembling of preexisting materials or of data that are selected, coordinated, or arranged in such a way that the resulting work as a whole constitutes an original work of authorship. The term "compilation" includes collective works.

A "computer program" is a set of statements or instructions to be used directly or indirectly in a computer in order to bring about a certain result.

"Copies" are material objects, other than phonorecords, in which a work is fixed by any method now known or later developed, and from which the work can be perceived, reproduced, or otherwise communicated, either directly or with the aid of a machine or device. The term "copies" includes the material object, other than a phonorecord, in which the work is first fixed.

"Copyright owner", with respect to any one of the exclusive rights comprised in a copyright, refers to the owner of that particular right.

A "Copyright Royalty Judge" is a Copyright Royalty Judge appointed under section 802 of this title, and includes any individual serving as an interim Copyright Royalty Judge under such section.

A work is "created" when it is fixed in a copy or phonorecord for the first time; where a work is prepared over a period of time, the portion of it that has been fixed at any particular time constitutes the work as of that time, and where the work has been prepared in different versions, each version constitutes a separate work.

A "derivative work" is a work based upon one or more preexisting works, such as a translation, musical arrangement, dramatization, fictionalization, motion picture version, sound recording, art reproduction, abridgment, condensation, or any other form in which a work may be recast, transformed, or adapted. A work consisting of editorial revisions, annotations, elaborations, or other modifications, which, as a whole, represent an original work of authorship, is a "derivative work".

A "device", "machine", or "process" is one now known or later developed. A "digital transmission" is a transmission in whole or in part in a digital or other nonanalog format.

To "display" a work means to show a copy of it, either directly or by means of a film, slide, television image, or any other device or process or, in the case of a motion picture or other audiovisual work, to show individual images nonsequentially.

An "establishment" is a store, shop, or any similar place of business open to the general public for the primary purpose of selling goods or services in which the majority of the gross square feet of space that is nonresidential is used for that purpose, and in which nondramatic musical works are performed publicly.

The term "financial gain" includes receipt, or expectation of receipt, of anything of value, including the receipt of other copyrighted works.

A work is "fixed" in a tangible medium of expression when its embodiment in a copy or phonorecord, by or under the authority of the author, is sufficiently permanent or stable to permit it to be perceived, reproduced, or otherwise communicated for a period of more than transitory duration. A work consisting of sounds, images, or both, that are being transmitted, is "fixed" for purposes of this title if a fixation of the work is being made simultaneously with its transmission.

A "food service or drinking establishment" is a restaurant, inn, bar, tavern, or any other similar place of business in which the public or patrons assemble for the primary purpose of being served food or drink, in which the majority of the gross square feet of space that is nonresidential is used for that purpose, and in which nondramatic musical works are performed publicly.

The "Geneva Phonograms Convention" is the Convention for the Protection of Producers of Phonograms Against Unauthorized Duplication of Their Phonograms, concluded at Geneva, Switzerland, on October 29, 1971.

The "gross square feet of space" of an establishment means the entire interior space of that establishment, and any adjoining outdoor space used to serve patrons, whether on a seasonal basis or otherwise.

The terms "including" and "such as" are illustrative and not limitative. An "international agreement" is—

(1) the Universal Copyright Convention;

(2) the Geneva Phonograms Convention; (3) the Berne Convention;

(4) the WTO Agreement;

(5) the WIPO Copyright Treaty;

(6) the WIPO Performances and Phonograms Treaty; and

(7) any other copyright treaty to which the United States is a party.

A "joint work" is a work prepared by two or more authors with the intention that their contributions be merged into inseparable or interdependent parts of a unitary whole.

"Literary works" are works, other than audiovisual works, expressed in words, numbers, or other verbal or numerical symbols or indicia, regardless of the nature of the material objects, such as books, periodicals, manuscripts, phonorecords, film, tapes, disks, or cards, in which they are embodied.

The term "motion picture exhibition facility'' means a movie theater, screening room, or other venue that is being used primarily for the exhibition of a copyrighted motion picture, if such exhibition is open to the public or is made to an assembled group of viewers outside of a normal circle of a family and its social acquaintances.

"Motion pictures" are audiovisual works consisting of a series of related images which, when shown in succession, impart an impression of motion, together with accompanying sounds, if any.

To "perform" a work means to recite, render, play, dance, or act it, either directly or by means of any device or process or, in the case of a motion picture or other audiovisual work, to show its images in any sequence or to make the sounds accompanying it audible.

A "performing rights society" is an association, corporation, or other entity that licenses the public performance of nondramatic musical works on behalf of copyright owners of such works, such as the American Society of Composers, Authors and Publishers (ASCAP), Broadcast Music, Inc. (BMI), and SESAC, Inc.

"Phonorecords" are material objects in which sounds, other than those accompanying a motion picture or other audiovisual work, are fixed by any method now known or later developed, and from which the sounds can be perceived, reproduced, or otherwise communicated, either directly or with the aid of a machine or device. The term "phonorecords" includes the material object in which the sounds are first fixed.

"Pictorial, graphic, and sculptural works" include two-dimensional and threedimensional works of fine, graphic, and applied art, photographs, prints and art reproductions, maps, globes, charts, diagrams, models, and technical drawings, including architectural plans. Such works shall include works of artistic craftsmanship insofar as their form but not their mechanical or utilitarian aspects are concerned; the design of a useful article, as defined in this section, shall be considered a pictorial, graphic, or sculptural work only if, and only to the extent that, such design incorporates pictorial, graphic, or sculptural features that can be identified separately from, and are capable of existing independently of, the utilitarian aspects of the article.

For purposes of section 513, a "proprietor" is an individual, corporation, partnership, or other entity, as the case may be, that owns an establishment or a food service or drinking establishment, except that no owner or operator of a radio or television station licensed by the Federal Communications Commission, cable system or satellite carrier, cable or satellite carrier service or programmer, provider of online services or network access or the operator of facilities therefor, telecommunications company, or any other such audio or audiovisual service or programmer now known or as may be developed in the future, commercial subscription music service, or owner or operator of any other transmission service, shall under any circumstances be deemed to be a proprietor.

A "pseudonymous work" is a work on the copies or phonorecords of which the author is identified under a fictitious name.

"Publication" is the distribution of copies or phonorecords of a work to the public by sale or other transfer of ownership, or by rental, lease, or lending. The offering to distribute copies or phonorecords to a group of persons for purposes of further distribution, public performance, or public display, constitutes publication. A public performance or display of a work does not of itself constitute publication.

To perform or display a work "publicly" means—

(1) to perform or display it at a place open to the public or at any place where a substantial number of persons outside of a normal circle of a family and its social acquaintances is gathered; or

(2) to transmit or otherwise communicate a performance or display of the work to a place specified by clause (1) or to the public, by means of any device or process, whether the members of the public capable of receiving the performance or display receive it in the same place or in separate places and at the same time or at different times.

"Registration", for purposes of sections 205(c)(2), 405, 406, 410(d), 411, 412, and 506(e), means a registration of a claim in the original or the renewed and extended term of copyright.

"Sound recordings" are works that result from the fixation of a series of musical, spoken, or other sounds, but not including the sounds accompanying a motion picture or other audiovisual work, regardless of the nature of the material objects, such as disks, tapes, or other phonorecords, in which they are embodied.

"State" includes the District of Columbia and the Commonwealth of Puerto Rico, and any territories to which this title is made applicable by an Act of Congress.

A "transfer of copyright ownership" is an assignment, mortgage, exclusive license, or any other conveyance, alienation, or hypothecation of a copyright or of any of the exclusive rights comprised in a copyright, whether or not it is limited in time or place of effect, but not including a nonexclusive license.

A "transmission program" is a body of material that, as an aggregate, has been produced for the sole purpose of transmission to the public in sequence and as a unit.

To "transmit" a performance or display is to communicate it by any device or process whereby images or sounds are received beyond the place from which they are sent.

A "treaty party" is a country or intergovernmental organization other than the United States that is a party to an international agreement.

The "United States", when used in a geographical sense, comprises the several States, the District of Columbia and the Commonwealth of Puerto Rico, and the organized territories under the jurisdiction of the United States Government.

For purposes of section 411, a work is a "United States work" only if— (1) in the case of a published work, the work is first published—

(A) in the United States;

(B) simultaneously in the United States and another treaty party or parties, whose law grants a term of copyright protection that is the same as or longer than the term provided in the United States;

(C) simultaneously in the United States and a foreign nation that is not a treaty party; or

(D) in a foreign nation that is not a treaty party, and all of the authors of the work are nationals, domiciliaries, or habitual residents of, or in the case of an audiovisual work legal entities with headquarters in, the United States;

(2) in the case of an unpublished work, all the authors of the work are nationals, domiciliaries, or habitual residents of the United States, or, in the case of an unpublished audiovisual work, all the authors are legal entities with headquarters in the United States; or

(3) in the case of a pictorial, graphic, or sculptural work incorporated in a building or structure, the building or structure is located in the United States.

A "useful article" is an article having an intrinsic utilitarian function that is not merely to portray the appearance of the article or to convey information. An article that is normally a part of a useful article is considered a "useful article".

The author's "widow" or "widower" is the author's surviving spouse under the law of the author's domicile at the time of his or her death, whether or not the spouse has later remarried.

The "WIPO Copyright Treaty" is the WIPO Copyright Treaty concluded at Geneva, Switzerland, on December 20, 1996.

The "WIPO Performances and Phonograms Treaty" is the WIPO Performances and Phonograms Treaty concluded at Geneva, Switzerland, on December 20, 1996.

A "work of visual art" is—

(1) a painting, drawing, print or sculpture, existing in a single copy, in a limited edition of 200 copies or fewer that are signed and consecutively numbered by the author, or, in the case of a sculpture, in multiple cast, carved, or fabricated sculptures of 200 or fewer that are consecutively numbered by the author and bear the signature or other identifying mark of the author; or

(2) a still photographic image produced for exhibition purposes only, existing in a single copy that is signed by the author, or in a limited edition of 200 copies or fewer that are signed and consecutively numbered by the author.

A work of visual art does not include—

(A) (i) any poster, map, globe, chart, technical drawing, diagram, model, applied art, motion picture or other audiovisual work, book, magazine, newspaper, periodical, data base, electronic information service, electronic publication, or similar publication;

(ii) any merchandising item or advertising, promotional, descriptive, covering, or packaging material or container;

(iii) any portion or part of any item described in clause (i) or (ii);

(B) any work made for hire; or

(C) any work not subject to copyright protection under this title. A "work of the United States Government" is a work prepared by an officer or employee of the United States Government as part of that person's official duties.

A "work made for hire" is—

(1) a work prepared by an employee within the scope of his or her employment; or

(2) a work specially ordered or commissioned for use as a contribution to a collective work, as a part of a motion picture or other audiovisual work, as a translation, as a supplementary work, as a compilation, as an instructional text, as a test, as answer material for a test, or as an atlas, if the parties expressly agree in a written instrument signed by them that the work shall be considered a work made for hire. For the purpose of the foregoing sentence, a "supplementary work" is a work prepared for publication as a secondary adjunct to a work by another author for the purpose of introducing, concluding, illustrating, explaining, revising, commenting upon, or assisting in the use of the other work, such as forewords, afterwords, pictorial illustrations, maps, charts, tables, editorial notes, musical arrangements, answer material for tests, bibliographies, appendixes, and indexes, and an "instructional text" is a literary, pictorial, or graphic work prepared for publication and with the purpose of use in systematic instructional activities.

In determining whether any work is eligible to be considered a work made for hire under paragraph (2), neither the amendment contained in section 1011(d) of the Intellectual Property and Communications Omnibus Reform Act of 1999, as enacted by section 1000(a)(9) of Public Law 106-113, nor the deletion of the words added by that amendment—

(A) shall be considered or otherwise given any legal significance, or

(B) shall be interpreted to indicate congressional approval or disapproval of, or acquiescence in, any judicial determination, by the courts or the Copyright Office. Paragraph (2) shall be interpreted as if both section 2(a)(1) of the Work Made for Hire and Copyright Corrections Act of 2000 and section 1011(d) of the Intellectual Property and Communications Omnibus Reform Act of 1999, as enacted by section 1000(a)(9) of Public Law 106-113, were never enacted, and without regard to any inaction or awareness by the Congress at any time of any judicial determinations.

The terms "WTO Agreement" and "WTO member country" have the meanings given those terms in paragraphs (9) and (10), respectively, of section 2 of the Uruguay Round Agreements Act.

§ 102 Subject Matter of Copyright: In General

(a) Copyright protection subsists, in accordance with this title, in original works of authorship fixed in any tangible medium of expression, now known or later developed, from which they can be perceived, reproduced, or otherwise communicated, either directly or with the aid of a machine or device. Works of authorship include the following categories:

(1) literary works;
(2) musical works, including any accompanying words;
(3) dramatic works, including any accompanying music;
(4) pantomimes and choreographic works;
(5) pictorial, graphic, and sculptural works;
(6) motion pictures and other audiovisual works;
(7) sound recordings; and
(8) architectural works.

(b) In no case does copyright protection for an original work of authorship extend to any idea, procedure, process, system, method of operation, concept, principle, or discovery, regardless of the form in which it is described, explained, illustrated, or embodied in such work.

§ 103 Subject Matter of Copyright: Compilations and Derivative Works

(a) The subject matter of copyright as specified by section 102 includes compilations and derivative works, but protection for a work employing preexisting material in which copyright subsists does not extend to any part of the work in which such material has been used unlawfully.

(b) The copyright in a compilation or derivative work extends only to the material contributed by the author of such work, as distinguished from the preexisting material employed in the work, and does not imply any exclusive right in the preexisting material. The copyright in such work is independent of, and does not affect or enlarge the scope, duration, ownership, or subsistence of, any copyright protection in the preexisting material.

§ 104 Subject Matter of Copyright: National Origin

(a) Unpublished Works.—The works specified by sections 102 and 103, while unpublished, are subject to protection under this title without regard to the nationality or domicile of the author.

(b) Published Works.—The works specified by sections 102 and 103, when published, are subject to protection under this title if—

(1) on the date of first publication, one or more of the authors is a national or domiciliary of the United States, or is a national, domiciliary, or sovereign authority of a treaty party, or is a stateless person, wherever that person may be domiciled; or

(2) the work is first published in the United States or in a foreign nation that, on the date of first publication, is a treaty party; or

(3) the work is a sound recording that was first fixed in a treaty party; or

(4) the work is a pictorial, graphic, or sculptural work that is incorporated in a building or other structure, or an architectural work that is embodied in a building and the building or structure is located in the United States or a treaty party; or

(5) the work is first published by the United Nations or any of its specialized agencies, or by the Organization of American States; or

(6) the work comes within the scope of a Presidential proclamation. Whenever the President finds that a particular foreign nation extends, to works by authors who are nationals or domiciliaries of the United States or to works that are first published in the United States, copyright protection on substantially the same basis as that on which the foreign nation extends protection to works of its own nationals and domiciliaries and works first published in that nation, the President may by proclamation extend protection under this title to works of which one or more of the authors is, on the date of first publication, a national, domiciliary, or sovereign authority of that nation, or which was first published in that nation. The President may revise, suspend, or revoke any such proclamation or impose any conditions or limitations on protection under a proclamation.

For purposes of paragraph (2), a work that is published in the United States or a treaty party within 30 days after publication in a foreign nation that is not a treaty party shall be considered to be first published in the United States or such treaty party, as the case may be.

(c) Effect of Berne Convention.—No right or interest in a work eligible for protection under this title may be claimed by virtue of, or in reliance upon, the provisions of the Berne Convention, or the adherence of the United States thereto. Any rights in a work eligible for protection under this title that derive from this title, other Federal or State statutes, or the common law, shall not be expanded or reduced by virtue of, or in reliance upon, the provisions of the Berne Convention, or the adherence of the United States thereto.

(d) Effect of Phonograms Treaties.—Notwithstanding the provisions of subsection (b), no works other than sound recordings shall be eligible for protection under this title solely by virtue of the adherence of the United States to the Geneva Phonograms Convention or the WIPO Performances and Phonograms Treaty.

§ 104A Copyright in Restored Works

(a) Automatic Protection and Term.—

(1) Term.—

(A) Copyright subsists, in accordance with this section, in restored works, and vests automatically on the date of restoration.

(B) Any work in which copyright is restored under this section shall subsist for the remainder of the term of copyright that the work would have otherwise been granted in the United States if the work never entered the public domain in the United States.

(2) Exception.—Any work in which the copyright was ever owned or administered by the Alien Property Custodian and in which the restored copyright would be owned by a government or instrumentality thereof, is not a restored work.

(b) Ownership of Restored Copyright.—A restored work vests initially in the author or initial rightholder of the work as determined by the law of the source country of the work.

(c) Filing of Notice of Intent to Enforce Restored Copyright Against Reliance Parties.—On or after the date of restoration, any person who owns a copyright in a restored work or an exclusive right therein may file with the Copyright Office a notice of intent to enforce that person's copyright or exclusive right or may serve such a notice directly on a reliance party. Acceptance of a notice by the Copyright Office is effective as to any reliance parties but shall not create a presumption of the validity of any of the facts stated therein. Service on a reliance party is effective as to that reliance party and any other reliance parties with actual knowledge of such service and of the contents of that notice.

(d) Remedies for Infringement of Restored Copyrights.—

(1) Enforcement of copyright in restored works in the absence of a reliance party.—As against any party who is not a reliance party, the remedies provided in chapter 5 of this title shall be available on or after the date of restoration of a restored copyright with respect to an act of infringement of the restored copyright that is commenced on or after the date of restoration.

(2) Enforcement of copyright in restored works as against reliance parties.—As against a reliance party, except to the extent provided in paragraphs (3) and (4), the remedies provided in chapter 5 of this title shall be available, with respect to an act of infringement of a restored copyright, on or after the date of restoration of the restored copyright if the requirements of either of the following subparagraphs are met:

(A)(i) The owner of the restored copyright (or such owner's agent) or the owner of an exclusive right therein (or such owner's agent) files with the Copyright Office, during the 24-month period beginning on the date of restoration, a notice of intent to enforce the restored copyright; and

(ii)(I) the act of infringement commenced after the end of the 12 month period beginning on the date of publication of the notice in the Federal Register;

(II) the act of infringement commenced before the end of the 12-month period described in subclause (I) and continued after the end of that 12-month period, in which case remedies shall be available only for infringement occurring after the end of that 12-month period; or

(III) copies or phonorecords of a work in which copyright has been restored under this section are made after publication of the notice of intent in the Federal Register.

(B) (i) The owner of the restored copyright (or such owner's agent) or the owner of an exclusive right therein (or such owner's agent) serves upon a reliance party a notice of intent to enforce a restored copyright; and

(ii) (I) the act of infringement commenced after the end of the 12 month period beginning on the date the notice of intent is received;

(II) the act of infringement commenced before the end of the 12 month period described in subclause (I) and continued after the end of that 12-month period, in which case remedies shall be available only for the infringement occurring after the end of that 12-month period; or

(III) copies or phonorecords of a work in which copyright has been restored under this section are made after receipt of the notice of intent.

In the event that notice is provided under both subparagraphs (A) and (B), the 12-month period referred to in such subparagraphs shall run from the earlier of publication or service of notice.

(3) Existing derivative works.—

(A) In the case of a derivative work that is based upon a restored work and is created—

(i) before the date of the enactment of the Uruguay Round Agreements Act, if the source country of the restored work is an eligible country on such date, or

(ii) before the date on which the source country of the restored work becomes an eligible country, if that country is not an eligible country on such date of enactment,

a reliance party may continue to exploit that derivative work for the duration of the restored copyright if the reliance party pays to the owner of the restored copyright reasonable compensation for conduct which would be subject to a remedy for infringement but for the provisions of this paragraph.

(B) In the absence of an agreement between the parties, the amount of such compensation shall be determined by an action in United States district court, and shall reflect any harm to the actual or potential market for or value of the restored work from the reliance party's continued exploitation of the work, as well as compensation for the relative contributions of expression of the author of the restored work and the reliance party to the derivative work.

(4) Commencement of infringement for reliance parties.—For purposes of section 412, in the case of reliance parties, infringement shall be deemed to have commenced before registration when acts which would have constituted

infringement had the restored work been subject to copyright were commenced before the date of restoration.

(e) Notices of Intent to Enforce a Restored Copyright.—

(1) Notices of intent filed with the Copyright Office.—

(A)(i) A notice of intent filed with the Copyright Office to enforce a restored copyright shall be signed by the owner of the restored copyright or the owner of an exclusive right therein, who files the notice under subsection (d)(2)(A)(i) (hereafter in this paragraph referred to as the "owner"), or by the owner's agent, shall identify the title of the restored work, and shall include an English translation of the title and any other alternative titles known to the owner by which the restored work may be identified, and an address and telephone number at which the owner may be contacted. If the notice is signed by an agent, the agency relationship must have been constituted in a writing signed by the owner before the filing of the notice. The Copyright Office may specifically require in regulations other information to be included in the notice, but failure to provide such other information shall not invalidate the notice or be a basis for refusal to list the restored work in the Federal Register.

(ii) If a work in which copyright is restored has no formal title, it shall be described in the notice of intent in detail sufficient to identify it.

(iii) Minor errors or omissions may be corrected by further notice at any time after the notice of intent is filed. Notices of corrections for such minor errors or omissions shall be accepted after the period established in subsection (d)(2)(A)(i). Notices shall be published in the Federal Register pursuant to subparagraph (B).

(B)(i) The Register of Copyrights shall publish in the Federal Register, commencing not later than 4 months after the date of restoration for a particular nation and every 4 months thereafter for a period of 2 years, lists identifying restored works and the ownership thereof if a notice of intent to enforce a restored copyright has been filed.

(ii) Not less than 1 list containing all notices of intent to enforce shall be maintained in the Public Information Office of the Copyright Office and shall be available for public inspection and copying during regular business hours pursuant to sections 705 and 708.

(C) The Register of Copyrights is authorized to fix reasonable fees based on the costs of receipt, processing, recording, and publication of notices of intent to enforce a restored copyright and corrections thereto.

(D)(i) Not later than 90 days before the date the Agreement on Trade-Related Aspects of Intellectual Property referred to in section 101(d)(15) of the Uruguay Round Agreements Act enters into force with respect to the United States, the Copyright Office shall issue and publish in the Federal Register regulations governing the filing under this subsection of notices of intent to enforce a restored copyright.

(ii) Such regulations shall permit owners of restored copyrights to file simultaneously for registration of the restored copyright.

(2) Notices of intent served on a reliance party.—

(A) Notices of intent to enforce a restored copyright may be served on a reliance party at any time after the date of restoration of the restored copyright.

(B) Notices of intent to enforce a restored copyright served on a reliance party shall be signed by the owner or the owner's agent, shall identify the restored work and the work in which the restored work is used, if any, in detail sufficient to identify them, and shall include an English translation of the title, any other alternative titles known to the owner by which the work may be identified, the use or uses to which the owner objects, and an address and telephone number at which the reliance party may contact the owner. If the notice is signed by an agent, the agency relationship must have been constituted in writing and signed by the owner before service of the notice.

(3) Effect of material false statements.—Any material false statement knowingly made with respect to any restored copyright identified in any notice of intent shall make void all claims and assertions made with respect to such restored copyright.

(f) Immunity from Warranty and Related Liability.—

(1) In general.—Any person who warrants, promises, or guarantees that a work does not violate an exclusive right granted in section 106 shall not be liable for legal, equitable, arbitral, or administrative relief if the warranty, promise, or guarantee is breached by virtue of the restoration of copyright under this section, if such warranty, promise, or guarantee is made before January 1, 1995.

(2) Performances.—No person shall be required to perform any act if such performance is made infringing by virtue of the restoration of copyright under the provisions of this section, if the obligation to perform was undertaken before January 1, 1995.

(g) Proclamation of Copyright Restoration.—Whenever the President finds that a particular foreign nation extends, to works by authors who are nationals or domiciliaries of the United States, restored copyright protection on substantially the same basis as provided under this section, the President may by proclamation extend restored protection provided under this section to any work—

(1) of which one or more of the authors is, on the date of first publication, a national, domiciliary, or sovereign authority of that nation; or

(2) which was first published in that nation.

The President may revise, suspend, or revoke any such proclamation or impose any conditions or limitations on protection under such a proclamation.

(h) Definitions.—For purposes of this section and section 109(a):

(1) The term "date of adherence or proclamation" means the earlier of the date on which a foreign nation which, as of the date the WTO Agreement enters into force with respect to the United States, is not a nation adhering to the Berne Convention or a WTO member country, becomes—

(A) a nation adhering to the Berne Convention;

(B) a WTO member country;

(C) a nation adhering to the WIPO Copyright Treaty;

(D) a nation adhering to the WIPO Performances and Phonograms Treaty; or

(E) subject to a Presidential proclamation under subsection (g).

(2) The "date of restoration" of a restored copyright is—

(A) January 1, 1996, if the source country of the restored work is a nation adhering to the Berne Convention or a WTO member country on such date, or

(B) the date of adherence or proclamation, in the case of any other source country of the restored work.

(3) The term "eligible country" means a nation, other than the United States, that—

(A) becomes a WTO member country after the date of the enactment of the Uruguay Round Agreements Act;

(B) on such date of enactment is, or after such date of enactment becomes, a nation adhering to the Berne Convention;

(C) adheres to the WIPO Copyright Treaty;

(D) adheres to the WIPO Performances and Phonograms Treaty; or

(E) after such date of enactment becomes subject to a proclamation under subsection (g).

(4) The term "reliance party" means any person who—

(A) with respect to a particular work, engages in acts, before the source country of that work becomes an eligible country, which would have violated section 106 if the restored work had been subject to copyright protection, and who, after the source country becomes an eligible country, continues to engage in such acts;

(B) before the source country of a particular work becomes an eligible country, makes or acquires 1 or more copies or phonorecords of that work; or

(C) as the result of the sale or other disposition of a derivative work covered under subsection (d)(3), or significant assets of a person described in subparagraph (A) or (B), is a successor, assignee, or licensee of that person.

(5) The term "restored copyright" means copyright in a restored work under this section.

(6) The term "restored work" means an original work of authorship that—

(A) is protected under subsection (a);

(B) is not in the public domain in its source country through expiration of term of protection;

(C) is in the public domain in the United States due to—

(i) noncompliance with formalities imposed at any time by United States copyright law, including failure of renewal, lack of proper notice, or failure to comply with any manufacturing requirements;

(ii) lack of subject matter protection in the case of sound recordings fixed before February 15, 1972; or

(iii) lack of national eligibility;

(D) has at least one author or rightholder who was, at the time the work was created, a national or domiciliary of an eligible country, and if published, was first published in an eligible country and not published in the United

States during the 30-day period following publication in such eligible country; and

(E) if the source country for the work is an eligible country solely by virtue of its adherence to the WIPO Performances and Phonograms Treaty, is a sound recording.

(7) The term "rightholder" means the person—

(A) who, with respect to a sound recording, first fixes a sound recording with authorization, or

(B) who has acquired rights from the person described in subparagraph (A) by means of any conveyance or by operation of law.

(8) The "source country" of a restored work is—

(A) a nation other than the United States;

(B) in the case of an unpublished work—

(i) the eligible country in which the author or rightholder is a national or domiciliary, or, if a restored work has more than 1 author or rightholder, of which the majority of foreign authors or rightholders are nationals or domiciliaries; or

(ii) if the majority of authors or rightholders are not foreign, the nation other than the United States which has the most significant contacts with the work; and

(C) in the case of a published work—

(i) the eligible country in which the work is first published, or

(ii) if the restored work is published on the same day in 2 or more eligible countries, the eligible country which has the most significant contacts with the work.

§ 105 Subject Matter of Copyright: United States Government Works

Copyright protection under this title is not available for any work of the United States Government, but the United States Government is not precluded from receiving and holding copyrights transferred to it by assignment, bequest, or otherwise.

§ 106 Exclusive Rights in Copyrighted Works

Subject to sections 107 through 122, the owner of copyright under this title has the exclusive rights to do and to authorize any of the following:

(1) to reproduce the copyrighted work in copies or phonorecords;

(2) to prepare derivative works based upon the copyrighted work;

(3) to distribute copies or phonorecords of the copyrighted work to the public by sale or other transfer of ownership, or by rental, lease, or lending;

(4) in the case of literary, musical, dramatic, and choreographic works, pantomimes, and motion pictures and other audiovisual works, to perform the copyrighted work publicly;

(5) in the case of literary, musical, dramatic, and choreographic works, pantomimes, and pictorial, graphic, or sculptural works, including the individual images of a motion picture or other audiovisual work, to display the copyrighted work publicly; and

(6) in the case of sound recordings, to perform the copyrighted work publicly by means of a digital audio transmission.

§ 106a Rights of Certain Authors to Attribution and Integrity

(a) Rights of Attribution and Integrity.—Subject to section 107 and independent of the exclusive rights provided in section 106, the author of a work of visual art—

(1) shall have the right—

(A) to claim authorship of that work, and

(B) to prevent the use of his or her name as the author of any work of visual art which he or she did not create;

(2) shall have the right to prevent the use of his or her name as the author of the work of visual art in the event of a distortion, mutilation, or other modification of the work which would be prejudicial to his or her honor or reputation; and

(3) subject to the limitations set forth in section 113(d), shall have the right—

(A) to prevent any intentional distortion, mutilation, or other modification of that work which would be prejudicial to his or her honor or reputation, and any intentional distortion, mutilation, or modification of that work is a violation of that right, and

(B) to prevent any destruction of a work of recognized stature, and any intentional or grossly negligent destruction of that work is a violation of that right.

(b) Scope and Exercise of Rights.—Only the author of a work of visual art has the rights conferred by subsection (a) in that work, whether or not the author is the copyright owner. The authors of a joint work of visual art are coowners of the rights conferred by subsection (a) in that work.

(c) Exceptions.—(1) The modification of a work of visual art which is the result of the passage of time or the inherent nature of the materials is not a distortion, mutilation, or other modification described in subsection (a)(3)(A).

(2) The modification of a work of visual art which is the result of conservation, or of the public presentation, including lighting and placement, of the work is not a destruction, distortion, mutilation, or other modification described in subsection (a)(3) unless the modification is caused by gross negligence.

(3) The rights described in paragraphs (1) and (2) of subsection (a) shall not apply to any reproduction, depiction, portrayal, or other use of a work in, upon, or in any connection with any item described in subparagraph (A) or (B) of the definition of "work of visual art" in section 101, and any such reproduction,

depiction, portrayal, or other use of a work is not a destruction, distortion, mutilation, or other modification described in paragraph (3) of subsection (a).

(d) Duration of Rights.—(1) With respect to works of visual art created on or after the effective date set forth in section 610(a) of the Visual Artists Rights Act of 1990, the rights conferred by subsection (a) shall endure for a term consisting of the life of the author.

(2) With respect to works of visual art created before the effective date set forth in section 610(a) of the Visual Artists Rights Act of 1990, but title to which has not, as of such effective date, been transferred from the author, the rights conferred by subsection (a) shall be coextensive with, and shall expire at the same time as, the rights conferred by section 106.

(3) In the case of a joint work prepared by two or more authors, the rights conferred by subsection (a) shall endure for a term consisting of the life of the last surviving author.

(4) All terms of the rights conferred by subsection (a) run to the end of the calendar year in which they would otherwise expire.

(e) Transfer and Waiver.—(1) The rights conferred by subsection (a) may not be transferred, but those rights may be waived if the author expressly agrees to such waiver in a written instrument signed by the author. Such instrument shall specifically identify the work, and uses of that work, to which the waiver applies, and the waiver shall apply only to the work and uses so identified. In the case of a joint work prepared by two or more authors, a waiver of rights under this paragraph made by one such author waives such rights for all such authors.

(2) Ownership of the rights conferred by subsection (a) with respect to a work of visual art is distinct from ownership of any copy of that work, or of a copyright or any exclusive right under a copyright in that work. Transfer of ownership of any copy of a work of visual art, or of a copyright or any exclusive right under a copyright, shall not constitute a waiver of the rights conferred by subsection (a). Except as may otherwise be agreed by the author in a written instrument signed by the author, a waiver of the rights conferred by subsection (a) with respect to a work of visual art shall not constitute a transfer of ownership of any copy of that work, or of ownership of a copyright or of any exclusive right under a copyright in that work.

§ 107 Limitations on Exclusive Rights: Fair Use

Notwithstanding the provisions of sections 106 and 106A, the fair use of a copyrighted work, including such use by reproduction in copies or phonorecords or by any other means specified by that section, for purposes such as criticism, comment, news reporting, teaching (including multiple copies for classroom use), scholarship, or research, is not an infringement of copyright. In determining whether the use made of a work in any particular case is a fair use the factors to be considered shall include—

(1) the purpose and character of the use, including whether such use is of a commercial nature or is for nonprofit educational purposes;

(2) the nature of the copyrighted work;

(3) the amount and substantiality of the portion used in relation to the copyrighted work as a whole; and

(4) the effect of the use upon the potential market for or value of the copyrighted work.

The fact that a work is unpublished shall not itself bar a finding of fair use if such finding is made upon consideration of all the above factors.

§ 108 Limitations on Exclusive Rights: Reproduction by Libraries and Archives

(a) Except as otherwise provided in this title and notwithstanding the provisions of section 106, it is not an infringement of copyright for a library or archives, or any of its employees acting within the scope of their employment, to reproduce no more than one copy or phonorecord of a work, except as provided in subsections (b) and (c), or to distribute such copy or phonorecord, under the conditions specified by this section, if—

(1) the reproduction or distribution is made without any purpose of direct or indirect commercial advantage;

(2) the collections of the library or archives are (i) open to the public, or (ii) available not only to researchers affiliated with the library or archives or with the institution of which it is a part, but also to other persons doing research in a specialized field; and

(3) the reproduction or distribution of the work includes a notice of copyright that appears on the copy or phonorecord that is reproduced under the provisions of this section, or includes a legend stating that the work may be protected by copyright if no such notice can be found on the copy or phonorecord that is reproduced under the provisions of this section.

(b) The rights of reproduction and distribution under this section apply to three copies or phonorecords of an unpublished work duplicated solely for purposes of preservation and security or for deposit for research use in another library or archives of the type described by clause (2) of subsection (a), if—

(1) the copy or phonorecord reproduced is currently in the collections of the library or archives; and

(2) any such copy or phonorecord that is reproduced in digital format is not otherwise distributed in that format and is not made available to the public in that format outside the premises of the library or archives.

(c) The right of reproduction under this section applies to three copies or phonorecords of a published work duplicated solely for the purpose of replacement of a copy or phonorecord that is damaged, deteriorating, lost, or stolen, or if the existing format in which the work is stored has become obsolete, if—

(1) the library or archives has, after a reasonable effort, determined that an unused replacement cannot be obtained at a fair price; and

(2) any such copy or phonorecord that is reproduced in digital format is not made available to the public in that format outside the premises of the library or archives in lawful possession of such copy.

For purposes of this subsection, a format shall be considered obsolete if the machine or device necessary to render perceptible a work stored in that format is no longer manufactured or is no longer reasonably available in the commercial marketplace.

(d) The rights of reproduction and distribution under this section apply to a copy, made from the collection of a library or archives where the user makes his or her request or from that of another library or archives, of no more than one article or other contribution to a copyrighted collection or periodical issue, or to a copy or phonorecord of a small part of any other copyrighted work, if—

(1) the copy or phonorecord becomes the property of the user, and the library or archives has had no notice that the copy or phonorecord would be used for any purpose other than private study, scholarship, or research; and

(2) the library or archives displays prominently, at the place where orders are accepted, and includes on its order form, a warning of copyright in accordance with requirements that the Register of Copyrights shall prescribe by regulation.

(e) The rights of reproduction and distribution under this section apply to the entire work, or to a substantial part of it, made from the collection of a library or archives where the user makes his or her request or from that of another library or archives, if the library or archives has first determined, on the basis of a reasonable investigation, that a copy or phonorecord of the copyrighted work cannot be obtained at a fair price, if—

(1) the copy or phonorecord becomes the property of the user, and the library or archives has had no notice that the copy or phonorecord would be used for any purpose other than private study, scholarship, or research; and

(2) the library or archives displays prominently, at the place where orders are accepted, and includes on its order form, a warning of copyright in accordance with requirements that the Register of Copyrights shall prescribe by regulation.

(f) Nothing in this section—

(1) shall be construed to impose liability for copyright infringement upon a library or archives or its employees for the unsupervised use of reproducing equipment located on its premises: *Provided,* That such equipment displays a notice that the making of a copy may be subject to the copyright law;

(2) excuses a person who uses such reproducing equipment or who requests a copy or phonorecord under subsection (d) from liability for copyright infringement for any such act, or for any later use of such copy or phonorecord, if it exceeds fair use as provided by section 107;

(3) shall be construed to limit the reproduction and distribution by lending of a limited number of copies and excerpts by a library or archives of an audiovisual news program, subject to clauses (1), (2), and (3) of subsection (a); or

(4) in any way affects the right of fair use as provided by section 107, or any contractual obligations assumed at any time by the library or archives when it obtained a copy or phonorecord of a work in its collections.

(g) The rights of reproduction and distribution under this section extend to the isolated and unrelated reproduction or distribution of a single copy or phonorecord of the same material on separate occasions, but do not extend to cases where the library or archives, or its employee—

(1) is aware or has substantial reason to believe that it is engaging in the related or concerted reproduction or distribution of multiple copies or phonorecords of the same material, whether made on one occasion or over a period of time, and whether intended for aggregate use by one or more individuals or for separate use by the individual members of a group; or

(2) engages in the systematic reproduction or distribution of single or multiple copies or phonorecords of material described in subsection (d): *Provided,* That nothing in this clause prevents a library or archives from participating in interlibrary arrangements that do not have, as their purpose or effect, that the library or archives receiving such copies or phonorecords for distribution does so in such aggregate quantities as to substitute for a subscription to or purchase of such work.

(h) (1) For purposes of this section, during the last 20 years of any term of copyright of a published work, a library or archives, including a nonprofit educational institution that functions as such, may reproduce, distribute, display, or perform in facsimile or digital form a copy or phonorecord of such work, or portions thereof, for purposes of preservation, scholarship, or research, if such library or archives has first determined, on the basis of a reasonable investigation, that none of the conditions set forth in subparagraphs (A), (B), and (C) of paragraph (2) apply.

(2) No reproduction, distribution, display, or performance is authorized under this subsection if—

(A) the work is subject to normal commercial exploitation;

(B) a copy or phonorecord of the work can be obtained at a reasonable price; or

(C) the copyright owner or its agent provides notice pursuant to regulations promulgated by the Register of Copyrights that either of the conditions set forth in subparagraphs (A) and (B) applies.

(3) The exemption provided in this subsection does not apply to any subsequent uses by users other than such library or archives.

(i) The rights of reproduction and distribution under this section do not apply to a musical work, a pictorial, graphic or sculptural work, or a motion picture or other audiovisual work other than an audiovisual work dealing with news, except that no such limitation shall apply with respect to rights granted by subsections

(b), (c), and (h), or with respect to pictorial or graphic works published as illustrations, diagrams, or similar adjuncts to works of which copies are reproduced or distributed in accordance with subsections (d) and (e).

§ 109 Limitations on Exclusive Rights: Effect of Transfer of Particular Copy or Phonorecord

(a) Notwithstanding the provisions of section 106(3), the owner of a particular copy or phonorecord lawfully made under this title, or any person authorized by such owner, is entitled, without the authority of the copyright owner, to sell or otherwise dispose of the possession of that copy or phonorecord. Notwithstanding the preceding sentence, copies or phonorecords of works subject to restored copyright under section 104A that are manufactured before the date of restoration of copyright or, with respect to reliance parties, before publication or service of notice under section 104A(e), may be sold or otherwise disposed of without the authorization of the owner of the restored copyright for purposes of direct or indirect commercial advantage only during the 12-month period beginning on—

(1) the date of the publication in the Federal Register of the notice of intent filed with the Copyright Office under section 104A(d)(2)(A), or

(2) the date of the receipt of actual notice served under section 104A(d)(2)(B),

whichever occurs first.

(b)(1)(A) Notwithstanding the provisions of subsection (a), unless authorized by the owners of copyright in the sound recording or the owner of copyright in a computer program (including any tape, disk, or other medium embodying such program), and in the case of a sound recording in the musical works embodied therein, neither the owner of a particular phonorecord nor any person in possession of a particular copy of a computer program (including any tape, disk, or other medium embodying such program), may, for the purposes of direct or indirect commercial advantage, dispose of, or authorize the disposal of, the possession of that phonorecord or computer program (including any tape, disk, or other medium embodying such program) by rental, lease, or lending, or by any other act or practice in the nature of rental, lease, or lending. Nothing in the preceding sentence shall apply to the rental, lease, or lending of a phonorecord for nonprofit purposes by a nonprofit library or nonprofit educational institution. The transfer of possession of a lawfully made copy of a computer program by a nonprofit educational institution to another nonprofit educational institution or to faculty, staff, and students does not constitute rental, lease, or lending for direct or indirect commercial purposes under this subsection.

(B) This subsection does not apply to—

(i) a computer program which is embodied in a machine or product and which cannot be copied during the ordinary operation or use of the machine or product; or

(ii) a computer program embodied in or used in conjunction with a limited purpose computer that is designed for playing video games and may be designed for other purposes.

(C) Nothing in this subsection affects any provision of chapter 9 of this title.

(2)(A) Nothing in this subsection shall apply to the lending of a computer program for nonprofit purposes by a nonprofit library, if each copy of a computer program which is lent by such library has affixed to the packaging containing the program a warning of copyright in accordance with requirements that the Register of Copyrights shall prescribe by regulation.

(B) Not later than three years after the date of the enactment of the Computer Software Rental Amendments Act of 1990, and at such times thereafter as the Register of Copyrights considers appropriate, the Register of Copyrights, after consultation with representatives of copyright owners and librarians, shall submit to the Congress a report stating whether this paragraph has achieved its intended purpose of maintaining the integrity of the copyright system while providing nonprofit libraries the capability to fulfill their function. Such report shall advise the Congress as to any information or recommendations that the Register of Copyrights considers necessary to carry out the purposes of this subsection.

(3) Nothing in this subsection shall affect any provision of the antitrust laws. For purposes of the preceding sentence, "antitrust laws" has the meaning given that term in the first section of the Clayton Act and includes section 5 of the Federal Trade Commission Act to the extent that section relates to unfair methods of competition.

(4) Any person who distributes a phonorecord or a copy of a computer program (including any tape, disk, or other medium embodying such program) in violation of paragraph (1) is an infringer of copyright under section 501 of this title and is subject to the remedies set forth in sections 502, 503, 504, and 505. Such violation shall not be a criminal offense under section 506 or cause such person to be subject to the criminal penalties set forth in section 2319 of title 18.

(c) Notwithstanding the provisions of section 106(5), the owner of a particular copy lawfully made under this title, or any person authorized by such owner, is entitled, without the authority of the copyright owner, to display that copy publicly, either directly or by the projection of no more than one image at a time, to viewers present at the place where the copy is located.

(d) The privileges prescribed by subsections (a) and (c) do not, unless authorized by the copyright owner, extend to any person who has acquired possession of the copy or phonorecord from the copyright owner, by rental, lease, loan, or otherwise, without acquiring ownership of it.

(e) Notwithstanding the provisions of sections 106(4) and 106(5), in the case of an electronic audiovisual game intended for use in coin-operated equipment, the owner of a particular copy of such a game lawfully made under this title, is entitled, without the authority of the copyright owner of the game, to publicly perform or display that game in coin-operated equipment, except that this subsection shall not apply to any work of authorship embodied in the audiovisual game if the copyright owner of the electronic audiovisual game is not also the copyright owner of the work of authorship.

§ 110 Limitations on Exclusive Rights: Exemption of Certain Performances and Displays

Notwithstanding the provisions of section 106, the following are not infringements of copyright:

(1) performance or display of a work by instructors or pupils in the course of face-to-face teaching activities of a nonprofit educational institution, in a classroom or similar place devoted to instruction, unless, in the case of a motion picture or other audiovisual work, the performance, or the display of individual images, is given by means of a copy that was not lawfully made under this title, and that the person responsible for the performance knew or had reason to believe was not lawfully made;

(2) except with respect to a work produced or marketed primarily for performance or display as part of mediated instructional activities transmitted via digital networks, or a performance or display that is given by means of a copy or phonorecord that is not lawfully made and acquired under this title, and the transmitting government body or accredited nonprofit educational institution knew or had reason to believe was not lawfully made and acquired, the performance of a nondramatic literary or musical work or reasonable and limited portions of any other work, or display of a work in an amount comparable to that which is typically displayed in the course of a live classroom session, by or in the course of a transmission, if—

(A) the performance or display is made by, at the direction of, or under the actual supervision of an instructor as an integral part of a class session offered as a regular part of the systematic mediated instructional activities of a governmental body or an accredited nonprofit educational institution;

(B) the performance or display is directly related and of material assistance to the teaching content of the transmission;

(C) the transmission is made solely for, and, to the extent technologically feasible, the reception of such transmission is limited to—

(i) students officially enrolled in the course for which the transmission is made; or

(ii) officers or employees of governmental bodies as a part of their official duties or employment; and

(D) the transmitting body or institution—

(i) institutes policies regarding copyright, provides informational materials to faculty, students, and relevant staff members that accurately describe, and promote compliance with, the laws of the United States relating to copyright, and provides notice to students that materials used in connection with the course may be subject to copyright protection; and

(ii) in the case of digital transmissions—

(I) applies technological measures that reasonably prevent—(aa) retention of the work in accessible form by recipients of the transmission from the transmitting body or institution for longer than the class session; and

(bb) unauthorized further dissemination of the work in accessible form by such recipients to others; and

(II) does not engage in conduct that could reasonably be expected to interfere with technological measures used by copyright owners to prevent such retention or unauthorized further dissemination;

(3) performance of a nondramatic literary or musical work or of a dramatico-musical work of a religious nature, or display of a work, in the course of services at a place of worship or other religious assembly;

(4) performance of a nondramatic literary or musical work otherwise than in a transmission to the public, without any purpose of direct or indirect commercial advantage and without payment of any fee or other compensation for the performance to any of its performers, promoters, or organizers, if—

(A) there is no direct or indirect admission charge; or

(B) the proceeds, after deducting the reasonable costs of producing the performance, are used exclusively for educational, religious, or charitable purposes and not for private financial gain, except where the copyright owner has served notice of objection to the performance under the following conditions:

(i) the notice shall be in writing and signed by the copyright owner or such owner's duly authorized agent; and

(ii) the notice shall be served on the person responsible for the performance at least seven days before the date of the performance, and shall state the reasons for the objection; and

(iii) the notice shall comply, in form, content, and manner of service, with requirements that the Register of Copyrights shall prescribe by regulation;

(5)(A) except as provided in subparagraph (B), communication of a transmission embodying a performance or display of a work by the public reception of the transmission on a single receiving apparatus of a kind commonly used in private homes, unless—

(i) a direct charge is made to see or hear the transmission; or

(ii) the transmission thus received is further transmitted to the public;

(B) communication by an establishment of a transmission or retransmission embodying a performance or display of a nondramatic musical work intended to be received by the general public, originated by a radio or television broadcast station licensed as such by the Federal Communications Commission, or, if an audiovisual transmission, by a cable system or satellite carrier, if—

(i) in the case of an establishment other than a food service or drinking establishment, either the establishment in which the communication occurs has less than 2,000 gross square feet of space (excluding space used for customer parking and for no other purpose), or the establishment in which the communication occurs has 2,000 or more gross square feet of space (excluding space used for customer parking and for no other purpose) and—

(I) if the performance is by audio means only, the performance is communicated by means of a total of not more than 6 loudspeakers, of which not more than 4 loudspeakers are located in any 1 room or adjoining outdoor space; or

(II) if the performance or display is by audiovisual means, any visual portion of the performance or display is communicated by means of a total of not more than 4 audiovisual devices, of which not more than 1 audiovisual device is located in any 1 room, and no such audiovisual device has a diagonal screen size greater than 55 inches, and any audio portion of the performance or display is communicated by means of a total of not more than 6 loudspeakers, of which not more than 4 loudspeakers are located in any 1 room or adjoining outdoor space;

(ii) in the case of a food service or drinking establishment, either the establishment in which the communication occurs has less than 3,750 gross square feet of space (excluding space used for customer parking and for no other purpose), or the establishment in which the communication occurs has 3,750 gross square feet of space or more (excluding space used for customer parking and for no other purpose) and—

(I) if the performance is by audio means only, the performance is communicated by means of a total of not more than 6 loudspeakers, of which not more than 4 loudspeakers are located in any 1 room or adjoining outdoor space; or

(II) if the performance or display is by audiovisual means, any visual portion of the performance or display is communicated by means of a total of not more than 4 audiovisual devices, of which not more than 1 audiovisual device is located in any 1 room, and no such audiovisual device has a diagonal screen size greater than 55 inches, and any audio portion of the performance or display is communicated by means of a total of not more than 6 loudspeakers, of which not more than 4 loudspeakers are located in any 1 room or adjoining outdoor space;

(iii) no direct charge is made to see or hear the transmission or retransmission;

(iv) the transmission or retransmission is not further transmitted beyond the establishment where it is received; and

(v) the transmission or retransmission is licensed by the copyright owner of the work so publicly performed or displayed;

(6) performance of a nondramatic musical work by a governmental body or a nonprofit agricultural or horticultural organization, in the course of an annual agricultural or horticultural fair or exhibition conducted by such body or organization; the exemption provided by this clause shall extend to any liability for copyright infringement that would otherwise be imposed on such body or organization, under doctrines of vicarious liability or related infringement, for a performance by a concessionaire, business establishment, or other person at such fair or exhibition, but shall not excuse any such person from liability for the performance;

(7) performance of a nondramatic musical work by a vending establishment open to the public at large without any direct or indirect admission charge, where the sole purpose of the performance is to promote the retail sale of

copies or phonorecords of the work, or of the audiovisual or other devices utilized in such performance, and the performance is not transmitted beyond the place where the establishment is located and is within the immediate area where the sale is occurring;

(8) performance of a nondramatic literary work, by or in the course of a transmission specifically designed for and primarily directed to blind or other handicapped persons who are unable to read normal printed material as a result of their handicap, or deaf or other handicapped persons who are unable to hear the aural signals accompanying a transmission of visual signals, if the performance is made without any purpose of direct or indirect commercial advantage and its transmission is made through the facilities of: (i) a governmental body; or (ii) a noncommercial educational broadcast station (as defined in section 397 of title 47); or (iii) a radio subcarrier authorization (as defined in 47 CFR 73.293–73.295 and 73.593–73.595); or (iv) a cable system (as defined in section 111 (f));

(9) performance on a single occasion of a dramatic literary work published at least ten years before the date of the performance, by or in the course of a transmission specifically designed for and primarily directed to blind or other handicapped persons who are unable to read normal printed material as a result of their handicap, if the performance is made without any purpose of direct or indirect commercial advantage and its transmission is made through the facilities of a radio subcarrier authorization referred to in clause (8) (iii), *Provided,* That the provisions of this clause shall not be applicable to more than one performance of the same work by the same performers or under the auspices of the same organization;

(10) notwithstanding paragraph (4), the following is not an infringement of copyright: performance of a nondramatic literary or musical work in the course of a social function which is organized and promoted by a nonprofit veterans' organization or a nonprofit fraternal organization to which the general public is not invited, but not including the invitees of the organizations, if the proceeds from the performance, after deducting the reasonable costs of producing the performance, are used exclusively for charitable purposes and not for financial gain. For purposes of this section the social functions of any college or university fraternity or sorority shall not be included unless the social function is held solely to raise funds for a specific charitable purpose; and

(11) the making imperceptible, by or at the direction of a member of a private household, of limited portions of audio or video content of a motion picture, during a performance in or transmitted to that household for private home viewing, from an authorized copy of the motion picture, or the creation or provision of a computer program or other technology that enables such making imperceptible and that is designed and marketed to be used, at the direction of a member of a private household, for such making imperceptible, if no fixed copy of the altered version of the motion picture is created by such computer program or other technology.

The exemptions provided under paragraph (5) shall not be taken into account in any administrative, judicial, or other governmental proceeding to set or adjust the royalties payable to copyright owners for the public performance or display of their works. Royalties payable to copyright owners for any public performance or display of their works other than such performances or displays as are exempted under paragraph (5) shall not be diminished in any respect as a result of such exemption.

In paragraph (2), the term "mediated instructional activities" with respect to the performance or display of a work by digital transmission under this section refers to activities that use such work as an integral part of the class experience, controlled by or under the actual supervision of the instructor and analogous to the type of performance or display that would take place in a live classroom setting. The term does not refer to activities that use, in 1 or more class sessions of a single course, such works as textbooks, course packs, or other material in any media, copies or phonorecords of which are typically purchased or acquired by the students in higher education for their independent use and retention or are typically purchased or acquired for elementary and secondary students for their possession and independent use.

For purposes of paragraph (2), accreditation—

(A) with respect to an institution providing post-secondary education, shall be as determined by a regional or national accrediting agency recognized by the Council on Higher Education Accreditation or the United States Department of Education; and

(B) with respect to an institution providing elementary or secondary education, shall be as recognized by the applicable state certification or licensing procedures.

For purposes of paragraph (2), no governmental body or accredited nonprofit educational institution shall be liable for infringement by reason of the transient or temporary storage of material carried out through the automatic technical process of a digital transmission of the performance or display of that material as authorized under paragraph (2). No such material stored on the system or network controlled or operated by the transmitting body or institution under this paragraph shall be maintained on such system or network in a manner ordinarily accessible to anyone other than anticipated recipients. No such copy shall be maintained on the system or network in a manner ordinarily accessible to such anticipated recipients for a longer period than is reasonably necessary to facilitate the transmissions for which it was made.

For purposes of paragraph (11), the term "making imperceptible" does not include the addition of audio or video content that is performed or displayed over or in place of existing content in a motion picture.

Nothing in paragraph (11) shall be construed to imply further rights under section 106 of this title, or to have any effect on defenses or limitations on rights granted under any other section of this title or under any other paragraph of this section.

§ 111 Limitations on Exclusive Rights: Secondary Transmissions of Broadcast Programming by Cable

(a) Certain Secondary Transmissions Exempted.—The secondary transmission of a performance or display of a work embodied in a primary transmission is not an infringement of copyright if—

(1) the secondary transmission is not made by a cable system, and consists entirely of the relaying, by the management of a hotel, apartment house, or similar establishment, of signals transmitted by a broadcast station licensed by the Federal Communications Commission, within the local service area of such station, to the private lodgings of guests or residents of such establishment, and no direct charge is made to see or hear the secondary transmission; or

(2) the secondary transmission is made solely for the purpose and under the conditions specified by paragraph (2) of section 110; or

(3) the secondary transmission is made by any carrier who has no direct or indirect control over the content or selection of the primary transmission or over the particular recipients of the secondary transmission, and whose activities with respect to the secondary transmission consist solely of providing wires, cables, or other communications channels for the use of others: *Provided,* That the provisions of this paragraph extend only to the activities of said carrier with respect to secondary transmissions and do not exempt from liability the activities of others with respect to their own primary or secondary transmissions;

(4) the secondary transmission is made by a satellite carrier pursuant to a statutory license under section 119 or section 122;

(5) the secondary transmission is not made by a cable system but is made by a governmental body, or other nonprofit organization, without any purpose of direct or indirect commercial advantage, and without charge to the recipients of the secondary transmission other than assessments necessary to defray the actual and reasonable costs of maintaining and operating the secondary transmission service.

(b) Secondary Transmission of Primary Transmission to Controlled Group.—Notwithstanding the provisions of subsections (a) and (c), the secondary transmission to the public of a performance or display of a work embodied in a primary transmission is actionable as an act of infringement under section 501, and is fully subject to the remedies provided by sections 502 through 506, if the primary transmission is not made for reception by the public at large but is controlled and limited to reception by particular members of the public: *Provided,* however, That such secondary transmission is not actionable as an act of infringement if—

(1) the primary transmission is made by a broadcast station licensed by the Federal Communications Commission; and

(2) the carriage of the signals comprising the secondary transmission is required under the rules, regulations, or authorizations of the Federal Communications Commission; and

(3) the signal of the primary transmitter is not altered or changed in any way by the secondary transmitter.

(c) Secondary Transmissions by Cable Systems.—

(1) Subject to the provisions of paragraphs (2), (3), and (4) of this subsection and section 114(d), secondary transmissions to the public by a cable system of a performance or display of a work embodied in a primary transmission made by a broadcast station licensed by the Federal Communications Commission or

by an appropriate governmental authority of Canada or Mexico shall be subject to statutory licensing upon compliance with the requirements of subsection (d) where the carriage of the signals comprising the secondary transmission is permissible under the rules, regulations, or authorizations of the Federal Communications Commission.

(2) Notwithstanding the provisions of paragraph (1) of this subsection, the willful or repeated secondary transmission to the public by a cable system of a primary transmission made by a broadcast station licensed by the Federal Communications Commission or by an appropriate governmental authority of Canada or Mexico and embodying a performance or display of a work is actionable as an act of infringement under section 501, and is fully subject to the remedies provided by sections 502 through 506, in the following cases:

(A) where the carriage of the signals comprising the secondary transmission is not permissible under the rules, regulations, or authorizations of the Federal Communications Commission; or

(B) where the cable system has not deposited the statement of account and royalty fee required by subsection (d).

(3) Notwithstanding the provisions of paragraph (1) of this subsection and subject to the provisions of subsection (e) of this section, the secondary transmission to the public by a cable system of a performance or display of a work embodied in a primary transmission made by a broadcast station licensed by the Federal Communications Commission or by an appropriate governmental authority of Canada or Mexico is actionable as an act of infringement under section 501, and is fully subject to the remedies provided by sections 502 through 506 and section 510, if the content of the particular program in which the performance or display is embodied, or any commercial advertising or station announcements transmitted by the primary transmitter during, or immediately before or after, the transmission of such program, is in any way willfully altered by the cable system through changes, deletions, or additions, except for the alteration, deletion, or substitution of commercial advertisements performed by those engaged in television commercial advertising market research: *Provided*, That the research company has obtained the prior consent of the advertiser who has purchased the original commercial advertisement, the television station broadcasting that commercial advertisement, and the cable system performing the secondary transmission: *And provided further*, That such commercial alteration, deletion, or substitution is not performed for the purpose of deriving income from the sale of that commercial time.

(4) Notwithstanding the provisions of paragraph (1) of this subsection, the secondary transmission to the public by a cable system of a performance or display of a work embodied in a primary transmission made by a broadcast station licensed by an appropriate governmental authority of Canada or Mexico is actionable as an act of infringement under section 501, and is fully subject to the remedies provided by sections 502 through 506, if (A) with respect to Canadian signals, the community of the cable system is located more than 150 miles from the United States-Canadian border and is also located south of the forty-second

parallel of latitude, or (B) with respect to Mexican signals, the secondary transmission is made by a cable system which received the primary transmission by means other than direct interception of a free space radio wave emitted by such broadcast television station, unless prior to April 15, 1976, such cable system was actually carrying, or was specifically authorized to carry, the signal of such foreign station on the system pursuant to the rules, regulations, or authorizations of the Federal Communications Commission.

(d) Statutory License for Secondary Transmissions by Cable Systems.—

(1) Statement of account and royalty fees.—Subject to paragraph (5), a cable system whose secondary transmissions have been subject to statutory licensing under subsection (c) shall, on a semiannual basis, deposit with the Register of Copyrights, in accordance with requirements that the Register shall prescribe by regulation the following—

(A) A statement of account, covering the six months next preceding, specifying the number of channels on which the cable system made secondary transmissions to its subscribers, the names and locations of all primary transmitters whose transmissions were further transmitted by the cable system, the total number of subscribers, the gross amounts paid to the cable system for the basic service of providing secondary transmissions of primary broadcast transmitters, and such other data as the Register of Copyrights may from time to time prescribe by regulation. In determining the total number of subscribers and the gross amounts paid to the cable system for the basic service of providing secondary transmissions of primary broadcast transmitters, the cable system shall not include subscribers and amounts collected from subscribers receiving secondary transmissions pursuant to section 119. Such statement shall also include a special statement of account covering any non-network television programming that was carried by the cable system in whole or in part beyond the local service area of the primary transmitter, under rules, regulations, or authorizations of the Federal Communications Commission permitting the substitution or addition of signals under certain circumstances, together with logs showing the times, dates, stations, and programs involved in such substituted or added carriage.

(B) Except in the case of a cable system whose royalty fee is specified in subparagraph (E) or (F), a total royalty fee payable to copyright owners pursuant to paragraph (3) for the period covered by the statement, computed on the basis of specified percentages of the gross receipts from subscribers to the cable service during such period for the basic service of providing secondary transmissions of primary broadcast transmitters, as follows:

(i) 1.064 percent of such gross receipts for the privilege of further transmitting, beyond the local service area of such primary transmitter, any non-network programming of a primary transmitter in whole or in part, such amount to be applied against the fee, if any, payable pursuant to clauses (ii) through (iv);

(ii) 1.064 percent of such gross receipts for the first distant signal equivalent;

(iii) 0.701 percent of such gross receipts for each of the second, third, and fourth distant signal equivalents; and

(iv) 0.330 percent of such gross receipts for the fifth distant signal equivalent and each distant signal equivalent thereafter.

(C) In computing amounts under clauses (ii) through (iv) of subparagraph (B)—

(i) any fraction of a distant signal equivalent shall be computed at its fractional value;

(ii) in the case of any cable system located partly within and partly outside of the local service area of a primary transmitter, gross receipts shall be limited to those gross receipts derived from subscribers located outside of the local service area of such primary transmitter; and

(iii) if a cable system provides a secondary transmission of a primary transmitter to some but not all communities served by that cable system—

(I) the gross receipts and the distant signal equivalent values for such secondary transmission shall be derived solely on the basis of the subscribers in those communities where the cable system provides such secondary transmission; and

(II) the total royalty fee for the period paid by such system shall not be less than the royalty fee calculated under subparagraph (B)(i) multiplied by the gross receipts from all subscribers to the system.

(D) A cable system that, on a statement submitted before the date of the enactment of the Satellite Television Extension and Localism Act of 2010, computed its royalty fee consistent with the methodology under subparagraph (C)(iii), or that amends a statement filed before such date of enactment to compute the royalty fee due using such methodology, shall not be subject to an action for infringement, or eligible for any royalty refund or offset, arising out of its use of such methodology on such statement.

(E) If the actual gross receipts paid by subscribers to a cable system for the period covered by the statement for the basic service of providing secondary transmissions of primary broadcast transmitters are $263,800 or less—

(i) gross receipts of the cable system for the purpose of this paragraph shall be computed by subtracting from such actual gross receipts the amount by which $263,800 exceeds such actual gross receipts, except that in no case shall a cable system's gross receipts be reduced to less than $10,400; and

(ii) the royalty fee payable under this paragraph to copyright owners pursuant to paragraph (3) shall be 0.5 percent, regardless of the number of distant signal equivalents, if any.

(F) If the actual gross receipts paid by subscribers to a cable system for the period covered by the statement for the basic service of providing secondary transmissions of primary broadcast transmitters are more than $263,800 but less than $527,600, the royalty fee payable under this paragraph to copyright owners pursuant to paragraph (3) shall be—

(i) 0.5 percent of any gross receipts up to $263,800, regardless of the number of distant signal equivalents, if any; and

(ii) 1 percent of any gross receipts in excess of $263,800, but less than $527,600, regardless of the number of distant signal equivalents, if any.

(G) A filing fee, as determined by the Register of Copyrights pursuant to section 708(a).

(2) Handling of fees.—The Register of Copyrights shall receive all fees (including the filing fee specified in paragraph (1)(G)) deposited under this section and, after deducting the reasonable costs incurred by the Copyright Office under this section, shall deposit the balance in the Treasury of the United States, in such manner as the Secretary of the Treasury directs. All funds held by the Secretary of the Treasury shall be invested in interest-bearing United States securities for later distribution with interest by the Librarian of Congress upon authorization by the Copyright Royalty Judges.

(3) Distribution of royalty fees to copyright owners.—The royalty fees thus deposited shall, in accordance with the procedures provided by clause (4), be distributed to those among the following copyright owners who claim that their works were the subject of secondary transmissions by cable systems during the relevant semiannual period:

(A) Any such owner whose work was included in a secondary transmission made by a cable system of a non-network television program in whole or in part beyond the local service area of the primary transmitter.

(B) Any such owner whose work was included in a secondary transmission identified in a special statement of account deposited under clause (1)(A).

(C) Any such owner whose work was included in non-network programming consisting exclusively of aural signals carried by a cable system in whole or in part beyond the local service area of the primary transmitter of such programs.

(4) Procedures for royalty fee distribution.—The royalty fees thus deposited shall be distributed in accordance with the following procedures:

(A) During the month of July in each year, every person claiming to be entitled to statutory license fees for secondary transmissions shall file a claim with the Copyright Royalty Judges, in accordance with requirements that the Copyright Royalty Judges shall prescribe by regulation. Notwithstanding any provisions of the antitrust laws, for purposes of this clause any claimants may agree among themselves as to the proportionate division of statutory licensing fees among them, may lump their claims together and file them jointly or as a single claim, or may designate a common agent to receive payment on their behalf.

(B) After the first day of August of each year, the Copyright Royalty Judges shall determine whether there exists a controversy concerning the distribution of royalty fees. If the Copyright Royalty Judges determine that no such controversy exists, the Copyright Royalty Judges shall authorize the Librarian of Congress to proceed to distribute such fees to the copyright owners entitled to receive them, or to their designated agents, subject to the deduction of reasonable administrative costs under this section. If the Copyright Royalty Judges find the existence of a controversy, the Copyright Royalty Judges shall, pursuant to chapter 8 of this title, conduct a proceeding to determine the distribution of royalty fees.

(C) During the pendency of any proceeding under this subsection, the Copyright Royalty Judges shall have the discretion to authorize the Librarian of Congress to proceed to distribute any amounts that are not in controversy.

(5) 3.75 percent rate and syndicated exclusivity surcharge not applicable to multicast streams.—The royalty rates specified in sections 256.2(c) and 256.2(d) of title 37, Code of Federal Regulations (commonly referred to as the "3.75 percent rate" and the "syndicated exclusivity surcharge", respectively), as in effect on the date of the enactment of the Satellite Television Extension and Localism Act of 2010, as such rates may be adjusted, or such sections redesignated, thereafter by the Copyright Royalty Judges, shall not apply to the secondary transmission of a multicast stream.

(6) Verification of accounts and fee payments.—The Register of Copyrights shall issue regulations to provide for the confidential verification by copyright owners whose works were embodied in the secondary transmissions of primary transmissions pursuant to this section of the information reported on the semiannual statements of account filed under this subsection for accounting periods beginning on or after January 1, 2010, in order that the auditor designated under subparagraph (A) is able to confirm the correctness of the calculations and royalty payments reported therein. The regulations shall—

(A) establish procedures for the designation of a qualified independent auditor—

(i) with exclusive authority to request verification of such a statement of account on behalf of all copyright owners whose works were the subject of secondary transmissions of primary transmissions by the cable system (that deposited the statement) during the accounting period covered by the statement; and

(ii) who is not an officer, employee, or agent of any such copyright owner for any purpose other than such audit;

(B) establish procedures for safeguarding all nonpublic financial and business information provided under this paragraph;

(C)(i) require a consultation period for the independent auditor to review its conclusions with a designee of the cable system;

(ii) establish a mechanism for the cable system to remedy any errors identified in the auditor's report and to cure any underpayment identified; and

(iii) provide an opportunity to remedy any disputed facts or conclusions;

(D) limit the frequency of requests for verification for a particular cable system and the number of audits that a multiple system operator can be required to undergo in a single year; and

(E) permit requests for verification of a statement of account to be made only within 3 years after the last day of the year in which the statement of account is filed.

(7) Acceptance of additional deposits.—Any royalty fee payments received by the Copyright Office from cable systems for the secondary transmission of primary transmissions that are in addition to the payments calculated and deposited in accordance with this subsection shall be deemed to have been deposited for the particular accounting period for which they are received and shall be distributed as specified under this subsection.

(e) Nonsimultaneous Secondary Transmissions by Cable Systems.—

(1) Notwithstanding those provisions of the subsection (f)(2) relating to nonsimultaneous secondary transmissions by a cable system, any such transmissions are actionable as an act of infringement under section 501, and are fully subject to the remedies provided by sections 502 through 506 and section 510, unless—

(A) the program on the videotape is transmitted no more than one time to the cable system's subscribers;

(B) the copyrighted program, episode, or motion picture videotape, including the commercials contained within such program, episode, or picture, is transmitted without deletion or editing;

(C) an owner or officer of the cable system

(i) prevents the duplication of the videotape while in the possession of the system,

(ii) prevents unauthorized duplication while in the possession of the facility making the videotape for the system if the system owns or controls the facility, or takes reasonable precautions to prevent such duplication if it does not own or control the facility,

(iii) takes adequate precautions to prevent duplication while the tape is being transported, and

(iv) subject to paragraph (2), erases or destroys, or causes the erasure or destruction of, the videotape;

(D) within forty-five days after the end of each calendar quarter, an owner or officer of the cable system executes an affidavit attesting

(i) to the steps and precautions taken to prevent duplication of the videotape, and

(ii) subject to paragraph (2), to the erasure or destruction of all videotapes made or used during such quarter;

(E) such owner or officer places or causes each such affidavit, and affidavits received pursuant to paragraph (2)(C), to be placed in a file, open to public inspection, at such system's main office in the community where the transmission is made or in the nearest community where such system maintains an office; and

(F) the nonsimultaneous transmission is one that the cable system would be authorized to transmit under the rules, regulations, and authorizations of the Federal Communications Commission in effect at the time of the nonsimultaneous transmission if the transmission had been made simultaneously, except that this subparagraph shall not apply to inadvertent or accidental transmissions.

(2) If a cable system transfers to any person a videotape of a program nonsimultaneously transmitted by it, such transfer is actionable as an act of infringement under section 501, and is fully subject to the remedies provided by sections

502 through 506, except that, pursuant to a written, nonprofit contract providing for the equitable sharing of the costs of such videotape and its transfer, a videotape nonsimultaneously transmitted by it, in accordance with paragraph (1), may be transferred by one cable system in Alaska to another system in Alaska, by one cable system in Hawaii permitted to make such nonsimultaneous transmissions to another such cable system in Hawaii, or by one cable system in Guam, the Northern Mariana Islands, or the Trust Territory of the Pacific Islands, to another cable system in any of those five entities, if—

(A) each such contract is available for public inspection in the offices of the cable systems involved, and a copy of such contract is filed, within thirty days after such contract is entered into, with the Copyright Office (which Office shall make each such contract available for public inspection); and

(B) the cable system to which the videotape is transferred complies with paragraph (1)(A), (B), (C)(i), (iii), and (iv), and (D) through (F);

(C) such system provides a copy of the affidavit required to be made in accordance with paragraph (1)(D) to each cable system making a previous nonsimultaneous transmission of the same videotape.

(3) This subsection shall not be construed to supersede the exclusivity protection provisions of any existing agreement, or any such agreement hereafter entered into, between a cable system and a television broadcast station in the area in which the cable system is located, or a network with which such station is affiliated.

(4) As used in this subsection, the term "videotape" means the reproduction of the images and sounds of a program or programs broadcast by a television broadcast station licensed by the Federal Communications Commission, regardless of the nature of the material objects, such as tapes or films, in which the reproduction is embodied.

(f) Definitions.—As used in this section, the following terms mean the following:

(1) Primary transmission.—A "primary transmission" is a transmission made to the public by a transmitting facility whose signals are being received and further transmitted by a secondary transmission service, regardless of where or when the performance or display was first transmitted. In the case of a television broadcast station, the primary stream and any multicast streams transmitted by the station constitute primary transmissions.

(2) Secondary transmission.—A "secondary transmission" is the further transmitting of a primary transmission simultaneously with the primary transmission, or nonsimultaneously with the primary transmission if by a cable system not located in whole or in part within the boundary of the fortyeight contiguous States, Hawaii, or Puerto Rico: *Provided, however,* That a nonsimultaneous further transmission by a cable system located in Hawaii of a primary transmission shall be deemed to be a secondary transmission if the carriage of the television broadcast signal comprising such further transmission is permissible under the rules, regulations, or authorizations of the Federal Communications Commission.

(3) Cable system.—A "cable system" is a facility, located in any State, territory, trust territory, or possession of the United States, that in whole or in part

receives signals transmitted or programs broadcast by one or more television broadcast stations licensed by the Federal Communications Commission, and makes secondary transmissions of such signals or programs by wires, cables, microwave, or other communications channels to subscribing members of the public who pay for such service. For purposes of determining the royalty fee under subsection (d)(1), two or more cable systems in contiguous communities under common ownership or control or operating from one headend shall be considered as one system.

(4) Local service area of a primary transmitter.—The "local service area of a primary transmitter", in the case of both the primary stream and any multicast streams transmitted by a primary transmitter that is a television broadcast station, comprises the area where such primary transmitter could have insisted upon its signal being retransmitted by a cable system pursuant to the rules, regulations, and authorizations of the Federal Communications Commission in effect on April 15, 1976, or such station's television market as defined in section 76.55(e) of title 47, Code of Federal Regulations (as in effect on September 18, 1993), or any modifications to such television market made, on or after September 18, 1993, pursuant to section 76.55(e) or 76.59 of title 47, Code of Federal Regulations or within the noise-limited contour as defined in 73.622(e)(1) of title 47, Code of Federal Regulations, or in the case of a television broadcast station licensed by an appropriate governmental authority of Canada or Mexico, the area in which it would be entitled to insist upon its signal being retransmitted if it were a television broadcast station subject to such rules, regulations, and authorizations. In the case of a low power television station, the "local service area of a primary transmitter" comprises the area within 35 miles of the transmitter site, except that in the case of such a station located in a standard metropolitan statistical area which has one of the 50 largest populations of all standard metropolitan statistical areas (based on the 1980 decennial census of population taken by the Secretary of Commerce), the number of miles shall be 20 miles. The "local service area of a primary transmitter", in the case of a radio broadcast station, comprises the primary service area of such station, pursuant to the rules and regulations of the Federal Communications Commission.

(5) Distant signal equivalent.—

(A) In general.—Except as provided under subparagraph (B), a "distant signal equivalent"—

(i) is the value assigned to the secondary transmission of any nonnetwork television programming carried by a cable system in whole or in part beyond the local service area of the primary transmitter of such programming; and

(ii) is computed by assigning a value of one to each primary stream and to each multicast stream (other than a simulcast) that is an independent station, and by assigning a value of one-quarter to each primary stream and to each multicast stream (other than a simulcast) that is a network station or a noncommercial educational station.

(B) Exceptions.—The values for independent, network, and noncommercial educational stations specified in subparagraph (A) are subject to the following:

(i) Where the rules and regulations of the Federal Communications Commission require a cable system to omit the further transmission of a particular program and such rules and regulations also permit the substitution of another program embodying a performance or display of a work in place of the omitted transmission, or where such rules and regulations in effect on the date of the enactment of the Copyright Act of 1976 permit a cable system, at its election, to effect such omission and substitution of a nonlive program or to carry additional programs not transmitted by primary transmitters within whose local service area the cable system is located, no value shall be assigned for the substituted or additional program.

(ii) Where the rules, regulations, or authorizations of the Federal Communications Commission in effect on the date of the enactment of the Copyright Act of 1976 permit a cable system, at its election, to omit the further transmission of a particular program and such rules, regulations, or authorizations also permit the substitution of another program embodying a performance or display of a work in place of the omitted transmission, the value assigned for the substituted or additional program shall be, in the case of a live program, the value of one full distant signal equivalent multiplied by a fraction that has as its numerator the number of days in the year in which such substitution occurs and as its denominator the number of days in the year.

(iii) In the case of the secondary transmission of a primary transmitter that is a television broadcast station pursuant to the late-night or specialty programming rules of the Federal Communications Commission, or the secondary transmission of a primary transmitter that is a television broadcast station on a part-time basis where full-time carriage is not possible because the cable system lacks the activated channel capacity to retransmit on a full-time basis all signals that it is authorized to carry, the values for independent, network, and noncommercial educational stations set forth in subparagraph (A), as the case may be, shall be multiplied by a fraction that is equal to the ratio of the broadcast hours of such primary transmitter retransmitted by the cable system to the total broadcast hours of the primary transmitter.

(iv) No value shall be assigned for the secondary transmission of the primary stream or any multicast streams of a primary transmitter that is a television broadcast station in any community that is within the local service area of the primary transmitter.

(6) Network Station.—

(A) Treatment of primary stream.—The term "network station" shall be applied to a primary stream of a television broadcast station that is owned or operated by, or affiliated with, one or more of the television networks in the United States providing nationwide transmissions, and that transmits a substantial part of the programming supplied by such networks for a substantial part of the primary stream's typical broadcast day.

(B) Treatment of Multicast Streams.—The term "network station" shall be applied to a multicast stream on which a television broadcast station transmits all or substantially all of the programming of an interconnected program service that—

(i) is owned or operated by, or affiliated with, one or more of the television networks described in subparagraph (A); and

(ii) offers programming on a regular basis for 15 or more hours per week to at least 25 of the affiliated television licensees of the interconnected program service in 10 or more States.

(7) Independent station.—The term "independent station" shall be applied to the primary stream or a multicast stream of a television broadcast station that is not a network station or a noncommercial educational station.

(8) Noncommercial educational station.—The term "noncommercial educational station" shall be applied to the primary stream or a multicast stream of a television broadcast station that is a noncommercial educational broadcast station as defined in section 397 of the Communications Act of 1934, as in effect on the date of the enactment of the Satellite Television Extension and Localism Act of 2010.

(9) Primary stream.—A "primary stream" is—

(A) the single digital stream of programming that, before June 12, 2009, was substantially duplicating the programming transmitted by the television broadcast station as an analog signal; or

(B) if there is no stream described in subparagraph (A), then the single digital stream of programming transmitted by the television broadcast station for the longest period of time.

(10) Primary transmitter.—A "primary transmitter" is a television or radio broadcast station licensed by the Federal Communications Commission, or by an appropriate governmental authority of Canada or Mexico, that makes primary transmissions to the public.

(11) Multicast stream.—A "multicast stream" is a digital stream of programming that is transmitted by a television broadcast station and is not the station's primary stream.

(12) Simulcast.—A "simulcast" is a multicast stream of a television broadcast station that duplicates the programming transmitted by the primary stream or another multicast stream of such station.

(13) Subscriber; subscribe.—

(A) Subscriber.—The term "subscriber" means a person or entity that receives a secondary transmission service from a cable system and pays a fee for the service, directly or indirectly, to the cable system.

(B) Subscribe.—The term "subscribe" means to elect to become a subscriber.

§ 112 Limitations on Exclusive Rights: Ephemeral Recordings

(a)(1) Notwithstanding the provisions of section 106, and except in the case of a motion picture or other audiovisual work, it is not an infringement of copyright for a transmitting organization entitled to transmit to the public a performance or display of a work, under a license, including a statutory license under section 114(f), or transfer of the copyright or under the limitations on exclusive rights in sound recordings specified by section 114(a) or for a transmitting organization that is a broadcast radio or television station licensed as such by the Federal Communications Commission and that makes a broadcast transmission of a performance of a sound recording in a digital format on a nonsubscription basis, to make no more than one copy or phonorecord of a particular transmission program embodying the performance or display, if—

(A) the copy or phonorecord is retained and used solely by the transmitting organization that made it, and no further copies or phonorecords are reproduced from it; and

(B) the copy or phonorecord is used solely for the transmitting organization's own transmissions within its local service area, or for purposes of archival preservation or security; and

(C) unless preserved exclusively for archival purposes, the copy or phonorecord is destroyed within six months from the date the transmission program was first transmitted to the public.

(2) In a case in which a transmitting organization entitled to make a copy or phonorecord under paragraph (1) in connection with the transmission to the public of a performance or display of a work is prevented from making such copy or phonorecord by reason of the application by the copyright owner of technical measures that prevent the reproduction of the work, the copyright owner shall make available to the transmitting organization the necessary means for permitting the making of such copy or phonorecord as permitted under that paragraph, if it is technologically feasible and economically reasonable for the copyright owner to do so. If the copyright owner fails to do so in a timely manner in light of the transmitting organization's reasonable business requirements, the transmitting organization shall not be liable for a violation of section 1201(a)(1) of this title for engaging in such activities as are necessary to make such copies or phonorecords as permitted under paragraph (1) of this subsection.

(b) Notwithstanding the provisions of section 106, it is not an infringement of copyright for a governmental body or other nonprofit organization entitled to transmit a performance or display of a work, under section 110(2) or under the limitations on exclusive rights in sound recordings specified by section 114(a), to make no more than thirty copies or phonorecords of a particular transmission program embodying the performance or display, if—

(1) no further copies or phonorecords are reproduced from the copies or phonorecords made under this clause; and

(2) except for one copy or phonorecord that may be preserved exclusively for archival purposes, the copies or phonorecords are destroyed within seven years from the date the transmission program was first transmitted to the public.

(c) Notwithstanding the provisions of section 106, it is not an infringement of copyright for a governmental body or other nonprofit organization to make for distribution no more than one copy or phonorecord, for each transmitting organization specified in clause (2) of this subsection, of a particular transmission program embodying a performance of a nondramatic musical work of a religious nature, or of a sound recording of such a musical work, if—

(1) there is no direct or indirect charge for making or distributing any such copies or phonorecords; and

(2) none of such copies or phonorecords is used for any performance other than a single transmission to the public by a transmitting organization entitled to transmit to the public a performance of the work under a license or transfer of the copyright; and

(3) except for one copy or phonorecord that may be preserved exclusively for archival purposes, the copies or phonorecords are all destroyed within one year from the date the transmission program was first transmitted to the public.

(d) Notwithstanding the provisions of section 106, it is not an infringement of copyright for a governmental body or other nonprofit organization entitled to transmit a performance of a work under section 110(8) to make no more than ten copies or phonorecords embodying the performance, or to permit the use of any such copy or phonorecord by any governmental body or nonprofit organization entitled to transmit a performance of a work under section 110(8), if—

(1) any such copy or phonorecord is retained and used solely by the organization that made it, or by a governmental body or nonprofit organization entitled to transmit a performance of a work under section 110(8), and no further copies or phonorecords are reproduced from it; and

(2) any such copy or phonorecord is used solely for transmissions authorized under section 110(8), or for purposes of archival preservation or security; and

(3) the governmental body or nonprofit organization permitting any use of any such copy or phonorecord by any governmental body or nonprofit organization under this subsection does not make any charge for such use.

(e) Statutory License.—(1) A transmitting organization entitled to transmit to the public a performance of a sound recording under the limitation on exclusive rights specified by section 114(d)(1)(C)(iv) or under a statutory license in accordance with section 114(f) is entitled to a statutory license, under the conditions specified by this subsection, to make no more than 1 phonorecord of the sound recording (unless the terms and conditions of the statutory license allow for more), if the following conditions are satisfied:

(A) The phonorecord is retained and used solely by the transmitting organization that made it, and no further phonorecords are reproduced from it.

(B) The phonorecord is used solely for the transmitting organization's own transmissions originating in the United States under a statutory license in accordance with section 114(f) or the limitation on exclusive rights specified by section 114(d)(1)(C)(iv).

(C) Unless preserved exclusively for purposes of archival preservation, the phonorecord is destroyed within 6 months from the date the sound recording was first transmitted to the public using the phonorecord.

(D) Phonorecords of the sound recording have been distributed to the public under the authority of the copyright owner or the copyright owner authorizes the transmitting entity to transmit the sound recording, and the transmitting entity makes the phonorecord under this subsection from a phonorecord lawfully made and acquired under the authority of the copyright owner.

(2) Notwithstanding any provision of the antitrust laws, any copyright owners of sound recordings and any transmitting organizations entitled to a statutory license under this subsection may negotiate and agree upon royalty rates and license terms and conditions for making phonorecords of such sound recordings under this section and the proportionate division of fees paid among copyright owners, and may designate common agents to negotiate, agree to, pay, or receive such royalty payments.

(3) Proceedings under chapter 8 shall determine reasonable rates and terms of royalty payments for the activities specified by paragraph (1) during the 5-year period beginning on January 1 of the second year following the year in which the proceedings are to be commenced, or such other period as the parties may agree. Such rates shall include a minimum fee for each type of service offered by transmitting organizations. Any copyright owners of sound recordings or any transmitting organizations entitled to a statutory license under this subsection may submit to the Copyright Royalty Judges licenses covering such activities with respect to such sound recordings. The parties to each proceeding shall bear their own costs.

(4) The schedule of reasonable rates and terms determined by the Copyright Royalty Judges shall, subject to paragraph (5), be binding on all copyright owners of sound recordings and transmitting organizations entitled to a statutory license under this subsection during the 5-year period specified in paragraph (3), or such other period as the parties may agree. Such rates shall include a minimum fee for each type of service offered by transmitting organizations. The Copyright Royalty Judges shall establish rates that most clearly represent the fees that would have been negotiated in the marketplace between a willing buyer and a willing seller. In determining such rates and terms, the Copyright Royalty Judges shall base their decision on economic, competitive, and programming information presented by the parties, including—

(A) whether use of the service may substitute for or may promote the sales of phonorecords or otherwise interferes with or enhances the copyright owner's traditional streams of revenue; and

(B) the relative roles of the copyright owner and the transmitting organization in the copyrighted work and the service made available to the public with respect to relative creative contribution, technological contribution, capital investment, cost, and risk.

In establishing such rates and terms, the Copyright Royalty Judges may consider the rates and terms under voluntary license agreements described in paragraphs (2) and (3). The Copyright Royalty Judges shall also establish requirements by which copyright owners may receive reasonable notice of the use of their sound recordings under this section, and under which records of such use shall be kept and made available by transmitting organizations entitled to obtain a statutory license under this subsection.

(5) License agreements voluntarily negotiated at any time between 1 or more copyright owners of sound recordings and 1 or more transmitting organizations entitled to obtain a statutory license under this subsection shall be given effect in lieu of any decision by the Librarian of Congress or determination by the Copyright Royalty Judges.

(6)(A) Any person who wishes to make a phonorecord of a sound recording under a statutory license in accordance with this subsection may do so without infringing the exclusive right of the copyright owner of the sound recording under section 106(1)—

(i) by complying with such notice requirements as the Copyright Royalty Judges shall prescribe by regulation and by paying royalty fees in accordance with this subsection; or

(ii) if such royalty fees have not been set, by agreeing to pay such royalty fees as shall be determined in accordance with this subsection.

(B) Any royalty payments in arrears shall be made on or before the 20th day of the month next succeeding the month in which the royalty fees are set.

(7) If a transmitting organization entitled to make a phonorecord under this subsection is prevented from making such phonorecord by reason of the application by the copyright owner of technical measures that prevent the reproduction of the sound recording, the copyright owner shall make available to the transmitting organization the necessary means for permitting the making of such phonorecord as permitted under this subsection, if it is technologically feasible and economically reasonable for the copyright owner to do so. If the copyright owner fails to do so in a timely manner in light of the transmitting organization's reasonable business requirements, the transmitting organization shall not be liable for a violation of section 1201(a)(1) of this title for engaging in such activities as are necessary to make such phonorecords as permitted under this subsection.

(8) Nothing in this subsection annuls, limits, impairs, or otherwise affects in any way the existence or value of any of the exclusive rights of the copyright owners in a sound recording, except as otherwise provided in this subsection, or in a musical work, including the exclusive rights to reproduce and distribute a sound recording or musical work, including by means of a digital phonorecord delivery, under section 106(1), 106(3), and 115, and the right to perform publicly a sound recording or musical work, including by means of a digital audio transmission, under sections 106(4) and 106(6).

(f)(1) Notwithstanding the provisions of section 106, and without limiting the application of subsection (b), it is not an infringement of copyright for a governmental body or other nonprofit educational institution entitled under section 110(2) to transmit a performance or display to make copies or phonorecords of a work that is in digital form and, solely to the extent permitted in paragraph (2), of a work that is in analog form, embodying the performance or display to be used for making transmissions authorized under section 110(2), if—

(A) such copies or phonorecords are retained and used solely by the body or institution that made them, and no further copies or phonorecords are reproduced from them, except as authorized under section 110(2); and

(B) such copies or phonorecords are used solely for transmissions authorized under section 110(2).

(2) This subsection does not authorize the conversion of print or other analog versions of works into digital formats, except that such conversion is permitted hereunder, only with respect to the amount of such works authorized to be performed or displayed under section 110(2), if—

(A) no digital version of the work is available to the institution; or

(B) the digital version of the work that is available to the institution is subject to technological protection measures that prevent its use for section 110(2).

(g) The transmission program embodied in a copy or phonorecord made under this section is not subject to protection as a derivative work under this title except with the express consent of the owners of copyright in the preexisting works employed in the program.

§ 113 Scope of Exclusive Rights in Pictorial, Graphic, and Sculptural Works

(a) Subject to the provisions of subsections (b) and (c) of this section, the exclusive right to reproduce a copyrighted pictorial, graphic, or sculptural work in copies under section 106 includes the right to reproduce the work in or on any kind of article, whether useful or otherwise.

(b) This title does not afford, to the owner of copyright in a work that portrays a useful article as such, any greater or lesser rights with respect to the making, distribution, or display of the useful article so portrayed than those afforded to such works under the law, whether title 17 or the common law or statutes of a State, in effect on December 31, 1977, as held applicable and construed by a court in an action brought under this title.

(c) In the case of a work lawfully reproduced in useful articles that have been offered for sale or other distribution to the public, copyright does not include any right to prevent the making, distribution, or display of pictures or photographs of such articles in connection with advertisements or commentaries related to the distribution or display of such articles, or in connection with news reports.

(d)(1) In a case in which—

(A) a work of visual art has been incorporated in or made part of a building in such a way that removing the work from the building will cause the destruction, distortion, mutilation, or other modification of the work as described in section 106A(a)(3), and

(B) the author consented to the installation of the work in the building either before the effective date set forth in section 610(a) of the Visual Artists Rights Act of 1990, or in a written instrument executed on or after such effective date that is signed by the owner of the building and the author and that specifies that installation of the work may subject the work to destruction, distortion, mutilation, or other modification, by reason of its removal, then the rights conferred by paragraphs (2) and (3) of section 106A(a) shall not apply.

(2) If the owner of a building wishes to remove a work of visual art which is a part of such building and which can be removed from the building without the destruction, distortion, mutilation, or other modification of the work as described in section 106A(a)(3), the author's rights under paragraphs (2) and (3) of section 106A(a) shall apply unless—

(A) the owner has made a diligent, good faith attempt without success to notify the author of the owner's intended action affecting the work of visual art, or

(B) the owner did provide such notice in writing and the person so notified failed, within 90 days after receiving such notice, either to remove the work or to pay for its removal.

For purposes of subparagraph (A), an owner shall be presumed to have made a diligent, good faith attempt to send notice if the owner sent such notice by registered mail to the author at the most recent address of the author that was recorded with the Register of Copyrights pursuant to paragraph (3). If the work is removed at the expense of the author, title to that copy of the work shall be deemed to be in the author.

(3) The Register of Copyrights shall establish a system of records whereby any author of a work of visual art that has been incorporated in or made part of a building, may record his or her identity and address with the Copyright Office. The Register shall also establish procedures under which any such author may update the information so recorded, and procedures under which owners of buildings may record with the Copyright Office evidence of their efforts to comply with this subsection.

§ 114 Scope of Exclusive Rights in Sound Recordings

(a) The exclusive rights of the owner of copyright in a sound recording are limited to the rights specified by clauses (1), (2), (3) and (6) of section 106, and do not include any right of performance under section 106(4).

(b) The exclusive right of the owner of copyright in a sound recording under clause (1) of section 106 is limited to the right to duplicate the sound recording in the form of phonorecords or copies that directly or indirectly recapture the actual sounds fixed in the recording. The exclusive right of the owner of copyright in a sound recording under clause (2) of section 106 is limited to the right to prepare a derivative work in which the actual sounds fixed in the sound recording are rearranged, remixed, or otherwise altered in sequence or quality. The exclusive rights of the owner of copyright in a sound recording under clauses (1) and (2) of section 106 do not extend to the making or duplication of another sound recording that consists entirely of an independent fixation of other sounds, even though such sounds imitate or simulate those in the copyrighted sound recording. The exclusive rights of the owner of copyright in a sound recording under clauses (1), (2), and (3) of section 106 do not apply to sound recordings included in educational television and radio programs (as defined in section 397 of title 47) distributed or transmitted by or through public broadcasting entities (as defined by section 118(f)): *Provided,* That copies or phonorecords of said programs are not commercially distributed by or through public broadcasting entities to the general public.

(c) This section does not limit or impair the exclusive right to perform publicly, by means of a phonorecord, any of the works specified by section 106(4).

(d) Limitations on Exclusive Right.—Notwithstanding the provisions of section 106(6)—

(1) Exempt transmissions and retransmissions.—The performance of a sound recording publicly by means of a digital audio transmission, other than as a part of an interactive service, is not an infringement of section 106(6) if the performance is part of—

(A) a nonsubscription broadcast transmission;

(B) a retransmission of a nonsubscription broadcast transmission: *Provided,* That, in the case of a retransmission of a radio station's broadcast transmission—

(i) the radio station's broadcast transmission is not willfully or repeatedly retransmitted more than a radius of 150 miles from the site of the radio broadcast transmitter, however—

(I) the 150 mile limitation under this clause shall not apply when a nonsubscription broadcast transmission by a radio station licensed by the Federal Communications Commission is retransmitted on a nonsubscription basis by a terrestrial broadcast station, terrestrial translator, or terrestrial repeater licensed by the Federal Communications Commission; and

(II) in the case of a subscription retransmission of a nonsubscription broadcast retransmission covered by subclause (I), the 150 mile radius shall be measured from the transmitter site of such broadcast retransmitter;

(ii) the retransmission is of radio station broadcast transmissions that are—

(I) obtained by the retransmitter over the air;

(II) not electronically processed by the retransmitter to deliver separate and discrete signals; and

(III) retransmitted only within the local communities served by the retransmitter;

(iii) the radio station's broadcast transmission was being retransmitted to cable systems (as defined in section 111(f)) by a satellite carrier on January 1, 1995, and that retransmission was being retransmitted by cable systems as a separate and discrete signal, and the satellite carrier obtains the radio station's broadcast transmission in an analog format: *Provided,* That the broadcast transmission being retransmitted may embody the programming of no more than one radio station; or

(iv) the radio station's broadcast transmission is made by a noncommercial educational broadcast station funded on or after January 1, 1995, under section 396(k) of the Communications Act of 1934 (47 U.S.C. 396(k)), consists solely of noncommercial educational and cultural radio programs, and the retransmission, whether or not simultaneous, is a nonsubscription terrestrial broadcast retransmission; or

(C) a transmission that comes within any of the following categories—

(i) a prior or simultaneous transmission incidental to an exempt transmission, such as a feed received by and then retransmitted by an exempt transmitter: *Provided,* That such incidental transmissions do not include any subscription transmission directly for reception by members of the public;

(ii) a transmission within a business establishment, confined to its premises or the immediately surrounding vicinity;

(iii) a retransmission by any retransmitter, including a multichannel video programming distributor as defined in section 602(12) of the Communications Act of 1934 (47 U.S.C. 522 (12)), of a transmission by a transmitter licensed to publicly perform the sound recording as a part of that transmission, if the retransmission is simultaneous with the licensed transmission and authorized by the transmitter; or

(iv) a transmission to a business establishment for use in the ordinary course of its business: *Provided,* That the business recipient does not retransmit the transmission outside of its premises or the immediately surrounding vicinity, and that the transmission does not exceed the sound recording performance complement. Nothing in this clause shall limit the scope of the exemption in clause (ii).

(2) Statutory licensing of certain transmissions.—The performance of a sound recording publicly by means of a subscription digital audio transmission not exempt under paragraph (1), an eligible nonsubscription transmission, or a transmission not exempt under paragraph (1) that is made by a preexisting satellite digital audio radio service shall be subject to statutory licensing, in accordance with subsection (f) if—

(A) (i) the transmission is not part of an interactive service;

(ii) except in the case of a transmission to a business establishment, the transmitting entity does not automatically and intentionally cause any device receiving the transmission to switch from one program channel to another; and

(iii) except as provided in section 1002(e), the transmission of the sound recording is accompanied, if technically feasible, by the information encoded in that sound recording, if any, by or under the authority of the copyright owner of that sound recording, that identifies the title of the sound recording, the featured recording artist who performs on the sound recording, and related information, including information concerning the underlying musical work and its writer;

(B) in the case of a subscription transmission not exempt under paragraph (1) that is made by a preexisting subscription service in the same transmission medium used by such service on July 31, 1998, or in the case of a transmission not exempt under paragraph (1) that is made by a preexisting satellite digital audio radio service—

(i) the transmission does not exceed the sound recording performance complement; and

(ii) the transmitting entity does not cause to be published by means of an advance program schedule or prior announcement the titles of the specific sound recordings or phonorecords embodying such sound recordings to be transmitted; and

(C) in the case of an eligible nonsubscription transmission or a subscription transmission not exempt under paragraph (1) that is made by a new subscription service or by a preexisting subscription service other than in the same transmission medium used by such service on July 31, 1998—

(i) the transmission does not exceed the sound recording performance complement, except that this requirement shall not apply in the case of a retransmission of a broadcast transmission if the retransmission is made by a transmitting entity that does not have the right or ability to control the programming of the broadcast station making the broadcast transmission, unless—

(I) the broadcast station makes broadcast transmissions—(aa) in digital format that regularly exceed the sound recording performance complement; or

(bb) in analog format, a substantial portion of which, on a weekly basis, exceed the sound recording performance complement; and

(II) the sound recording copyright owner or its representative has notified the transmitting entity in writing that broadcast transmissions of the copyright owner's sound recordings exceed the sound recording performance complement as provided in this clause;

(ii) the transmitting entity does not cause to be published, or induce or facilitate the publication, by means of an advance program schedule or prior announcement, the titles of the specific sound recordings to be transmitted, the phonorecords embodying such sound recordings, or, other than for illustrative purposes, the names of the featured recording artists, except that this clause does not disqualify a transmitting entity that makes a prior announcement that a particular artist will be featured within an unspecified future time period, and in the case of a retransmission of a broadcast transmission by a transmitting entity that does not have the right or ability to control the programming of the broadcast transmission, the requirement of this clause shall not apply to a prior oral announcement by the broadcast station, or to an advance program schedule published, induced, or facilitated by the broadcast station, if the transmitting entity does not have actual knowledge and has not received written notice from the copyright owner or its representative that the broadcast station publishes or induces or facilitates the publication of such advance program schedule, or if such advance program schedule is a schedule of classical music programming published by the broadcast station in the same manner as published by that broadcast station on or before September 30, 1998;

(iii) the transmission—

(I) is not part of an archived program of less than 5 hours duration;

(II) is not part of an archived program of 5 hours or greater in duration that is made available for a period exceeding 2 weeks;

(III) is not part of a continuous program which is of less than 3 hours duration; or

(IV) is not part of an identifiable program in which performances of sound recordings are rendered in a predetermined order, other than an archived or continuous program, that is transmitted at—

(aa) more than 3 times in any 2-week period that have been publicly announced in advance, in the case of a program of less than 1 hour in duration, or

(bb) more than 4 times in any 2-week period that have been publicly announced in advance, in the case of a program of 1 hour or more in duration, except that the requirement of this subclause shall not apply in the case of a retransmission of a broadcast transmission by a transmitting entity that does not have the right or ability to control the programming of the broadcast transmission, unless the transmitting entity is given notice in writing by the copyright owner of the sound recording that the broadcast station makes broadcast transmissions that regularly violate such requirement;

(iv) the transmitting entity does not knowingly perform the sound recording, as part of a service that offers transmissions of visual images contemporaneously with transmissions of sound recordings, in a manner that is likely to cause confusion, to cause mistake, or to deceive, as to the affiliation, connection, or association of the copyright owner or featured recording artist with the transmitting entity or a particular product or service advertised by the transmitting entity, or as to the origin, sponsorship, or approval by the copyright owner or featured recording artist of the activities of the transmitting entity other than the performance of the sound recording itself;

(v) the transmitting entity cooperates to prevent, to the extent feasible without imposing substantial costs or burdens, a transmission recipient or any other person or entity from automatically scanning the transmitting entity's transmissions alone or together with transmissions by other transmitting entities in order to select a particular sound recording to be transmitted to the transmission recipient, except that the requirement of this clause shall not apply to a satellite digital audio service that is in operation, or that is licensed by the Federal Communications Commission, on or before July 31, 1998;

(vi) the transmitting entity takes no affirmative steps to cause or induce the making of a phonorecord by the transmission recipient, and if the technology used by the transmitting entity enables the transmitting entity to limit the making by the transmission recipient of phonorecords of the transmission directly in a digital format, the transmitting entity sets such technology to limit such making of phonorecords to the extent permitted by such technology;

(vii) phonorecords of the sound recording have been distributed to the public under the authority of the copyright owner or the copyright owner authorizes the transmitting entity to transmit the sound recording, and the transmitting entity makes the transmission from a phonorecord lawfully made under the authority of the copyright owner, except that the requirement of this clause shall not apply to a retransmission of a broadcast transmission by a transmitting entity that does not have the right or ability to control the programming of the broadcast transmission, unless the transmitting entity is given notice in writing by the copyright owner of the sound recording that the broadcast station makes broadcast transmissions that regularly violate such requirement;

(viii) the transmitting entity accommodates and does not interfere with the transmission of technical measures that are widely used by sound recording copyright owners to identify or protect copyrighted works, and that are technically feasible of being transmitted by the transmitting entity without imposing substantial costs on the transmitting entity or resulting in perceptible aural or visual degradation of the digital signal, except that the requirement of this clause shall not apply to a satellite digital audio service that is in operation, or that is licensed under the authority of the Federal Communications Commission, on or before July 31, 1998, to the extent that such service has designed, developed, or made commitments to procure equipment or technology that is not compatible with such technical measures before such technical measures are widely adopted by sound recording copyright owners; and

(ix) the transmitting entity identifies in textual data the sound recording during, but not before, the time it is performed, including the title of the sound recording, the title of the phonorecord embodying such sound recording, if any, and the featured recording artist, in a manner to permit it to be displayed to the transmission recipient by the device or technology intended for receiving the service provided by the transmitting entity, except that the obligation in this clause shall not take effect until 1 year after the date of the enactment of the Digital Millennium Copyright Act and shall not apply in the case of a retransmission of a broadcast transmission by a transmitting entity that does not have the right or ability to control the programming of the broadcast transmission, or in the case in which devices or technology intended for receiving the service provided by the transmitting entity that have the capability to display such textual data are not common in the marketplace.

(3) Licenses for transmissions by interactive services.—

(A) No interactive service shall be granted an exclusive license under section 106(6) for the performance of a sound recording publicly by means of digital audio transmission for a period in excess of 12 months, except that with respect to an exclusive license granted to an interactive service by a licensor that holds the copyright to 1,000 or fewer sound recordings, the period of such license shall not exceed 24 months: Provided, however, That the grantee of such exclusive license shall be ineligible to receive another exclusive license for the performance of that sound recording for a period of 13 months from the expiration of the prior exclusive license.

(B) The limitation set forth in subparagraph (A) of this paragraph shall not apply if—

(i) the licensor has granted and there remain in effect licenses under section 106(6) for the public performance of sound recordings by means of digital audio transmission by at least 5 different interactive services; Provided, however, That each such license must be for a minimum of 10 percent of the copyrighted sound recordings owned by the licensor that have been licensed to interactive services, but in no event less than 50 sound recordings; or

(ii) the exclusive license is granted to perform publicly up to 45 seconds of a sound recording and the sole purpose of the performance is to promote the distribution or performance of that sound recording.

(C) Notwithstanding the grant of an exclusive or nonexclusive license of the right of public performance under section 106(6), an interactive service may not publicly perform a sound recording unless a license has been granted for the public performance of any copyrighted musical work contained in the sound recording: Provided, That such license to publicly perform the copyrighted musical work may be granted either by a performing rights society representing the copyright owner or by the copyright owner.

(D) The performance of a sound recording by means of a retransmission of a digital audio transmission is not an infringement of section 106(6) if—

(i) the retransmission is of a transmission by an interactive service licensed to publicly perform the sound recording to a particular member of the public as part of that transmission; and

(ii) the retransmission is simultaneous with the licensed transmission, authorized by the transmitter, and limited to that particular member of the public intended by the interactive service to be the recipient of the transmission.

(E) For the purposes of this paragraph—

(i) a "licensor" shall include the licensing entity and any other entity under any material degree of common ownership, management, or control that owns copyrights in sound recordings; and

(ii) a "performing rights society" is an association or corporation that licenses the public performance of nondramatic musical works on behalf of the copyright owner, such as the American Society of Composers, Authors and Publishers, Broadcast Music, Inc., and SESAC, Inc.

(4) Rights not otherwise limited.—

(A) Except as expressly provided in this section, this section does not limit or impair the exclusive right to perform a sound recording publicly by means of a digital audio transmission under section 106(6).

(B) Nothing in this section annuls or limits in any way—

(i) the exclusive right to publicly perform a musical work, including by means of a digital audio transmission, under section 106(4);

(ii) the exclusive rights in a sound recording or the musical work embodied therein under sections 106(1), 106(2) and 106(3); or

(iii) any other rights under any other clause of section 106, or remedies available under this title as such rights or remedies exist either before or after the date of enactment of the Digital Performance Right in Sound Recordings Act of 1995.

(C) Any limitations in this section on the exclusive right under section 106(6) apply only to the exclusive right under section 106(6) and not to any other exclusive rights under section 106. Nothing in this section shall be construed to annul, limit, impair or otherwise affect in any way the ability of the owner of a copyright in a sound recording to exercise the rights under sections 106(1), 106(2) and 106(3), or to obtain the remedies available under this title pursuant to such rights, as such rights and remedies exist either before or after the date of enactment of the Digital Performance Right in Sound Recordings Act of 1995.

(e) Authority for Negotiations.—

(1) Notwithstanding any provision of the antitrust laws, in negotiating statutory licenses in accordance with subsection (f), any copyright owners of sound recordings and any entities performing sound recordings affected by this section may negotiate and agree upon the royalty rates and license terms and conditions for the performance of such sound recordings and the proportionate division of fees paid among copyright owners, and may designate common agents on a nonexclusive basis to negotiate, agree to, pay, or receive payments.

(2) For licenses granted under section 106(6), other than statutory licenses, such as for performances by interactive services or performances that exceed the sound recording performance complement—

(A) copyright owners of sound recordings affected by this section may designate common agents to act on their behalf to grant licenses and receive and remit royalty payments: *Provided,* That each copyright owner shall establish the royalty rates and material license terms and conditions unilaterally, that is, not in agreement, combination, or concert with other copyright owners of sound recordings; and

(B) entities performing sound recordings affected by this section may designate common agents to act on their behalf to obtain licenses and collect and pay royalty fees: *Provided,* That each entity performing sound recordings shall determine the royalty rates and material license terms and conditions unilaterally, that is, not in agreement, combination, or concert with other entities performing sound recordings.

(f) Licenses for Certain Nonexempt Transmissions.

(1)(A) Proceedings under chapter 8 shall determine reasonable rates and terms of royalty payments for subscription transmissions by preexisting subscription services and transmissions by preexisting satellite digital audio radio services specified by subsection (d)(2) during the 5-year period beginning on January 1 of the second year following the year in which the proceedings are to be commenced, except in the case of a different transitional period provided under section 6(b)(3) of the Copyright Royalty and Distribution Reform Act of 2004, or such other period as the parties may agree. Such terms and rates shall distinguish among the different types of digital audio transmission services then in operation. Any copyright owners of sound recordings, preexisting subscription services, or preexisting satellite digital audio radio services may submit to the Copyright Royalty Judges licenses covering such subscription transmissions with respect to such sound recordings. The parties to each proceeding shall bear their own costs.

(B) The schedule of reasonable rates and terms determined by the Copyright Royalty Judges shall, subject to paragraph (3), be binding on all copyright owners of sound recordings and entities performing sound recordings affected by this paragraph during the 5-year period specified in subparagraph (A), a transitional period provided under section 6(b)(3) of the Copyright Royalty and Distribution Reform Act of 2004, or such other period as the parties may agree. In establishing rates and terms for preexisting subscription services and preexisting satellite digital audio radio services, in addition to the objectives set forth in section 801(b)(1), the Copyright Royalty Judges may consider the rates and terms for comparable types of subscription digital audio transmission services and comparable circumstances under voluntary license agreements described in subparagraph (A).

(C) The procedures under subparagraphs (A) and (B) also shall be initiated pursuant to a petition filed by any copyright owners of sound recordings, any preexisting subscription services, or any preexisting satellite digital audio radio services indicating that a new type of subscription digital audio transmission service on which sound recordings are performed is or is about to become operational, for the purpose of determining reasonable terms and rates of royalty payments with respect to such new type of transmission service for the period beginning with the inception of such new type of service and ending on the date on which the royalty rates and terms for subscription digital audio transmission services most recently determined under subparagraph (A) or (B) and chapter 8 expire, or such other period as the parties may agree.

(2)(A) Proceedings under chapter 8 shall determine reasonable rates and terms of royalty payments for public performances of sound recordings by means of eligible nonsubscription transmission services and new subscription services specified by subsection (d)(2) during the 5-year period beginning on January 1 of the second year following the year in which the proceedings are to be commenced, except in the case of a different transitional period provided under section 6(b)(3) of the Copyright Royalty and Distribution Reform Act of 2004, or such other period as the parties may agree. Such rates and terms shall distinguish among the different types of eligible nonsubscription transmission services and new subscription services then in operation and shall include a minimum fee for each such type of service. Any copyright owners of sound recordings or any entities performing sound recordings affected by this paragraph may submit to the Copyright Royalty Judges licenses covering such eligible nonsubscription transmissions and new subscription services with respect to such sound recordings. The parties to each proceeding shall bear their own costs.

(B) The schedule of reasonable rates and terms determined by the Copyright Royalty Judges shall, subject to paragraph (3), be binding on all copyright owners of sound recordings and entities performing sound recordings affected by this paragraph during the 5-year period specified in subparagraph (A), a transitional period provided under section 6(b)(3) of the Copyright Royalty and Distribution Act of 2004, or such other period as the parties may agree. Such rates and terms shall distinguish among the different types of eligible nonsubscription transmission services then in operation and shall include a minimum fee for each such type of service, such differences to be based on criteria including, but not limited to, the quantity and nature of the use of sound recordings and the degree to which use of the service may substitute for or may promote the purchase of phonorecords by consumers. In establishing rates and terms for transmissions by eligible nonsubscription services and new subscription services, the Copyright Royalty Judges shall establish rates and terms that most clearly represent the rates and terms that would have been negotiated in the marketplace between a willing buyer and a willing seller. In determining such rates and terms, the Copyright Royalty Judges shall base their decision on economic, competitive and programming information presented by the parties, including—

(i) whether use of the service may substitute for or may promote the sales of phonorecords or otherwise may interfere with or may enhance the sound recording copyright owner's other streams of revenue from its sound recordings; and

(ii) the relative roles of the copyright owner and the transmitting entity in the copyrighted work and the service made available to the public with respect to relative creative contribution, technological contribution, capital investment, cost, and risk.

In establishing such rates and terms, the Copyright Royalty Judges may consider the rates and terms for comparable types of digital audio transmission services and comparable circumstances under voluntary license agreements described in subparagraph (A).

(C) The procedures under subparagraphs (A) and (B) shall also be initiated pursuant to a petition filed by any copyright owners of sound recordings or any eligible nonsubscription service or new subscription service indicating that a new type of eligible nonsubscription service or new subscription service on which sound recordings are performed is or is about to become operational, for the purpose of determining reasonable terms and rates of royalty payments with respect to such new type of service for the period beginning with the inception of such new type of service and ending on the date on which the royalty rates and terms for eligible nonsubscription services and new subscription services, as the case may be, most recently determined under subparagraph (A) or (B) and chapter 8 expire, or such other period as the parties may agree.

(3) License agreements voluntarily negotiated at any time between 1 or more copyright owners of sound recordings and 1 or more entities performing sound recordings shall be given effect in lieu of any decision by the Librarian of Congress or determination by the Copyright Royalty Judges.

(4)(A) The Copyright Royalty Judges shall also establish requirements by which copyright owners may receive reasonable notice of the use of their sound recordings under this section, and under which records of such use shall be kept and made available by entities performing sound recordings. The notice and recordkeeping rules in effect on the day before the effective date of the Copyright Royalty and Distribution Reform Act of 2004 shall remain in effect unless and until new regulations are promulgated by the Copyright Royalty Judges. If new regulations are promulgated under this subparagraph, the Copyright Royalty Judges shall take into account the substance and effect of the rules in effect on the day before the effective date of the Copyright Royalty and Distribution Reform Act of 2004 and shall, to the extent practicable, avoid significant disruption of the functions of any designated agent authorized to collect and distribute royalty fees.

(B) Any person who wishes to perform a sound recording publicly by means of a transmission eligible for statutory licensing under this subsection may do so without infringing the exclusive right of the copyright owner of the sound recording—

(i) by complying with such notice requirements as the Copyright Royalty Judges shall prescribe by regulation and by paying royalty fees in accordance with this subsection; or

(ii) if such royalty fees have not been set, by agreeing to pay such royalty fees as shall be determined in accordance with this subsection.

(C) Any royalty payments in arrears shall be made on or before the twentieth day of the month next succeeding the month in which the royalty fees are set.

(5)(A) Notwithstanding section 112(e) and the other provisions of this subsection, the receiving agent may enter into agreements for the reproduction and performance of sound recordings under section 112(e) and this section by any 1 or more commercial webcasters or noncommercial webcasters for a period of not more than 11 years beginning on January 1, 2005, that, once published in the Federal Register pursuant to subparagraph (B), shall be binding on all copyright owners of sound recordings and other persons entitled to payment under this section, in lieu of any determination by the Copyright Royalty Judges. Any such agreement for commercial webcasters may include provisions for payment of royalties on the basis of a percentage of revenue or expenses, or both, and include a minimum fee. Any such agreement may include other terms and conditions, including requirements by which copyright owners may receive notice of the use of their sound recordings and under which records of such use shall be kept and made available by commercial webcasters or noncommercial webcasters. The receiving agent shall be under no obligation to negotiate any such agreement. The receiving agent shall have no obligation to any copyright owner of sound recordings or any other person entitled to payment under this section in negotiating any such agreement, and no liability to any copyright owner of sound recordings or any other person entitled to payment under this section for having entered into such agreement.

(B) The Copyright Office shall cause to be published in the Federal Register any agreement entered into pursuant to subparagraph (A). Such publication shall include a statement containing the substance of subparagraph (C). Such agreements shall not be included in the Code of Federal Regulations. Thereafter, the terms of such agreement shall be available, as an option, to any commercial webcaster or noncommercial webcaster meeting the eligibility conditions of such agreement.

(C) Neither subparagraph (A) nor any provisions of any agreement entered into pursuant to subparagraph (A), including any rate structure, fees, terms, conditions, or notice and recordkeeping requirements set forth therein, shall be admissible as evidence or otherwise taken into account in any administrative, judicial, or other government proceeding involving the setting or adjustment of the royalties payable for the public performance or reproduction in ephemeral phonorecords or copies of sound recordings, the determination of terms or conditions related thereto, or the establishment of notice or recordkeeping requirements by the Copyright Royalty Judges under paragraph (4) or section 112(e)(4). It is the intent of Congress that any royalty rates, rate structure, definitions, terms, conditions, or notice and recordkeeping requirements, included in such agreements shall be considered as a compromise motivated by the unique business, economic and political circumstances of webcasters, copyright owners, and performers rather than as matters that would have been negotiated in the marketplace between a willing buyer and a willing seller, or otherwise meet the objectives set forth in section 801(b). This subparagraph shall not apply to the extent that the receiving agent and a webcaster that is party to an agreement entered into pursuant to subparagraph (A) expressly authorize the submission of the agreement in a proceeding under this subsection.

(D) Nothing in the Webcaster Settlement Act of 2008, the Webcaster Settlement Act of 2009, or any agreement entered into pursuant to subparagraph (A) shall be taken into account by the United States Court of Appeals for the District of Columbia Circuit in its review of the determination by the Copyright Royalty Judges of May 1, 2007, of rates and terms for the digital performance of sound recordings and ephemeral recordings, pursuant to sections 112 and 114.

(E) As used in this paragraph—

(i) the term "noncommercial webcaster" means a webcaster that—

(I) is exempt from taxation under section 501 of the Internal Revenue Code of 1986 (26 U.S.C. 501);

(II) has applied in good faith to the Internal Revenue Service for exemption from taxation under section 501 of the Internal Revenue Code and has a commercially reasonable expectation that such exemption shall be granted; or

(III) is operated by a State or possession or any governmental entity or subordinate thereof, or by the United States or District of Columbia, for exclusively public purposes;

(ii) the term "receiving agent" shall have the meaning given that term in section 261.2 of title 37, Code of Federal Regulations, as published in the Federal Register on July 8, 2002; and

(iii) the term "webcaster" means a person or entity that has obtained a compulsory license under section 112 or 114 and the implementing regulations therefor.

(F) The authority to make settlements pursuant to subparagraph (A) shall expire at 11:59 p.m. Eastern time on the 30th day after the date of the enactment of the Webcaster Settlement Act of 2009.

(g) Proceeds from Licensing of Transmissions.—

(1) Except in the case of a transmission licensed under a statutory license in accordance with subsection (f) of this section—

(A) a featured recording artist who performs on a sound recording that has been licensed for a transmission shall be entitled to receive payments from the copyright owner of the sound recording in accordance with the terms of the artist's contract; and

(B) a nonfeatured recording artist who performs on a sound recording that has been licensed for a transmission shall be entitled to receive payments from the copyright owner of the sound recording in accordance with the terms of the nonfeatured recording artist's applicable contract or other applicable agreement.

(2) An agent designated to distribute receipts from the licensing of transmissions in accordance with subsection (f) shall distribute such receipts as follows:

(A) 50 percent of the receipts shall be paid to the copyright owner of the exclusive right under section 106(6) of this title to publicly perform a sound recording by means of a digital audio transmission.

(B) 2½ percent of the receipts shall be deposited in an escrow account managed by an independent administrator jointly appointed by copyright owners of sound recordings and the American Federation of Musicians (or any successor entity) to be distributed to nonfeatured musicians (whether or not members of the American Federation of Musicians) who have performed on sound recordings.

(C) 2½ percent of the receipts shall be deposited in an escrow account managed by an independent administrator jointly appointed by copyright owners of sound recordings and the American Federation of Television and Radio Artists (or any successor entity) to be distributed to nonfeatured vocalists (whether or not members of the American Federation of Television and Radio Artists) who have performed on sound recordings.

(D) 45 percent of the receipts shall be paid, on a per sound recording basis, to the recording artist or artists featured on such sound recording (or the persons conveying rights in the artists' performance in the sound recordings).

(3) A nonprofit agent designated to distribute receipts from the licensing of transmissions in accordance with subsection (f) may deduct from any of its receipts, prior to the distribution of such receipts to any person or entity entitled thereto other than copyright owners and performers who have elected to receive royalties from another designated agent and have notified such nonprofit agent in writing of such election, the reasonable costs of such agent incurred after November 1, 1995, in—

(A) the administration of the collection, distribution, and calculation of the royalties;

(B) the settlement of disputes relating to the collection and calculation of the royalties; and

(C) the licensing and enforcement of rights with respect to the making of ephemeral recordings and performances subject to licensing under section 112 and this section, including those incurred in participating in negotiations or arbitration proceedings under section 112 and this section, except that all costs incurred relating to the section 112 ephemeral recordings right may only be deducted from the royalties received pursuant to section 112.

(4) Notwithstanding paragraph (3), any designated agent designated to distribute receipts from the licensing of transmissions in accordance with subsection (f) may deduct from any of its receipts, prior to the distribution of such receipts, the reasonable costs identified in paragraph (3) of such agent incurred after November 1, 1995, with respect to such copyright owners and performers who have entered with such agent a contractual relationship that specifies that such costs may be deducted from such royalty receipts.

(h) Licensing to Affiliates.—

(1) If the copyright owner of a sound recording licenses an affiliated entity the right to publicly perform a sound recording by means of a digital audio transmission under section 106(6), the copyright owner shall make the licensed sound recording available under section 106(6) on no less favorable terms and conditions to all bona fide entities that offer similar services, except that, if there are material differences in the scope of the requested license with respect to the type of service, the particular sound recordings licensed, the frequency of use, the number of subscribers served, or the duration, then the copyright owner may establish different terms and conditions for such other services.

(2) The limitation set forth in paragraph (1) of this subsection shall not apply in the case where the copyright owner of a sound recording licenses—

(A) an interactive service; or

(B) an entity to perform publicly up to 45 seconds of the sound recording and the sole purpose of the performance is to promote the distribution or performance of that sound recording.

(i) No Effect on Royalties for Underlying Works.—License fees payable for the public performance of sound recordings under section 106(6) shall not be taken into account in any administrative, judicial, or other governmental proceeding to set or adjust the royalties payable to copyright owners of musical works for the public performance of their works. It is the intent of Congress that royalties payable to copyright owners of musical works for the public performance of their works shall not be diminished in any respect as a result of the rights granted by section 106(6).

(j) Definitions.—As used in this section, the following terms have the following meanings:

(1) An "affiliated entity" is an entity engaging in digital audio transmissions covered by section 106(6), other than an interactive service, in which the licensor has any direct or indirect partnership or any ownership interest amounting to 5 percent or more of the outstanding voting or nonvoting stock.

(2) An "archived program" is a predetermined program that is available repeatedly on the demand of the transmission recipient and that is performed in the same order from the beginning, except that an archived program shall not include a recorded event or broadcast transmission that makes no more than an incidental use of sound recordings, as long as such recorded event or broadcast transmission does not contain an entire sound recording or feature a particular sound recording.

(3) A "broadcast" transmission is a transmission made by a terrestrial broadcast station licensed as such by the Federal Communications Commission.

(4) A "continuous program" is a predetermined program that is continuously performed in the same order and that is accessed at a point in the program that is beyond the control of the transmission recipient.

(5) A "digital audio transmission" is a digital transmission as defined in section 101, that embodies the transmission of a sound recording. This term does not include the transmission of any audiovisual work.

(6) An "eligible nonsubscription transmission" is a noninteractive nonsubscription digital audio transmission not exempt under subsection (d)(1) that is made as part of a service that provides audio programming consisting, in whole or in part, of performances of sound recordings, including retransmissions of broadcast transmissions, if the primary purpose of the service is to provide to the public such audio or other entertainment programming, and the primary purpose of the service is not to sell, advertise, or promote particular products or services other than sound recordings, live concerts, or other music-related events.

(7) An "interactive service" is one that enables a member of the public to receive a transmission of a program specially created for the recipient, or on request, a transmission of a particular sound recording, whether or not as part of a program, which is selected by or on behalf of the recipient. The ability of individuals to request that particular sound recordings be performed for reception by the public at large, or in the case of a subscription service, by all subscribers of the service, does not make a service interactive, if the programming on each channel of the service does not substantially consist of sound recordings that are performed within 1 hour of the request or at a time designated by either the transmitting entity or the individual making such request. If an entity offers both interactive and noninteractive services (either concurrently or at different

times), the noninteractive component shall not be treated as part of an interactive service.

(8) A "new subscription service" is a service that performs sound recordings by means of noninteractive subscription digital audio transmissions and that is not a preexisting subscription service or a preexisting satellite digital audio radio service.

(9) A "nonsubscription" transmission is any transmission that is not a subscription transmission.

(10) A "preexisting satellite digital audio radio service" is a subscription satellite digital audio radio service provided pursuant to a satellite digital audio radio service license issued by the Federal Communications Commission on or before July 31, 1998, and any renewal of such license to the extent of the scope of the original license, and may include a limited number of sample channels representative of the subscription service that are made available on a nonsubscription basis in order to promote the subscription service.

(11) A "preexisting subscription service" is a service that performs sound recordings by means of noninteractive audio-only subscription digital audio transmissions, which was in existence and was making such transmissions to the public for a fee on or before July 31, 1998, and may include a limited number of sample channels representative of the subscription service that are made available on a nonsubscription basis in order to promote the subscription service.

(12) A "retransmission" is a further transmission of an initial transmission, and includes any further retransmission of the same transmission. Except as provided in this section, a transmission qualifies as a "retransmission" only if it is simultaneous with the initial transmission. Nothing in this definition shall be construed to exempt a transmission that fails to satisfy a separate element required to qualify for an exemption under section 114(d)(1).

(13) The "sound recording performance complement" is the transmission during any 3-hour period, on a particular channel used by a transmitting entity, of no more than—

(A) 3 different selections of sound recordings from any one phonorecord lawfully distributed for public performance or sale in the United States, if no more than 2 such selections are transmitted consecutively; or

(B) 4 different selections of sound recordings—

(i) by the same featured recording artist; or

(ii) from any set or compilation of phonorecords lawfully distributed together as a unit for public performance or sale in the United States,

if no more than three such selections are transmitted consecutively: *Provided*, That the transmission of selections in excess of the numerical limits provided for in clauses (A) and (B) from multiple phonorecords shall nonetheless qualify as a sound recording performance complement if the programming of the multiple phonorecords was not willfully intended to avoid the numerical limitations prescribed in such clauses.

(14) A "subscription" transmission is a transmission that is controlled and limited to particular recipients, and for which consideration is required to be

paid or otherwise given by or on behalf of the recipient to receive the transmission or a package of transmissions including the transmission.

(15) A "transmission" is either an initial transmission or a retransmission.

§ 115 Scope of Exclusive Rights in Nondramatic Musical Works: Compulsory License for Making and Distributing Phonorecords

In the case of nondramatic musical works, the exclusive rights provided by clauses (1) and (3) of section 106, to make and to distribute phonorecords of such works, are subject to compulsory licensing under the conditions specified by this section.

(a) Availability and Scope of Compulsory License.—

(1) When phonorecords of a nondramatic musical work have been distributed to the public in the United States under the authority of the copyright owner, any other person, including those who make phonorecords or digital phonorecord deliveries, may, by complying with the provisions of this section, obtain a compulsory license to make and distribute phonorecords of the work. A person may obtain a compulsory license only if his or her primary purpose in making phonorecords is to distribute them to the public for private use, including by means of a digital phonorecord delivery. A person may not obtain a compulsory license for use of the work in the making of phonorecords duplicating a sound recording fixed by another, unless:

(i) such sound recording was fixed lawfully; and

(ii) the making of the phonorecords was authorized by the owner of copyright in the sound recording or, if the sound recording was fixed before February 15, 1972, by any person who fixed the sound recording pursuant to an express license from the owner of the copyright in the musical work or pursuant to a valid compulsory license for use of such work in a sound recording.

(2) A compulsory license includes the privilege of making a musical arrangement of the work to the extent necessary to conform it to the style or manner of interpretation of the performance involved, but the arrangement shall not change the basic melody or fundamental character of the work, and shall not be subject to protection as a derivative work under this title, except with the express consent of the copyright owner.

(b) Notice of Intention to Obtain Compulsory License.—

(1) Any person who wishes to obtain a compulsory license under this section shall, before or within thirty days after making, and before distributing any phonorecords of the work, serve notice of intention to do so on the copyright owner. If the registration or other public records of the Copyright Office do not identify the copyright owner and include an address at which notice can be served, it shall be sufficient to file the notice of intention in the Copyright Office. The notice shall comply, in form, content, and manner of service, with requirements that the Register of Copyrights shall prescribe by regulation.

(2) Failure to serve or file the notice required by clause (1) forecloses the possibility of a compulsory license and, in the absence of a negotiated license, renders the making and distribution of phonorecords actionable as acts of infringement under section 501 and fully subject to the remedies provided by sections 502 through 506 and 509.

(c) Royalty Payable under Compulsory License.—

(1) To be entitled to receive royalties under a compulsory license, the copyright owner must be identified in the registration or other public records of the Copyright Office. The owner is entitled to royalties for phonorecords made and distributed after being so identified, but is not entitled to recover for any phonorecords previously made and distributed.

(2) Except as provided by clause (1), the royalty under a compulsory license shall be payable for every phonorecord made and distributed in accordance with the license. For this purpose, and other than as provided in paragraph (3), a phonorecord is considered "distributed" if the person exercising the compulsory license has voluntarily and permanently parted with its possession. With respect to each work embodied in the phonorecord, the royalty shall be either two and three-fourths cents, or one-half of one cent per minute of playing time or fraction thereof, whichever amount is larger.

(3) (A) A compulsory license under this section includes the right of the compulsory licensee to distribute or authorize the distribution of a phonorecord of a nondramatic musical work by means of a digital transmission which constitutes a digital phonorecord delivery, regardless of whether the digital transmission is also a public performance of the sound recording under section 106(6) of this title or of any nondramatic musical work embodied therein under section 106(4) of this title. For every digital phonorecord delivery by or under the authority of the compulsory licensee—

(i) on or before December 31, 1997, the royalty payable by the compulsory licensee shall be the royalty prescribed under paragraph (2) and chapter 8 of this title; and

(ii) on or after January 1, 1998, the royalty payable by the compulsory licensee shall be the royalty prescribed under subparagraphs (B) through (E) and chapter 8 of this title.

(B) Notwithstanding any provision of the antitrust laws, any copyright owners of nondramatic musical works and any persons entitled to obtain a compulsory license under subsection (a)(1) may negotiate and agree upon the terms and rates of royalty payments under this section and the proportionate division of fees paid among copyright owners, and may designate common agents on a nonexclusive basis to negotiate, agree to, pay or receive such royalty payments. Such authority to negotiate the terms and rates of royalty payments includes, but is not limited to, the authority to negotiate the year during which the royalty rates prescribed under this subparagraph and subparagraphs (C) through (E) and chapter 8 of this title shall next be determined.

(C) Proceedings under chapter 8 shall determine reasonable rates and terms of royalty payments for the activities specified by this section during the period beginning with the effective date of such rates and terms, but not earlier than January 1 of the second year following the year in which the petition requesting the proceeding is filed, and ending on the effective date of successor rates and terms, or such other period as the parties may agree. Such terms and rates shall distinguish between (i) digital phonorecord deliveries where the reproduction or distribution of a phonorecord is incidental to the transmission which constitutes the digital phonorecord delivery, and (ii) digital phonorecord deliveries in general. Any copyright owners of nondramatic musical works and any persons entitled to obtain a compulsory license under subsection (a)(1) may submit to the Copyright Royalty Judges licenses covering such activities. The parties to each proceeding shall bear their own costs.

(D) The schedule of reasonable rates and terms determined by the Copyright Royalty Judges shall, subject to subparagraph (E), be binding on all copyright owners of nondramatic musical works and persons entitled to obtain a compulsory license under subsection (a)(1) during the period specified in subparagraph (C), such other period as may be determined pursuant to subparagraphs (B) and (C), or such other period as the parties may agree. Such terms and rates shall distinguish between (i) digital phonorecord deliveries where the reproduction or distribution of a phonorecord is incidental to the transmission which constitutes the digital phonorecord delivery, and (ii) digital phonorecord deliveries in general. In addition to the objectives set forth in section 801(b)(1), in establishing such rates and terms, the Copyright Royalty Judges may consider rates and terms under voluntary license agreements described in subparagraphs (B) and (C). The royalty rates payable for a compulsory license for a digital phonorecord delivery under this section shall be established de novo and no precedential effect shall be given to the amount of the royalty payable by a compulsory licensee for digital phonorecord deliveries on or before December 31, 1997. The Copyright Royalty Judges shall also establish requirements by which copyright owners may receive reasonable notice of the use of their works under this section, and under which records of such use shall be kept and made available by persons making digital phonorecord deliveries.

(E)(i) License agreements voluntarily negotiated at any time between one or more copyright owners of nondramatic musical works and one or more persons entitled to obtain a compulsory license under subsection (a)(1) shall be given effect in lieu of any determination by the Librarian of Congress and Copyright Royalty Judges. Subject to clause (ii), the royalty rates determined pursuant to subparagraph (C) and (D) shall be given effect as to digital phonorecord deliveries in lieu of any contrary royalty rates specified in a contract pursuant to which a recording artist who is the author of a nondramatic musical work grants a license under that person's exclusive rights in the musical work under paragraphs (1) and (3) of section 106 or commits another person to grant a license in that musical work under paragraphs (1) and (3) of section 106, to a person desiring to fix in a tangible medium of expression a sound recording embodying the musical work.

(ii) The second sentence of clause (i) shall not apply to—

(I) a contract entered into on or before June 22, 1995 and not modified thereafter for the purpose of reducing the royalty rates determined pursuant to subparagraph (C) and (D) or of increasing the number of musical works within the scope of the contract covered by the reduced rates, except if a contract entered into on or before June 22, 1995, is modified thereafter for the purpose of increasing the number of musical works within the scope of the contract, any contrary royalty rates specified in the contract shall be given effect in lieu of royalty rates determined pursuant to subparagraph (C) and (D) for the number of musical works within the scope of the contract as of June 22, 1995; and

(II) a contract entered into after the date that the sound recording is fixed in a tangible medium of expression substantially in a form intended for commercial release, if at the time the contract is entered into, the recording artist retains the right to grant licenses as to the musical work under paragraphs (1) and (3) of section 106.

(F) Except as provided in section 1002(e) of this title, a digital phonorecord delivery licensed under this paragraph shall be accompanied by the information encoded in the sound recording, if any, by or under the authority of the copyright owner of that sound recording, that identifies the title of the sound recording, the featured recording artist who performs on the sound recording, and related information, including information concerning the underlying musical work and its writer.

(G)(i) A digital phonorecord delivery of a sound recording is actionable as an act of infringement under section 501, and is fully subject to the remedies provided by sections 502 through 506, unless—

(I) the digital phonorecord delivery has been authorized by the copyright owner of the sound recording; and

(II) the owner of the copyright in the sound recording or the entity making the digital phonorecord delivery has obtained a compulsory license under this section or has otherwise been authorized by the copyright owner of the musical work to distribute or authorize the distribution, by means of a digital phonorecord delivery, of each musical work embodied in the sound recording.

(ii) Any cause of action under this subparagraph shall be in addition to those available to the owner of the copyright in the nondramatic musical work under subsection (c)(6) and section 106(4) and the owner of the copyright in the sound recording under section 106(6).

(H) The liability of the copyright owner of a sound recording for infringement of the copyright in a nondramatic musical work embodied in the sound recording shall be determined in accordance with applicable law, except that the owner of a copyright in a sound recording shall not be liable for a digital phonorecord delivery by a third party if the owner of the copyright in the sound recording does not license the distribution of a phonorecord of the nondramatic musical work.

(I) Nothing in section 1008 shall be construed to prevent the exercise of the rights and remedies allowed by this paragraph, paragraph (6), and chapter 5 in the event of a digital phonorecord delivery, except that no action alleging infringement of copyright may be brought under this title against a manufacturer, importer or distributor of a digital audio recording device, a digital audio recording medium, an analog recording device, or an analog recording medium, or against a consumer, based on the actions described in such section.

(J) Nothing in this section annuls or limits

(i) the exclusive right to publicly perform a sound recording or the musical work embodied therein, including by means of a digital transmission, under sections 106(4) and 106(6), (ii) except for compulsory licensing under the conditions specified by this section, the exclusive rights to reproduce and distribute the sound recording and the musical work embodied therein under sections 106(1) and 106(3), including by means of a digital phonorecord delivery, or (iii) any other rights under any other provision of section 106, or remedies available under this title, as such rights or remedies exist either before or after the date of enactment of the Digital Performance Right in Sound Recordings Act of 1995.

(K) The provisions of this section concerning digital phonorecord deliveries shall not apply to any exempt transmissions or retransmissions under section 114(d)(1). The exemptions created in section 114(d)(1) do not expand or reduce the rights of copyright owners under section 106(1) through (5) with respect to such transmissions and retransmissions.

(4) A compulsory license under this section includes the right of the maker of a phonorecord of a nondramatic musical work under subsection (a)(1) to distribute or authorize distribution of such phonorecord by rental, lease, or lending (or by acts or practices in the nature of rental, lease, or lending). In addition to any royalty payable under clause (2) and chapter 8 of this title, a royalty shall be payable by the compulsory licensee for every act of distribution of a phonorecord by or in the nature of rental, lease, or lending, by or under the authority of the compulsory licensee. With respect to each nondramatic musical work embodied in the phonorecord, the royalty shall be a proportion of the revenue received by the compulsory licensee from every such act of distribution of the phonorecord under this clause equal to the proportion of the revenue received by the compulsory licensee from distribution of the phonorecord under clause (2) that is payable by a compulsory licensee under that clause and under chapter 8. The Register of Copyrights shall issue regulations to carry out the purpose of this clause.

(5) Royalty payments shall be made on or before the twentieth day of each month and shall include all royalties for the month next preceding. Each monthly payment shall be made under oath and shall comply with requirements that the Register of Copyrights shall prescribe by regulation. The Register shall also prescribe regulations under which detailed cumulative annual statements of account, certified by a certified public accountant, shall be filed for every compulsory license under this section. The regulations covering both the monthly and the annual statements of account shall prescribe the form, content, and manner

of certification with respect to the number of records made and the number of records distributed.

(6) If the copyright owner does not receive the monthly payment and the monthly and annual statements of account when due, the owner may give written notice to the licensee that, unless the default is remedied within thirty days from the date of the notice, the compulsory license will be automatically terminated. Such termination renders either the making or the distribution, or both, of all phonorecords for which the royalty has not been paid, actionable as acts of infringement under section 501 and fully subject to the remedies provided by sections 502 through 506.

(d) Definition.—As used in this section, the following term has the following meaning: A "digital phonorecord delivery" is each individual delivery of a phonorecord by digital transmission of a sound recording which results in a specifically identifiable reproduction by or for any transmission recipient of a phonorecord of that sound recording, regardless of whether the digital transmission is also a public performance of the sound recording or any nondramatic musical work embodied therein. A digital phonorecord delivery does not result from a real-time, non-interactive subscription transmission of a sound recording where no reproduction of the sound recording or the musical work embodied therein is made from the inception of the transmission through to its receipt by the transmission recipient in order to make the sound recording audible.

§ 116 Negotiated Licenses for Public Performances by Means of Coin-Operated Phonorecord Players

(a) Applicability of Section.—This section applies to any nondramatic musical work embodied in a phonorecord.

(b) Negotiated Licenses.—

(1) Authority for negotiations.—Any owners of copyright in works to which this section applies and any operators of coin-operated phonorecord players may negotiate and agree upon the terms and rates of royalty payments for the performance of such works and the proportionate division of fees paid among copyright owners, and may designate common agents to negotiate, agree to, pay, or receive such royalty payments.

(2) Chapter 8 proceeding.—Parties not subject to such a negotiation may have the terms and rates and the division of fees described in paragraph

(1) determined in a proceeding in accordance with the provisions of chapter 8.

(c) License Agreements Superior to Determinations by Copyright Royalty Judges.—License agreements between one or more copyright owners and one or more operators of coin-operated phonorecord players, which are negotiated in accordance with subsection (b), shall be given effect in lieu of any otherwise applicable determination by the Copyright Royalty Judges.

(d) Definitions.—As used in this section, the following terms mean the following:

(1) A "coin-operated phonorecord player" is a machine or device that—

(A) is employed solely for the performance of nondramatic musical works by means of phonorecords upon being activated by the insertion of coins, currency, tokens, or other monetary units or their equivalent;

(B) is located in an establishment making no direct or indirect charge for admission;

(C) is accompanied by a list which is comprised of the titles of all the musical works available for performance on it, and is affixed to the phonorecord player or posted in the establishment in a prominent position where it can be readily examined by the public; and

(D) affords a choice of works available for performance and permits the choice to be made by the patrons of the establishment in which it is located.

(2) An "operator" is any person who, alone or jointly with others—

(A) owns a coin-operated phonorecord player;

(B) has the power to make a coin-operated phonorecord player available for placement in an establishment for purposes of public performance; or

(C) has the power to exercise primary control over the selection of the musical works made available for public performance on a coin-operated phonorecord player.

§ 117 Limitations on Exclusive Rights: Computer Programs

(a) Making of Additional Copy or Adaptation by Owner of Copy.—Notwithstanding the provisions of section 106, it is not an infringement for the owner of a copy of a computer program to make or authorize the making of another copy or adaptation of that computer program provided:

(1) that such a new copy or adaptation is created as an essential step in the utilization of the computer program in conjunction with a machine and that it is used in no other manner, or

(2) that such new copy or adaptation is for archival purposes only and that all archival copies are destroyed in the event that continued possession of the computer program should cease to be rightful.

(b) Lease, Sale, or Other Transfer of Additional Copy or Adaptation.—Any exact copies prepared in accordance with the provisions of this section may be leased, sold, or otherwise transferred, along with the copy from which such copies were prepared, only as part of the lease, sale, or other transfer of all rights in the program. Adaptations so prepared may be transferred only with the authorization of the copyright owner.

(c) Machine Maintenance or Repair.—Notwithstanding the provisions of section 106, it is not an infringement for the owner or lessee of a machine to make or authorize the making of a copy of a computer program if such copy is made solely by virtue of the activation of a machine that lawfully contains an authorized copy of the computer program, for purposes only of maintenance or repair of that machine, if—

(1) such new copy is used in no other manner and is destroyed immediately after the maintenance or repair is completed; and

(2) with respect to any computer program or part thereof that is not necessary for that machine to be activated, such program or part thereof is not accessed or used other than to make such new copy by virtue of the activation of the machine.

(d) Definitions.—For purposes of this section—

(1) the "maintenance" of a machine is the servicing of the machine in order to make it work in accordance with its original specifications and any changes to those specifications authorized for that machine; and

(2) the "repair" of a machine is the restoring of the machine to the state of working in accordance with its original specifications and any changes to those specifications authorized for that machine.

§ 118 Scope of Exclusive Rights: Use of Certain Works in Connection with Noncommercial Broadcasting

(a) The exclusive rights provided by section 106 shall, with respect to the works specified by subsection (b) and the activities specified by subsection (d), be subject to the conditions and limitations prescribed by this section.

(b) Notwithstanding any provision of the antitrust laws, any owners of copyright in published nondramatic musical works and published pictorial, graphic, and sculptural works and any public broadcasting entities, respectively, may negotiate and agree upon the terms and rates of royalty payments and the proportionate division of fees paid among various copyright owners, and may designate common agents to negotiate, agree to, pay, or receive payments.

(1) Any owner of copyright in a work specified in this subsection or any public broadcasting entity may submit to the Copyright Royalty Judges proposed licenses covering such activities with respect to such works.

(2) License agreements voluntarily negotiated at any time between one or more copyright owners and one or more public broadcasting entities shall be given effect in lieu of any determination by the Librarian of Congress or the Copyright Royalty Judges, if copies of such agreements are filed with the Copyright Royalty Judges within 30 days of execution in accordance with regulations that the Copyright Royalty Judges shall issue.

(3) Voluntary negotiation proceedings initiated pursuant to a petition filed under section 804(a) for the purpose of determining a schedule of terms and rates of royalty payments by public broadcasting entities to owners of copyright in works specified by this subsection and the proportionate division of fees paid among various copyright owners shall cover the 5-year period beginning on January 1 of the second year following the year in which the petition is filed. The parties to each negotiation proceeding shall bear their own costs.

(4) In the absence of license agreements negotiated under paragraph (2) or (3), the Copyright Royalty Judges shall, pursuant to chapter 8, conduct a proceeding to determine and publish in the Federal Register a schedule of rates and terms which, subject to paragraph (2), shall be binding on all owners of copyright in works specified by this subsection and public broadcasting entities, regardless

of whether such copyright owners have submitted proposals to the Copyright Royalty Judges. In establishing such rates and terms the Copyright Royalty Judges may consider the rates for comparable circumstances under voluntary license agreements negotiated as provided in paragraph (2) or (3). The Copyright Royalty Judges shall also establish requirements by which copyright owners may receive reasonable notice of the use of their works under this section, and under which records of such use shall be kept by public broadcasting entities.

(c) Subject to the terms of any voluntary license agreements that have been negotiated as provided by subsection (b) (2) or (3), a public broadcasting entity may, upon compliance with the provisions of this section, including the rates and terms established by the Copyright Royalty Judges under subsection (b)(4), engage in the following activities with respect to published nondramatic musical works and published pictorial, graphic, and sculptural works:

(1) performance or display of a work by or in the course of a transmission made by a noncommercial educational broadcast station referred to in subsection (f); and

(2) production of a transmission program, reproduction of copies or phonorecords of such a transmission program, and distribution of such copies or phonorecords, where such production, reproduction, or distribution is made by a nonprofit institution or organization solely for the purpose of transmissions specified in paragraph (1); and

(3) the making of reproductions by a governmental body or a nonprofit institution of a transmission program simultaneously with its transmission as specified in paragraph (1), and the performance or display of the contents of such program under the conditions specified by paragraph (1) of section 110, but only if the reproductions are used for performances or displays for a period of no more than seven days from the date of the transmission specified in paragraph (1), and are destroyed before or at the end of such period. No person supplying, in accordance with paragraph (2), a reproduction of a transmission program to governmental bodies or nonprofit institutions under this paragraph shall have any liability as a result of failure of such body or institution to destroy such reproduction: *Provided*, That it shall have notified such body or institution of the requirement for such destruction pursuant to this paragraph: *And provided further*, That if such body or institution itself fails to destroy such reproduction it shall be deemed to have infringed.

(d) Except as expressly provided in this subsection, this section shall have no applicability to works other than those specified in subsection (b). Owners of copyright in nondramatic literary works and public broadcasting entities may, during the course of voluntary negotiations, agree among themselves, respectively, as to the terms and rates of royalty payments without liability under the antitrust laws. Any such terms and rates of royalty payments shall be effective upon filing with the Copyright Royalty Judges, in accordance with regulations that the Copyright Royalty Judges shall prescribe as provided in section 803(b)(6).

(e) Nothing in this section shall be construed to permit, beyond the limits of fair use as provided by section 107, the unauthorized dramatization of a nondramatic musical work, the production of a transmission program drawn to any substantial extent from a published compilation of pictorial, graphic, or sculptural works, or the unauthorized use of any portion of an audiovisual work.

(f) As used in this section, the term "public broadcasting entity" means a noncommercial educational broadcast station as defined in section 397 of title 47 and any nonprofit institution or organization engaged in the activities described in paragraph (2) of subsection (c).

§ 119 Limitations on Exclusive Rights: Secondary Transmissions of Distant Television Programming by Satellite

(a) Secondary Transmissions by Satellite Carriers.—

(1) Non-network Stations.—Subject to the provisions of paragraphs (4), (5), and (7) of this subsection and section 114(d), secondary transmissions of a performance or display of a work embodied in a primary transmission made by a non-network station shall be subject to statutory licensing under this section if the secondary transmission is made by a satellite carrier to the public for private home viewing or for viewing in a commercial establishment, with regard to secondary transmissions the satellite carrier is in compliance with the rules, regulations, or authorizations of the Federal Communications Commission governing the carriage of television broadcast station signals, and the carrier makes a direct or indirect charge for each retransmission service to each subscriber receiving the secondary transmission or to a distributor that has contracted with the carrier for direct or indirect delivery of the secondary transmission to the public for private home viewing or for viewing in a commercial establishment.

(2) Network stations.—

(A) In general.—Subject to the provisions of subparagraph (B) of this paragraph and paragraphs (4), (5), (6), and (7) of this subsection and section 114(d), secondary transmissions of a performance or display of a work embodied in a primary transmission made by a network station shall be subject to statutory licensing under this section if the secondary transmission is made by a satellite carrier to the public for private home viewing, with regard to secondary transmissions the satellite carrier is in compliance with the rules, regulations, or authorizations of the Federal Communications Commission governing the carriage of television broadcast station signals, and the carrier makes a direct or indirect charge for such retransmission service to each subscriber receiving the secondary transmission.

(B) Secondary transmissions to unserved households.—

(i) In general.—The statutory license provided for in subparagraph (A) shall be limited to secondary transmissions of the signals of no more than two network stations in a single day for each television network to persons who reside in unserved households.

(ii) Accurate determinations of eligibility.—

(I) Accurate predictive model.—In determining presumptively whether a person resides in an unserved household under subsection (d)(10)(A), a court shall rely on the Individual Location Longley Rice model set forth by the Federal Communications Commission in Docket No. 98-201, as that model may be amended by the Commission over time under section 339(c)(3) of the Communications Act of 1934 to increase the accuracy of that model.

(II) Accurate measurements.—For purposes of site measurements to determine whether a person resides in an unserved household under subsection (d)(10)(A), a court shall rely on section 339(c)(4) of the Communications Act of 1934.

(iii) C-band exemption to unserved households.—

(I) In general.—The limitations of clause (i) shall not apply to any secondary transmissions by C-band services of network stations that a subscriber to C-band service received before any termination of such secondary transmissions before October 31, 1999.

(II) Definition.—In this clause, the term "C-band service" means a service that is licensed by the Federal Communications Commission and operates in the Fixed Satellite Service under part 25 of title 47, Code of Federal Regulations.

(III) Accurate predictive model with respect to digital signals.—Notwithstanding subclause (I), in determining presumptively whether a person resides in an unserved household under subsection (d)(10)(A) with respect to digital signals, a court shall rely on a predictive model set forth by the Federal Communications Commission pursuant to a rulemaking as provided in section 339(c)(3) of the Communications Act of 1934 (47 U.S.C. 339(c)(3)), as that model may be amended by the Commission over time under such section to increase the accuracy of that model. Until such time as the Commission sets forth such model, a court shall rely on the predictive model as recommended by the Commission with respect to digital signals in its Report to Congress in ET Docket No. 05–182, FCC 05–199 (released December 9, 2005).

(C) Submission of subscriber lists to networks.—

(i) Initial lists.—A satellite carrier that makes secondary transmissions of a primary transmission made by a network station pursuant to subparagraph (A) shall, not later than 90 days after commencing such secondary transmissions, submit to the network that owns or is affiliated with the network station a list identifying (by name and address, including street or rural route number, city, State, and 9-digit zip code) all subscribers to which the satellite carrier makes secondary transmissions of that primary transmission to subscribers in unserved households.

(ii) Monthly lists.—After the submission of the initial lists under clause (i), the satellite carrier shall, not later than the 15th of each month, submit to the network a list, aggregated by designated market area, identifying (by name and address, including street or rural route number, city, State, and 9-digit zip code) any persons who have been added or dropped as subscribers under clause (i) since the last submission under this subparagraph.

(iii) Use of subscriber information.—Subscriber information submitted by a satellite carrier under this subparagraph may be used only for purposes of monitoring compliance by the satellite carrier with this subsection.

(iv) Applicability.—The submission requirements of this subparagraph shall apply to a satellite carrier only if the network to which the submissions are to be made places on file with the Register of Copyrights a document identifying the name and address of the person to whom such submissions are to be made. The Register shall maintain for public inspection a file of all such documents.

(3) Statutory license where retransmissions into local market available.—

(A) Rules for subscribers to signals under subsection (e).—

(i) For those receiving distant signals.—In the case of a subscriber of a satellite carrier who is eligible to receive the secondary transmission of the primary transmission of a network station solely by reason of subsection (e) (in this subparagraph referred to as a "distant signal"), and who, as of October 1, 2004, is receiving the distant signal of that network station, the following shall apply:

(I) In a case in which the satellite carrier makes available to the subscriber the secondary transmission of the primary transmission of a local network station affiliated with the same television network pursuant to the statutory license under section 122, the statutory license under paragraph (2) shall apply only to secondary transmissions by that satellite carrier to that subscriber of the distant signal of a station affiliated with the same television network—

(aa) if, within 60 days after receiving the notice of the satellite carrier under section 338(h)(1) of the Communications Act of 1934, the subscriber elects to retain the distant signal; but

(bb) only until such time as the subscriber elects to receive such local signal.

(II) Notwithstanding subclause (I), the statutory license under paragraph (2) shall not apply with respect to any subscriber who is eligible to receive the distant signal of a television network station solely by reason of subsection (e), unless the satellite carrier, within 60 days after the date of the enactment of the Satellite Home Viewer Extension and Reauthorization Act of 2004, submits to that television network a list, aggregated by designated market area (as defined in section 122(j)(2)(C)), that—

(aa) identifies that subscriber by name and address (street or rural route number, city, State, and zip code) and specifies the distant signals received by the subscriber; and

(bb) states, to the best of the satellite carrier's knowledge and belief, after having made diligent and good faith inquiries, that the subscriber is eligible under subsection (e) to receive the distant signals.

(ii) For those not receiving distant signals.—In the case of any subscriber of a satellite carrier who is eligible to receive the distant signal of a network station solely by reason of subsection (e) and who did not receive a distant signal of a station affiliated with the same network on October 1, 2004, the statutory license under paragraph (2) shall not apply to secondary transmissions by that satellite carrier to that subscriber of the distant signal of a station affiliated with the same network.

(B) Rules for lawful subscribers as of date of enactment of 2010 act.—In the case of a subscriber of a satellite carrier who, on the day before the date of the enactment of the Satellite Television Extension and Localism Act of 2010, was lawfully receiving the secondary transmission of the primary transmission of a network station under the statutory license under paragraph (2) (in this subparagraph referred to as the distant signal), other than subscribers to whom subparagraph (A) applies, the statutory license under paragraph (2) shall apply to secondary transmissions by that satellite carrier to that subscriber of the distant signal of a station affiliated with the same television network, and the subscriber's household shall continue to be considered to be an unserved household with respect to such network, until such time as the subscriber elects to terminate such secondary transmissions, whether or not the subscriber elects to subscribe to receive the secondary transmission of the primary transmission of a local network station affiliated with the same network pursuant to the statutory license under section 122.

(C) Future applicability.—

(i) When local signal available at time of subscription.—The statutory license under paragraph (2) shall not apply to the secondary transmission by a satellite carrier of the primary transmission of a network station to a person who is not a subscriber lawfully receiving such secondary transmission as of the date of the enactment of the Satellite Television Extension and Localism Act of 2010 and, at the time such person seeks to subscribe to receive such secondary transmission, resides in a local market where the satellite carrier makes available to that person the secondary transmission of the primary transmission of a local network station affiliated with the same network pursuant to the statutory license under section 122.

(ii) When local signal available after subscription.—In the case of a subscriber who lawfully subscribes to and receives the secondary transmission by a satellite carrier of the primary transmission of a network station under the statutory license under paragraph (2) (in this clause referred to as the "distant signal") on or after the date of the enactment of the Satellite Television Extension and Localism Act of 2010, the statutory license under paragraph (2) shall apply to secondary transmissions by that satellite carrier to that subscriber of the distant signal of a station affiliated with the same television network, and the subscriber's household shall continue to be considered to be an unserved household with respect to such network, until such time as the subscriber elects to terminate such secondary transmissions, but only if such subscriber subscribes to the secondary transmission of the primary transmission of a local network station affiliated with the same network within 60 days after the satellite carrier makes available to the subscriber such secondary transmission of the primary transmission of such local network station.

(D) Other provisions not affected.—This paragraph shall not affect the applicability of the statutory license to secondary transmissions to unserved households included under paragraph (11).

(E) Waiver.—A subscriber who is denied the secondary transmission of a network station under subparagraph (B) or (C) may request a waiver from such denial by submitting a request, through the subscriber's satellite carrier, to the network station in the local market affiliated with the same network where the subscriber is located. The network station shall accept or reject the subscriber's request for a waiver within 30 days after receipt of the request. If the network station fails to accept or reject the subscriber's request for a waiver within that 30-day period, that network station shall be deemed to agree to the waiver request. Unless specifically stated by the network station, a waiver that was granted before the date of the enactment of the Satellite Home Viewer Extension and Reauthorization Act of 2004 under section 339(c)(2) of the Communications Act of 1934 shall not constitute a waiver for purposes of this subparagraph.

(F) Available defined.—For purposes of this paragraph, a satellite carrier makes available a secondary transmission of the primary transmission of a local station to a subscriber or person if the satellite carrier offers that secondary transmission to other subscribers who reside in the same 9-digit zip code as that subscriber or person.

(4) Noncompliance with reporting and payment requirements.—Notwithstanding the provisions of paragraphs (1) and (2), the willful or repeated secondary transmission to the public by a satellite carrier of a primary transmission made by a non-network station or a network station and embodying a performance or display of a work is actionable as an act of infringement under section 501, and is fully subject to the remedies provided by sections 502 through 506, where the satellite carrier has not deposited the statement of account and royalty fee required by subsection (b), or has failed to make the submissions to networks required by paragraph (2)(C).

(5) Willful alterations.—Notwithstanding the provisions of paragraphs (1) and (2), the secondary transmission to the public by a satellite carrier of a performance or display of a work embodied in a primary transmission made by a non-network station or a network station is actionable as an act of infringement under section 501, and is fully subject to the remedies provided by sections 502 through 506 and section 510, if the content of the particular program in which the performance or display is embodied, or any commercial advertising or station announcement transmitted by the primary transmitter during, or immediately before or after, the transmission of such program, is in any way willfully altered by the satellite carrier through changes, deletions, or additions, or is combined with programming from any other broadcast signal.

(6) Violation of territorial restrictions on statutory license for network stations.—

(A) Individual violations.—The willful or repeated secondary transmission by a satellite carrier of a primary transmission made by a network station and embodying a performance or display of a work to a subscriber who is not eligible to receive the transmission under this section is actionable as an act of infringement under section 501 and is fully subject to the remedies provided by sections 502 through 506, except that—

(i) no damages shall be awarded for such act of infringement if the satellite carrier took corrective action by promptly withdrawing service from the ineligible subscriber, and

(ii) any statutory damages shall not exceed $250 for such subscriber for each month during which the violation occurred.

(B) Pattern of violations.—If a satellite carrier engages in a willful or repeated pattern or practice of delivering a primary transmission made by a network station and embodying a performance or display of a work to subscribers who are not eligible to receive the transmission under this section, then in addition to the remedies set forth in subparagraph (A)—

(i) if the pattern or practice has been carried out on a substantially nationwide basis, the court shall order a permanent injunction barring the secondary transmission by the satellite carrier, for private home viewing, of the primary transmissions of any primary network station affiliated with the same network, and the court may order statutory damages of not to exceed $2,500,000 for each 3-month period during which the pattern or practice was carried out; and

(ii) if the pattern or practice has been carried out on a local or regional basis, the court shall order a permanent injunction barring the secondary transmission, for private home viewing in that locality or region, by the satellite carrier of the primary transmissions of any primary network station affiliated with the same network, and the court may order statutory damages of not to exceed $2,500,000 for each 6-month period during which the pattern or practice was carried out.

The court shall direct one half of any statutory damages ordered under clause (i) to be deposited with the Register of Copyrights for distribution to copyright owners pursuant to subsection (b). The Copyright Royalty Judges shall issue regulations establishing procedures for distributing such funds, on a proportional basis, to copyright owners whose works were included in the secondary transmissions that were the subject of the statutory damages.

(C) Previous subscribers excluded.—Subparagraphs (A) and (B) do not apply to secondary transmissions by a satellite carrier to persons who subscribed to receive such secondary transmissions from the satellite carrier or a distributor before November 16, 1988.

(D) Burden of proof.—In any action brought under this paragraph, the satellite carrier shall have the burden of proving that its secondary transmission of a primary transmission by a network station is to a subscriber who is eligible to receive the secondary transmission under this section.

(E) Exception.—The secondary transmission by a satellite carrier of a performance or display of a work embodied in a primary transmission made by a network station to subscribers who do not reside in unserved households shall not be an act of infringement if—

(i) the station on May 1, 1991, was retransmitted by a satellite carrier and was not on that date owned or operated by or affiliated with a television network that offered interconnected program service on a regular basis for 15 or more hours per week to at least 25 affiliated television licensees in 10 or more States;

(ii) as of July 1, 1998, such station was retransmitted by a satellite carrier under the statutory license of this section; and

(iii) the station is not owned or operated by or affiliated with a television network that, as of January 1, 1995, offered interconnected program service on a regular basis for 15 or more hours per week to at least 25 affiliated television licensees in 10 or more States.

(7) Discrimination by a satellite carrier.—Notwithstanding the provisions of paragraph (1), the willful or repeated secondary transmission to the public by a satellite carrier of a performance or display of a work embodied in a primary transmission made by a non-network station or a network station is actionable as an act of infringement under section 501, and is fully subject to the remedies provided by sections 502 through 506, if the satellite carrier unlawfully discriminates against a distributor.

(8) Geographic limitation on secondary transmissions.—The statutory license created by this section shall apply only to secondary transmissions to households located in the United States.

(9) Loser pays for signal intensity measurement; recovery of measurement costs in a civil action.—In any civil action filed relating to the eligibility of subscribing households as unserved households—

(A) a network station challenging such eligibility shall, within 60 days after receipt of the measurement results and a statement of such costs, reimburse the satellite carrier for any signal intensity measurement that is conducted by that carrier in response to a challenge by the network station and that establishes the household is an unserved household; and

(B) a satellite carrier shall, within 60 days after receipt of the measurement results and a statement of such costs, reimburse the network station challenging such eligibility for any signal intensity measurement that is conducted by that station and that establishes the household is not an unserved household.

(10) Inability to conduct measurement.—If a network station makes a reasonable attempt to conduct a site measurement of its signal at a subscriber's household and is denied access for the purpose of conducting the measurement, and is otherwise unable to conduct a measurement, the satellite carrier shall within 60 days notice thereof, terminate service of the station's network to that household.

(11) Service to recreational vehicles and commercial trucks.—

(A) Exemption.—

(i) In general.—For purposes of this subsection, and subject to clauses (ii) and (iii), the term "unserved household" shall include—

(I) recreational vehicles as defined in regulations of the Secretary of Housing and Urban Development under section 3282.8 of title 24, Code of Federal Regulations; and

(II) commercial trucks that qualify as commercial motor vehicles under regulations of the Secretary of Transportation under section 383.5 of title 49, Code of Federal Regulations.

(ii) Limitation.—Clause (i) shall apply only to a recreational vehicle or commercial truck if any satellite carrier that proposes to make a secondary transmission of a network station to the operator of such a recreational vehicle or commercial truck complies with the documentation requirements under subparagraphs (B) and (C).

(iii) Exclusion.—For purposes of this subparagraph, the terms "recreational vehicle" and "commercial truck" shall not include any fixed dwelling, whether a mobile home or otherwise.

(B) Documentation requirements.—A recreational vehicle or commercial truck shall be deemed to be an unserved household beginning 10 days after the relevant satellite carrier provides to the network that owns or is affiliated with the network station that will be secondarily transmitted to the recreational vehicle or commercial truck the following documents:

(i) Declaration.—A signed declaration by the operator of the recreational vehicle or commercial truck that the satellite dish is permanently attached to the recreational vehicle or commercial truck, and will not be used to receive satellite programming at any fixed dwelling.

(ii) Registration.—In the case of a recreational vehicle, a copy of the current State vehicle registration for the recreational vehicle.

(iii) Registration and license.—In the case of a commercial truck, a copy of—

(I) the current State vehicle registration for the truck; and

(II) a copy of a valid, current commercial driver's license, as defined in regulations of the Secretary of Transportation under section 383 of title 49, Code of Federal Regulations, issued to the operator.

(C) Updated documentation requirements.—If a satellite carrier wishes to continue to make secondary transmissions to a recreational vehicle or commercial truck for more than a 2-year period, that carrier shall provide each network, upon request, with updated documentation in the form described under subparagraph (B) during the 90 days before expiration of that 2-year period.

(12) Statutory license contingent on compliance with fcc rules and remedial steps.—Notwithstanding any other provision of this section, the willful or repeated secondary transmission to the public by a satellite carrier of a primary transmission embodying a performance or display of a work made by a broadcast station licensed by the Federal Communications Commission is actionable as an act of infringement under section 501, and is fully subject to the remedies provided by sections 502 through 506, if, at the time of such transmission, the satellite carrier is not in compliance with the rules, regulations, and authorizations of the Federal Communications Commission concerning the carriage of television broadcast station signals.

(13) Waivers.—A subscriber who is denied the secondary transmission of a signal of a network station under subsection (a)(2)(B) may request a waiver from

such denial by submitting a request, through the subscriber's satellite carrier, to the network station asserting that the secondary transmission is prohibited. The network station shall accept or reject a subscriber's request for a waiver within 30 days after receipt of the request. If a television network station fails to accept or reject a subscriber's request for a waiver within the 30-day period after receipt of the request, that station shall be deemed to agree to the waiver request and have filed such written waiver. Unless specifically stated by the network station, a waiver that was granted before the date of the enactment of the Satellite Home Viewer Extension and Reauthorization Act of 2004 under section 339(c)(2) of the Communications Act of 1934, and that was in effect on such date of enactment, shall constitute a waiver for purposes of this paragraph.

(14) Restricted transmission of out-of-state distant network signals into certain markets.—

(A) Out-of-state network affiliates.—Notwithstanding any other provision of this title, the statutory license in this subsection and subsection (b) shall not apply to any secondary transmission of the primary transmission of a network station located outside of the State of Alaska to any subscriber in that State to whom the secondary transmission of the primary transmission of a television station located in that State is made available by the satellite carrier pursuant to section 122.

(B) Exception.—The limitation in subparagraph (A) shall not apply to the secondary transmission of the primary transmission of a digital signal of a network station located outside of the State of Alaska if at the time that the secondary transmission is made, no television station licensed to a community in the State and affiliated with the same network makes primary transmissions of a digital signal.

(b) Deposit of Statements and Fees; Verification Procedures.—

(1) Deposits with the register of copyrights.—A satellite carrier whose secondary transmissions are subject to statutory licensing under subsection (a) shall, on a semiannual basis, deposit with the Register of Copyrights, in accordance with requirements that the Register shall prescribe by regulation—

(A) a statement of account, covering the preceding 6-month period, specifying the names and locations of all non-network and network stations whose signals were retransmitted, at any time during that period, to subscribers as described in subsections (a)(1) and (a)(2), the total number of subscribers that received such retransmissions, and such other data as the Register of Copyrights may from time to time prescribe by regulation;

(B) a royalty fee payable to copyright owners pursuant to paragraph (4) for that 6-month period, computed by multiplying the total number of subscribers receiving each secondary transmission of a primary stream or multicast stream of each non-network station or network station during each calendar year month by the appropriate rate in effect under this subsection; and

(C) a filing fee, as determined by the Register of Copyrights pursuant to section 708(a).

(2) Verification of accounts and fee payments.—The Register of Copyrights shall issue regulations to permit interested parties to verify and audit the statements of account and royalty fees submitted by satellite carriers under this subsection.

(3) Investment of fees.—The Register of Copyrights shall receive all fees (including the filing fee specified in paragraph (1)(c)) deposited under this section and, after deducting the reasonable costs incurred by the Copyright Office under this section (other than the costs deducted under paragraph (5)), shall deposit the balance in the Treasury of the United States, in such manner as the Secretary of the Treasury directs. All funds held by the Secretary of the Treasury shall be invested in interest-bearing securities of the United States for later distribution with interest by the Librarian of Congress as provided by this title.

(4) Persons to whom fees are distributed.—The royalty fees deposited under paragraph (3) shall, in accordance with the procedures provided by paragraph (5), be distributed to those copyright owners whose works were included in a secondary transmission made by a satellite carrier during the applicable 6-month accounting period and who file a claim with the Copyright Royalty Judges under paragraph (5).

(5) Procedures for distribution.—The royalty fees deposited under paragraph (3) shall be distributed in accordance with the following procedures:

(A) Filing of claims for fees.—During the month of July in each year, each person claiming to be entitled to statutory license fees for secondary transmissions shall file a claim with the Copyright Royalty Judges, in accordance with requirements that the Copyright Royalty Judges shall prescribe by regulation. For purposes of this paragraph, any claimants may agree among themselves as to the proportionate division of statutory license fees among them, may lump their claims together and file them jointly or as a single claim, or may designate a common agent to receive payment on their behalf.

(B) Determination of controversy; distributions.—After the first day of August of each year, the Copyright Royalty Judges shall determine whether there exists a controversy concerning the distribution of royalty fees. If the Copyright Royalty Judges determine that no such controversy exists, the Copyright Royalty Judges shall authorize the Librarian of Congress to proceed to distribute such fees to the copyright owners entitled to receive them, or to their designated agents, subject to the deduction of reasonable administrative costs under this section. If the Copyright Royalty Judges find the existence of a controversy, the Copyright Royalty Judges shall, pursuant to chapter 8 of this title, conduct a proceeding to determine the distribution of royalty fees.

(C) Withholding of fees during controversy.—During the pendency of any proceeding under this subsection, the Copyright Royalty Judges shall have the discretion to authorize the Librarian of Congress to proceed to distribute any amounts that are not in controversy.

(c) Adjustment of Royalty Fees.—

(1) Applicability and determination of royalty fees for signals.—

(A) Initial fee.—The appropriate fee for purposes of determining the royalty fee under subsection (b)(1)(B) for the secondary transmission of the primary transmissions of network stations and non-network stations shall be the appropriate fee set forth in part 258 of title 37, Code of Federal Regulations, as in effect on July 1, 2009, as modified under this paragraph.

(B) Fee set by voluntary negotiation.—On or before June 1, 2010, the Copyright Royalty Judges shall cause to be published in the Federal Register of the initiation of voluntary negotiation proceedings for the purpose of determining the royalty fee to be paid by satellite carriers for the secondary transmission of the primary transmission of network stations and non-network stations under subsection (b)(1)(B).

(C) Negotiations.—Satellite carriers, distributors, and copyright owners entitled to royalty fees under this section shall negotiate in good faith in an effort to reach a voluntary agreement or agreements for the payment of royalty fees. Any such satellite carriers, distributors and copyright owners may at any time negotiate and agree to the royalty fee, and may designate common agents to negotiate, agree to, or pay such fees. If the parties fail to identify common agents, the Copyright Royalty Judges shall do so, after requesting recommendations from the parties to the negotiation proceeding. The parties to each negotiation proceeding shall bear the cost thereof.

(D) Agreements binding on parties; filing of agreements; public notice.—(i) Voluntary agreements; filing.—Voluntary agreements negotiated at any time in accordance with this paragraph shall be binding upon all satellite carriers, distributors, and copyright owners that are parties thereto. Copies of such agreements shall be filed with the Copyright Office within 30 days after execution in accordance with regulations that the Register of Copyrights shall prescribe.

(ii) Procedure for adoption of fees.—

(I) Publication of notice.—Within 10 days after publication in the Federal Register of a notice of the initiation of voluntary negotiation proceedings, parties who have reached a voluntary agreement may request that the royalty fees in that agreement be applied to all satellite carriers, distributors, and copyright owners without convening a proceeding under subparagraph (F).

(II) Public notice of fees.—Upon receiving a request under subclause (I), the Copyright Royalty Judges shall immediately provide public notice of the royalty fees from the voluntary agreement and afford parties an opportunity to state that they object to those fees.

(III) Adoption of fees.—The Copyright Royalty Judges shall adopt the royalty fees from the voluntary agreement for all satellite carriers, distributors, and copyright owners without convening the proceeding under subparagraph (F) unless a party with an intent to participate in that proceeding and a significant interest in the outcome of that proceeding objects under subclause (II).

(E) Period agreement is in effect.—The obligation to pay the royalty fees established under a voluntary agreement which has been filed with the Copyright Royalty Judges in accordance with this paragraph shall become effective on the date specified in the agreement, and shall remain in effect until

December 31, 2014, in accordance with the terms of the agreement, whichever is later.

(F) Fee set by copyright royalty judges proceeding.—

(i) Notice of initiation of the proceeding.—On or before September 1, 2010, the Copyright Royalty Judges shall cause notice to be published in the Federal Register of the initiation of a proceeding for the purpose of determining the royalty fees to be paid for the secondary transmission of the primary transmissions of network stations and non-network stations under subsection (b)(1)(B) by satellite carriers and distributors—

(I) in the absence of a voluntary agreement filed in accordance with subparagraph (D) that establishes royalty fees to be paid by all satellite carriers and distributors; or

(II) if an objection to the fees from a voluntary agreement submitted for adoption by the Copyright Royalty Judges to apply to all satellite carriers, distributors, and copyright owners is received under subparagraph (D) from a party with an intent to participate in the proceeding and a significant interest in the outcome of that proceeding.

Such proceeding shall be conducted under chapter 8.

(ii) Establishment of royalty fees.—In determining royalty fees under this subparagraph, the Copyright Royalty Judges shall establish fees for the secondary transmissions of the primary transmissions of network stations and non-network stations that most clearly represent the fair market value of secondary transmissions, except that the Copyright Royalty Judges shall adjust royalty fees to account for the obligations of the parties under any applicable voluntary agreement filed with the Copyright Royalty Judges in accordance with subparagraph (D). In determining the fair market value, the Judges shall base their decision on economic, competitive, and programming information presented by the parties, including—

(I) the competitive environment in which such programming is distributed, the cost of similar signals in similar private and compulsory license marketplaces, and any special features and conditions of the retransmission marketplace;

(II) the economic impact of such fees on copyright owners and satellite carriers; and

(III) the impact on the continued availability of secondary transmissions to the public.

(iii) Effective date for decision of copyright royalty judges.—The obligation to pay the royalty fees established under a determination that is made by the Copyright Royalty Judges in a proceeding under this paragraph shall be effective as of January 1, 2010.

(iv) Persons subject to royalty fees.—The royalty fees referred to in clause (iii) shall be binding on all satellite carriers, distributors and copyright owners, who are not party to a voluntary agreement filed with the Copyright Office under subparagraph (D).

(2) Annual royalty fee adjustment.—Effective January 1 of each year, the royalty fee payable under subsection (b)(1)(B) for the secondary transmission

of the primary transmissions of network stations and non-network stations shall be adjusted by the Copyright Royalty Judges to reflect any changes occurring in the cost of living as determined by the most recent Consumer Price Index (for all consumers and for all items) published by the Secretary of Labor before December 1 of the preceding year. Notification of the adjusted fees shall be published in the Federal Register at least 25 days before January 1.

(d) Definitions.—As used in this section—

(1) Distributor.—The term "distributor" means an entity that contracts to distribute secondary transmissions from a satellite carrier and, either as a single channel or in a package with other programming, provides the secondary transmission either directly to individual subscribers or indirectly through other program distribution entities in accordance with the provisions of this section.

(2) Network station.—The term "network station" means—

(A) a television station licensed by the Federal Communications Commission, including any translator station or terrestrial satellite station that rebroadcasts all or substantially all of the programming broadcast by a network station, that is owned or operated by, or affiliated with, one or more of the television networks in the United States that offer an interconnected program service on a regular basis for 15 or more hours per week to at least 25 of its affiliated television licensees in 10 or more States; or

(B) a noncommercial educational broadcast station (as defined in section 397 of the Communications Act of 1934).

(3) Primary network station.—The term "primary network station" means a network station that broadcasts or rebroadcasts the basic programming service of a particular national network.

(4) Primary transmission.—The term "primary transmission" has the meaning given that term in section 111(f) of this title.

(5) Private home viewing.—The term "private home viewing" means the viewing, for private use in a household by means of satellite reception equipment that is operated by an individual in that household and that serves only such household, of a secondary transmission delivered by a satellite carrier of a primary transmission of a television station licensed by the Federal Communications Commission.

(6) Satellite carrier.—The term "satellite carrier" means an entity that uses the facilities of a satellite or satellite service licensed by the Federal Communications Commission and operates in the Fixed-Satellite Service under part 25 of title 47, Code of Federal Regulations, or the Direct Broadcast Satellite Service under part 100 of title 47, Code of Federal Regulations to establish and operate a channel of communications for point-to-multipoint distribution of television station signals, and that owns or leases a capacity or service on a satellite in order to provide such point-to-multipoint distribution, except to the extent that such entity provides such distribution pursuant to tariff under the Communications Act of 1934, other than for private home viewing pursuant to this section.

(7) Secondary transmission.—The term "secondary transmission" has the meaning given that term in section 111(f) of this title.

(8) Subscriber; subscribe.—

(A) Subscriber.—The term "subscriber" means a person or entity that receives a secondary transmission service from a satellite carrier and pays a fee for the service, directly or indirectly, to the satellite carrier or to a distributor.

(B) Subscribe.—The term "subscribe" means to elect to become a subscriber.

(9) Non-network station.—The term "non-network station" means a television station, other than a network station, licensed by the Federal Communications Commission, that is secondarily transmitted by a satellite carrier.

(10) Unserved household.—The term "unserved household", with respect to a particular television network, means a household that—

(A) cannot receive, through the use of an antenna, an over-the-air signal containing the primary stream, or, on or after the qualifying date, the multicast stream, originating in that household's local market and affiliated with that network of—

(i) if the signal originates as an analog signal, Grade B intensity as defined by the Federal Communications Commission in section 73.683(a) of title 47, Code of Federal Regulations, as in effect on January 1, 1999; or

(ii) if the signal originates as a digital signal, intensity defined in the values for the digital television noise-limited service contour, as defined in regulations issued by the Federal Communications Commission (section 73.622(e) of title 47, Code of Federal Regulations), as such regulations may be amended from time to time;

(B) is subject to a waiver that meets the standards of subsection (a)(13) whether or not the waiver was granted before the date of the enactment of the Satellite Television Extension and Localism Act of 2010;

(C) is a subscriber to whom subsection (e) applies;

(D) is a subscriber to whom subsection (a)(11) applies; or

(E) is a subscriber to whom the exemption under subsection (a)(2)(B)(iii) applies.

(11) Local market.—The term "local market" has the meaning given such term under section 122(j).

(12) Commercial establishment.—The term "commercial establishment"—

(A) means an establishment used for commercial purposes, such as a bar, restaurant, private office, fitness club, oil rig, retail store, bank or other financial institution, supermarket, automobile or boat dealership, or any other establishment with a common business area; and

(B) does not include a multi-unit permanent or temporary dwelling where private home viewing occurs, such as a hotel, dormitory, hospital, apartment, condominium, or prison.

(13) Qualifying date.—The term "qualifying date", for purposes of paragraph (10)(A), means—

(A) October 1, 2010, for multicast streams that exist on March 31, 2010; and

(B) January 1, 2011, for all other multicast streams.

(14) Multicast stream.—The term "multicast stream" means a digital stream containing programming and program-related material affiliated with a television network, other than the primary stream.

(15) Primary stream.—The term "primary stream" means—

(A) the single digital stream of programming as to which a television broadcast station has the right to mandatory carriage with a satellite carrier under the rules of the Federal Communications Commission in effect on July 1, 2009; or

(B) if there is no stream described in subparagraph (A), then either—

(i) the single digital stream of programming associated with the network last transmitted by the station as an analog signal; or

(ii) if there is no stream described in clause (i), then the single digital stream of programming affiliated with the network that, as of July 1, 2009, had been offered by the television broadcast station for the longest period of time.

(e) Moratorium on Copyright Liability.—Until December 31, 201462, a subscriber who does not receive a signal of Grade A intensity (as defined in the regulations of the Federal Communications Commission under section 73.683(a) of title 47, Code of Federal Regulations, as in effect on January 1, 1999, or predicted by the Federal Communications Commission using the Individual Location Longley-Rice methodology described by the Federal Communications Commission in Docket No. 98-201) of a local network television broadcast station shall remain eligible to receive signals of network stations affiliated with the same network, if that subscriber had satellite service of such network signal terminated after July 11, 1998, and before October 31, 1999, as required by this section, or received such service on October 31, 1999.

(f) Expedited Consideration by Justice Department of Voluntary Agreements to Provide Satellite Secondary Transmissions to Local Markets.—

(1) In general.—In a case in which no satellite carrier makes available, to subscribers located in a local market, as defined in section 122(j)(2), the secondary transmission into that market of a primary transmission of one or more television broadcast stations licensed by the Federal Communications Commission, and two or more satellite carriers request a business review letter in accordance with section 50.6 of title 28, Code of Federal Regulations (as in effect on July 7, 2004), in order to assess the legality under the antitrust laws of proposed business conduct to make or carry out an agreement to provide such secondary transmission into such local market, the appropriate official of the Department of Justice shall respond to the request no later than 90 days after the date on which the request is received.

(2) Definition.—For purposes of this subsection, the term "antitrust laws"—

(A) has the meaning given that term in subsection (a) of the first section of the Clayton Act (15 U.S.C. 12(a)), except that such term includes section 5 of the Federal Trade Commission Act (15 U.S.C. 45) to the extent such section 5 applies to unfair methods of competition; and

(B) includes any State law similar to the laws referred to in paragraph (1).

(g) Certain Waivers Granted to Providers of Local-into-Local Service to All DMAs.—

(1) Injunction waiver.—A court that issued an injunction pursuant to subsection (a)(7)(B) before the date of the enactment of this subsection shall waive such injunction if the court recognizes the entity against which the injunction was issued as a qualified carrier.

(2) Limited temporary waiver.—

(A) In general.—Upon a request made by a satellite carrier, a court that issued an injunction against such carrier under subsection (a)(7)(B) before the date of the enactment of this subsection shall waive such injunction with respect to the statutory license provided under subsection (a)(2) to the extent necessary to allow such carrier to make secondary transmissions of primary transmissions made by a network station to unserved households located in short markets in which such carrier was not providing local service pursuant to the license under section 122 as of December 31, 2009.

(B) Expiration of temporary waiver.—A temporary waiver of an injunction under subparagraph (A) shall expire after the end of the 120 day period beginning on the date such temporary waiver is issued unless extended for good cause by the court making the temporary waiver.

(C) Failure to provide local-into-local service to all dmas.—

(i) Failure to act reasonably in good faith.—If the court issuing a temporary waiver under subparagraph (A) determines that the satellite carrier that made the request for such waiver has failed to act reasonably or has failed to make a good faith effort to provide local-intolocal service to all DMAs, such failure—

(I) is actionable as an act of infringement under section 501 and the court may in its discretion impose the remedies provided for in sections 502 through 506 and subsection (a)(6)(B) of this section; and

(II) shall result in the termination of the waiver issued under subparagraph (A).

(ii) Failure to provide local-into-local service.—If the court issuing a temporary waiver under subparagraph (A) determines that the satellite carrier that made the request for such waiver has failed to provide local-into-local service to all DMAs, but determines that the carrier acted reasonably and in good faith, the court may in its discretion impose financial penalties that reflect—

(I) the degree of control the carrier had over the circumstances that resulted in the failure;

(II) the quality of the carrier's efforts to remedy the failure; and

(III) the severity and duration of any service interruption.

(D) Single temp0rary waiver available.—An entity may only receive one temporary waiver under this paragraph.

(E) Short market defined.—For purposes of this paragraph, the term "short market" means a local market in which programming of one or more of the four most widely viewed television networks nationwide as measured on the date of the enactment of this subsection is not offered on the primary stream transmitted by any local television broadcast station.

(3) Establishment of qualified carrier recognition.—

(A) Statement of eligibility.—An entity seeking to be recognized as a qualified carrier under this subsection shall file a statement of eligibility with the court that imposed the injunction. A statement of eligibility must include—

(i) an affidavit that the entity is providing local-into-local service to all DMAs;

(ii) a motion for a waiver of the injunction;

(iii) a motion that the court appoint a special master under Rule 53 of the Federal Rules of Civil Procedure;

(iv) an agreement by the carrier to pay all expenses incurred by the special master under paragraph (4)(B)(ii); and

(v) a certification issued pursuant to section 342(a) of Communications Act of 1934.

(B) Grant of recognition as a qualified carrier.—Upon receipt of a statement of eligibility, the court shall recognize the entity as a qualified carrier and issue the waiver under paragraph (1). Upon motion pursuant to subparagraph (A)(iii), the court shall appoint a special master to conduct the examination and provide a report to the court as provided in paragraph (4)(B).

(C) Voluntary termination.—At any time, an entity recognized as a qualified carrier may file a statement of voluntary termination with the court certifying that it no longer wishes to be recognized as a qualified carrier. Upon receipt of such statement, the court shall reinstate the injunction waived under paragraph (1).

(D) Loss of recognition prevents future recognition.—No entity may be recognized as a qualified carrier if such entity had previously been recognized as a qualified carrier and subsequently lost such recognition or voluntarily terminated such recognition under subparagraph (C).

(4) Qualified carrier obligations and compliance.—

(A) Continuing obligations.—

(i) In general.—An entity recognized as a qualified carrier shall continue to provide local-into-local service to all DMAs.

(ii) Cooperation with compliance examination.—An entity recognized as a qualified carrier shall fully cooperate with the special master appointed by the court under paragraph (3)(B) in an examination set forth in subparagraph (B).

(B) Qualified carrier compliance examination.—

(i) Examination and report.—A special master appointed by the court under paragraph (3)(B) shall conduct an examination of, and file a report on, the qualified carrier's compliance with the royalty payment and household eligibility requirements of the license under this section. The report shall address the qualified carrier's conduct during the period beginning on the date on which the qualified carrier is recognized as such under paragraph (3)(B) and ending on April 30, 2012.

(ii) Records of qualified carrier.—Beginning on the date that is one year after the date on which the qualified carrier is recognized as such under paragraph (3)(B), but not later than December 1, 2011, the qualified carrier shall provide the special master with all records that the special master considers to be directly pertinent to the following requirements under this section:

(I) Proper calculation and payment of royalties under the statutory license under this section.

(II) Provision of service under this license to eligible subscribers only.

(iii) Submission of report.—The special master shall file the report required by clause (i) not later than July 24, 2012, with the court referred to in paragraph (1) that issued the injunction, and the court shall transmit a copy of the report to the Register of Copyrights, the Committees on the Judiciary and on Energy and Commerce of the House of Representatives, and the Committees on the Judiciary and on Commerce, Science, and Transportation of the Senate.

(iv) Evidence of infringement.—The special master shall include in the report a statement of whether the examination by the special master indicated that there is substantial evidence that a copyright holder could bring a successful action under this section against the qualified carrier for infringement.

(v) Subsequent examination.—If the special master's report includes a statement that its examination indicated the existence of substantial evidence that a copyright holder could bring a successful action under this section against the qualified carrier for infringement, the special master shall, not later than 6 months after the report under clause

(i) is filed, initiate another examination of the qualified carrier's compliance with the royalty payment and household eligibility requirements of the license under this section since the last report was filed under clause

(iii). The special master shall file a report on the results of the examination conducted under this clause with the court referred to in paragraph

(1) that issued the injunction, and the court shall transmit a copy to the Register of Copyrights, the Committees on the Judiciary and on Energy and Commerce of the House of Representatives, and the Committees on the Judiciary and on Commerce, Science, and Transportation of the Senate. The report shall include a statement described in clause (iv).

(vi) Compliance.—Upon motion filed by an aggrieved copyright owner, the court recognizing an entity as a qualified carrier shall terminate such designation upon finding that the entity has failed to cooperate with an examination required by this subparagraph.

(vii) Oversight.—During the period of time that the special master is conducting an examination under this subparagraph, the Comptroller General shall monitor the degree to which the entity seeking to be recognized or recognized as a qualified carrier under paragraph (3) is complying with the special master's examination. The qualified carrier shall make available to the Comptroller General all records and individuals that the Comptroller General considers necessary to meet the Comptroller General's obligations under this clause. The Comptroller General shall report the results of the monitoring required by this clause to the Committees on the Judiciary and on Energy and Commerce of the House of Representatives and the Committees on the Judiciary and on Commerce, Science, and Transportation of the Senate at intervals of not less than six months during such period.

(C) Affirmation.—A qualified carrier shall file an affidavit with the district court and the Register of Copyrights 30 months after such status was granted stating that, to the best of the affiant's knowledge, it is in compliance with the requirements for a qualified carrier. The qualified carrier shall attach to its affidavit copies of all reports or orders issued by the court, the special master, and the Comptroller General.

(D) Compliance determination.—Upon the motion of an aggrieved television broadcast station, the court recognizing an entity as a qualified carrier may make a determination of whether the entity is providing local-into-local service to all DMAs.

(E) Pleading requirement.—In any motion brought under subparagraph (D), the party making such motion shall specify one or more designated market areas (as such term is defined in section 122(j)(2)(C)) for which the failure to provide service is being alleged, and, for each such designated market area, shall plead with particularity the circumstances of the alleged failure.

(F) Burden of proof.—In any proceeding to make a determination under subparagraph (D), and with respect to a designated market area for which failure to provide service is alleged, the entity recognized as a qualified carrier shall have the burden of proving that the entity provided local-into-local service with a good quality satellite signal to at least 90 percent of the households in such designated market area (based on the most recent census data released by the United States Census Bureau) at the time and place alleged.

(5) Failure to provide service.—

(A) Penalties.—If the court recognizing an entity as a qualified carrier finds that such entity has willfully failed to provide local-into-local service to all DMAs, such finding shall result in the loss of recognition of the entity as a qualified carrier and the termination of the waiver provided under paragraph (1), and the court may, in its discretion—

(i) treat such failure as an act of infringement under section 501, and subject such infringement to the remedies provided for in sections 502 through 506 and subsection (a)(6)(B) of this section; and

(ii) impose a fine of not less than $250,000 and not more than $5,000,000.

(B) Exception for nonwillful violation.—If the court determines that the failure to provide local-into-local service to all DMAs is nonwillful, the court may in its discretion impose financial penalties for noncompliance that reflect—

(i) the degree of control the entity had over the circumstances that resulted in the failure;

(ii) the quality of the entity's efforts to remedy the failure and restore service; and

(iii) the severity and duration of any service interruption.

(6) Penalties for violation of license.—A court that finds, under subsection (a)(6)(A), that an entity recognized as a qualified carrier has willfully made a secondary transmission of a primary transmission made by a network station and embodying a performance or display of a work to a subscriber who is not eligible to receive the transmission under this section shall reinstate the injunction waived under paragraph (1), and the court may order statutory damages of not more than $2,500,000.

(7) Local-into-local service to all dmas defined.—For purposes of this subsection:

(A) In general.—An entity provides "local-into-local service to all DMAs" if the entity provides local service in all designated market areas (as such term is defined in section 122(j)(2)(C)) pursuant to the license under section 122.

(B) Household coverage.—For purposes of subparagraph (A), an entity that makes available local-into-local service with a good quality satellite signal to at least 90 percent of the households in a designated market area based on the most recent census data released by the United States Census Bureau shall be considered to be providing local service to such designated market area.

(C) Good quality satellite signal defined.—The term "good quality satellite signal" has the meaning given such term under section 342(e)(2) of Communications Act of 1934.

§ 120 Scope of Exclusive Rights in Architectural Works

(a) Pictorial Representations Permitted.—The copyright in an architectural work that has been constructed does not include the right to prevent the making, distributing, or public display of pictures, paintings, photographs, or other pictorial representations of the work, if the building in which the work is embodied is located in or ordinarily visible from a public place.

(b) Alterations to and Destruction of Buildings.—Notwithstanding the provisions of section 106(2), the owners of a building embodying an architectural work may, without the consent of the author or copyright owner of the architectural work, make or authorize the making of alterations to such building, and destroy or authorize the destruction of such building.

§ 121 Limitations on Exclusive Rights: Reproduction for Blind or Other People with Disabilities

(a) Notwithstanding the provisions of section 106, it is not an infringement of copyright for an authorized entity to reproduce or to distribute copies or phonorecords of a previously published, nondramatic literary work if such copies or phonorecords are reproduced or distributed in specialized formats exclusively for use by blind or other persons with disabilities.

(b)(1) Copies or phonorecords to which this section applies shall—

(A) not be reproduced or distributed in a format other than a specialized format exclusively for use by blind or other persons with disabilities;

(B) bear a notice that any further reproduction or distribution in a format other than a specialized format is an infringement; and

(C) include a copyright notice identifying the copyright owner and the date of the original publication.

(2) The provisions of this subsection shall not apply to standardized, secure, or norm-referenced tests and related testing material, or to computer programs, except the portions thereof that are in conventional human language (including descriptions of pictorial works) and displayed to users in the ordinary course of using the computer programs.

(c) Notwithstanding the provisions of section 106, it is not an infringement of copyright for a publisher of print instructional materials for use in elementary or secondary schools to create and distribute to the National Instructional Materials Access Center copies of the electronic files described in sections 612(a)(23)(C), 613(a)(6), and section 674(e) of the Individuals with Disabilities Education Act that contain the contents of print instructional materials using the National Instructional Material Accessibility Standard (as defined in section 674(e)(3) of that Act), if—

(1) the inclusion of the contents of such print instructional materials is required by any State educational agency or local educational agency;

(2) the publisher had the right to publish such print instructional materials in print formats; and

(3) such copies are used solely for reproduction or distribution of the contents of such print instructional materials in specialized formats.

(d) For purposes of this section, the term—

(1) "authorized entity" means a nonprofit organization or a governmental agency that has a primary mission to provide specialized services relating to training, education, or adaptive reading or information access needs of blind or other persons with disabilities;

(2) "blind or other persons with disabilities" means individuals who are eligible or who may qualify in accordance with the Act entitled "An Act to provide books for the adult blind", approved March 3, 1931 (2 U.S.C. 135a; 46 Stat. 1487) to receive books and other publications produced in specialized formats;

(3) "print instructional materials" has the meaning given under section 674(e)(3)(C) of the Individuals with Disabilities Education Act; and

(4) "specialized formats" means—

(A) braille, audio, or digital text which is exclusively for use by blind or other persons with disabilities; and

(B) with respect to print instructional materials, includes large print formats when such materials are distributed exclusively for use by blind or other persons with disabilities.

§ 122 Limitations on Exclusive Rights: Secondary Transmissions of Local Television Programming by Satellite

(a) Secondary Transmissions into Local Markets.—

(1) Secondary transmissions of television broadcast stations within a local market.—A secondary transmission of a performance or display of a work embodied in a primary transmission of a television broadcast station into the station's local market shall be subject to statutory licensing under this section if—

(A) the secondary transmission is made by a satellite carrier to the public;

(B) with regard to secondary transmissions, the satellite carrier is in compliance with the rules, regulations, or authorizations of the Federal Communications Commission governing the carriage of television broadcast station signals; and

(C) the satellite carrier makes a direct or indirect charge for the secondary transmission to—

(i) each subscriber receiving the secondary transmission; or

(ii) a distributor that has contracted with the satellite carrier for direct or indirect delivery of the secondary transmission to the public.

(2) Significantly viewed stations.—

(A) In general.—A secondary transmission of a performance or display of a work embodied in a primary transmission of a television broadcast station to subscribers who receive secondary transmissions of primary transmissions under paragraph (1) shall be subject to statutory licensing under this paragraph if the secondary transmission is of the primary transmission of a network station or a nonnetwork station to a subscriber who resides outside the station's local market but within a community in which the signal has been determined by the Federal Communications Commission to be significantly viewed in such community, pursuant to the rules, regulations, and authorizations of the Federal Communications Commission in effect on April 15, 1976, applicable to determining with respect to a cable system whether signals are significantly viewed in a community.

(B) Waiver.—A subscriber who is denied the secondary transmission of the primary transmission of a network station or a non-network station under subparagraph (A) may request a waiver from such denial by submitting a request, through the subscriber's satellite carrier, to the network station or non-network station in the local market affiliated with the same network or non-network where the subscriber is located. The network station or non-network station shall accept or reject the subscriber's request for a waiver within 30 days after receipt of the request. If the network station or non-network station fails to accept or reject the subscriber's request for a waiver within that

30-day period, that network station or non-network station shall be deemed to agree to the waiver request.

(3) Secondary transmission of low power programming.—

(A) In general.—Subject to subparagraphs (B) and (C), a secondary transmission of a performance or display of a work embodied in a primary transmission of a television broadcast station to subscribers who receive secondary transmissions of primary transmissions under paragraph (1) shall be subject to statutory licensing under this paragraph if the secondary transmission is of the primary transmission of a television broadcast station that is licensed as a low power television station, to a subscriber who resides within the same designated market area as the station that originates the transmission.

(B) No applicability to repeaters and translators.—Secondary transmissions provided for in subparagraph (A) shall not apply to any low power television station that retransmits the programs and signals of another television station for more than 2 hours each day.

(C) No impact on other secondary transmissions obligations.—A satellite carrier that makes secondary transmissions of a primary transmission of a low power television station under a statutory license provided under this section is not required, by reason of such secondary transmissions, to make any other secondary transmissions.

(4) Special exceptions.—A secondary transmission of a performance or display of a work embodied in a primary transmission of a television broadcast station to subscribers who receive secondary transmissions of primary transmissions under paragraph (1) shall, if the secondary transmission is made by a satellite carrier that complies with the requirements of paragraph (1), be subject to statutory licensing under this paragraph as follows:

(A) States with single full-power network station.—In a State in which there is licensed by the Federal Communications Commission a single full-power station that was a network station on January 1, 1995, the statutory license provided for in this paragraph shall apply to the secondary transmission by a satellite carrier of the primary transmission of that station to any subscriber in a community that is located within that State and that is not within the first 50 television markets as listed in the regulations of the Commission as in effect on such date (47 C.F.R. 76.51).

(B) States with all network stations and non-network stations in same local market.—In a State in which all network stations and non-network stations licensed by the Federal Communications Commission within that State as of January 1, 1995, are assigned to the same local market and that local market does not encompass all counties of that State, the statutory license provided under this paragraph shall apply to the secondary transmission by a satellite carrier of the primary transmissions of such station to all subscribers in the State who reside in a local market that is within the first 50 major television markets as listed in the regulations of the Commission as in effect on such date (section 76.51 of title 47, Code of Federal Regulations).

(C) Additional stations.—In the case of that State in which are located 4 counties that—

(i) on January 1, 2004, were in local markets principally comprised of counties in another State, and

(ii) had a combined total of 41,340 television households, according to the U.S. Television Household Estimates by Nielsen Media Research for 2004, the statutory license provided under this paragraph shall apply to secondary transmissions by a satellite carrier to subscribers in any such county of the primary transmissions of any network station located in that State, if the satellite carrier was making such secondary transmissions to any subscribers in that county on January 1, 2004.

(D) Certain additional stations.—If 2 adjacent counties in a single State are in a local market comprised principally of counties located in another State, the statutory license provided for in this paragraph shall apply to the secondary transmission by a satellite carrier to subscribers in those 2 counties of the primary transmissions of any network station located in the capital of the State in which such 2 counties are located, if—

(i) the 2 counties are located in a local market that is in the top 100 markets for the year 2003 according to Nielsen Media Research; and

(ii) the total number of television households in the 2 counties combined did not exceed 10,000 for the year 2003 according to Nielsen Media Research.

(E) Networks of noncommercial educational broadcast stations.—In the case of a system of three or more noncommercial educational broadcast stations licensed to a single State, public agency, or political, educational, or special purpose subdivision of a State, the statutory license provided for in this paragraph shall apply to the secondary transmission of the primary transmission of such system to any subscriber in any county or county equivalent within such State, if such subscriber is located in a designated market area that is not otherwise eligible to receive the secondary transmission of the primary transmission of a noncommercial educational broadcast station located within the State pursuant to paragraph (1).

(5) Applicability of royalty rates and procedures.—The royalty rates and procedures under section 119(b) shall apply to the secondary transmissions to which the statutory license under paragraph (4) applies.

(b) Reporting Requirements.—

(1) Initial lists.—A satellite carrier that makes secondary transmissions of a primary transmission made by a network station under subsection (a) shall, within 90 days after commencing such secondary transmissions, submit to the network that owns or is affiliated with the network station—

(A) a list identifying (by name in alphabetical order and street address, including county and 9-digit zip code) all subscribers to which the satellite carrier makes secondary transmissions of that primary transmission under subsection (a);

(B) a separate list, aggregated by designated market area (by name and address, including street or rural route number, city, State, and 9-digit zip code), which shall indicate those subscribers being served pursuant to paragraph (2) of subsection (a).

(2) Subsequent lists.—After the list is submitted under paragraph (1), the satellite carrier shall, on the 15th of each month, submit to the network—

(A) a list identifying (by name in alphabetical order and street address, including county and 9-digit zip code) any subscribers who have been added or dropped as subscribers since the last submission under this subsection; and

(B) a separate list, aggregated by designated market area (by name and street address, including street or rural route number, city, State, and 9-digit zip code), identifying those subscribers whose service pursuant to paragraph (2) of subsection (a) has been added or dropped since the last submission under this subsection.

(3) Use of subscriber information.—Subscriber information submitted by a satellite carrier under this subsection may be used only for the purposes of monitoring compliance by the satellite carrier with this section.

(4) Requirements of networks.—The submission requirements of this subsection shall apply to a satellite carrier only if the network to which the submissions are to be made places on file with the Register of Copyrights a document identifying the name and address of the person to whom such submissions are to be made. The Register of Copyrights shall maintain for public inspection a file of all such documents.

(c) No Royalty Fee Required for Certain Secondary Transmissions.—A satellite carrier whose secondary transmissions are subject to statutory licensing under paragraphs (1), (2), and (3) of subsection (a) shall have no royalty obligation for such secondary transmissions.

(d) Noncompliance with Reporting and Regulatory Requirements.—Notwithstanding subsection (a), the willful or repeated secondary transmission to the public by a satellite carrier into the local market of a television broadcast station of a primary transmission embodying a performance or display of a work made by that television broadcast station is actionable as an act of infringement under section 501, and is fully subject to the remedies provided under sections 502 through 506, if the satellite carrier has not complied with the reporting requirements of subsection (b) or with the rules, regulations, and authorizations of the Federal Communications Commission concerning the carriage of television broadcast signals.

(e) Willful Alterations.—Notwithstanding subsection (a), the secondary transmission to the public by a satellite carrier into the local market of a television broadcast station of a performance or display of a work embodied in a primary transmission made by that television broadcast station is actionable as an act of infringement under section 501, and is fully subject to the remedies provided by sections 502 through 506 and section 510, if the content of the particular program in which the performance or display is embodied, or any commercial advertising or station announcement transmitted by the primary transmitter during, or immediately before or after, the transmission of such program, is in any way willfully altered by the satellite carrier through changes, deletions, or additions, or is combined with programming from any other broadcast signal.

(f) Violation of Territorial Restrictions on Statutory License for Television Broadcast Stations.—

(1) Individual violations.—The willful or repeated secondary transmission to the public by a satellite carrier of a primary transmission embodying a performance or display of a work made by a television broadcast station to a subscriber who does not reside in that station's local market, and is not subject to statutory licensing under section 119, subject to statutory licensing by reason of paragraph (2)(A), (3), or (4) of subsection (a), or subject to a private licensing agreement, is actionable as an act of infringement under section 501 and is fully subject to the remedies provided by sections 502 through 506 and 509, except that—

(A) no damages shall be awarded for such act of infringement if the satellite carrier took corrective action by promptly withdrawing service from the ineligible subscriber; and

(B) any statutory damages shall not exceed $250 for such subscriber for each month during which the violation occurred.

(2) Pattern of violations.—If a satellite carrier engages in a willful or repeated pattern or practice of secondarily transmitting to the public a primary transmission embodying a performance or display of a work made by a television broadcast station to subscribers who do not reside in that station's local market, and are not subject to statutory licensing under section 119, subject to statutory licensing by reason of paragraph (2)(A), (3), or (4) of subsection (a), or subject to a private licensing agreement, then in addition to the remedies under paragraph (1)—

(A) if the pattern or practice has been carried out on a substantially nationwide basis, the court—

(i) shall order a permanent injunction barring the secondary transmission by the satellite carrier of the primary transmissions of that television broadcast station (and if such television broadcast station is a network station, all other television broadcast stations affiliated with such network); and

(ii) may order statutory damages not exceeding $2,500,000 for each 6-month period during which the pattern or practice was carried out; and

(B) if the pattern or practice has been carried out on a local or regional basis with respect to more than one television broadcast station, the court—

(i) shall order a permanent injunction barring the secondary transmission in that locality or region by the satellite carrier of the primary transmissions of any television broadcast station; and

(ii) may order statutory damages not exceeding $2,500,000 for each 6-month period during which the pattern or practice was carried out.

(g) Burden of Proof.—In any action brought under subsection (f), the satellite carrier shall have the burden of proving that its secondary transmission of a primary transmission by a television broadcast station is made only to subscribers located within that station's local market or subscribers being served in compliance with section 119, paragraph (2)(A), (3), or (4) of subsection (a), or a private licensing agreement.

(h) Geographic Limitations on Secondary Transmissions.—The statutory license created by this section shall apply to secondary transmissions to locations in the United States.

(i) Exclusivity with Respect to Secondary Transmissions of Broadcast Stations by Satellite to Members of the Public.—No provision of section 111 or any other law (other than this section and section 119) shall be construed to contain any authorization, exemption, or license through which secondary transmissions by satellite carriers of programming contained in a primary transmission made by a television broadcast station may be made without obtaining the consent of the copyright owner.

(j) Definitions.—In this section—

(1) Distributor.—The term "distributor" means an entity that contracts to distribute secondary transmissions from a satellite carrier and, either as a single channel or in a package with other programming, provides the secondary transmission either directly to individual subscribers or indirectly through other program distribution entities.

(2) Local market.—

(A) In general.—The term "local market", in the case of both commercial and noncommercial television broadcast stations, means the designated market area in which a station is located, and—

(i) in the case of a commercial television broadcast station, all commercial television broadcast stations licensed to a community within the same designated market area are within the same local market; and

(ii) in the case of a noncommercial educational television broadcast station, the market includes any station that is licensed to a community within the same designated market area as the noncommercial educational television broadcast station.

(B) County of license.—In addition to the area described in subparagraph (A), a station's local market includes the county in which the station's community of license is located.

(C) Designated market area.—For purposes of subparagraph (A), the term "designated market area" means a designated market area, as determined by Nielsen Media Research and published in the 1999–2000 Nielsen Station Index Directory and Nielsen Station Index United States Television Household Estimates or any successor publication.

(D) Certain areas outside of any designated market area.—Any census area, borough, or other area in the State of Alaska that is outside of a designated market area, as determined by Nielsen Media Research, shall be deemed to be part of one of the local markets in the State of Alaska. A satellite carrier may determine which local market in the State of Alaska will be deemed to be the relevant local market in connection with each subscriber in such census area, borough, or other area.

(3) Low power television station.—The term "low power television station" means a low power TV station as defined in section 74.701(f) of title 47, Code of Federal Regulations, as in effect on June 1, 2004. For purposes of this paragraph, the term "low power television station" includes a low power television station

that has been accorded primary status as a Class A television licensee under section 73.6001(a) of title 47, Code of Federal Regulations.

(4) Network station; non-network station; satellite carrier; secondary transmission.—The terms "network station", "non-network station", "satellite carrier", and "secondary transmission" have the meanings given such terms under section 119(d).

(5) Noncommercial educational broadcast station.—The term "noncommercial educational broadcast station" means a television broadcast station that is a noncommercial educational broadcast station as defined in section 397 of the Communications Act of 1934, as in effect on the date of the enactment of the Satellite Television Extension and Localism Act of 2010.

(6) Subscriber.—The term "subscriber" means a person or entity that receives a secondary transmission service from a satellite carrier and pays a fee for the service, directly or indirectly, to the satellite carrier or to a distributor.

(7) Television broadcast station.—The term "television broadcast station"—

(A) means an over-the-air, commercial or noncommercial television broadcast station licensed by the Federal Communications Commission under subpart E of part 73 of title 47, Code of Federal Regulations, except that such term does not include a low-power or translator television station; and

(B) includes a television broadcast station licensed by an appropriate governmental authority of Canada or Mexico if the station broadcasts primarily in the English language and is a network station as defined in section 119(d)(2)(A).

Chapter 1 Notes

1. In 1980, section 117 was amended in its entirety and given a new title. However, the table of sections was not changed to reflect the new title. Pub. L. No. 96-517, 94 Stat. 3015, 3028. In 1997, a technical amendment made that change. Pub. L. No. 105-80, 111 Stat. 1529, 1534.

The Satellite Television Extension and Localism Act of 2010 Act amended the title for section 111 by adding "of broadcast programming by cable." Pub. L. No. 111-175, 124 Stat. 1218, 1231. It amended the title for section 119 by substituting "distant television programming by satellite" for "superstations and network stations for private home viewing." *Id.*, 124 Stat. 1219. The Act also amended the title for section 122 by deleting "by satellite carriers within local markets" from the title and inserting "of local television programming by satellite." *Id.*, 124 Stat. 1227.

2. The Audio Home Recording Act of 1992 amended section 101 by inserting "Except as otherwise provided in this title," at the beginning of the first sentence. Pub. L. No. 102-563, 106 Stat. 4237, 4248.

The Berne Convention Implementation Act of 1988 amended section 101 by adding a definition for "Berne Convention work." Pub. L. No. 100-568, 102 Stat. 2853, 2854. In 1990, the Architectural Works Copyright Protection Act amended the definition of "Berne Convention work" by adding paragraph (5). Pub. L. No. 101-650, 104 Stat. 5089, 5133. The WIPO Copyright and Performances and Phonograms Treaties Implementation Act of 1998 deleted the definition of "Berne Convention work" from section 101. Pub. L. No. 105-304, 112 Stat. 2860, 2861. The definition of "Berne Convention work," as deleted, is contained in Appendix N.

3. In 1990, the Architectural Works Copyright Protection Act amended section 101 by adding the definition for "architectural work." Pub. L. No. 101-650, 104 Stat. 5089, 5133. That Act states that the definition is applicable to "any architectural work that, on the date of the enactment of this Act, is unconstructed and embodied in unpublished plans or drawings, except that protection for such architectural work under title 17, *United States Code*, by virtue of the amendments made by this title, shall terminate on December 31, 2002, unless the work is constructed by that date."

4. The Berne Convention Implementation Act of 1988 amended section 101 by adding the definition of "Berne Convention." Pub. L. No. 100-568, 102 Stat. 2853, 2854.

5. In 1980, the definition of "computer program" was added to section 101 and placed at the end. Pub. L. No. 96-517, 94 Stat. 3015, 3028. The Intellectual Property and High Technology Technical Amendments Act of 2002 amended section 101 by moving the definition for computer program from the end of section 101 to be in alphabetical order, after "compilation." Pub. L. No. 107-273, 116 Stat. 1758, 1909.

6. The Copyright Royalty and Distribution Reform Act of 2004 amended section 101 by adding the definition for "Copyright Royalty Judge." That Act inserted the definition in the wrong alphabetical order, placing it after "copies," instead of "copyright owner." Pub. L. No. 108-419, 118 Stat. 2341, 2361. A technical correction in the Copyright Cleanup, Clarification, and Corrections Act of 2010 put it after the definition for copyright owner. Pub. L. No. 111 295, 124 Stat. 3180, 3181.

7. The Digital Performance Right in Sound Recordings Act of 1995 amended section 101 by adding the definition of "digital transmission." Pub. L. No.104-39, 109 Stat. 336, 348.

8. The Fairness in Music Licensing Act of 1998 amended section 101 by adding the definition of "establishment." Pub. L. No. 105-298, 112 Stat. 2827, 2833.

9. In 1997, the No Electronic Theft (NET) Act amended section 101 by adding the definition for "financial gain." Pub. L. No. 105-147, 111 Stat. 2678.

10. The Fairness in Music Licensing Act of 1998 amended section 101 by adding the definition for "food service or drinking establishment." It inserted the definition in the wrong alphabetical order, placing it after "establishment" instead of "fixed." Pub. L. No. 105-298, 112 Stat. 2827, 2833. A technical correction in the Copyright Cleanup, Clarification, and Corrections Act of 2010 placed it after the definition for fixed. Pub. L. No. 111-295, 124 Stat. 3180, 3181.

11. The WIPO Copyright and Performances and Phonograms Treaties Implementation Act of 1998 amended section 101 by adding the definition of "Geneva Phonograms Convention." Pub. L. No. 105-304, 112 Stat. 2860, 2861.

12. The Fairness in Music Licensing Act of 1998 amended section 101 by adding the definition of "gross square feet of space." Pub. L. No. 105-298, 112 Stat. 2827, 2833.

13. The WIPO Copyright and Performances and Phonograms Treaties Implementation Act of 1998 requires that paragraph (5) of the definition of "international agreement" take effect upon entry into force of the WIPO Copyright Treaty with respect to the United States, which occurred March 6, 2002. Pub. L. No. 105-304, 112 Stat. 2860, 2877.

14. The WIPO Copyright and Performances and Phonograms Treaties Implementation Act of 1998 requires that paragraph (6) of the definition of "international agreement" take effect upon entry into force of the WIPO Performances and Phonograms Treaty with respect to the United States, which occurred May 20, 2002. Pub. L. No. 105-304, 112 Stat. 2860, 2877.

15. The WIPO Copyright and Performances and Phonograms Treaties Implementation Act of 1998 amended section 101 by adding the definition of "international agreement." Pub. L. No. 105-304, 112 Stat. 2860, 2861.

16. The Artists' Rights and Theft Prevention Act of 2005 amended section 101 by adding the definition for "motion picture exhibition facility." That Act inserted the definition in the wrong alphabetical order, placing it after "motion pictures," instead of before. Pub. L. No. 109-9, 119 Stat. 218, 220. A technical correction in the Copyright Cleanup, Clarification, and Corrections Act of 2010 put it after the definition for literary works. Pub. L. No. 111-295, 124 Stat. 3180, 3181.

17. The Fairness in Music Licensing Act of 1998 amended section 101 by adding the definition of "performing rights society." Pub. L. No. 105-298, 112 Stat. 2827, 2833.

18. The Berne Convention Implementation Act of 1988 amended the definition of "Pictorial, graphic, and sculptural works" by inserting "diagrams, models, and technical drawings, including architectural plans" in the first sentence, in lieu of "technical drawings, diagrams, and models." Pub. L. No. 100-568, 102 Stat. 2853, 2854.

19. The Fairness in Music Licensing Act of 1998 amended section 101 by adding the definition of "proprietor." Pub. L. No. 105-298, 112 Stat. 2827, 2833. In 1999, a technical amendment added the phrase "For purposes of section 513,", to the beginning of the definition of "proprietor." Pub. L. No. 106-44, 113 Stat. 221, 222.

20. The Copyright Renewal Act of 1992 amended section 101 by adding the definition of "registration." Pub. L. No. 102-307, 106 Stat. 264, 266.

21. The WIPO Copyright and Performances and Phonograms Treaties Implementation Act of 1998 amended section 101 by adding the definition of "treaty party." Pub. L. No. 105-304, 112 Stat. 2860, 2861.

22. The Berne Convention Implementation Act of 1988 amended section 101 by adding the definition of "country of origin" of a Berne Convention work, for purposes of section 411. Pub. L. No. 100-568, 102 Stat. 2853, 2854. The WIPO Copyright and Performances and Phonograms Treaties Implementation Act of 1998 amended that definition by changing it to a definition for "United States work," for purposes of section 411. Pub. L. No. 105-304, 112 Stat. 2860, 2861. In 1999, a technical amendment moved the definition of "United States work" to place it in alphabetical order, after the definition for "United States." Pub. L. No. 106-44, 113 Stat. 221, 222.

23. The WIPO Copyright and Performances and Phonograms Treaties Implementation Act of 1998 amended section 101 by adding the definition of "WIPO Copyright Treaty." Pub. L. No. 105-304, 112 Stat. 2860, 2861. That definition is required to take effect upon entry into force of the WIPO Copyright Treaty with respect to the United States, which occurred March 6, 2002. Pub. L. No. 105-304, 112 Stat. 2860, 2877.

24. The WIPO Copyright and Performances and Phonograms Treaties Implementation Act of 1998 amended section 101 by adding the definition of "WIPO Performances and Phonograms Treaty." Pub. L. No. 105-304, 112 Stat. 2860, 2862. That definition is required to take effect upon entry into force of the WIPO Performances and Phonograms Treaty with respect to the United States, which occurred May 20, 2002. Pub. L. No. 105-304, 112 Stat. 2860, 2877.

25. The Visual Artists Rights Act of 1990 amended section 101 by adding the definition of "work of visual art." Pub. L. No. 101-650, 104 Stat. 5089, 5128.

26. The Satellite Home Viewer Improvement Act of 1999 amended the definition of "a work made for hire" by inserting "as a sound recording" after "audiovisual work." Pub. L. No. 106-113, 113 Stat. 1501, app. I at 1501A-544. The Work Made for Hire and Copyright Corrections Act of 2000 amended the definition of "work made for hire" by deleting "as a sound recording" after "audiovisual work." Pub. L. No. 106-379, 114 Stat. 1444. The Act also added a second paragraph to part (2) of that definition. *Id.* These changes are effective retroactively, as of November 29, 1999.

27. The WIPO Copyright and Performances and Phonograms Treaties Implementation Act of 1998 amended section 101 by adding the definitions of "WTO Agreement" and "WTO member country," thereby transferring those definitions to section 101 from section 104A. Pub. L. No. 105-304, 112 Stat. 2860, 2862. See also note 31, *infra.*

28. In 1990, the Architectural Works Copyright Protection Act amended subsection 102(a) by adding at the end thereof paragraph (8). Pub. L. No. 101-650, 104 Stat. 5089, 5133.

29. The Berne Convention Implementation Act of 1988 amended section 104(b) by redesignating paragraph (4) as paragraph (5), by inserting after paragraph (3) a new paragraph (4), and by adding subsection (c) at the end. Pub. L. No. 100-568, 102 Stat. 2853, 2855. The WIPO Copyright and Performances and Phonograms Treaties Implementation Act of 1998 amended section 104 as follows: 1) by amending subsection (b) to redesignate paragraphs (3) and (5) as (5) and (6), respectively, and by adding a new paragraph (3); 2) by amending section 104(b), throughout; and 3) by adding section 104(d). Pub. L. No. 105-304, 112 Stat. 2860, 2862.

30. The WIPO Copyright and Performances and Phonograms Treaties Implementation Act of 1998 requires that subsection (d), regarding the effect of phonograms treaties, take effect upon entry into force of the WIPO Performances and Phonograms Treaty with respect to the United States, which occurred May 20, 2002. Pub. L. No. 105-304, 112 Stat. 2860, 2877.

31. In 1993, the North American Free Trade Agreement Implementation Act added section 104A. Pub. L. No. 103-182, 107 Stat. 2057, 2115. In 1994, the Uruguay Round Agreements Act amended section 104A in its entirety with an amendment in

the nature of a substitute. Pub. L. No. 103-465, 108 Stat. 4809, 4976. On November 13, 1997, section 104A was amended by replacing subsection (d)(3)(A), by striking the last sentence of subsection (e)(1)(B)(ii), and by rewriting paragraphs (2) and (3) of subsection (h). Pub. L. No. 105-80, 111 Stat. 1529, 1530. The WIPO Copyright and Performances and Phonograms Treaties Implementation Act of 1998 amended section 104A by rewriting paragraphs (1) and (3) of subsection (h); by adding subparagraph (E) to subsection (h)(6); and by amending subsection (h)(8)(B)(i). Pub. L. No. 105-304, 112 Stat. 2860, 2862. That Act also deleted paragraph (9), thereby transferring the definitions for "WTO Agreement" and "WTO member country" from section 104A to section 101. Pub. L. No. 105-304, 112 Stat. 2860, 2863. See also note 27, *supra.*

32. The WIPO Copyright and Performances and Phonograms Treaties Implementation Act of 1998 requires that subparagraph (C) of the definition of "date of adherence or proclamation" take effect upon entry into force of the WIPO Copyright Treaty with respect to the United States, which occurred March 6, 2002. Pub. L. No. 105-304, 112 Stat. 2860, 2877.

33. The WIPO Copyright and Performances and Phonograms Treaties Implementation Act of 1998 requires that subparagraph (D) of the definition of "date of adherence or proclamation" take effect upon entry into force of the WIPO Performances and Phonograms Treaty with respect to the United States, which occurred May 20, 2002. Pub. L. No. 105-304, 112 Stat. 2860, 2877.

34. The WIPO Copyright and Performances and Phonograms Treaties Implementation Act of 1998 requires that subparagraph (C) of the definition of "eligible country" take effect upon entry into force of the WIPO Copyright Treaty with respect to the United States, which occurred March 6, 2002. Pub. L. No. 105-304, 112 Stat. 2860, 2877.

35. The WIPO Copyright and Performances and Phonograms Treaties Implementation Act of 1998 requires that subparagraph (D) of the definition of "eligible country" take effect upon entry into force of the WIPO Performance and Phonograms Treaty with respect to the United States, which occurred May 20, 2002. Pub. L. No. 105-304, 112 Stat. 2860, 2877.

36. The WIPO Copyright and Performances and Phonograms Treaties Implementation Act of 1998 requires that subparagraph (E) of the definition of "restored work" take effect upon entry into force of the WIPO Performances and Phonograms Treaty with respect to the United States, which occurred May 20, 2002. Pub. L. No. 105-304, 112 Stat. 2860, 2877.

37. In 1968, the Standard Reference Data Act provided an exception to section 105, Pub. L. No. 90-396, 82 Stat. 339. Section 6 of that act amended title 15 of the *United States Code* by authorizing the Secretary of Commerce, at 15 *U.S.C.* 290e, to secure copyright and renewal thereof on behalf of the United States as author or proprietor "in all or any part of any standard reference data which he prepares or makes available under this chapter," and to "authorize the reproduction and publication thereof by others." See also section 105(f) of the Transitional and Supplementary Provisions of the Copyright Act of 1976, in Appendix A. Pub. L. No. 94-553, 90 Stat. 2541.

Concerning the liability of the United States Government for copyright infringement, also see 28 *U.S.C.* 1498. Title 28 of the *United States Code* is entitled "Judiciary and Judicial Procedure," included in the appendices to this volume.

38. The Digital Performance Right in Sound Recordings Act of 1995 amended section 106 by adding paragraph (6). Pub. L. No. 104-39, 109 Stat. 336. In 1999, a technical amendment substituted "121" for "120." Pub. L. No. 106-44, 113 Stat. 221, 222. The Intellectual Property and High Technology Technical Amendments Act of 2002 amended section 106 by substituting sections "107 through 122" for "107 through 121." Pub. L. No. 107-273, 116 Stat. 1758, 1909.

39. The Visual Artists Rights Act of 1990 added section 106A. Pub. L. No. 101-650, 104 Stat. 5089, 5128. The Act states that, generally, section 106A is to take effect 6 months after its date of enactment, December 1, 1990, and that the rights created by section 106A shall apply to (1) works created before such effective date but title to which has not, as of such effective date, been transferred from the author and (2) works created on or after such effective date, but shall not apply to any destruction, distortion, mutilation, or other modification (as described in section 106A(a)(3)) of any work which occurred before such effective date. See also, note 3, chapter 3.

40. The Visual Artists Rights Act of 1990 amended section 107 by adding the reference to section 106A. Pub. L. No. 101-650, 104 Stat. 5089, 5132. In 1992, section 107 was also amended to add the last sentence. Pub. L. No. 102-492, 106 Stat. 3145.

41. The Copyright Amendments Act of 1992 amended section 108 by repealing subsection (i) in its entirety. Pub. L. No. 102-307, 106 Stat. 264, 272. In 1998, the Sonny Bono Copyright Term Extension Act amended section 108 by redesignating subsection (h) as (i) and adding a new subsection (h). Pub. L. No. 105-298, 112 Stat. 2827, 2829. Also in 1998, the Digital Millennium Copyright Act amended section 108 by making changes in subsections (a), (b), and (c). Pub. L. No. 105-304, 112 Stat. 2860, 2889. In 2005, the Preservation of Orphan Works Act amended subsection 108(i) by adding a reference to subsection (h). It substituted "(b), (c), and (h)" for "(b) and (c)." Pub. L. No. 109-9, 119 Stat. 218, 226, 227.

42. The Record Rental Amendment of 1984 amended section 109 by redesignating subsections (b) and (c) as subsections (c) and (d), respectively, and by inserting a new subsection (b) after subsection (a). Pub. L. No. 98-450, 98 Stat. 1727. Section 4(b) of the Act states that the provisions of section 109(b), as added by section 2 of the Act, "shall not affect the right of an owner of a particular phonorecord of a sound recording, who acquired such ownership before [October 4, 1984], to dispose of the possession of that particular phonorecord on or after such date of enactment in any manner permitted by section 109 of title 17, *United States Code*, as in effect on the day before the date of the enactment of this Act." Pub. L. No. 98-450, 98 Stat. 1727, 1728. Section 4(c) of the Act also states that the amendments "shall not apply to rentals, leasings, lendings (or acts or practices in the nature of rentals, leasings, or lendings) occurring after the date which is 13 years after [October 4, 1984]." In 1988, the Record Rental Amendment Act of 1984 was amended to extend the time period in section 4(c) from 5 years to 13 years. Pub. L. No. 100-617, 102 Stat. 3194. In 1993, the North American Free Trade Agreement Implementation

Act repealed section 4(c) of the Record Rental Amendment of 1984. Pub. L. No. 103-182, 107 Stat. 2057, 2114. Also in 1988, technical amendments to section 109(d) inserted "(c)" in lieu of "(b)" and substituted "copyright" in lieu of "coyright." Pub. L. No. 100-617, 102 Stat. 3194.

The Computer Software Rental Amendments Act of 1990 amended section 109(b) as follows: 1) paragraphs (2) and (3) were redesignated as paragraphs (3) and (4), respectively; 2) paragraph (1) was struck out and new paragraphs (1) and (2) were inserted in lieu thereof; and 3) paragraph (4), as redesignated, was amended in its entirety with a new paragraph (4) inserted in lieu thereof. Pub. L. No. 101-650, 104 Stat. 5089, 5134. The Act states that section 109(b), as amended, "shall not affect the right of a person in possession of a particular copy of a computer program, who acquired such copy before the date of the enactment of this Act, to dispose of the possession of that copy on or after such date of enactment in any manner permitted by section 109 of title 17, *United States Code*, as in effect on the day before such date of enactment." The Act also states that the amendments made to section 109(b) "shall not apply to rentals, leasings, or lendings (or acts or practices in the nature of rentals, leasings, or lendings) occurring on or after October 1, 1997." However, this limitation, which is set forth in the first sentence of section 804 (c) of the Computer Software Rental Amendments Act of 1990, at 104 Stat. 5136, was subsequently deleted in 1994 by the Uruguay Round Agreements Act. Pub. L. No. 103-465, 108 Stat. 4809, 4974.

The Computer Software Rental Amendments Act of 1990 also amended section 109 by adding at the end thereof subsection (e). Pub. L. No. 101-650, 104 Stat. 5089, 5135. That Act states that the provisions contained in the new subsection (e) shall take effect 1 year after its date of enactment. It was enacted on December 1, 1990. The Act also states that such amendments so made "shall not apply to public performances or displays that occur on or after October 1, 1995."

In 1994, the Uruguay Round Agreements Act amended section 109(a) by adding the second sentence, which begins with "Notwithstanding the preceding sentence." Pub. L. No. 103-465, 108 Stat. 4809, 4981.

The Prioritizing Resources and Organization for Intellectual Property Act of 2008 amended part (4) of subsection 109(b) by removing the reference to section 509 (which was repealed). Pub. L. No. 110-403, 122 Stat. 4256, 4264.

43. In 1988, the Extension of Record Rental Amendment amended section 110 by adding paragraph (10). Pub. L. No. 97-366, 96 Stat. 1759. In 1997, the Technical Corrections to the Satellite Home Viewer Act amended section 110 by inserting a semicolon in lieu of the period at the end of paragraph (8); by inserting "; and" in lieu of the period at the end of paragraph (9); and by inserting "(4)" in lieu of "4 above" in paragraph (10). Pub. L. No. 105-80, 111 Stat. 1529, 1534. The Fairness in Music Licensing Act of 1998 amended section 110, in paragraph 5, by adding subparagraph (B) and by making conforming amendments to subparagraph (A); by adding the phrase "or of the audiovisual or other devices utilized in such performance" to paragraph 7; and by adding the last paragraph to section 110 that begins "The exemptions provided under paragraph (5)." Pub. L. No. 105-298, 112 Stat. 2827, 2830. In 1999, a technical amendment made corrections to conform paragraph designations that were affected by amendments previously made by the Fairness in Music Licensing Act

of 1998. Pub. L. No. 106-44, 113 Stat. 221. The Technology, Education, and Copyright Harmonization Act of 2002 amended section 110 by substituting new language for paragraph 110(2) and by adding all the language at the end of section 110 that concerns paragraph 110(2). Pub. L. No. 107-273, 116 Stat. 1758, 1910.

The Family Movie Act of 2005 amended section 110 by adding paragraph (11) and by adding a new paragraph at the end of that section. Pub. L. No. 109-9, 119 Stat. 218, 223.

44. Throughout section 111 the date of enactment of the Satellite Television and Localism Act of 2010 (STELA) is incorporated by reference. STELA was enacted on May 27, 2010. Pub. L. No. 111-175, 124 Stat. 1218. There are also references to the date of enactment for the Copyright Act of 1976, which is October 19, 1976. Pub. L. No. 94-553, 90 Stat. 2541.

In 1986, section 111(d) was amended by striking out paragraph (1) and by redesignating paragraphs (2), (3), (4), and (5) as paragraphs (1), (2), (3), and (4), respectively. Pub. L. 99-397, 100 Stat. 848. Also, in 1986, section 111(f) was amended by substituting "subsection (d)(1)" for "subsection (d)(2)" in the last sentence of the definition of "secondary transmission" and by adding a new sentence after the first sentence in the definition of "local service area of a primary transmitter." Pub. L. No. 99-397, 100 Stat. 848.

The Satellite Home Viewer Act of 1988 amended subsection 111(a) by striking "or" at the end of paragraph (3), by redesignating paragraph (4) as paragraph (5), and by inserting a new paragraph (4). Pub. L. No. 100-667, 102 Stat. 3935, 3949. That Act also amended section (d)(1)(A) by adding the second sentence, which begins with "In determining the total number." *Id.* The Copyright Royalty Tribunal Reform Act of 1993 amended section 111(d) by substituting "Librarian of Congress" for "Copyright Royalty Tribunal" where appropriate, by inserting a new sentence in lieu of the second and third sentences of paragraph (2), and, in paragraph (4), by amending subparagraph (B) in its entirety. Pub. L. No. 103-198, 107 Stat. 2304, 2311.

The Satellite Home Viewer Act of 1994 amended section 111(f) by inserting "microwave" after "wires, cables," in the paragraph relating to the definition of "cable system" and by inserting new matter after "April 15, 1976," in the paragraph relating to the definition of "local service area of a primary transmitter." Pub. L. No. 103-369, 108 Stat. 3477, 3480. That Act provides that the amendment "relating to the definition of the local service area of a primary transmitter, shall take effect on July 1, 1994." *Id.*

In 1995, the Digital Performance in Sound Recordings Act amended section 111(c)(1) by inserting "and section 114(d)" in the first sentence, after "of this subsection." Pub. L. No. 104-39, 109 Stat. 336, 348.

The Satellite Home Viewer Improvement Act of 1999 amended section 111 by substituting "statutory" for "compulsory" and "programming" for "programing," wherever they appeared. Pub. L. No. 106-113, 113 Stat. 1501, app. I at 1501A-543. The Act also amended sections 111(a) and (b) by inserting "performance or display of a work embodied in a primary transmission" in lieu of "primary transmission embodying a performance or display of a work." It amended paragraph (1) of section 111(c) by inserting "a performance or display of a work embodied in" after "by a cable system of " and by striking "and embodying a performance or display of a work." It amended subparagraphs (3) and (4) of section 111(a) by inserting "a performance or display of a work embodied in a primary transmission" in lieu of "a primary transmission" and by striking "and embodying a performance or display of a work." *Id.*

The Copyright Royalty and Distribution Reform Act of 2004 made amendments to subsection 111(d) to conform it to revised chapter 8, substituting "Copyright Royalty Judges" for "Librarian of Congress" where appropriate, along with making other conforming amendments. Pub. L. No. 108-419, 118 Stat. 2341, 2361. The Satellite Home Viewer Extension and Reauthorization Act of 2004 amended section 111 by deleting "for private home viewing" in subsections (a)(4) and (d)(1)(A). Pub. L. No. 108-447, 118 Stat. 2809, 3393, 3406.

In 2006, the Copyright Royalty Judges Program Technical Corrections Act amended section 111(d)(2) by substituting "upon authorization by the Copyright Royalty Judges for everything in the second sentence after "Librarian of Congress"; by substituting new text for the second sentence of (4)(B); by making a technical correction in the last sentence of (4)(B) to change "finds" to "find"; and by revising (4)(C) in its entirety. Pub. L. No. 109-303, 120 Stat. 1478, 1481.

The Prioritizing Resources and Organization for Intellectual Property Act of 2008 amended section 111 by deleting all references to section 509 (which was repealed). Pub. L. No. 110-403, 122 Stat. 4256, 4264.

The Satellite Television Extension and Localism Act of 2010 amended subsection 111(d)(1) by deleting subparagraphs (B) through (D) and replacing them with new subparagraphs (B) through (G); by amending paragraph 111(d)(2) to insert the parenthetical language in the first sentence; and by adding new paragraphs (5) through (7). Pub. L. No. 111-175, 124 Stat. 1218, 1232-35. It amended subsection 111(f) by numbering the previously undesignated paragraphs and adding subtitles to them; by substituting new language for the definition for primary transmissions; by amending the definition for local service area of a primary transmitter; by substituting new provisions to define distant signal equivalent, network station, independent station, and noncommercial educational station; and by adding definitions for primary stream, primary transmitter, multicast stream, simulcast, subscriber, and subscribe. *Id.*, at 124 Stat. 1235-38. The Act also added "of broadcast programming by cable" to the section title and made perfecting, technical, and clarifying amendments throughout section 111. *Id.*

45. Royalty rates specified by the compulsory licensing provisions of this section are subject to adjustment by copyright royalty judges appointed by the Librarian of Congress in accordance with the provisions of chapter 8 of title 17 of the *United States Code*, as amended by the Copyright Royalty and Distribution Reform Act of 2004, Pub. L. No. 108-419, 118 Stat. 2341. See chapter 8, *infra.* Regulations for adjusting royalty rates may be found in subchapter B of chapter 11, title 37, *Code of Federal Regulations.*

46. In 1998, the Digital Millennium Copyright Act amended section 112 by redesignating subsection (a) as subsection (a)(1); by redesignating former sections (a)(1), (a)(2), and (a)(3) as subsections (a)(1)(A), (a)(1)(B), and (a)(1)(C), respectively; by adding subsection (a)(2); and by amending the language in new subsection (a)(1). Pub. L. No. 105-304, 112 Stat. 2860, 2888. The Digital

Millennium Copyright Act also amended section 112 by redesignating subsection (e) as subsection (f) and adding a new subsection (e). Pub. L. No. 105-304, 112 Stat. 2860, 2899. In 1999, a technical amendment to section 112(e) redesignated paragraphs

(3) through (10) as (2) through (9) and corrected the paragraph references throughout that section to conform to those redesignations. Pub. L. No. 106-44, 113 Stat. 221. The Technology, Education, and Copyright Harmonization Act of 2002 amended section 112 by redesignating subsection 112(f) as 112(g) and adding a new paragraph (f). Pub. L. No. 107-273, 116 Stat. 1758, 1912.

The Copyright Royalty and Distribution Reform Act of 2004 amended subsection 112(e) to conform it to revised chapter 8, by substituting new language for the first sentences of paragraphs (3) and (4); by deleting paragraph (6) and renumbering paragraphs (7) through (9) as (6) through (8); by changing references to the "Librarian of Congress" to "Copyright Royalty Judges," with corresponding grammatical changes, throughout; and by striking references to negotiations in paragraphs (3) and (4) along with making other conforming amendments. Pub. L. No. 108-419, 118 Stat. 2341, 2361.

47. The Visual Artists Rights Act of 1990 amended section 113 by adding subsection (d) at the end thereof. Pub. L. No. 101-650, 104 Stat. 5089, 5130.

48. The Digital Performance Right in Sound Recordings Act of 1995 amended section 114 as follows: 1) in subsection (a), by striking "and (3)" and inserting in lieu thereof "(3) and (6)"; 2) in subsection (b) in the first sentence, by striking "phonorecords, or of copies of motion pictures and other audiovisual works," and inserting "phonorecords or copies"; and 3) by striking subsection (d) and inserting in lieu thereof new subsections (d), (e), (f), (g), (h), (i), and (j). Pub. L. No. 104-39, 109 Stat. 336. In 1997, subsection 114(f) was amended in paragraph (1) by inserting all the text that appears after "December 31, 2000" and in paragraph (2) by striking "and publish in the Federal Register." Pub. L. No. 105-80, 111 Stat. 1529, 1531.

In 1998, the Digital Millennium Copyright Act amended section 114(d) by replacing paragraphs (1)(A) and (2) with amendments in the nature of substitutes. Pub. L. No. 105-304, 112 Stat. 2860, 2890. That Act also amended section 114(f) by revising the title; by redesignating paragraph (1) as paragraph (1)(A); by adding paragraph (1)(B) in lieu of paragraphs (2), (3), (4), and (5); and by amending the language in newly designated paragraph (1)(A), including revising the effective date from December 31, 2000, to December 31, 2001. Pub. L. No. 105-304, 112 Stat. 2860, 2894. The Digital Millennium Copyright Act also amended subsection 114(g) by substituting "transmission" in lieu of "subscription transmission," wherever it appears and, in the first sentence in paragraph (g)(1), by substituting "transmission licensed under a statutory license" in lieu of "subscription transmission licensed." Pub. L. No. 105-304, 112 Stat. 2860, 2897. That Act also amended subsection 114(j) by redesignating paragraphs (2), (3), (5), (6), (7), and (8) as (3), (5), (9), (12), (13), and (14), respectively; by amending paragraphs (4) and (9) in their entirety and redesignating them as paragraphs (7) and (15), respectively; and by adding new definitions, including, paragraph (2) defining "archived program," paragraph (4) defining "continuous program," paragraph (6) defining "eligible nonsubscription transmission," paragraph (8) defining "new subscription service," paragraph (10) defining "preexisting satellite digital audio radio service," and paragraph (11) defining "preexisting subscription service." Pub. L. No. 105-304, 112 Stat. 2860, 2897.

The Small Webcaster Settlement Act of 2002 amended section 114 by adding paragraph (5) to subsection 114(f), by amending paragraph 114(g)(2), and by adding paragraph 114(g) (3). Pub. L. No. 107-321, 116 Stat. 2780, 2781 and 2784.

The Copyright Royalty and Distribution Reform Act of 2004 amended subsection 114(f) to conform it to revised chapter 8, by substituting new language for the first sentences of subparagraphs (1)(A), (1)(B), (2)(A), and (2)(B); by substituting new language for subparagraphs

(1)(C) and (2)(C); by changing references to the "Librarian of Congress" in paragraphs (1), (2), (3), and (4) to "Copyright Royalty Judges," and making corresponding grammatical changes; by striking references to negotiations in paragraphs (1), (2), (3), and (4) and replacing with corresponding grammatical changes and conforming language; and by adding new language at the end of subparagraph (4)(A). Pub. L. No. 108-419, 118 Stat. 2341, 2362-2364.

In 2006, the Copyright Royalty Judges Program Technical Corrections Act amended section 114 by substituting new text after "proceedings are to be commenced" in the first sentence in (f)(1)(A); by amending (2)(A) in its entirety; and by inserting "described in" in the last sentence of (2)(B) which repeats an amendment already made by the Copyright Royalty and Distribution Reform Act of 2004, Pub. L. No. 108 419, 118 Stat. 2341, 2364. Pub. L. No. 109-303, 120 Stat. 1478, 1481-82.

The Webcaster Settlement Act of 2008 amended part (5) of subsection 114(f) by deleting "small" wherever it appears before "commercial webcasters." Pub. L. No. 110-435, 122 Stat. 4974. The Webcaster Settlement Act of 2008 also amended subpart (5)(A) by substituting "for a period of not more than 11 years beginning on January 1, 2005" for "during the period beginning on October 28, 1998, and ending on December 31, 2004," by substituting "the Copyright Royalty Judges" in place of references to the copyright arbitration royalty panel and the Librarian of Congress and by changing "shall" to "may" at the beginning of the second sentence. *Id.* It amended subpart (5)(C) by adding a sentence at the end that states, "This subparagraph shall not apply to the extent that the receiving agent and a webcaster that is party to an agreement entered into pursuant to subparagraph (A) expressly authorize the submission of the agreement in a proceeding under this subsection." *Id.* It amended subpart (5)(D) by changing the reference in the first sentence from "the Small Webcaster Settlement Act of 2002" to "the Webcaster Settlement Act of 2008" and by substituting "Copyright Royalty Judges of May 1, 2007" for "Librarian of Congress of July 8, 2002." *Id.* It also amended subpart (5)(F) by substituting "February 15, 2009" for "December 15, 2002, except with respect to noncommercial webcasters for whom the authority shall expire May 31, 2003." *Id.*

The Webcaster Settlement Act of 2009 amended section 114 by adding a reference to itself in the first sentence of subpart (f)(5)(D); by deleting this text: "to make eligible nonsubscription transmissions and ephemeral recordings" at the end of subpart (f)(5)(E)(iii); and, in subpart (f)(5)(F), by changing the expiration to

11:59 p.m. on July 30, 2009, which is 30 days after the Webcaster Settlement Act of 2009 was enacted on June 30, 2009. Pub. L. No. 111-36, 123 Stat. 1926.

A technical correction in the Copyright Cleanup, Clarification, and Corrections Act of 2010 amended section 114(b) to change the reference from "section 118(g)" to "section 118(f)." Pub. L. No. 111-295, 124 Stat. 3180, 3181. It also amended subparagraph 114(f)(2)(C) by substituting "eligible nonsubscription services and new subscription services" for "preexisting subscription digital audio transmission services or preexisting satellite digital radio audio services." *Id.*

49. The Copyright Royalty and Distribution Reform Act of 2004 did not conform the pronoun "its" to "their" when it substituted "Copyright Royalty Judges" for "copyright arbitration royalty panel." Pub. L. No. 108-419, 118 Stat. 2341, 2362. A technical amendment in the Copyright Cleanup, Clarification, and Corrections Act of 2010 corrected this error. Pub. L. No. 111-295, 124 Stat. 3180, 3181.

50. The Record Rental Amendment of 1984 amended section 115 by redesignating paragraphs (3) and (4) of subsection (c) as paragraphs (4) and (5), respectively, and by adding a new paragraph (3). Pub. L. No. 98-450, 98 Stat. 1727.

The Digital Performance Right in Sound Recordings Act of 1995 amended section 115 as follows: 1) in the first sentence of subsection (a)(1), by striking "any other person" and inserting in lieu thereof "any other person, including those who make phonorecords or digital phonorecord deliveries,"; 2) in the second sentence of the same subsection, by inserting before the period "including by means of a digital phonorecord delivery"; 3) in the second sentence of subsection (c)(2), by inserting "and other than as provided in paragraph (3)," after "For this purpose,"; 4) by redesignating paragraphs (3), (4), and (5) of subsection (c) as paragraphs (4), (5), and (6), respectively, and by inserting after paragraph (2) a new paragraph (3); and (5) by adding after subsection (c) a new subsection (d). Pub. L. No. 104-39, 109 Stat. 336, 344.

In 1997, section 115 was amended by striking "and publish in the Federal Register" in subparagraph 115(c)(3)(D). Pub. L. No. 105-80, 111 Stat. 1529, 1531. The same legislation also amended section 115(c)(3)(E) by replacing the phrases "sections 106(1) and (3)" and "sections 106(1) and 106(3)" with "paragraphs (1) and (3) of section 106." Pub. L. No. 105-80, 111 Stat. 1529, 1534.

The Copyright Royalty and Distribution Reform Act of 2004 amended paragraph 115(c)(3) to conform it to revised chapter 8, by substituting new language for the first sentences of subparagraphs (3)(C) and (3)(D); by changing references to the "Librarian of Congress" in subparagraphs (3)(C), (3)(D), and (3)(E) to "Copyright Royalty Judges," with corresponding grammatical changes; by striking references to negotiations in subparagraphs (3)(C) and (D) and making corresponding grammatical changes and conforming language; by deleting subparagraph (F); and by redesignating paragraphs (G) through (L) as paragraphs (F) through (K) with corresponding technical changes in subparagraphs (A), (B), and (E) to conform references to the subparagraphs subject to that redesignation. Pub. L. No. 108-419, 118 Stat. 2341, 2364–2365. The Copyright Royalty and Distribution Act of 2004 also amended the first sentence of subparagraph 115(c)(3)(B) by substituting "section" for "paragraph" and by inserting "on a nonexclusive basis" after "common agents."

Id. at 2364. It also amended subparagraph 115(c)(3)(E) by inserting "as to digital phonorecord deliveries" after "shall be given effect." *Id.* at 2365.

In 2006, the Copyright Royalty Judges Program Technical Corrections Act amended 115(c)(3)(B) by inserting "this subparagraph and subparagraphs (C) through (E)" in lieu of "subparagraphs (B) through (F)." The Act also amended the third sentence of 115(c)(3)(D) by inserting "in subparagraphs (B) and (C)" after "described"; and 115(c)(3)(E)(i) and (ii)(I) by substituting "(C) and (D)" for "(C) or (D)" wherever it appears. Pub. L. No. 109-303, 120 Stat. 1478, 1482.

The Prioritizing Resources and Organization for Intellectual Property Act of 2008 amended subparts (3)(G)(i) and (6) of subsection 115(c) by removing the reference to section 509 (which was repealed). Pub. L. No. 110-403, 122 Stat. 4256, 4264.

51. See note 45, *supra.*

52. There was a minor drafting error for this text in the Prioritizing Resources and Organization for Intellectual Property Act of 2008. Pub. L. No. 110-403, 122 Stat. 4256, 5264. To delete the reference to "and section 509," it deleted "and 509," thereby, technically, not including the word "section" in the amendment. A technical amendment in the Copyright Cleanup, Clarification, and Corrections Act of 2010 corrected this error. Pub. L. No. 111-295, 124 Stat. 3180, 3181.

53. The Berne Convention Implementation Act of 1988 added section 116A. Pub. L. No. 100-568, 102 Stat. 2853, 2855. The Copyright Royalty Tribunal Reform Act of 1993 redesignated section 116A as section 116; repealed the preexisting section 116; in the redesignated section 116, struck subsections (b), (e), (f), and (g), and redesignated subsections (c) and (d) as subsections (b) and (c), respectively; and substituted, where appropriate, "Librarian of Congress" or "copyright arbitration royalty panel" for "Copyright Royalty Tribunal." Pub. L. No. 103-198, 107 Stat. 2304, 2309. In 1997, section 116 was amended by rewriting subsection

(b)(2) and by adding a new subsection (d). Pub. L. No. 105-80, 111 Stat. 1529, 1531. The Copyright Royalty and Distribution Reform Act of 2004 amended section 116 to conform it to revised chapter 8, by substituting new language for subsection (b)(2); by changing the title of subsection (c) to "License Agreements Superior to Determinations by Copyright Royalty Judges" from "License Agreements Superior to Copyright Arbitration Royalty Panel Determinations"; and, in subsection (c), by striking "copyright arbitration royalty panel" and replacing it with "Copyright Royalty Judges." Pub. L. No. 108-419, 118 Stat. 2341, 2365.

54. In 1980, section 117 was amended in its entirety. Pub. L. No. 96-517, 94 Stat. 3015, 3028. In 1998, the Computer Maintenance Competition Assurance Act amended section 117 by inserting headings for subsections (a) and (b) and by adding subsections (c) and (d). Pub. L. No. 105-304, 112 Stat. 2860, 2887.

55. The Copyright Royalty Tribunal Reform Act of 1993 amended section 118 by striking the first two sentences of subsection (b), by substituting a new first sentence in paragraph (3), and by making general conforming amendments throughout. Pub. L. 103-198, 107 Stat. 2304, 2309. In 1999, a technical amendment deleted paragraph (2) from section 118(e). Pub. L. No. 106-44, 113 Stat. 221, 222. The Intellectual Property and High Technology Technical Amendments Act of 2002

amended section 118 by deleting "to it" in the second sentence in subsection (b)(1). Pub. L. No. 107-273, 116 Stat. 1758, 1909.

The Copyright Royalty and Distribution Reform Act of 2004 amended section 118 to conform it to revised chapter 8, by deleting the last sentence in paragraph (b)(1); by revising paragraph (b)(2) by rewriting the end of sentence after "determination by the"; by substituting new language for the first sentence of paragraph (3), thereby, creating new paragraphs (3) and (4); by deleting subsection (c) and redesignating sections (d) through (g) as (c) through (f) with a corresponding technical change in section (f) to refer to "subsection (c)" instead of "subsection (d)"; and by changing references to the "Librarian of Congress" in subsections (b) and (d), as redesignated, to "Copyright Royalty Judges," with corresponding grammatical or procedural changes. Pub. L. No. 108-419, 118 Stat. 2341, 2365–2366. The Copyright Royalty and Distribution Act of 2004 also amended section 116 in text that became the second sentence of new subparagraph (4), see *supra*, by inserting "or (3)" after "paragraph 2". *Id.* at 2366. It further amended subsection 116(c) by inserting "or (3)" after "provided by subsection (b)(2)" and by inserting "to the extent that they were accepted by the Librarian of Congress" after "under subsection (b)(3)" and before the comma. *Id.*

In 2006, the Copyright Royalty Judges Program Technical Corrections Act amended subsection 118(b)(3) by inserting "owners of copyright in works" in lieu of "copyright owners in works"; by amending the first sentence of (c); and by substituting "(f)" for "(g)" in (c)(1). Pub. L. No. 109-303, 120 Stat. 1478, 1482.

The Prioritizing Resources and Organization for Intellectual Property Act of 2008 amended subparts (6), (7)(A), (8) and (13) of subsection 119(a) by removing the reference to section 509 (which was repealed). Pub. L. No. 110-403, 122 Stat. 4256, 4264.

56. Throughout section 119 the date of enactment of the Satellite Television and Localism Act of 2010 (STELA) is incorporated by reference. STELA was enacted on May 27, 2010. Pub. L. No. 111-175, 124 Stat. 1218.

The Satellite Home Viewer Act of 1988 added section 119. Pub. L. No. 100-667, 102 Stat. 3935, 3949. The Copyright Royalty Tribunal Reform Act of 1993 amended subsections (b) and (c) of section 119 by substituting "Librarian of Congress" in lieu of "Copyright Royalty Tribunal" wherever it appeared and by making related conforming amendments. Pub. L. No. 103-198, 107 Stat. 2304, 2310. The Copyright Royalty Tribunal Reform Act of 1993 also amended paragraph (c)(3) by deleting subparagraphs (B), (C), (E), and (F) and by redesignating subparagraph (D) as (B), (G) as (C), and (H) as (D). The redesignated subparagraph (C) was amended in its entirety and paragraph (c)(4) was deleted. *Id.*

The Satellite Home Viewer Act of 1994 further amended section 119. Pub. L. No. 103-369, 108 Stat. 3477. In 1997, technical corrections and clarifications were made to the Satellite Home Viewer Act of 1994. Pub. L. No. 105-80, 111 Stat. 1529. Those two acts amended section 119 as follows: 1) by deleting or replacing obsolete effective dates; 2) in subsection (a)(5), by adding subparagraph (D); 3) in subsection (a), by adding paragraphs (8), (9), and (10); 4) in subsection (b)(1)(B), by adjusting the royalty rate for retransmitted superstations; 5) in subsection (c)(3), by replacing subparagraph (B) with an amendment in the nature of a substitute;

6) in subsections (d)(2) and (d)(6), by modifying the definition of "network station" and "satellite carrier"; and 7) in subsection (d), by adding paragraph 11 to define "local market."

Pursuant to section 4 of the Satellite Home Viewer Act of 1994, the changes made by that Act to section 119 of the *United States Code* ceased to be effective on December 31, 1999. Pub. L. No. 103-369, 108 Stat. 3477, 3481. However, section 1003 of the Satellite Home Viewer Improvement Act of 1999 extended that date to December 31, 2004. Pub. L. No. 106-113, 113 Stat. 1501, app. I at 1501A-527.

The Digital Performance Right in Sound Recordings Act of 1995 amended section 119 in the first sentence of subsections (a)(1) and (a)(2)(A), respectively, by inserting the words "and section 114(d)" after "of this subsection." Pub. L. No. 104-39, 109 Stat. 336, 348. In 1999, a technical amendment substituted "network station's" for "network's stations" in section 119(a)(8)(C)(ii). Pub. L. No. 106-44, 113 Stat. 221, 222.

The Satellite Home Viewer Improvement Act of 1999 amended section 119(a)(1) as follows: 1) by inserting "and PBS satellite feed" after "Superstations" in the paragraph heading; 2) by inserting "performance or display of a work embodied in a primary transmission made by a superstation or by the Public Broadcasting Service satellite feed" in lieu of "primary transmission made by a superstation and embodying a performance or display of a work," (see this note, *infra*); and 3) by adding the last sentence, which begins "In the case of the Public Broadcasting Service." Pub. L. No. 106-113, 113 Stat. 1501, app. I at 1501A-530 and 543. The Act states that these amendments shall be effective as of July 1, 1999, except for a portion of the second item, starting with "performance or display" through "superstation." Pub. L. No. 106-113, 113 Stat. 1501, app. I at 1501A-544. The Act also amended section 119(a) by inserting the phrase "with regard to secondary transmissions the satellite carrier is in compliance with the rules, regulations, or authorization of the Federal Communications Commission governing the carriage of television broadcast stations signals" in paragraphs (1) and (2) and by inserting into paragraph (2), "a performance or display of a work embodied in a primary transmission made by a network station" in lieu of "programming contained in a primary transmission made by a network station and embodying a performance or display of a work." *Id.* at 1501A-531 and 544. The Act amended section 119(a)(2) by substituting new language for paragraph (B) and, in paragraph (C), by deleting "currently" after "the satellite carrier" near the end of the first sentence. *Id.* at 1501A-528 and 544. It also amended section 119(a)(4) by inserting "a performance or display of a work embodied in" after "by a satellite carrier of" and by deleting "and embodying a performance or display of a work." *Id.* at 1501A-544. The Satellite Home Viewer Improvement Act of 1999 further amended section 119(a) by adding subparagraph (E) to paragraph (5). *Id.* at 1501A-528. It amended section 19(a)(6) by inserting "performance or display of a work embodied in" after "by a satellite carrier of" and by deleting "and embodying a performance or display of a work." *Id.* The Act also amended section 119(a) by adding paragraphs (11) and (12). *Id.* at 1501A-529 and 531.

The Satellite Home Viewer Improvement Act of 1999 amended section 119(b)(1) by inserting "or the Public Broadcasting Service satellite feed" into subparagraph

(B). *Id.* at 1501A-530. The Act amended section 119(c) by adding a new paragraph (4). *Id.* at 1501A-527. The Act amended section 119(d) by substituting new language for paragraphs (9) through (11) and by adding paragraph (12). *Id.* at 1501A-527, 530, and 531. The Act substituted new language for section 119(e). *Id.* at 1501A-529.

The Intellectual Property and High Technology Technical Amendments Act of 2002 amended section 119(a)(6) by substituting "of a performance" for "of performance." Pub. L. No. 107-273, 116 Stat. 1758, 1909. The Act also amended section 119(b)(1)(A) by substituting "retransmitted" and "retransmissions" for "transmitted" and "transmitted," respectively, in paragraph (1)(A). *Id.*

The Copyright Royalty and Distribution Reform Act of 2004 amended section 119 to conform it to revised chapter 8 by changing references to the "Librarian of Congress" in subsections (b) and (c) to "Copyright Royalty Judges," with corresponding grammatical adjustments and procedural references; by substituting new language for subparagraphs (b)(4)(B) and (C); by deleting the term "arbitration" wherever it appears with "proceedings," along with corresponding grammatical adjustments; by amending the title of subparagraph 119(c)(3)(C) to insert "Determination under Chapter 8" in lieu of "Decision of Arbitration Panel or Order of Librarian"; and, also, in subparagraph (c)(3)(C), by substituting new language for clauses (i) and (ii). Pub. L. No. 108-419, 118 Stat. 2341, 2364-2365.

The Satellite Home Viewer Extension and Reauthorization Act of 2004 amended paragraph 119(a)(1) by deleting "and PBS satellite feed" from the title; by deleting "or by the Public Broadcasting Service satellite feed" from the first sentence; by deleting the last sentence, which concerned Public Broadcasting Service satellite feed; by inserting "or for viewing in a commercial establishment" after "for private home viewing"; and by substituting "subscriber" for "household." Pub. L. No. 108-447, 118 Stat. 2809, 3393, 3394 and 3406. It amended subparagraph 119(a)(2)(B) by inserting at the end of clause (i) "The limitation in this clause shall not apply to secondary transmissions under paragraph (3)." *Id.* at 3397. It amended subsection (C) in its entirety by substituting new language. *Id.* at 3394. It amended paragraph 119(a)(5), which is now renumbered as 119(a)(7), in the first sentence of subparagraph (A), by inserting "who is not eligible to receive the transmission under this section" in lieu of "who does not reside in an unserved household" and, in the first sentence of subparagraph (B), by making the same change but using "are" instead of "is"; and, in subparagraph (D), by substituting "is to a subscriber who is eligible to receive the secondary transmission under this section" in lieu of "is for private home viewing to an unserved household." *Id.* at 3404. The Act further amended subsection 119(a) by adding new paragraphs, redesignated as paragraphs (3) and (4); by deleting paragraph eight; by renumbering the paragraphs affected by those changes; and by revising the references to old paragraph numbers, accordingly, in paragraphs (1) and (2), to be the new numbers as redesignated. *Id.* at 3394, 3396, and 3397. The Act further amended subsection 119(a) by adding at the end three new paragraphs, designated as new paragraphs (14), (15) and (16). *Id.* at 3400, 3404 and 3408.

The Satellite Home Viewer Extension and Reauthorization Act of 2004 amended the title of subsection 119(b) by deleting "for Private Home Viewing." *Id.* at 3406. It also amended subparagraph 119(b)(A) and paragraph 119(b)(3) by deleting "for private home viewing." *Id.* The Act amended subparagraph (119)(b)(1)(B) in its entirety by substituting new language. *Id.* at 3400. It added a new paragraph at the end of paragraph 119(b)(1). *Id.* at 3401.

The Satellite Home Viewer Extension and Reauthorization Act of 2004 amended subsection 119(c) in its entirety. *Id.* The Satellite Home Viewer Extension and Reauthorization Act of 2004 amended paragraph 119(d)(1) by deleting "for private home viewing" after "individual subscribers" and by adding at the end "in accordance with the provisions of this section." *Id.* at 3406. It amended paragraph 119(d)(2)(A) by substituting, at the beginning of the first sentence, "a television station licensed by the Federal Communications Commission" in lieu of "a television broadcast station." *Id.* The Act amended paragraph 119(d)(8) by substituting "or entity that" in lieu of "who"; by deleting "for private home viewing"; and by inserting at the end "in accordance with the provisions of this section." *Id.* It amended subparagraph 119(d)(10)(D) by changing "(a)(11)" to "(a)(12)". *Id.* at 3405. It amended in their entireties paragraph 119(d)(9), subparagraph (119)(d)(10)(B) and paragraphs 119(d)(11) and (12). *Id.* at 3405 and 3406.

The Satellite Home Viewer Extension and Reauthorization Act of 2004 amended subsection 119(e) by changing the date at the beginning of the sentence from "December 31, 2004" to "December 31, 2009". *Id.* at 3394. The Satellite Home Viewer Extension and Reauthorization Act of 2004 amended section 119 by adding a new subsection (f). *Id.* at 3394.

In 2006, the Copyright Royalty Judges Program Technical Corrections Act amended section 119 by revising the second sentence of (b)(4)(B); by amending (b)(4)(C) in its entirety; and by making a technical correction to substitute "arbitration" for "arbitrary" in (c)(1)(F)(i). Pub. L. No. 109-303, 120 Stat. 1478, 1482–83. The Prioritizing Resources and Organization for Intellectual Property Act of 2008 amended subparts (6), (7)(A), (8) and (13) of subsection 119(a) by removing the reference to section 509 (which was repealed). Pub. L. No. 110-403, 122 Stat. 4256, 4264.

Besides the changes described below, the Satellite Television Extension and Localism Act of 2010 made perfecting, technical, and clarifying amendments throughout section 119. Pub. L. No. 111-175, 124 Stat. 1218. It also substituted "non-network station" for "superstation" throughout section 119, revised the title to substitute "distant television programming by satellite" for "superstations and network stations for private home viewing" and added a new subsection (g) at the end. *Id.*, at 124 Stat. 1219, 1233, 1239-44.

In addition, the Satellite Television Extension and Localism Act of 2010 amended subsection 119(a) by deleting the last sentence in subparagraph (a)(2)(B)(i); by deleting paragraph (a)(2)(C); by revising subparagraphs (a)(2)(C)(i) and (ii) in the newly designated paragraph (C); by deleting subsection (a)(3); in the newly designated paragraph (3), by revising subparagraphs (a)(3)(B) and (C) and deleting "analog" wherever it appeared; in newly designated subsection (a)(6), by increasing the statutory damages in (6)(A)(ii) and (6)(B)(ii) from $5 and $250,000 to $250 and $2,500,000, respectively, and, in (6)(B)(i), by substituting "$2,500,000 for each 3-month period" for "$250,000 for each 6 month period;" by adding the last paragraph of subsection (a)(6) and by deleting subsection (a)(15). Pub. L. No. 111-175, 124 Stat. 1218, 1219, 1223-28.

Furthermore, the Satellite Television Extension and Localism Act of 2010 amended the title of subsection 119(b) by deleting "Statutory License for Secondary Transmissions" and inserting "Deposit of Statements and Fees; Verification Procedures. *Id.*, 124 Stat. 1220. It amended subsection 119(b) by revising subparagraph (b)(1)(B), adding new subparagraph (b)(1)(C), and deleting the last sentence of (b)(1); by adding a new subsection (b)(2); and by inserting "(including the filing fee specified in paragraph (1)(C))" into the first sentence of subsection (b)(3). *Id.*, at 124 Stat. 1220, 1223-24.

Public Law No. 111-175 amended subsection 119(c) throughout by deleting the references to "analog" and to "arbitration" and by substituting "Copyright Royalty Judges" for "Librarian of Congress." *Id.*, at 124 Stat. 1220, 21. It also amended subsection 119(c) by changing the reference in paragraph (c)(1)(A) from July 1, 2004, to July 1, 2009, to designate the regulation that establishes royalty fees and by changing the deadline in paragraph (c)(1)(A) from January 2, 2004, to January 2, 2010, for publishing notice in the *Federal Register* of initiation of voluntary negotiations. *Id.*, at 124 Stat. 1220. The Act further amended subsection 119(c) by adding subtitles for the subparagraphs in paragraph (c)(1)(D); by substituting "Copyright Royalty Judges" for "Copyright Office" in paragraph (c)(1)(E); by amending the subtitle for paragraph (c)(1)(F) to substitute "Copyright Royalty Judges Proceeding" for "Compulsory Arbitration;" and by revising subparagraphs (c)(1)(F)(ii) and (iii). *Id.*, at 124 Stat. 1221. The Act revised subsection 119(c)(2). *Id.*, at 124 Stat. 1222.

Public Law No. 111-175 amended subsection 119(d) by revising the definitions for subscriber and subscribe at (d)(8), for unserved household at (d)(10) and for local market at (d)(11). *Id.*, at 124 Stat. 1219, 1222-23. It deleted the definition for low power television station at (d)(12) and added definitions for qualifying date in newly designated paragraph (d)(13), multicast stream at (d)(14), and primary stream at (d)(15). *Id.*, at 124 Stat. 1219-20, 1223. The Act amended subsection 119(e) to extend the moratorium on copyright liability for satellite subscribers from December 31, 20o9, to December 31, 2014. See note 62.

A technical correction in the Copyright Cleanup, Clarification, and Corrections Act of 2010 amended subparagraph 119(g)(4)(B)(vi) to substitute "an examination" for "the examinations." Pub. L. No. 111-295, 124 Stat. 3180, 3181.

57. The Satellite Home Viewer Improvement Act of 1999 amended section 119(a)(1) by deleting "primary transmission made by a superstation and embodying a performance or display of a work" and inserting in its place "performance or display of a work embodied in a primary transmission made by a superstation." Pub. L. No. 106-113, 113 Stat. 1501, app. I at 1501A-543. This amendatory language did not take into account a prior amendment that had inserted "or by the Public Broadcasting Service satellite feed" after "superstation" into the phrase quoted above that was deleted. Pub. L. No. 106-113, 113 Stat. 1501, app. I at 1501A-530. There was no mention of the phrase "or by the Public Broadcasting Service satellite feed" in that second amendment. The Intellectual Property and High Technology Technical Amendments Act of 2002 clarified these provisions. Pub. L. No. 107-273, 116 Stat.

1758, 1908. The Act deleted the first change and amended the second to clarify that the amended language should read, "performance or display of a work embodied in a primary transmission made by a superstation or by the Public Broadcasting Service satellite feed." *Id.* The Prioritizing Resources and Organization for Intellectual Property Act of 2008 amended subparts (6), (7)(A), (8) and (13) of subsection 119(a) by removing the reference to section 509 (which was repealed). Pub. L. No. 110-403, 122 Stat. 4256, 4264.

58. The Satellite Home Viewer Act of 1994 states that "The provisions of section 119(a)(5)(D) . . . relating to the burden of proof of satellite carriers, shall take effect on January 1, 1997, with respect to civil actions relating to the eligibility of subscribers who subscribed to service as an unserved household before the date of the enactment of this Act [, October 18, 1994]." Pub. L. No. 103-369, 108 Stat. 3477, 3481.

59. The Intellectual Property and High Technology Technical Amendments Act of 2002 made a technical correction to insert the word "a" before "performance." Pub. L. No. 107-273, 116 Stat. 1758, 1909.

60. The Satellite Home Viewer Improvement Act of 1999 stated that section 119(a), "as amended by section 1005(e)" of the same Act, was amended to add a new paragraph (12). Pub. L. No. 106-113, 113 Stat. 1501, app. I at 1501A-531. The Intellectual Property and High Technology Technical Amendments Act of 2002 made a technical correction to clarify that the amendment was to section 119(a) as amended by "section 1005(d)" of the Satellite Home Viewer Improvement Act of 1999 rather than "section 1005(e)." Pub. L. No. 107-273, 116 Stat. 1758, 1908.

61. Section 119(c)(1)(E) establishes the time period during which the obligation to pay royalty fees established under a voluntary agreement that is filed with the Copyright Royalty Judges is in effect. The December 31, 2009, expiration date was extended several times. The 2010 Department of Defense Appropriations Act extended the date to February 28, 2010. Pub. L. No. 111-118, 123 Stat. 3409, 3469. The Temporary Extension Act of 2010 extended it to March 28, 2010. Pub. L. No. 111-144, 124 Stat. 42, 47. The Satellite Television Extension Act of 2010 extended it to April 30, 2010. Pub. L. No. 111-151, 124 Stat. 1027. The Continuing Extension Act of 2010 extended the date to May 31, 2010. Pub. L. No. 111-157, 124 Stat. 1116, 1119. Finally, the Satellite Television Extension and Localism Act of 2010 extended it to December 31, 2014. Pub. L. No. 111-175, 124 Stat. 1218, 1221.

62. Subsection 119(e) establishes a moratorium on copyright liability for satellite subscribers that was due to expire on December 31, 2009, but was extended several times. The 2010 Department of Defense Appropriations Act extended the date to February 28, 2010. Pub. L. No. 111-118, 123 Stat. 3409, 3469. The Temporary Extension Act of 2010 extended it to March 28, 2010. Pub. L. No. 111-144, 124 Stat. 42, 47. The Satellite Television Extension Act of 2010 extended it to April 30, 2010. Pub. L. No. 111-151, 124 Stat. 1027. The Continuing Extension Act of 2010 extended the date to May 31, 2010. Pub. L. No. 111-157, 124 Stat. 1116, 1119. Finally, the Satellite Television Extension and Localism Act of 2010 extended it to December 31, 2014. Pub. L. No. 111-175, 124 Stat. 1218, 1221.

63. In 1990, the Architectural Works Copyright Protection Act added section 120. Pub. L. No. 101-650, 104 Stat. 5089, 5133. The effective date provision of the

Act states that its amendments apply to any work created on or after the date it was enacted, which was December 1, 1990. It also states that the amendments apply to "any architectural work that, on [December 1, 1990], is unconstructed and embodied in unpublished plans or drawings, except that protection for such architectural work under title 17, *United States Code*, by virtue of the amendments made by [the Act], shall terminate on December 31, 2002, unless the work is constructed by that date." *Id.*, 104 Stat. 5089, 5134.

64. The Legislative Branch Appropriations Act, 1997, added section 121. Pub. L. No. 104-197, 110 Stat. 2394, 2416. The Work Made for Hire and Copyright Corrections Act of 2000 amended section 121 by substituting "section 106" for "sections 106 and 710." Pub. L. No. 106-379, 114 Stat. 1444, 1445.

The Individuals with Disabilities Education Improvement Act of 2004 amended section 121 by amending paragraph (c)(3) in its entirety; by adding a new paragraph (c)(4); by redesignating subsection (c) as (d); and by adding a new subsection (c). Pub. L. No. 108-446, 118 Stat. 2647, 2807.

65. Throughout section 122 the date of enactment of the Satellite Television and Localism Act of 2010 (STELA) is incorporated by reference. STELA was enacted on May 27, 2010. Pub. L. No. 111-175, 124 Stat. 1218.

The Satellite Home Viewer Improvement Act of 1999 added section 122. Pub. L. No. 106-113, 113 Stat. 1501, app. I at 1501A-523. The Act states that section 122 shall be effective as of November 29, 1999. Pub. L. No. 106-113, 113 Stat. 1501, app. I at 1501A-544.

The Satellite Home Viewer Extension and Reauthorization Act of 2004 amended section 122 by adding a subparagraph (D) to paragraph (j)(2). Pub. L. No. 108-447, 118 Stat. 2809, 3393, 3409.

The Prioritizing Resources and Organization for Intellectual Property Act of 2008 amended subsections 122(a) and (b) by removing the reference to section 509 (which was repealed). Pub. L. No. 110-403, 122 Stat. 4256, 4264.

The Satellite Television Extension and Localism Act of 2010 amended section 122 by deleting "by satellite carriers within local markets" from the title and replacing that text with "of local television programming by satellite;" by revising subsection (a); by deleting the second half of the first sentence in paragraph (b)(1) and adding subparagraphs (A) and (B); by revising subparagraph (b)(2) to add paragraphs (A) and (B); and by inserting, "For Certain Secondary Transmissions" at the end of the title for subsection (c) and adding "paragraphs (1), (2), and (3) of" to the text. Pub. L. No. 111-175, 124 Stat. 1218, 1227-30. The Act amended subsection 122(f) by inserting "subject to statutory licensing by reason of paragraph (2)(A), (3), or (4) of subsection (a), or subject to" into paragraphs (1) and (2); by increasing the statutory damages in subparagraph (1)(B) from $5 to $250; by increasing the statutory damages in subparts (2)(A)(ii) and (B)(ii) from $250,000 to $2,500,000; and by inserting "paragraph (2)(A), (3), or (4) of subsection (a)" into subsection (g). *Id.*, at 124 Stat. 1230. It also amended subsection (j) by inserting new paragraphs (3) and (5); by revising newly designated paragraph (6); and by inserting "non-network station" into the title of newly designated paragraph (4) and its text. *Id.*, at 124 Stat. 1231.

Chapter 2 Copyright Ownership and Transfer

§ 201 Ownership of Copyright

(a) Initial Ownership.—Copyright in a work protected under this title vests initially in the author or authors of the work. The authors of a joint work are coowners of copyright in the work.

(b) Works Made for Hire.—In the case of a work made for hire, the employer or other person for whom the work was prepared is considered the author for purposes of this title, and, unless the parties have expressly agreed otherwise in a written instrument signed by them, owns all of the rights comprised in the copyright.

(c) Contributions to Collective Works.—Copyright in each separate contribution to a collective work is distinct from copyright in the collective work as a whole, and vests initially in the author of the contribution. In the absence of an express transfer of the copyright or of any rights under it, the owner of copyright in the collective work is presumed to have acquired only the privilege of reproducing and distributing the contribution as part of that particular collective work, any revision of that collective work, and any later collective work in the same series.

(d) Transfer of Ownership.—

(1) The ownership of a copyright may be transferred in whole or in part by any means of conveyance or by operation of law, and may be bequeathed by will or pass as personal property by the applicable laws of intestate succession.

(2) Any of the exclusive rights comprised in a copyright, including any subdivision of any of the rights specified by section 106, may be transferred as provided by clause (1) and owned separately. The owner of any particular exclusive right is entitled, to the extent of that right, to all of the protection and remedies accorded to the copyright owner by this title.

(e) Involuntary Transfer.—When an individual author's ownership of a copyright, or of any of the exclusive rights under a copyright, has not previously been transferred voluntarily by that individual author, no action by any governmental body or other official or organization purporting to seize, expropriate, transfer, or exercise rights of ownership with respect to the copyright, or any of the exclusive rights under a copyright, shall be given effect under this title, except as provided under title 11.

§ 202 Ownership of Copyright as Distinct from Ownership of Material Object

Ownership of a copyright, or of any of the exclusive rights under a copyright, is distinct from ownership of any material object in which the work is embodied. Transfer of ownership of any material object, including the copy or phonorecord in which the work is first fixed, does not of itself convey any rights in the copyrighted work embodied in the object; nor, in the absence of an agreement, does transfer of ownership of a copyright or of any exclusive rights under a copyright convey property rights in any material object.

§ 203 Termination of Transfers and Licenses Granted by the Author

(a) Conditions for Termination.—In the case of any work other than a work made for hire, the exclusive or nonexclusive grant of a transfer or license of copyright or of any right under a copyright, executed by the author on or after January 1, 1978, otherwise than by will, is subject to termination under the following conditions:

(1) In the case of a grant executed by one author, termination of the grant may be effected by that author or, if the author is dead, by the person or persons who, under clause (2) of this subsection, own and are entitled to exercise a total of more than one-half of that author's termination interest. In the case of a grant executed by two or more authors of a joint work, termination of the grant may be effected by a majority of the authors who executed it; if any of such authors is dead, the termination interest of any such author may be exercised as a unit by the person or persons who, under clause (2) of this subsection, own and are entitled to exercise a total of more than one-half of that author's interest.

(2) Where an author is dead, his or her termination interest is owned, and may be exercised, as follows:

(A) The widow or widower owns the author's entire termination interest unless there are any surviving children or grandchildren of the author, in which case the widow or widower owns one-half of the author's interest.

(B) The author's surviving children, and the surviving children of any dead child of the author, own the author's entire termination interest unless there is a widow or widower, in which case the ownership of one-half of the author's interest is divided among them.

(C) The rights of the author's children and grandchildren are in all cases divided among them and exercised on a per stirpes basis according to the number of such author's children represented; the share of the children of a dead child in a termination interest can be exercised only by the action of a majority of them.

(D) In the event that the author's widow or widower, children, and grandchildren are not living, the author's executor, administrator, personal representative, or trustee shall own the author's entire termination interest.

(3) Termination of the grant may be effected at any time during a period of five years beginning at the end of thirty-five years from the date of execution of the grant; or, if the grant covers the right of publication of the work, the period begins at the end of thirty-five years from the date of publication of the work under the grant or at the end of forty years from the date of execution of the grant, whichever term ends earlier.

(4) The termination shall be effected by serving an advance notice in writing, signed by the number and proportion of owners of termination interests required under clauses (1) and (2) of this subsection, or by their duly authorized agents, upon the grantee or the grantee's successor in title.

(A) The notice shall state the effective date of the termination, which shall fall within the five-year period specified by clause (3) of this subsection, and the notice shall be served not less than two or more than ten years before that date. A copy of the notice shall be recorded in the Copyright Office before the effective date of termination, as a condition to its taking effect.

(B) The notice shall comply, in form, content, and manner of service, with requirements that the Register of Copyrights shall prescribe by regulation.

(5) Termination of the grant may be effected notwithstanding any agreement to the contrary, including an agreement to make a will or to make any future grant.

(b) Effect of Termination.—Upon the effective date of termination, all rights under this title that were covered by the terminated grants revert to the author, authors, and other persons owning termination interests under clauses (1) and (2) of subsection (a), including those owners who did not join in signing the notice of termination under clause (4) of subsection (a), but with the following limitations:

(1) A derivative work prepared under authority of the grant before its termination may continue to be utilized under the terms of the grant after its termination, but this privilege does not extend to the preparation after the termination of other derivative works based upon the copyrighted work covered by the terminated grant.

(2) The future rights that will revert upon termination of the grant become vested on the date the notice of termination has been served as provided by clause (4) of subsection (a). The rights vest in the author, authors, and other persons named in, and in the proportionate shares provided by, clauses (1) and (2) of subsection (a).

(3) Subject to the provisions of clause (4) of this subsection, a further grant, or agreement to make a further grant, of any right covered by a terminated grant is valid only if it is signed by the same number and proportion of the owners, in whom the right has vested under clause (2) of this subsection, as are required to terminate the grant under clauses (1) and (2) of subsection (a). Such further grant or agreement is effective with respect to all of the persons in whom the right it covers has vested under clause (2) of this subsection, including those who did not join in signing it. If any person dies after rights under a terminated grant have vested in him or her, that person's legal representatives, legatees, or heirs at law represent him or her for purposes of this clause.

(4) A further grant, or agreement to make a further grant, of any right covered by a terminated grant is valid only if it is made after the effective date of the termination. As an exception, however, an agreement for such a further grant may be made between the persons provided by clause (3) of this subsection and the original grantee or such grantee's successor in title, after the notice of termination has been served as provided by clause (4) of subsection (a).

(5) Termination of a grant under this section affects only those rights covered by the grants that arise under this title, and in no way affects rights arising under any other Federal, State, or foreign laws.

(6) Unless and until termination is effected under this section, the grant, if it does not provide otherwise, continues in effect for the term of copyright provided by this title.

§ 204 Execution of Transfers of Copyright Ownership

(a) A transfer of copyright ownership, other than by operation of law, is not valid unless an instrument of conveyance, or a note or memorandum of the transfer, is in writing and signed by the owner of the rights conveyed or such owner's duly authorized agent.

(b) A certificate of acknowledgment is not required for the validity of a transfer, but is prima facie evidence of the execution of the transfer if—

(1) in the case of a transfer executed in the United States, the certificate is issued by a person authorized to administer oaths within the United States; or

(2) in the case of a transfer executed in a foreign country, the certificate is issued by a diplomatic or consular officer of the United States, or by a person authorized to administer oaths whose authority is proved by a certificate of such an officer.

§ 205 Recordation of Transfers and Other Documents

(a) Conditions for Recordation.—Any transfer of copyright ownership or other document pertaining to a copyright may be recorded in the Copyright Office if the document filed for recordation bears the actual signature of the person who executed it, or if it is accompanied by a sworn or official certification that it is a true copy of the original, signed document. A sworn or official certification may be submitted to the Copyright Office electronically, pursuant to regulations established by the Register of Copyrights.

(b) Certificate of Recordation.—The Register of Copyrights shall, upon receipt of a document as provided by subsection (a) and of the fee provided by section 708, record the document and return it with a certificate of recordation.

(c) Recordation as Constructive Notice.—Recordation of a document in the Copyright Office gives all persons constructive notice of the facts stated in the recorded document, but only if—

(1) the document, or material attached to it, specifically identifies the work to which it pertains so that, after the document is indexed by the Register of Copyrights, it would be revealed by a reasonable search under the title or registration number of the work; and

(2) registration has been made for the work.

(d) Priority between Conflicting Transfers.—As between two conflicting transfers, the one executed first prevails if it is recorded, in the manner required to give constructive notice under subsection (c), within one month after its execution in the United States or within two months after its execution outside the United States, or at any time before recordation in such manner of the later transfer. Otherwise the later transfer prevails if recorded first in such manner, and if taken in good faith, for valuable consideration or on the basis of a binding promise to pay royalties, and without notice of the earlier transfer.

(e) Priority between Conflicting Transfer of Ownership and Nonexclusive License.—A nonexclusive license, whether recorded or not, prevails over a conflicting transfer of copyright ownership if the license is evidenced by a written instrument signed by the owner of the rights licensed or such owner's duly authorized agent, and if

(1) the license was taken before execution of the transfer; or

(2) the license was taken in good faith before recordation of the transfer and without notice of it.

Chapter 2 Notes

1. In 1978, section 201(e) was amended by deleting the period at the end and adding ", except as provided under title 11." See note 2, *infra*.

2. Title 11 of the *United States Code* is entitled "Bankruptcy."

3. In 1998, the Sonny Bono Copyright Term Extension Act amended section 203 by deleting "by his widow or her widower and his or her grandchildren" from the first sentence in paragraph (2) of subsection (a) and by adding subparagraph (D) to paragraph (2). Pub. L. No. 105-298, 112 Stat. 2827, 2829.

4. The Berne Convention Implementation Act of 1988 amended section 205 by deleting subsection (d) and redesignating subsections (e) and (f) as subsections (d) and (e), respectively. Pub. L. No. 100-568, 102 Stat. 2853, 2857. The Copyright Cleanup, Clarification, and Corrections Act of 2010 amended section 205(a) to add the last sentence. Pub. L. No. 111-295, 124 Stat. 3180.

Chapter 3 Duration of Copyright

§ 301 Preemption with Respect to Other Laws

(a) On and after January 1, 1978, all legal or equitable rights that are equivalent to any of the exclusive rights within the general scope of copyright as specified by section 106 in works of authorship that are fixed in a tangible medium of expression and come within the subject matter of copyright as specified by sections 102 and 103, whether created before or after that date and whether published or unpublished, are governed exclusively by this title. Thereafter, no person is entitled to any such right or equivalent right in any such work under the common law or statutes of any State.

(b) Nothing in this title annuls or limits any rights or remedies under the common law or statutes of any State with respect to—

(1) subject matter that does not come within the subject matter of copyright as specified by sections 102 and 103, including works of authorship not fixed in any tangible medium of expression; or

(2) any cause of action arising from undertakings commenced before January 1, 1978;

(3) activities violating legal or equitable rights that are not equivalent to any of the exclusive rights within the general scope of copyright as specified by section 106; or

(4) State and local landmarks, historic preservation, zoning, or building codes, relating to architectural works protected under section 102(a)(8).

(c) With respect to sound recordings fixed before February 15, 1972, any rights or remedies under the common law or statutes of any State shall not be annulled or limited by this title until February 15, 2067. The preemptive provisions of subsection (a) shall apply to any such rights and remedies pertaining to any cause of action arising from undertakings commenced on and after February 15, 2067. Notwithstanding the provisions of section 303, no sound recording fixed before February 15, 1972, shall be subject to copyright under this title before, on, or after February 15, 2067.

(d) Nothing in this title annuls or limits any rights or remedies under any other Federal statute.

(e) The scope of Federal preemption under this section is not affected by the adherence of the United States to the Berne Convention or the satisfaction of obligations of the United States thereunder.

(f)(1) On or after the effective date set forth in section 610(a) of the Visual Artists Rights Act of 1990, all legal or equitable rights that are equivalent to any of the rights conferred by section 106A with respect to works of visual art to which the rights conferred by section 106A apply are governed exclusively by section 106A and section 113(d) and the provisions of this title relating to such sections. Thereafter, no person is entitled to any such right or equivalent right in any work of visual art under the common law or statutes of any State.

(2) Nothing in paragraph (1) annuls or limits any rights or remedies under the common law or statutes of any State with respect to—

(A) any cause of action from undertakings commenced before the effective date set forth in section 610(a) of the Visual Artists Rights Act of 1990;

(B) activities violating legal or equitable rights that are not equivalent to any of the rights conferred by section 106A with respect to works of visual art; or

(C) activities violating legal or equitable rights which extend beyond the life of the author.

§ 302 Duration of Copyright: Works Created on or After January 1, 1978

(a) In General.—Copyright in a work created on or after January 1, 1978, subsists from its creation and, except as provided by the following subsections, endures for a term consisting of the life of the author and 70 years after the author's death.

(b) Joint Works.—In the case of a joint work prepared by two or more authors who did not work for hire, the copyright endures for a term consisting of the life of the last surviving author and 70 years after such last surviving author's death.

(c) Anonymous Works, Pseudonymous Works, and Works Made for Hire.—In the case of an anonymous work, a pseudonymous work, or a work made for hire, the copyright endures for a term of 95 years from the year of its first publication, or a term of 120 years from the year of its creation, whichever expires first. If, before the end of such term, the identity of one or more of the authors of an anonymous or pseudonymous work is revealed in the records of a registration made for that work under subsections (a) or (d) of section 408, or in the records provided by this subsection, the copyright in the work endures for the term specified by subsection (a) or (b), based on the life of the author or authors whose identity has been revealed. Any person having an interest in the copyright in an anonymous or pseudonymous work may at any time record, in records to be maintained by the Copyright Office for that purpose, a statement identifying one or more authors of the work; the statement shall also identify the person filing it, the nature of that person's interest, the source of the information recorded, and the particular work affected, and shall comply in form and content with requirements that the Register of Copyrights shall prescribe by regulation.

(d) Records Relating to Death of Authors.—Any person having an interest in a copyright may at any time record in the Copyright Office a statement of the date of death of the author of the copyrighted work, or a statement that the author is still living on a particular date. The statement shall identify the person filing it, the nature of that person's interest, and the source of the information recorded, and shall comply in form and content with requirements that the Register of Copyrights shall prescribe by regulation. The Register shall maintain current records of information relating to the death of authors of copyrighted works, based on such recorded statements and, to the extent the Register considers practicable, on data contained in any of the records of the Copyright Office or in other reference sources.

(e) Presumption as to Author's Death.—After a period of 95 years from the year of first publication of a work, or a period of 120 years from the year of its creation, whichever expires first, any person who obtains from the Copyright Office a certified report that the records provided by subsection (d) disclose nothing to indicate that the author of the work is living, or died less than 70 years before, is entitled to the benefit of a presumption that the author has been dead for at least 70 years. Reliance in good faith upon this presumption shall be a complete defense to any action for infringement under this title.

§ 303 Duration of Copyright: Works Created but Not Published or Copyrighted Before January 1, 1978

(a) Copyright in a work created before January 1, 1978, but not theretofore in the public domain or copyrighted, subsists from January 1, 1978, and endures for the term provided by section 302. In no case, however, shall the term of copyright in such a work expire before December 31, 2002; and, if the work is published on or before December 31, 2002, the term of copyright shall not expire before December 31, 2047.

(b) The distribution before January 1, 1978, of a phonorecord shall not for any purpose constitute a publication of any musical work, dramatic work, or literary work embodied therein.

§ 304 Duration of Copyright: Subsisting Copyrights

(a) Copyrights in Their First Term on January 1, 1978.—

(1) (A) Any copyright, in the first term of which is subsisting on January 1, 1978, shall endure for 28 years from the date it was originally secured.

(B) In the case of—

(i) any posthumous work or of any periodical, cyclopedic, or other composite work upon which the copyright was originally secured by the proprietor thereof, or

(ii) any work copyrighted by a corporate body (otherwise than as assignee or licensee of the individual author) or by an employer for whom such work is made for hire, the proprietor of such copyright shall be entitled to a renewal and extension of the copyright in such work for the further term of 67 years.

(C) In the case of any other copyrighted work, including a contribution by an individual author to a periodical or to a cyclopedic or other composite work—

(i) the author of such work, if the author is still living,

(ii) the widow, widower, or children of the author, if the author is not living,

(iii) the author's executors, if such author, widow, widower, or children are not living, or

(iv) the author's next of kin, in the absence of a will of the author, shall be entitled to a renewal and extension of the copyright in such work for a further term of 67 years.

(2)(A) At the expiration of the original term of copyright in a work specified in paragraph (1)(B) of this subsection, the copyright shall endure for a renewed and extended further term of 67 years, which—

(i) if an application to register a claim to such further term has been made to the Copyright Office within 1 year before the expiration of the original term of copyright, and the claim is registered, shall vest, upon the beginning of such further term, in the proprietor of the copyright who is entitled to claim the renewal of copyright at the time the application is made; or

(ii) if no such application is made or the claim pursuant to such application is not registered, shall vest, upon the beginning of such further term, in the person or entity that was the proprietor of the copyright as of the last day of the original term of copyright.

(B) At the expiration of the original term of copyright in a work specified in paragraph (1)(C) of this subsection, the copyright shall endure for a renewed and extended further term of 67 years, which—

(i) if an application to register a claim to such further term has been made to the Copyright Office within 1 year before the expiration of the original term of copyright, and the claim is registered, shall vest, upon the beginning of such further term, in any person who is entitled under paragraph (1)(C) to the renewal and extension of the copyright at the time the application is made; or

(ii) if no such application is made or the claim pursuant to such application is not registered, shall vest, upon the beginning of such further term, in any person entitled under paragraph (1)(C), as of the last day of the original term of copyright, to the renewal and extension of the copyright.

(3)(A) An application to register a claim to the renewed and extended term of copyright in a work may be made to the Copyright Office—

(i) within 1 year before the expiration of the original term of copyright by any person entitled under paragraph (1)(B) or (C) to such further term of 67 years; and

(ii) at any time during the renewed and extended term by any person in whom such further term vested, under paragraph (2)(A) or (B), or by any successor or assign of such person, if the application is made in the name of such person.

(B) Such an application is not a condition of the renewal and extension of the copyright in a work for a further term of 67 years.

(4)(A) If an application to register a claim to the renewed and extended term of copyright in a work is not made within 1 year before the expiration of the original term of copyright in a work, or if the claim pursuant to such application is not registered, then a derivative work prepared under authority of a grant of a transfer or license of the copyright that is made before the expiration of the original term of copyright may continue to be used under the terms of the grant during the renewed and extended term of copyright without infringing the copyright, except that such use does not extend to the preparation during such renewed and extended term of other derivative works based upon the copyrighted work covered by such grant.

(B) If an application to register a claim to the renewed and extended term of copyright in a work is made within 1 year before its expiration, and the claim is registered, the certificate of such registration shall constitute prima facie evidence as to the validity of the copyright during its renewed and extended term and of the facts stated in the certificate. The evidentiary weight to be accorded the certificates of a registration of a renewed and extended term of copyright made after the end of that 1-year period shall be within the discretion of the court.

(b) Copyrights in Their Renewal Term at the Time of the Effective Date of the Sonny Bono Copyright Term Extension Act—Any copyright still in its renewal term at the time that the Sonny Bono Copyright Term Extension Act becomes effective shall have a copyright term of 95 years from the date copyright was originally secured.

(c) Termination of Transfers and Licenses Covering Extended Renewal Term.—In the case of any copyright subsisting in either its first or renewal term on January 1, 1978, other than a copyright in a work made for hire, the exclusive or nonexclusive grant of a transfer or license of the renewal copyright or any right under it, executed before January 1, 1978, by any of the persons designated by subsection (a)(1)(C) of this section, otherwise than by will, is subject to termination under the following conditions:

(1) In the case of a grant executed by a person or persons other than the author, termination of the grant may be effected by the surviving person or persons who executed it. In the case of a grant executed by one or more of the authors of the work, termination of the grant may be effected, to the extent of a particular author's share in the ownership of the renewal copyright, by the author who executed it or, if such author is dead, by the person or persons who, under clause (2) of this subsection, own and are entitled to exercise a total of more than one-half of that author's termination interest.

(2) Where an author is dead, his or her termination interest is owned, and may be exercised, as follows:

(A) The widow or widower owns the author's entire termination interest unless there are any surviving children or grandchildren of the author, in which case the widow or widower owns one-half of the author's interest.

(B) The author's surviving children, and the surviving children of any dead child of the author, own the author's entire termination interest unless there is a widow or widower, in which case the ownership of one-half of the author's interest is divided among them.

(C) The rights of the author's children and grandchildren are in all cases divided among them and exercised on a per stirpes basis according to the number of such author's children represented; the share of the children of a dead child in a termination interest can be exercised only by the action of a majority of them.

(D) In the event that the author's widow or widower, children, and grandchildren are not living, the author's executor, administrator, personal representative, or trustee shall own the author's entire termination interest.

(3) Termination of the grant may be effected at any time during a period of five years beginning at the end of fifty-six years from the date copyright was originally secured, or beginning on January 1, 1978, whichever is later.

(4) The termination shall be effected by serving an advance notice in writing upon the grantee or the grantee's successor in title. In the case of a grant executed by a person or persons other than the author, the notice shall be signed by all of those entitled to terminate the grant under clause (1) of this subsection, or by their duly authorized agents. In the case of a grant executed by one or more of the authors of the work, the notice as to any one author's share shall be signed by that author or his or her duly authorized agent or, if that author is dead, by the number and proportion of the owners of his or her termination interest required under clauses (1) and (2) of this subsection, or by their duly authorized agents.

(A) The notice shall state the effective date of the termination, which shall fall within the five-year period specified by clause (3) of this subsection, or, in the case of a termination under subsection (d), within the five-year period specified by subsection (d)(2), and the notice shall be served not less than two or more than ten years before that date. A copy of the notice shall be recorded in the Copyright Office before the effective date of termination, as a condition to its taking effect.

(B) The notice shall comply, in form, content, and manner of service, with requirements that the Register of Copyrights shall prescribe by regulation.

(5) Termination of the grant may be effected notwithstanding any agreement to the contrary, including an agreement to make a will or to make any future grant.

(6) In the case of a grant executed by a person or persons other than the author, all rights under this title that were covered by the terminated grant revert, upon the effective date of termination, to all of those entitled to terminate the grant under clause (1) of this subsection. In the case of a grant executed by one or more of the authors of the work, all of a particular author's rights under this title that were covered by the terminated grant revert, upon the effective date of termination, to that author or, if that author is dead, to the persons owning his or her termination interest under clause (2) of this subsection, including those owners who did not join in signing the notice of termination under clause (4) of this subsection. In all cases the reversion of rights is subject to the following limitations:

(A) A derivative work prepared under authority of the grant before its termination may continue to be utilized under the terms of the grant after its termination, but this privilege does not extend to the preparation after the termination of other derivative works based upon the copyrighted work covered by the terminated grant.

(B) The future rights that will revert upon termination of the grant become vested on the date the notice of termination has been served as provided by clause (4) of this subsection.

(C) Where the author's rights revert to two or more persons under clause (2) of this subsection, they shall vest in those persons in the proportionate shares provided by that clause. In such a case, and subject to the provisions of subclause (D) of this clause, a further grant, or agreement to make a further grant, of a particular author's share with respect to any right covered by a terminated grant is valid only if it is signed by the same number and proportion of the owners, in whom the right has vested under this clause, as are required to terminate the grant under clause (2) of this subsection. Such further grant or agreement is effective with respect to all of the persons in whom the right it covers has vested under this subclause, including those who did not join in signing it. If any person dies after rights under a terminated grant have vested in him or her, that person's legal representatives, legatees, or heirs at law represent him or her for purposes of this subclause.

(D) A further grant, or agreement to make a further grant, of any right covered by a terminated grant is valid only if it is made after the effective date of the termination. As an exception, however, an agreement for such a further grant may be made between the author or any of the persons provided by the first sentence of clause (6) of this subsection, or between the persons provided by subclause (C) of this clause, and the original grantee or such grantee's successor in title, after the notice of termination has been served as provided by clause (4) of this subsection.

(E) Termination of a grant under this subsection affects only those rights covered by the grant that arise under this title, and in no way affects rights arising under any other Federal, State, or foreign laws.

(F) Unless and until termination is effected under this subsection, the grant, if it does not provide otherwise, continues in effect for the remainder of the extended renewal term.

(d) Termination Rights Provided in Subsection (c) Which Have Expired on or before the Effective Date of the Sonny Bono Copyright Term Extension Act.—In the case of any copyright other than a work made for hire, subsisting in its renewal term on the effective date of the Sonny Bono Copyright Term Extension Act for which the termination right provided in subsection (c) has expired by such date, where the author or owner of the termination right has not previously exercised such termination right, the exclusive or nonexclusive grant of a transfer or license of the renewal copyright or any right under it, executed before January 1, 1978, by any of the persons designated in subsection (a)(1)(C) of this section, other than by will, is subject to termination under the following conditions:

(1) The conditions specified in subsections (c) (1), (2), (4), (5), and (6) of this section apply to terminations of the last 20 years of copyright term as provided by the amendments made by the Sonny Bono Copyright Term Extension Act.

(2) Termination of the grant may be effected at any time during a period of 5 years beginning at the end of 75 years from the date copyright was originally secured.

§ 305 Duration of Copyright: Terminal Date

All terms of copyright provided by sections 302 through 304 run to the end of the calendar year in which they would otherwise expire.

Chapter 3 Notes

1. Private Law 92-60, 85 Stat. 857, effective December 15, 1971, states that: [A]ny provision of law to the contrary notwithstanding, copyright is hereby granted to the trustees under the will of Mary Baker Eddy, their successors, and assigns, in the work "Science and Health with Key to the Scriptures" (entitled also in some editions "Science and Health" or "Science and Health; with a Key to the Scriptures"), by Mary Baker Eddy, including all editions thereof in English and translation heretofore published, or hereafter published by or on behalf of said trustees, their successors or assigns, for a term of seventy-five years from the effective date of this Act or from the date of first publication, whichever is later.

But *cf. United Christian Scientists* v. *Christian Science Board of Directors, First Church of Christ, Scientist,* 829 F.2d 1152, 4 USPQ2d 1177 (D.C. Cir. 1987) (holding Priv. L. 92-60, 85 Stat. 857, to be unconstitutional because it violates the Establishment Clause).

2. The Berne Convention Implementation Act of 1988 amended section 301 by adding at the end thereof subsection (e). Pub. L. No. 100-568, 102 Stat. 2853, 2857. In 1990, the Architectural Works Copyright Protection Act amended section 301(b) by adding at the end thereof paragraph (4). Pub. L. No. 101-650, 104 Stat. 5133, 5134. The Visual Artists Rights Act of 1990 amended section 301 by adding at the end thereof subsection (f). Pub. L. No. 101650, 104 Stat. 5089, 5131. In 1998, the Sonny Bono Copyright Term Extension Act amended section 301 by changing "February 15, 2047" to "February 15, 2067" each place it appeared in subsection (c). Pub. L. No. 105-298, 112 Stat. 2827.

3. The Visual Artists Rights Act of 1990, which added subsection (f), states, "Subject to subsection (b) and except as provided in subsection (c), this title and the amendments made by this title take effect 6 months after the date of the enactment of this Act," that is, 6 months after December 1, 1990. Pub. L. No. 101-650, 104 Stat. 5089, 5132. See also note 39, chapter 1.

4. In 1998, the Sonny Bono Copyright Term Extension Act amended section 302 by substituting "70" for "fifty," "95" for "seventy-five," and "120" for "one hundred" each place they appeared. Pub. L. No. 105-298, 112 Stat. 2827. This change was effective October 27, 1998. *Id.*

5. In 1997, section 303 was amended by adding subsection (b). Pub. L. No. 105-80, 111 Stat. 1529, 1534. In 1998, the Sonny Bono Copyright Term Extension Act amended section 303 by substituting "December 31, 2047" for "December 31, 2027." Pub. L. No. 105-298, 112 Stat. 2827. The Copyright Cleanup, Clarification, and Corrections Act of 2010 amended section 303(b) to substitute "any musical work, dramatic work, or literary work" for "the musical work." Pub. L. No. 111-295, 124 Stat. 3180, 3181.

6. The Copyright Renewal Act of 1992 amended section 304 by substituting a new subsection (a) and by making a conforming amendment in the matter preceding paragraph (1) of subsection (c). Pub. L. No. 102-307, 106 Stat. 264. The Act, as amended by the Sonny Bono Copyright Term Extension Act, states that the renewal and extension of a copyright for a further term of 67 years "shall have the same effect with respect to any grant, before the effective date of the Sonny Bono Copyright Term Extension Act [October 27, 1998], of a transfer or license of the further term as did the renewal of a copyright before the effective date of the Sonny Bono Copyright Term Extension Act [October 27, 1998] under the law in effect at the time of such grant." The Act also states that the 1992 amendments "shall apply only to those copyrights secured between January 1, 1964, and December 31, 1977. Copyrights secured before January 1, 1964, shall be governed by the provisions of section 304(a) of title 17, United States Code, as in effect on the day before . . . [enactment on June 26, 1992], except each reference to forty-seven years in such provisions shall be deemed to be 67 years." Pub. L. No. 102-307, 106 Stat. 264, 266, as amended by the Sonny Bono Copyright Term Extension Act, Pub. L. No. 105-298, 112 Stat. 2827, 2828.

In 1998, the Sonny Bono Copyright Term Extension Act amended section 304 by substituting "67" for "47" wherever it appeared in subsection (a), by substituting a new subsection (b), and by adding subsection (d) at the end thereof. Pub. L. No. 105-298, 112 Stat. 2827. That Act also amended subsection 304(c) by deleting "by his widow or her widower and his or her children or grandchildren" from the first sentence of paragraph (2), by adding subparagraph (D) at the end of paragraph (2) and by inserting "or, in the case of a termination under subsection (d), within the five-year period specified by subsection (d)(2)," into the first sentence of subparagraph (4)(A). *Id.*

7. In 1998, the Sonny Bono Copyright Term Extension Act amendment to subsection 304(b) completely deleted the previous language that was originally part of the 1976 Copyright Act. Pub. L. No. 105-298, 112 Stat. 2827. That earlier statutory language continues to be relevant for calculating the term of protection for copyrights commencing between September 19, 1906, and December 31, 1949. The 1976 Copyright Act extended the terms for those copyrights by 20 years, provided they were in their renewal term between December 31, 1976, and December 31, 1977. The deleted language states:

The duration of any copyright, the renewal term of which is subsisting at any time between December 31, 1976, and December 31, 1977, inclusive, or for which renewal registration is made between December 31, 1976, and December 31, 1977, inclusive, is extended to endure for a term of seventy-five years from the date copyright was originally secured.

The effective date of this provision was October 19, 1976. That effective date provision is contained in Appendix A, herein, as section 102 of the Transitional and Supplementary Provisions of the Copyright Act of 1976. Copyright Act of 1976, Pub. L. No. 94-553, 90 Stat. 2541, 2598.

In addition, prior to the 1976 Copyright Act, Congress enacted a series of nine interim extensions for works whose copyright protection began between September 19, 1906, and December 31, 1918, if they were in their renewal terms. Without these interim extensions, copyrights commencing during that time period would have otherwise expired after 56 years, at the end of their renewal terms, between September 19, 1962, and December 31, 1976. The nine Acts authorizing the interim extensions are as follows, in chronological order:

Pub. L. No. 87-668, 76 Stat. 555 (extending copyrights from September 19, 1962, to December 31, 1965)

Pub. L. No. 89-142, 79 Stat. 581 (extending copyrights to December 31, 1967)

Pub. L. No. 90-141, 81 Stat. 464 (extending copyrights to December 31, 1968)

Pub. L. No. 90-416, 82 Stat. 397 (extending copyrights to December 31, 1969)

Pub. L. No. 91-147, 83 Stat. 360 (extending copyrights to December 31, 1970)

Pub. L. No. 91-555, 84 Stat. 1441 (extending copyrights to December 31, 1971)

Pub. L. No. 92-170, 85 Stat. 490 (extending copyrights to December 31, 1972)

Pub. L. No. 92-566, 86 Stat. 1181 (extending copyrights to December 31, 1974)

Pub. L. No. 93-573, 88 Stat. 1873 (extending copyrights to December 31, 1976)

8. The effective date of the Sonny Bono Copyright Term Extension Act is October 27, 1998.

9. See note 8, *supra.*

Chapter 4 Copyright Notice, Deposit, and Registration

§ 401 Notice of Copyright: Visually Perceptible Copies

(a) General Provisions.—Whenever a work protected under this title is published in the United States or elsewhere by authority of the copyright owner, a notice of copyright as provided by this section may be placed on publicly distributed copies from which the work can be visually perceived, either directly or with the aid of a machine or device.

(b) Form of Notice.—If a notice appears on the copies, it shall consist of the following three elements:

(1) the symbol © (the letter C in a circle), or the word "Copyright", or the abbreviation "Copr."; and

(2) the year of first publication of the work; in the case of compilations or derivative works incorporating previously published material, the year date of first publication of the compilation or derivative work is sufficient. The year date may be omitted where a pictorial, graphic, or sculptural work, with accompanying text matter, if any, is reproduced in or on greeting cards, postcards, stationery, jewelry, dolls, toys, or any useful articles; and

(3) the name of the owner of copyright in the work, or an abbreviation by which the name can be recognized, or a generally known alternative designation of the owner.

(c) Position of Notice.—The notice shall be affixed to the copies in such manner and location as to give reasonable notice of the claim of copyright. The Register of Copyrights shall prescribe by regulation, as examples, specific methods of affixation and positions of the notice on various types of works that will satisfy this requirement, but these specifications shall not be considered exhaustive.

(d) Evidentiary Weight of Notice.—If a notice of copyright in the form and position specified by this section appears on the published copy or copies to which a defendant in a copyright infringement suit had access, then no weight shall be given to such a defendant's interposition of a defense based on innocent infringement in mitigation of actual or statutory damages, except as provided in the last sentence of section 504(c)(2).

§ 402 Notice of Copyright: Phonorecords of Sound Recordings

(a) General Provisions.—Whenever a sound recording protected under this title is published in the United States or elsewhere by authority of the copyright owner, a notice of copyright as provided by this section may be placed on publicly distributed phonorecords of the sound recording.

(b) Form of Notice.—If a notice appears on the phonorecords, it shall consist of the following three elements:

(1) the symbol ℗ (the letter P in a circle); and

(2) the year of first publication of the sound recording; and

(3) the name of the owner of copyright in the sound recording, or an abbreviation by which the name can be recognized, or a generally known alternative designation of the owner; if the producer of the sound recording is named on the phonorecord labels or containers, and if no other name appears in conjunction with the notice, the producer's name shall be considered a part of the notice.

(c) Position of Notice.—The notice shall be placed on the surface of the phonorecord, or on the phonorecord label or container, in such manner and location as to give reasonable notice of the claim of copyright.

(d) Evidentiary Weight of Notice.—If a notice of copyright in the form and position specified by this section appears on the published phonorecord or phonorecords to which a defendant in a copyright infringement suit had access, then no weight shall be given to such a defendant's interposition of a defense based on innocent infringement in mitigation of actual or statutory damages, except as provided in the last sentence of section 504(c)(2).

§ 403 Notice of Copyright: Publications Incorporating United States Government Works

Sections 401(d) and 402(d) shall not apply to a work published in copies or phonorecords consisting predominantly of one or more works of the United States Government unless the notice of copyright appearing on the published copies or phonorecords to which a defendant in the copyright infringement suit had access includes a statement identifying, either affirmatively or negatively, those portions of the copies or phonorecords embodying any work or works protected under this title.

§ 404 Notice of Copyright: Contributions to Collective Works

(a) A separate contribution to a collective work may bear its own notice of copyright, as provided by sections 401 through 403. However, a single notice applicable to the collective work as a whole is sufficient to invoke the provisions of section 401(d) or 402(d), as applicable with respect to the separate contributions it contains (not including advertisements inserted on behalf of persons other than the owner of copyright in the collective work), regardless of the ownership of copyright in the contributions and whether or not they have been previously published.

(b) With respect to copies and phonorecords publicly distributed by authority of the copyright owner before the effective date of the Berne Convention Implementation Act of 1988, where the person named in a single notice applicable to a collective work as a whole is not the owner of copyright in a separate contribution that does not bear its own notice, the case is governed by the provisions of section 406(a).

§ 405 Notice of Copyright: Omission of Notice on Certain Copies and Phonorecords

(a) Effect of Omission on Copyright.—With respect to copies and phonorecords publicly distributed by authority of the copyright owner before the effective date of the Berne Convention Implementation Act of 1988, the omission of the copyright notice described in sections 401 through 403 from copies or phonorecords publicly distributed by authority of the copyright owner does not invalidate the copyright in a work if—

(1) the notice has been omitted from no more than a relatively small number of copies or phonorecords distributed to the public; or

(2) registration for the work has been made before or is made within five years after the publication without notice, and a reasonable effort is made to add notice to all copies or phonorecords that are distributed to the public in the United States after the omission has been discovered; or

(3) the notice has been omitted in violation of an express requirement in writing that, as a condition of the copyright owner's authorization of the public distribution of copies or phonorecords, they bear the prescribed notice.

(b) Effect of Omission on Innocent Infringers.—Any person who innocently infringes a copyright, in reliance upon an authorized copy or phonorecord from which the copyright notice has been omitted and which was publicly distributed by authority of the copyright owner before the effective date of the Berne Convention Implementation Act of 1988, incurs no liability for actual or statutory damages under section 504 for any infringing acts committed before receiving actual notice that registration for the work has been made under section 408, if such person proves that he or she was misled by the omission of notice. In a suit for infringement in such a case the court may allow or disallow recovery of any of the infringer's profits attributable to the infringement, and may enjoin the continuation of the infringing undertaking or may require, as a condition for permitting the continuation of the infringing undertaking, that the infringer pay the copyright owner a reasonable license fee in an amount and on terms fixed by the court.

(c) Removal of Notice.—Protection under this title is not affected by the removal, destruction, or obliteration of the notice, without the authorization of the copyright owner, from any publicly distributed copies or phonorecords.

§ 406 Notice of Copyright: Error in Name or Date on Certain Copies and Phonorecords

(a) Error in Name.—With respect to copies and phonorecords publicly distributed by authority of the copyright owner before the effective date of the Berne Convention Implementation Act of 1988, where the person named in the copyright notice on copies or phonorecords publicly distributed by authority of the copyright owner is not the owner of copyright, the validity and ownership of the copyright are not affected. In such a case, however, any person who innocently begins an undertaking that infringes the copyright has a complete defense to any action for such infringement if such person proves that he or she was misled by the notice and began the undertaking in good faith under a purported transfer or license from the person named therein, unless before the undertaking was begun—

(1) registration for the work had been made in the name of the owner of copyright; or

(2) a document executed by the person named in the notice and showing the ownership of the copyright had been recorded.

The person named in the notice is liable to account to the copyright owner for all receipts from transfers or licenses purportedly made under the copyright by the person named in the notice.

(b) Error in Date.—When the year date in the notice on copies or phonorecords distributed before the effective date of the Berne Convention Implementation Act of 1988 by authority of the copyright owner is earlier than the year in which publication first occurred, any period computed from the year of first publication under section 302 is to be computed from the year in the notice. Where the year date is more than one year later than the year in which publication first occurred, the work is considered to have been published without any notice and is governed by the provisions of section 405.

(c) Omission of Name or Date.—Where copies or phonorecords publicly distributed before the effective date of the Berne Convention Implementation Act of 1988 by authority of the copyright owner contain no name or no date that could reasonably be considered a part of the notice, the work is considered to have been published without any notice and is governed by the provisions of section 405 as in effect on the day before the effective date of the Berne Convention Implementation Act of 1988.

§ 407 Deposit of Copies or Phonorecords for Library of Congress

(a) Except as provided by subsection (c), and subject to the provisions of subsection (e), the owner of copyright or of the exclusive right of publication in a work published in the United States shall deposit, within three months after the date of such publication—

(1) two complete copies of the best edition; or

(2) if the work is a sound recording, two complete phonorecords of the best edition, together with any printed or other visually perceptible material published with such phonorecords.

Neither the deposit requirements of this subsection nor the acquisition provisions of subsection (e) are conditions of copyright protection.

(b) The required copies or phonorecords shall be deposited in the Copyright Office for the use or disposition of the Library of Congress. The Register of Copyrights shall, when requested by the depositor and upon payment of the fee prescribed by section 708, issue a receipt for the deposit.

(c) The Register of Copyrights may by regulation exempt any categories of material from the deposit requirements of this section, or require deposit of only one copy or phonorecord with respect to any categories. Such regulations shall provide either for complete exemption from the deposit requirements of this section, or for alternative forms of deposit aimed at providing a satisfactory archival record of a work without imposing practical or financial hardships on the depositor, where the individual author is the owner of copyright in a pictorial, graphic, or sculptural work and (i) less than five copies of the work have been published, or (ii) the work has been published in a limited edition consisting of numbered copies, the monetary value of which would make the mandatory deposit of two copies of the best edition of the work burdensome, unfair, or unreasonable.

(d) At any time after publication of a work as provided by subsection (a), the Register of Copyrights may make written demand for the required deposit on any of the persons obligated to make the deposit under subsection (a). Unless deposit is made within three months after the demand is received, the person or persons on whom the demand was made are liable—

(1) to a fine of not more than $250 for each work; and

(2) to pay into a specially designated fund in the Library of Congress the total retail price of the copies or phonorecords demanded, or, if no retail price has been fixed, the reasonable cost to the Library of Congress of acquiring them; and

(3) to pay a fine of $2,500, in addition to any fine or liability imposed under clauses (1) and (2), if such person willfully or repeatedly fails or refuses to comply with such a demand.

(e) With respect to transmission programs that have been fixed and transmitted to the public in the United States but have not been published, the Register of Copyrights shall, after consulting with the Librarian of Congress and other interested organizations and officials, establish regulations governing the acquisition, through deposit or otherwise, of copies or phonorecords of such programs for the collections of the Library of Congress.

(1) The Librarian of Congress shall be permitted, under the standards and conditions set forth in such regulations, to make a fixation of a transmission program directly from a transmission to the public, and to reproduce one copy or phonorecord from such fixation for archival purposes.

(2) Such regulations shall also provide standards and procedures by which the Register of Copyrights may make written demand, upon the owner of the right of transmission in the United States, for the deposit of a copy or phonorecord of a specific transmission program. Such deposit may, at the option of the owner of the right of transmission in the United States, be accomplished by gift, by loan for purposes of reproduction, or by sale at a price not to exceed the cost of reproducing and supplying the copy or phonorecord. The regulations established under this clause shall provide reasonable periods of not less than three months for compliance with a demand, and shall allow for extensions of such periods and adjustments in the scope of the demand or the methods for fulfilling it, as reasonably warranted by the circumstances. Willful failure or refusal to comply with the conditions prescribed by such regulations shall subject the owner of the right of transmission in the United States to liability for an amount, not to exceed the cost of reproducing and supplying the copy or phonorecord in question, to be paid into a specially designated fund in the Library of Congress.

(3) Nothing in this subsection shall be construed to require the making or retention, for purposes of deposit, of any copy or phonorecord of an unpublished transmission program, the transmission of which occurs before the receipt of a specific written demand as provided by clause (2).

(4) No activity undertaken in compliance with regulations prescribed under clauses (1) and (2) of this subsection shall result in liability if intended solely to assist in the acquisition of copies or phonorecords under this subsection.

§ 408 Copyright Registration in General

(a) Registration Permissive.—At any time during the subsistence of the first term of copyright in any published or unpublished work in which the copyright was secured before January 1, 1978, and during the subsistence of any copyright secured on or after that date, the owner of copyright or of any exclusive right in the work may obtain registration of the copyright claim by delivering to the Copyright Office the deposit specified by this section, together with the application and fee specified by sections 409 and 708. Such registration is not a condition of copyright protection.

(b) Deposit for Copyright Registration.—Except as provided by subsection (c), the material deposited for registration shall include—

(1) in the case of an unpublished work, one complete copy or phonorecord;

(2) in the case of a published work, two complete copies or phonorecords of the best edition;

(3) in the case of a work first published outside the United States, one complete copy or phonorecord as so published;

(4) in the case of a contribution to a collective work, one complete copy or phonorecord of the best edition of the collective work.

Copies or phonorecords deposited for the Library of Congress under section 407 may be used to satisfy the deposit provisions of this section, if they are accompanied by the prescribed application and fee, and by any additional identifying material that the Register may, by regulation, require. The Register shall also prescribe regulations establishing requirements under which copies or phonorecords acquired for the Library of Congress under subsection (e) of section 407, otherwise than by deposit, may be used to satisfy the deposit provisions of this section.

(c) Administrative Classification and Optional Deposit—

(1) The Register of Copyrights is authorized to specify by regulation the administrative classes into which works are to be placed for purposes of deposit and registration, and the nature of the copies or phonorecords to be deposited in the various classes specified. The regulations may require or permit, for particular classes, the deposit of identifying material instead of copies or phonorecords, the deposit of only one copy or phonorecord where two would normally be required, or a single registration for a group of related works. This administrative classification of works has no significance with respect to the subject matter of copyright or the exclusive rights provided by this title.

(2) Without prejudice to the general authority provided under clause (1), the Register of Copyrights shall establish regulations specifically permitting a single registration for a group of works by the same individual author, all first published as contributions to periodicals, including newspapers, within a twelve-month period, on the basis of a single deposit, application, and registration fee, under the following conditions—

(A) if the deposit consists of one copy of the entire issue of the periodical, or of the entire section in the case of a newspaper, in which each contribution was first published; and

(B) if the application identifies each work separately, including the periodical containing it and its date of first publication.

(3) As an alternative to separate renewal registrations under subsection (a) of section 304, a single renewal registration may be made for a group of works by the same individual author, all first published as contributions to periodicals, including newspapers, upon the filing of a single application and fee, under all of the following conditions:

(A) the renewal claimant or claimants, and the basis of claim or claims under section 304(a), is the same for each of the works; and

(B) the works were all copyrighted upon their first publication, either through separate copyright notice and registration or by virtue of a general copyright notice in the periodical issue as a whole; and

(C) the renewal application and fee are received not more than twentyeight or less than twenty-seven years after the thirty-first day of December of the calendar year in which all of the works were first published; and

(D) the renewal application identifies each work separately, including the periodical containing it and its date of first publication.

(d) Corrections and Amplifications.—The Register may also establish, by regulation, formal procedures for the filing of an application for supplementary registration, to correct an error in a copyright registration or to amplify the information given in a registration. Such application shall be accompanied by the fee provided by section 708, and shall clearly identify the registration to be corrected or amplified. The information contained in a supplementary registration augments but does not supersede that contained in the earlier registration.

(e) Published Edition of Previously Registered Work.—Registration for the first published edition of a work previously registered in unpublished form may be made even though the work as published is substantially the same as the unpublished version.

(f) Preregistration of Works Being Prepared for Commercial Distribution.—

(1) Rulemaking.—Not later than 180 days after the date of enactment of this subsection, the Register of Copyrights shall issue regulations to establish procedures for preregistration of a work that is being prepared for commercial distribution and has not been published.

(2) Class of works.—The regulations established under paragraph (1) shall permit preregistration for any work that is in a class of works that the Register determines has had a history of infringement prior to authorized commercial distribution.

(3) Application for registration.—Not later than 3 months after the first publication of a work preregistered under this subsection, the applicant shall submit to the Copyright Office–

(A) an application for registration of the work;

(B) a deposit; and

(C) the applicable fee.

(4) Effect of untimely application.—An action under this chapter for infringement of a work preregistered under this subsection, in a case in which the infringement commenced no later than 2 months after the first publication of

the work, shall be dismissed if the items described in paragraph (3) are not submitted to the Copyright Office in proper form within the earlier of—

(A) 3 months after the first publication of the work; or

(B) 1 month after the copyright owner has learned of the infringement.

§ 409 Application for Copyright Registration

The application for copyright registration shall be made on a form prescribed by the Register of Copyrights and shall include—

(1) the name and address of the copyright claimant;

(2) in the case of a work other than an anonymous or pseudonymous work, the name and nationality or domicile of the author or authors, and, if one or more of the authors is dead, the dates of their deaths;

(3) if the work is anonymous or pseudonymous, the nationality or domicile of the author or authors;

(4) in the case of a work made for hire, a statement to this effect;

(5) if the copyright claimant is not the author, a brief statement of how the claimant obtained ownership of the copyright;

(6) the title of the work, together with any previous or alternative titles under which the work can be identified;

(7) the year in which creation of the work was completed;

(8) if the work has been published, the date and nation of its first publication;

(9) in the case of a compilation or derivative work, an identification of any preexisting work or works that it is based on or incorporates, and a brief, general statement of the additional material covered by the copyright claim being registered; and

(10) any other information regarded by the Register of Copyrights as bearing upon the preparation or identification of the work or the existence, ownership, or duration of the copyright.

If an application is submitted for the renewed and extended term provided for in section 304(a)(3)(A) and an original term registration has not been made, the Register may request information with respect to the existence, ownership, or duration of the copyright for the original term.

§ 410 Registration of Claim and Issuance of Certificate

(a) When, after examination, the Register of Copyrights determines that, in accordance with the provisions of this title, the material deposited constitutes copyrightable subject matter and that the other legal and formal requirements of this title have been met, the Register shall register the claim and issue to the applicant a certificate of registration under the seal of the Copyright Office. The certificate shall contain the information given in the application, together with the number and effective date of the registration.

(b) In any case in which the Register of Copyrights determines that, in accordance with the provisions of this title, the material deposited does not constitute copyrightable subject matter or that the claim is invalid for any other reason, the Register shall refuse registration and shall notify the applicant in writing of the reasons for such refusal.

(c) In any judicial proceedings the certificate of a registration made before or within five years after first publication of the work shall constitute prima facie evidence of the validity of the copyright and of the facts stated in the certificate. The evidentiary weight to be accorded the certificate of a registration made thereafter shall be within the discretion of the court.

(d) The effective date of a copyright registration is the day on which an application, deposit, and fee, which are later determined by the Register of Copyrights or by a court of competent jurisdiction to be acceptable for registration, have all been received in the Copyright Office.

§ 411 Registration and Civil Infringement Actions

(a) Except for an action brought for a violation of the rights of the author under section 106A(a), and subject to the provisions of subsection (b), no civil action for infringement of the copyright in any United States work shall be instituted until preregistration or registration of the copyright claim has been made in accordance with this title. In any case, however, where the deposit, application, and fee required for registration have been delivered to the Copyright Office in proper form and registration has been refused, the applicant is entitled to institute a civil action for infringement if notice thereof, with a copy of the complaint, is served on the Register of Copyrights. The Register may, at his or her option, become a party to the action with respect to the issue of registrability of the copyright claim by entering an appearance within sixty days after such service, but the Register's failure to become a party shall not deprive the court of jurisdiction to determine that issue.

(b)(1) A certificate of registration satisfies the requirements of this section and section 412, regardless of whether the certificate contains any inaccurate information, unless—

(A) the inaccurate information was included on the application for copyright registration with knowledge that it was inaccurate; and

(B) the inaccuracy of the information, if known, would have caused the Register of Copyrights to refuse registration.

(2) In any case in which inaccurate information described under paragraph (1) is alleged, the court shall request the Register of Copyrights to advise the court whether the inaccurate information, if known, would have caused the Register of Copyrights to refuse registration.

(3) Nothing in this subsection shall affect any rights, obligations, or requirements of a person related to information contained in a registration certificate, except for the institution of and remedies in infringement actions under this section and section 412.

(c) In the case of a work consisting of sounds, images, or both, the first fixation of which is made simultaneously with its transmission, the copyright owner may, either before or after such fixation takes place, institute an action for infringement under section 501, fully subject to the remedies provided by sections 502 through 505 and section 510, if, in accordance with requirements that the Register of Copyrights shall prescribe by regulation, the copyright owner—

(1) serves notice upon the infringer, not less than 48 hours before such fixation, identifying the work and the specific time and source of its first transmission, and declaring an intention to secure copyright in the work; and

(2) makes registration for the work, if required by subsection (a), within three months after its first transmission.

§ 412 Registration as Prerequisite to Certain Remedies for Infringement

In any action under this title, other than an action brought for a violation of the rights of the author under section 106A(a), an action for infringement of the copyright of a work that has been preregistered under section 408(f) before the commencement of the infringement and that has an effective date of registration not later than the earlier of 3 months after the first publication of the work or 1 month after the copyright owner has learned of the infringement, or an action instituted under section 411(c), no award of statutory damages or of attorney's fees, as provided by sections 504 and 505, shall be made for—

(1) any infringement of copyright in an unpublished work commenced before the effective date of its registration; or

(2) any infringement of copyright commenced after first publication of the work and before the effective date of its registration, unless such registration is made within three months after the first publication of the work.

Chapter 4 Notes

1. The Prioritizing Resources and Organization for Intellectual Property Act of 2008 amended the title for section 411 by adding "civil" to it, so that the new title is "Registration and civil infringement actions." Pub. L. No. 110-403, 122 Stat. 4256, 4258.

2. The Berne Convention Implementation Act of 1988 amended section 401 as follows: 1) in subsection (a), by changing the heading to "General Provisions" and by inserting "may be placed on" in lieu of "shall be placed on all"; 2) in subsection (b), by inserting "If a notice appears on the copies, it" in lieu of "The notice appearing on the copies"; and 3) by adding subsection (d). Pub. L. No. 100-568, 102 Stat. 2853, 2857.

3. The Berne Convention Implementation Act of 1988 amended section 402 as follows: 1) in subsection (a), by changing the heading to "General Provisions" and by inserting "may be placed on" in lieu of "shall be placed on all"; 2) in subsection

(b), by inserting "If a notice appears on the phonorecords, it" in lieu of "The notice appearing on the phonorecords"; and 3) by adding subsection (d). Pub. L. No. 100-568, 102 Stat. 2853, 2857.

4. The Berne Convention Implementation Act of 1988 amended section 403 in its entirety. Pub. L. No. 100-568, 102 Stat. 2853, 2858.

5. The Berne Convention Implementation Act of 1988 amended section 404 as follows: 1) in the second sentence of subsection (a), by inserting "to invoke the provisions of section 401(d) or 402(d), as applicable" in lieu of "to satisfy the requirements of sections 401 through 403" and 2) in subsection (b), by inserting "With respect to copies and phonorecords publicly distributed by authority of the copyright owner before the effective date of the Berne Convention Implementation Act of 1988," at the beginning of the sentence. Pub. L. No. 100-568, 102 Stat. 2853, 2858.

6. The Berne Convention Implementation Act of 1988 amended section 405 as follows: 1) in subsection (a), by inserting "With respect to copies and phonorecords publicly distributed by authority of the copyright owner before the effective date of the Berne Convention Implementation Act of 1988, the omission of the copyright notice described in" at the beginning of the first sentence, in lieu of "The omission of the copyright notice prescribed by"; 2) in subsection (b), by inserting after "omitted," in the first sentence, "and which was publicly distributed by authority of the copyright owner before the effective date of the Berne Convention Implementation Act of 1988"; and 3) by amending the section heading to add "on certain copies and phonorecords" at the end thereof. Pub. L. No. 100-568, 102 Stat. 2853, 2858.

7. The Berne Convention Implementation Act of 1988 amended section 406 as follows: 1) in subsection (a), by inserting "With respect to copies and phonorecords publicly distributed by authority of the copyright owner before the effective date of the Berne Convention Implementation Act of 1988," at the beginning of the first sentence; 2) in subsection (b), by inserting "before the effective date of the Berne Convention Implementation Act of 1988" after "distributed"; 3) in subsection (c), by inserting "before the effective date of the Berne Convention Implementation Act of 1988" after "publicly distributed" and by inserting "as in effect on the day before the effective date of the Berne Convention Implementation Act of 1988" after "405"; and 4) by amending the section heading to add "on certain copies and phonorecords" at the end thereof. Pub. L. No. 100-568, 102 Stat. 2853, 2858.

8. The Berne Convention Implementation Act of 1988 amended section 407 by striking out the words "with notice of copyright" in subsection (a). Pub. L. No. 100-568, 102 Stat. 2853, 2859.

9. The Berne Convention Implementation Act of 1988 amended section 408 by deleting "Subject to the provisions of section 405(a)," at the beginning of the second sentence of subsection (a). Pub. L. No. 100-568, 102 Stat. 2853, 2859. That Act also amended section 408(c)(2) by inserting "the following conditions:" in lieu of "all of the following conditions" and by striking subparagraph (A) and by redesignating subparagraphs (B) and (C) as subparagraphs

(A) and (B), respectively. *Id.* The Copyright Renewal Act of 1992 amended section 408 by revising the first sentence of subsection (a), preceding the words "the owner of copyright or of any exclusive right." Pub. L. No. 102-307, 106 Stat. 264, 266.

The Artists' Rights and Theft Prevention Act of 2005 amended section 408 by adding a new subsection (f). Pub. L. No. 109-9, 119 Stat. 218, 221.

10. The Copyright Renewal Act of 1992 amended section 409 by adding the last sentence. Pub. L. No. 102-307, 106 Stat. 264, 266. The Copyright Cleanup, Clarification, and Corrections Act of 2010 amended section 409 to delete subsection (10) and redesignate subsection (11) as (10). Pub. L. No. 111-295, 124 Stat. 3180.

11. The Berne Convention Implementation Act of 1988 amended section 411 as follows: 1) in subsection (a), by inserting "Except for actions for infringement of copyright in Berne Convention works whose country of origin is not the United States, and" before "subject"; 2) in paragraph (b)(2), by inserting ", if required by subsection (a)," after "work"; and 3) by inserting "and infringement actions" in the heading, in lieu of "as prerequisite to infringement suit." Pub. L. No. 100-568, 102 Stat. 2853, 2859.

The Visual Artists Rights Act of 1990 amended section 411(a) by inserting "and an action brought for a violation of the rights of the author under section 106A(a)" after "United States." Pub. L. No. 101-650, 104 Stat. 5089, 5131. In 1997, section 411(b)(1) was amended in its entirety. Pub. L. No. 105-80, 111 Stat. 1529, 1532.

The WIPO Copyright and Performances and Phonograms Treaties Implementation Act of 1998 amended the first sentence in section 411(a) by deleting "actions for infringement of copyright in Berne Convention works whose country of origin is not the United and" and by inserting "United States" after "no action for infringement of the copyright in any." Pub. L. No. 105-304, 112 Stat. 2860, 2863.

The Artists' Rights and Theft Prevention Act of 2005 amended subsection 411(a) by inserting "preregistration" in the first sentence, after "shall be instituted until" in the first sentence. Pub. L. No. 109-9, 119 Stat. 218, 222.

The Prioritizing Resources and Organization for Intellectual Property Act of 2008 amended the title for section 411 by inserting "civil," so that the new title is "Registration and civil infringement actions." Pub. L. No. 110-403, 122 Stat. 4256, 4257. It also amended subsection (a) to insert "civil" before "action" in the first and second sentences. *Id.* It redesignated subsection (b) as subsection (c) and added a new subsection (b). *Id.* at 4257-58. The Act also amended the newly designated subsection (c) to remove the reference to section 509 (which was repealed). *Id.* at 4257-58.

12. The Visual Artists Rights Act of 1990 amended section 412 by inserting "an action brought for a violation of the rights of the author under section 106A(a) or" after "other than." Pub. L. No. 101-650, 104 Stat. 5089, 5131.

The Artists' Rights and Theft Prevention Act of 2005 amended subsection 412 by inserting the clause that follows "section 106A(a)," in the text preceding subparagraph (1). Pub. L. No. 109-9, 119 Stat. 218, 222.

The Prioritizing Resources and Organization for Intellectual Property Act of 2008 amended section 412 by making a conforming amendment to substitute "subsection 411(b)" for "subsection 411(c)." Pub. L. No. 110-403, 122 Stat. 4256, 4258.

Chapter 5 Copyright Infringement and Remedies

§ 501 Infringement of Copyright

(a) Anyone who violates any of the exclusive rights of the copyright owner as provided by sections 106 through 122 or of the author as provided in section 106A(a), or who imports copies or phonorecords into the United States in violation of section 602, is an infringer of the copyright or right of the author, as the case may be. For purposes of this chapter (other than section 506), any reference to copyright shall be deemed to include the rights conferred by section 106A(a). As used in this subsection, the term "anyone" includes any State, any instrumentality of a State, and any officer or employee of a State or instrumentality of a State acting in his or her official capacity. Any State, and any such instrumentality, officer, or employee, shall be subject to the provisions of this title in the same manner and to the same extent as any nongovernmental entity.

(b) The legal or beneficial owner of an exclusive right under a copyright is entitled, subject to the requirements of section 411, to institute an action for any infringement of that particular right committed while he or she is the owner of it. The court may require such owner to serve written notice of the action with a copy of the complaint upon any person shown, by the records of the Copyright Office or otherwise, to have or claim an interest in the copyright, and shall require that such notice be served upon any person whose interest is likely to be affected by a decision in the case. The court may require the joinder, and shall permit the intervention, of any person having or claiming an interest in the copyright.

(c) For any secondary transmission by a cable system that embodies a performance or a display of a work which is actionable as an act of infringement under subsection (c) of section 111, a television broadcast station holding a copyright or other license to transmit or perform the same version of that work shall, for purposes of subsection (b) of this section, be treated as a legal or beneficial owner if such secondary transmission occurs within the local service area of that television station.

(d) For any secondary transmission by a cable system that is actionable as an act of infringement pursuant to section 111(c)(3), the following shall also have standing to sue: (i) the primary transmitter whose transmission has been altered by the cable system; and (ii) any broadcast station within whose local service area the secondary transmission occurs.

(e) With respect to any secondary transmission that is made by a satellite carrier of a performance or display of a work embodied in a primary transmission and is actionable as an act of infringement under section 119(a)(5), a network station holding a copyright or other license to transmit or perform the same version of that work shall, for purposes of subsection (b) of this section, be treated as a legal or beneficial owner if such secondary transmission occurs within the local service area of that station.

(f)(1) With respect to any secondary transmission that is made by a satellite carrier of a performance or display of a work embodied in a primary transmission and is actionable as an act of infringement under section 122, a television broadcast station holding a copyright or other license to transmit or perform the same version of that work shall, for purposes of subsection (b) of this section, be treated as a legal or beneficial owner if such secondary transmission occurs within the local market of that station.

(2) A television broadcast station may file a civil action against any satellite carrier that has refused to carry television broadcast signals, as required under section 122(a)(2), to enforce that television broadcast station's rights under section 338(a) of the Communications Act of 1934.

§ 502 Remedies for Infringement: Injunctions

(a) Any court having jurisdiction of a civil action arising under this title may, subject to the provisions of section 1498 of title 28, grant temporary and final injunctions on such terms as it may deem reasonable to prevent or restrain infringement of a copyright.

(b) Any such injunction may be served anywhere in the United States on the person enjoined; it shall be operative throughout the United States and shall be enforceable, by proceedings in contempt or otherwise, by any United States court having jurisdiction of that person. The clerk of the court granting the injunction shall, when requested by any other court in which enforcement of the injunction is sought, transmit promptly to the other court a certified copy of all the papers in the case on file in such clerk's office.

§ 503 Remedies for Infringement: Impounding and Disposition of Infringing Articles

(a)(1) At any time while an action under this title is pending, the court may order the impounding, on such terms as it may deem reasonable—

(A) of all copies or phonorecords claimed to have been made or used in violation of the exclusive right of the copyright owner;

(B) of all plates, molds, matrices, masters, tapes, film negatives, or other articles by means of which such copies or phonorecords may be reproduced; and

(C) of records documenting the manufacture, sale, or receipt of things involved in any such violation, provided that any records seized under this subparagraph shall be taken into the custody of the court.

(2) For impoundments of records ordered under paragraph (1)(C), the court shall enter an appropriate protective order with respect to discovery and use of any records or information that has been impounded. The protective order shall provide for appropriate procedures to ensure that confidential, private, proprietary, or privileged information contained in such records is not improperly disclosed or used.

(3) The relevant provisions of paragraphs (2) through (11) of section 34(d) of the Trademark Act (15 U.S.C. 1116(d)(2) through (11)) shall extend to any impoundment of records ordered under paragraph (1)(C) that is based upon an ex parte application, notwithstanding the provisions of rule 65 of the Federal Rules of Civil Procedure. Any references in paragraphs (2) through (11) of section 34(d) of the Trademark Act to section 32 of such Act shall be read as references to section 501 of this title, and references to use of a counterfeit mark in connection with the sale, offering for sale, or distribution of goods or services shall be read as references to infringement of a copyright.

(b) As part of a final judgment or decree, the court may order the destruction or other reasonable disposition of all copies or phonorecords found to have been made or used in violation of the copyright owner's exclusive rights, and of all plates, molds, matrices, masters, tapes, film negatives, or other articles by means of which such copies or phonorecords may be reproduced.

§ 504 Remedies for Infringement: Damages and Profits

(a) In General.—Except as otherwise provided by this title, an infringer of copyright is liable for either—

(1) the copyright owner's actual damages and any additional profits of the infringer, as provided by subsection (b); or

(2) statutory damages, as provided by subsection (c).

(b) Actual Damages and Profits.—The copyright owner is entitled to recover the actual damages suffered by him or her as a result of the infringement, and any profits of the infringer that are attributable to the infringement and are not taken into account in computing the actual damages. In establishing the infringer's profits, the copyright owner is required to present proof only of the infringer's gross revenue, and the infringer is required to prove his or her deductible expenses and the elements of profit attributable to factors other than the copyrighted work.

(c) Statutory Damages.—

(1) Except as provided by clause (2) of this subsection, the copyright owner may elect, at any time before final judgment is rendered, to recover, instead of actual damages and profits, an award of statutory damages for all infringements involved in the action, with respect to any one work, for which any one infringer is liable individually, or for which any two or more infringers are liable jointly and severally, in a sum of not less than $750 or more than $30,000 as the court considers just. For the purposes of this subsection, all the parts of a compilation or derivative work constitute one work.

(2) In a case where the copyright owner sustains the burden of proving, and the court finds, that infringement was committed willfully, the court in its discretion may increase the award of statutory damages to a sum of not more than $150,000. In a case where the infringer sustains the burden of proving, and the court finds, that such infringer was not aware and had no reason to believe that his or her acts constituted an infringement of copyright, the court in its discretion may reduce the award of statutory damages to a sum of not less than $200. The court shall remit statutory damages in any case where an infringer believed and had reasonable grounds for believing that his or her use of the copyrighted work was a fair use under section 107, if the infringer was: (i) an employee or agent of a nonprofit educational institution, library, or archives acting within the scope of his or her employment who, or such institution, library, or archives itself, which infringed by reproducing the work in copies or phonorecords; or (ii) a public broadcasting entity which or a person who, as a regular part of the nonprofit activities of a public broadcasting entity (as defined in section 118(f)) infringed by performing a published nondramatic literary work or by reproducing a transmission program embodying a performance of such a work.

(3) (A) In a case of infringement, it shall be a rebuttable presumption that the infringement was committed willfully for purposes of determining relief if the violator, or a person acting in concert with the violator, knowingly provided or knowingly caused to be provided materially false contact information to a domain name registrar, domain name registry, or other domain name registration authority in registering, maintaining, or renewing a domain name used in connection with the infringement.

(B) Nothing in this paragraph limits what may be considered willful infringement under this subsection.

(C) For purposes of this paragraph, the term "domain name" has the meaning given that term in section 45 of the Act entitled "An Act to provide for the registration and protection of trademarks used in commerce, to carry out the provisions of certain international conventions, and for other purposes" approved July 5, 1946 (commonly referred to as the "Trademark Act of 1946"; 15 U.S.C. 1127).

(d) Additional Damages in Certain Cases.—In any case in which the court finds that a defendant proprietor of an establishment who claims as a defense that its activities were exempt under section 110(5) did not have reasonable grounds to believe that its use of a copyrighted work was exempt under such section, the plaintiff shall be entitled to, in addition to any award of damages under this section, an additional award of two times the amount of the license fee that the proprietor of the establishment concerned should have paid the plaintiff for such use during the preceding period of up to 3 years.

§ 505 Remedies for Infringement: Costs and Attorney's Fees

In any civil action under this title, the court in its discretion may allow the recovery of full costs by or against any party other than the United States or an officer thereof. Except as otherwise provided by this title, the court may also award a reasonable attorney's fee to the prevailing party as part of the costs.

§ 506 Criminal Offenses

(a) Criminal Infringement.—

(1) In general.—Any person who willfully infringes a copyright shall be punished as provided under section 2319 of title 18, if the infringement was committed—

(A) for purposes of commercial advantage or private financial gain;

(B) by the reproduction or distribution, including by electronic means, during any 180-day period, of 1 or more copies or phonorecords of 1 or more copyrighted works, which have a total retail value of more than $1,000; or

(C) by the distribution of a work being prepared for commercial distribution, by making it available on a computer network accessible to members of the public, if such person knew or should have known that the work was intended for commercial distribution.

(2) Evidence.—For purposes of this subsection, evidence of reproduction or distribution of a copyrighted work, by itself, shall not be sufficient to establish willful infringement of a copyright.

(3) Definition.—In this subsection, the term "work being prepared for commercial distribution" means—

(A) a computer program, a musical work, a motion picture or other audiovisual work, or a sound recording, if, at the time of unauthorized distribution—

(i) the copyright owner has a reasonable expectation of commercial distribution; and

(ii) the copies or phonorecords of the work have not been commercially distributed; or

(B) a motion picture, if, at the time of unauthorized distribution, the motion picture—

(i) has been made available for viewing in a motion picture exhibition facility; and

(ii) has not been made available in copies for sale to the general public in the United States in a format intended to permit viewing outside a motion picture exhibition facility.

(b) Forfeiture, Destruction, and Restitution.—Forfeiture, destruction, and restitution relating to this section shall be subject to section 2323 of title 18, to the extent provided in that section, in addition to any other similar remedies provided by law.

(c) Fraudulent Copyright Notice.—Any person who, with fraudulent intent, places on any article a notice of copyright or words of the same purport that such person knows to be false, or who, with fraudulent intent, publicly distributes or imports for public distribution any article bearing such notice or words that such person knows to be false, shall be fined not more than $2,500.

(d) Fraudulent Removal of Copyright Notice.—Any person who, with fraudulent intent, removes or alters any notice of copyright appearing on a copy of a copyrighted work shall be fined not more than $2,500.

(e) False Representation.—Any person who knowingly makes a false representation of a material fact in the application for copyright registration provided for by section 409, or in any written statement filed in connection with the application, shall be fined not more than $2,500.

(f) Rights of Attribution and Integrity.—Nothing in this section applies to infringement of the rights conferred by section 106A(a).

§ 507 Limitations on Actions

(a) Criminal Proceedings.—Except as expressly provided otherwise in this title, no criminal proceeding shall be maintained under the provisions of this title unless it is commenced within 5 years after the cause of action arose.

(b) Civil Actions.—No civil action shall be maintained under the provisions of this title unless it is commenced within three years after the claim accrued.

§ 508 Notification of Filing and Determination of Actions

(a) Within one month after the filing of any action under this title, the clerks of the courts of the United States shall send written notification to the Register of Copyrights setting forth, as far as is shown by the papers filed in the court, the names and addresses of the parties and the title, author, and registration number of each work involved in the action. If any other copyrighted work is later included in the action by amendment, answer, or other pleading, the clerk shall also send a notification concerning it to the Register within one month after the pleading is filed.

(b) Within one month after any final order or judgment is issued in the case, the clerk of the court shall notify the Register of it, sending with the notification a copy of the order or judgment together with the written opinion, if any, of the court.

(c) Upon receiving the notifications specified in this section, the Register shall make them a part of the public records of the Copyright Office.

§ 509 [Repealed]

§ 510 Remedies for Alteration of Programming by Cable Systems

(a) In any action filed pursuant to section 111(c)(3), the following remedies shall be available:

(1) Where an action is brought by a party identified in subsections (b) or (c) of section 501, the remedies provided by sections 502 through 505, and the remedy provided by subsection (b) of this section; and

(2) When an action is brought by a party identified in subsection (d) of section 501, the remedies provided by sections 502 and 505, together with any actual damages suffered by such party as a result of the infringement, and the remedy provided by subsection (b) of this section.

(b) In any action filed pursuant to section 111(c)(3), the court may decree that, for a period not to exceed thirty days, the cable system shall be deprived of the benefit of a statutory license for one or more distant signals carried by such cable system.

§ 511 Liability of States, Instrumentalities of States, and State Officials for Infringement of Copyright

(a) In General.—Any State, any instrumentality of a State, and any officer or employee of a State or instrumentality of a State acting in his or her official capacity, shall not be immune, under the Eleventh Amendment of the Constitution of the United States or under any other doctrine of sovereign immunity, from suit in Federal Court by any person, including any governmental or nongovernmental entity, for a violation of any of the exclusive rights of a copyright owner provided by sections 106 through 122, for importing copies of phonorecords in violation of section 602, or for any other violation under this title.

(b) Remedies.—In a suit described in subsection (a) for a violation described in that subsection, remedies (including remedies both at law and in equity) are available for the violation to the same extent as such remedies are available for such a violation in a suit against any public or private entity other than a State, instrumentality of a State, or officer or employee of a State acting in his or her official capacity. Such remedies include impounding and disposition of infringing articles under section 503, actual damages and profits and statutory damages under section 504, costs and attorney's fees under section 505, and the remedies provided in section 510.

§ 512 Limitations on Liability Relating to Material Online

(a) Transitory Digital Network Communications.—A service provider shall not be liable for monetary relief, or, except as provided in subsection (j), for injunctive or other equitable relief, for infringement of copyright by reason of the provider's transmitting, routing, or providing connections for, material through a system or network controlled or operated by or for the service provider, or by reason of the intermediate and transient storage of that material in the course of such transmitting, routing, or providing connections, if—

(1) the transmission of the material was initiated by or at the direction of a person other than the service provider;

(2) the transmission, routing, provision of connections, or storage is carried out through an automatic technical process without selection of the material by the service provider;

(3) the service provider does not select the recipients of the material except as an automatic response to the request of another person;

(4) no copy of the material made by the service provider in the course of such intermediate or transient storage is maintained on the system or network in a manner ordinarily accessible to anyone other than anticipated recipients, and no such copy is maintained on the system or network in a manner ordinarily accessible to such anticipated recipients for a longer period than is reasonably necessary for the transmission, routing, or provision of connections; and

(5) the material is transmitted through the system or network without modification of its content.

(b) System Caching.—

(1) Limitation on liability.—A service provider shall not be liable for monetary relief, or, except as provided in subsection (j), for injunctive or other equitable relief, for infringement of copyright by reason of the intermediate and temporary storage of material on a system or network controlled or operated by or for the service provider in a case in which—

(A) the material is made available online by a person other than the service provider;

(B) the material is transmitted from the person described in subparagraph (A) through the system or network to a person other than the person described in subparagraph (A) at the direction of that other person; and

(C) the storage is carried out through an automatic technical process for the purpose of making the material available to users of the system or network who, after the material is transmitted as described in subparagraph

(B), request access to the material from the person described in subparagraph (A), if the conditions set forth in paragraph (2) are met.

(2) Conditions.—The conditions referred to in paragraph (1) are that—

(A) the material described in paragraph (1) is transmitted to the subsequent users described in paragraph (1)(C) without modification to its content from the manner in which the material was transmitted from the person described in paragraph (1)(A);

(B) the service provider described in paragraph (1) complies with rules concerning the refreshing, reloading, or other updating of the material when specified by the person making the material available online in accordance with a generally accepted industry standard data communications protocol for the system or network through which that person makes the material available, except that this subparagraph applies only if those rules are not used by the person described in paragraph (1)(A) to prevent or unreasonably impair the intermediate storage to which this subsection applies;

(C) the service provider does not interfere with the ability of technology associated with the material to return to the person described in paragraph (1)(A) the information that would have been available to that person if the material had been obtained by the subsequent users described in paragraph (1)(C) directly from that person, except that this subparagraph applies only if that technology—

(i) does not significantly interfere with the performance of the provider's system or network or with the intermediate storage of the material;

(ii) is consistent with generally accepted industry standard communications protocols; and

(iii) does not extract information from the provider's system or network other than the information that would have been available to the person described in paragraph (1)(A) if the subsequent users had gained access to the material directly from that person;

(D) if the person described in paragraph (1)(A) has in effect a condition that a person must meet prior to having access to the material, such as a condition based on payment of a fee or provision of a password or other information, the service provider permits access to the stored material in significant part only to users of its system or network that have met those conditions and only in accordance with those conditions; and

(E) if the person described in paragraph (1)(A) makes that material available online without the authorization of the copyright owner of the material, the service provider responds expeditiously to remove, or disable access to, the material that is claimed to be infringing upon notification of claimed infringement as described in subsection (c)(3), except that this subparagraph applies only if—

(i) the material has previously been removed from the originating site or access to it has been disabled, or a court has ordered that the material be removed from the originating site or that access to the material on the originating site be disabled; and

(ii) the party giving the notification includes in the notification a statement confirming that the material has been removed from the originating site or access to it has been disabled or that a court has ordered that the material be removed from the originating site or that access to the material on the originating site be disabled.

(c) Information Residing on Systems or Networks at Direction of Users.—

(1) In general.—A service provider shall not be liable for monetary relief, or, except as provided in subsection (j), for injunctive or other equitable relief, for infringement of copyright by reason of the storage at the direction of a user of material that resides on a system or network controlled or operated by or for the service provider, if the service provider—

(A)(i) does not have actual knowledge that the material or an activity using the material on the system or network is infringing;

(ii) in the absence of such actual knowledge, is not aware of facts or circumstances from which infringing activity is apparent; or

(iii) upon obtaining such knowledge or awareness, acts expeditiously to remove, or disable access to, the material;

(B) does not receive a financial benefit directly attributable to the infringing activity, in a case in which the service provider has the right and ability to control such activity; and

(C) upon notification of claimed infringement as described in paragraph (3), responds expeditiously to remove, or disable access to, the material that is claimed to be infringing or to be the subject of infringing activity.

(2) Designated agent.—The limitations on liability established in this subsection apply to a service provider only if the service provider has designated an agent to receive notifications of claimed infringement described in paragraph (3), by making available through its service, including on its website in a location accessible to the public, and by providing to the Copyright Office, substantially the following information:

(A) the name, address, phone number, and electronic mail address of the agent.

(B) other contact information which the Register of Copyrights may deem appropriate.

The Register of Copyrights shall maintain a current directory of agents available to the public for inspection, including through the Internet, and may require payment of a fee by service providers to cover the costs of maintaining the directory.

(3) Elements of notification.—

(A) To be effective under this subsection, a notification of claimed infringement must be a written communication provided to the designated agent of a service provider that includes substantially the following:

(i) A physical or electronic signature of a person authorized to act on behalf of the owner of an exclusive right that is allegedly infringed.

(ii) Identification of the copyrighted work claimed to have been infringed, or, if multiple copyrighted works at a single online site are covered by a single notification, a representative list of such works at that site.

(iii) Identification of the material that is claimed to be infringing or to be the subject of infringing activity and that is to be removed or access to which is to be disabled, and information reasonably sufficient to permit the service provider to locate the material.

(iv) Information reasonably sufficient to permit the service provider to contact the complaining party, such as an address, telephone number, and, if available, an electronic mail address at which the complaining party may be contacted.

(v) A statement that the complaining party has a good faith belief that use of the material in the manner complained of is not authorized by the copyright owner, its agent, or the law.

(vi) A statement that the information in the notification is accurate, and under penalty of perjury, that the complaining party is authorized to act on behalf of the owner of an exclusive right that is allegedly infringed.

(B)(i) Subject to clause (ii), a notification from a copyright owner or from a person authorized to act on behalf of the copyright owner that fails to comply substantially with the provisions of subparagraph (A) shall not be considered under paragraph (1)(A) in determining whether a service provider has actual knowledge or is aware of facts or circumstances from which infringing activity is apparent.

(ii) In a case in which the notification that is provided to the service provider's designated agent fails to comply substantially with all the provisions of subparagraph (A) but substantially complies with clauses (ii), (iii), and (iv) of subparagraph (A), clause (i) of this subparagraph applies only if the service provider promptly attempts to contact the person making the notification or takes other reasonable steps to assist in the receipt of notification that substantially complies with all the provisions of subparagraph (A).

(d) Information Location Tools.—A service provider shall not be liable for monetary relief, or, except as provided in subsection (j), for injunctive or other equitable relief, for infringement of copyright by reason of the provider referring or linking users to an online location containing infringing material or infringing activity, by using information location tools, including a directory, index, reference, pointer, or hypertext link, if the service provider—

(1) (A) does not have actual knowledge that the material or activity is infringing;

(B) in the absence of such actual knowledge, is not aware of facts or circumstances from which infringing activity is apparent; or

(C) upon obtaining such knowledge or awareness, acts expeditiously to remove, or disable access to, the material;

(2) does not receive a financial benefit directly attributable to the infringing activity, in a case in which the service provider has the right and ability to control such activity; and

(3) upon notification of claimed infringement as described in subsection (c)(3), responds expeditiously to remove, or disable access to, the material that is claimed to be infringing or to be the subject of infringing activity, except that, for purposes of this paragraph, the information described in subsection (c)(3)(A)(iii) shall be identification of the reference or link, to material or activity claimed to be infringing, that is to be removed or access to which is to be disabled, and information reasonably sufficient to permit the service provider to locate that reference or link.

(e) Limitation on Liability of Nonprofit Educational Institutions.—

(1) When a public or other nonprofit institution of higher education is a service provider, and when a faculty member or graduate student who is an employee of such institution is performing a teaching or research function, for the purposes of subsections (a) and (b) such faculty member or graduate student shall be considered to be a person other than the institution, and for the purposes of subsections (c) and (d) such faculty member's or graduate student's knowledge or awareness of his or her infringing activities shall not be attributed to the institution, if—

(A) such faculty member's or graduate student's infringing activities do not involve the provision of online access to instructional materials that are or were required or recommended, within the preceding 3-year period, for a course taught at the institution by such faculty member or graduate student;

(B) the institution has not, within the preceding 3-year period, received more than 2 notifications described in subsection (c)(3) of claimed infringement by such faculty member or graduate student, and such notifications of claimed infringement were not actionable under subsection (f); and

(C) the institution provides to all users of its system or network informational materials that accurately describe, and promote compliance with, the laws of the United States relating to copyright.

(2) For the purposes of this subsection, the limitations on injunctive relief contained in subsections (j)(2) and (j)(3), but not those in (j)(1), shall apply.

(f) Misrepresentations.—Any person who knowingly materially misrepresents under this section—

(1) that material or activity is infringing, or

(2) that material or activity was removed or disabled by mistake or misidentification, shall be liable for any damages, including costs and attorneys' fees, incurred by the alleged infringer, by any copyright owner or copyright owner's authorized licensee, or by a service provider, who is injured by such misrepresentation, as the result of the service provider relying upon such misrepresentation in removing or disabling access to the material or activity claimed to be infringing, or in replacing the removed material or ceasing to disable access to it.

(g) Replacement of Removed or Disabled Material and Limitation on Other Liability.—

(1) No liability for taking down generally.—Subject to paragraph (2), a service provider shall not be liable to any person for any claim based on the service provider's good faith disabling of access to, or removal of, material or activity claimed to be infringing or based on facts or circumstances from which infringing activity is apparent, regardless of whether the material or activity is ultimately determined to be infringing.

(2) Exception.—Paragraph (1) shall not apply with respect to material residing at the direction of a subscriber of the service provider on a system or network controlled or operated by or for the service provider that is removed, or to which access is disabled by the service provider, pursuant to a notice provided under subsection (c)(1)(C), unless the service provider—

(A) takes reasonable steps promptly to notify the subscriber that it has removed or disabled access to the material;

(B) upon receipt of a counter notification described in paragraph (3), promptly provides the person who provided the notification under subsection (c)(1)(C) with a copy of the counter notification, and informs that person that it will replace the removed material or cease disabling access to it in 10 business days; and

(C) replaces the removed material and ceases disabling access to it not less than 10, nor more than 14, business days following receipt of the counter notice, unless its designated agent first receives notice from the person who submitted the notification under subsection (c)(1)(C) that such person has filed an action seeking a court order to restrain the subscriber from engaging in infringing activity relating to the material on the service provider's system or network.

(3) Contents of counter notification.—To be effective under this subsection, a counter notification must be a written communication provided to the service provider's designated agent that includes substantially the following:

(A) A physical or electronic signature of the subscriber.

(B) Identification of the material that has been removed or to which access has been disabled and the location at which the material appeared before it was removed or access to it was disabled.

(C) A statement under penalty of perjury that the subscriber has a good faith belief that the material was removed or disabled as a result of mistake or misidentification of the material to be removed or disabled.

(D) The subscriber's name, address, and telephone number, and a statement that the subscriber consents to the jurisdiction of Federal District Court for the judicial district in which the address is located, or if the subscriber's address is outside of the United States, for any judicial district in which the service provider may be found, and that the subscriber will accept service of process from the person who provided notification under subsection (c)(1)(C) or an agent of such person.

(4) Limitation on other liability.—A service provider's compliance with paragraph (2) shall not subject the service provider to liability for copyright infringement with respect to the material identified in the notice provided under subsection (c)(1)(C).

(h) Subpoena to Identify Infringer.—

(1) Request.—A copyright owner or a person authorized to act on the owner's behalf may request the clerk of any United States district court to issue a subpoena to a service provider for identification of an alleged infringer in accordance with this subsection.

(2) Contents of request.—The request may be made by filing with the clerk—

(A) a copy of a notification described in subsection (c)(3)(A);

(B) a proposed subpoena; and

(C) a sworn declaration to the effect that the purpose for which the subpoena is sought is to obtain the identity of an alleged infringer and that such information will only be used for the purpose of protecting rights under this title.

(3) Contents of subpoena.—The subpoena shall authorize and order the service provider receiving the notification and the subpoena to expeditiously disclose to the copyright owner or person authorized by the copyright owner information sufficient to identify the alleged infringer of the material described in the notification to the extent such information is available to the service provider.

(4) Basis for granting subpoena.—If the notification filed satisfies the provisions of subsection (c)(3)(A), the proposed subpoena is in proper form, and the accompanying declaration is properly executed, the clerk shall expeditiously

issue and sign the proposed subpoena and return it to the requester for delivery to the service provider.

(5) Actions of service provider receiving subpoena.—Upon receipt of the issued subpoena, either accompanying or subsequent to the receipt of a notification described in subsection (c)(3)(A), the service provider shall expeditiously disclose to the copyright owner or person authorized by the copyright owner the information required by the subpoena, notwithstanding any other provision of law and regardless of whether the service provider responds to the notification.

(6) Rules applicable to subpoena.—Unless otherwise provided by this section or by applicable rules of the court, the procedure for issuance and delivery of the subpoena, and the remedies for noncompliance with the subpoena, shall be governed to the greatest extent practicable by those provisions of the Federal Rules of Civil Procedure governing the issuance, service, and enforcement of a subpoena *duces tecum.*

(i) Conditions for Eligibility.—

(1) Accommodation of technology.—The limitations on liability established by this section shall apply to a service provider only if the service provider—

(A) has adopted and reasonably implemented, and informs subscribers and account holders of the service provider's system or network of, a policy that provides for the termination in appropriate circumstances of subscribers and account holders of the service provider's system or network who are repeat infringers; and

(B) accommodates and does not interfere with standard technical measures.

(2) Definition.—As used in this subsection, the term "standard technical measures" means technical measures that are used by copyright owners to identify or protect copyrighted works and—

(A) have been developed pursuant to a broad consensus of copyright owners and service providers in an open, fair, voluntary, multi-industry standards process;

(B) are available to any person on reasonable and nondiscriminatory terms; and

(C) do not impose substantial costs on service providers or substantial burdens on their systems or networks.

(j) Injunctions.—The following rules shall apply in the case of any application for an injunction under section 502 against a service provider that is not subject to monetary remedies under this section:

(1) Scope of relief.—(A) With respect to conduct other than that which qualifies for the limitation on remedies set forth in subsection (a), the court may grant injunctive relief with respect to a service provider only in one or more of the following forms:

(i) An order restraining the service provider from providing access to infringing material or activity residing at a particular online site on the provider's system or network.

(ii) An order restraining the service provider from providing access to a subscriber or account holder of the service provider's system or network who is engaging in infringing activity and is identified in the order, by terminating the accounts of the subscriber or account holder that are specified in the order.

(iii) Such other injunctive relief as the court may consider necessary to prevent or restrain infringement of copyrighted material specified in the order of the court at a particular online location, if such relief is the least burdensome to the service provider among the forms of relief comparably effective for that purpose.

(B) If the service provider qualifies for the limitation on remedies described in subsection (a), the court may only grant injunctive relief in one or both of the following forms:

(i) An order restraining the service provider from providing access to a subscriber or account holder of the service provider's system or network who is using the provider's service to engage in infringing activity and is identified in the order, by terminating the accounts of the subscriber or account holder that are specified in the order.

(ii) An order restraining the service provider from providing access, by taking reasonable steps specified in the order to block access, to a specific, identified, online location outside the United States.

(2) Considerations.—The court, in considering the relevant criteria for injunctive relief under applicable law, shall consider—

(A) whether such an injunction, either alone or in combination with other such injunctions issued against the same service provider under this subsection, would significantly burden either the provider or the operation of the provider's system or network;

(B) the magnitude of the harm likely to be suffered by the copyright owner in the digital network environment if steps are not taken to prevent or restrain the infringement;

(C) whether implementation of such an injunction would be technically feasible and effective, and would not interfere with access to noninfringing material at other online locations; and

(D) whether other less burdensome and comparably effective means of preventing or restraining access to the infringing material are available.

(3) Notice and ex parte orders.—Injunctive relief under this subsection shall be available only after notice to the service provider and an opportunity for the service provider to appear are provided, except for orders ensuring the preservation of evidence or other orders having no material adverse effect on the operation of the service provider's communications network.

(k) Definitions.—

(1) Service provider.—(A) As used in subsection (a), the term "service provider" means an entity offering the transmission, routing, or providing of connections for digital online communications, between or among points specified by a user, of material of the user's choosing, without modification to the content of the material as sent or received.

(B) As used in this section, other than subsection (a), the term "service provider" means a provider of online services or network access, or the operator of facilities therefor, and includes an entity described in subparagraph (A).

(2) Monetary relief.—As used in this section, the term "monetary relief" means damages, costs, attorneys' fees, and any other form of monetary payment.

(l) Other Defenses Not Affected.—The failure of a service provider's conduct to qualify for limitation of liability under this section shall not bear adversely upon the consideration of a defense by the service provider that the service provider's conduct is not infringing under this title or any other defense.

(m) Protection of Privacy.—Nothing in this section shall be construed to condition the applicability of subsections (a) through (d) on—

(1) a service provider monitoring its service or affirmatively seeking facts indicating infringing activity, except to the extent consistent with a standard technical measure complying with the provisions of subsection (i); or

(2) a service provider gaining access to, removing, or disabling access to material in cases in which such conduct is prohibited by law.

(n) Construction.—Subsections (a), (b), (c), and (d) describe separate and distinct functions for purposes of applying this section. Whether a service provider qualifies for the limitation on liability in any one of those subsections shall be based solely on the criteria in that subsection, and shall not affect a determination of whether that service provider qualifies for the limitations on liability under any other such subsection.

§ 513 Determination of Reasonable License Fees for Individual Proprietors

In the case of any performing rights society subject to a consent decree which provides for the determination of reasonable license rates or fees to be charged by the performing rights society, notwithstanding the provisions of that consent decree, an individual proprietor who owns or operates fewer than 7 nonpublicly traded establishments in which nondramatic musical works are performed publicly and who claims that any license agreement offered by that performing rights society is unreasonable in its license rate or fee as to that individual proprietor, shall be entitled to determination of a reasonable license rate or fee as follows:

(1) The individual proprietor may commence such proceeding for determination of a reasonable license rate or fee by filing an application in the applicable district court under paragraph (2) that a rate disagreement exists and by serving a copy of the application on the performing rights society. Such proceeding shall commence in the applicable district court within 90 days after the service of such copy, except that such 90-day requirement shall be subject to the administrative requirements of the court.

(2) The proceeding under paragraph (1) shall be held, at the individual proprietor's election, in the judicial district of the district court with jurisdiction over the applicable consent decree or in that place of holding court of a district

court that is the seat of the Federal circuit (other than the Court of Appeals for the Federal Circuit) in which the proprietor's establishment is located.

(3) Such proceeding shall be held before the judge of the court with jurisdiction over the consent decree governing the performing rights society. At the discretion of the court, the proceeding shall be held before a special master or magistrate judge appointed by such judge. Should that consent decree provide for the appointment of an advisor or advisors to the court for any purpose, any such advisor shall be the special master so named by the court.

(4) In any such proceeding, the industry rate shall be presumed to have been reasonable at the time it was agreed to or determined by the court. Such presumption shall in no way affect a determination of whether the rate is being correctly applied to the individual proprietor.

(5) Pending the completion of such proceeding, the individual proprietor shall have the right to perform publicly the copyrighted musical compositions in the repertoire of the performing rights society by paying an interim license rate or fee into an interest bearing escrow account with the clerk of the court, subject to retroactive adjustment when a final rate or fee has been determined, in an amount equal to the industry rate, or, in the absence of an industry rate, the amount of the most recent license rate or fee agreed to by the parties.

(6) Any decision rendered in such proceeding by a special master or magistrate judge named under paragraph (3) shall be reviewed by the judge of the court with jurisdiction over the consent decree governing the performing rights society. Such proceeding, including such review, shall be concluded within 6 months after its commencement.

(7) Any such final determination shall be binding only as to the individual proprietor commencing the proceeding, and shall not be applicable to any other proprietor or any other performing rights society, and the performing rights society shall be relieved of any obligation of nondiscrimination among similarly situated music users that may be imposed by the consent decree governing its operations.

(8) An individual proprietor may not bring more than one proceeding provided for in this section for the determination of a reasonable license rate or fee under any license agreement with respect to any one performing rights society.

(9) For purposes of this section, the term "industry rate" means the license fee a performing rights society has agreed to with, or which has been determined by the court for, a significant segment of the music user industry to which the individual proprietor belongs.

Chapter 5 Notes

1. See note 8, *infra.*

2. In 1998, two sections 512 were enacted into law. First, on October 17, 1998, the Fairness in Music Licensing Act of 1998 was enacted. This Act amended chapter 5 to add section 512 entitled "Determination of reasonable license fees for individual

proprietors." Pub. L. No. 105-298, 112 Stat. 2827, 2831. Second, on October 28, 1998, the Online Copyright Infringement Liability Limitation Act was enacted. This Act amended chapter 5 to add section 512 entitled "Limitations on liability relating to material online." Pub. L. No. 105-304, 112 Stat. 2860, 2877. Consequently, in 1999, a technical correction was enacted to redesignate the section 512 that was entitled "Determination of reasonable license fees for individual proprietors" as section 513. Also, the table of sections was amended to reflect that change. Pub. L. No. 106-44, 113 Stat. 221. See also note 12, *infra.*

3. The Berne Convention Implementation Act of 1988 amended section 501(b) by striking out "sections 205(d) and 411" and inserting in lieu thereof "section 411." Pub. L. No. 100-568, 102 Stat. 2853, 2860. The Satellite Home Viewer Act of 1988 amended section 501 by adding subsection (e). Pub. L. No. 100-667, 102 Stat. 3935, 3957.

In 1990, the Copyright Remedy Clarification Act amended section 501(a) by adding the last two sentences. Pub. L. No. 101-553, 104 Stat. 2749. The Visual Artists Rights Act of 1990 also amended section 501(a) as follows: 1) by inserting "or of the author as provided in section 106A(a)" after "118" and 2) by striking out "copyright." and inserting in lieu thereof "copyright or right of the author, as the case may be. For purposes of this chapter (other than section 506), any reference to copyright shall be deemed to include the rights conferred by section 106A(a)." Pub. L. No. 101-650, 104 Stat. 5089, 5131.

In 1999, a technical correction amended the first sentence in subsection 501(a) by inserting "121" in lieu of "118." Pub. L. No. 106-44, 113 Stat. 221, 222. The Satellite Home Viewer Improvement Act of 1999 amended section 501 by adding a subsection (f) and, in subsection

(e), by inserting "performance or display of a work embodied in a primary transmission" in lieu of "primary transmission embodying the performance or display of a work." Pub. L. No. 106-113, 113 Stat. 1501, app. I at 1501A-527 and 544. The Satellite Home Viewer Improvement Act of 1999 states that section 501(f) shall be effective as of July 1, 1999. Pub. L. No. 106-113, 113 Stat. 1501, app. I at 1501A-544.

The Intellectual Property and High Technology Technical Amendments Act of 2002 amended section 501(a) by substituting sections "106 through 122" for "106 through 121." Pub. L. No. 107-273, 116 Stat. 1758, 1909.

4. The Prioritizing Resources and Organization for Intellectual Property Act of 2008 amended section 503 by revising subsection (a) in its entirety. Pub. L. No. 110-403, 122 Stat. 4256, 4258. A technical correction in the Copyright Cleanup, Clarification, and Corrections Act of 2010 amended section 503(a)(1)(B) to substitute "copies or phonorecords" for "copies of phonorecords." Pub. L. No. 111-295, 124 Stat. 3180, 3181.

5. The Berne Convention Implementation Act of 1988 amended section 504(c) as follows: 1) in paragraph (1), by inserting "$500" in lieu of "$250" and by inserting "$20,000" in lieu of "$10,000" and 2) in paragraph (2), by inserting "$100,000" in lieu of "$50,000" and by inserting "$200" in lieu of "$100." Pub. L. No. 100-568, 102 Stat. 2853, 2860. The Digital Theft Deterrence and Copyright Damages Improvement Act of 1999 amended section 504(c), in paragraph (1), by substituting "$750"

for "$500" and "$30,000" for "$20,000" and, in paragraph (2), by substituting "$150,000" for "$100,000." Pub. L. No. 106-160, 113 Stat. 1774.

The Fraudulent Online Identity Sanctions Act of 2004 amended section 504(c) by adding a new subparagraph (3). Pub. L. No. 108-482, 118 Stat. 3912, 3916. The Copyright Cleanup, Clarification, and Corrections Act of 2010 amended paragraph 504(c)(2) to change the reference from "subsection (g) of section 118" to "section 118(f)." Pub. L. No. 111-295, 124 Stat. 3180, 3181.

6. The Piracy and Counterfeiting Amendments Act of 1982 amended section 506 by substituting a new subsection (a). Pub. L. No. 97-180, 96 Stat. 91, 93. The Visual Artists Rights Act of 1990 amended section 506 by adding subsection (f). Pub. L. No.101-650, 104 Stat. 5089, 5131. In 1997, the No Electronic Theft (NET) Act again amended section 506 by amending subsection (a) in its entirety. Pub. L. No. 105-147, 111 Stat. 2678. That Act also directed the United States Sentencing Commission to "ensure that the applicable guideline range for a defendant convicted of a crime against intellectual property . . . is sufficiently stringent to deter such a crime" and to "ensure that the guidelines provide for consideration of the retail value and quantity of the items with respect to which the crime against intellectual property was committed." Pub. L. No. 105-147, 111 Stat. 2678, 2680. See also note 2 in Appendix H.

The Artists' Rights and Theft Prevention Act of 2005 amended subsection 506(a) in its entirety. Pub. L. No. 109-9, 119 Stat. 218, 220.

The Prioritizing Resources and Organization for Intellectual Property Act of 2008 amended section 506 by revising subsection (b) in its entirety. Pub. L. No. 110-403, 122 Stat. 4256, 4260.

7. In 1997, the No Electronic Theft (NET) Act amended section 507(a) by inserting "5" in lieu of "three." Pub. L. No. 105-147, 111 Stat. 2678.

8. The Prioritizing Resources and Organization for Intellectual Property Act of 2008 repealed section 509. Pub. L. No. 110-403, 122 Stat. 4256, 4260. In lieu of this provision, refer to section 2323, chapter 113 of title 18, *United States Code*, entitled, "Forfeiture, Destruction and Restitution." Section 2323 is included in Appendix H to this volume.

9. The Satellite Home Viewer Improvement Act of 1999 amended the heading for section 510 by substituting "programming" for "programing" and, in subsection (b), by substituting "statutory" for "compulsory." Pub. L. No. 106-113, 113 Stat. 1501, app. I at 1501A-543.

10. In 1990, the Copyright Remedy Clarification Act added section 511. Pub. L. No. 101553, 104 Stat. 2749. In 1999, a technical correction amended subsection 511(a) by inserting "121" in lieu of "119." Pub. L. No. 106-44, 113 Stat. 221, 222. The Intellectual Property and High Technology Technical Amendments Act of 2002 amended section 511(a) by substituting sections "106 through 122" for "106 through 121." Pub. L. No. 107-273, 116 Stat. 1758, 1909.

11. In 1998, the Online Copyright Infringement Liability Limitation Act added section 512. Pub. L. No. 105-304, 112 Stat. 2860, 2877. In 1999, a technical correction deleted the heading for paragraph (2) of section 512(e), which was "Injunctions." Pub. L. No. 106-44, 113 Stat. 221, 222. The Copyright Cleanup, Clarification, and Corrections Act of 2010 amended the last sentence in paragraph 512(c)(2)(B)

by deleting "in both electronic and hard copy formats" after "through the Internet." Pub. L. No. 111-295, 124 Stat. 3180, 3180.

12. The Fairness in Music Licensing Act of 1998 added section 513. Pub. L. No. 105-298, 112 Stat. 2827, 2831. This section was originally designated as section 512. However, because two sections 512 had been enacted into law in 1998, a technical amendment redesignated this as section 513. Pub. L. No. 106-44, 113 Stat. 221. See also note 2, *supra.*

Chapter 6 Importation and Exportation

§ 601 [Repealed]

§ 602 Infringing Importation or Exportation of Copies or Phonorecords

(a) Infringing Importation or Exportation.—

(1) Importation.—Importation into the United States, without the authority of the owner of copyright under this title, of copies or phonorecords of a work that have been acquired outside the United States is an infringement of the exclusive right to distribute copies or phonorecords under section 106, actionable under section 501.

(2) Importation or exportation of infringing items.—Importation into the United States or exportation from the United States, without the authority of the owner of copyright under this title, of copies or phonorecords, the making of which either constituted an infringement of copyright, or which would have constituted an infringement of copyright if this title had been applicable, is an infringement of the exclusive right to distribute copies or phonorecords under section 106, actionable under sections 501 and 506.

(3) Exceptions.—This subsection does not apply to—

(A) importation or exportation of copies or phonorecords under the authority or for the use of the Government of the United States or of any State or political subdivision of a State, but not including copies or phonorecords for use in schools, or copies of any audiovisual work imported for purposes other than archival use;

(B) importation or exportation, for the private use of the importer or exporter and not for distribution, by any person with respect to no more than one copy or phonorecord of any one work at any one time, or by any person arriving from outside the United States or departing from the United States with respect to copies or phonorecords forming part of such person's personal baggage; or

(C) importation by or for an organization operated for scholarly, educational, or religious purposes and not for private gain, with respect to no more than one copy of an audiovisual work solely for its archival purposes, and no more than five copies or phonorecords of any other work for its library lending or archival purposes, unless the importation of such copies or phonorecords is part of an activity consisting of systematic reproduction or distribution, engaged in by such organization in violation of the provisions of section 108(g)(2).

(b) Import Prohibition.—In a case where the making of the copies or phonorecords would have constituted an infringement of copyright if this title had been applicable, their importation is prohibited. In a case where the copies or phonorecords were lawfully made, the United States Customs and Border Protection Service has no authority to prevent their importation. In either case, the Secretary of the Treasury is authorized to prescribe, by regulation, a procedure under which any person claiming an interest in the copyright in a particular work may, upon payment of a specified fee, be entitled to notification by the United States Customs and Border Protection Service of the importation of articles that appear to be copies or phonorecords of the work.

§ 603 Importation Prohibitions: Enforcement and Disposition of Excluded Articles

(a) The Secretary of the Treasury and the United States Postal Service shall separately or jointly make regulations for the enforcement of the provisions of this title prohibiting importation.

(b) These regulations may require, as a condition for the exclusion of articles under section 602—

(1) that the person seeking exclusion obtain a court order enjoining importation of the articles; or

(2) that the person seeking exclusion furnish proof, of a specified nature and in accordance with prescribed procedures, that the copyright in which such person claims an interest is valid and that the importation would violate the prohibition in section 602; the person seeking exclusion may also be required to post a surety bond for any injury that may result if the detention or exclusion of the articles proves to be unjustified.

(c) Articles imported in violation of the importation prohibitions of this title are subject to seizure and forfeiture in the same manner as property imported in violation of the customs revenue laws. Forfeited articles shall be destroyed as directed by the Secretary of the Treasury or the court, as the case may be.

Chapter 6 Notes

1. The Prioritizing Resources and Organization for Intellectual Property Act of 2008 amended the title of Chapter 6 to add "Exportation" to it, so that now the title is "Manufacturing Requirements, Importation, and Exportation." Pub. L. No. 110-403, 122 Stat. 4256, 4259. The Copyright Cleanup, Clarification, and Corrections

Act of 2010 amended chapter six by repealing section 601, entitled "Manufacture, importation, and public distribution of certain copies." Pub. L. No. 111-295, 124 Stat. 3180. It also deleted "Manufacturing Requirements" from the beginning of the title for chapter six. *Id.*

2. See note 4, *infra.*

3. The Prioritizing Resources and Organization for Intellectual Property Act of 2008 amended the title of section 602 in the table of contents by adding "Exportation" to it, so that the new title is "Infringing importation or exportation of copies or phonorecords." Pub. L. No. 110-403, 122 Stat. 4256, 4260.

4. In 1982, section 601(a) was amended in the first sentence by substituting "1986" for "1982." Pub. L. No. 97-215, 96 Stat. 178. The Prioritizing Resources and Organization for Intellectual Property Act of 2008 amended the first sentence of subpart 601(b)(2) by inserting "and Border Protection" after "United States Customs." Pub. L. No. 110-403, 122 Stat. 4256, 4260. The Copyright Cleanup, Clarification, and Corrections Act of 2010 repealed section 601. Pub. L. No. 111-295, 124 Stat. 3180.

5. The Prioritizing Resources and Organization for Intellectual Property Act of 2008 amended the title of section 602 by adding "Exportation" to it, so the new title is "Infringing importation or exportation of copies or phonorecords." Pub. L. No. 110-403, 122 Stat. 4256, 4260. The Act also amended section 602(a) by dividing it into paragraphs (1), (2) and (3); by adding paragraph (2); and by adding references to exports throughout. *Id.* at 425960. It further amended subsection 602(b) by inserting the subtitle, "Import Prohibition," at the beginning and adding references to the United Stated Customs and Border Protection Service. *Id.* at 4260. The Copyright Cleanup, Clarification, and Corrections Act of 2010 made a conforming amendment to subsection 602(b) to eliminate the reference to section 601, which the Act had repealed, by deleting at the end of the second sentence "unless the provisions of section 601 are applicable." Pub. L. No. 111-295, 124 Stat. 3180, 3181.

6. The Anticounterfeiting Consumer Protection Act of 1996 amended the last sentence of section 603(c) by deleting the semicolon and all text immediately following the words "as the case may be." Pub. L. No. 104-153, 110 Stat. 1386, 1388.

Chapter 7 Copyright Office

§ 701 The Copyright Office: General Responsibilities and Organization

(a) All administrative functions and duties under this title, except as otherwise specified, are the responsibility of the Register of Copyrights as director of the Copyright Office of the Library of Congress. The Register of Copyrights, together with the subordinate officers and employees of the Copyright Office, shall be appointed by the Librarian of Congress, and shall act under the Librarian's general direction and supervision.

(b) In addition to the functions and duties set out elsewhere in this chapter, the Register of Copyrights shall perform the following functions:

(1) Advise Congress on national and international issues relating to copyright, other matters arising under this title, and related matters.

(2) Provide information and assistance to Federal departments and agencies and the Judiciary on national and international issues relating to copyright, other matters arising under this title, and related matters.

(3) Participate in meetings of international intergovernmental organizations and meetings with foreign government officials relating to copyright, other matters arising under this title, and related matters, including as a member of United States delegations as authorized by the appropriate Executive branch authority.

(4) Conduct studies and programs regarding copyright, other matters arising under this title, and related matters, the administration of the Copyright Office, or any function vested in the Copyright Office by law, including educational programs conducted cooperatively with foreign intellectual property offices and international intergovernmental organizations.

(5) Perform such other functions as Congress may direct, or as may be appropriate in furtherance of the functions and duties specifically set forth in this title.

(c) The Register of Copyrights shall adopt a seal to be used on and after January 1, 1978, to authenticate all certified documents issued by the Copyright Office.

(d) The Register of Copyrights shall make an annual report to the Librarian of Congress of the work and accomplishments of the Copyright Office during the previous fiscal year. The annual report of the Register of Copyrights shall be published separately and as a part of the annual report of the Librarian of Congress.

(e) Except as provided by section 706(b) and the regulations issued thereunder, all actions taken by the Register of Copyrights under this title are subject to the provisions of the Administrative Procedure Act of June 11, 1946, as amended (c. 324, 60 Stat. 237, title 5, United States Code, Chapter 5, Subchapter II and Chapter 7).

(f) The Register of Copyrights shall be compensated at the rate of pay in effect for level III of the Executive Schedule under section 5314 of title 5.3. The Librarian of Congress shall establish not more than four positions for Associate Registers of Copyrights, in accordance with the recommendations of the Register of Copyrights. The Librarian shall make appointments to such positions after consultation with the Register of Copyrights. Each Associate Register of Copyrights shall be paid at a rate not to exceed the maximum annual rate of basic pay payable for GS-18 of the General Schedule under section 5332 of title 5.

§ 702 Copyright Office Regulations

The Register of Copyrights is authorized to establish regulations not inconsistent with law for the administration of the functions and duties made the responsibility of the Register under this title. All regulations established by the Register under this title are subject to the approval of the Librarian of Congress.

§ 703 Effective Date of Actions in Copyright Office

In any case in which time limits are prescribed under this title for the performance of an action in the Copyright Office, and in which the last day of the prescribed period falls on a Saturday, Sunday, holiday, or other nonbusiness day within the District of Columbia or the Federal Government, the action may be taken on the next succeeding business day, and is effective as of the date when the period expired.

§ 704 Retention and Disposition of Articles Deposited in Copyright Office

(a) Upon their deposit in the Copyright Office under sections 407 and 408, all copies, phonorecords, and identifying material, including those deposited in connection with claims that have been refused registration, are the property of the United States Government.

(b) In the case of published works, all copies, phonorecords, and identifying material deposited are available to the Library of Congress for its collections, or for exchange or transfer to any other library. In the case of unpublished works, the Library is entitled, under regulations that the Register of Copyrights shall prescribe, to select any deposits for its collections or for transfer to the National Archives of the United States or to a Federal records center, as defined in section 2901 of title 44.

(c) The Register of Copyrights is authorized, for specific or general categories of works, to make a facsimile reproduction of all or any part of the material deposited under section 408, and to make such reproduction a part of the Copyright Office records of the registration, before transferring such material to the Library of Congress as provided by subsection (b), or before destroying or otherwise disposing of such material as provided by subsection (d).

(d) Deposits not selected by the Library under subsection (b), or identifying portions or reproductions of them, shall be retained under the control of the Copyright Office, including retention in Government storage facilities, for the longest period considered practicable and desirable by the Register of Copyrights and the

Librarian of Congress. After that period it is within the joint discretion of the Register and the Librarian to order their destruction or other disposition; but, in the case of unpublished works, no deposit shall be knowingly or intentionally destroyed or otherwise disposed of during its term of copyright unless a facsimile reproduction of the entire deposit has been made a part of the Copyright Office records as provided by subsection (c).

(e) The depositor of copies, phonorecords, or identifying material under section 408, or the copyright owner of record, may request retention, under the control of the Copyright Office, of one or more of such articles for the full term of copyright in the work. The Register of Copyrights shall prescribe, by regulation, the conditions under which such requests are to be made and granted, and shall fix the fee to be charged under section 708(a) if the request is granted.

§ 705 Copyright Office Records: Preparation, Maintenance, Public Inspection, and Searching

(a) The Register of Copyrights shall ensure that records of deposits, registrations, recordations, and other actions taken under this title are maintained, and that indexes of such records are prepared.

(b) Such records and indexes, as well as the articles deposited in connection with completed copyright registrations and retained under the control of the Copyright Office, shall be open to public inspection.

(c) Upon request and payment of the fee specified by section 708, the Copyright Office shall make a search of its public records, indexes, and deposits, and shall furnish a report of the information they disclose with respect to any particular deposits, registrations, or recorded documents.

§ 706 Copies of Copyright Office Records

(a) Copies may be made of any public records or indexes of the Copyright Office; additional certificates of copyright registration and copies of any public records or indexes may be furnished upon request and payment of the fees specified by section 708.

(b) Copies or reproductions of deposited articles retained under the control of the Copyright Office shall be authorized or furnished only under the conditions specified by the Copyright Office regulations.

§ 707 Copyright Office Forms and Publications

(a) Catalog of Copyright Entries.—The Register of Copyrights shall compile and publish at periodic intervals catalogs of all copyright registrations. These catalogs shall be divided into parts in accordance with the various classes of works, and the Register has discretion to determine, on the basis of practicability and usefulness, the form and frequency of publication of each particular part.

(b) Other Publications.—The Register shall furnish, free of charge upon request, application forms for copyright registration and general informational material in connection with the functions of the Copyright Office. The Register also has the authority to publish compilations of information, bibliographies, and other material he or she considers to be of value to the public.

(c) Distribution of Publications.—All publications of the Copyright Office shall be furnished to depository libraries as specified under section 1905 of title 44, and, aside from those furnished free of charge, shall be offered for sale to the public at prices based on the cost of reproduction and distribution.

§ 708 Copyright Office Fees

(a) Fees.—Fees shall be paid to the Register of Copyrights—

(1) on filing each application under section 408 for registration of a copyright claim or for a supplementary registration, including the issuance of a certificate of registration if registration is made;

(2) on filing each application for registration of a claim for renewal of a subsisting copyright under section 304(a), including the issuance of a certificate of registration if registration is made;

(3) for the issuance of a receipt for a deposit under section 407;

(4) for the recordation, as provided by section 205, of a transfer of copyright ownership or other document;

(5) for the filing, under section 115(b), of a notice of intention to obtain a compulsory license;

(6) for the recordation, under section 302(c), of a statement revealing the identity of an author of an anonymous or pseudonymous work, or for the recordation, under section 302(d), of a statement relating to the death of an author;

(7) for the issuance, under section 706, of an additional certificate of registration;

(8) for the issuance of any other certification;

(9) for the making and reporting of a search as provided by section 705, and for any related services;

(10) on filing a statement of account based on secondary transmissions of primary transmissions pursuant to section 119 or 122; and

(11) on filing a statement of account based on secondary transmissions of primary transmissions pursuant to section 111.

The Register is authorized to fix fees for other services, including the cost of preparing copies of Copyright Office records, whether or not such copies are certified, based on the cost of providing the service. Fees established under paragraphs (10) and (11) shall be reasonable and may not exceed one-half of the cost necessary to cover reasonable expenses incurred by the Copyright Office for the collection and administration of the statements of account and any royalty fees deposited with such statements.

(b) Adjustment of Fees.—The Register of Copyrights may, by regulation, adjust the fees for the services specified in paragraphs (1) through (9) of subsection (a) in the following manner:

(1) The Register shall conduct a study of the costs incurred by the Copyright Office for the registration of claims, the recordation of documents, and the provision of services. The study shall also consider the timing of any adjustment in fees and the authority to use such fees consistent with the budget.

(2) The Register may, on the basis of the study under paragraph (1), and subject to paragraph (5), adjust fees to not more than that necessary to cover the reasonable costs incurred by the Copyright Office for the services described in paragraph (1), plus a reasonable inflation adjustment to account for any estimated increase in costs.

(3) Any fee established under paragraph (2) shall be rounded off to the nearest dollar, or for a fee less than $12, rounded off to the nearest 50 cents.

(4) Fees established under this subsection shall be fair and equitable and give due consideration to the objectives of the copyright system.

(5) If the Register determines under paragraph (2) that fees should be adjusted, the Register shall prepare a proposed fee schedule and submit the schedule with the accompanying economic analysis to the Congress. The fees proposed by the Register may be instituted after the end of 120 days after the schedule is submitted to the Congress unless, within that 120-day period, a law is enacted stating in substance that the Congress does not approve the schedule.

(c) The fees prescribed by or under this section are applicable to the United States Government and any of its agencies, employees, or officers, but the Register of Copyrights has discretion to waive the requirement of this subsection in occasional or isolated cases involving relatively small amounts.

(d) (1) Except as provided in paragraph (2), all fees received under this section shall be deposited by the Register of Copyrights in the Treasury of the United States and shall be credited to the appropriations for necessary expenses of the Copyright Office. Such fees that are collected shall remain available until expended. The Register may, in accordance with regulations that he or she shall prescribe, refund any sum paid by mistake or in excess of the fee required by this section.

(2) In the case of fees deposited against future services, the Register of Copyrights shall request the Secretary of the Treasury to invest in interestbearing securities in the United States Treasury any portion of the fees that, as determined by the Register, is not required to meet current deposit account demands. Funds from such portion of fees shall be invested in securities that permit funds to be available to the Copyright Office at all times if they are determined to be necessary to meet current deposit account demands. Such investments shall be in public debt securities with maturities suitable to the needs of the Copyright Office, as determined by the Register of Copyrights, and bearing interest at rates determined by the Secretary of the Treasury, taking into consideration current market yields on outstanding marketable obligations of the United States of comparable maturities.

(3) The income on such investments shall be deposited in the Treasury of the United States and shall be credited to the appropriations for necessary expenses of the Copyright Office.

§ 709 Delay in Delivery Caused by Disruption of Postal or Other Services

In any case in which the Register of Copyrights determines, on the basis of such evidence as the Register may by regulation require, that a deposit, application, fee, or any other material to be delivered to the Copyright Office by a particular date, would have been received in the Copyright Office in due time except for a general disruption or suspension of postal or other transportation or communications services, the actual receipt of such material in the Copyright Office within one month after the date on which the Register determines that the disruption or suspension of such services has terminated, shall be considered timely.

Chapter 7 Notes

1. The Work Made for Hire and Copyright Corrections Act of 2000 amended the table of sections for chapter 7 by deleting section 710, entitled, "Reproduction for use of the blind and physically handicapped: Voluntary licensing forms and procedures." Pub. L. No. 106-379, 114 Stat. 1444, 1445.

2. The Copyright Fees and Technical Amendments Act of 1989 amended section 701 by adding subsection (e). Pub. L. No. 101-319, 104 Stat. 290. In 1998, the Digital Millennium Copyright Act amended section 701 by adding a new subsection (b), redesignating former subsections (b) through (e) as (c) through (f) respectively, and, in the new subsection (f), by substituting "III" for "IV" and "5314" for "5315." Pub. L. No. 105-304, 112 Stat. 2860, 2887.

3. Title 5 of the *United States Code* is entitled "Government Organization and Employees."

4. Copyright Office regulations are published in the *Federal Register* and in title 37, chapter II, of the *Code of Federal Regulations.*

5. The Copyright Cleanup, Clarification, and Corrections Act of 2010 amended the last sentence of section 704(e) to substitute "section 708(a)" for "section 708(a)(10)." Pub. L. No. 111-295, 124 Stat. 3180, 3181.

6. The Work Made for Hire and Copyright Corrections Act of 2000 amended section 705 by rewriting paragraph (a). Pub. L. No. 106-379, 114 Stat. 1444, 1445.

7. The Copyright Fees and Technical Amendments Act of 1989 amended section 708 by substituting a new subsection (a), by redesignating subsections (b) and (c) as subsections (c) and (d), respectively, and by adding a new subsection (b). Pub. L. No. 101-318, 104 Stat. 287. The Act states that these amendments "shall take effect 6 months after the date of the enactment of this Act" and shall apply to:

(A) claims to original, supplementary, and renewal copyright received for registration, and to items received for recordation in the Copyright Office, on or after such effective date, and

(B) other requests for services received on or after such effective date, or received before such effective date for services not yet rendered as of such date.

With respect to prior claims, the Act states that claims to original, supplementary, and renewal copyright received for registration and items received for recordation in acceptable form in the Copyright Office before the above mentioned effective date, and requests for services which are rendered before such effective date "shall be governed by section 708 of title 17, United States Code, as in effect before such effective date." Pub. L. No. 101-318, 104 Stat. 287, 288.

The Copyright Renewal Act of 1992 amended paragraph (2) of section 708(a) by striking the words "in its first term" and by substituting "$20" in lieu of "$12." Pub. L. No. 102-307, 106 Stat. 264, 266.

In 1997, section 708 was amended by rewriting subsections (b) and (d) in their entirety. Pub. L. No. 105-80, 111 Stat. 1529, 1532.

The Work Made for Hire and Copyright Corrections Act of 2000 amended section 708 by rewriting subsection (a), by substituting new language for the first sentence in subsection (b) and by substituting "adjustment" for "increase" in paragraph (b)(1), the word "adjust" for "increase" in paragraph (b)(2) and the word "adjusted" for "increased" in paragraph (b)(5). Pub. L. No. 106-379, 114 Stat. 1444, 1445. The Act also stated that "The fees under section 708(a) of title 17, United States Code, on the date of the enactment of this Act shall be the fees in effect under section 708(a) of such title on the day before such date of enactment."

The Satellite Television Extension and Localism Act of 2010 amended subsection 708(a) by adding new paragraphs (10) and (11) and by adding the last sentence. Pub. L. No. 111-175, 124 Stat. 1218, 1244-45.

8. The current fees may be found in the *Code of Federal Regulations,* at 37 CFR §201.3, as authorized by Pub. L. No. 105-80, 111 Stat. 1529, 1532. In Pub. L. No. 105-80, Congress amended section 708(b) to require that the Register of Copyrights establish fees by regulation rather than by codifying them in title 17, *United States Code,* as was previously done.

Chapter 8 Proceedings by Copyright Royalty Judges

§ 801 Copyright Royalty Judges; Appointment and Functions

(a) Appointment.—The Librarian of Congress shall appoint 3 full-time Copyright Royalty Judges, and shall appoint 1 of the 3 as the Chief Copyright Royalty Judge. The Librarian shall make appointments to such positions after consultation with the Register of Copyrights.

(b) Functions.—Subject to the provisions of this chapter, the functions of the Copyright Royalty Judges shall be as follows:

(1) To make determinations and adjustments of reasonable terms and rates of royalty payments as provided in sections 112(e), 114, 115, 116, 118, 119, and

1004. The rates applicable under sections 114(f)(1)(B), 115, and 116 shall be calculated to achieve the following objectives:

(A) To maximize the availability of creative works to the public.

(B) To afford the copyright owner a fair return for his or her creative work and the copyright user a fair income under existing economic conditions.

(C) To reflect the relative roles of the copyright owner and the copyright user in the product made available to the public with respect to relative creative contribution, technological contribution, capital investment, cost, risk, and contribution to the opening of new markets for creative expression and media for their communication.

(D) To minimize any disruptive impact on the structure of the industries involved and on generally prevailing industry practices.

(2) To make determinations concerning the adjustment of the copyright royalty rates under section 111 solely in accordance with the following provisions:

(A) The rates established by section 111(d)(1)(B) may be adjusted to reflect—

(i) national monetary inflation or deflation; or

(ii) changes in the average rates charged cable subscribers for the basic service of providing secondary transmissions to maintain the real constant dollar level of the royalty fee per subscriber which existed as of the date of October 19, 1976,

except that—

(I) if the average rates charged cable system subscribers for the basic service of providing secondary transmissions are changed so that the average rates exceed national monetary inflation, no change in the rates established by section 111(d)(1)(B) shall be permitted; and

(II) no increase in the royalty fee shall be permitted based on any reduction in the average number of distant signal equivalents per subscriber. The Copyright Royalty Judges may consider all factors relating to the maintenance of such level of payments, including, as an extenuating factor, whether the industry has been restrained by subscriber rate regulating authorities from increasing the rates for the basic service of providing secondary transmissions.

(B) In the event that the rules and regulations of the Federal Communications Commission are amended at any time after April 15, 1976, to permit the carriage by cable systems of additional television broadcast signals beyond the local service area of the primary transmitters of such signals, the royalty rates established by section 111(d)(1)(B) may be adjusted to ensure that the rates for the additional distant signal equivalents resulting from such carriage are reasonable in the light of the changes effected by the amendment to such rules and regulations. In determining the reasonableness of rates proposed following an amendment of Federal Communications Commission rules and regulations, the Copyright Royalty Judges shall consider, among other factors, the economic impact on copyright owners and users; except that no adjustment in royalty rates shall be made under this subparagraph with respect to any distant signal equivalent or fraction thereof represented by—

(i) carriage of any signal permitted under the rules and regulations of the Federal Communications Commission in effect on April 15, 1976, or the carriage of a signal of the same type (that is, independent, network, or noncommercial educational) substituted for such permitted signal; or

(ii) a television broadcast signal first carried after April 15, 1976, pursuant to an individual waiver of the rules and regulations of the Federal Communications Commission, as such rules and regulations were in effect on April 15, 1976.

(C) In the event of any change in the rules and regulations of the Federal Communications Commission with respect to syndicated and sports program exclusivity after April 15, 1976, the rates established by section 111(d)(1)(B) may be adjusted to assure that such rates are reasonable in light of the changes to such rules and regulations, but any such adjustment shall apply only to the affected television broadcast signals carried on those systems affected by the change.

(D) The gross receipts limitations established by section 111(d)(1)(C) and (D) shall be adjusted to reflect national monetary inflation or deflation or changes in the average rates charged cable system subscribers for the basic service of providing secondary transmissions to maintain the real constant dollar value of the exemption provided by such section, and the royalty rate specified therein shall not be subject to adjustment.

(3)(A) To authorize the distribution, under sections 111, 119, and 1007, of those royalty fees collected under sections 111, 119, and 1005, as the case may be, to the extent that the Copyright Royalty Judges have found that the distribution of such fees is not subject to controversy.

(B) In cases where the Copyright Royalty Judges determine that controversy exists, the Copyright Royalty Judges shall determine the distribution of such fees, including partial distributions, in accordance with section 111, 119, or 1007, as the case may be.

(C) Notwithstanding section 804(b)(8), the Copyright Royalty Judges, at any time after the filing of claims under section 111, 119, or 1007, may, upon motion of one or more of the claimants and after publication in the Federal Register of a request for responses to the motion from interested claimants, make a partial distribution of such fees, if, based upon all responses received during the 30day period beginning on the date of such publication, the Copyright Royalty Judges conclude that no claimant entitled to receive such fees has stated a reasonable objection to the partial distribution, and all such claimants—

(i) agree to the partial distribution;

(ii) sign an agreement obligating them to return any excess amounts to the extent necessary to comply with the final determination on the distribution of the fees made under subparagraph (B);

(iii) file the agreement with the Copyright Royalty Judges; and

(iv) agree that such funds are available for distribution.

(D) The Copyright Royalty Judges and any other officer or employee acting in good faith in distributing funds under subparagraph (C) shall not be held liable for the payment of any excess fees under subparagraph (C). The Copyright Royalty Judges shall, at the time the final determination is made, calculate any such excess amounts.

(4) To accept or reject royalty claims filed under sections 111, 119, and 1007, on the basis of timeliness or the failure to establish the basis for a claim.

(5) To accept or reject rate adjustment petitions as provided in section 804 and petitions to participate as provided in section 803(b) (1) and (2).

(6) To determine the status of a digital audio recording device or a digital audio interface device under sections 1002 and 1003, as provided in section 1010.

(7) (A) To adopt as a basis for statutory terms and rates or as a basis for the distribution of statutory royalty payments, an agreement concerning such matters reached among some or all of the participants in a proceeding at any time during the proceeding, except that—

(i) the Copyright Royalty Judges shall provide to those that would be bound by the terms, rates, or other determination set by any agreement in a proceeding to determine royalty rates an opportunity to comment on the agreement and shall provide to participants in the proceeding under section 803(b)(2) that would be bound by the terms, rates, or other determination set by the agreement an opportunity to comment on the agreement and object to its adoption as a basis for statutory terms and rates; and

(ii) the Copyright Royalty Judges may decline to adopt the agreement as a basis for statutory terms and rates for participants that are not parties to the agreement, if any participant described in clause (i) objects to the agreement and the Copyright Royalty Judges conclude, based on the record before them if one exists, that the agreement does not provide a reasonable basis for setting statutory terms or rates.

(B) License agreements voluntarily negotiated pursuant to section 112(e)(5), 114(f)(3), 115(c)(3)(E)(i), 116(c), or 118(b)(2) that do not result in statutory terms and rates shall not be subject to clauses (i) and (ii) of subparagraph (A).

(C) Interested parties may negotiate and agree to, and the Copyright Royalty Judges may adopt, an agreement that specifies as terms notice and recordkeeping requirements that apply in lieu of those that would otherwise apply under regulations.

(8) To perform other duties, as assigned by the Register of Copyrights within the Library of Congress, except as provided in section 802(g), at times when Copyright Royalty Judges are not engaged in performing the other duties set forth in this section.

(c) Rulings.—The Copyright Royalty Judges may make any necessary procedural or evidentiary rulings in any proceeding under this chapter and may, before commencing a proceeding under this chapter, make any such rulings that would apply to the proceedings conducted by the Copyright Royalty Judges.

(d) Administrative Support.—The Librarian of Congress shall provide the Copyright Royalty Judges with the necessary administrative services related to proceedings under this chapter.

(e) Location in Library of Congress.—The offices of the Copyright Royalty Judges and staff shall be in the Library of Congress.

(f) Effective Date of Actions.—On and after the date of the enactment of the Copyright Royalty and Distribution Reform Act of 2004, in any case in which time limits are prescribed under this title for performance of an action with or by the Copyright Royalty Judges, and in which the last day of the prescribed period falls on a Saturday, Sunday, holiday, or other nonbusiness day within the District of Columbia or the Federal Government, the action may be taken on the next succeeding business day, and is effective as of the date when the period expired.

§ 802 Copyright Royalty Judgeships; Staff

(a) Qualifications of Copyright Royalty Judges.—

(1) In general.—Each Copyright Royalty Judge shall be an attorney who has at least 7 years of legal experience. The Chief Copyright Royalty Judge shall have at least 5 years of experience in adjudications, arbitrations, or court trials. Of the other 2 Copyright Royalty Judges, 1 shall have significant knowledge of copyright law, and the other shall have significant knowledge of economics. An individual may serve as a Copyright Royalty Judge only if the individual is free of any financial conflict of interest under subsection (h).

(2) Definition.—In this subsection, the term "adjudication" has the meaning given that term in section 551 of title 5, but does not include mediation.

(b) Staff.—The Chief Copyright Royalty Judge shall hire 3 full-time staff members to assist the Copyright Royalty Judges in performing their functions.

(c) Terms.—The individual first appointed as the Chief Copyright Royalty Judge shall be appointed to a term of 6 years, and of the remaining individuals first appointed as Copyright Royalty Judges, 1 shall be appointed to a term of 4 years, and the other shall be appointed to a term of 2 years. Thereafter, the terms of succeeding Copyright Royalty Judges shall each be 6 years. An individual serving as a Copyright Royalty Judge may be reappointed to subsequent terms. The term of a Copyright Royalty Judge shall begin when the term of the predecessor of that Copyright Royalty Judge ends. When the term of office of a Copyright Royalty Judge ends, the individual serving that term may continue to serve until a successor is selected.

(d) Vacancies or Incapacity.—

(1) Vacancies.—If a vacancy should occur in the position of Copyright Royalty Judge, the Librarian of Congress shall act expeditiously to fill the vacancy, and may appoint an interim Copyright Royalty Judge to serve until another Copyright Royalty Judge is appointed under this section. An individual appointed to fill the vacancy occurring before the expiration of the term for which the predecessor of that individual was appointed shall be appointed for the remainder of that term.

(2) Incapacity.—In the case in which a Copyright Royalty Judge is temporarily unable to perform his or her duties, the Librarian of Congress may appoint

an interim Copyright Royalty Judge to perform such duties during the period of such incapacity.

(e) Compensation.—

(1) Judges.—The Chief Copyright Royalty Judge shall receive compensation at the rate of basic pay payable for level AL-1 for administrative law judges pursuant to section 5372(b) of title 5, and each of the other two Copyright Royalty Judges shall receive compensation at the rate of basic pay payable for level AL-2 for administrative law judges pursuant to such section. The compensation of the Copyright Royalty Judges shall not be subject to any regulations adopted by the Office of Personnel Management pursuant to its authority under section 5376(b)(1) of title 5.

(2) Staff members.—Of the staff members appointed under subsection (b)—

(A) the rate of pay of 1 staff member shall be not more than the basic rate of pay payable for level 10 of GS-15 of the General Schedule;

(B) the rate of pay of 1 staff member shall be not less than the basic rate of pay payable for GS-13 of the General Schedule and not more than the basic rate of pay payable for level 10 of GS-14 of such Schedule; and

(C) the rate of pay for the third staff member shall be not less than the basic rate of pay payable for GS-8 of the General Schedule and not more than the basic rate of pay payable for level 10 of GS-11 of such Schedule.

(3) Locality pay.—All rates of pay referred to under this subsection shall include locality pay.

(f) Independence of Copyright Royalty Judge.—

(1) In making determinations.—

(A) In general.—(i) Subject to subparagraph (B) and clause (ii) of this subparagraph, the Copyright Royalty Judges shall have full independence in making determinations concerning adjustments and determinations of copyright royalty rates and terms, the distribution of copyright royalties, the acceptance or rejection of royalty claims, rate adjustment petitions, and petitions to participate, and in issuing other rulings under this title, except that the Copyright Royalty Judges may consult with the Register of Copyrights on any matter other than a question of fact.

(ii) One or more Copyright Royalty Judges may, or by motion to the Copyright Royalty Judges, any participant in a proceeding may, request from the Register of Copyrights an interpretation of any material questions of substantive law that relate to the construction of provisions of this title and arise in the course of the proceeding. Any request for a written interpretation shall be in writing and on the record, and reasonable provision shall be made to permit participants in the proceeding to comment on the material questions of substantive law in a manner that minimizes duplication and delay. Except as provided in subparagraph (B), the Register of Copyrights shall deliver to the Copyright Royalty Judges a written response within 14 days after the receipt of all briefs and comments from the participants. The Copyright Royalty Judges shall apply the legal interpretation embodied in the response of the Register of Copyrights if it is timely delivered, and the response shall be included in the record that accompanies the final determination. The authority under this clause shall not be construed to authorize the Register of Copyrights to provide an interpretation of questions of procedure before the Copyright Royalty Judges, the ultimate adjustments and determinations of copyright royalty rates and terms, the ultimate distribution of copyright royalties, or the acceptance or rejection of royalty claims, rate adjustment petitions, or petitions to participate in a proceeding.

(B) Novel questions.—(i) In any case in which a novel material question of substantive law concerning an interpretation of those provisions of this title that are the subject of the proceeding is presented, the Copyright Royalty Judges shall request a decision of the Register of Copyrights, in writing, to resolve such novel question. Reasonable provision shall be made for comment on such request by the participants in the proceeding, in such a way as to minimize duplication and delay. The Register of Copyrights shall transmit his or her decision to the Copyright Royalty Judges within 30 days after the Register of Copyrights receives all of the briefs or comments of the participants. Such decision shall be in writing and included by the Copyright Royalty Judges in the record that accompanies their final determination. If such a decision is timely delivered to the Copyright Royalty Judges, the Copyright Royalty Judges shall apply the legal determinations embodied in the decision of the Register of Copyrights in resolving material questions of substantive law.

(ii) In clause (i), a "novel question of law" is a question of law that has not been determined in prior decisions, determinations, and rulings described in section 803(a).

(C) Consultation.—Notwithstanding the provisions of subparagraph (A), the Copyright Royalty Judges shall consult with the Register of Copyrights with respect to any determination or ruling that would require that any act be performed by the Copyright Office, and any such determination or ruling shall not be binding upon the Register of Copyrights.

(D) Review of legal conclusions by the register of copyrights.—The Register of Copyrights may review for legal error the resolution by the Copyright Royalty Judges of a material question of substantive law under this title that underlies or is contained in a final determination of the Copyright Royalty Judges. If the Register of Copyrights concludes, after taking into consideration the views of the participants in the proceeding, that any resolution reached by the Copyright Royalty Judges was in material error, the Register of Copyrights shall issue a written decision correcting such legal error, which shall be made part of the record of the proceeding. The Register of Copyrights shall issue such written decision not later than 60 days after the date on which the final determination by the Copyright Royalty Judges is issued. Additionally, the Register of Copyrights shall cause to be published in the Federal Register such written decision, together with a specific identification of the legal conclusion of the Copyright Royalty Judges that is determined to be erroneous. As to conclusions of substantive law involving an interpretation of the statutory provisions of this title, the decision of the Register of Copyrights shall be binding as precedent upon the Copyright Royalty Judges in subsequent proceedings under this chapter. When a decision has been rendered pursuant to this subparagraph, the Register of Copyrights may, on the basis of and in accordance with such decision, intervene as of right in any appeal of a final determination of the Copyright Royalty Judges pursuant to section 803(d) in the United States Court of Appeals for the District of Columbia Circuit. If, prior to intervening in such an appeal, the Register of Copyrights gives notification to, and undertakes to consult with, the Attorney General with respect to such intervention, and the Attorney General fails, within a reasonable period after receiving such notification, to intervene in such appeal, the Register of Copyrights may intervene in such appeal in his or her own name by any attorney designated by the Register of Copyrights for such purpose. Intervention by the Register of Copyrights in his or her own name shall not preclude the Attorney General from intervening on behalf of the United States in such an appeal as may be otherwise provided or required by law.

(E) Effect on judicial review.—Nothing in this section shall be interpreted to alter the standard applied by a court in reviewing legal determinations involving an interpretation or construction of the provisions of this title or to affect the extent to which any construction or interpretation of the provisions of this title shall be accorded deference by a reviewing court.

(2) Performance appraisals.—

(A) In general.—Notwithstanding any other provision of law or any regulation of the Library of Congress, and subject to subparagraph (B), the Copyright Royalty Judges shall not receive performance appraisals.

(B) Relating to sanction or removal.—To the extent that the Librarian of Congress adopts regulations under subsection (h) relating to the sanction or removal of a Copyright Royalty Judge and such regulations require documentation to establish the cause of such sanction or removal, the Copyright Royalty Judge may receive an appraisal related specifically to the cause of the sanction or removal.

(g) Inconsistent Duties Barred.—No Copyright Royalty Judge may undertake duties that conflict with his or her duties and responsibilities as a Copyright Royalty Judge.

(h) Standards of Conduct.—The Librarian of Congress shall adopt regulations regarding the standards of conduct, including financial conflict of interest and restrictions against ex parte communications, which shall govern the Copyright Royalty Judges and the proceedings under this chapter.

(i) Removal or Sanction.—The Librarian of Congress may sanction or remove a Copyright Royalty Judge for violation of the standards of conduct adopted under subsection (h), misconduct, neglect of duty, or any disqualifying physical or mental disability. Any sanction or removal may be made only after notice and opportunity for a hearing, but the Librarian of Congress may suspend the Copyright Royalty Judge during the pendency of such hearing. The Librarian shall appoint an interim Copyright Royalty Judge during the period of any such suspension.

§ 803 Proceedings of Copyright Royalty Judges

(a) Proceedings.—

(1) In general.—The Copyright Royalty Judges shall act in accordance with this title, and to the extent not inconsistent with this title, in accordance with subchapter II of chapter 5 of title 5, in carrying out the purposes set forth in section 801. The Copyright Royalty Judges shall act in accordance with regulations issued by the Copyright Royalty Judges and the Librarian of Congress, and on the basis of a written record, prior determinations and interpretations of the Copyright Royalty Tribunal, Librarian of Congress, the Register of Copyrights, copyright arbitration royalty panels (to the extent those determinations are not inconsistent with a decision of the Librarian of Congress or the Register of Copyrights), and the Copyright Royalty Judges (to the extent those determinations are not inconsistent with a decision of the Register of Copyrights that was timely delivered to the Copyright Royalty Judges pursuant to section 802(f)(1)(A) or (B), or with a decision of the Register of Copyrights pursuant to section 802(f)(1)(D)), under this chapter, and decisions of the court of appeals under this chapter before, on, or after the effective date of the Copyright Royalty and Distribution Reform Act of 2004.

(2) Judges acting as panel and individually.—The Copyright Royalty Judges shall preside over hearings in proceedings under this chapter en banc. The Chief Copyright Royalty Judge may designate a Copyright Royalty Judge to preside individually over such collateral and administrative proceedings, and over such proceedings under paragraphs (1) through (5) of subsection (b), as the Chief Judge considers appropriate.

(3) Determinations.—Final determinations of the Copyright Royalty Judges in proceedings under this chapter shall be made by majority vote. A Copyright Royalty Judge dissenting from the majority on any determination under this chapter may issue his or her dissenting opinion, which shall be included with the determination.

(b) Procedures.—

(1) Initiation.—

(A) Call for petitions to participate.—(i) The Copyright Royalty Judges shall cause to be published in the Federal Register notice of commencement of proceedings under this chapter, calling for the filing of petitions to participate in a proceeding under this chapter for the purpose of making the relevant determination under section 111, 112, 114, 115, 116, 118, 119, 1004, or 1007, as the case may be—

(I) promptly upon a determination made under section 804(a);

(II) by no later than January 5 of a year specified in paragraph (2) of section 804(b) for the commencement of proceedings;

(III) by no later than January 5 of a year specified in subparagraph (A) or (B) of paragraph (3) of section 804(b) for the commencement of proceedings, or as otherwise provided in subparagraph (A) or (C) of such paragraph for the commencement of proceedings;

(IV) as provided under section 804(b)(8); or

(V) by no later than January 5 of a year specified in any other provision of section 804(b) for the filing of petitions for the commencement of proceedings, if a petition has not been filed by that date, except that the publication of notice requirement shall not apply in the case of proceedings under section 111 that are scheduled to commence in 2005.

(ii) Petitions to participate shall be filed by no later than 30 days after publication of notice of commencement of a proceeding under clause (i), except that the Copyright Royalty Judges may, for substantial good cause shown and if there is no prejudice to the participants that have already filed petitions, accept late petitions to participate at any time up to the date that is 90 days before the date on which participants in the proceeding are to file their written direct statements. Notwithstanding the preceding sentence, petitioners whose petitions are filed more than 30 days after publication of notice of commencement of a proceeding are not eligible to object to a settlement reached during the voluntary negotiation period under paragraph (3), and any objection filed by such a petitioner shall not be taken into account by the Copyright Royalty Judges.

(B) Petitions to participate.—Each petition to participate in a proceeding shall describe the petitioner's interest in the subject matter of the proceeding. Parties with similar interests may file a single petition to participate.

(2) Participation in general.—Subject to paragraph (4), a person may participate in a proceeding under this chapter, including through the submission of briefs or other information, only if—

(A) that person has filed a petition to participate in accordance with paragraph (1) (either individually or as a group under paragraph (1)(B));

(B) the Copyright Royalty Judges have not determined that the petition to participate is facially invalid;

(C) the Copyright Royalty Judges have not determined, sua sponte or on the motion of another participant in the proceeding, that the person lacks a significant interest in the proceeding; and

(D) the petition to participate is accompanied by either—

(i) in a proceeding to determine royalty rates, a filing fee of $150; or

(ii) in a proceeding to determine distribution of royalty fees—

(I) a filing fee of $150; or

(II) a statement that the petitioner (individually or as a group) will not seek a distribution of more than $1000, in which case the amount distributed to the petitioner shall not exceed $1000.

(3) Voluntary negotiation period.—

(A) Commencement of proceedings.—

(i) Rate adjustment proceeding.—Promptly after the date for filing of petitions to participate in a proceeding, the Copyright Royalty Judges shall make available to all participants in the proceeding a list of such participants and shall initiate a voluntary negotiation period among the participants.

(ii) Distribution proceeding.—Promptly after the date for filing of petitions to participate in a proceeding to determine the distribution of royalties, the Copyright Royalty Judges shall make available to all participants in the proceeding a list of such participants. The initiation of a voluntary negotiation period among the participants shall be set at a time determined by the Copyright Royalty Judges.

(B) Length of proceedings.—The voluntary negotiation period initiated under subparagraph (A) shall be 3 months.

(C) Determination of subsequent proceedings.—At the close of the voluntary negotiation proceedings, the Copyright Royalty Judges shall, if further proceedings under this chapter are necessary, determine whether and to what extent paragraphs (4) and (5) will apply to the parties.

(4) Small claims procedure in distribution proceedings.—

(A) In general.—If, in a proceeding under this chapter to determine the distribution of royalties, the contested amount of a claim is $10,000 or less, the Copyright Royalty Judges shall decide the controversy on the basis of the filing of the written direct statement by the participant, the response by any opposing participant, and 1 additional response by each such party.

(B) Bad faith inflation of claim.—If the Copyright Royalty Judges determine that a participant asserts in bad faith an amount in controversy in excess of $10,000 for the purpose of avoiding a determination under the procedure set forth in subparagraph (A), the Copyright Royalty Judges shall impose a fine on that participant in an amount not to exceed the difference between the actual amount distributed and the amount asserted by the participant.

(5) Paper proceedings.—The Copyright Royalty Judges in proceedings under this chapter may decide, sua sponte or upon motion of a participant, to determine issues on the basis of the filing of the written direct statement by the participant, the response by any opposing participant, and one additional response by each such participant. Prior to making such decision to proceed on such a paper record only, the Copyright Royalty Judges shall offer to all parties to the proceeding the opportunity to comment on the decision. The procedure under this paragraph—

(A) shall be applied in cases in which there is no genuine issue of material fact, there is no need for evidentiary hearings, and all participants in the proceeding agree in writing to the procedure; and

(B) may be applied under such other circumstances as the Copyright Royalty Judges consider appropriate.

(6) Regulations.—

(A) In general.—The Copyright Royalty Judges may issue regulations to carry out their functions under this title. All regulations issued by the Copyright Royalty Judges are subject to the approval of the Librarian of Congress and are subject to judicial review pursuant to chapter 7 of title 5, except as set forth in subsection (d). Not later than 120 days after Copyright Royalty Judges or interim Copyright Royalty Judges, as the case may be, are first appointed after the enactment of the Copyright Royalty and Distribution Reform Act of 2004, such judges shall issue regulations to govern proceedings under this chapter.

(B) Interim regulations.—Until regulations are adopted under subparagraph (A), the Copyright Royalty Judges shall apply the regulations in effect under this chapter on the day before the effective date of the Copyright Royalty and Distribution Reform Act of 2004, to the extent such regulations are not inconsistent with this chapter, except that functions carried out under such regulations by the Librarian of Congress, the Register of Copyrights, or copyright arbitration royalty panels that, as of such date of enactment, are to be carried out by the Copyright Royalty Judges under this chapter, shall be carried out by the Copyright Royalty Judges under such regulations.

(C) Requirements.—Regulations issued under subparagraph (A) shall include the following:

(i) The written direct statements and written rebuttal statements of all participants in a proceeding under paragraph (2) shall be filed by a date specified by the Copyright Royalty Judges, which, in the case of written direct statements, may be not earlier than 4 months, and not later than 5 months, after the end of the voluntary negotiation period under paragraph (3). Notwithstanding the preceding sentence, the Copyright Royalty Judges may allow a participant in a proceeding to file an amended written direct statement based on new information received during the discovery process, within 15 days after the end of the discovery period specified in clause (iv).

(ii) (I) Following the submission to the Copyright Royalty Judges of written direct statements and written rebuttal statements by the participants in a proceeding under paragraph (2), the Copyright Royalty Judges, after taking into consideration the views of the participants in the proceeding, shall determine a schedule for conducting and completing discovery.

(II) In this chapter, the term "written direct statements" means witness statements, testimony, and exhibits to be presented in the proceedings, and such other information that is necessary to establish terms and rates, or the distribution of royalty payments, as the case may be, as set forth in regulations issued by the Copyright Royalty Judges.

(iii) Hearsay may be admitted in proceedings under this chapter to the extent deemed appropriate by the Copyright Royalty Judges.

(iv) Discovery in connection with written direct statements shall be permitted for a period of 60 days, except for discovery ordered by the Copyright Royalty Judges in connection with the resolution of motions, orders, and disputes pending at the end of such period. The Copyright Royalty Judges may order a discovery schedule in connection with written rebuttal statements.

(v) Any participant under paragraph (2) in a proceeding under this chapter to determine royalty rates may request of an opposing participant nonprivileged documents directly related to the written direct statement or written rebuttal statement of that participant. Any objection to such a request shall be resolved by a motion or request to compel production made to the Copyright Royalty Judges in accordance with regulations adopted by the Copyright Royalty Judges. Each motion or request to compel discovery shall be determined by the Copyright Royalty Judges, or by a Copyright Royalty Judge when permitted under subsection (a)(2). Upon such motion, the Copyright Royalty Judges may order discovery pursuant to regulations established under this paragraph.

(vi) (I) Any participant under paragraph (2) in a proceeding under this chapter to determine royalty rates may, by means of written motion or on the record, request of an opposing participant or witness other relevant information and materials if, absent the discovery sought, the Copyright Royalty Judges' resolution of the proceeding would be substantially impaired. In determining whether discovery will be granted under this clause, the Copyright Royalty Judges may consider—

(aa) whether the burden or expense of producing the requested information or materials outweighs the likely benefit, taking into account the needs and resources of the participants, the importance of the issues at stake, and the probative value of the requested information or materials in resolving such issues;

(bb) whether the requested information or materials would be unreasonably cumulative or duplicative, or are obtainable from another source that is more convenient, less burdensome, or less expensive; and

(cc) whether the participant seeking discovery has had ample opportunity by discovery in the proceeding or by other means to obtain the information sought.

(II) This clause shall not apply to any proceeding scheduled to commence after December 31, 2010.

(vii) In a proceeding under this chapter to determine royalty rates, the participants entitled to receive royalties shall collectively be permitted to take no more than 10 depositions and secure responses to no more than 25 interrogatories, and the participants obligated to pay royalties shall collectively be permitted to take no more than 10 depositions and secure responses to no more than 25 interrogatories. The Copyright Royalty Judges shall resolve any disputes among similarly aligned participants to allocate the number of depositions or interrogatories permitted under this clause.

(viii) The rules and practices in effect on the day before the effective date of the Copyright Royalty and Distribution Reform Act of 2004, relating to discovery in proceedings under this chapter to determine the distribution of royalty fees, shall continue to apply to such proceedings on and after such effective date.

(ix) In proceedings to determine royalty rates, the Copyright Royalty Judges may issue a subpoena commanding a participant or witness to appear and give testimony, or to produce and permit inspection of documents or tangible things, if the Copyright Royalty Judges' resolution of the proceeding would be substantially impaired by the absence of such testimony or production of documents or tangible things. Such subpoena shall specify with reasonable particularity the materials to be produced or the scope and nature of the required testimony. Nothing in this clause shall preclude the Copyright Royalty Judges from requesting the production by a nonparticipant of information or materials relevant to the resolution by the Copyright Royalty Judges of a material issue of fact.

(x) The Copyright Royalty Judges shall order a settlement conference among the participants in the proceeding to facilitate the presentation of offers of settlement among the participants. The settlement conference shall be held during a 21-day period following the 60-day discovery period specified in clause (iv) and shall take place outside the presence of the Copyright Royalty Judges.

(xi) No evidence, including exhibits, may be submitted in the written direct statement or written rebuttal statement of a participant without a sponsoring witness, except where the Copyright Royalty Judges have taken official notice, or in the case of incorporation by reference of past records, or for good cause shown.

(c) Determination of Copyright Royalty Judges.—

(1) Timing.—The Copyright Royalty Judges shall issue their determination in a proceeding not later than 11 months after the conclusion of the 21-day settlement conference period under subsection (b)(6)(C)(x), but, in the case of a proceeding to determine successors to rates or terms that expire on a specified date, in no event later than 15 days before the expiration of the then current statutory rates and terms.

(2) Rehearings.—

(A) In general.—The Copyright Royalty Judges may, in exceptional cases, upon motion of a participant in a proceeding under subsection (b)(2), order a rehearing, after the determination in the proceeding is issued under paragraph (1), on such matters as the Copyright Royalty Judges determine to be appropriate.

(B) Timing for filing motion.—Any motion for a rehearing under subparagraph (A) may only be filed within 15 days after the date on which the Copyright Royalty Judges deliver to the participants in the proceeding their initial determination.

(C) Participation by opposing party not required.—In any case in which a rehearing is ordered, any opposing party shall not be required to participate in the rehearing, except that nonparticipation may give rise to the limitations with respect to judicial review provided for in subsection (d)(1).

(D) No negative inference.—No negative inference shall be drawn from lack of participation in a rehearing.

(E) Continuity of rates and terms.—(i) If the decision of the Copyright Royalty Judges on any motion for a rehearing is not rendered before the expiration of the statutory rates and terms that were previously in effect, in the case of a proceeding to determine successors to rates and terms that expire on a specified date, then—

(I) the initial determination of the Copyright Royalty Judges that is the subject of the rehearing motion shall be effective as of the day following the date on which the rates and terms that were previously in effect expire; and

(II) in the case of a proceeding under section 114(f)(1)(C) or 114(f)(2)(C), royalty rates and terms shall, for purposes of section 114(f)(4)(B), be deemed to have been set at those rates and terms contained in the initial determination of the Copyright Royalty Judges that is the subject of the rehearing motion, as of the date of that determination.

(ii) The pendency of a motion for a rehearing under this paragraph shall not relieve persons obligated to make royalty payments who would be affected by the determination on that motion from providing the statements of account and any reports of use, to the extent required, and paying the royalties required under the relevant determination or regulations.

(iii) Notwithstanding clause (ii), whenever royalties described in clause (ii) are paid to a person other than the Copyright Office, the entity designated by the Copyright Royalty Judges to which such royalties are paid by the copyright user (and any successor thereto) shall, within 60 days after the motion for rehearing is resolved or, if the motion is granted, within 60 days after the rehearing is concluded, return any excess amounts previously paid to the extent necessary to comply with the final determination of royalty rates by the Copyright Royalty Judges. Any underpayment of royalties resulting from a rehearing shall be paid within the same period.

(3) Contents of determination.—A determination of the Copyright Royalty Judges shall be supported by the written record and shall set forth the findings of fact relied on by the Copyright Royalty Judges. Among other terms adopted in a determination, the Copyright Royalty Judges may specify notice and recordkeeping requirements of users of the copyrights at issue that apply in lieu of those that would otherwise apply under regulations.

(4) Continuing jurisdiction.—The Copyright Royalty Judges may issue an amendment to a written determination to correct any technical or clerical errors in the determination or to modify the terms, but not the rates, of royalty payments in response to unforeseen circumstances that would frustrate the proper implementation of such determination. Such amendment shall be set forth in a written addendum to the determination that shall be distributed to the participants of the proceeding and shall be published in the Federal Register.

(5) Protective order.—The Copyright Royalty Judges may issue such orders as may be appropriate to protect confidential information, including orders excluding confidential information from the record of the determination that is published or made available to the public, except that any terms or rates of royalty payments or distributions may not be excluded.

(6) Publication of determination.—By no later than the end of the 60-day period provided in section 802(f)(1)(D), the Librarian of Congress shall cause the determination, and any corrections thereto, to be published in the Federal Register. The Librarian of Congress shall also publicize the determination and corrections in such other manner as the Librarian considers appropriate, including, but not limited to, publication on the Internet. The Librarian of Congress shall also make the determination, corrections, and the accompanying record available for public inspection and copying.

(7) Late payment.—A determination of the Copyright Royalty Judges may include terms with respect to late payment, but in no way shall such terms prevent the copyright holder from asserting other rights or remedies provided under this title.

(d) Judicial Review.—

(1) Appeal.—Any determination of the Copyright Royalty Judges under subsection (c) may, within 30 days after the publication of the determination in the Federal Register, be appealed, to the United States Court of Appeals for the District of Columbia Circuit, by any aggrieved participant in the proceeding under subsection (b)(2) who fully participated in the proceeding and who would be bound by the determination. Any participant that did not participate in a rehearing may not raise any issue that was the subject of that rehearing at any stage of judicial review of the hearing determination. If no appeal is brought within that 30-day period, the determination of the Copyright Royalty Judges shall be final, and the royalty fee or determination with respect to the distribution of fees, as the case may be, shall take effect as set forth in paragraph (2).

(2) Effect of rates.—

(A) Expiration on specified date.—When this title provides that the royalty rates and terms that were previously in effect are to expire on a specified date, any adjustment or determination by the Copyright Royalty Judges of successor rates and terms for an ensuing statutory license period shall be effective as of the day following the date of expiration of the rates and terms that were previously in effect, even if the determination of the Copyright Royalty Judges is rendered on a later date. A licensee shall be obligated to continue making payments under the rates and terms previously in effect until such time as rates and terms for the successor period are established. Whenever royalties pursuant to this section are paid to a person other than the Copyright Office, the entity designated by the Copyright Royalty Judges to which such royalties are paid by the copyright user (and any successor thereto) shall, within 60 days after the final determination of the Copyright Royalty Judges establishing rates and terms for a successor period or the exhaustion of all rehearings or appeals of such determination, if any, return any excess amounts previously paid to the extent necessary to comply with the final determination of royalty rates. Any underpayment of royalties by a copyright user shall be paid to the entity designated by the Copyright Royalty Judges within the same period.

(B) Other cases.—In cases where rates and terms have not, prior to the inception of an activity, been established for that particular activity under the relevant license, such rates and terms shall be retroactive to the inception of activity under the relevant license covered by such rates and terms. In other cases where rates and terms do not expire on a specified date, successor rates and terms shall take effect on the first day of the second month that begins after the publication of the determination of the Copyright Royalty Judges in the Federal Register, except as otherwise provided in this title, or by the Copyright Royalty Judges, or as agreed by the participants in a proceeding that would be bound by the rates and terms. Except as otherwise provided in this title, the rates and terms, to the extent applicable, shall remain in effect until such successor rates and terms become effective.

(C) Obligation to make payments.—

(i) The pendency of an appeal under this subsection shall not relieve persons obligated to make royalty payments under section 111, 112, 114, 115, 116, 118, 119, or 1003, who would be affected by the determination on appeal, from–

(I) providing the applicable statements of account and report of use; and

(II) paying the royalties required under the relevant determination or regulations.

(ii) Notwithstanding clause (i), whenever royalties described in clause (i) are paid to a person other than the Copyright Office, the entity designated by the Copyright Royalty Judges to which such royalties are paid by the copyright user (and any successor thereto) shall, within 60 days after the final resolution of the appeal, return any excess amounts previously paid (and interest thereon, if ordered pursuant to paragraph (3)) to the extent necessary to comply with the final determination of royalty rates on appeal. Any underpayment of royalties resulting from an appeal (and interest thereon, if ordered pursuant to paragraph (3)) shall be paid within the same period.

(3) Jurisdiction of court.—Section 706 of title 5 shall apply with respect to review by the court of appeals under this subsection. If the court modifies or vacates a determination of the Copyright Royalty Judges, the court may enter its own determination with respect to the amount or distribution of royalty fees and costs, and order the repayment of any excess fees, the payment of any underpaid fees, and the payment of interest pertaining respectively thereto, in accordance with its final judgment. The court may also vacate the determination of the Copyright Royalty Judges and remand the case to the Copyright Royalty Judges for further proceedings in accordance with subsection (a).

(e) Administrative Matters.—

(1) Deduction of costs of Library of Congress and Copyright Office from filing fees.—

(A) Deduction from filing fees.—The Librarian of Congress may, to the extent not otherwise provided under this title, deduct from the filing fees collected under subsection (b) for a particular proceeding under this chapter the reasonable costs incurred by the Librarian of Congress, the Copyright Office, and the Copyright Royalty Judges in conducting that proceeding, other than the salaries of the Copyright Royalty Judges and the 3 staff members appointed under section 802(b).

(B) Authorization of appropriations.—There are authorized to be appropriated such sums as may be necessary to pay the costs incurred under this chapter not covered by the filing fees collected under subsection (b). All funds made available pursuant to this subparagraph shall remain available until expended.

(2) Positions required for administration of compulsory licensing.—Section 307 of the Legislative Branch Appropriations Act, 1994, shall not apply to employee positions in the Library of Congress that are required to be filled in order to carry out section 111, 112, 114, 115, 116, 118, or 119 or chapter 10.

§ 804 Institution of Proceedings

(a) Filing of Petition.—With respect to proceedings referred to in paragraphs (1) and (2) of section 801(b) concerning the determination or adjustment of royalty rates as provided in sections 111, 112, 114, 115, 116, 118, 119, and 1004, during the calendar years specified in the schedule set forth in subsection (b), any owner or user of a copyrighted work whose royalty rates are specified by this title, or are established under this chapter before or after the enactment of the Copyright Royalty and Distribution Reform Act of 2004, may file a petition with the Copyright Royalty Judges declaring that the petitioner requests a determination or adjustment of the rate. The Copyright Royalty Judges shall make a determination as to whether the petitioner has such a significant interest in the royalty rate in which a determination or adjustment is requested. If the Copyright Royalty Judges determine that the petitioner has such a significant interest, the Copyright Royalty Judges shall cause notice of this determination, with the reasons for such determination, to be published in the Federal Register, together with the notice of commencement of proceedings under this chapter. With respect to proceedings under paragraph (1) of section 801(b) concerning the determination or adjustment of royalty rates as provided in sections 112 and 114, during the calendar years specified in the schedule set forth in subsection (b), the Copyright Royalty Judges shall cause notice of commencement of proceedings under this chapter to be published in the Federal Register as provided in section 803(b)(1)(A).

(b) Timing of Proceedings.—

(1) Section 111 proceedings.—(A) A petition described in subsection (a) to initiate proceedings under section 801(b)(2) concerning the adjustment of royalty rates under section 111 to which subparagraph (A) or (D) of section 801(b)(2) applies may be filed during the year 2015 and in each subsequent fifth calendar year.

(B) In order to initiate proceedings under section 801(b)(2) concerning the adjustment of royalty rates under section 111 to which subparagraph (B) or (C) of section 801(b)(2) applies, within 12 months after an event described in either of those subsections, any owner or user of a copyrighted work whose royalty rates are specified by section 111, or by a rate established under this chapter before or after the enactment of the Copyright Royalty and Distribution Reform Act of 2004, may file a petition with the Copyright Royalty Judges declaring that the petitioner requests an adjustment of the rate. The Copyright Royalty Judges shall then proceed as set forth in subsection (a) of this section. Any change in royalty rates made under this chapter pursuant to this subparagraph may be reconsidered in the year 2015, and each fifth calendar year thereafter, in accordance with the provisions in section 801(b)(2)(B) or (C), as the case may be. A petition for adjustment of rates established by section 111(d)(1)(B) as a result of a change in the rules and regulations of the Federal Communications Commission shall set forth the change on which the petition is based.

(C) Any adjustment of royalty rates under section 111 shall take effect as of the first accounting period commencing after the publication of the determination of the Copyright Royalty Judges in the Federal Register, or on such other date as is specified in that determination.

(2) Certain section 112 proceedings.—Proceedings under this chapter shall be commenced in the year 2007 to determine reasonable terms and rates of royalty payments for the activities described in section 112(e)(1) relating to the limitation on exclusive rights specified by section 114(d)(1)(C)(iv), to become effective on January 1, 2009. Such proceedings shall be repeated in each subsequent fifth calendar year.

(3) Section 114 and corresponding 112 proceedings.—

(A) For eligible nonsubscription services and new subscription services.—Proceedings under this chapter shall be commenced as soon as practicable after the date of enactment of the Copyright Royalty and Distribution Reform Act of 2004 to determine reasonable terms and rates of royalty payments under sections 114 and 112 for the activities of eligible nonsubscription transmission services and new subscription services, to be effective for the period beginning on January 1, 2006, and ending on December 31, 2010. Such proceedings shall next be commenced in January 2009 to determine reasonable terms and rates of royalty payments, to become effective on January 1, 2011. Thereafter, such proceedings shall be repeated in each subsequent fifth calendar year.

(B) For preexisting subscription and satellite digital audio radio services.—Proceedings under this chapter shall be commenced in January 2006 to determine reasonable terms and rates of royalty payments under sections 114 and 112 for the activities of preexisting subscription services, to be effective during the period beginning on January 1, 2008, and ending on December 31, 2012, and preexisting satellite digital audio radio services, to be effective during the period beginning on January 1, 2007, and ending on December 31, 2012. Such proceedings shall next be commenced in 2011 to determine reasonable terms and rates of royalty payments, to become effective on January 1, 2013. Thereafter, such proceedings shall be repeated in each subsequent fifth calendar year.

(C)(i) Notwithstanding any other provision of this chapter, this subparagraph shall govern proceedings commenced pursuant to section 114(f)(1)(C) and 114(f)(2)(C) concerning new types of services.

(ii) Not later than 30 days after a petition to determine rates and terms for a new type of service is filed by any copyright owner of sound recordings, or such new type of service, indicating that such new type of service is or is about to become operational, the Copyright Royalty Judges shall issue a notice for a proceeding to determine rates and terms for such service.

(iii) The proceeding shall follow the schedule set forth in subsections (b), (c), and (d) of section 803, except that—

(I) the determination shall be issued by not later than 24 months after the publication of the notice under clause (ii); and

(II) the decision shall take effect as provided in subsections (c)(2) and (d)(2) of section 803 and section 114(f)(4)(B)(ii) and (C).

(iv) The rates and terms shall remain in effect for the period set forth in section 114(f)(1)(C) or 114(f)(2)(C), as the case may be.

(4) Section 115 proceedings.—A petition described in subsection (a) to initiate proceedings under section 801(b)(1) concerning the adjustment or determination of royalty rates as provided in section 115 may be filed in the year 2006 and in each subsequent fifth calendar year, or at such other times as the parties have agreed under section 115(c)(3) (B) and (C).

(5) Section 116 proceedings.—(A) A petition described in subsection (a) to initiate proceedings under section 801(b) concerning the determination of royalty rates and terms as provided in section 116 may be filed at any time within 1 year after negotiated licenses authorized by section 116 are terminated or expire and are not replaced by subsequent agreements.

(B) If a negotiated license authorized by section 116 is terminated or expires and is not replaced by another such license agreement which provides permission to use a quantity of musical works not substantially smaller than the quantity of such works performed on coin-operated phonorecord players during the 1-year period ending March 1, 1989, the Copyright Royalty Judges shall, upon petition filed under paragraph (1) within 1 year after such termination or expiration, commence a proceeding to promptly establish an interim royalty rate or rates for the public performance by means of a coin-operated phonorecord player of nondramatic musical works embodied in phonorecords which had been subject to the terminated or expired negotiated license agreement. Such rate or rates shall be the same as the last such rate or rates and shall remain in force until the conclusion of proceedings by the Copyright Royalty Judges, in accordance with section 803, to adjust the royalty rates applicable to such works, or until superseded by a new negotiated license agreement, as provided in section 116(b).

(6) Section 118 proceedings.—A petition described in subsection (a) to initiate proceedings under section 801(b)(1) concerning the determination of reasonable terms and rates of royalty payments as provided in section 118 may be filed in the year 2006 and in each subsequent fifth calendar year.

(7) Section 1004 proceedings.—A petition described in subsection (a) to initiate proceedings under section 801(b)(1) concerning the adjustment of reasonable royalty rates under section 1004 may be filed as provided in section 1004(a)(3).

(8) Proceedings concerning distribution of royalty fees.—With respect to proceedings under section 801(b)(3) concerning the distribution of royalty fees in certain circumstances under section 111, 119, or 1007, the Copyright Royalty Judges shall, upon a determination that a controversy exists concerning such distribution, cause to be published in the Federal Register notice of commencement of proceedings under this chapter.

§ 805 General Rule for Voluntarily Negotiated Agreements

Any rates or terms under this title that—

(1) are agreed to by participants to a proceeding under section 803(b)(3),

(2) are adopted by the Copyright Royalty Judges as part of a determination under this chapter, and

(3) are in effect for a period shorter than would otherwise apply under a determination pursuant to this chapter,

shall remain in effect for such period of time as would otherwise apply under such determination, except that the Copyright Royalty Judges shall adjust the rates pursuant to the voluntary negotiations to reflect national monetary inflation during the additional period the rates remain in effect.

Chapter 8 Notes

1. The Copyright Royalty and Distribution Reform Act of 2004 amended chapter 8 in its entirety. Pub. L. No. 108-419, 118 Stat. 2341.

In 2006, the Copyright Royalty Judges Program Technical Corrections Act amended chapter 8 throughout. Pub. L. No. 109-303, 120 Stat. 1478. Section 6 of that Act states, "Except as provided under subsection (b), this Act and the amendments made by this Act shall be effective as if included in the Copyright Royalty and Distribution Reform Act of 2004." *Id.* at 1483.

2. In 2006, the Copyright Royalty Judges Program Technical Corrections Act amended section 801 by inserting a comma after "119" in the first sentence of subsection (b)(1) and by adding a new subsection (f) at the end. Pub. L. No. 109-303, 120 Stat. 1478. It also amended the language in 803(b)(3)(C) that preceded (i) and substituted "the" for "such" in (i). *Id.* at 1483.

3. In 2006, the Copyright Royalty Judges Program Technical Corrections Act amended 802(f)(1)(A)(i) by substituting "subparagraph (B) and clause (ii) of this subparagraph" for "clause (ii) of this subparagraph and subparagraph (B),"; by amending (f)(1)(A)(ii) in its entirety; and by inserting a comma after "undertakes to consult with" in the seventh sentence of (f)(1)(D). Pub. L. No. 109-303, 120 Stat. 1478-79.

4. The Satellite Home Viewer Extension and Reauthorization Act of 2004 amended section 803(b)(1)(A)(i)(V) by inserting at the end, "except that in the case of proceedings under section 111 that are scheduled to commence in 2005, such notice may not be published." Pub. L. No. 108-447, 118 Stat. 2809, 3393, 3409.

In 2006, the Copyright Royalty Judges Program Technical Corrections Act amended paragraph 803(a)(1) by inserting a new sentence at the beginning and by amending the second sentence. Pub. L. No. 109-303, 120 Stat. 1478, 1479. It amended (b)(1)(A)(i)(V) by inserting "the publication of notice requirement shall not apply" prior to "in the case of" and deleting from the end of the sentence "such notice may not be published." *Id.* It amended (b)(2)(A) by deleting from the end "together with a filing fee of $150" and by adding a new clause (D). *Id.* at 1479-80. It amended (b)(3)(A) by changing the heading and adding the text for (ii). *Id.* at 1480. It amended (b)(4)(A) by deleting the last sentence. *Id.* It amended the first sentence of (b)(6)(C)(i) by inserting "and written rebuttal statements" after "direct statements" and inserting "in the case of written direct statements" after "Copyright Royalty Judge." *Id.* It entirely amended (b)(6)(C)(ii)(I), iv, and x. *Id.* It amended (c)(2)(B) by deleting "concerning rates and terms" at the end of the sentence; (c)(4) by deleting "with the approval of the Register of Copyrights" in the first sentence after "Copyright Royalty Judges;" and (c) (7) by making a technical correction to add "the" before "Copyright Royalty Judges." *Id.* It amended (d)(2)C)(i)(I) by inserting "applicable" before "statements of account" and deleting "any" before "reports of use." *Id.* at 1481. It amended (d)(3) by inserting a new sentence at the beginning and by deleting "pursuant to section 706 of title 5" at the beginning of what was previously the first sentence, now the second sentence. *Id.*

The Copyright Cleanup, Clarification, and Corrections Act of 2010 amended subparagraph 803(b)(6)(A) by revising the second sentence in its entirety. Pub. L. No. 111-295, 124 Stat. 3180, 3181.

5. In 2006, the Copyright Royalty Judges Program Technical Corrections Act amended paragraph 804(b)(1)(B) by substituting "801(b)(2)(B) or (C)" for "801(b)(3)(B) or (C)" in the third sentence. Pub. L. No. 109-303, 120 Stat. 1478, 1481. It amended (b)(3)(A) by substituting "date of enactment" for "effective date" and (b)(3)(C)(i) and (ii) by making technical corrections to correct grammatical errors. *Id.*

The Satellite Television Extension and Localism Act of 2010 amended paragraphs 804(b)(1)(A) and (B) by substituting "2015" for "2005." Pub. L. No. 111-175, 124 Stat. 1218, 1238.

Chapter 9 Protection of Semiconductor Chip Products

§ 901 Definitions

(a) As used in this chapter—

(1) a "semiconductor chip product" is the final or intermediate form of any product—

(A) having two or more layers of metallic, insulating, or semiconductor material, deposited or otherwise placed on, or etched away or otherwise removed from, a piece of semiconductor material in accordance with a predetermined pattern; and

(B) intended to perform electronic circuitry functions;

(2) a "mask work" is a series of related images, however fixed or encoded—

(A) having or representing the predetermined, three-dimensional pattern of metallic, insulating, or semiconductor material present or removed from the layers of a semiconductor chip product; and

(B) in which series the relation of the images to one another is that each image has the pattern of the surface of one form of the semiconductor chip product;

(3) a mask work is "fixed" in a semiconductor chip product when its embodiment in the product is sufficiently permanent or stable to permit the mask work to be perceived or reproduced from the product for a period of more than transitory duration;

(4) to "distribute" means to sell, or to lease, bail, or otherwise transfer, or to offer to sell, lease, bail, or otherwise transfer;

(5) to "commercially exploit" a mask work is to distribute to the public for commercial purposes a semiconductor chip product embodying the mask work; except that such term includes an offer to sell or transfer a semiconductor chip product only when the offer is in writing and occurs after the mask work is fixed in the semiconductor chip product;

(6) the "owner" of a mask work is the person who created the mask work, the legal representative of that person if that person is deceased or under a legal incapacity, or a party to whom all the rights under this chapter of such person or representative are transferred in accordance with section 903(b); except that, in the case of a work made within the scope of a person's employment, the owner is the employer for whom the person created the mask work or a party to whom all the rights under this chapter of the employer are transferred in accordance with section 903(b);

(7) an "innocent purchaser" is a person who purchases a semiconductor chip product in good faith and without having notice of protection with respect to the semiconductor chip product;

(8) having "notice of protection" means having actual knowledge that, or reasonable grounds to believe that, a mask work is protected under this chapter; and

(9) an "infringing semiconductor chip product" is a semiconductor chip product which is made, imported, or distributed in violation of the exclusive rights of the owner of a mask work under this chapter.

(b) For purposes of this chapter, the distribution or importation of a product incorporating a semiconductor chip product as a part thereof is a distribution or importation of that semiconductor chip product.

§ 902 Subject Matter of Protection

(a)(1) Subject to the provisions of subsection (b), a mask work fixed in a semiconductor chip product, by or under the authority of the owner of the mask work, is eligible for protection under this chapter if—

(A) on the date on which the mask work is registered under section 908, or is first commercially exploited anywhere in the world, whichever occurs first, the owner of the mask work is (i) a national or domiciliary of the United States, (ii) a national, domiciliary, or sovereign authority of a foreign nation that is a party to a treaty affording protection to mask works to which the United States is also a party, or (iii) a stateless person, wherever that person may be domiciled;

(B) the mask work is first commercially exploited in the United States; or

(C) the mask work comes within the scope of a Presidential proclamation issued under paragraph (2).

(2) Whenever the President finds that a foreign nation extends, to mask works of owners who are nationals or domiciliaries of the United States protection (A) on substantially the same basis as that on which the foreign nation extends protection to mask works of its own nationals and domiciliaries and mask works first commercially exploited in that nation, or (B) on substantially the same basis as provided in this chapter, the President may by proclamation extend protection under this chapter to mask works (i) of owners who are, on the date on which the mask works are registered under section 908, or the date on which the mask works are first commercially exploited anywhere in the world, whichever occurs first, nationals, domiciliaries, or sovereign authorities of that nation, or (ii) which are first commercially exploited in that nation. The President may revise, suspend, or revoke any such proclamation or impose any conditions or limitations on protection extended under any such proclamation.

(b) Protection under this chapter shall not be available for a mask work that—

(1) is not original; or

(2) consists of designs that are staple, commonplace, or familiar in the semiconductor industry, or variations of such designs, combined in a way that, considered as a whole, is not original.

(c) In no case does protection under this chapter for a mask work extend to any idea, procedure, process, system, method of operation, concept, principle, or discovery, regardless of the form in which it is described, explained, illustrated, or embodied in such work.

§ 903 Ownership, Transfer, Licensing, and Recordation

(a) The exclusive rights in a mask work subject to protection under this chapter belong to the owner of the mask work.

(b) The owner of the exclusive rights in a mask work may transfer all of those rights, or license all or less than all of those rights, by any written instrument signed by such owner or a duly authorized agent of the owner. Such rights may be transferred or licensed by operation of law, may be bequeathed by will, and may pass as personal property by the applicable laws of intestate succession.

(c)(1) Any document pertaining to a mask work may be recorded in the Copyright Office if the document filed for recordation bears the actual signature of the person who executed it, or if it is accompanied by a sworn or official certification that it is a true copy of the original, signed document. The Register of Copyrights shall, upon receipt of the document and the fee specified pursuant to section 908(d), record the document and return it with a certificate of recordation. The recordation of any transfer or license under this paragraph gives all persons constructive notice of the facts stated in the recorded document concerning the transfer or license.

(2) In any case in which conflicting transfers of the exclusive rights in a mask work are made, the transfer first executed shall be void as against a subsequent transfer which is made for a valuable consideration and without notice of the first transfer, unless the first transfer is recorded in accordance with paragraph

(1) within three months after the date on which it is executed, but in no case later than the day before the date of such subsequent transfer.

(d) Mask works prepared by an officer or employee of the United States Government as part of that person's official duties are not protected under this chapter, but the United States Government is not precluded from receiving and holding exclusive rights in mask works transferred to the Government under subsection (b).

§ 904 Duration of Protection

(a) The protection provided for a mask work under this chapter shall commence on the date on which the mask work is registered under section 908, or the date on which the mask work is first commercially exploited anywhere in the world, whichever occurs first.

(b) Subject to subsection (c) and the provisions of this chapter, the protection provided under this chapter to a mask work shall end ten years after the date on which such protection commences under subsection (a).

(c) All terms of protection provided in this section shall run to the end of the calendar year in which they would otherwise expire.

§ 905 Exclusive Rights in Mask Works

The owner of a mask work provided protection under this chapter has the exclusive rights to do and to authorize any of the following:

(1) to reproduce the mask work by optical, electronic, or any other means;

(2) to import or distribute a semiconductor chip product in which the mask work is embodied; and

(3) to induce or knowingly to cause another person to do any of the acts described in paragraphs (1) and (2).

§ 906 Limitation on Exclusive Rights: Reverse Engineering; First Sale

(a) Notwithstanding the provisions of section 905, it is not an infringement of the exclusive rights of the owner of a mask work for—

(1) a person to reproduce the mask work solely for the purpose of teaching, analyzing, or evaluating the concepts or techniques embodied in the mask work or the circuitry, logic flow, or organization of components used in the mask work; or

(2) a person who performs the analysis or evaluation described in paragraph (1) to incorporate the results of such conduct in an original mask work which is made to be distributed.

(b) Notwithstanding the provisions of section 905(2), the owner of a particular semiconductor chip product made by the owner of the mask work, or by any person authorized by the owner of the mask work, may import, distribute, or otherwise dispose of or use, but not reproduce, that particular semiconductor chip product without the authority of the owner of the mask work.

§ 907 Limitation on Exclusive Rights: Innocent Infringement

(a) Notwithstanding any other provision of this chapter, an innocent purchaser of an infringing semiconductor chip product—

(1) shall incur no liability under this chapter with respect to the importation or distribution of units of the infringing semiconductor chip product that occurs before the innocent purchaser has notice of protection with respect to the mask work embodied in the semiconductor chip product; and

(2) shall be liable only for a reasonable royalty on each unit of the infringing semiconductor chip product that the innocent purchaser imports or distributes after having notice of protection with respect to the mask work embodied in the semiconductor chip product.

(b) The amount of the royalty referred to in subsection (a)(2) shall be determined by the court in a civil action for infringement unless the parties resolve the issue by voluntary negotiation, mediation, or binding arbitration.

(c) The immunity of an innocent purchaser from liability referred to in subsection (a)(1) and the limitation of remedies with respect to an innocent purchaser referred to in subsection (a)(2) shall extend to any person who directly or indirectly purchases an infringing semiconductor chip product from an innocent purchaser.

(d) The provisions of subsections (a), (b), and (c) apply only with respect to those units of an infringing semiconductor chip product that an innocent purchaser purchased before having notice of protection with respect to the mask work embodied in the semiconductor chip product.

§ 908 Registration of Claims of Protection

(a) The owner of a mask work may apply to the Register of Copyrights for registration of a claim of protection in a mask work. Protection of a mask work under this chapter shall terminate if application for registration of a claim of protection in the mask work is not made as provided in this chapter within two years after the date on which the mask work is first commercially exploited anywhere in the world.

(b) The Register of Copyrights shall be responsible for all administrative functions and duties under this chapter. Except for section 708, the provisions of chapter 7 of this title relating to the general responsibilities, organization, regulatory authority, actions, records, and publications of the Copyright Office shall apply to this chapter, except that the Register of Copyrights may make such changes as may be necessary in applying those provisions to this chapter.

(c) The application for registration of a mask work shall be made on a form prescribed by the Register of Copyrights. Such form may require any information regarded by the Register as bearing upon the preparation or identification of the mask work, the existence or duration of protection of the mask work under this chapter, or ownership of the mask work. The application shall be accompanied by the fee set pursuant to subsection (d) and the identifying material specified pursuant to such subsection.

(d) The Register of Copyrights shall by regulation set reasonable fees for the filing of applications to register claims of protection in mask works under this chapter, and for other services relating to the administration of this chapter or the rights under this chapter, taking into consideration the cost of providing those services, the benefits of a public record, and statutory fee schedules under this title. The Register shall also specify the identifying material to be deposited in connection with the claim for registration.

(e) If the Register of Copyrights, after examining an application for registration, determines, in accordance with the provisions of this chapter, that the application relates to a mask work which is entitled to protection under this chapter, then the Register shall register the claim of protection and issue to the applicant a certificate of registration of the claim of protection under the seal of the Copyright Office. The effective date of registration of a claim of protection shall be the date on which an application, deposit of identifying material, and fee, which are determined by the Register of Copyrights or by a court of competent jurisdiction to be acceptable for registration of the claim, have all been received in the Copyright Office.

(f) In any action for infringement under this chapter, the certificate of registration of a mask work shall constitute prima facie evidence (1) of the facts stated in the certificate, and (2) that the applicant issued the certificate has met the requirements of this chapter, and the regulations issued under this chapter, with respect to the registration of claims.

(g) Any applicant for registration under this section who is dissatisfied with the refusal of the Register of Copyrights to issue a certificate of registration under this section may seek judicial review of that refusal by bringing an action for such review in an appropriate United States district court not later than sixty days after the refusal. The provisions of chapter 7 of title 5 shall apply to such judicial review. The failure of the Register of Copyrights to issue a certificate of registration within four months after an application for registration is filed shall be deemed to be a refusal to issue a certificate of registration for purposes of this subsection and section 910(b)(2), except that, upon a showing of good cause, the district court may shorten such four-month period.

§ 909 Mask Work Notice

(a) The owner of a mask work provided protection under this chapter may affix notice to the mask work, and to masks and semiconductor chip products embodying the mask work, in such manner and location as to give reasonable notice of such protection. The Register of Copyrights shall prescribe by regulation, as examples, specific methods of affixation and positions of notice for purposes of this section, but these specifications shall not be considered exhaustive. The affixation of such notice is not a condition of protection under this chapter, but shall constitute prima facie evidence of notice of protection.

(b) The notice referred to in subsection (a) shall consist of—

(1) the words "mask work", the symbol *M*, or the symbol Ⓜ (the letter M in a circle); and

(2) the name of the owner or owners of the mask work or an abbreviation by which the name is recognized or is generally known.

§ 910 Enforcement of Exclusive Rights

(a) Except as otherwise provided in this chapter, any person who violates any of the exclusive rights of the owner of a mask work under this chapter, by conduct in or affecting commerce, shall be liable as an infringer of such rights. As used in this subsection, the term "any person" includes any State, any instrumentality of a State, and any officer or employee of a State or instrumentality of a State acting in his or her official capacity. Any State, and any such instrumentality, officer, or employee, shall be subject to the provisions of this chapter in the same manner and to the same extent as any nongovernmental entity.

(b) (1) The owner of a mask work protected under this chapter, or the exclusive licensee of all rights under this chapter with respect to the mask work, shall, after a certificate of registration of a claim of protection in that mask work has been issued under section 908, be entitled to institute a civil action for any infringement with respect to the mask work which is committed after the commencement of protection of the mask work under section 904(a).

(2) In any case in which an application for registration of a claim of protection in a mask work and the required deposit of identifying material and fee have been received in the Copyright Office in proper form and registration of the mask work has been refused, the applicant is entitled to institute a civil action for infringement under this chapter with respect to the mask work if notice of the action, together with a copy of the complaint, is served on the Register of Copyrights, in accordance with the Federal Rules of Civil Procedure. The Register may, at his or her option, become a party to the action with respect to the issue of whether the claim of protection is eligible for registration by entering an appearance within sixty days after such service, but the failure of the Register to become a party to the action shall not deprive the court of jurisdiction to determine that issue.

(c)(1) The Secretary of the Treasury and the United States Postal Service shall separately or jointly issue regulations for the enforcement of the rights set forth in section 905 with respect to importation. These regulations may require, as a condition for the exclusion of articles from the United States, that the person seeking exclusion take any one or more of the following actions:

(A) Obtain a court order enjoining, or an order of the International Trade Commission under section 337 of the Tariff Act of 1930 excluding, importation of the articles.

(B) Furnish proof that the mask work involved is protected under this chapter and that the importation of the articles would infringe the rights in the mask work under this chapter.

(C) Post a surety bond for any injury that may result if the detention or exclusion of the articles proves to be unjustified.

(2) Articles imported in violation of the rights set forth in section 905 are subject to seizure and forfeiture in the same manner as property imported in violation of the customs laws. Any such forfeited articles shall be destroyed as directed by the Secretary of the Treasury or the court, as the case may be, except that the articles may be returned to the country of export whenever it is shown to the satisfaction of the Secretary of the Treasury that the importer had no reasonable grounds for believing that his or her acts constituted a violation of the law.

§ 911 Civil Actions

(a) Any court having jurisdiction of a civil action arising under this chapter may grant temporary restraining orders, preliminary injunctions, and permanent injunctions on such terms as the court may deem reasonable to prevent or restrain infringement of the exclusive rights in a mask work under this chapter.

(b) Upon finding an infringer liable, to a person entitled under section 910(b)(1) to institute a civil action, for an infringement of any exclusive right under this chapter, the court shall award such person actual damages suffered by the person as a result of the infringement. The court shall also award such person the infringer's profits that are attributable to the infringement and are not taken into account in computing the award of actual damages. In establishing the infringer's profits, such person is required to present proof only of the infringer's gross revenue, and the infringer is required to prove his or her deductible expenses and the elements of profit attributable to factors other than the mask work.

(c) At any time before final judgment is rendered, a person entitled to institute a civil action for infringement may elect, instead of actual damages and profits as provided by subsection (b), an award of statutory damages for all infringements involved in the action, with respect to any one mask work for which any one infringer is liable individually, or for which any two or more infringers are liable jointly and severally, in an amount not more than $250,000 as the court considers just.

(d) An action for infringement under this chapter shall be barred unless the action is commenced within three years after the claim accrues.

(e)(1) At any time while an action for infringement of the exclusive rights in a mask work under this chapter is pending, the court may order the impounding, on such terms as it may deem reasonable, of all semiconductor chip products, and any drawings, tapes, masks, or other products by means of which such products may be reproduced, that are claimed to have been made, imported, or used in violation of those exclusive rights. Insofar as practicable, applications for orders under this paragraph shall be heard and determined in the same manner as an application for a temporary restraining order or preliminary injunction.

(2) As part of a final judgment or decree, the court may order the destruction or other disposition of any infringing semiconductor chip products, and any masks, tapes, or other articles by means of which such products may be reproduced.

(f) In any civil action arising under this chapter, the court in its discretion may allow the recovery of full costs, including reasonable attorneys' fees, to the prevailing party.

(g)(1) Any State, any instrumentality of a State, and any officer or employee of a State or instrumentality of a State acting in his or her official capacity, shall not be immune, under the Eleventh Amendment of the Constitution of the United States or under any other doctrine of sovereign immunity, from suit in Federal court by any person, including any governmental or nongovernmental entity, for a violation of any of the exclusive rights of the owner of a mask work under this chapter, or for any other violation under this chapter.

(2) In a suit described in paragraph (1) for a violation described in that paragraph, remedies (including remedies both at law and in equity) are available for the violation to the same extent as such remedies are available for such a violation in a suit against any public or private entity other than a State, instrumentality of a State, or officer or employee of a State acting in his or her official capacity. Such remedies include actual damages and profits under subsection (b), statutory damages under subsection (c), impounding and disposition of infringing articles under subsection (e), and costs and attorney's fees under subsection (f).

§ 912 Relation to Other Laws

(a) Nothing in this chapter shall affect any right or remedy held by any person under chapters 1 through 8 or 10 of this title, or under title 35.

(b) Except as provided in section 908(b) of this title, references to "this title" or "title 17" in chapters 1 through 8 or 10 of this title shall be deemed not to apply to this chapter.

(c) The provisions of this chapter shall preempt the laws of any State to the extent those laws provide any rights or remedies with respect to a mask work which are equivalent to those rights or remedies provided by this chapter, except that such preemption shall be effective only with respect to actions filed on or after January 1, 1986.

(d) Notwithstanding subsection (c), nothing in this chapter shall detract from any rights of a mask work owner, whether under Federal law (exclusive of this chapter) or under the common law or the statutes of a State, heretofore or hereafter declared or enacted, with respect to any mask work first commercially exploited before July 1, 1983.

§ 913 Transitional Provisions

(a) No application for registration under section 908 may be filed, and no civil action under section 910 or other enforcement proceeding under this chapter may be instituted, until sixty days after the date of the enactment of this chapter.

(b) No monetary relief under section 911 may be granted with respect to any conduct that occurred before the date of the enactment of this chapter, except as provided in subsection (d).

(c) Subject to subsection (a), the provisions of this chapter apply to all mask works that are first commercially exploited or are registered under this chapter, or both, on or after the date of the enactment of this chapter.

(d)(1) Subject to subsection (a), protection is available under this chapter to any mask work that was first commercially exploited on or after July 1, 1983, and before the date of the enactment of this chapter, if a claim of protection in the mask work is registered in the Copyright Office before July 1, 1985, under section 908.

(2) In the case of any mask work described in paragraph (1) that is provided protection under this chapter, infringing semiconductor chip product units manufactured before the date of the enactment of this chapter may, without liability under sections 910 and 911, be imported into or distributed in the United States, or both, until two years after the date of registration of the mask work under section 908, but only if the importer or distributor, as the case may be, first pays or offers to pay the reasonable royalty referred to in section 907(a)(2) to the mask work owner, on all such units imported or distributed, or both, after the date of the enactment of this chapter.

(3) In the event that a person imports or distributes infringing semiconductor chip product units described in paragraph (2) of this subsection without first paying or offering to pay the reasonable royalty specified in such paragraph, or if the person refuses or fails to make such payment, the mask work owner shall be entitled to the relief provided in sections 910 and 911.

§ 914 International Transitional Provisions

(a) Notwithstanding the conditions set forth in subparagraphs (A) and (C) of section 902(a)(1) with respect to the availability of protection under this chapter to nationals, domiciliaries, and sovereign authorities of a foreign nation, the Secretary of Commerce may, upon the petition of any person, or upon the Secretary's own motion, issue an order extending protection under this chapter to such foreign nationals, domiciliaries, and sovereign authorities if the Secretary finds—

(1) that the foreign nation is making good faith efforts and reasonable progress toward—

(A) entering into a treaty described in section 902(a)(1)(A); or

(B) enacting or implementing legislation that would be in compliance with subparagraph (A) or (B) of section 902(a)(2); and

(2) that the nationals, domiciliaries, and sovereign authorities of the foreign nation, and persons controlled by them, are not engaged in the misappropriation, or unauthorized distribution or commercial exploitation, of mask works; and

(3) that issuing the order would promote the purposes of this chapter and international comity with respect to the protection of mask works.

(b) While an order under subsection (a) is in effect with respect to a foreign nation, no application for registration of a claim for protection in a mask work under this chapter may be denied solely because the owner of the mask work is a national, domiciliary, or sovereign authority of that foreign nation, or solely because the mask work was first commercially exploited in that foreign nation.

(c) Any order issued by the Secretary of Commerce under subsection (a) shall be effective for such a period as the Secretary designates in the order, except that no such order may be effective after that date on which the authority of the Secretary of Commerce terminates under subsection (e). The effective date of any such order shall also be designated in the order. In the case of an order issued upon the petition of a person, such effective date may be no earlier than the date on which the Secretary receives such petition.

(d)(1) Any order issued under this section shall terminate if—

(A) the Secretary of Commerce finds that any of the conditions set forth in paragraphs (1), (2), and (3) of subsection (a) no longer exist; or

(B) mask works of nationals, domiciliaries, and sovereign authorities of that foreign nation or mask works first commercially exploited in that foreign nation become eligible for protection under subparagraph (A) or

(C) of section 902(a)(1).

(2) Upon the termination or expiration of an order issued under this section, registrations of claims of protection in mask works made pursuant to that order shall remain valid for the period specified in section 904.

(e) The authority of the Secretary of Commerce under this section shall commence on the date of the enactment of this chapter, and shall terminate on July 1, 1995.

(f) (1) The Secretary of Commerce shall promptly notify the Register of Copyrights and the Committees on the Judiciary of the Senate and the House of Representatives of the issuance or termination of any order under this section, together with a statement of the reasons for such action. The Secretary shall also publish such notification and statement of reasons in the Federal Register.

(2) Two years after the date of the enactment of this chapter, the Secretary of Commerce, in consultation with the Register of Copyrights, shall transmit to the Committees on the Judiciary of the Senate and the House of Representatives a report on the actions taken under this section and on the current status of international recognition of mask work protection. The report shall include such recommendation for modifications of the protection accorded under this chapter to mask works owned by nationals, domiciliaries, or sovereign authorities of

foreign nations as the Secretary, in consultation with the Register of Copyrights, considers would promote the purposes of this chapter and international comity with respect to mask work protection. Not later than July 1, 1994, the Secretary of Commerce, in consultation with the Register of Copyrights, shall transmit to the Committees on the Judiciary of the Senate and the House of Representatives a report updating the matters contained in the report transmitted under the preceding sentence.

Chapter 9 Notes

1. In 1984, the Semiconductor Chip Protection Act amended title 17 of the *United States Code* to add a new chapter 9 entitled "Protection of Semiconductor Chip Products." Pub. L. No. 98-620, 98 Stat. 3335, 3347.

2. In 1997, the heading for section 903 in the table of sections was changed from "Ownership and Transfer" to "Ownership, transfer, licensure, and recordation." Pub. L. No. 105-80, 111 Stat. 1529, 1535. The Intellectual Property and High Technology Technical Amendments Act of 2002 amended the heading for section 903 in the table of sections for chapter 9 by substituting "licensing" for "licensure." Pub. L. No. 107-273, 116 Stat. 1758, 1910.

3. In 1987, section 902 was amended by adding the last sentence in subsection (a)(2). Pub. L. No. 100-159, 101 Stat. 899, 900.

4. In 1997, section 909 was amended by correcting misspellings in subsection (b)(1). Pub. L. No. 105-80, 111 Stat. 1529, 1535.

5. In 1990, the Copyright Remedy Clarification Act amended section 910 by adding the last two sentences to subsection (a). Pub. L. No. 101-553, 104 Stat. 2749, 2750. In 1997, a technical correction amended section 910(a) by capitalizing the first word of the second sentence. Pub. L. No. 105-80, 111 Stat. 1529 1535.

6. In 1990, the Copyright Remedy Clarification Act amended section 911 by adding subsection (g). Pub. L. No. 101-553, 104 Stat. 2749, 2750.

7. In 1988, the Judicial Improvements and Access to Justice Act amended section 912 by deleting subsection (d) and redesignating subsection (e) as subsection (d). Pub. L. No. 100-702, 102 Stat. 4642, 4672. The Audio Home Recording Act of 1992 amended section 912 by inserting "or 10" after "8" in subsections (a) and (b). Pub. L. No. 102-563, 106 Stat. 4237, 4248.

8. In 1987, section 914 was amended in subsection (e) by inserting "on July 1, 1991" in lieu of "three years after such date of enactment" and by adding the last sentence to subsection (f)(2). Pub. L. No. 100-159, 101 Stat. 899. The Semiconductor International Protection Extension Act of 1991 amended section 914 by inserting "or implementing" after "enacting" in the first sentence of subsection (a)(1)(B), by changing the date in subsection (e) to "July 1, 1995" and by changing the date in the last sentence of subsection (f)(2) to "July 1, 1994." Pub. L. No. 102-64, 105 Stat. 320.

On July 1, 1995, section 914 expired as required by subsection (e). It was rendered largely unnecessary upon the entry into force on January 1, 1995, of the

Agreement on Trade-Related Aspects of Intellectual Property Rights (TRIPs) (Annex 1C to the World Trade Organization (WTO) Agreement). Part II, section 6, of TRIPs protects semiconductor chip products and was the basis for Presidential Proclamation No. 6780, March 23, 1995, under section 902(a)(2) extending protection to all present and future WTO members (146 countries as of April 4, 2003), as of January 1, 1996. See Appendix M. For a discussion of Congressional findings regarding extending protection to semiconductor chip products of foreign entities, see Pub. L. No. 100-159, 101 Stat. 899, and the Semiconductor International Protection Extension Act of 1991, Pub. L. No. 102-64, 105 Stat. 320.

Chapter 10 Digital Audio Recording Devices and Media

Subchapter A—Definitions

§ 1001 Definitions

As used in this chapter, the following terms have the following meanings:

(1) A "digital audio copied recording" is a reproduction in a digital recording format of a digital musical recording, whether that reproduction is made directly from another digital musical recording or indirectly from a transmission.

(2) A "digital audio interface device" is any machine or device that is designed specifically to communicate digital audio information and related interface data to a digital audio recording device through a nonprofessional interface.

(3) A "digital audio recording device" is any machine or device of a type commonly distributed to individuals for use by individuals, whether or not included with or as part of some other machine or device, the digital recording function of which is designed or marketed for the primary purpose of, and that is capable of, making a digital audio copied recording for private use, except for—

(A) professional model products, and

(B) dictation machines, answering machines, and other audio recording equipment that is designed and marketed primarily for the creation of sound recordings resulting from the fixation of nonmusical sounds.

(4)(A) A "digital audio recording medium" is any material object in a form commonly distributed for use by individuals, that is primarily marketed or most commonly used by consumers for the purpose of making digital audio copied recordings by use of a digital audio recording device.

(B) Such term does not include any material object—

(i) that embodies a sound recording at the time it is first distributed by the importer or manufacturer; or

(ii) that is primarily marketed and most commonly used by consumers either for the purpose of making copies of motion pictures or other audiovisual works or for the purpose of making copies of nonmusical literary works, including computer programs or data bases.

(5)(A) A "digital musical recording" is a material object—

(i) in which are fixed, in a digital recording format, only sounds, and material, statements, or instructions incidental to those fixed sounds, if any, and

(ii) from which the sounds and material can be perceived, reproduced, or otherwise communicated, either directly or with the aid of a machine or device.

(B) A "digital musical recording" does not include a material object—

(i) in which the fixed sounds consist entirely of spoken word recordings, or

(ii) in which one or more computer programs are fixed, except that a digital musical recording may contain statements or instructions constituting the fixed sounds and incidental material, and statements or instructions to be used directly or indirectly in order to bring about the perception, reproduction, or communication of the fixed sounds and incidental material.

(C) For purposes of this paragraph—

(i) a "spoken word recording" is a sound recording in which are fixed only a series of spoken words, except that the spoken words may be accompanied by incidental musical or other sounds, and

(ii) the term "incidental" means related to and relatively minor by comparison.

(6) "Distribute" means to sell, lease, or assign a product to consumers in the United States, or to sell, lease, or assign a product in the United States for ultimate transfer to consumers in the United States.

(7) An "interested copyright party" is—

(A) the owner of the exclusive right under section 106(1) of this title to reproduce a sound recording of a musical work that has been embodied in a digital musical recording or analog musical recording lawfully made under this title that has been distributed;

(B) the legal or beneficial owner of, or the person that controls, the right to reproduce in a digital musical recording or analog musical recording a musical work that has been embodied in a digital musical recording or analog musical recording lawfully made under this title that has been distributed;

(C) a featured recording artist who performs on a sound recording that has been distributed; or

(D) any association or other organization—

(i) representing persons specified in subparagraph (A), (B), or (C), or

(ii) engaged in licensing rights in musical works to music users on behalf of writers and publishers.

(8) To "manufacture" means to produce or assemble a product in the United States. A "manufacturer" is a person who manufactures.

(9) A "music publisher" is a person that is authorized to license the reproduction of a particular musical work in a sound recording.

(10) A "professional model product" is an audio recording device that is designed, manufactured, marketed, and intended for use by recording professionals in the ordinary course of a lawful business, in accordance with such requirements as the Secretary of Commerce shall establish by regulation.

(11) The term "serial copying" means the duplication in a digital format of a copyrighted musical work or sound recording from a digital reproduction of a digital musical recording. The term "digital reproduction of a digital musical recording" does not include a digital musical recording as distributed, by authority of the copyright owner, for ultimate sale to consumers.

(12) The "transfer price" of a digital audio recording device or a digital audio recording medium—

(A) is, subject to subparagraph (B)—

(i) in the case of an imported product, the actual entered value at United States Customs (exclusive of any freight, insurance, and applicable duty), and

(ii) in the case of a domestic product, the manufacturer's transfer price (FOB the manufacturer, and exclusive of any direct sales taxes or excise taxes incurred in connection with the sale); and

(B) shall, in a case in which the transferor and transferee are related entities or within a single entity, not be less than a reasonable arms-length price under the principles of the regulations adopted pursuant to section 482 of the Internal Revenue Code of 1986, or any successor provision to such section.

(13) A "writer" is the composer or lyricist of a particular musical work.

Subchapter B—Copying Controls

§ 1002 Incorporation of Copying Controls

(a) Prohibition on Importation, Manufacture, and Distribution.—No person shall import, manufacture, or distribute any digital audio recording device or digital audio interface device that does not conform to—

(1) the Serial Copy Management System;

(2) a system that has the same functional characteristics as the Serial Copy Management System and requires that copyright and generation status information be accurately sent, received, and acted upon between devices using the

system's method of serial copying regulation and devices using the Serial Copy Management System; or

(3) any other system certified by the Secretary of Commerce as prohibiting unauthorized serial copying.

(b) Development of Verification Procedure.—The Secretary of Commerce shall establish a procedure to verify, upon the petition of an interested party, that a system meets the standards set forth in subsection (a)(2).

(c) Prohibition on Circumvention of the System.—No person shall import, manufacture, or distribute any device, or offer or perform any service, the primary purpose or effect of which is to avoid, bypass, remove, deactivate, or otherwise circumvent any program or circuit which implements, in whole or in part, a system described in subsection (a).

(d) Encoding of Information on Digital Musical Recordings.—

(1) Prohibition on encoding inaccurate information.—No person shall encode a digital musical recording of a sound recording with inaccurate information relating to the category code, copyright status, or generation status of the source material for the recording.

(2) Encoding of copyright status not required.—Nothing in this chapter requires any person engaged in the importation or manufacture of digital musical recordings to encode any such digital musical recording with respect to its copyright status.

(e) Information Accompanying Transmission in Digital Format.—Any person who transmits or otherwise communicates to the public any sound recording in digital format is not required under this chapter to transmit or otherwise communicate the information relating to the copyright status of the sound recording. Any such person who does transmit or otherwise communicate such copyright status information shall transmit or communicate such information accurately.

Subchapter C—Royalty Payments

§ 1003 Obligation to Make Royalty Payments

(a) Prohibition on Importation and Manufacture.—No person shall import into and distribute, or manufacture and distribute, any digital audio recording device or digital audio recording medium unless such person records the notice specified by this section and subsequently deposits the statements of account and applicable royalty payments for such device or medium specified in section 1004.

(b) Filing of Notice.—The importer or manufacturer of any digital audio recording device or digital audio recording medium, within a product category or utilizing a technology with respect to which such manufacturer or importer has not previously filed a notice under this subsection, shall file with the Register of Copyrights a notice with respect to such device or medium, in such form and content as the Register shall prescribe by regulation.

(c) Filing of Quarterly and Annual Statements of Account.—

(1) Generally.—Any importer or manufacturer that distributes any digital audio recording device or digital audio recording medium that it manufactured or imported shall file with the Register of Copyrights, in such form and content as the Register shall prescribe by regulation, such quarterly and annual statements of account with respect to such distribution as the Register shall prescribe by regulation.

(2) Certification, verification, and confidentiality.—Each such statement shall be certified as accurate by an authorized officer or principal of the importer or manufacturer. The Register shall issue regulations to provide for the verification and audit of such statements and to protect the confidentiality of the information contained in such statements. Such regulations shall provide for the disclosure, in confidence, of such statements to interested copyright parties.

(3) Royalty payments.—Each such statement shall be accompanied by the royalty payments specified in section 1004.

§ 1004 Royalty Payments

(a) Digital Audio Recording Devices.—

(1) Amount of payment.—The royalty payment due under section 1003 for each digital audio recording device imported into and distributed in the United States, or manufactured and distributed in the United States, shall be 2 percent of the transfer price. Only the first person to manufacture and distribute or import and distribute such device shall be required to pay the royalty with respect to such device.

(2) Calculation for devices distributed with other devices.—With respect to a digital audio recording device first distributed in combination with one or more devices, either as a physically integrated unit or as separate components, the royalty payment shall be calculated as follows:

(A) If the digital audio recording device and such other devices are part of a physically integrated unit, the royalty payment shall be based on the transfer price of the unit, but shall be reduced by any royalty payment made on any digital audio recording device included within the unit that was not first distributed in combination with the unit.

(B) If the digital audio recording device is not part of a physically integrated unit and substantially similar devices have been distributed separately at any time during the preceding 4 calendar quarters, the royalty payment shall be based on the average transfer price of such devices during those 4 quarters.

(C) If the digital audio recording device is not part of a physically integrated unit and substantially similar devices have not been distributed separately at any time during the preceding 4 calendar quarters, the royalty payment shall be based on a constructed price reflecting the proportional value of such device to the combination as a whole.

(3) Limits on royalties.—Notwithstanding paragraph (1) or (2), the amount of the royalty payment for each digital audio recording device shall not be less

than $1 nor more than the royalty maximum. The royalty maximum shall be $8 per device, except that in the case of a physically integrated unit containing more than 1 digital audio recording device, the royalty maximum for such unit shall be $12. During the 6th year after the effective date of this chapter, and not more than once each year thereafter, any interested copyright party may petition the Copyright Royalty Judges to increase the royalty maximum and, if more than 20 percent of the royalty payments are at the relevant royalty maximum, the Copyright Royalty Judges shall prospectively increase such royalty maximum with the goal of having no more than 10 percent of such payments at the new royalty maximum; however the amount of any such increase as a percentage of the royalty maximum shall in no event exceed the percentage increase in the Consumer Price Index during the period under review.

(b) Digital Audio Recording Media.—The royalty payment due under section 1003 for each digital audio recording medium imported into and distributed in the United States, or manufactured and distributed in the United States, shall be 3 percent of the transfer price. Only the first person to manufacture and distribute or import and distribute such medium shall be required to pay the royalty with respect to such medium.

§ 1005 Deposit of Royalty Payments and Deduction of Expenses

The Register of Copyrights shall receive all royalty payments deposited under this chapter and, after deducting the reasonable costs incurred by the Copyright Office under this chapter, shall deposit the balance in the Treasury of the United States as offsetting receipts, in such manner as the Secretary of the Treasury directs. All funds held by the Secretary of the Treasury shall be invested in interestbearing United States securities for later distribution with interest under section 1007. The Register may, in the Register's discretion, 4 years after the close of any calendar year, close out the royalty payments account for that calendar year, and may treat any funds remaining in such account and any subsequent deposits that would otherwise be attributable to that calendar year as attributable to the succeeding calendar year.

§ 1006 Entitlement to Royalty Payments

(a) Interested Copyright Parties.—The royalty payments deposited pursuant to section 1005 shall, in accordance with the procedures specified in section 1007, be distributed to any interested copyright party—

(1) whose musical work or sound recording has been—

(A) embodied in a digital musical recording or an analog musical recording lawfully made under this title that has been distributed, and

(B) distributed in the form of digital musical recordings or analog musical recordings or disseminated to the public in transmissions, during the period to which such payments pertain; and

(2) who has filed a claim under section 1007.

(b) Allocation of Royalty Payments to Groups.—The royalty payments shall be divided into 2 funds as follows:

(1) The sound recordings fund.—$66\frac{2}{3}$ percent of the royalty payments shall be allocated to the Sound Recordings Fund. $2\frac{5}{8}$ percent of the royalty payments allocated to the Sound Recordings Fund shall be placed in an escrow account managed by an independent administrator jointly appointed by the interested copyright parties described in section 1001(7)(A) and the American Federation of Musicians (or any successor entity) to be distributed to nonfeatured musicians (whether or not members of the American Federation of Musicians or any successor entity) who have performed on sound recordings distributed in the United States. $1\frac{3}{8}$ percent of the royalty payments allocated to the Sound Recordings Fund shall be placed in an escrow account managed by an independent administrator jointly appointed by the interested copyright parties described in section 1001(7)(A) and the American Federation of Television and Radio Artists (or any successor entity) to be distributed to nonfeatured vocalists (whether or not members of the American Federation of Television and Radio Artists or any successor entity) who have performed on sound recordings distributed in the United States. 40 percent of the remaining royalty payments in the Sound Recordings Fund shall be distributed to the interested copyright parties described in section 1001(7)(C), and 60 percent of such remaining royalty payments shall be distributed to the interested copyright parties described in section 1001(7)(A).

(2) The Musical Works Fund.—

(A) $33\frac{1}{3}$ percent of the royalty payments shall be allocated to the Musical Works Fund for distribution to interested copyright parties described in section 1001(7)(B).

(B)(i) Music publishers shall be entitled to 50 percent of the royalty payments allocated to the Musical Works Fund.

(ii) Writers shall be entitled to the other 50 percent of the royalty payments allocated to the Musical Works Fund.

(c) Allocation of Royalty Payments Within Groups.—If all interested copyright parties within a group specified in subsection (b) do not agree on a voluntary proposal for the distribution of the royalty payments within each group, the Copyright Royalty Judges shall, pursuant to the procedures specified under section 1007(c), allocate royalty payments under this section based on the extent to which, during the relevant period—

(1) for the Sound Recordings Fund, each sound recording was distributed in the form of digital musical recordings or analog musical recordings; and

(2) for the Musical Works Fund, each musical work was distributed in the form of digital musical recordings or analog musical recordings or disseminated to the public in transmissions.

§ 1007 Procedures for Distributing Royalty Payments

(a) Filing of Claims and Negotiations.—

(1) Filing of claims.—During the first 2 months of each calendar year, every interested copyright party seeking to receive royalty payments to which such party is entitled under section 1006 shall file with the Copyright Royalty Judges a claim for payments collected during the preceding year in such form and manner as the Copyright Royalty Judges shall prescribe by regulation.

(2) Negotiations.—Notwithstanding any provision of the antitrust laws, for purposes of this section interested copyright parties within each group specified in section 1006(b) may agree among themselves to the proportionate division of royalty payments, may lump their claims together and file them jointly or as a single claim, or may designate a common agent, including any organization described in section 1001(7)(D), to negotiate or receive payment on their behalf; except that no agreement under this subsection may modify the allocation of royalties specified in section 1006(b).

(b) Distribution of Payments in the Absence of a Dispute.—After the period established for the filing of claims under subsection (a), in each year, the Copyright Royalty Judges shall determine whether there exists a controversy concerning the distribution of royalty payments under section 1006(c). If the Copyright Royalty Judges determine that no such controversy exists, the Copyright Royalty Judges shall, within 30 days after such determination, authorize the distribution of the royalty payments as set forth in the agreements regarding the distribution of royalty payments entered into pursuant to subsection (a). The Librarian of Congress shall, before such royalty payments are distributed, deduct the reasonable administrative costs incurred under this section.

(c) Resolution of Disputes.—If the Copyright Royalty Judges find the existence of a controversy, the Copyright Royalty Judges shall, pursuant to chapter 8 of this title, conduct a proceeding to determine the distribution of royalty payments. During the pendency of such a proceeding, the Copyright Royalty Judges shall withhold from distribution an amount sufficient to satisfy all claims with respect to which a controversy exists, but shall, to the extent feasible, authorize the distribution of any amounts that are not in controversy. The Librarian of Congress shall, before such royalty payments are distributed, deduct the reasonable administrative costs incurred under this section.

Subchapter D—Prohibition on Certain Infringement Actions, Remedies, and Arbitration

§ 1008 Prohibition on Certain Infringement Actions

No action may be brought under this title alleging infringement of copyright based on the manufacture, importation, or distribution of a digital audio recording device, a digital audio recording medium, an analog recording device, or an analog recording medium, or based on the noncommercial use by a consumer of such a device or medium for making digital musical recordings or analog musical recordings.

§ 1009 Civil Remedies

(a) Civil Actions.—Any interested copyright party injured by a violation of section 1002 or 1003 may bring a civil action in an appropriate United States district court against any person for such violation.

(b) Other Civil Actions.—Any person injured by a violation of this chapter may bring a civil action in an appropriate United States district court for actual damages incurred as a result of such violation.

(c) Powers of the Court.—In an action brought under subsection (a), the court—

(1) may grant temporary and permanent injunctions on such terms as it deems reasonable to prevent or restrain such violation;

(2) in the case of a violation of section 1002, or in the case of an injury resulting from a failure to make royalty payments required by section 1003, shall award damages under subsection (d);

(3) in its discretion may allow the recovery of costs by or against any party other than the United States or an officer thereof; and

(4) in its discretion may award a reasonable attorney's fee to the prevailing party.

(d) Award of Damages.—

(1) Damages for section 1002 or 1003 violations.—

(A) Actual damages.—

(i) In an action brought under subsection (a), if the court finds that a violation of section 1002 or 1003 has occurred, the court shall award to the complaining party its actual damages if the complaining party elects such damages at any time before final judgment is entered.

(ii) In the case of section 1003, actual damages shall constitute the royalty payments that should have been paid under section 1004 and deposited under section 1005. In such a case, the court, in its discretion, may award an additional amount of not to exceed 50 percent of the actual damages.

(B) Statutory damages for section 1002 violations.—

(i) Device.—A complaining party may recover an award of statutory damages for each violation of section 1002(a) or (c) in the sum of not more than $2,500 per device involved in such violation or per device on which a service prohibited by section 1002(c) has been performed, as the court considers just.

(ii) Digital musical recording.—A complaining party may recover an award of statutory damages for each violation of section 1002(d) in the sum of not more than $25 per digital musical recording involved in such violation, as the court considers just.

(iii) Transmission.—A complaining party may recover an award of damages for each transmission or communication that violates section 1002(e) in the sum of not more than $10,000, as the court considers just.

(2) Repeated violations.—In any case in which the court finds that a person has violated section 1002 or 1003 within 3 years after a final judgment against

that person for another such violation was entered, the court may increase the award of damages to not more than double the amounts that would otherwise be awarded under paragraph (1), as the court considers just.

(3) Innocent violations of section 1002.—The court in its discretion may reduce the total award of damages against a person violating section 1002 to a sum of not less than $250 in any case in which the court finds that the violator was not aware and had no reason to believe that its acts constituted a violation of section 1002.

(e) Payment of Damages.—Any award of damages under subsection (d) shall be deposited with the Register pursuant to section 1005 for distribution to interested copyright parties as though such funds were royalty payments made pursuant to section 1003.

(f) Impounding of Articles.—At any time while an action under subsection (a) is pending, the court may order the impounding, on such terms as it deems reasonable, of any digital audio recording device, digital musical recording, or device specified in section 1002(c) that is in the custody or control of the alleged violator and that the court has reasonable cause to believe does not comply with, or was involved in a violation of, section 1002.

(g) Remedial Modification and Destruction of Articles.—In an action brought under subsection (a), the court may, as part of a final judgment or decree finding a violation of section 1002, order the remedial modification or the destruction of any digital audio recording device, digital musical recording, or device specified in section 1002(c) that—

(1) does not comply with, or was involved in a violation of, section 1002, and

(2) is in the custody or control of the violator or has been impounded under subsection (f).

§ 1010 Determination of Certain Disputes

(a) Scope of Determination.—Before the date of first distribution in the United States of a digital audio recording device or a digital audio interface device, any party manufacturing, importing, or distributing such device, and any interested copyright party may mutually agree to petition the Copyright Royalty Judges to determine whether such device is subject to section 1002, or the basis on which royalty payments for such device are to be made under section 1003.

(b) Initiation of Proceedings.—The parties under subsection (a) shall file the petition with the Copyright Royalty Judges requesting the commencement of a proceeding. Within 2 weeks after receiving such a petition, the Chief Copyright Royalty Judge shall cause notice to be published in the Federal Register of the initiation of the proceeding.

(c) Stay of Judicial Proceedings.—Any civil action brought under section 1009 against a party to a proceeding under this section shall, on application of one of the parties to the proceeding, be stayed until completion of the proceeding.

(d) Proceeding.—The Copyright Royalty Judges shall conduct a proceeding with respect to the matter concerned, in accordance with such procedures as the Copyright Royalty Judges may adopt. The Copyright Royalty Judges shall act on the basis of a fully documented written record. Any party to the proceeding may submit relevant information and proposals to the Copyright Royalty Judges. The parties to the proceeding shall each bear their respective costs of participation.

(e) Judicial Review.—Any determination of the Copyright Royalty Judges under subsection (d) may be appealed, by a party to the proceeding, in accordance with section 803(d) of this title. The pendency of an appeal under this subsection shall not stay the determination of the Copyright Royalty Judges. If the court modifies the determination of the Copyright Royalty Judges, the court shall have jurisdiction to enter its own decision in accordance with its final judgment. The court may further vacate the determination of the Copyright Royalty Judges and remand the case for proceedings as provided in this section.

Chapter 10 Notes

1. The Audio Home Recording Act of 1992 added chapter 10, entitled "Digital Audio Recording Devices and Media," to title 17. Pub. L. No. 102-563, 106 Stat. 4237.

2. The Copyright Royalty Tribunal Reform Act of 1993 amended section 1004(a)(3) by substituting "Librarian of Congress" in lieu of "Copyright Royalty Tribunal," where appropriate. Pub. L. No. 103-198, 107 Stat. 2304, 2312.

The Copyright Royalty and Distribution Reform Act of 2004 amended paragraph 1004(a)(3) by substituting "Copyright Royalty Judges" in lieu of "Librarian of Congress," wherever it appeared. Pub. L. No. 108-419, 118 Stat. 2341, 2368.

3. The Copyright Royalty Tribunal Reform Act of 1993 amended section 1005 by striking the last sentence which began "The Register shall submit to the Copyright Royalty Tribunal." Pub. L. No. 103-198, 107 Stat. 2304, 2312.

4. The Copyright Royalty Tribunal Reform Act of 1993 amended section 1006(c) by substituting "Librarian of Congress" in lieu of "Copyright Royalty Tribunal," where appropriate. Pub. L. No. 103-198, 107 Stat. 2304, 2312. In 1997, section 1006(b)(1) was amended to insert "Federation of Television" in lieu of "Federation Television" wherever it appeared. Pub. L. No. 105-80, 111 Stat. 1529, 1535.

The Copyright Royalty and Distribution Reform Act of 2004 amended subsection 1006(c) by substituting "Copyright Royalty Judges" for "Librarian of Congress shall convene a copyright arbitration royalty panel which" in matter preceding paragraph (1). Pub. L. No. 108-419, 118 Stat. 2341, 2368.

5. The Copyright Royalty Tribunal Reform Act of 1993 amended section 1007 by substituting "Librarian of Congress" in lieu of "Copyright Royalty Tribunal" or "Tribunal," where appropriate, by amending the first sentence in subsection (c) and by inserting "the reasonable administrative costs incurred by the Librarian" in the last sentence of subsection (c), in lieu of "its reasonable administrative costs." Pub. L. No. 103-198, 107 Stat. 2304, 2312.

In 1997, section 1007 was amended, in subsection (a)(1), by inserting "calendar year 1992" in lieu of "the calendar year in which this chapter takes effect" and, in subsection (b), by inserting "1992" in lieu of "the year in which this section takes effect," and also in subsection (b), by inserting "After" in lieu of "Within 30 days after." Pub. L. No. 105-80, 111 Stat. 1529, 1534 and 1535.

The Copyright Royalty and Distribution Reform Act of 2004 Act amended paragraph 1007(a)(1) and subsections (b) and (c) in their entirety. Pub. L. No. 108-419, 118 Stat. 2341, 2368.

In 2006, the Copyright Royalty Judges Program Technical Corrections Act amended subsections 1007(b) and (c) by making technical and conforming amendments to correct references to the Copyright Royalty Board and deleting "Librarian of Congress," where appropriate. Pub. L. No. 109-303, 120 Stat. 1478, 1483.

6. The Copyright Royalty Tribunal Reform Act of 1993 amended section 1010 by substituting "Librarian of Congress" in lieu of "Copyright Royalty Tribunal" or "Tribunal," where appropriate, and by inserting "Librarian's" in lieu of "its." Pub. L. No. 103-198, 107 Stat. 2304, 2312. That Act, which established copyright arbitration royalty panels, states that "[a]ll royalty rates and all determinations with respect to the proportionate division of compulsory license fees among copyright claimants, whether made by the Copyright Royalty Tribunal, or by voluntary agreement, before the effective date set forth in subsection (a) [December 17, 1993] shall remain in effect until modified by voluntary agreement or pursuant to the amendments made by this Act." Pub. L. No. 103-198, 107 Stat. 2304, 2313.

The Copyright Royalty and Distribution Reform Act of 2004 Act amended section 1010 in its entirety. Pub. L. No. 108-419, 118 Stat. 2341, 2368.

Chapter 11 Sound Recordings and Music Videos

§ 1101 Unauthorized Fixation and Trafficking in Sound Recordings and Music Videos

(a) Unauthorized Acts.—Anyone who, without the consent of the performer or performers involved—

(1) fixes the sounds or sounds and images of a live musical performance in a copy or phonorecord, or reproduces copies or phonorecords of such a performance from an unauthorized fixation,

(2) transmits or otherwise communicates to the public the sounds or sounds and images of a live musical performance, or

(3) distributes or offers to distribute, sells or offers to sell, rents or offers to rent, or traffics in any copy or phonorecord fixed as described in paragraph (1), regardless of whether the fixations occurred in the United States,

shall be subject to the remedies provided in sections 502 through 505, to the same extent as an infringer of copyright.

(b) Definition.—As used in this section, the term "traffic in" means transport, transfer, or otherwise dispose of, to another, as consideration for anything of value, or make or obtain control of with intent to transport, transfer, or dispose of.

(c) Applicability.—This section shall apply to any act or acts that occur on or after the date of the enactment of the Uruguay Round Agreements Act.

(d) State Law Not Preempted.—Nothing in this section may be construed to annul or limit any rights or remedies under the common law or statutes of any State.

Chapter 11 Note

1. In 1994, the Uruguay Round Agreements Act added chapter 11, entitled "Sound Recordings and Music Videos," to title 17. Pub. L. No. 103-465, 108 Stat. 4809, 4974.

Chapter 12 Copyright Protection and Management Systems

§ 1201 Circumvention of Copyright Protection Systems

(a) Violations Regarding Circumvention of Technological Measures.—(1)(A) No person shall circumvent a technological measure that effectively controls access to a work protected under this title. The prohibition contained in the preceding sentence shall take effect at the end of the 2-year period beginning on the date of the enactment of this chapter.

(B) The prohibition contained in subparagraph (A) shall not apply to persons who are users of a copyrighted work which is in a particular class of works, if such persons are, or are likely to be in the succeeding 3-year period, adversely affected by virtue of such prohibition in their ability to make noninfringing uses of that particular class of works under this title, as determined under subparagraph (C).

(C) During the 2-year period described in subparagraph (A), and during each succeeding 3-year period, the Librarian of Congress, upon the recommendation of the Register of Copyrights, who shall consult with the Assistant Secretary for Communications and Information of the Department of Commerce and report and comment on his or her views in making such recommendation, shall make the determination in a rulemaking proceeding for purposes of subparagraph (B) of whether persons who are users of a copyrighted work are, or are likely to be in the succeeding 3-year period, adversely affected by the prohibition under subparagraph (A) in their ability to make noninfringing uses under this title of a particular class of copyrighted works. In conducting such rulemaking, the Librarian shall examine—

(i) the availability for use of copyrighted works;

(ii) the availability for use of works for nonprofit archival, preservation, and educational purposes;

(iii) the impact that the prohibition on the circumvention of technological measures applied to copyrighted works has on criticism, comment, news reporting, teaching, scholarship, or research;

(iv) the effect of circumvention of technological measures on the market for or value of copyrighted works; and

(v) such other factors as the Librarian considers appropriate.

(D) The Librarian shall publish any class of copyrighted works for which the Librarian has determined, pursuant to the rulemaking conducted under subparagraph (C), that noninfringing uses by persons who are users of a copyrighted work are, or are likely to be, adversely affected, and the prohibition contained in subparagraph (A) shall not apply to such users with respect to such class of works for the ensuing 3-year period.

(E) Neither the exception under subparagraph (B) from the applicability of the prohibition contained in subparagraph (A), nor any determination made in a rulemaking conducted under subparagraph (C), may be used as a defense in any action to enforce any provision of this title other than this paragraph.

(2) No person shall manufacture, import, offer to the public, provide, or otherwise traffic in any technology, product, service, device, component, or part thereof, that—

(A) is primarily designed or produced for the purpose of circumventing a technological measure that effectively controls access to a work protected under this title;

(B) has only limited commercially significant purpose or use other than to circumvent a technological measure that effectively controls access to a work protected under this title; or

(C) is marketed by that person or another acting in concert with that person with that person's knowledge for use in circumventing a technological measure that effectively controls access to a work protected under this title.

(3) As used in this subsection—

(A) to "circumvent a technological measure" means to descramble a scrambled work, to decrypt an encrypted work, or otherwise to avoid, bypass, remove, deactivate, or impair a technological measure, without the authority of the copyright owner; and

(B) a technological measure "effectively controls access to a work" if the measure, in the ordinary course of its operation, requires the application of information, or a process or a treatment, with the authority of the copyright owner, to gain access to the work.

(b) Additional Violations.—(1) No person shall manufacture, import, offer to the public, provide, or otherwise traffic in any technology, product, service, device, component, or part thereof, that—

(A) is primarily designed or produced for the purpose of circumventing protection afforded by a technological measure that effectively protects a right of a copyright owner under this title in a work or a portion thereof;

(B) has only limited commercially significant purpose or use other than to circumvent protection afforded by a technological measure that effectively protects a right of a copyright owner under this title in a work or a portion thereof; or

(C) is marketed by that person or another acting in concert with that person with that person's knowledge for use in circumventing protection afforded by a technological measure that effectively protects a right of a copyright owner under this title in a work or a portion thereof.

(2) As used in this subsection—

(A) to "circumvent protection afforded by a technological measure" means avoiding, bypassing, removing, deactivating, or otherwise impairing a technological measure; and

(B) a technological measure "effectively protects a right of a copyright owner under this title" if the measure, in the ordinary course of its operation, prevents, restricts, or otherwise limits the exercise of a right of a copyright owner under this title.

(c) Other Rights, Etc., Not Affected.—(1) Nothing in this section shall affect rights, remedies, limitations, or defenses to copyright infringement, including fair use, under this title.

(2) Nothing in this section shall enlarge or diminish vicarious or contributory liability for copyright infringement in connection with any technology, product, service, device, component, or part thereof.

(3) Nothing in this section shall require that the design of, or design and selection of parts and components for, a consumer electronics, telecommunications, or computing product provide for a response to any particular technological measure, so long as such part or component, or the product in which such part or component is integrated, does not otherwise fall within the prohibitions of subsection (a)(2) or (b)(1).

(4) Nothing in this section shall enlarge or diminish any rights of free speech or the press for activities using consumer electronics, telecommunications, or computing products.

(d) Exemption for Nonprofit Libraries, Archives, and Educational Institutions.—(1) A nonprofit library, archives, or educational institution which gains access to a commercially exploited copyrighted work solely in order to make a good faith determination of whether to acquire a copy of that work for the sole purpose of engaging in conduct permitted under this title shall not be in violation of subsection (a)(1)(A). A copy of a work to which access has been gained under this paragraph—

(A) may not be retained longer than necessary to make such good faith determination; and

(B) may not be used for any other purpose.

(2) The exemption made available under paragraph (1) shall only apply with respect to a work when an identical copy of that work is not reasonably available in another form.

(3) A nonprofit library, archives, or educational institution that willfully for the purpose of commercial advantage or financial gain violates paragraph (1)—

(A) shall, for the first offense, be subject to the civil remedies under section 1203; and

(B) shall, for repeated or subsequent offenses, in addition to the civil remedies under section 1203, forfeit the exemption provided under paragraph (1).

(4) This subsection may not be used as a defense to a claim under subsection (a)(2) or (b), nor may this subsection permit a nonprofit library, archives, or educational institution to manufacture, import, offer to the public, provide, or otherwise traffic in any technology, product, service, component, or part thereof, which circumvents a technological measure.

(5) In order for a library or archives to qualify for the exemption under this subsection, the collections of that library or archives shall be—

(A) open to the public; or

(B) available not only to researchers affiliated with the library or archives or with the institution of which it is a part, but also to other persons doing research in a specialized field.

(e) Law Enforcement, Intelligence, and Other Government Activities.—This section does not prohibit any lawfully authorized investigative, protective, information security, or intelligence activity of an officer, agent, or employee of the United States, a State, or a political subdivision of a State, or a person acting pursuant to a contract with the United States, a State, or a political subdivision of a State. For purposes of this subsection, the term "information security" means activities carried out in order to identify and address the vulnerabilities of a government computer, computer system, or computer network.

(f) Reverse Engineering.—(1) Notwithstanding the provisions of subsection (a)(1)(A), a person who has lawfully obtained the right to use a copy of a computer program may circumvent a technological measure that effectively controls access to a particular portion of that program for the sole purpose of identifying and analyzing those elements of the program that are necessary to achieve interoperability of an independently created computer program with other programs, and that have not previously been readily available to the person engaging in the circumvention, to the extent any such acts of identification and analysis do not constitute infringement under this title.

(2) Notwithstanding the provisions of subsections (a)(2) and (b), a person may develop and employ technological means to circumvent a technological measure, or to circumvent protection afforded by a technological measure, in order to enable the identification and analysis under paragraph (1), or for the purpose of enabling interoperability of an independently created computer program with other programs, if such means are necessary to achieve such interoperability, to the extent that doing so does not constitute infringement under this title.

(3) The information acquired through the acts permitted under paragraph (1), and the means permitted under paragraph (2), may be made available to others if the person referred to in paragraph (1) or (2), as the case may be,

provides such information or means solely for the purpose of enabling interoperability of an independently created computer program with other programs, and to the extent that doing so does not constitute infringement under this title or violate applicable law other than this section.

(4) For purposes of this subsection, the term "interoperability" means the ability of computer programs to exchange information, and of such programs mutually to use the information which has been exchanged.

(g) Encryption Research.—

(1) Definitions.—For purposes of this subsection—

(A) the term "encryption research" means activities necessary to identify and analyze flaws and vulnerabilities of encryption technologies applied to copyrighted works, if these activities are conducted to advance the state of knowledge in the field of encryption technology or to assist in the development of encryption products; and

(B) the term "encryption technology" means the scrambling and descrambling of information using mathematical formulas or algorithms.

(2) Permissible acts of encryption research.—Notwithstanding the provisions of subsection (a)(1)(A), it is not a violation of that subsection for a person to circumvent a technological measure as applied to a copy, phonorecord, performance, or display of a published work in the course of an act of good faith encryption research if—

(A) the person lawfully obtained the encrypted copy, phonorecord, performance, or display of the published work;

(B) such act is necessary to conduct such encryption research;

(C) the person made a good faith effort to obtain authorization before the circumvention; and

(D) such act does not constitute infringement under this title or a violation of applicable law other than this section, including section 1030 of title 18 and those provisions of title 18 amended by the Computer Fraud and Abuse Act of 1986.

(3) Factors in determining exemption.—In determining whether a person qualifies for the exemption under paragraph (2), the factors to be considered shall include—

(A) whether the information derived from the encryption research was disseminated, and if so, whether it was disseminated in a manner reasonably calculated to advance the state of knowledge or development of encryption technology, versus whether it was disseminated in a manner that facilitates infringement under this title or a violation of applicable law other than this section, including a violation of privacy or breach of security;

(B) whether the person is engaged in a legitimate course of study, is employed, or is appropriately trained or experienced, in the field of encryption technology; and

(C) whether the person provides the copyright owner of the work to which the technological measure is applied with notice of the findings and documentation of the research, and the time when such notice is provided.

(4) Use of technological means for research activities.—Notwithstanding the provisions of subsection (a)(2), it is not a violation of that subsection for a person to—

(A) develop and employ technological means to circumvent a technological measure for the sole purpose of that person performing the acts of good faith encryption research described in paragraph (2); and

(B) provide the technological means to another person with whom he or she is working collaboratively for the purpose of conducting the acts of good faith encryption research described in paragraph (2) or for the purpose of having that other person verify his or her acts of good faith encryption research described in paragraph (2).

(5) Report to Congress.—Not later than 1 year after the date of the enactment of this chapter, the Register of Copyrights and the Assistant Secretary for Communications and Information of the Department of Commerce shall jointly report to the Congress on the effect this subsection has had on—

(A) encryption research and the development of encryption technology;

(B) the adequacy and effectiveness of technological measures designed to protect copyrighted works; and

(C) protection of copyright owners against the unauthorized access to their encrypted copyrighted works.

The report shall include legislative recommendations, if any.

(h) Exceptions Regarding Minors.—In applying subsection (a) to a component or part, the court may consider the necessity for its intended and actual incorporation in a technology, product, service, or device, which—

(1) does not itself violate the provisions of this title; and

(2) has the sole purpose to prevent the access of minors to material on the Internet.

(i) Protection of Personally Identifying Information.—

(1) Circumvention permitted.—Notwithstanding the provisions of subsection (a)(1)(A), it is not a violation of that subsection for a person to circumvent a technological measure that effectively controls access to a work protected under this title, if—

(A) the technological measure, or the work it protects, contains the capability of collecting or disseminating personally identifying information reflecting the online activities of a natural person who seeks to gain access to the work protected;

(B) in the normal course of its operation, the technological measure, or the work it protects, collects or disseminates personally identifying information about the person who seeks to gain access to the work protected, without providing conspicuous notice of such collection or dissemination to such person, and without providing such person with the capability to prevent or restrict such collection or dissemination;

(C) the act of circumvention has the sole effect of identifying and disabling the capability described in subparagraph (A), and has no other effect on the ability of any person to gain access to any work; and

(D) the act of circumvention is carried out solely for the purpose of preventing the collection or dissemination of personally identifying information about a natural person who seeks to gain access to the work protected, and is not in violation of any other law.

(2) Inapplicability to certain technological measures.—This subsection does not apply to a technological measure, or a work it protects, that does not collect or disseminate personally identifying information and that is disclosed to a user as not having or using such capability.

(j) Security Testing.—

(1) Definition.—For purposes of this subsection, the term "security testing" means accessing a computer, computer system, or computer network, solely for the purpose of good faith testing, investigating, or correcting, a security flaw or vulnerability, with the authorization of the owner or operator of such computer, computer system, or computer network.

(2) Permissible acts of security testing.—Notwithstanding the provisions of subsection (a)(1)(A), it is not a violation of that subsection for a person to engage in an act of security testing, if such act does not constitute infringement under this title or a violation of applicable law other than this section, including section 1030 of title 18 and those provisions of title 18 amended by the Computer Fraud and Abuse Act of 1986.

(3) Factors in determining exemption.—In determining whether a person qualifies for the exemption under paragraph (2), the factors to be considered shall include—

(A) whether the information derived from the security testing was used solely to promote the security of the owner or operator of such computer, computer system or computer network, or shared directly with the developer of such computer, computer system, or computer network; and

(B) whether the information derived from the security testing was used or maintained in a manner that does not facilitate infringement under this title or a violation of applicable law other than this section, including a violation of privacy or breach of security.

(4) Use of technological means for security testing.—Notwithstanding the provisions of subsection (a)(2), it is not a violation of that subsection for a person to develop, produce, distribute or employ technological means for the sole purpose of performing the acts of security testing described in subsection (2), provided such technological means does not otherwise violate section (a)(2).

(k) Certain Analog Devices and Certain Technological Measures.—

(1) Certain analog devices.—

(A) Effective 18 months after the date of the enactment of this chapter, no person shall manufacture, import, offer to the public, provide or otherwise traffic in any—

(i) VHS format analog video cassette recorder unless such recorder conforms to the automatic gain control copy control technology;

(ii) 8mm format analog video cassette camcorder unless such camcorder conforms to the automatic gain control technology;

(iii) Beta format analog video cassette recorder, unless such recorder conforms to the automatic gain control copy control technology, except that this requirement shall not apply until there are 1,000 Beta format analog video cassette recorders sold in the United States in any one calendar year after the date of the enactment of this chapter;

(iv) 8mm format analog video cassette recorder that is not an analog video cassette camcorder, unless such recorder conforms to the automatic gain control copy control technology, except that this requirement shall not apply until there are 20,000 such recorders sold in the United States in any one calendar year after the date of the enactment of this chapter; or

(v) analog video cassette recorder that records using an NTSC format video input and that is not otherwise covered under clauses (i) through

(iv), unless such device conforms to the automatic gain control copy control technology.

(B) Effective on the date of the enactment of this chapter, no person shall manufacture, import, offer to the public, provide or otherwise traffic in—

(i) any VHS format analog video cassette recorder or any 8mm format analog video cassette recorder if the design of the model of such recorder has been modified after such date of enactment so that a model of recorder that previously conformed to the automatic gain control copy control technology no longer conforms to such technology; or

(ii) any VHS format analog video cassette recorder, or any 8mm format analog video cassette recorder that is not an 8mm analog video cassette camcorder, if the design of the model of such recorder has been modified after such date of enactment so that a model of recorder that previously conformed to the four-line colorstripe copy control technology no longer conforms to such technology.

Manufacturers that have not previously manufactured or sold a VHS format analog video cassette recorder, or an 8mm format analog cassette recorder, shall be required to conform to the four-line colorstripe copy control technology in the initial model of any such recorder manufactured after the date of the enactment of this chapter, and thereafter to continue conforming to the four-line colorstripe copy control technology. For purposes of this subparagraph, an analog video cassette recorder "conforms to" the four-line colorstripe copy control technology if it records a signal that, when played back by the playback function of that recorder in the normal viewing mode, exhibits, on a reference display device, a display containing distracting visible lines through portions of the viewable picture.

(2) Certain encoding restrictions.—No person shall apply the automatic gain control copy control technology or colorstripe copy control technology to prevent or limit consumer copying except such copying—

(A) of a single transmission, or specified group of transmissions, of live events or of audiovisual works for which a member of the public has exercised choice in selecting the transmissions, including the content of the transmissions or the time of receipt of such transmissions, or both, and as to which such member is charged a separate fee for each such transmission or specified group of transmissions;

(B) from a copy of a transmission of a live event or an audiovisual work if such transmission is provided by a channel or service where payment is made by a member of the public for such channel or service in the form of a subscription fee that entitles the member of the public to receive all of the programming contained in such channel or service;

(C) from a physical medium containing one or more prerecorded audiovisual works; or

(D) from a copy of a transmission described in subparagraph (A) or from a copy made from a physical medium described in subparagraph (C).

In the event that a transmission meets both the conditions set forth in subparagraph (A) and those set forth in subparagraph (B), the transmission shall be treated as a transmission described in subparagraph (A).

(3) Inapplicability.—This subsection shall not—

(A) require any analog video cassette camcorder to conform to the automatic gain control copy control technology with respect to any video signal received through a camera lens;

(B) apply to the manufacture, importation, offer for sale, provision of, or other trafficking in, any professional analog video cassette recorder; or

(C) apply to the offer for sale or provision of, or other trafficking in, any previously owned analog video cassette recorder, if such recorder was legally manufactured and sold when new and not subsequently modified in violation of paragraph (1)(B).

(4) Definitions.—For purposes of this subsection:

(A) An "analog video cassette recorder" means a device that records, or a device that includes a function that records, on electromagnetic tape in an analog format the electronic impulses produced by the video and audio portions of a television program, motion picture, or other form of audiovisual work.

(B) An "analog video cassette camcorder" means an analog video cassette recorder that contains a recording function that operates through a camera lens and through a video input that may be connected with a television or other video playback device.

(C) An analog video cassette recorder "conforms" to the automatic gain control copy control technology if it—

(i) detects one or more of the elements of such technology and does not record the motion picture or transmission protected by such technology; or

(ii) records a signal that, when played back, exhibits a meaningfully distorted or degraded display.

(D) The term "professional analog video cassette recorder" means an analog video cassette recorder that is designed, manufactured, marketed, and intended for use by a person who regularly employs such a device for a lawful business or industrial use, including making, performing, displaying, distributing, or transmitting copies of motion pictures on a commercial scale.

(E) The terms "VHS format," "8mm format," "Beta format," "automatic gain control copy control technology," "colorstripe copy control technology," "four-line version of the colorstripe copy control technology," and "NTSC" have the meanings that are commonly understood in the consumer electronics and motion picture industries as of the date of the enactment of this chapter.

(5) Violations.—Any violation of paragraph (1) of this subsection shall be treated as a violation of subsection (b)(1) of this section. Any violation of paragraph (2) of this subsection shall be deemed an "act of circumvention" for the purposes of section 1203(c)(3)(A) of this chapter.

§ 1202 Integrity of Copyright Management Information

(a) False Copyright Management Information.—No person shall knowingly and with the intent to induce, enable, facilitate, or conceal infringement—

(1) provide copyright management information that is false, or

(2) distribute or import for distribution copyright management information that is false.

(b) Removal or Alteration of Copyright Management Information.—No person shall, without the authority of the copyright owner or the law—

(1) intentionally remove or alter any copyright management information,

(2) distribute or import for distribution copyright management information knowing that the copyright management information has been removed or altered without authority of the copyright owner or the law, or

(3) distribute, import for distribution, or publicly perform works, copies of works, or phonorecords, knowing that copyright management information has been removed or altered without authority of the copyright owner or the law, knowing, or, with respect to civil remedies under section 1203, having reasonable grounds to know, that it will induce, enable, facilitate, or conceal an infringement of any right under this title.

(c) Definition.—As used in this section, the term "copyright management information" means any of the following information conveyed in connection with copies or phonorecords of a work or performances or displays of a work, including in digital form, except that such term does not include any personally identifying information about a user of a work or of a copy, phonorecord, performance, or display of a work:

(1) The title and other information identifying the work, including the information set forth on a notice of copyright.

(2) The name of, and other identifying information about, the author of a work.

(3) The name of, and other identifying information about, the copyright owner of the work, including the information set forth in a notice of copyright.

(4) With the exception of public performances of works by radio and television broadcast stations, the name of, and other identifying information about, a performer whose performance is fixed in a work other than an audiovisual work.

(5) With the exception of public performances of works by radio and television broadcast stations, in the case of an audiovisual work, the name of, and other identifying information about, a writer, performer, or director who is credited in the audiovisual work.

(6) Terms and conditions for use of the work.

(7) Identifying numbers or symbols referring to such information or links to such information.

(8) Such other information as the Register of Copyrights may prescribe by regulation, except that the Register of Copyrights may not require the provision of any information concerning the user of a copyrighted work.

(d) Law Enforcement, Intelligence, and Other Government Activities.—This section does not prohibit any lawfully authorized investigative, protective, information security, or intelligence activity of an officer, agent, or employee of the United States, a State, or a political subdivision of a State, or a person acting pursuant to a contract with the United States, a State, or a political subdivision of a State. For purposes of this subsection, the term "information security" means activities carried out in order to identify and address the vulnerabilities of a government computer, computer system, or computer network.

(e) Limitations on Liability.—

(1) Analog transmissions.—In the case of an analog transmission, a person who is making transmissions in its capacity as a broadcast station, or as a cable system, or someone who provides programming to such station or system, shall not be liable for a violation of subsection (b) if—

(A) avoiding the activity that constitutes such violation is not technically feasible or would create an undue financial hardship on such person; and

(B) such person did not intend, by engaging in such activity, to induce, enable, facilitate, or conceal infringement of a right under this title.

(2) Digital transmissions.—

(A) If a digital transmission standard for the placement of copyright management information for a category of works is set in a voluntary, consensus standard-setting process involving a representative cross-section of broadcast stations or cable systems and copyright owners of a category of works that are intended for public performance by such stations or systems, a person identified in paragraph (1) shall not be liable for a violation of subsection (b) with respect to the particular copyright management information addressed by such standard if—

(i) the placement of such information by someone other than such person is not in accordance with such standard; and

(ii) the activity that constitutes such violation is not intended to induce, enable, facilitate, or conceal infringement of a right under this title.

(B) Until a digital transmission standard has been set pursuant to subparagraph (A) with respect to the placement of copyright management information for a category of works, a person identified in paragraph (1) shall not be liable for a violation of subsection (b) with respect to such copyright management information, if the activity that constitutes such violation is not intended to induce, enable, facilitate, or conceal infringement of a right under this title, and if—

(i) the transmission of such information by such person would result in a perceptible visual or aural degradation of the digital signal; or

(ii) the transmission of such information by such person would conflict with—

(I) an applicable government regulation relating to transmission of information in a digital signal;

(II) an applicable industry-wide standard relating to the transmission of information in a digital signal that was adopted by a voluntary consensus standards body prior to the effective date of this chapter; or

(III) an applicable industry-wide standard relating to the transmission of information in a digital signal that was adopted in a voluntary, consensus standards-setting process open to participation by a representative cross-section of broadcast stations or cable systems and copyright owners of a category of works that are intended for public performance by such stations or systems.

(3) Definitions.—As used in this subsection—

(A) the term "broadcast station" has the meaning given that term in section 3 of the Communications Act of 1934 (47 U.S.C. 153); and

(B) the term "cable system" has the meaning given that term in section 602 of the Communications Act of 1934 (47 U.S.C. 522).

§ 1203 Civil Remedies

(a) Civil Actions.—Any person injured by a violation of section 1201 or 1202 may bring a civil action in an appropriate United States district court for such violation.

(b) Powers of the Court.—In an action brought under subsection (a), the court—

(1) may grant temporary and permanent injunctions on such terms as it deems reasonable to prevent or restrain a violation, but in no event shall impose a prior restraint on free speech or the press protected under the 1st amendment to the Constitution;

(2) at any time while an action is pending, may order the impounding, on such terms as it deems reasonable, of any device or product that is in the custody or control of the alleged violator and that the court has reasonable cause to believe was involved in a violation;

(3) may award damages under subsection (c);

(4) in its discretion may allow the recovery of costs by or against any party other than the United States or an officer thereof;

(5) in its discretion may award reasonable attorney's fees to the prevailing party; and

(6) may, as part of a final judgment or decree finding a violation, order the remedial modification or the destruction of any device or product involved in the

violation that is in the custody or control of the violator or has been impounded under paragraph (2).

(c) Award of Damages.—

(1) In general.—Except as otherwise provided in this title, a person committing a violation of section 1201 or 1202 is liable for either—

(A) the actual damages and any additional profits of the violator, as provided in paragraph (2), or

(B) statutory damages, as provided in paragraph (3).

(2) Actual damages.—The court shall award to the complaining party the actual damages suffered by the party as a result of the violation, and any profits of the violator that are attributable to the violation and are not taken into account in computing the actual damages, if the complaining party elects such damages at any time before final judgment is entered.

(3) Statutory damages.—(A) At any time before final judgment is entered, a complaining party may elect to recover an award of statutory damages for each violation of section 1201 in the sum of not less than $200 or more than $2,500 per act of circumvention, device, product, component, offer, or performance of service, as the court considers just.

(B) At any time before final judgment is entered, a complaining party may elect to recover an award of statutory damages for each violation of section 1202 in the sum of not less than $2,500 or more than $25,000.

(4) Repeated violations.—In any case in which the injured party sustains the burden of proving, and the court finds, that a person has violated section 1201 or 1202 within three years after a final judgment was entered against the person for another such violation, the court may increase the award of damages up to triple the amount that would otherwise be awarded, as the court considers just.

(5) Innocent violations.—

(A) In general.—The court in its discretion may reduce or remit the total award of damages in any case in which the violator sustains the burden of proving, and the court finds, that the violator was not aware and had no reason to believe that its acts constituted a violation.

(B) Nonprofit library, archives, educational institutions, or public broadcasting entities.—

(i) Definition.—In this subparagraph, the term "public broadcasting entity" has the meaning given such term under section 118(f).

(ii) In general.—In the case of a nonprofit library, archives, educational institution, or public broadcasting entity, the court shall remit damages in any case in which the library, archives, educational institution, or public broadcasting entity sustains the burden of proving, and the court finds, that the library, archives, educational institution, or public broadcasting entity was not aware and had no reason to believe that its acts constituted a violation.

§ 1204 Criminal Offenses and Penalties

(a) In General.—Any person who violates section 1201 or 1202 willfully and for purposes of commercial advantage or private financial gain—

(1) shall be fined not more than $500,000 or imprisoned for not more than 5 years, or both, for the first offense; and

(2) shall be fined not more than $1,000,000 or imprisoned for not more than 10 years, or both, for any subsequent offense.

(b) Limitation for Nonprofit Library, Archives, Educational Institution, or Public Broadcasting Entity.—Subsection (a) shall not apply to a nonprofit library, archives, educational institution, or public broadcasting entity (as defined under section 118(f)).

(c) Statute of Limitations.—No criminal proceeding shall be brought under this section unless such proceeding is commenced within five years after the cause of action arose.

§ 1205 Savings Clause

Nothing in this chapter abrogates, diminishes, or weakens the provisions of, nor provides any defense or element of mitigation in a criminal prosecution or civil action under, any Federal or State law that prevents the violation of the privacy of an individual in connection with the individual's use of the Internet.

Chapter 12 Notes

1. The WIPO Copyright and Performances and Phonograms Treaties Implementation Act of 1998 added chapter 12, entitled "Copyright Protection and Management Systems," to title 17. Pub. L. No. 105-304, 112 Stat. 2860, 2863. The WIPO Copyright and Performances and Phonograms Treaties Implementation Act of 1998 is title I of the Digital Millennium Copyright Act. Pub. L. No. 105-304, 112 Stat. 2860.

2. The Satellite Home Viewer Improvement Act of 1999 amended section 1201(a)(1)(C) by deleting "on the record." Pub. L. No. 106-113, 113 Stat. 1501, app. I at 1501A-594.

3. In 1999, section 1202 was amended by inserting "category of works" for "category or works," in subsection (e)(2)(B). Pub. L. No. 106-44, 113 Stat. 221, 222.

4. The Satellite Home Viewer Improvement Act of 1999 amended section 1203(c)(5)(B) in its entirety. Pub. L. No. 106-113, 113 Stat. 1501, app. I at 1501A-593. The Copyright Cleanup, Clarification, and Corrections Act of 2010 amended subparagraph 1203(c)(5)(B)(i) to substitute "section 118(f)" for "section 118(g)." Pub. L. No. 111-295, 124 Stat. 3180, 3181.

5. The Satellite Home Viewer Improvement Act of 1999 amended section 1204(b) in its entirety. Pub. L. No. 106-113, 113 Stat. 1501, app. I at 1501A-593. The Copyright Cleanup, Clarification, and Corrections Act of 2010 amended subsection 1204(b) to substitute "section 118(f)" for "section 118(g)." Pub. L. No. 111-295, 124 Stat. 3180, 3181.

Chapter 13 Protection of Original Designs

§ 1301 Designs Protected

(a) Designs Protected.—

(1) In general.—The designer or other owner of an original design of a useful article which makes the article attractive or distinctive in appearance to the purchasing or using public may secure the protection provided by this chapter upon complying with and subject to this chapter.

(2) Vessel features.—The design of a vessel hull, deck, or combination of a hull and deck, including a plug or mold, is subject to protection under this chapter, notwithstanding section 1302(4).

(3) Exceptions.—Department of Defense rights in a registered design under this chapter, including the right to build to such registered design, shall be determined solely by operation of section 2320 of title 10 or by the instrument under which the design was developed for the United States Government.

(b) Definitions.—For the purpose of this chapter, the following terms have the following meanings:

(1) A design is "original" if it is the result of the designer's creative endeavor that provides a distinguishable variation over prior work pertaining to similar articles which is more than merely trivial and has not been copied from another source.

(2) A "useful article" is a vessel hull or deck, including a plug or mold, which in normal use has an intrinsic utilitarian function that is not merely to portray the appearance of the article or to convey information. An article which normally is part of a useful article shall be deemed to be a useful article.

(3) A "vessel" is a craft—

(A) that is designed and capable of independently steering a course on or through water through its own means of propulsion; and

(B) that is designed and capable of carrying and transporting one or more passengers.

(4) A "hull" is the exterior frame or body of a vessel, exclusive of the deck, superstructure, masts, sails, yards, rigging, hardware, fixtures, and other attachments.

(5) A "plug" means a device or model used to make a mold for the purpose of exact duplication, regardless of whether the device or model has an intrinsic utilitarian function that is not only to portray the appearance of the product or to convey information.

(6) A "mold" means a matrix or form in which a substance for material is used, regardless of whether the matrix or form has an intrinsic utilitarian function that is not only to portray the appearance of the product or to convey information.

(7) A "deck" is the horizontal surface of a vessel that covers the hull, including exterior cabin and cockpit surfaces and exclusive of masts, sails, yards, rigging, hardware, fixtures, and other attachments.

§ 1302 Designs Not Subject to Protection

Protection under this chapter shall not be available for a design that is—

(1) not original;

(2) staple or commonplace, such as a standard geometric figure, a familiar symbol, an emblem, or a motif, or another shape, pattern, or configuration which has become standard, common, prevalent, or ordinary;

(3) different from a design excluded by paragraph (2) only in insignificant details or in elements which are variants commonly used in the relevant trades;

(4) dictated solely by a utilitarian function of the article that embodies it; or

(5) embodied in a useful article that was made public by the designer or owner in the United States or a foreign country more than 2 years before the date of the application for registration under this chapter.

§ 1303 Revisions, Adaptations, and Rearrangements

Protection for a design under this chapter shall be available notwithstanding the employment in the design of subject matter excluded from protection under section 1302 if the design is a substantial revision, adaptation, or rearrangement of such subject matter. Such protection shall be independent of any subsisting protection in subject matter employed in the design, and shall not be construed as securing any right to subject matter excluded from protection under this chapter or as extending any subsisting protection under this chapter.

§ 1304 Commencement of Protection

The protection provided for a design under this chapter shall commence upon the earlier of the date of publication of the registration under section 1313(a) or the date the design is first made public as defined by section 1310(b).

§ 1305 Term of Protection

(a) In General.—Subject to subsection (b), the protection provided under this chapter for a design shall continue for a term of 10 years beginning on the date of the commencement of protection under section 1304.

(b) Expiration.—All terms of protection provided in this section shall run to the end of the calendar year in which they would otherwise expire.

(c) Termination of Rights.—Upon expiration or termination of protection in a particular design under this chapter, all rights under this chapter in the design shall terminate, regardless of the number of different articles in which the design may have been used during the term of its protection.

§ 1306 Design Notice

(a) Contents of Design Notice.—

(1) Whenever any design for which protection is sought under this chapter is made public under section 1310(b), the owner of the design shall, subject to the provisions of section 1307, mark it or have it marked legibly with a design notice consisting of—

(A) the words "Protected Design", the abbreviation "Prot'd Des.", or the letter "D" within a circle, Ⓓ, or the symbol "*D*";

(B) the year of the date on which protection for the design commenced; and

(C) the name of the owner, an abbreviation by which the name can be recognized, or a generally accepted alternative designation of the owner. Any distinctive identification of the owner may be used for purposes of subparagraph (C) if it has been recorded by the Administrator before the design marked with such identification is registered.

(2) After registration, the registration number may be used instead of the elements specified in subparagraphs (B) and (C) of paragraph (1).

(b) Location of Notice.—The design notice shall be so located and applied as to give reasonable notice of design protection while the useful article embodying the design is passing through its normal channels of commerce.

(c) Subsequent Removal of Notice.—When the owner of a design has complied with the provisions of this section, protection under this chapter shall not be affected by the removal, destruction, or obliteration by others of the design notice on an article.

§ 1307 Effect of Omission of Notice

(a) Actions with Notice.—Except as provided in subsection (b), the omission of the notice prescribed in section 1306 shall not cause loss of the protection under this chapter or prevent recovery for infringement under this chapter against any person who, after receiving written notice of the design protection, begins an undertaking leading to infringement under this chapter.

(b) Actions without Notice.—The omission of the notice prescribed in section 1306 shall prevent any recovery under section 1323 against a person who began an undertaking leading to infringement under this chapter before receiving written notice of the design protection. No injunction shall be issued under this chapter with respect to such undertaking unless the owner of the design reimburses that person for any reasonable expenditure or contractual obligation in connection with such undertaking that was incurred before receiving written notice of the design protection, as the court in its discretion directs. The burden of providing written notice of design protection shall be on the owner of the design.

§ 1308 Exclusive Rights

The owner of a design protected under this chapter has the exclusive right to—

(1) make, have made, or import, for sale or for use in trade, any useful article embodying that design; and

(2) sell or distribute for sale or for use in trade any useful article embodying that design.

§ 1309 Infringement

(a) Acts of Infringement.—Except as provided in subsection (b), it shall be infringement of the exclusive rights in a design protected under this chapter for any person, without the consent of the owner of the design, within the United States and during the term of such protection, to—

(1) make, have made, or import, for sale or for use in trade, any infringing article as defined in subsection (e); or

(2) sell or distribute for sale or for use in trade any such infringing article.

(b) Acts of Sellers and Distributors.—A seller or distributor of an infringing article who did not make or import the article shall be deemed to have infringed on a design protected under this chapter only if that person—

(1) induced or acted in collusion with a manufacturer to make, or an importer to import such article, except that merely purchasing or giving an order to purchase such article in the ordinary course of business shall not of itself constitute such inducement or collusion; or

(2) refused or failed, upon the request of the owner of the design, to make a prompt and full disclosure of that person's source of such article, and that person orders or reorders such article after receiving notice by registered or certified mail of the protection subsisting in the design.

(c) Acts without Knowledge.—It shall not be infringement under this section to make, have made, import, sell, or distribute, any article embodying a design which was created without knowledge that a design was protected under this chapter and was copied from such protected design.

(d) Acts in Ordinary Course of Business.—A person who incorporates into that person's product of manufacture an infringing article acquired from others in the ordinary course of business, or who, without knowledge of the protected design embodied in an infringing article, makes or processes the infringing article for the account of another person in the ordinary course of business, shall not be deemed to have infringed the rights in that design under this chapter except under a condition contained in paragraph (1) or (2) of subsection (b). Accepting an order or reorder from the source of the infringing article shall be deemed ordering or reordering within the meaning of subsection (b)(2).

(e) Infringing Article Defined.—As used in this section, an "infringing article" is any article the design of which has been copied from a design protected under this chapter, without the consent of the owner of the protected design. An infringing article is not an illustration or picture of a protected design in an advertisement, book, periodical, newspaper, photograph, broadcast, motion picture, or similar medium. A design shall not be deemed to have been copied from a protected design if it is original and not substantially similar in appearance to a protected design.

(f) Establishing Originality.—The party to any action or proceeding under this chapter who alleges rights under this chapter in a design shall have the burden of establishing the design's originality whenever the opposing party introduces an earlier work which is identical to such design, or so similar as to make prima facie showing that such design was copied from such work.

(g) Reproduction for Teaching or Analysis.—It is not an infringement of the exclusive rights of a design owner for a person to reproduce the design in a useful article or in any other form solely for the purpose of teaching, analyzing, or evaluating the appearance, concepts, or techniques embodied in the design, or the function of the useful article embodying the design.

§ 1310 Application for Registration

(a) Time Limit for Application for Registration.—Protection under this chapter shall be lost if application for registration of the design is not made within 2 years after the date on which the design is first made public.

(b) When Design Is Made Public.—A design is made public when an existing useful article embodying the design is anywhere publicly exhibited, publicly distributed, or offered for sale or sold to the public by the owner of the design or with the owner's consent.

(c) Application by Owner of Design.—Application for registration may be made by the owner of the design.

(d) Contents of Application.—The application for registration shall be made to the Administrator and shall state—

(1) the name and address of the designer or designers of the design;

(2) the name and address of the owner if different from the designer;

(3) the specific name of the useful article embodying the design;

(4) the date, if any, that the design was first made public, if such date was earlier than the date of the application;

(5) affirmation that the design has been fixed in a useful article; and

(6) such other information as may be required by the Administrator. The application for registration may include a description setting forth the salient features of the design, but the absence of such a description shall not prevent registration under this chapter.

(e) Sworn Statement.—The application for registration shall be accompanied by a statement under oath by the applicant or the applicant's duly authorized agent or representative, setting forth, to the best of the applicant's knowledge and belief—

(1) that the design is original and was created by the designer or designers named in the application;

(2) that the design has not previously been registered on behalf of the applicant or the applicant's predecessor in title; and

(3) that the applicant is the person entitled to protection and to registration under this chapter.

If the design has been made public with the design notice prescribed in section 1306, the statement shall also describe the exact form and position of the design notice.

(f) Effect of Errors.—(1) Error in any statement or assertion as to the utility of the useful article named in the application under this section, the design of which is sought to be registered, shall not affect the protection secured under this chapter.

(2) Errors in omitting a joint designer or in naming an alleged joint designer shall not affect the validity of the registration, or the actual ownership or the protection of the design, unless it is shown that the error occurred with deceptive intent.

(g) Design Made in Scope of Employment.—In a case in which the design was made within the regular scope of the designer's employment and individual authorship of the design is difficult or impossible to ascribe and the application so states, the name and address of the employer for whom the design was made may be stated instead of that of the individual designer.

(h) Pictorial Representation of Design.—The application for registration shall be accompanied by two copies of a drawing or other pictorial representation of the useful article embodying the design, having one or more views, adequate to show the design, in a form and style suitable for reproduction, which shall be deemed a part of the application.

(i) Design in More than One Useful Article.—If the distinguishing elements of a design are in substantially the same form in different useful articles, the design shall be protected as to all such useful articles when protected as to one of them, but not more than one registration shall be required for the design.

(j) Application for More than One Design.—More than one design may be included in the same application under such conditions as may be prescribed by the Administrator. For each design included in an application the fee prescribed for a single design shall be paid.

§ 1311 Benefit of Earlier Filing Date in Foreign Country

An application for registration of a design filed in the United States by any person who has, or whose legal representative or predecessor or successor in title has, previously filed an application for registration of the same design in a foreign country which extends to designs of owners who are citizens of the United States, or to applications filed under this chapter, similar protection to that provided under this chapter shall have that same effect as if filed in the United States on the date on which the application was first filed in such foreign country, if the application in the United States is filed within 6 months after the earliest date on which any such foreign application was filed.

§ 1312 Oaths and Acknowledgments

(a) In General.—Oaths and acknowledgments required by this chapter—

(1) may be made—

(A) before any person in the United States authorized by law to administer oaths; or

(B) when made in a foreign country, before any diplomatic or consular officer of the United States authorized to administer oaths, or before any official authorized to administer oaths in the foreign country concerned, whose authority shall be proved by a certificate of a diplomatic or consular officer of the United States; and

(2) shall be valid if they comply with the laws of the State or country where made.

(b) Written Declaration in Lieu of Oath.—(1) The Administrator may by rule prescribe that any document which is to be filed under this chapter in the Office of the Administrator and which is required by any law, rule, or other regulation to be under oath, may be subscribed to by a written declaration in such form as the Administrator may prescribe, and such declaration shall be in lieu of the oath otherwise required.

(2) Whenever a written declaration under paragraph (1) is used, the document containing the declaration shall state that willful false statements are punishable by fine or imprisonment, or both, pursuant to section 1001 of title 18, and may jeopardize the validity of the application or document or a registration resulting therefrom.

§ 1313 Examination of Application and Issue or Refusal of Registration

(a) Determination of Registrability of Design; Registration.—Upon the filing of an application for registration in proper form under section 1310, and upon payment of the fee prescribed under section 1316, the Administrator shall determine whether or not the application relates to a design which on its face appears to be subject to protection under this chapter, and, if so, the Register shall register the design. Registration under this subsection shall be announced by publication. The date of registration shall be the date of publication.

(b) Refusal to Register; Reconsideration.—If, in the judgment of the Administrator, the application for registration relates to a design which on its face is not subject to protection under this chapter, the Administrator shall send to the applicant a notice of refusal to register and the grounds for the refusal. Within 3 months after the date on which the notice of refusal is sent, the applicant may, by written request, seek reconsideration of the application. After consideration of such a request, the Administrator shall either register the design or send to the applicant a notice of final refusal to register.

(c) Application to Cancel Registration.—Any person who believes he or she is or will be damaged by a registration under this chapter may, upon payment of the prescribed fee, apply to the Administrator at any time to cancel the registration on the ground that the design is not subject to protection under this chapter, stating the reasons for the request. Upon receipt of an application for cancellation, the Administrator shall send to the owner of the design, as shown in the records of the Office of the Administrator, a notice of the application, and the owner shall have a period of 3 months after the date on which such notice is mailed in which to present arguments to the Administrator for support of the validity of the registration. The Administrator shall also have the authority to establish, by regulation, conditions under which the opposing parties may appear and be heard in support of their arguments. If, after the periods provided for the presentation of arguments have expired, the Administrator determines that the applicant for cancellation has established that the design is not subject to protection under this chapter, the Administrator shall order the registration stricken from the record. Cancellation under this subsection shall be announced by publication, and notice of the Administrator's final determination with respect to any application for cancellation shall be sent to the applicant and to the owner of record. Costs of the cancellation procedure under this subsection shall be borne by the nonprevailing party or parties, and the Administrator shall have the authority to assess and collect such costs.

§ 1314 Certification of Registration

Certificates of registration shall be issued in the name of the United States under the seal of the Office of the Administrator and shall be recorded in the official records of the Office. The certificate shall state the name of the useful article, the date of filing of the application, the date of registration, and the date the design was made public, if earlier than the date of filing of the application, and shall contain a reproduction of the drawing or other pictorial representation of the design. If a description of the salient features of the design appears in the application, the description shall also appear in the certificate. A certificate of registration shall be admitted in any court as prima facie evidence of the facts stated in the certificate.

§ 1315 Publication of Announcements and Indexes

(a) Publications of the Administrator.—The Administrator shall publish lists and indexes of registered designs and cancellations of designs and may also publish the drawings or other pictorial representations of registered designs for sale or other distribution.

(b) File of Representatives of Registered Designs.—The Administrator shall establish and maintain a file of the drawings or other pictorial representations of registered designs. The file shall be available for use by the public under such conditions as the Administrator may prescribe.

§ 1316 Fees

The Administrator shall by regulation set reasonable fees for the filing of applications to register designs under this chapter and for other services relating to the administration of this chapter, taking into consideration the cost of providing these services and the benefit of a public record.

§ 1317 Regulations

The Administrator may establish regulations for the administration of this chapter.

§ 1318 Copies of Records

Upon payment of the prescribed fee, any person may obtain a certified copy of any official record of the Office of the Administrator that relates to this chapter. That copy shall be admissible in evidence with the same effect as the original.

§ 1319 Correction of Errors in Certificates

The Administrator may, by a certificate of correction under seal, correct any error in a registration incurred through the fault of the Office, or, upon payment of the required fee, any error of a clerical or typographical nature occurring in good faith but not through the fault of the Office. Such registration, together with the certificate, shall thereafter have the same effect as if it had been originally issued in such corrected form.

§ 1320 Ownership and Transfer

(a) Property Right in Design.—The property right in a design subject to protection under this chapter shall vest in the designer, the legal representatives of a deceased designer or of one under legal incapacity, the employer for whom the designer created the design in the case of a design made within the regular scope of the designer's employment, or a person to whom the rights of the designer or of such employer have been transferred. The person in whom the property right is vested shall be considered the owner of the design.

(b) Transfer of Property Right.—The property right in a registered design, or a design for which an application for registration has been or may be filed, may be assigned, granted, conveyed, or mortgaged by an instrument in writing, signed by the owner, or may be bequeathed by will.

(c) Oath or Acknowledgment of Transfer.—An oath or acknowledgment under section 1312 shall be prima facie evidence of the execution of an assignment, grant, conveyance, or mortgage under subsection (b).

(d) Recordation of Transfer.—An assignment, grant, conveyance, or mortgage under subsection (b) shall be void as against any subsequent purchaser or mortgagee for a valuable consideration, unless it is recorded in the Office of the Administrator within 3 months after its date of execution or before the date of such subsequent purchase or mortgage.

§ 1321 Remedy for Infringement

(a) In General.—The owner of a design is entitled, after issuance of a certificate of registration of the design under this chapter, to institute an action for any infringement of the design.

(b) Review of Refusal to Register.—(1) Subject to paragraph (2), the owner of a design may seek judicial review of a final refusal of the Administrator to register the design under this chapter by bringing a civil action, and may in the same action, if the court adjudges the design subject to protection under this chapter, enforce the rights in that design under this chapter.

(2) The owner of a design may seek judicial review under this section if—

(A) the owner has previously duly filed and prosecuted to final refusal an application in proper form for registration of the design;

(B) the owner causes a copy of the complaint in the action to be delivered to the Administrator within 10 days after the commencement of the action; and

(C) the defendant has committed acts in respect to the design which would constitute infringement with respect to a design protected under this chapter.

(c) Administrator as Party to Action.—The Administrator may, at the Administrator's option, become a party to the action with respect to the issue of registrability of the design claim by entering an appearance within 60 days after being served with the complaint, but the failure of the Administrator to become a party shall not deprive the court of jurisdiction to determine that issue.

(d) Use of Arbitration to Resolve Dispute.—The parties to an infringement dispute under this chapter, within such time as may be specified by the Administrator by regulation, may determine the dispute, or any aspect of the dispute, by arbitration. Arbitration shall be governed by title 9. The parties shall give notice of any arbitration award to the Administrator, and such award shall, as between the parties to the arbitration, be dispositive of the issues to which it relates. The arbitration award shall be unenforceable until such notice is given. Nothing in this subsection shall preclude the Administrator from determining whether a design is subject to registration in a cancellation proceeding under section 1313(c).

§ 1322 Injunctions

(a) In General.—A court having jurisdiction over actions under this chapter may grant injunctions in accordance with the principles of equity to prevent infringement of a design under this chapter, including, in its discretion, prompt relief by temporary restraining orders and preliminary injunctions.

(b) Damages for Injunctive Relief Wrongfully Obtained.—A seller or distributor who suffers damage by reason of injunctive relief wrongfully obtained under this section has a cause of action against the applicant for such injunctive relief and may recover such relief as may be appropriate, including damages for lost profits, cost of materials, loss of good will, and punitive damages in instances where the injunctive relief was sought in bad faith, and, unless the court finds extenuating circumstances, reasonable attorney's fees.

§ 1323 Recovery for Infringement

(a) Damages.—Upon a finding for the claimant in an action for infringement under this chapter, the court shall award the claimant damages adequate to compensate for the infringement. In addition, the court may increase the damages to such amount, not exceeding $50,000 or $1 per copy, whichever is greater, as the court determines to be just. The damages awarded shall constitute compensation and not a penalty. The court may receive expert testimony as an aid to the determination of damages.

(b) Infringer's Profits.—As an alternative to the remedies provided in subsection (a), the court may award the claimant the infringer's profits resulting from the sale of the copies if the court finds that the infringer's sales are reasonably related to the use of the claimant's design. In such a case, the claimant shall be required to prove only the amount of the infringer's sales and the infringer shall be required to prove its expenses against such sales.

(c) Statute of Limitations.—No recovery under subsection (a) or (b) shall be had for any infringement committed more than 3 years before the date on which the complaint is filed.

(d) Attorney's Fees.—In an action for infringement under this chapter, the court may award reasonable attorney's fees to the prevailing party.

(e) Disposition of Infringing and Other Articles.—The court may order that all infringing articles, and any plates, molds, patterns, models, or other means specifically adapted for making the articles, be delivered up for destruction or other disposition as the court may direct.

§ 1324 Power of Court over Registration

In any action involving the protection of a design under this chapter, the court, when appropriate, may order registration of a design under this chapter or the cancellation of such a registration. Any such order shall be certified by the court to the Administrator, who shall make an appropriate entry upon the record.

§ 1325 Liability for Action on Registration Fraudulently Obtained

Any person who brings an action for infringement knowing that registration of the design was obtained by a false or fraudulent representation materially affecting the rights under this chapter, shall be liable in the sum of $10,000, or such part of that amount as the court may determine. That amount shall be to compensate the defendant and shall be charged against the plaintiff and paid to the defendant, in addition to such costs and attorney's fees of the defendant as may be assessed by the court.

§ 1326 Penalty for False Marking

(a) In General.—Whoever, for the purpose of deceiving the public, marks upon, applies to, or uses in advertising in connection with an article made, used, distributed, or sold, a design which is not protected under this chapter, a design notice specified in section 1306, or any other words or symbols importing that the design is protected under this chapter, knowing that the design is not so protected, shall pay a civil fine of not more than $500 for each such offense.

(b) Suit by Private Persons.—Any person may sue for the penalty established by subsection (a), in which event one-half of the penalty shall be awarded to the person suing and the remainder shall be awarded to the United States.

§ 1327 Penalty for False Representation

Whoever knowingly makes a false representation materially affecting the rights obtainable under this chapter for the purpose of obtaining registration of a design under this chapter shall pay a penalty of not less than $500 and not more than $1,000, and any rights or privileges that individual may have in the design under this chapter shall be forfeited.

§ 1328 Enforcement by Treasury and Postal Service

(a) Regulations.—The Secretary of the Treasury and the United States Postal Service shall separately or jointly issue regulations for the enforcement of the rights set forth in section 1308 with respect to importation. Such regulations may require, as a condition for the exclusion of articles from the United States, that the person seeking exclusion take any one or more of the following actions:

(1) Obtain a court order enjoining, or an order of the International Trade Commission under section 337 of the Tariff Act of 1930 excluding, importation of the articles.

(2) Furnish proof that the design involved is protected under this chapter and that the importation of the articles would infringe the rights in the design under this chapter.

(3) Post a surety bond for any injury that may result if the detention or exclusion of the articles proves to be unjustified.

(b) Seizure and Forfeiture.—Articles imported in violation of the rights set forth in section 1308 are subject to seizure and forfeiture in the same manner as property imported in violation of the customs laws. Any such forfeited articles shall be destroyed as directed by the Secretary of the Treasury or the court, as the case may be, except that the articles may be returned to the country of export whenever it is shown to the satisfaction of the Secretary of the Treasury that the importer had no reasonable grounds for believing that his or her acts constituted a violation of the law.

§ 1329 Relation to Design Patent Law

The issuance of a design patent under title 35, United States Code, for an original design for an article of manufacture shall terminate any protection of the original design under this chapter.

§ 1330 Common Law and Other Rights Unaffected

Nothing in this chapter shall annul or limit—

(1) common law or other rights or remedies, if any, available to or held by any person with respect to a design which has not been registered under this chapter; or

(2) any right under the trademark laws or any right protected against unfair competition.

§ 1331 Administrator; Office of the Administrator

In this chapter, the "Administrator" is the Register of Copyrights, and the "Office of the Administrator" and the "Office" refer to the Copyright Office of the Library of Congress.

§ 1332 No Retroactive Effect

Protection under this chapter shall not be available for any design that has been made public under section 1310(b) before the effective date of this chapter.

Chapter 13 Notes

1. In 1998, the Vessel Hull Design Protection Act added chapter 13, entitled "Protection of Original Designs," to title 17. Pub. L. No. 105-304, 112 Stat. 2860, 2905. The Vessel Hull Design Protection Act is title V of the Digital Millennium Copyright Act, Pub. L. No. 105-304, 112 Stat. 2860, 2905.

2. The Satellite Home Viewer Improvement Act of 1999 amended section 1301(b)(3) in its entirety. Pub. L. No. 106-113, 113 Stat. 1501, app. I at 1501A-593.

The Vessel Hull Design Protection Amendments of 2008 amended subsection 1301(a) by revising the definition of vessel hulls in subpart (2) and by providing an exception for the U.S. Department of Defense in a new subpart (3). Pub. L. No. 110-434, 122 Stat. 4972. It also amended subsection 301(b) by revising the definitions of useful article in subpart (2) and hull in subpart (4) and by adding a definition for deck in new subpart (7). *Id.*

3. In 1999, section 1302(5) was amended to substitute "2 years" in lieu of "1 year." Pub. L. No. 106-44, 113 Stat. 221, 222.

4. The Satellite Home Viewer Improvement Act of 1999 amended section 1313(c) by adding at the end thereof the last sentence, which begins "Costs of the cancellation procedure." Pub. L. No. 106-113, 113 Stat. 1501, app. I at 1501A-594.

5. In 1999, section 1320 was amended to change the spelling in the heading of subsection

(c) from "acknowledgement" to "acknowledgment." Pub. L. No. 106-44, 113 Stat. 221, 222.

6. The effective date of chapter 13 is October 28, 1998. See section 505 of the Digital Millennium Copyright Act, which appears in Appendix B.

Trademark Act of 1946, as Amended

PUBLIC LAW 79-489, CHAPTER 540, APPROVED JULY 5, 1946; 60 STAT. 427

The headings used for sections and subsections or paragraphs in the following reprint of the Act are not part of the Act but have been added for convenience in using this reprint. Prior trademark statutes may be found in Title 15, Chapter 3, of the U.S. Code and in the Statutes at Large. The present Act forms Chapter 22 of Title 15 of the U.S. Code and the U.S. Code citations have been placed at the beginning of each section and subsection. Included are an amendment to repeal § 40 by the Act of May 24, 1949, 63 Stat. 109; and amendments made to § 7(a) by the Act of August 17, 1950, Public Law 81-710, 64 Stat. 459; to § 21 by the Act of July 19, 1952, Public Law 82-593, 66 Stat. 792; to §§ 17, 20, 21, 24 and 31 by the Act of August 8, 1958, Public Law 85- 609, 72 Stat. 540; to § 44(d) by the Act of October 3, 1961, Public Law 87-333, 75 Stat. 748; to §§ 1(a)(1), 2(d), 6, 7(a), 7(d), 7(e), 7(f), 9, 10, 12(a), 12(c), 13, 14, 15, 16, 21, 23, 24, 29, 30, 32, 33, 35, 44(b), 44(e) and 45 by the Act of October 9, 1962, Public Law 87-772, 76 Stat. 769; to § 31 by the Act of July 24, 1965, Public Law 89-83, 79 Stat. 260; to § 29 and the terms "Patent Office" and "Commissioner of Patents" by the Act of January 2, 1975, Public Law 93-596, 88 Stat. 1949; to §§ 13, 21 and 35 by the Act of January 2, 1975, Public Law 93-600, 88 Stat. 1955; to § 42 by the Act of October 3, 1978, Public Law 95-410, 92 Stat. 903; to § 17 by the Act of October 15, 1980, Public Law 96-455, 94 Stat. 2024; to § 31 by the Act of December 12, 1980, Public Law 96-517, 94 Stat. 3018; to §§ 21 and 39 by the Act of April 2, 1982, Public Law 97-164, 96 Stat. 25; to §§ 8(a), 8(b), 11, 13, 14, 15, 16 and 31 by the Act of August 27, 1982, Public Law 97-247, 96 Stat. 317; to § 31(a) by the Act of September 8, 1982, Public Law 97-256, 96 Stat. 816; to add § 39(a) by the Act of October 12, 1982, Public Law 97-296, 96 Stat. 1316; to §§ 34, 35 and 36 by the Act of October 12, 1984, Public Law 98-473, 98 Stat. 2179; to §§ 14(c), 21 and 45 by the Act of November 8, 1984, Public Law 98-620, 98 Stat. 3335; to §§ 1, 2(d), 2(e), 2(f), 3, 4, 5, 6(b), 7, 8(a), 9(a), 9(c), 10, 12(a), 13, 14, 15, 16, 18, 19, 21, 23, 24, 26, 27, 29, 30, 32(2), 33, 34, 35(a), 36, 39, 43(a), 44 and 45 and to add § 51 by the Act of November 16, 1988, Public Law 100-667, 102 Stat. 3935; to § 31(a) by the

Act of December 10, 1991, Public Law 102-204, 105 Stat. 1640; to § 32(1), to add § 40, and to §§ 43(a) and 45 by the Act of October 27, 1992, Public Law 102- 542, 106 Stat. 3567; to §§ 2(e), 2(f) and 23(a) by the Act of December 8, 1993, Public Law 103- 182, 107 Stat. 2057; to §§ 2(a) and 45 by the Act of December 8, 1994, Public Law 103-465, 108 Stat. 4982; to §§ 43 and 45 by the Act of January 16, 1996, Public Law 104-98, 109 Stat. 985; to §§ 34(d)(9) and 35 by Public Law No. 104-153, 110 Stat. 1386, July 2, 1996; to §§ 1(d)(1), 2, 7(a), 14, 23(c), 26, 31, 32(1), 33(b), 39(a) and 42 by Public Law 105-330, 112 Stat. 3069, effective October 30, 1998; to §§ 2, 10, 13(a), 14, 24, 32, 34(a), 35(a), 36, 40, 43(a), 43(c)(2) and 45 by Public Law 106-43, 113 Stat. 218, effective August 5, 1999; to §§ 1(a), 1(b), 12(b), 8, 9, 10, 44(d) and 44(e) by Public Law 105-330, 112 Stat. 3064, effective October 30, 1999; to §§ 1, 2, 6, 7, 8, 9, 10, 12, 13, 15, 16, 17, 18, 21, 23, 24, 30, 31, 32, 34, 35, 37, 41, 43, 44, 45, and 47(b) by Public Law 106-113, 113 Stat. 1501A, Nov. 29, 1999; to add Title XII, by Public Law 107-273, 116 Stat. 1758, effective Nov. 2, 2003; by Public Law 108-482, 118 Stat. 3916, effective Dec. 23, 2004; by Public Law 109-312, 120 Stat. 1730, Oct. 6, 2006.

TITLE I - THE PRINCIPAL REGISTER

§ 1 (15 U.S.C. § 1051). Application for registration; verification

(a)(1) The owner of a trademark used in commerce may request registration of its trademark on the principal register hereby established by paying the prescribed fee and filing in the Patent and Trademark Office an application and a verified statement, in such form as may be prescribed by the Director, and such number of specimens or facsimiles of the mark as used as may be required by the Director.

(2) The application shall include specification of the applicant's domicile and citizenship, the date of the applicant's first use of the mark, the date of the applicant's first use of the mark in commerce, the goods in connection with which the mark is used, and a drawing of the mark.

(3) The statement shall be verified by the applicant and specify that—

(A) the person making the verification believes that he or she, or the juristic person in whose behalf he or she makes the verification, to be the owner of the mark sought to be registered;

(B) to the best of the verifier's knowledge and belief, the facts recited in the application are accurate;

(C) the mark is in use in commerce; and

(D) to the best of the verifier's knowledge and belief, no other person has the right to use such mark in commerce either in the identical form thereof or in such near resemblance thereto as to be likely, when used on or in connection with the goods of such other person, to cause confusion, or to cause mistake, or to deceive, except that, in the case of every application claiming concurrent use, the applicant shall—

(i) state exceptions to the claim of exclusive use; and

(ii) shall specify, to the extent of the verifier's knowledge—

(I) any concurrent use by others;

(II) the goods on or in connection with which and the areas in which each concurrent use exists;

(III) the periods of each use; and

(IV) the goods and area for which the applicant desires registration.

(4) The applicant shall comply with such rules or regulations as may be prescribed by the Director. The Director shall promulgate rules prescribing the requirements for the application and for obtaining a filing date herein.

(b)(1) A person who has a bona fide intention, under circumstances showing the good faith of such person, to use a trademark in commerce may request registration of its trademark on the principal register hereby established by paying the prescribed fee and filing in the Patent and Trademark Office an application and a verified statement, in such form as may be prescribed by the Director.

(2) The application shall include specification of the applicant's domicile and citizenship, the goods in connection with which the applicant has a bona fide intention to use the mark, and a drawing of the mark.

(3) The statement shall be verified by the applicant and specify—

(A) that the person making the verification believes that he or she, or the juristic person in whose behalf he or she makes the verification, to be entitled to use the mark in commerce;

(B) the applicant's bona fide intention to use the mark in commerce;

(C) that, to the best of the verifier's knowledge and belief, the facts recited in the application are accurate; and

(D) that, to the best of the verifier's knowledge and belief, no other person has the right to use such mark in commerce either in the identical form thereof or in such near resemblance thereto as to be likely, when used on or in connection with the goods of such other person, to cause confusion, or to cause mistake, or to deceive. Except for applications filed pursuant to section 1126 of this title, no mark shall be registered until the applicant has met the requirements of subsections (c) and (d) of this section.

(4) The applicant shall comply with such rules or regulations as may be prescribed by the Director. The Director shall promulgate rules prescribing the requirements for the application and for obtaining a filing date herein.

(c) At any time during examination of an application filed under subsection (b) of this section, an applicant who has made use of the mark in commerce may claim the benefits of such use for purposes of this chapter, by amending his or her application to bring it into conformity with the requirements of subsection (a) of this section.

(d) (1) Within six months after the date on which the notice of allowance with respect to a mark is issued under section 1063(b)(2) of this title to an applicant under subsection (b) of this section, the applicant shall file in the Patent and Trademark Office, together with such number of specimens or facsimiles of the mark as used in commerce as may be required by the Director and payment of the prescribed fee, a verified statement that the mark is in use in commerce and specifying the date of the applicant's first use of the mark in commerce and those goods or services specified in the notice of allowance on or in connection with which the mark is used in commerce. Subject to examination and acceptance of the statement of use, the mark shall be registered in the Patent and Trademark Office, a certificate of registration shall be issued for those goods or services recited in the statement of use for which the mark is entitled to registration, and notice of registration shall be published in the Official Gazette of the Patent and Trademark Office. Such examination may include an examination of the factors set forth in subsections (a) through (e) of section 1052 of this title. The notice of registration shall specify the goods or services for which the mark is registered.

(2) The Director shall extend, for one additional 6-month period, the time for filing the statement of use under paragraph (1), upon written request of the applicant before the expiration of the 6-month period provided in paragraph (1). In addition to an extension under the preceding sentence, the Director may, upon a showing of good cause by the applicant, further extend the time for filing the statement of use under paragraph (1) for periods aggregating not more than 24 months, pursuant to written request of the applicant made before the expiration of the last extension granted under this paragraph. Any request for an extension under this paragraph shall be accompanied by a verified statement that the applicant has a continued bona fide intention to use the mark in commerce and specifying those goods or services identified in the notice of allowance on or in connection with which the applicant has a continued bona fide intention to use the mark in commerce. Any request for an extension under this paragraph shall be accompanied by payment of the prescribed fee. The Director shall issue regulations setting forth guidelines for determining what constitutes good cause for purposes of this paragraph.

(3) The Director shall notify any applicant who files a statement of use of the acceptance or refusal thereof and, if the statement of use is refused, the reasons for the refusal. An applicant may amend the statement of use.

(4) The failure to timely file a verified statement of use under paragraph (1) or an extension request under paragraph (2) shall result in abandonment of the application, unless it can be shown to the satisfaction of the Director that the delay in responding was unintentional, in which case the time for filing may be extended, but for a period not to exceed the period specified in paragraphs (1) and (2) for filing a statement of use.

(e) If the applicant is not domiciled in the United States the applicant may designate, by a document filed in the United States Patent and Trademark Office, the name and address of a person resident in the United States on whom may be served notices or process in proceedings affecting the mark. Such notices or process may be served upon the person so designated by leaving with that person or mailing to that person a copy thereof at the address specified in the last designation so filed. If the person so designated cannot be found at the address given in the last designation, or if the registrant does not designate by a document filed in the United States Patent and Trademark Office the name and address of a person resident in the United States on whom may be served notices or process in proceedings affecting the mark, such notices or process may be served on the Director.

(Amended Oct. 9, 1962, 76 Stat. 769; Jan. 2, 1975, 88 Stat. 1949; Nov. 16, 1988, 102 Stat. 3935; Oct. 30, 1998, 112 Stat. 3064; Nov. 29, 1999, 113 Stat. 1501A; Nov. 2, 2002, 116 Stat. 1906.)

§ 2 (15 U.S.C. § 1052). Trademarks registrable on the principal register; concurrent registration

No trademark by which the goods of the applicant may be distinguished from the goods of others shall be refused registration on the principal register on account of its nature unless it—

(a) Consists of or comprises immoral, deceptive, or scandalous matter; or matter which may disparage or falsely suggest a connection with persons, living or dead, institutions, beliefs, or national symbols, or bring them into contempt, or disrepute; or a geographical indication which, when used on or in connection with wines or spirits, identifies a place other than the origin of the goods and is first used on or in connection with wines or spirits by the applicant on or after one year after the date on which the WTO Agreement (as defined in section 3501(9) of title 19) enters into force with respect to the United States.

(b) Consists of or comprises the flag or coat of arms or other insignia of the United States, or of any State or municipality, or of any foreign nation, or any simulation thereof.

(c) Consists of or comprises a name, portrait, or signature identifying a particular living individual except by his written consent, or the name, signature, or portrait of a deceased President of the United States during the life of his widow, if any, except by the written consent of the widow.

(d) Consists of or comprises a mark which so resembles a mark registered in the Patent and Trademark Office, or a mark or trade name previously used in the United States by another and not abandoned, as to be likely, when used on or in connection with the goods of the applicant, to cause confusion, or to cause mistake, or to deceive: *Provided*, That if the Director determines that confusion, mistake, or deception is not likely to result from the continued use by more than one person of the same or similar marks under conditions and limitations as to the mode or place of use of the marks or the goods on or in connection with which such marks are used, concurrent registrations may be issued to such persons when they have become entitled to use such marks as a result of their concurrent lawful use in commerce prior to (1) the earliest of the filing dates of the applications pending or of any registration issued under this chapter; (2) July 5, 1947, in the case of registrations previously issued under the Act of March 3, 1881, or February 20, 1905, and continuing in full force and effect on that date; or (3) July 5, 1947, in the case of applications filed under the Act of February 20, 1905, and registered after July 5, 1947. Use prior to the filing date of any pending application or a registration shall not be required when the owner of such application or registration consents to the grant of a concurrent registration to the applicant. Concurrent registrations may also be issued by the Director when a court of competent jurisdiction has finally determined that more than one person is entitled to use the same or similar marks in commerce. In issuing concurrent registrations, the Director shall prescribe conditions and limitations as to the mode or place of use of the mark or the goods on or in connection with which such mark is registered to the respective persons.

(e) Consists of a mark which, (1) when used on or in connection with the goods of the applicant is merely descriptive or deceptively misdescriptive of them, (2) when used on or in connection with the goods of the applicant is primarily geographically descriptive of them, except as indications of regional origin may be registrable under section 1054 of this title, (3) when used on or in connection with the goods of the applicant is primarily geographically deceptively misdescriptive of them, (4) is primarily merely a surname, or (5) comprises any matter that, as a whole, is functional.

(f) Except as expressly excluded in subsections (a), (b), (c), (d), (e)(3), and (e)(5) of this section, nothing herein shall prevent the registration of a mark used by the applicant which has become distinctive of the applicant's goods in commerce. The Director may accept as prima facie evidence that the mark has become distinctive, as used on or in connection with the applicant's goods in commerce, proof of substantially exclusive and continuous use thereof as a mark by the applicant in commerce for the five years before the date on which the claim of distinctiveness is made. Nothing in this section shall prevent the registration of a mark which, when used on or in connection with the goods of the applicant, is primarily geographically deceptively misdescriptive of them, and which became distinctive of the applicant's goods in commerce before the date of the enactment of the North American Free Trade Agreement Implementation Act. A mark which would be likely to cause dilution by blurring or dilution by tarnishment under section 43(c), may be refused registration only pursuant to a proceeding brought under section 13. A registration for a mark which would be likely to cause dilution by blurring or dilution by tarnishment under section 1125(c) of this title, may be canceled pursuant to a proceeding brought under either section 1064 of this title or section 1092 of this titleA registration for a mark which would be likely to cause dilution by blurring or dilution by tarnishment under section 43(c), may be canceled pursuant to a proceeding brought under either section 14 or section 24

(Amended Oct. 9, 1962, 76 Stat. 769; Jan. 2, 1975, 88 Stat. 1949; Nov. 16, 1988, 102 Stat. 3037; Dec. 8, 1993, 107 Stat. 2057; Dec. 8, 1994, 108 Stat. 4982; Oct. 30, 1998, 112 Stat. 3069; Aug. 5, 1999, 113 Stat. 218; Nov. 29, 1999, 113 Stat. 1501A-583; Oct. 6, 2006, 120 Stat. 1730.)

§ 3 (15 U.S.C. § 1053). Service marks registrable

Subject to the provisions relating to the registration of trademarks, so far as they are applicable, service marks shall be registrable, in the same manner and with the same effect as are trademarks, and when registered they shall be entitled to the protection provided in this chapter herein in the case of trademarks. Applications and procedure under this section shall conform as nearly as practicable to those prescribed for the registration of trademarks.

(Amended Nov. 16, 1988, 102 Stat. 3938.)

§ 4 (15 U.S.C. § 1054). Collective marks and certification marks registrable

Subject to the provisions relating to the registration of trademarks, so far as they are applicable, collective and certification marks, including indications of regional origin, shall be registrable under this chapter, in the same manner and with the same effect as are trademarks, by persons, and nations, States, municipalities, and the like, exercising legitimate control over the use of the marks sought to be registered, even though not possessing an industrial or commercial establishment, and when registered they shall be entitled to the protection provided in this chapter in the case of trademarks, except in the case of certification marks when used so as to represent falsely that the owner or a user thereof makes or sells the goods or performs the services on or in connection with which such mark is used. Applications and procedure under this section shall conform as nearly as practicable to those prescribed for the registration of trademarks.

(Amended Nov. 16, 1988, 102 Stat. 3938.)

§ 5 (15 U.S.C. § 1055). Use by related companies

Where a registered mark or a mark sought to be registered is or may be used legitimately by related companies, such use shall inure to the benefit of the registrant or applicant for registration, and such use shall not affect the validity of such mark or of its registration, provided such mark is not used in such manner as to deceive the public. If first use of a mark by a person is controlled by the registrant or applicant for registration of the mark with respect to the nature and quality of the goods or services, such first use shall inure to the benefit of the registrant or applicant, as the case may be.

(Amended Nov. 16, 1988, 102 Stat. 3938.)

§ 6 (15 U.S.C. § 1056). Disclaimers

(a) The Director may require the applicant to disclaim an unregistrable component of a mark otherwise registrable. An applicant may voluntarily disclaim a component of a mark sought to be registered.

(b) No disclaimer, including those made under subsection (e) of section 1057 of this title, shall prejudice or affect the applicant's or registrant's rights then existing or thereafter arising in the disclaimed matter, or his right of registration on another application if the disclaimed matter be or shall have become distinctive of his goods or services.

(Amended Oct. 9, 1962, 76 Stat. 769; Nov. 16, 1988, 102 Stat. 3938; Nov. 29, 1999, 113 Stat. 1501A-583.)

§ 7 (15 U.S.C. § 1057). Certificates of registration

(a) *Issuance and form.* Certificates of registration of marks registered upon the principal register shall be issued in the name of the United States of America, under the seal of the United States Patent and Trademark Office, and shall be signed by the Director or have his signature placed thereon, and a record thereof shall be kept in the United States Patent and Trademark Office. The registration shall reproduce the mark, and state that the mark is registered on the principal register under this chapter, the date of the first use of the mark, the date of the first use of the mark in commerce, the particular goods or services for which it is registered, the number and date of the registration, the term thereof, the date on which the application for registration was received in the United States Patent and Trademark Office, and any conditions and limitations that may be imposed in the registration.

(b) *Certificate as prima facie evidence.* A certificate of registration of a mark upon the principal register provided by this chapter shall be prima facie evidence of the validity of the registered mark and of the registration of the mark, of the owner's ownership of the mark, and of the owner's exclusive right to use the registered mark in commerce on or in connection with the goods or services specified in the certificate, subject to any conditions or limitations stated in the certificate.

(c) *Application to register mark considered constructive use.* Contingent on the registration of a mark on the principal register provided by this chapter, the filing of the application to register such mark shall constitute constructive use of the mark, conferring a right of priority, nationwide in effect, on or in connection with the goods or services specified in the registration against any other person except for a person whose mark has not been abandoned and who, prior to such filing—

(1) has used the mark;

(2) has filed an application to register the mark which is pending or has resulted in registration of the mark; or

(3) has filed a foreign application to register the mark on the basis of which he or she has acquired a right of priority, and timely files an application under section 1126(d) to register the 125mark which is pending or has resulted in registration of the mark.

(d) *Issuance to assignee.* A certificate of registration of a mark may be issued to the assignee of the applicant, but the assignment must first be recorded in the United States Patent and Trademark Office. In case of change of ownership the Director shall, at the request of the owner and upon a proper showing and the payment of the prescribed fee, issue to such assignee a new certificate of registration of the said mark in the name of such assignee, and for the unexpired part of the original period.

(e) *Surrender, cancellation, or amendment by owner.* Upon application of the owner the Director may permit any registration to be surrendered for cancellation, and upon cancellation appropriate entry shall be made in the records of the United States Patent and Trademark Office. Upon application of the owner and payment of the prescribed fee, the Director for good cause may permit any registration to be amended or to be disclaimed in part: *Provided,* That the amendment or disclaimer does not alter materially the character of the mark. Appropriate entry shall be made in the records of the United States Patent and Trademark Office and upon the certificate of registration.

(f) *Copies of United States Patent and Trademark Office records as evidence.* Copies of any records, books, papers, or drawings belonging to the United States Patent and Trademark Office relating to marks, and copies of registrations, when authenticated by the seal of the United States Patent and Trademark Office and certified by the Director, or in his name by an employee of the Office duly designated by the Director, shall be evidence in all cases wherein the originals would be evidence; and any person making application therefor and paying the prescribed fee shall have such copies.

(g) *Correction of Patent and Trademark Office mistake.* Whenever a material mistake in a registration, incurred through the fault of the United States Patent and Trademark Office, is clearly disclosed by the records of the Office a certificate stating the fact and nature of such mistake shall be issued without charge and recorded and a printed copy thereof shall be attached to each printed copy of the registration and such corrected registration shall thereafter have the same effect as if the same had been originally issued in such corrected form, or in the discretion of the Director a new certificate of registration may be issued without charge. All certificates of correction heretofore issued in accordance with the rules of the United States Patent and Trademark Office and the registrations to which they are attached shall have the same force and effect as if such certificates and their issue had been specifically authorized by statute.

(h) *Correction of applicant's mistake.* Whenever a mistake has been made in a registration and a showing has been made that such mistake occurred in good faith through the fault of the applicant, the Director is authorized to issue a certificate of correction or, in his discretion, a new certificate upon the payment of the prescribed fee: *Provided,* That the correction does not involve such changes in the registration as to require republication of the mark.

(Amended Aug. 17, 1950, 64 Stat. 459; Oct. 9, 1962, 76 Stat. 769; Jan. 2, 1975, 88 Stat. 1949; Nov. 16, 1988, 102 Stat. 3938; Oct. 30, 1998, 112 Stat. 3069; Nov. 29, 1999, 113 Stat. 1501A-583; Mar. 17, 2010, 124 Stat. 66.)

§ 8 (15 U.S.C. § 1058). Duration, affidavits and fees

(a) *Time Periods for Required Affidavits.* Each registration shall remain in force for 10 years, except that the registration of any mark shall be canceled by the Director unless the owner of the registration files in the United States Patent and Trademark Office affidavits that meet the requirements of subsection (b), within the following time periods:

(1) Within the 1-year period immediately preceding the expiration of 6 years following the date of registration under this Act or the date of the publication under section 12(c).

(2) Within the 1-year period immediately preceding the expiration of 10 years following the date of registration, and each successive 10-year period following the date of registration.

(3) The owner may file the affidavit required under this section within the 6-month grace period immediately following the expiration of the periods established in paragraphs (1) and (2), together with the fee described in subsection (b) and the additional grace period surcharge prescribed by the Director.

(b) *Requirements for Affidavit.* The affidavit referred to in subsection (a) shall—

(1)(A) state that the mark is in use in commerce;

(B) set forth the goods and services recited in the registration on or in connection with which the mark is in use in commerce;

(C) be accompanied by such number of specimens or facsimiles showing current use of the mark in commerce as may be required by the Director; and

(D) be accompanied by the fee prescribed by the Director; or

(2)(A) set forth the goods and services recited in the registration on or in connection with which the mark is not in use in commerce;

(B) include a showing that any nonuse is due to special circumstances which excuse such nonuse and is not due to any intention to abandon the mark; and

(C) be accompanied by the fee prescribed by the Director.

(c) *Deficient Affidavit.* If any submission filed within the period set forth in subsection (a) is deficient, including that the affidavit was not filed in the name of the owner of the registration, the deficiency may be corrected after the statutory time period, within the time prescribed after notification of the deficiency. Such submission shall be accompanied by the additional deficiency surcharge prescribed by the Director.

(d) *Notice of Requirement.* Special notice of the requirement for such affidavit shall be attached to each certificate of registration and notice of publication under section 12(c).

(e) *Notification of Acceptance or Refusal.* The Director shall notify any owner who files any affidavit required by this section of the Director's acceptance or refusal thereof and, in the case of a refusal, the reasons therefor.

(f) *Designation of Resident for Service of Process and Notices.* If the owner is not domiciled in the United States, the owner may designate, by a document filed in the United States Patent and Trademark Office, the name and address of a person resident in the United States on whom may be served notices or process in proceedings affecting the mark. Such notices or process may be served upon the person so designated by leaving with that person or mailing to that person a copy thereof at the address specified in the last designation so filed. If the person so designated cannot be found at the last designated address, or if the owner does not designate by a document filed in the United States Patent and Trademark Office the name and address of a person resident in the United States on whom may be served notices or process in proceedings affecting the mark, such notices or process may be served on the Director.

(Jan. 2, 1975, 88 Stat. 1949; Aug. 27, 1982, 96 Stat. 320; Nov. 16, 1988, 102 Stat. 3939; Oct. 30, 1998, 112 Stat. 3066; Nov. 29, 1999, 113 Stat. 1536, 1501A-583; Nov. 2, 2002, 116 Stat. 1906; Mar. 17, 2010, 124 Stat. 66.)

§ 9 (15 U.S.C. § 1059). Renewal of registration

(a) Subject to the provisions of section 1058 of this title, each registration may be renewed for periods of 10 years at the end of each successive 10-year period following the date of registration upon payment of the prescribed fee and the filing of a written application, in such form as may be prescribed by the Director. Such application may be made at any time within 1 year before the end of each successive 10-year period for which the registration was issued or renewed, or it may be made within a grace period of 6 months after the end of each successive 10-year period, upon payment of a fee and surcharge prescribed therefor. If any application filed under this section is deficient, the deficiency may be corrected within the time prescribed after notification of the deficiency, upon payment of a surcharge prescribed therefor.

(b) If the Director refuses to renew the registration, the Director shall notify the registrant of the Director's refusal and the reasons therefor.

(c) If the registrant is not domiciled in the United States the registrant may designate, by a document filed in the United States Patent and Trademark Office, the name and address of a person resident in the United States on whom may be served notices or process in proceedings affecting the mark. Such notices or process may be served upon the person so designated by leaving with that person or mailing to that person a copy thereof at the address specified in the last designation so filed. If the person so designated cannot be found at the address given in the last designation, or if the registrant does not designate by a document filed in the United States Patent and Trademark Office the name and address of a person resident in the United States on whom may be served notices or process in proceedings affecting the mark, such notices or process may be served on the Director.

(Amended Oct. 9, 1962, 76 Stat. 770; Nov. 16, 1988, 102 Stat. 3939; Oct. 30, 1998, 112 Stat. 3067; Nov. 29, 1999, 113 Stat. 1501A-583; Nov. 2, 2002, 116 Stat. 1907.)

§ 10 (15 U.S.C. § 1060). Assignment

(a)(1) A registered mark or a mark for which an application to register has been filed shall be assignable with the good will of the business in which the mark is used, or with that part of the good will of the business connected with the use of and symbolized by the mark. Notwithstanding the preceding sentence, no application to register a mark under section 1051(b) of this title shall be assignable prior to the filing of an amendment under section 1051(c) of this title to bring the application into conformity with section 1051(a) of this title or the filing of the verified statement of use under section 1051(d) of this title, except for an assignment to a successor to the business of the applicant, or portion thereof, to which the mark pertains, if that business is ongoing and existing.

(2) In any assignment authorized by this section, it shall not be necessary to include the good will of the business connected with the use of and symbolized

by any other mark used in the business or by the name or style under which the business is conducted.

(3) Assignments shall be by instruments in writing duly executed. Acknowledgment shall be prima facie evidence of the execution of an assignment, and when the prescribed information reporting the assignment is recorded in the United States Patent and Trademark Office, the record shall be prima facie evidence of execution.

(4) An assignment shall be void against any subsequent purchaser for valuable consideration without notice, unless the prescribed information reporting the assignment is recorded in the United States Patent and Trademark Office within 3 months after the date of the assignment or prior to the subsequent purchase.

(5) The United States Patent and Trademark Office shall maintain a record of information on assignments, in such form as may be prescribed by the Director.

(b) An assignee not domiciled in the United States may designate by a document filed in the United States Patent and Trademark Office the name and address of a person resident in the United States on whom may be served notices or process in proceedings affecting the mark. Such notices or process may be served upon the person so designated by leaving with that person or mailing to that person a copy thereof at the address specified in the last designation so filed. If the person so designated cannot be found at the address given in the last designation, or if the assignee does not designate by a document filed in the United States Patent and Trademark Office the name and address of a person resident in the United States on whom may be served notices or process in proceedings affecting the mark, such notices or process may be served upon the Director.

(Amended Oct. 9, 1962, 76 Stat. 770; Jan. 2, 1975, 88 Stat. 1949; Nov. 16, 1988, 102 Stat. 3937; Aug. 5, 1999, 113 Stat. 218; Oct. 30, 1998, 112 Stat. 3068, effective Oct. 30, 1999; Aug. 5, 1999, 113 Stat. 218; Nov. 29, 1999, 113 Stat. 1501A-583; Nov. 2, 2002, 116 Stat. 1907.)

§ 11 (15 U.S.C. § 1061). Acknowledgments and verifications

Acknowledgments and verifications required hereunder under this chapter may be made before any person within the United States authorized by law to administer oaths, or, when made in a foreign country, before any diplomatic or consular officer of the United States or before any official authorized to administer oaths in the foreign country concerned whose authority shall be proved by a certificate of a diplomatic or consular officer of the United States, or apostille of an official designated by a foreign country which, by treaty or convention, accords like effect to apostilles of designated officials in the United States, and shall be valid if they comply with the laws of the state or country where made.

(Amended Aug. 27, 1982, 96 Stat. 321.)

§ 12 (15 U.S.C. § 1062). Publication

(a) Upon the filing of an application for registration and payment of the prescribed fee, the Director shall refer the application to the examiner in charge of the registration of marks, who shall cause an examination to be made and, if on such examination it shall appear that the applicant is entitled to registration, or would be entitled to registration upon the acceptance of the statement of use required by section 1051(d) of this title, the Director shall cause the mark to be published in the Official Gazette of the Patent and Trademark Office: *Provided,* That in the case of an applicant claiming concurrent use, or in the case of an application to be placed in an interference as provided for in section 1066 of this title, the mark, if otherwise registrable, may be published subject to the determination of the rights of the parties to such proceedings.

(b) If the applicant is found not entitled to registration, the examiner shall advise the applicant thereof and of the reason therefor. The applicant shall have a period of six months in which to reply or amend his application, which shall then be reexamined. This procedure may be repeated until (1) the examiner finally refuses registration of the mark or (2) the applicant fails for a period of six months to reply or amend or appeal, whereupon the application shall be deemed to have been abandoned, unless it can be shown to the satisfaction of the Director that the delay in responding was unintentional, whereupon such time may be extended.

(c) A registrant of a mark registered under the provision of the Act of March 3, 1881, or the Act of February 20, 1905, may, at any time prior to the expiration of the registration thereof, upon the payment of the prescribed fee file with the Director an affidavit setting forth those goods stated in the registration on which said mark is in use in commerce and that the registrant claims the benefits of this chapter for said mark. The Director shall publish notice thereof with a reproduction of said mark in the Official Gazette, and notify the registrant of such publication and of the requirement for the affidavit of use or nonuse as provided for in subsection (b) of section 1058 of this title. Marks published under this subsection shall not be subject to the provisions of section 1063 of this title.

(Amended Oct. 9, 1962, 76 Stat. 770; Jan. 2, 1975, 88 Stat. 1949; Nov. 16, 1988, 102 Stat. 3940; Oct. 30, 1998, 112 Stat. 3066; Nov. 29, 1999, 113 Stat. 1501A-583.)

§ 13 (15 U.S.C. § 1063). Opposition

(a) Any person who believes that he would be damaged by the registration of a mark upon the principal register, including the registration of any mark which would be likely to cause dilution by blurring or dilution by tarnishment under section 1125(c) of this title, may, upon payment of the prescribed fee, file an opposition in the Patent and Trademark Office, stating the grounds therefor, within thirty days after the publication under subsection (a) of section 1062 of this title of the mark sought to be registered. Upon written request prior to the expiration of the thirty-day period, the time for filing opposition shall be extended for an additional thirty days, and further extensions of time for filing opposition may be granted by the Director for good cause when requested prior to the expiration of an extension. The Director shall notify the applicant of each extension of the time for filing opposition. An opposition may be amended under such conditions as may be prescribed by the Director.

(b) Unless registration is successfully opposed—

(1) a mark entitled to registration on the principal register based on an application filed under section 1051(a) of this title or pursuant to section 1126 shall be registered in the Patent and Trademark Office, a certificate of registration shall be issued, and notice of the registration shall be published in the Official Gazette of the Patent and Trademark Office; or

(2) a notice of allowance shall be issued to the applicant if the applicant applied for registration under section 1051(b) of this title.

(Amended Oct. 9, 1962, 76 Stat. 771; Jan. 2, 1975, 88 Stat. 1949; Jan. 2, 1975, 88 Stat. 1955; Aug. 27, 1982, 96 Stat. 320; Nov. 16, 1988, 102 Stat. 3940; Aug. 5, 1999, 113 Stat. 218; Nov. 29, 1999, 113 Stat. 1501A-583; Oct. 6, 2006, 120 Stat. 1730.)

§ 14 (15 U.S.C. § 1064). Cancellation

A petition to cancel a registration of a mark, stating the grounds relied upon, may, upon payment of the prescribed fee, be filed as follows by any person who believes that he is or will be damaged, including as a result of a likelihood of dilution by blurring or dilution by tarnishment under section 1125(c) of this title, by the registration of a mark on the principal register established by this chapter, or under the Act of March 3, 1881, or the Act of February 20, 1905:

(1) Within five years from the date of the registration of the mark under this chapter.

(2) Within five years from the date of publication under section 1062(c) of this title of a mark registered under the Act of March 3, 1881, or the Act of February 20, 1905.

(3) At any time if the registered mark becomes the generic name for the goods or services, or a portion thereof, for which it is registered, or is functional, or has been abandoned, or its registration was obtained fraudulently or contrary to the provisions of section 1054 of this title or of subsection (a), (b), or (c) of section 1052 of this title for a registration under this chapter, or contrary to similar prohibitory provisions of such said prior Acts for a registration under such Acts, or if the registered mark is being used by, or with the permission of, the registrant so as to misrepresent the source of the goods or services on or in connection with which the mark is used. If the registered mark becomes the generic name for less than all of the goods or services for which it is registered, a petition to cancel the registration for only those goods or services may be filed. A registered mark shall not be deemed to be the generic name of goods or services solely because such mark is also used as a name of or to identify a unique product or service. The primary significance of the registered mark to the relevant public

rather than purchaser motivation shall be the test for determining whether the registered mark has become the generic name of goods or services on or in connection with which it has been used.

(4) At any time if the mark is registered under the Act of March 3, 1881, or the Act of February 20, 1905, and has not been published under the provisions of subsection (c) of section 1062 of this title.

(5) At any time in the case of a certification mark on the ground that the registrant (A) does not control, or is not able legitimately to exercise control over, the use of such mark, or (B) engages in the production or marketing of any goods or services to which the certification mark is applied, or (C) permits the use of the certification mark for purposes other than to certify, or (D) discriminately refuses to certify or to continue to certify the goods or services of any person who maintains the standards or conditions which such mark certifies:

Provided, That the Federal Trade Commission may apply to cancel on the grounds specified in paragraphs (3) and (5) of this section any mark registered on the principal register established by this chapter, and the prescribed fee shall not be required. Nothing in paragraph (5) shall be deemed to prohibit the registrant from using its certification mark in advertising or promoting recognition of the certification program or of the goods or services meeting the certification standards of the registrant. Such uses of the certification mark shall not be grounds for cancellation under paragraph (5), so long as the registrant does not itself produce, manufacture, or sell any of the certified goods or services to which its identical certification mark is applied.

(Amended Oct. 9, 1962, 76 Stat. 771; Aug. 27, 1982, 96 Stat. 320; Nov. 8, 1984, 98 Stat. 3335; Nov. 16, 1988, 102 Stat. 3940; Oct. 30, 1998, 112 Stat. 3069; Aug. 5, 1999, 113 Stat. 218; Oct. 6, 2006, 120 Stat. 1730.)

§ 15 (15 U.S.C. § 1065). Incontestability of right to use mark under certain conditions

Except on a ground for which application to cancel may be filed at anytime under paragraphs (3) and (5) of section 1064 of this title, and except to the extent, if any, to which the use of a mark registered on the principal register infringes a valid right acquired under the law of any State or Territory by use of a mark or trade name continuing from a date prior to the date of registration under this chapter of such registered mark, the right of the owner to use such registered mark in commerce for the goods or services on or in connection with which such registered mark has been in continuous use for five consecutive years subsequent to the date of such registration and is still in use in commerce, shall be incontestable: *Provided,* That—

(1) there has been no final decision adverse to the owner's claim of ownership of such mark for such goods or services, or to the owner's right to register the same or to keep the same on the register; and

(2) there is no proceeding involving said rights pending in the United States Patent and Trademark Office or in a court and not finally disposed of; and

(3) an affidavit is filed with the Director within one year after the expiration of any such five-year period setting forth those goods or services stated in the registration on or in connection with which such mark has been in continuous use for such five consecutive years and is still in use in commerce, and the other matters specified in paragraphs (1) and (2) of this section; and

(4) no incontestable right shall be acquired in a mark which is the generic name for the goods or services or a portion thereof, for which it is registered.

Subject to the conditions above specified in this section, the incontestable right with reference to a mark registered under this chapter shall apply to a mark registered under the Act of March 3, 1881, or the Act of February 20, 1905, upon the filing of the required affidavit with the Director within one year after the expiration of any period of five consecutive years after the date of publication of a mark under the provisions of subsection (c) of section 1062 of this title.

The Director shall notify any registrant who files the above-prescribed affidavit of the filing thereof.

(Amended Oct. 9, 1962, 76 Stat. 771; Jan. 2, 1975, 88 Stat. 1949; Aug. 27, 1982, 96 Stat. 320; Nov. 16, 1988, 102 Stat. 3941; Nov. 29, 1999, 113 Stat. 1501A-583; Mar. 17, 2010, 124 Stat. 66.)

§ 16 (15 U.S.C. § 1066). Interference

Upon petition showing extraordinary circumstances, the Director may declare that an interference exists when application is made for the registration of a mark which so resembles a mark previously registered by another, or for the registration of which another has previously made application, as to be likely when used on or in connection with the goods or services of the applicant to cause confusion or mistake or to deceive. No interference shall be declared between an application and the registration of a mark the right to the use of which has become incontestable.

(Amended Oct. 9, 1962, 76 Stat. 771; Aug. 27, 1982, 96 Stat. 321; Nov. 16, 1988, 102 Stat. 3941; Nov. 29, 1999, 113 Stat. 1501A-583.)

§ 17 (15 U.S.C. § 1067). Interference, opposition, and proceedings for concurrent use registration or for cancellation; notice; Trademark Trial and Appeal Board

(a) In every case of interference, opposition to registration, application to register as a lawful concurrent user, or application to cancel the registration of a mark, the Director shall give notice to all parties and shall direct a Trademark Trial and Appeal Board to determine and decide the respective rights of registration.

(b) The Trademark Trial and Appeal Board shall include the Director, Deputy Director of the United States Patent and Trademark Office, the Commissioner for Patents, the Commissioner for Trademarks, and administrative trademark judges who are appointed by the Director.

(c) Authority of the Secretary. The Secretary of Commerce may, in his or her discretion, deem the appointment of an administrative trademark judge who, before the date of the enactment of this subsection, August 12, 2008, held office pursuant to an appointment by the Director to take effect on the date on which the Director initially appointed the administrative trademark judge.

(d) Defense to Challenge of Appointment. It shall be a defense to a challenge to the appointment of an administrative trademark judge on the basis of the judge's having been originally appointed by the Director that the administrative trademark judge so appointed was acting as a de facto officer.

(Amended Aug. 8, 1958, 72 Stat. 540; Jan. 2, 1975, 88 Stat. 1949; 94 Stat. 2024; Nov. 29, 1999, 113 Stat. 1536, 1501A-580; Nov. 2, 2002, 116 Stat. 1902; Aug. 12, 2008, 122 Stat. 3014.)

§ 18 (15 U.S.C. § 1068). Refusal, cancellation, or restriction of registration; concurrent use

In such proceedings the Director may refuse to register the opposed mark, may cancel the registration, in whole or in part, may modify the application or registration by limiting the goods or services specified therein, may otherwise restrict or rectify with respect to the register the registration of a registered mark, may refuse to register any or all of several interfering marks, or may register the mark or marks for the person or persons entitled thereto, as the rights of the parties under this chapter may be established in the proceedings: Provided, That in the case of the registration of any mark based on concurrent use, the Director shall determine and fix the conditions and limitations provided for in subsection (d) of section 1052 of this title. However, no final judgment shall be entered in favor of an applicant under section 1051(b) of this title before the mark is registered, if such applicant cannot prevail without establishing constructive use pursuant to section 1057(c) of this title.

(Amended Nov. 16, 1988, 102 Stat. 3941; Nov. 29, 1999, 113 Stat. 1501A-583.)

§ 19 (15 U.S.C. § 1069). Applicability, in inter partes proceeding, of equitable principles of laches, estoppel and acquiescence

In all inter partes proceedings equitable principles of laches, estoppel, and acquiescence, where applicable, may be considered and applied.

(Amended Jan. 2, 1975, 88 Stat. 1949; Nov. 16, 1988, 102 Stat. 3941.)

§ 20 (15 U.S.C. § 1070). Appeal from examiner to Trademark Trial and Appeal Board

An appeal may be taken to the Trademark Trial and Appeal Board from any final decision of the examiner in charge of the registration of marks upon the payment of the prescribed fee.

(Amended Aug. 8, 1958, 72 Stat. 540.)

§ 21 (15 U.S.C. § 1071). Appeal to courts

(a) *Persons entitled to appeal; United States Court of Appeals for the Federal Circuit; waiver of civil action; election of civil action by adverse party; procedure.*

(1) An applicant for registration of a mark, party to an interference proceeding, party to an opposition proceeding, party to an application to register as a lawful concurrent user, party to a cancellation proceeding, a registrant who has filed an affidavit as provided in section 1058 or section 71 of this title, or an applicant for renewal, who is dissatisfied with the decision of the Director or Trademark Trial and Appeal Board, may appeal to the United States Court of Appeals for the Federal Circuit thereby waiving his right to proceed under subsection (b) of this section: *Provided,* That such appeal shall be dismissed if any adverse party to the proceeding, other than the Director, shall, within twenty days after the appellant has filed notice of appeal according to paragraph (2) of this subsection, files notice with the Director that he elects to have all further proceedings conducted as provided in subsection (b) of this section. Thereupon the appellant shall have thirty days thereafter within which to file a civil action under subsection (b) of this section, in default of which the decision appealed from shall govern the further proceedings in the case.

(2) When an appeal is taken to the United States Court of Appeals for the Federal Circuit, the appellant shall file in the United States Patent and Trademark Office a written notice of appeal directed to the Director, within such time after the date of the decision from which the appeal is taken as the Director prescribes, but in no case less than 60 days after that date.

(3) The Director shall transmit to the United States Court of Appeals for the Federal Circuit a certified list of the documents comprising the record in the United States Patent and Trademark Office. The court may request that the Director forward the original or certified copies of such documents during pendency of the appeal. In an ex parte case, the Director shall submit to that court a brief explaining the grounds for the decision of the United States Patent and Trademark Office, addressing all the issues involved in the appeal. The court shall, before hearing an appeal, give notice of the time and place of the hearing to the Director and the parties in the appeal.

(4) The United States Court of Appeals for the Federal Circuit shall review the decision from which the appeal is taken on the record before the United States Patent and Trademark Office. Upon its determination the court shall issue its mandate and opinion to the Director, which shall be entered of record in the United States Patent and Trademark Office and shall govern the further proceedings in the case. However, no final judgment shall be entered in favor of an applicant under section 1051(b) of this title before the mark is registered, if such applicant cannot prevail without establishing constructive use pursuant to section 1057(c) of this title.

(b) *Civil action; persons entitled to; jurisdiction of court; status of Director; procedure.*

(1) Whenever a person authorized by subsection (a) of this section to appeal to the United States Court of Appeals for the Federal Circuit is dissatisfied with the decision of the Director or Trademark Trial and Appeal Board, said person

may, unless appeal has been taken to said United States Court of Appeals for the Federal Circuit, have remedy by a civil action if commenced within such time after such decision, not less than sixty days, as the Director appoints or as provided in subsection (a) of this section. The court may adjudge that an applicant is entitled to a registration upon the application involved, that a registration involved should be canceled, or such other matter as the issues in the proceeding require, as the facts in the case may appear. Such adjudication shall authorize the Director to take any necessary action, upon compliance with the requirements of law. However, no final judgment shall be entered in favor of an applicant under section 1051(b) of this title before the mark is registered, if such applicant cannot prevail without establishing constructive use pursuant to section 1057(c) of this title.

(2) The Director shall not be made a party to an inter partes proceeding under this subsection, but he shall be notified of the filing of the complaint by the clerk of the court in which it is filed and shall have the right to intervene in the action.

(3) In any case where there is no adverse party, a copy of the complaint shall be served on the Director, and, unless the court finds the expenses to be unreasonable, all the expenses of the proceeding shall be paid by the party bringing the case, whether the final decision is in favor of such party or not. In suits brought hereunder, the record in the United States Patent and Trademark Office shall be admitted on motion of any party, upon such terms and conditions as to costs, expenses, and the further cross-examination of the witnesses as the court imposes, without prejudice to the right of any party to take further testimony. The testimony and exhibits of the record in the United States Patent and Trademark Office, when admitted, shall have the same effect as if originally taken and produced in the suit.

(4) Where there is an adverse party, such suit may be instituted against the party in interest as shown by the records of the United States Patent and Trademark Office at the time of the decision complained of, but any party in interest may become a party to the action. If there are adverse parties residing in a plurality of districts not embraced within the same State, or an adverse party residing in a foreign country, the United States District Court for the Eastern District of Virginia shall have jurisdiction and may issue summons against the adverse parties directed to the marshal of any district in which any adverse party resides. Summons against adverse parties residing in foreign countries may be served by publication or otherwise as the court directs.

(Amended July 19, 1952, 66 Stat. 814; Aug. 8, 1958, 72 Stat. 540; Oct. 9, 1962, 76 Stat. 771; Jan. 2, 1975, 88 Stat. 1949; Jan. 2, 1975, 88 Stat. 1955; Apr. 2, 1982, 96 Stat. 49; Nov. 8, 1984, 98 Stat. 3363; Nov. 16, 1988, 102 Stat. 3942; Nov. 29, 1999, 113 Stat. 1501A-583; Mar. 17, 2010, 124 Stat. 66; Sept. 16, 2011, 125 Stat. 316.)

§ 22 (15 U.S.C. § 1072). Registration as notice

Registration of a mark on the principal register provided by this chapter or under the Act of March 3, 1881, or the Act of February 20, 1905, shall be constructive notice of the registrant's claim of ownership thereof.

TITLE II - THE SUPPLEMENTAL REGISTER

§ 23 (15 U.S.C. § 1091). Filing and registration for foreign use

(a) In addition to the principal register, the Director shall keep a continuation of the register provided in paragraph (b) of section 1 of the Act of March 19, 1920, entitled "An Act to give effect to certain provisions of the convention for the protection of trademarks and commercial names, made and signed in the city of Buenos Aires, in the Argentine Republic, August 20, 1910, and for other purposes," to be called the supplemental register. All marks capable of distinguishing applicant's goods or services and not registrable on the principal register herein provided, except those declared to be unregistrable under subsections (a), (b), (c), (d), and (e)(3) of section 1052 of this title, which are in lawful use in commerce by the owner thereof, on or in connection with any goods or services may be registered on the supplemental register upon the payment of the prescribed fee and compliance with the provisions of subsections (a) and (e) of section 1051 of this title so far as they are applicable. Nothing in this section shall prevent the registration on the supplemental register of a mark, capable of distinguishing the applicant's goods or services and not registrable on the principal register under this chapter, that is declared to be unregistrable under section 1052(e)(3) of this title, if such mark has been in lawful use in commerce by the owner thereof, on or in connection with any goods or services, since before December 8, 1993.

(b) Upon the filing of an application for registration on the supplemental register and payment of the prescribed fee the Director shall refer the application to the examiner in charge of the registration of marks, who shall cause an examination to be made and if on such examination it shall appear that the applicant is entitled to registration, the registration shall be granted. If the applicant is found not entitled to registration the provisions of subsection (b) of section 1062 of this title shall apply.

(c) For the purposes of registration on the supplemental register, a mark may consist of any trademark, symbol, label, package, configuration of goods, name, word, slogan, phrase, surname, geographical name, numeral, device, any matter that as a whole is not functional, or any combination of any of the foregoing, but such mark must be capable of distinguishing the applicant's goods or services.

(Amended Oct. 9, 1962, 76 Stat. 773; Nov. 16, 1988, 102 Stat. 3942; Dec. 8, 1993, 107 Stat. 2057; Oct. 30, 1998, 112 Stat. 3069; Nov. 29, 1999, 113 Stat. 1501A-583; Nov. 2, 2002, 116 Stat. 1908.)

§ 24 (15 U.S.C. § 1092). Cancellation

Marks for the supplemental register shall not be published for or be subject to opposition, but shall be published on registration in the Official Gazette of the Patent and Trademark Office. Whenever any person believes that such person is or will be damaged by the registration of a mark on the supplemental register–

(1) for which the effective filing date is after the date on which such person's mark became famous and which would be likely to cause dilution by blurring or dilution by tarnishment under section 1125(c) of this title; or

(2) on grounds other than dilution by blurring or dilution by tarnishment, such person may at any time, upon payment of the prescribed fee and the filing of a petition stating the ground therefor, apply to the Director to cancel such registration.

The Director shall refer such application to the Trademark Trial and Appeal Board, which shall give notice thereof to the registrant. If it is found after a hearing before the Board that the registrant is not entitled to registration, or that the mark has been abandoned, the registration shall be canceled by the Director. However, no final judgment shall be entered in favor of an applicant under section 1051(b) of this title before the mark is registered, if such applicant cannot prevail without establishing constructive use pursuant to section 1057(c).

(Amended Aug. 8, 1958, 72 Stat. 540; Oct. 9, 1962, 76 Stat. 773; Jan. 2, 1975, 88 Stat. 1949; Nov. 16, 1988, 102 Stat. 3943; Aug. 5, 1999, 113 Stat. 218; Nov. 29, 1999, 113 Stat. 1501A-583; Oct. 6, 2006, 120 Stat. 1730.)

§ 25 (15 U.S.C. § 1093). Supplemental registration certificate

The certificates of registration for marks registered on the supplemental register shall be conspicuously different from certificates issued for marks registered on the principal register.

§ 26 (15 U.S.C. § 1094). General provisions

The provisions of this chapter shall govern so far as applicable applications for registration and registrations on the supplemental register as well as those on the principal register, but applications for and registrations on the supplemental register shall not be subject to or receive the advantages of sections 1051(b), 1052(e), 1052(f), 1057(b), 1057(c), 1062(a), 1063 to 1068, inclusive, 1072, 1115 and 1124 of this title.

(Amended Nov. 16, 1988, 102 Stat. 3943; Oct. 30, 1998, 112 Stat. 3069.)

§ 27 (15 U.S.C. § 1095). Principal registration not precluded by supplemental registration

Registration of a mark on the supplemental register, or under the Act of March 19, 1920, shall not preclude registration by the registrant on the principal register established by this chapter. Registration of a mark on the supplemental register shall not constitute an admission that the mark has not acquired distinctiveness.

(Amended Nov. 16, 1988, 102 Stat. 3943.)

§ 28 (15 U.S.C. § 1096). Department of Treasury; supplemental registration not filed

Registration on the supplemental register or under the Act of March 19, 1920, shall not be filed in the Department of the Treasury or be used to stop importations.

TITLE III - NOTICE OF REGISTRATION

§ 29 (15 U.S.C. § 1111). Notice of registration; display with mark; recovery of profits and damages in infringement suit

Notwithstanding the provisions of section 1072 of this title, a registrant of a mark registered in the Patent and Trademark Office, may give notice that his mark is registered by displaying with the mark the words "Registered in U.S. Patent and Trademark Office" or "Reg. U.S. Pat. & Tm. Off." or the letter R enclosed within a circle, thus ®; and in any suit for infringement under this chapter by such a registrant failing to give such notice of registration, no profits and no damages shall be recovered under the provisions of this chapter unless the defendant had actual notice of the registration.

(Amended Oct. 9, 1962, 76 Stat. 773; Jan. 2, 1975, 88 Stat. 1949; Nov. 16, 1988, 102 Stat. 1343.)

**Note: The amendment of the wording of this term by Public Law 93-596 became effective on January 2, 1975. However, the amendment provides that any registrant may continue to give notice of his registration in accordance with § 29 of the Trademark Act of 1946, as amended Oct. 9, 1962, as an alternative to notice in accordance with § 29 of the Trademark Act as amended by Public Law 93-596, regardless of whether his mark was registered before or after January 2, 1975.*

TITLE IV - CLASSIFICATION

§ 30 (15 U.S.C. § 1112). Classification of goods and services; registration in plurality of classes

The Director may establish a classification of goods and services, for convenience of Patent and Trademark Office administration, but not to limit or extend the applicant's or registrant's rights. The applicant may apply to register a mark for any or all of the goods or services on or in connection with which he or she is using or has a bona fide intention to use the mark in commerce: Provided, That if the Director by regulation permits the filing of an application for the registration of a mark for goods or services which fall within a plurality of classes, a fee equaling the sum of the fees for filing an application in each class shall be paid, and the Director may issue a single certificate of registration for such mark.

(Amended Oct. 9, 1962, 76 Stat. 773; Jan. 2, 1975, 88 Stat. 1949; Nov. 16, 1988, 102 Stat. 3943; Nov. 29, 1999, 113 Stat. 1501A-583.)

TITLE V - FEES AND CHARGES

§ 31 (15 U.S.C. § 1113). Fees

(a) The Director shall establish fees for the filing and processing of an application for the registration of a trademark or other mark and for all other services performed by and materials furnished by the Patent and Trademark Office related to trademarks and other marks. Fees established under this subsection may be adjusted by the Director once each year to reflect, in the aggregate, any fluctuations during the preceding 12 months in the Consumer Price Index, as determined by the Secretary of Labor. Changes of less than 1 percent may be ignored. No fee established under this section shall take effect until at least 30 days after notice of the fee has been published in the Federal Register and in the Official Gazette of the Patent and Trademark Office.

(b) The Director may waive the payment of any fee for any service or material related to trademarks or other marks in connection with an occasional request made by a department or agency of the Government, or any officer thereof. The Indian Arts and Crafts Board will not be charged any fee to register Government trademarks of genuineness and quality for Indian products or for products of particular Indian tribes and groups.

(Amended Aug. 8, 1958, 72 Stat. 540; July 24, 1965, 79 Stat. 260; Jan. 2, 1975, 88 Stat. 1949; Dec. 12, 1980, 94 Stat. 3018; Aug. 27, 1982, 96 Stat. 319; Sept. 8, 1982, 96 Stat. 256; Dec. 10, 1991, 105 Stat. 1640; Oct. 30, 1998, 112 Stat. 3069; Nov. 29, 1999, 113 Stat. 1501A-554.)

TITLE VI - REMEDIES

§ 32 (15 U.S.C. § 1114). Remedies; infringement; innocent infringers

(1) Any person who shall, without the consent of the registrant—

(a) use in commerce any reproduction, counterfeit, copy, or colorable imitation of a registered mark in connection with the sale, offering for sale, distribution, or advertising of any goods or services on or in connection with which such use is likely to cause confusion, or to cause mistake, or to deceive; or

(b) reproduce, counterfeit, copy or colorably imitate a registered mark and apply such reproduction, counterfeit, copy or colorable imitation to labels, signs, prints, packages, wrappers, receptacles or advertisements intended to be used in commerce upon or in connection with the sale, offering for sale, distribution, or advertising of goods or services on or in connection with which such use is likely to cause confusion, or to cause mistake, or to deceive,

shall be liable in a civil action by the registrant for the remedies hereinafter provided. Under subsection (b) hereof, the registrant shall not be entitled to recover profits or damages unless the acts have been committed with knowledge that such imitation is intended to be used to cause confusion, or to cause mistake, or to deceive.

As used in this paragraph, the term "any person" includes the United States, all agencies and instrumentalities thereof, and all individuals, firms, corporations, or other persons acting for the United States and with the authorization and consent of the United States, and any State, any instrumentality of a State, and any officer or employee of a State or instrumentality of a State acting in his or her official capacity. The United States, all agencies and instrumentalities thereof, and all individuals, firms, corporations, other persons acting for the United States and with the authorization and consent of the United States, and any State, and any such instrumentality, officer, or employee, shall be subject to the provisions of this chapter in the same manner and to the same extent as any nongovernmental entity.

(2) Notwithstanding any other provision of this chapter, the remedies given to the owner of a right infringed under this chapter or to a person bringing an action under section 1125(a) or (d) of this title shall be limited as follows:

(A) Where an infringer or violator is engaged solely in the business of printing the mark or violating matter for others and establishes that he or she was an innocent infringer or innocent violator, the owner of the right infringed or person bringing the action under section 1125(a) of this title shall be entitled as against such infringer or violator only to an injunction against future printing.

(B) Where the infringement or violation complained of is contained in or is part of paid advertising matter in a newspaper, magazine, or other similar periodical or in an electronic communication as defined in section 2510(12) of title 18, United States Code, the remedies of the owner of the right infringed or person bringing the action under section 1125(a) of this title as against the publisher or distributor of such newspaper, magazine, or other similar periodical or electronic communication shall be limited to an injunction against the presentation of such advertising matter in future issues of such newspapers, magazines, or other similar periodicals or in future transmissions of such electronic communications. The limitations of this subparagraph shall apply only to innocent infringers and innocent violators.

(C) Injunctive relief shall not be available to the owner of the right infringed or person bringing the action under section 1125(a) of this title with respect to an issue of a newspaper, magazine, or other similar periodical or an electronic communication containing infringing matter or violating matter where restraining the dissemination of such infringing matter or violating matter in any particular issue of such periodical or in an electronic communication would delay the delivery of such issue or transmission of such electronic communication after the regular time for such delivery or transmission, and such delay would be due to the method by which publication and distribution of such periodical or transmission of such electronic communication is customarily conducted in accordance with sound business practice, and not due to any method or device adopted to evade this section or to prevent or delay the issuance of an injunction or restraining order with respect to such infringing matter or violating matter.

(D)(i)(I) A domain name registrar, a domain name registry, or other domain name registration authority that takes any action described under clause (ii) affecting a domain name shall not be liable for monetary relief or, except as provided in subclause (II), for injunctive relief, to any person for such action, regardless of whether the domain name is finally determined to infringe or dilute the mark.

(II) A domain name registrar, domain name registry, or other domain name registration authority described in subclause (I) may be subject to injunctive relief only if such registrar, registry, or other registration authority has—

(aa) not expeditiously deposited with a court, in which an action has been filed regarding the disposition of the domain name, documents sufficient for the court to establish the court's control and authority regarding the disposition of the registration and use of the domain name;

(bb) transferred, suspended, or otherwise modified the domain name during the pendency of the action, except upon order of the court; or

(cc) willfully failed to comply with any such court order.

(ii) An action referred to under clause (i)(I) is any action of refusing to register, removing from registration, transferring, temporarily disabling, or permanently canceling a domain name—

(I) in compliance with a court order under section 43(d); or

(II) in the implementation of a reasonable policy by such registrar, registry, or authority prohibiting the registration of a domain name that is identical to, confusingly similar to, or dilutive of another's mark.

(iii) A domain name registrar, a domain name registry, or other domain name registration authority shall not be liable for damages under this section for the registration or maintenance of a domain name for another absent a showing of bad faith intent to profit from such registration or maintenance of the domain name.

(iv) If a registrar, registry, or other registration authority takes an action described under clause (ii) based on a knowing and material misrepresentation by any other person that a domain name is identical to, confusingly similar to, or dilutive of a mark, the person making the knowing and material misrepresentation shall be liable for any damages, including costs and attorney's fees, incurred by the domain name registrant as a result of such action. The court may also grant injunctive relief to the domain name registrant, including the reactivation of the domain name or the transfer of the domain name to the domain name registrant.

(v) A domain name registrant whose domain name has been suspended, disabled, or transferred under a policy described under clause (ii)(II) may, upon notice to the mark owner, file a civil action to establish that the registration or use of the domain name by such registrant is not unlawful under this chapter. The court may grant injunctive relief to the domain name registrant, including the reactivation of the domain name or transfer of the domain name to the domain name registrant.

(E) As used in this paragraph—(i) the term "violator" means a person who violates section 1125(a) of this title; and

(ii) the term "violating matter" means matter that is the subject of a violation under section 1125(a) of this title.

(Amended Oct. 9, 1962, 76 Stat. 773; Nov. 16, 1988, 102 Stat. 3943; Oct. 27, 1992, 106 Stat. 3567; Oct. 30, 1998, 112 Stat. 3069; Aug. 5, 1999, 113 Stat. 218; Nov. 29, 1999, 113 Stat. 1501A-549.)

§ 33 (15 U.S.C. § 1115). Registration as evidence of right to exclusive use; defenses

(a) Any registration issued under the Act of March 3, 1881, or the Act of February 20, 1905, or of a mark registered on the principal register provided by this chapter and owned by a party to an action shall be admissible in evidence and shall be prima facie evidence of the validity of the registered mark and of the registration of the mark, of the registrant's ownership of the mark, and of the registrant's exclusive right to use the registered mark in commerce on or in connection with the goods or services specified in the registration subject to any conditions or limitations stated therein, but shall not preclude another person from proving any legal or equitable defense or defect, including those set forth in subsection (b) of this section, which might have been asserted if such mark had not been registered.

(b) To the extent that the right to use the registered mark has become incontestable under section 1065 of this title, the registration shall be conclusive evidence of the validity of the registered mark and of the registration of the mark, of the registrant's ownership of the mark, and of the registrant's exclusive right to use the registered mark in commerce. Such conclusive evidence shall relate to the exclusive right to use the mark on or in connection with the goods or services specified in the affidavit filed under the provisions of section 1065 of this title, or in the renewal application filed under the provisions of section 1059 of this title if the goods or services specified in the renewal are fewer in number, subject to any conditions or limitations in the registration or in such affidavit or renewal application. Such conclusive evidence of the right to use the registered mark shall be subject to proof of infringement as defined in section 1114 of this title, and shall be subject to the following defenses or defects:

(1) That the registration or the incontestable right to use the mark was obtained fraudulently; or

(2) That the mark has been abandoned by the registrant; or

(3) That the registered mark is being used, by or with the permission of the registrant or a person in privity with the registrant, so as to misrepresent the source of the goods or services on or in connection with which the mark is used; or

(4) That the use of the name, term, or device charged to be an infringement is a use, otherwise than as a mark, of the party's individual name in his own business, or of the individual name of anyone in privity with such party, or of a term or device which is descriptive of and used fairly and in good faith only to describe the goods or services of such party, or their geographic origin; or

(5) That the mark whose use by a party is charged as an infringement was adopted without knowledge of the registrant's prior use and has been continuously used by such party or those in privity with him from a date prior to (A) the date of constructive use of the mark established pursuant to section 1057(c) of this title, (B) the registration of the mark under this Act if the application for registration is filed before the effective date of the Trademark Law Revision Act of 1988, or (C) publication of the registered mark under subsection (c) of section 1062 of this title: Provided, however, That this defense or defect shall apply only for the area in which such continuous prior use is proved; or

(6) That the mark whose use is charged as an infringement was registered and used prior to the registration under this Act or publication under subsection (c) of section 1062 of this title of the registered mark of the registrant, and not abandoned: Provided, however, That this defense or defect shall apply only for the area in which the mark was used prior to such registration or such publication of the registrant's mark; or

(7) That the mark has been or is being used to violate the antitrust laws of the United States; or

(8) That the mark is functional; or

(9) That equitable principles, including laches, estoppel, and acquiescence, are applicable.

(Amended Oct. 9, 1962, 76 Stat. 774; Nov. 16, 1988, 102 Stat. 3944; Oct. 30, 1998, 112 Stat. 3069; Nov. 2, 2002, 116 Stat. 1908.)

§ 34 (15 U.S.C. § 1116). Injunctions; enforcement; notice of filing suit given Director

(a) The several courts vested with jurisdiction of civil actions arising under this chapter shall have power to grant injunctions, according to the principles of equity and upon such terms as the court may deem reasonable, to prevent the violation of any right of the registrant of a mark registered in the Patent and Trademark Office or to prevent a violation under subsection (a), (c), or (d) of section 1125 of this title. Any such injunction may include a provision directing the defendant to file with the court and serve on the plaintiff within thirty days after the service on the defendant of such injunction, or such extended period as the court may direct, a report in writing under oath setting forth in detail the manner and form in which the defendant has complied with the injunction. Any such injunction granted upon hearing, after notice to the defendant, by any district court of the United States, may be served on the parties against whom such injunction is granted anywhere in the United States where they may be found, and shall be operative and may be enforced by proceedings to punish for contempt, or otherwise, by the court by which such injunction was granted, or by any other United States district court in whose jurisdiction the defendant may be found.

(b) The said courts shall have jurisdiction to enforce said injunction, as herein provided, as fully as if the injunction had been granted by the district court in which it is sought to be enforced. The clerk of the court or judge granting the injunction shall, when required to do so by the court before which application to enforce said injunction is made, transfer without delay to said court a certified copy of all papers on file in his office upon which said injunction was granted.

(c) It shall be the duty of the clerks of such courts within one month after the filing of any action, suit, or proceeding involving a mark registered under the provisions of this chapter to give notice thereof in writing to the Director setting forth in order so far as known the names and addresses of the litigants and the designating number or numbers of the registration or registrations upon which the action, suit, or proceeding has been brought, and in the event any other registration be subsequently included in the action, suit, or proceeding by amendment, answer, or other pleading, the clerk shall give like notice thereof to the Director, and within one month after the judgment is entered or an appeal is taken, the clerk of the court shall give notice thereof to the Director, and it shall be the duty of the Director on receipt of such notice forthwith to endorse the same upon the file wrapper of the said registration or registrations and to incorporate the same as a part of the contents of said file wrapper.

(d) (1) (A) In the case of a civil action arising under section 1114(1)(a) of this title or section 220506 of title 36, United States Code, with respect to a violation that consists of using a counterfeit mark in connection with the sale, offering for sale, or distribution of goods or services, the court may, upon ex parte application, grant an order under subsection (a) of this section pursuant to this subsection providing for the seizure of goods and counterfeit marks involved in such violation and the means of making such marks, and records documenting the manufacture, sale, or receipt of things involved in such violation.

(B) As used in this subsection the term "counterfeit mark" means—

(i) a counterfeit of a mark that is registered on the principal register in the United States Patent and Trademark Office for such goods or services sold, offered for sale, or distributed and that is in use, whether or not the person against whom relief is sought knew such mark was so registered; or

(ii) a spurious designation that is identical with, or substantially indistinguishable from, a designation as to which the remedies of this Act are made available by reason of section 220506 of title 36, United States Code;

but such term does not include any mark or designation used on or in connection with goods or services of which the manufacturer or producer was, at the time of the manufacture or production in question authorized to use the mark or designation for the type of goods or services so manufactured or produced, by the holder of the right to use such mark or designation.

(2) The court shall not receive an application under this subsection unless the applicant has given such notice of the application as is reasonable under the circumstances to the United States attorney for the judicial district in which such order is sought. Such attorney may participate in the proceedings arising under such application if such proceedings may affect evidence of an offense against the United States. The court may deny such application if the court determines that the public interest in a potential prosecution so requires.

(3) The application for an order under this subsection shall—

(A) be based on an affidavit or the verified complaint establishing facts sufficient to support the findings of fact and conclusions of law required for such order; and

(B) contain the additional information required by paragraph (5) of this subsection to be set forth in such order.

(4) The court shall not grant such an application unless—

(A) the person obtaining an order under this subsection provides the security determined adequate by the court for the payment of such damages as any person may be entitled to recover as a result of a wrongful seizure or wrongful attempted seizure under this subsection; and

(B) the court finds that it clearly appears from specific facts that—

(i) an order other than an ex parte seizure order is not adequate to achieve the purposes of section 1114 of this title;

(ii) the applicant has not publicized the requested seizure;

(iii) the applicant is likely to succeed in showing that the person against whom seizure would be ordered used a counterfeit mark in connection with the sale, offering for sale, or distribution of goods or services;

(iv) an immediate and irreparable injury will occur if such seizure is not ordered;

(v) the matter to be seized will be located at the place identified in the application;

(vi) the harm to the applicant of denying the application outweighs the harm to the legitimate interests of the person against whom seizure would be ordered of granting the application; and

(vii) the person against whom seizure would be ordered, or persons acting in concert with such person, would destroy, move, hide, or otherwise make such matter inaccessible to the court, if the applicant were to proceed on notice to such person.

(5) An order under this subsection shall set forth—

(A) the findings of fact and conclusions of law required for the order;

(B) a particular description of the matter to be seized, and a description of each place at which such matter is to be seized;

(C) the time period, which shall end not later than seven days after the date on which such order is issued, during which the seizure is to be made;

(D) the amount of security required to be provided under this subsection; and

(E) a date for the hearing required under paragraph (10) of this subsection.

(6) The court shall take appropriate action to protect the person against whom an order under this subsection is directed from publicity, by or at the behest of the plaintiff, about such order and any seizure under such order.

(7) Any materials seized under this subsection shall be taken into the custody of the court. For seizures made under this section, the court shall enter an appropriate protective order with respect to discovery and use of any records or information that has been seized. The protective order shall provide for appropriate procedures to ensure that confidential, private, proprietary, or privileged information contained in such records is not improperly disclosed or used.

(8) An order under this subsection, together with the supporting documents, shall be sealed until the person against whom the order is directed has an opportunity to contest such order, except that any person against whom such order is issued shall have access to such order and supporting documents after the seizure has been carried out.

(9) The court shall order that service of a copy of the order under this subsection shall be made by a Federal law enforcement officer (such as a United States marshal or an officer or agent of the United States Customs Service, Secret Service, Federal Bureau of Investigation, or Post Office) or may be made by a State or local law enforcement officer, who, upon making service, shall carry out the seizure under the order. The court shall issue orders, when appropriate, to protect the defendant from undue damage from the disclosure of trade secrets or other confidential information during the course of the seizure, including, when appropriate, orders restricting the access of the applicant (or any agent or employee of the applicant) to such secrets or information.

(10) (A) The court shall hold a hearing, unless waived by all the parties, on the date set by the court in the order of seizure. That date shall be not sooner than ten days after the order is issued and not later than fifteen days after the order is issued, unless the applicant for the order shows good cause for another date or unless the party against whom such order is directed consents to another date for such hearing. At such hearing the party obtaining the order shall have the burden to prove that the facts supporting findings of fact and conclusions of law necessary to support such order are still in effect. If that party fails to meet that burden, the seizure order shall be dissolved or modified appropriately.

(B) In connection with a hearing under this paragraph, the court may make such orders modifying the time limits for discovery under the Rules of Civil Procedure as may be necessary to prevent the frustration of the purposes of such hearing.

(11) A person who suffers damage by reason of wrongful seizure under this subsection has a cause of action against the applicant for the order under which such seizure was made, and shall be entitled to recover such relief as may be appropriate, including damages for lost profits, cost of materials, loss of goodwill, and punitive damages in instances where the seizure was sought in bad faith, and, unless the court finds extenuating circumstances, to recover a reasonable attorney's fee. The court in its discretion may award prejudgment interest on relief recovered under this paragraph, at an annual interest rate established under section 6621(a)(2) of the Internal Revenue Code of 1986, commencing on the date of service of the claimant's pleading setting forth the claim under this paragraph and ending on the date such recovery is granted, or for such shorter time as the court deems appropriate.

(Amended Jan. 2, 1975, 88 Stat. 1949; Oct. 12, 1984, 98 Stat. 2179; Nov. 16, 1988, 102 Stat. 3945; July 2, 1996, 110 Stat. 1386; Aug. 5, 1999, 113 Stat. 218; Nov. 29, 1999, 113 Stat. 1501A-548; Nov. 2, 2002, 116 Stat. 1908; Oct. 13, 2008, 122 Stat. 4258)

§ 35 (15 U.S.C. § 1117). Recovery of profits, damages, and costs

(a) When a violation of any right of the registrant of a mark registered in the Patent and Trademark Office, a violation under section 1125(a) or (d) of this title or a willful violation under section 1125(c) of this title, shall have been established in any civil action arising under this Act, the plaintiff shall be entitled, subject to the provisions of sections 1111 and 1114 and subject to the principles of equity, to recover (1) defendant's profits, (2) any damages sustained by the plaintiff, and (3) the costs of the action. The court shall assess such profits and damages or cause the same to be assessed under its direction. In assessing profits the plaintiff shall be required to prove defendant's sale only; defendant must prove all elements of cost or deduction claimed. In assessing damages the court may enter judgment, according to the circumstances of the case, for any sum above the amount found as actual damages, not exceeding three times such amount. If the court shall find that the amount of the recovery based on profits is either inadequate or excessive the court may in its discretion enter judgment for such sum as the court shall find to be just, according to the circumstances of the case. Such sum in either of the above circumstances shall constitute compensation and not a penalty. The court in exceptional cases may award reasonable attorney fees to the prevailing party.

(b) In assessing damages under subsection (a) for any violation of section 32(1)(a) of this Act or section 220506 of title 36, United States Code, in a case involving use of a counterfeit mark or designation (as defined in section 34(d) of this Act), the court shall, unless the court finds extenuating circumstances, enter judgment for three times such profits or damages, whichever amount is greater, together with a reasonable attorney's fee, if the violation consists of–

(1) intentionally using a mark or designation, knowing such mark or designation is a counterfeit mark (as defined in section 34(d) of this Act), in connection with the sale, offering for sale, or distribution of goods or services; or

(2) providing goods or services necessary to the commission of a violation specified in paragraph (1), with the intent that the recipient of the goods or services would put the goods or services to use in committing the violation.

In such a case, the court may award prejudgment interest on such amount at an annual interest rate established under section 6621(a)(2) of the Internal Revenue Code of 1986, beginning on the date of the service of the claimant's pleadings setting forth the claim for such entry of judgment and ending on the date such entry is made, or for such shorter time as the court considers appropriate.

(c) In a case involving the use of a counterfeit mark (as defined in section 1116(d) of this title) in connection with the sale, offering for sale, or distribution of goods or services, the plaintiff may elect, at any time before final judgment is rendered by the trial court, to recover, instead of actual damages and profits under subsection (a) of this section, an award of statutory damages for any such use in connection with the sale, offering for sale, or distribution of goods or services in the amount of -

(1) not less than $1,000 or more than $200,000 per counterfeit mark per type of goods or services sold, offered for sale, or distributed, as the court considers just; or

(2) if the court finds that the use of the counterfeit mark was willful, not more than $2,000,000 per counterfeit mark per type of goods or services sold, offered for sale, or distributed, as the court considers just.

(d) In a case involving a violation of section 1125(d)(1) of this title, the plaintiff may elect, at any time before final judgment is rendered by the trial court, to recover, instead of actual damages and profits, an award of statutory damages in the amount of not less than $1,000 and not more than $100,000 per domain name, as the court considers just.

(e) In the case of a violation referred to in this section, it shall be a rebuttable presumption that the violation is willful for purposes of determining relief if the violator, or a person acting in concert with the violator, knowingly provided or knowingly caused to be provided materially false contact information to a domain name registrar, domain name registry, or other domain name registration authority in registering, maintaining, or renewing a domain name used in connection with the violation. Nothing in this subsection limits what may be considered a willful violation under this section.

(Amended Oct. 9, 1962, 76 Stat. 774; Jan. 2, 1975, 88 Stat. 1949; Jan. 2, 1975, 88 Stat. 1955; Oct. 12, 1984, 98 Stat. 2182; Nov. 16, 1988, 102 Stat. 3945; July 2, 1996, 110 Stat. 1386; Aug. 5, 1999, 113 Stat. 218; Nov. 29, 1999, 113 Stat. 1501A-549; Nov. 2, 2002, 116 Stat. 1906, 1908; Dec. 23, 2004, 118 Stat. 3916; Oct. 13, 2008, 122 Stat. 4259.)

§ 36 (15 U.S.C. § 1118). Destruction of infringing articles

In any action arising under this chapter, in which a violation of any right of the registrant of a mark registered in the Patent and Trademark Office, a violation under section 1125(a) of this title, or a willful violation under section 1125(c) of this title, shall have been established, the court may order that all labels, signs, prints, packages, wrappers, receptacles, and advertisements in the possession of the defendant, bearing the registered mark or, in the case of a violation of section 1125(a) of this title or a willful violation under section 1125(c) of this title, the word, term, name, symbol, device, combination thereof, designation, description, or representation that is the subject of the violation, or any reproduction, counterfeit, copy, or colorable imitation thereof, and all plates, molds, matrices, and other means of making the same, shall be delivered up and destroyed. The party seeking an order under this section for destruction of articles seized under section 1161(d) of this title shall give ten days' notice to the United States attorney for the judicial district in which such order is sought (unless good cause is shown for lesser notice) and such United States attorney may, if such destruction may affect evidence of an offense against the United States, seek a hearing on such destruction or participate in any hearing otherwise to be held with respect to such destruction.

(Amended Jan. 2, 1975, 88 Stat. 1949; Oct. 12, 1984, 98 Stat. 2182; Nov. 16, 1988, 102 Stat. 3945; Aug. 5, 1999, 113 Stat. 218.)

§ 37 (15 U.S.C. § 1119). Power of court over registration; certification of decrees and orders

In any action involving a registered mark the court may determine the right to registration, order the cancellation of registrations, in whole or in part, restore cancelled registrations, and otherwise rectify the register with respect to the registrations of any party to the action. Decrees and orders shall be certified by the court to the Director, who shall make appropriate entry upon the records of the Patent and Trademark Office, and shall be controlled thereby.

(Amended Jan. 2, 1975, 88 Stat. 1949; Nov. 29, 1999, 113 Stat. 1501A-583.)

§ 38 (15 U.S.C. § 1120). Fraud; civil liability

Any person who shall procure registration in the Patent and Trademark Office of a mark by a false or fraudulent declaration or representation, oral or in writing, or by any false means, shall be liable in a civil action by any person injured thereby for any damages sustained in consequence thereof.

(Amended Jan. 2, 1975, 88 Stat. 1949.)

§ 39 (15 U.S.C. § 1121). Jurisdiction of Federal courts; State, local, and other agency requirements

(a) The district and territorial courts of the United States shall have original jurisdiction, the courts of appeal of the United States (other than the United States Court of Appeals for the Federal Circuit) and the United States Court of Appeals for the District of Columbia shall have appellate jurisdiction, of all actions arising under this chapter, without regard to the amount in controversy or to diversity or lack of diversity of the citizenship of the parties.

(b) No State or other jurisdiction of the United States or any political subdivision or any agency thereof may require alteration of a registered mark, or require that additional trademarks, service marks, trade names, or corporate names that may be associated with or incorporated into the registered mark be displayed in the mark in a manner differing from the display of such additional trademarks, service marks, trade names, or corporate names contemplated by the registered mark as exhibited in the certificate of registration issued by the United States Patent and Trademark Office.

(Amended Apr. 2, 1982, 96 Stat. 46; Nov. 16, 1988, 102 Stat. 3946; Oct. 30, 1998, 112 Stat. 3069.)

§ 40 (15 U.S.C. § 1122). Liability of States, instrumentalities of States and State officials

(a) WAIVER OF SOVEREIGN IMMUNITY BY THE UNITED STATES—The United States, all agencies and instrumentalities thereof, and all individuals, firms, corporations, other persons acting for the United States and with the authorization and consent of the United States, shall not be immune from suit in Federal or State court by any person, including any governmental or nongovernmental entity, for any violation under this chapter.

(b) WAIVER OF SOVEREIGN IMMUNITY BY STATES- Any State, instrumentality of a State or any officer or employee of a State or instrumentality of a State acting in his or her official capacity, shall not be immune, under the eleventh amendment of the Constitution of the United States or under any other doctrine of sovereign immunity, from suit in Federal court by any person, including any governmental or nongovernmental entity for any violation under this chapter.

(c) In a suit described in subsection (a) or (b) for a violation described therein, remedies (including remedies both at law and in equity) are available for the violation to the same extent as such remedies are available for such a violation in a suit against any person other than the United States or any agency or instrumentality thereof, or any individual, firm, corporation, or other person acting for the United States and with authorization and consent of the United States, or a State, instrumentality of a State, or officer or employee of a State or instrumentality of a State acting in his or her official capacity. Such remedies include injunctive relief under section 1116 of this title, actual damages, profits, costs and attorney's fees under section 1117 f this title, destruction of infringing articles under section 1118 f this title, the remedies provided for under sections 1114, 1119, 1120, 1124 and 1125 of this title, and for any other remedies provided under this chapter.

(Added Oct. 27, 1992, 106 Stat. 3567; amended Aug. 5, 1999, 113 Stat. 218.)

§ 41 (15 U.S.C. § 1123). Rules and regulations

The Director shall make rules and regulations, not inconsistent with law, for the conduct of proceedings in the Patent and Trademark Office under this chapter.

(Amended Jan. 2, 1975, 88 Stat. 1949; Nov. 29, 1999, 113 Stat. 1501A-583.)

TITLE VII - IMPORTATION FORBIDDEN OF GOODS BEARING INFRINGING MARKS OR NAMES

§ 42 (15 U.S.C. § 1124). Importation of goods bearing infringing marks or names forbidden

Except as provided in subsection (d) of section 526 of the Tariff Act of 1930, no article of imported merchandise which shall copy or simulate the name of any domestic manufacture, or manufacturer, or trader, or of any manufacturer or trader located in any foreign country which, by treaty, convention, or law affords similar privileges to citizens of the United States, or which shall copy or simulate a trademark registered in accordance with the provisions of this chapter or shall bear a name or mark calculated to induce the public to believe that the article is manufactured in the United States, or that it is manufactured in any foreign country or locality other than the country or locality in which it is in fact manufactured, shall be admitted to entry at any customhouse of the United States; and, in order to aid the officers of the customs in enforcing this prohibition, any domestic manufacturer or trader, and any foreign manufacturer or trader, who is entitled under the provisions of a treaty, convention, declaration, or agreement between the United States and any foreign country to the advantages afforded by law to citizens of the United States in respect to trademarks and commercial names, may require his name and residence, and the name of the locality in which his goods are manufactured, and a copy of the certificate of registration of his trademark, issued in accordance with the provisions of this chapter, to be recorded in books which shall be kept for this purpose in the Department of the Treasury, under such regulations as the Secretary of the Treasury shall prescribe, and may furnish to the Department facsimiles of his name, the name of the locality in which his goods are manufactured, or of his registered trademark, and thereupon the Secretary of the Treasury shall cause one or more copies of the same to be transmitted to each collector or other proper officer of customs.

(Amended Oct. 3, 1978, 92 Stat. 903; Oct. 30, 1998, 112 Stat. 3069.)

Note: Copies of regulations referred to in this section may be obtained from the Department of the Treasury.

TITLE VIII - FALSE DESIGNATIONS OF ORIGIN, FALSE DESCRIPTIONS. AND DILUTION FORBIDDEN

§ 43 (15 U.S.C. § 1125). False designations of origin; false description or representation

(a) (1) Any person who, on or in connection with any goods or services, or any container for goods, uses in commerce any word, term, name, symbol, or device, or any combination thereof, or any false designation of origin, false or misleading description of fact, or false or misleading representation of fact, which—

(A) is likely to cause confusion, or to cause mistake, or to deceive as to the affiliation, connection, or association of such person with another person, or as to the origin, sponsorship, or approval of his or her goods, services, or commercial activities by another person, or

(B) in commercial advertising or promotion, misrepresents the nature, characteristics, qualities, or geographic origin of his or her or another person's goods, services, or commercial activities,

shall be liable in a civil action by any person who believes that he or she is or is likely to be damaged by such act.

(2) As used in this subsection, the term "any person" includes any State, instrumentality of a State or employee of a State or instrumentality of a State acting in his or her official capacity. Any State, and any such instrumentality, officer, or employee, shall be subject to the provisions of this chapter in the same manner and to the same extent as any nongovernmental entity.

(3) In a civil action for trade dress infringement under this chapter for trade dress not registered on the principal register, the person who asserts trade dress protection has the burden of proving that the matter sought to be protected is not functional.

(b) Any goods marked or labeled in contravention of the provisions of this section shall not be imported into the United States or admitted to entry at any customhouse of the United States. The owner, importer, or consignee of goods refused entry at any customhouse under this section may have any recourse by protest or appeal that is given under the customs revenue laws or may have the remedy given by this chapter in cases involving goods refused entry or seized.

(c) Dilution by Blurring; Dilution by Tarnishment.—

(1) Injunctive relief.—Subject to the principles of equity, the owner of a famous mark that is distinctive, inherently or through acquired distinctiveness, shall be entitled to an injunction against another person who, at any time after the owner's mark has become famous, commences use of a mark or trade name in commerce that is likely to cause dilution by blurring or dilution by tarnishment of the famous mark, regardless of the presence or absence of actual or likely confusion, of competition, or of actual economic injury.

(2) Definitions.—(A) For purposes of paragraph (1), a mark is famous if it is widely recognized by the general consuming public of the United States as a designation of source of the goods or services of the mark's owner. In determining whether a mark possesses the requisite degree of recognition, the court may consider all relevant factors, including the following:

(i) The duration, extent, and geographic reach of advertising and publicity of the mark, whether advertised or publicized by the owner or third parties.

(ii) The amount, volume, and geographic extent of sales of goods or services offered under the mark.

(iii) The extent of actual recognition of the mark.

(iv) Whether the mark was registered under the Act of March 3, 1881, or the Act of February 20, 1905, or on the principal register.

(B) For purposes of paragraph (1), 'dilution by blurring' is association arising from the similarity between a mark or trade name and a famous mark that impairs the distinctiveness of the famous mark. In determining whether a mark or trade name is likely to cause dilution by blurring, the court may consider all relevant factors, including the following:

(i) The degree of similarity between the mark or trade name and the famous mark.

(ii) The degree of inherent or acquired distinctiveness of the famous mark.

(iii) The extent to which the owner of the famous mark is engaging in substantially exclusive use of the mark.

(iv) The degree of recognition of the famous mark.

(v) Whether the user of the mark or trade name intended to create an association with the famous mark.

(vi) Any actual association between the mark or trade name and the famous mark.

(C) For purposes of paragraph (1), 'dilution by tarnishment' is association arising from the similarity between a mark or trade name and a famous mark that harms the reputation of the famous mark.

(3) Exclusions.—The following shall not be actionable as dilution by blurring or dilution by tarnishment under this subsection:

(A) Any fair use, including a nominative or descriptive fair use, or facilitation of such fair use, of a famous mark by another person other than as a designation of source for the person's own goods or services, including use in connection with—

(i) advertising or promotion that permits consumers to compare goods or services; or

(ii) identifying and parodying, criticizing, or commenting upon the famous mark owner or the goods or services of the famous mark owner.

(B) All forms of news reporting and news commentary.

(C) Any noncommercial use of a mark.

(4) Burden of proof.—In a civil action for trade dress dilution under this Act for trade dress not registered on the principal register, the person who asserts trade dress protection has the burden of proving that—

(A) the claimed trade dress, taken as a whole, is not functional and is famous; and

(B) if the claimed trade dress includes any mark or marks registered on the principal register, the unregistered matter, taken as a whole, is famous separate and apart from any fame of such registered marks.

(5) Additional remedies.—In an action brought under this subsection, the owner of the famous mark shall be entitled to injunctive relief as set forth in section 34. The owner of the famous mark shall also be entitled to the remedies set forth in sections 35(a) and 36, subject to the discretion of the court and the principles of equity if—

(A) the mark or trade name that is likely to cause dilution by blurring or dilution by tarnishment was first used in commerce by the person against whom the injunction is sought after the date of enactment of the Trademark Dilution Revision Act of 2006; and

(B) in a claim arising under this subsection—

(i) by reason of dilution by blurring, the person against whom the injunction is sought willfully intended to trade on the recognition of the famous mark; or

(ii) by reason of dilution by tarnishment, the person against whom the injunction is sought willfully intended to harm the reputation of the famous mark.

(6) Ownership of valid registration a complete bar to action.—The ownership by a person of a valid registration under the Act of March 3, 1881, or the Act of February 20, 1905, or on the principal register under this Act shall be a complete bar to an action against that person, with respect to that mark, that—

(A) (i) is brought by another person under the common law or a statute of a State; and

(Bii) (i) seeks to prevent dilution by blurring or dilution by tarnishment; or

(iiB) asserts any claim of actual or likely damage or harm to the distinctiveness or reputation of a mark, label, or form of advertisement.

(7) Savings clause.—Nothing in this subsection shall be construed to impair, modify, or supersede the applicability of the patent laws of the United States.

(d) (1) (A) A person shall be liable in a civil action by the owner of a mark, including a personal name which is protected as a mark under this section, if, without regard to the goods or services of the parties, that person—

(i) has a bad faith intent to profit from that mark, including a personal name which is protected as a mark under this section; and

(ii) registers, traffics in, or uses a domain name that—

(I) in the case of a mark that is distinctive at the time of registration of the domain name, is identical or confusingly similar to that mark;

(II) in the case of a famous mark that is famous at the time of registration of the domain name, is identical or confusingly similar to or dilutive of that mark; or

(III) is a trademark, word, or name protected by reason of section 706 of title 18, United States Code, or section 220506 of title 36.

(B) (i) In determining whether a person has a bad faith intent described under subparagraph (A), a court may consider factors such as, but not limited to—

(I) the trademark or other intellectual property rights of the person, if any, in the domain name;

(II) the extent to which the domain name consists of the legal name of the person or a name that is otherwise commonly used to identify that person;

(III) the person's prior use, if any, of the domain name in connection with the bona fide offering of any goods or services;

(IV) the person's bona fide noncommercial or fair use of the mark in a site accessible under the domain name;

(V) the person's intent to divert consumers from the mark owner's online location to a site accessible under the domain name that could harm the goodwill represented by the mark, either for commercial gain or with the intent to tarnish or disparage the mark, by creating a likelihood of confusion as to the source, sponsorship, affiliation, or endorsement of the site;

(VI) the person's offer to transfer, sell, or otherwise assign the domain name to the mark owner or any third party for financial gain without having used, or having an intent to use, the domain name in the bona fide offering of any goods or services, or the person's prior conduct indicating a pattern of such conduct;

(VII) the person's provision of material and misleading false contact information when applying for the registration of the domain name, the person's intentional failure to maintain accurate contact information, or the person's prior conduct indicating a pattern of such conduct;

(VIII) the person's registration or acquisition of multiple domain names which the person knows are identical or confusingly similar to marks of others that are distinctive at the time of registration of such domain names, or dilutive of famous marks of others that are famous at the time of registration of such domain names, without regard to the goods or services of the parties; and

(IX) the extent to which the mark incorporated in the person's domain name registration is or is not distinctive and famous within the meaning of subsection (c)(1) of this section.

(ii) Bad faith intent described under subparagraph (A) shall not be found in any case in which the court determines that the person believed and had reasonable grounds to believe that the use of the domain name was a fair use or otherwise lawful.

(C) In any civil action involving the registration, trafficking, or use of a domain name under this paragraph, a court may order the forfeiture or cancellation of the domain name or the transfer of the domain name to the owner of the mark.

(D) A person shall be liable for using a domain name under subparagraph (A) only if that person is the domain name registrant or that registrant's authorized licensee.

(E) As used in this paragraph, the term "traffics in" refers to transactions that include, but are not limited to, sales, purchases, loans, pledges, licenses, exchanges of currency, and any other transfer for consideration or receipt in exchange for consideration.

(2)(A) The owner of a mark may file an in rem civil action against a domain name in the judicial district in which the domain name registrar, domain name registry, or other domain name authority that registered or assigned the domain name is located if—

(i) the domain name violates any right of the owner of a mark registered in the Patent and Trademark Office, or protected under subsection (a) or (c) of this section; and

(ii) the court finds that the owner—

(I) is not able to obtain in personam jurisdiction over a person who would have been a defendant in a civil action under paragraph (1); or

(II) through due diligence was not able to find a person who would have been a defendant in a civil action under paragraph (1) by—

(aa) sending a notice of the alleged violation and intent to proceed under this paragraph to the registrant of the domain name at the postal and e-mail address provided by the registrant to the registrar; and

(bb) publishing notice of the action as the court may direct promptly after filing the action.

(B) The actions under subparagraph (A)(ii) shall constitute service of process.

(C) In an in rem action under this paragraph, a domain name shall be deemed to have its situs in the judicial district in which—

(i) the domain name registrar, registry, or other domain name authority that registered or assigned the domain name is located; or

(ii) documents sufficient to establish control and authority regarding the disposition of the registration and use of the domain name are deposited with the court.

(D) (i) The remedies in an in rem action under this paragraph shall be limited to a court order for the forfeiture or cancellation of the domain name or the transfer of the domain name to the owner of the mark. Upon receipt of written notification of a filed, stamped copy of a complaint filed by the owner of a mark in a United States district court under this paragraph, the domain name registrar, domain name registry, or other domain name authority shall—

(I) expeditiously deposit with the court documents sufficient to establish the court's control and authority regarding the disposition of the registration and use of the domain name to the court; and

(II) not transfer, suspend, or otherwise modify the domain name during the pendency of the action, except upon order of the court.

(ii) The domain name registrar or registry or other domain name authority shall not be liable for injunctive or monetary relief under this paragraph except in the case of bad faith or reckless disregard, which includes a willful failure to comply with any such court order.

(3) The civil action established under paragraph (1) and the in rem action established under paragraph (2), and any remedy available under either such action, shall be in addition to any other civil action or remedy otherwise applicable.

(4) The in rem jurisdiction established under paragraph (2) shall be in addition to any other jurisdiction that otherwise exists, whether in rem or in personam.

(Amended Nov. 16, 1988, 102 Stat. 3946; Oct. 27, 1992, 106 Stat. 3567; Jan. 16, 1996, 109 Stat. 985; Aug. 5, 1999, 113 Stat. 218; Nov. 29, 1999, 113 Stat. 1501A-545; Oct. 6, 2006, 120 Stat. 1730.)

TITLE IX - INTERNATIONAL CONVENTIONS

§ 44 (15 U.S.C. § 1126). International conventions; register of marks

(a) The Director shall keep a register of all marks communicated to him by the international bureaus provided for by the conventions for the protection of industrial property, trademarks, trade and commercial names, and the repression of unfair competition to which the United States is or may become a party, and upon the payment of the fees required by such conventions and the fees required in this Act may place the marks so communicated upon such register. This register shall show a facsimile of the mark or trade or commercial name; the name, citizenship, and address of the registrant; the number, date, and place of the first registration of the mark, including the dates on which application for such registration was filed and granted and the term of such registration; a list of goods or services to which the mark is applied as shown by the registration in the country of origin, and such other data as may be useful concerning the mark. This register shall be a continuation of the register provided in section 1(a) of the Act of March 19, 1920.

Note: The United States is not at present a party to the parts of the international conventions providing for international bureaus for the registration or communication of trademarks.

(b) Any person whose country of origin is a party to any convention or treaty relating to trademarks, trade or commercial names, or the repression of unfair competition, to which the United States is also a party, or extends reciprocal rights to nationals of the United States by law, shall be entitled to the benefits of this section under the conditions expressed herein to the extent necessary to give effect to any provision of such convention, treaty or reciprocal law, in addition to the rights to which any owner of a mark is otherwise entitled by this chapter.

Notes: International Convention for the Protection of Industrial Property of 1883 (Paris); revised at Washington in 1911, 204 O.G. 1011, July 21, 1914 (37 Stat. 1645; Treaty Series 579); at the Hague in 1925, 407 O.G. 298, June 9, 1931 (47 Stat. 1789; Treaty Series 834; 2 Bevans 524); at London in 1934, 613 O.G. 23, August 3, 1948 (53 Stat. 1748; Treaty Series 941; 3 Bevans 223); at Lisbon in 1958, 775 O.G. 321, February 13, 1962 (53 Stat. 1748; 13 U.S.T. 1; TIAS 9431); and at Stockholm July 14, 1967, 852 O.G. 511, July 16, 1968 (21 U.S.T. 1583; TIAS 6923). A list of the member countries together with an indication of the latest Act by which each country is bound and the date from which each is considered to be bound appears annually (January issue) in "Industrial Property," a monthly review of the World Intellectual Property Organization (WIPO), Geneva, Switzerland.

General Inter-American Convention for Trade-Mark and Commercial Protection (Pan-American Trade-Mark Convention) of 1929, 46 Stat. 2907; Pan-American Trade-Mark Convention of 1923, 44 Stat. 2494; Pan-American Trade-Mark Convention of 1910, 39 Stat. 1675.

List of the States which are parties to the above conventions may be found in "Treaties in Force," a list of treaties and other international agreements in force on the first day of January of each year, compiled annually by the Office of the Legal Adviser, U.S. Department of State.

(c) No registration of a mark in the United States by a person described in subsection (b) of this section shall be granted until such mark has been registered in the country of origin of the applicant, unless the applicant alleges use in commerce. For the purposes of this section, the country of origin of the applicant is the country in which he has a bona fide and effective industrial or commercial establishment, or if he has not such an establishment the country in which he is domiciled, or if he has not a domicile in any of the countries described in subsection (b) of this section, the country of which he is a national.

(d) An application for registration of a mark under sections 1051, 1053, 1054, or 1091 of this title or under subsection (e) of this section filed by a person described in subsection (b) of this section who has previously duly filed an application for registration of the same mark in one of the countries described in subsection (b) of this section shall be accorded the same force and effect as would be accorded to the same application if filed in the United States on the same date on which the application was first filed in such foreign country: *Provided,* That—

(1) the application in the United States is filed within 6 months from the date on which the application was first filed in the foreign country;

(2) the application conforms as nearly as practicable to the requirements of this chapter, including a statement that the applicant has a bona fide intention to use the mark in commerce;

(3) the rights acquired by third parties before the date of the filing of the first application in the foreign country shall in no way be affected by a registration obtained on an application filed under this subsection;

(4) nothing in this subsection shall entitle the owner of a registration granted under this section to sue for acts committed prior to the date on which his mark was registered in this country unless the registration is based on use in commerce.

In like manner and subject to the same conditions and requirements, the right provided in this section may be based upon a subsequent regularly filed application in the same foreign country, instead of the first filed foreign application: *Provided,* That any foreign application filed prior to such subsequent application has been withdrawn, abandoned, or otherwise disposed of, without having been laid open to public inspection and without leaving any rights outstanding, and has not served, nor thereafter shall serve, as a basis for claiming a right of priority.

Note: Section 3 of Public Law 333, 87th Cong., approved October 3, 1961, 75 Stat. 748, the provision which added the last paragraph above provided that "This Act shall take effect on the date when the Convention of Paris for the Protection of Industrial Property of March 20, 1883, as revised at Lisbon October 31, 1958, comes into force with respect to the United States and is applied only to applications thereafter filed in the United States by persons entitled to the benefit of said convention, as revised at the time of such filing." This provision became effective Jan. 4, 1962.

(e) A mark duly registered in the country of origin of the foreign applicant may be registered on the principal register if eligible, otherwise on the supplemental register herein provided. Such applicant shall submit, within such time period as may be prescribed by the Director, a true copy, a photocopy, a certification, or a certified copy of the registration in the country of origin of the applicant. The application must state the applicant's bona fide intention to use the mark in commerce, but use in commerce shall not be required prior to registration.

(f) The registration of a mark under the provisions of subsections (c), (d), and (e) of this section by a person described in subsection (b) of this section shall be independent of the registration in the country of origin and the duration, validity, or transfer in the United States of such registration shall be governed by the provisions of this chapter.

(g) Trade names or commercial names of persons described in subsection (b) of this section shall be protected without the obligation of filing or registration whether or not they form parts of marks.

(h) Any person designated in subsection (b) of this section as entitled to the benefits and subject to the provisions of this chapter shall be entitled to effective protection against unfair competition, and the remedies provided herein for infringement of marks shall be available so far as they may be appropriate in repressing acts of unfair competition.

(i) Citizens or residents of the United States shall have the same benefits as are granted by this section to persons described in subsection (b) of this section.

(Amended Oct. 3, 1961, 75 Stat. 748; Oct. 9, 1962, 76 Stat. 774; Nov. 16, 1988, 102 Stat. 3946; Oct. 30, 1998, 112 Stat. 3068; Nov. 29, 1999, 113 Stat. 1501A-583; Nov. 2, 2002, 116 Stat. 1908.)

TITLE X - CONSTRUCTION AND DEFINITIONS

§ 45 (15 U.S.C. § 1127).

In the construction of this chapter, unless the contrary is plainly apparent from the context—

United States. The United States includes and embraces all territory which is under its jurisdiction and control.

Commerce. The word "commerce" means all commerce which may lawfully be regulated by Congress.

Principal Register, Supplemental Register. The term "principal register" refers to the register provided for by sections 1051 to 1072 of this title, and the term "supplemental register" refers to the register provided for by sections 1091 to 1096 of this title.

Person, juristic person. The term "person" and any other word or term used to designate the applicant or other entitled to a benefit or privilege or rendered liable under the provisions of this Act includes a juristic person as well as a natural person. The term "juristic person" includes a firm, corporation, union, association, or other organization capable of suing and being sued in a court of law.

The term "person" also includes the United States, any agency or instrumentality thereof, or any individual, firm, or corporation acting for the United States and with the authorization and consent of the United States. The United States, any agency or instrumentality thereof, and any individual, firm, or corporation acting for the United States and with the authorization and consent of the United States, shall be subject to the provisions of this chapter in the same manner and to the same extent as any nongovernmental entity.

The term "person" also includes any State, any instrumentality of a State, and any officer or employee of a State or instrumentality of a State acting in his or her official capacity. Any State, and any such instrumentality, officer, or employee, shall be subject to the provisions of this chapter in the same manner and to the same extent as any non-governmental entity.

Applicant, registrant. The terms "applicant" and "registrant" embrace the legal representatives, predecessors, successors and assigns of such applicant or registrant.

Director. The term "Director" means the Under Secretary of Commerce for Intellectual Property and Director of the United States Patent and Trademark Office.

Related company. The term "related company" means any person whose use of a mark is controlled by the owner of the mark with respect to the nature and quality of the goods or services on or in connection with which the mark is used.

Trade name, commercial name. The terms "trade name" and "commercial name" mean any name used by a person to identify his or her business or vocation.

Trademark. The term "trademark" includes any word, name, symbol, or device, or any combination thereof—

(1) used by a person, or

(2) which a person has a bona fide intention to use in commerce and applies to register on the principal register established by this chapter,

to identify and distinguish his or her goods, including a unique product, from those manufactured or sold by others and to indicate the source of the goods, even if that source is unknown.

Service mark. The term "service mark" means any word, name, symbol, or device, or any combination thereof—

(1) used by a person, or

(2) which a person has a bona fide intention to use in commerce and applies to register on the principal register established by this chapter,

to identify and distinguish the services of one person, including a unique service, from the services of others and to indicate the source of the services, even if that source is unknown. Titles, character names, and other distinctive features of radio or television programs may be registered as service marks notwithstanding that they, or the programs, may advertise the goods of the sponsor.

Certification mark. The term "certification mark" means any word, name, symbol, or device, or any combination thereof—

(1) used by a person other than its owner, or

(2) which its owner has a bona fide intention to permit a person other than the owner to use in commerce and files an application to register on the principal register established by this chapter,

to certify regional or other origin, material, mode of manufacture, quality, accuracy, or other characteristics of such person's goods or services or that the work or labor on the goods or services was performed by members of a union or other organization.

Collective mark. The term "collective mark" means a trademark or service mark—

(1) used by the members of a cooperative, an association, or other collective group or organization, or

(2) which such cooperative, association, or other collective group or organization has a bona fide intention to use in commerce and applies to register on the principal register established by this chapter,

and includes marks indicating membership in a union, an association, or other organization.

Mark. The term "mark" includes any trademark, service mark, collective mark, or certification mark.

Use in commerce. The term "use in commerce" means the bona fide use of a mark in the ordinary course of trade, and not made merely to reserve a right in a mark. For purposes of this chapter, a mark shall be deemed to be in use in commerce—

(1) on goods when—

(A) it is placed in any manner on the goods or their containers or the displays associated therewith or on the tags or labels affixed thereto, or if the nature of the goods makes such placement impracticable, then on documents associated with the goods or their sale, and

(B) the goods are sold or transported in commerce, and

(2) on services when it is used or displayed in the sale or advertising of services and the services are rendered in commerce, or the services are rendered in more than one State or in the United States and a foreign country and the person rendering the services is engaged in commerce in connection with the services.

Abandonment of mark. A mark shall be deemed to be "abandoned" if either of the following occurs:

(1) When its use has been discontinued with intent not to resume such use. Intent not to resume may be inferred from circumstances. Nonuse for 3 consecutive years shall be prima facie evidence of abandonment. "Use" of a mark means the bona fide use of such mark made in the ordinary course of trade, and not made merely to reserve a right in a mark.

(2) When any course of conduct of the owner, including acts of omission as well as commission, causes the mark to become the generic name for the goods or services on or in connection with which it is used or otherwise to lose its significance as a mark. Purchaser motivation shall not be a test for determining abandonment under this paragraph.

Colorable imitation. The term "colorable imitation" includes any mark which so resembles a registered mark as to be likely to cause confusion or mistake or to deceive.

Registered mark. The term "registered mark" means a mark registered in the United States Patent and Trademark Office under this chapter or under the Act of March 3, 1881, or the Act of February 20, 1905, or the Act of March 19, 1920. The phrase "marks registered in the Patent and Trademark Office" means registered marks.

Prior acts. The term "Act of March 3, 1881," "Act of February 20, 1905," or "Act of March 19, 1920," means the respective Act as amended.

Counterfeit. A "counterfeit" is a spurious mark which is identical with, or substantially indistinguishable from, a registered mark.

Domain name. The term "domain name" means any alphanumeric designation which is registered with or assigned by any domain name registrar, domain name registry, or other domain name registration authority as part of an electronic address on the Internet.

Internet. The term "Internet" has the meaning given that term in section 230(f)(1) of the Communications Act of 1934 (47 U.S.C. 230(f)(1)).

Singular and plural. Words used in the singular include the plural and vice versa.

Intent of Chapter. The intent of this chapter is to regulate commerce within the control of Congress by making actionable the deceptive and misleading use of marks in such commerce; to protect registered marks used in such commerce from interference by State, or territorial legislation; to protect persons engaged in such commerce against unfair competition; to prevent fraud and deception in such commerce by the use of reproductions, copies, counterfeits, or colorable imitations of registered marks; and to provide rights and remedies stipulated by treaties and conventions respecting trademarks, trade names, and unfair competition entered into between the United States and foreign nations.

(Amended Oct. 9, 1962, 76 Stat. 774; Jan. 2, 1975, 88 Stat. 1949; Nov. 8, 1984, 98 Stat. 3335, Nov. 16, 1988, 102 Stat. 3946; Oct. 27, 1992, 106 Stat. 3568; Dec. 8, 1994, 108 Stat. 4981; Jan. 16, 1996, 109 Stat. 986; Aug. 5, 1999, 113 Stat. 219-220; Nov. 29, 1999, 113 Stat. 1501A-550, 1501A-583; Oct. 6, 2006, 120 Stat. 1730.)

TITLE XI - REPEAL OF PREVIOUS ACTS

§ 46(a) (15 U.S.C. § 1051 note). Time of taking effect - Repeal of prior acts

This Act shall be in force and take effect one year from its enactment, but except as otherwise herein specifically provided shall not affect any suit, proceeding, or appeal then pending. All Acts and parts of Acts inconsistent herewith are hereby repealed effective one year from the enactment hereof, including the following Acts insofar as they are inconsistent herewith: The Act of Congress approved March 3, 1881, entitled "An Act to authorize the registration of trademarks and protect the same"; the Act approved August 5, 1882, entitled "An Act relating to the registration of trademarks"; the Act of February 20, 1905 (U.S.C., title 15, secs. 81 to 109, inclusive), entitled "An Act to authorize the registration of trademarks used in commerce with foreign nations or among the several States or with Indian tribes, and to protect the same", and the amendments thereto by the Acts of May 4, 1906 (U.S.C., title 15, secs. 131 and 132; 34 Stat. 169), March 2, 1907 (34 Stat. 1251, 1252), February 18, 1909 (35 Stat. 627, 628), February 18, 1911 (36 Stat. 918), January 8, 1913 (37 Stat. 649), June 7, 1924 (43 Stat. 647), March 4, 1925 (43 Stat. 1268, 1269), April 11, 1930 (46 Stat. 155), June 10, 1938 (Public, Numbered 586, Seventy-fifth Congress, ch. 332, third session); the Act of March 19, 1920 (U.S.C., title 15, secs. 121 to 128, inclusive), entitled "an Act to give effect to certain provisions of the convention for the protection of trademarks and commercial names made and signed in the city of Buenos Aires, in the Argentine Republic, August 20, 1910, and for other purposes", and the amendments thereto, including the Act of June 10, 1938 (Public, Numbered 586, Seventy-fifth Congress, ch. 332, third session): *Provided,* That this repeal shall not affect the validity of registrations granted or applied for under any of said Acts prior to the effective date of this Act, or rights or remedies thereunder except as provided in sections 8, 12, 14, 15, and 47 of this Act; but nothing contained in this Act shall be construed as limiting, restricting, modifying, or repealing any statute in force on the effective date of this Act which does not relate to trademarks, or as restricting or increasing the authority of any Federal departments or regulatory agency except as may be specifically provided in this Act.

§ 46(b) (15 U.S.C. § 1051 note). Existing registrations under prior acts

Acts of 1881 and 1905. Registrations now existing under the Act of March 3, 1881, or the Act of February 20, 1905, shall continue in full force and effect for the unexpired terms thereof and may be renewed under the provisions of section 9 of this Act. Such registrations and the renewals thereof shall be subject to and shall be entitled to the benefits of the provisions of this Act to the same extent and with the same force and effect as though registered on the principal register established by this Act except as limited in sections 8, 12, 14, and 15 of this Act. Marks registered under the "10-year proviso" of section 5 of the Act of February 20, 1905, as amended, shall be deemed to have become distinctive of the registrant's goods in commerce under paragraph (f) of section 2 of this Act and may be renewed under section 9 hereof as marks coming within said paragraph.

Act of 1920. Registrations now existing under the Act of March 19, 1920, shall expire 6 months after the effective date of this Act, or twenty years from the dates of their registrations, whichever date is later. Such registrations shall be subject to and entitled to the benefits of the provisions of this Act relating to marks registered on the supplemental register established by this Act, and may not be renewed unless renewal is required to support foreign registrations. In that event renewal may be effected on the supplemental register under the provisions of section 9 of this Act.

Subject to registration under this Act. Marks registered under previous Acts may, if eligible, also be registered under this Act.

§ 47(a) (15 U.S.C. § 1051 note). Applications pending on effective date of Act

All applications for registration pending in the Patent and Trademark Office at the effective date of this Act may be amended, if practicable, to bring them under the provisions of this Act. The prosecution of such applications so amended and the grant of registrations thereon shall be proceeded with in accordance with the provisions of this Act. If such amendments are not made, the prosecution of said applications shall be proceeded with and registrations thereon granted in accordance with the Acts under which said applications were filed, and said Acts are hereby continued in force to this extent for this purpose only, notwithstanding the foregoing general repeal thereof.

§ 47(b) (15 U.S.C. § 1051 note). Appeals pending on effective date of Act

In any case in which an appeal is pending before the United States Court of Customs and Patent Appeals or any United States Circuit Court of Appeals or the United States Court of Appeals for the District of Columbia or the United States Supreme Court at the effective date of this Act, the court, if it be of the opinion that the provisions of this Act are applicable to the subject matter of the appeal, may apply such provision or may remand the case to the Director or to the district court for the taking of additional evidence or a new trial or for reconsideration of the decision on the record as made, as the appellate court may deem proper.

(Amended Nov. 29, 1999, 113 Stat. 1501A-583.)

§ 48 (15 U.S.C. § 1051 note). Prior acts not repealed

Section 4 of the Act of January 5, 1905 (U.S.C., title 36, section 4), as amended, entitled "An Act to incorporate the National Red Cross," and section 7 of the Act of June 15, 1916 (U.S.C., title 36, section 27), entitled "An Act to incorporate the Boy Scouts of America, and for other purposes," and the Act of June 20, 1936 (U.S.C., title 22, section 248), entitled "An Act to prohibit the commercial use of the coat of arms of the Swiss Confederation," are not repealed or affected by this Act.

Note: The first and third of the laws referred to in this section have been repealed and replaced by §§ 706 and 708, respectively, of U.S.C., Title 18, Crimes and Criminal Procedure, enacted June 25, 1948, effective September 1, 1948.

§ 49 (15 U.S.C. § 1051 note). Preservation of existing rights

Nothing herein shall adversely affect the rights or the enforcement of rights in marks acquired in good faith prior to the effective date of this Act.

§ 50 (15 U.S.C. § 1051 note). Severability

If any provision of this Act or the application of such provision to any person or circumstance is held invalid, the remainder of the Act shall not be affected thereby.

§ 51 (15 U.S.C. § 1058 note). Applications pending on effective date of the Trademark Law Revision Act of 1988

All certificates of registration based upon applications for registration pending in the Patent and Trademark Office on the effective date of the Trademark Law Revision Act of 1988 shall remain in force for a period of 10 years.

TITLE XII—THE MADRID PROTOCOL

Sec. 60 (15 U.S.C. § 1141). Definitions

In this subchapter:

(1) *Basic application.*—The term 'basic application' means the application for the registration of a mark that has been filed with an Office of a Contracting

Party and that constitutes the basis for an application for the international registration of that mark.

(2) *Basic registration.*—The term 'basic registration' means the registration of a mark that has been granted by an Office of a Contracting Party and that constitutes the basis for an application for the international registration of that mark.

(3) *Contracting party.*—The term 'Contracting Party' means any country or inter-governmental organization that is a party to the Madrid Protocol.

(4) *Date of recordal.*—The term 'date of recordal' means the date on which a request for extension of protection, filed after an international registration is granted, is recorded on the International Register.

(5) *Declaration of bona fide intention to use the mark in commerce.*—The term 'declaration of bona fide intention to use the mark in commerce' means a declaration that is signed by the applicant for, or holder of, an international registration who is seeking extension of protection of a mark to the United States and that contains a statement that—

(A) the applicant or holder has a bona fide intention to use the mark in commerce;

(B) the person making the declaration believes himself or herself, or the firm, corporation, or association in whose behalf he or she makes the declaration, to be entitled to use the mark in commerce; and

(C) no other person, firm, corporation, or association, to the best of his or her knowledge and belief, has the right to use such mark in commerce either in the identical form of the mark or in such near resemblance to the mark as to be likely, when used on or in connection with the goods of such other person, firm, corporation, or association, to cause confusion, mistake, or deception.

(6) *Extension of protection.*—The term 'extension of protection' means the protection resulting from an international registration that extends to the United States at the request of the holder of the international registration, in accordance with the Madrid Protocol.

(7) *Holder of an international registration.*—A 'holder' of an international registration is the natural or juristic person in whose name the international registration is recorded on the International Register.

(8) *International application.*—The term 'international application' means an application for international registration that is filed under the Madrid Protocol.

(9) *International bureau.*—The term 'International Bureau' means the International Bureau of the World Intellectual Property Organization.

(10) *International register.*—The term 'International Register' means the official collection of data concerning international registrations maintained by the International Bureau that the Madrid Protocol or its implementing regulations require or permit to be recorded.

(11) *International registration.*—The term 'international registration' means the registration of a mark granted under the Madrid Protocol.

(12) *International registration date.*—The term 'international registration date' means the date assigned to the international registration by the International Bureau.

(13) *Madrid protocol.*—The term 'Madrid Protocol' means the Protocol Relating to the Madrid Agreement Concerning the International Registration of Marks, adopted at Madrid, Spain, on June 27, 1989.

(14) *Notification of refusal.*—The term 'notification of refusal' means the notice sent by the United States Patent and Trademark Office to the International Bureau declaring that an extension of protection cannot be granted.

(15) *Office of a contracting party.*—The term 'Office of a Contracting Party' means—

(A) the office, or governmental entity, of a Contracting Party that is responsible for the registration of marks; or

(B) the common office, or governmental entity, of more than 1 Contracting Party that is responsible for the registration of marks and is so recognized by the International Bureau.

(16) *Office of origin.*—The term 'office of origin' means the Office of a Contracting Party with which a basic application was filed or by which a basic registration was granted.

(17) *Opposition period.*—The term 'opposition period' means the time allowed for filing an opposition in the United States Patent and Trademark Office, including any extension of time granted under section 13.

[Added Nov. 2, 2003, 116 Stat. 1758, 1913]

Sec. 61 (15 U.S.C. § 1141a). International applications based on United States applications or registrations

(a) *In General.*—The owner of a basic application pending before the United States Patent and Trademark Office, or the owner of a basic registration granted by the United States Patent and Trademark Office may file an international application by submitting to the United States Patent and Trademark Office a written application in such form, together with such fees, as may be prescribed by the Director.

(b) *Qualified Owners.*—A qualified owner, under subsection (a) of this section, shall—

(1) be a national of the United States;

(2) be domiciled in the United States; or

(3) have a real and effective industrial or commercial establishment in the United States.

[Added Nov. 2, 2003, 116 Stat. 1758, 1915]

Sec. 62 (15 U.S.C. § 1141b). Certification of the international application

(a) *Certification Procedure.*—Upon the filing of an application for international registration and payment of the prescribed fees, the Director shall examine the international application for the purpose of certifying that the information contained in the international application corresponds to the information contained in the basic application or basic registration at the time of the certification.

(b) *Transmittal.*—Upon examination and certification of the international application, the Director shall transmit the international application to the International Bureau.

[Added Nov. 2, 2003, 116 Stat. 1758, 1915]

Sec. 63 (15 U.S.C. § 1141c). Restriction, abandonment, cancellation, or expiration of a basic application or basic registration

With respect to an international application transmitted to the International Bureau under section 1141b of this title, the Director shall notify the International Bureau whenever the basic application or basic registration which is the basis for the international application has been restricted, abandoned, or canceled, or has expired, with respect to some or all of the goods and services listed in the international registration—

(1) within 5 years after the international registration date; or

(2) more than 5 years after the international registration date if the restriction, abandonment, or cancellation of the basic application or basic registration resulted from an action that began before the end of that 5-year period.

[Added Nov. 2, 2003, 116 Stat. 1758, 1915]

Sec. 64 (15 U.S.C. § 1141d). Request for extension of protection subsequent to International registration

The holder of an international registration that is based upon a basic application filed with the United States Patent and Trademark Office or a basic registration granted by the Patent and Trademark Office may request an extension of protection of its international registration by filing such a request—

(1) directly with the International Bureau; or

(2) with the United States Patent and Trademark Office for transmittal to the International Bureau, if the request is in such form, and contains such transmittal fee, as may be prescribed by the Director.

[Added Nov. 2, 2003, 116 Stat. 1758, 1916]

Sec. 65 (15 U.S.C. § 1141e). Extension of protection of an international registration to the United States under the Madrid Protocol

(a) *In General.*—Subject to the provisions of section 68, the holder of an international registration shall be entitled to the benefits of extension of protection of that international registration to the United States to the extent necessary to give effect to any provision of the Madrid Protocol.

(b) *If the United States Is Office of Origin.*—Where the United States Patent and Trademark Office is the office of origin for a trademark application or registration, any international registration based on such application or registration cannot be used to obtain the benefits of the Madrid Protocol in the United States.

[Added Nov. 2, 2003, 116 Stat. 1758, 1916]

Sec. 66 (15 U.S.C. § 1141f). Effect of filing a request for extension of protection of an international registration to the United States

(a) *Requirement for Request for Extension of Protection.*—A request for extension of protection of an international registration to the United States that the International Bureau transmits to the United States Patent and Trademark Office shall be deemed to be properly filed in the United States if such request, when received by the International Bureau, has attached to it a declaration of bona fide intention to use the mark in commerce that is verified by the applicant for, or holder of, the international registration.

(b) *Effect of Proper Filing.*—Unless extension of protection is refused under section 681141h of this title, the proper filing of the request for extension of protection under subsection (a) shall constitute constructive use of the mark, conferring the same rights as those specified in section 1057(c) of this title, as of the earliest of the following:

(1) The international registration date, if the request for extension of protection was filed in the international application.

(2) The date of recordal of the request for extension of protection, if the request for extension of protection was made after the international registration date.

(3) The date of priority claimed pursuant to section 1141g of this title.

[Added Nov. 2, 2003, 116 Stat. 1758, 1916]

Sec. 67 (15 U.S.C. § 1141g). Right of priority for request for extension of protection to the United States.

The holder of an international registration with a request for an extension of protection to the United States shall be entitled to claim a date of priority based on a right of priority within the meaning of Article 4 of the Paris Convention for the Protection of Industrial Property if—

(1) the request for extension of protection contains a claim of priority; and

(2) the date of international registration or the date of the recordal of the request for extension of protection to the United States is not later than 6 months after the date of the first regular national filing (within the meaning of Article 4(A)(3) of the Paris Convention for the Protection of Industrial Property) or a subsequent application (within the meaning of Article 4(C)(4) of the Paris Convention for the Protection of Industrial Property).

[Added Nov. 2, 2003, 116 Stat. 1758, 1917]

Sec. 68 (15 U.S.C. § 1141h). Examination of and opposition to request for extension of protection; notification of refusal

(a) *Examination and Opposition.*—

(1) A request for extension of protection described in section 1141f(a) of this title shall be examined as an application for registration on the Principal

Register under this Act, and if on such examination it appears that the applicant is entitled to extension of protection under this title, the Director shall cause the mark to be published in the Official Gazette of the United States Patent and Trademark Office.

(2) Subject to the provisions of subsection (c) of this section, a request for extension of protection under this title shall be subject to opposition under section 1063 of this title.

(3) Extension of protection shall not be refused on the ground that the mark has not been used in commerce.

(4) Extension of protection shall be refused to any mark not registrable on the Principal Register.

(b) *Notification of Refusal.*—If, a request for extension of protection is refused under subsection (a) of this section, the Director shall declare in a notification of refusal (as provided in subsection (c) of this section) that the extension of protection cannot be granted, together with a statement of all grounds on which the refusal was based.

(c) *Notice to International Bureau.*—

(1) Within 18 months after the date on which the International Bureau transmits to the Patent and Trademark Office a notification of a request for extension of protection, the Director shall transmit to the International Bureau any of the following that applies to such request:

(A) A notification of refusal based on an examination of the request for extension of protection.

(B) A notification of refusal based on the filing of an opposition to the request.

(C) A notification of the possibility that an opposition to the request may be filed after the end of that 18-month period.

(2) If the Director has sent a notification of the possibility of opposition under paragraph (1)(C), the Director shall, if applicable, transmit to the International Bureau a notification of refusal on the basis of the opposition, together with a statement of all the grounds for the opposition, within 7 months after the beginning of the opposition period or within 1 month after the end of the opposition period, whichever is earlier.

(3) If a notification of refusal of a request for extension of protection is transmitted under paragraph (1) or (2), no grounds for refusal of such request other than those set forth in such notification may be transmitted to the International Bureau by the Director after the expiration of the time periods set forth in paragraph (1) or (2), as the case may be.

(4) If a notification specified in paragraph (1) or (2) is not sent to the International Bureau within the time period set forth in such paragraph, with respect to a request for extension of protection, the request for extension of protection shall not be refused and the Director shall issue a certificate of extension of protection pursuant to the request.

(d) *Designation of Agent for Service of Process.*—In responding to a notification of refusal with respect to a mark, the holder of the international registration of the mark may designate, by a document filed in the United States Patent and Trademark

Office, the name and address of a person residing in the United States on whom notices or process in proceedings affecting the mark may be served. Such notices or process may be served upon the person designated by leaving with that person, or mailing to that person, a copy thereof at the address specified in the last designation filed. If the person designated cannot be found at the address given in the last designation, or if the holder does not designate by a document filed in the United States Patent and Trademark Office the name and address of a person residing in the United States for service of notices or process in proceedings affecting the mark, the notice or process may be served on the Director.

[Added Nov. 2, 2003, 116 Stat. 1758, 1917]

Sec. 69 (15 U.S.C. § 1141i). Effect of extension of protection.

(a) *Issuance of Extension of Protection.*—Unless a request for extension of protection is refused under section 1141h of this title, the Director shall issue a certificate of extension of protection pursuant to the request and shall cause notice of such certificate of extension of protection to be published in the Official Gazette of the United States Patent and Trademark Office.

(b) *Effect of Extension of Protection.*—From the date on which a certificate of extension of protection is issued under subsection (a) of this section—

(1) such extension of protection shall have the same effect and validity as a registration on the Principal Register; and

(2) the holder of the international registration shall have the same rights and remedies as the owner of a registration on the Principal Register.

[Added Nov. 2, 2003, 116 Stat. 1758, 1918]

Sec. 70 (15 U.S.C. § 1141j). Dependence of extension of protection to the United States on the underlying international registration

(a) *Effect of Cancellation of International Registration.*—If the International Bureau notifies the United States Patent and Trademark Office of the cancellation of an international registration with respect to some or all of the goods and services listed in the international registration, the Director shall cancel any extension of protection to the United States with respect to such goods and services as of the date on which the international registration was canceled.

(b) *Effect of Failure To Renew International Registration.*—If the International Bureau does not renew an international registration, the corresponding extension of protection to the United States shall cease to be valid as of the date of the expiration of the international registration.

(c) *Transformation of an Extension of Protection Into a United States Application.*—The holder of an international registration canceled in whole or in part by the International Bureau at the request of the office of origin, under article 6(4) of the Madrid Protocol, may file an application, under section 1051 or 1126 of this title, for the registration of the same mark for any of the goods and services to which the cancellation applies that were covered by an extension of protection to the United States based on that international registration. Such an application shall be treated as if it had been filed on the international registration date or the date of recordal of the request for extension of protection with the International Bureau, whichever date applies, and, if the extension of protection enjoyed priority under section 1141g of this title, shall enjoy the same priority. Such an application shall be entitled to the benefits conferred by this subsection only if the application is filed not later than 3 months after the date on which the international registration was canceled, in whole or in part, and only if the application complies with all the requirements of this Act which apply to any application filed pursuant to section 1051 or 1126 of this title.

[Added Nov. 2, 2003, 116 Stat. 1758, 1918]

Sec. 71 (15 U.S.C. § 1141k). Duration, affidavits and fees

(a) *Time Periods for Required Affidavits.* Each extension of protection for which a certificate has been issued under section 69 shall remain in force for the term of the international registration upon which it is based, except that the extension of protection of any mark shall be canceled by the Director unless the holder of the international registration files in the United States Patent and Trademark Office affidavits that meet the requirements of subsection (b), within the following time periods:

(1) Within the 1-year period immediately preceding the expiration of 6 years following the date of issuance of the certificate of extension of protection.

(2) Within the 1-year period immediately preceding the expiration of 10 years following the date of issuance of the certificate of extension of protection, and each successive 10-year period following the date of issuance of the certificate of extension of protection.

(3) The holder may file the affidavit required under this section within a grace period of 6 months after the end of the applicable time period established in paragraph (1) or (2), together with the fee described in subsection (b) and the additional grace period surcharge prescribed by the Director.

(b) *Requirements for Affidavit.* The affidavit referred to in subsection (a) shall—

(1) (A) state that the mark is in use in commerce;

(B) set forth the goods and services recited in the extension of protection on or in connection with which the mark is in use in commerce;

(C) be accompanied by such number of specimens or facsimiles showing current use of the mark in commerce as may be required by the Director; and

(D) be accompanied by the fee prescribed by the Director; or

(2) (A) set forth the goods and services recited in the extension of protection on or in connection with which the mark is not in use in commerce;

(B) include a showing that any nonuse is due to special circumstances which excuse such nonuse and is not due to any intention to abandon the mark; and

(C) be accompanied by the fee prescribed by the Director.

(c) *Deficient Affidavit.* If any submission filed within the period set forth in subsection (a) is deficient, including that the affidavit was not filed in the name of the holder of the international registration, the deficiency may be corrected after the statutory time period, within the time prescribed after notification of the deficiency. Such submission shall be accompanied by the additional deficiency surcharge prescribed by the Director.

(d) *Notice of Requirement.* Special notice of the requirement for such affidavit shall be attached to each certificate of extension of protection.

(e) *Notification of Acceptance or Refusal.* The Director shall notify the holder of the international registration who files any affidavit required by this section of the Director's acceptance or refusal thereof and, in the case of a refusal, the reasons therefor.

(f) *Designation of Resident for Service of Process and Notices.* If the holder of the international registration of the mark is not domiciled in the United States, the holder may designate, by a document filed in the United States Patent and Trademark Office, the name and address of a person resident in the United States on whom may be served notices or process in proceedings affecting the mark. Such notices or process may be served upon the person so designated by leaving with that person or mailing to that person a copy thereof at the address specified in the last designation so filed. If the person so designated cannot be found at the last designated address, or if the holder does not designate by a document filed in the United States Patent and Trademark Office the name and address of a person resident in the United States on whom may be served notices or process in proceedings affecting the mark, such notices or process may be served on the Director.

(Amended Mar. 17, 2010, 124 Stat. 66.)

Sec. 72 (15 U.S.C. § 1141*l*). Assignment of an extension of protection

An extension of protection may be assigned, together with the goodwill associated with the mark, only to a person who is a national of, is domiciled in, or has a bona fide and effective industrial or commercial establishment either in a country that is a Contracting Party or in a country that is a member of an intergovernmental organization that is a Contracting Party.

[Added Nov. 2, 2003, 116 Stat. 1758, 1920]

Sec. 73 (15 U.S.C. § 1141m). Incontestability

The period of continuous use prescribed under section 1065 of this title for a mark covered by an extension of protection issued under this title may begin no earlier than the date on which the Director issues the certificate of the extension of protection under section 1141i of this title, except as provided in section 1141n of this title.

[Added Nov. 2, 2003, 116 Stat. 1758, 1920]

Sec. 74 (15 U.S.C. § 1141n). Rights of extension of protection

When a United States registration and a subsequently issued certificate of extension of protection to the United States are owned by the same person, identify the same mark, and list the same goods or services, the extension of protection shall have the same rights that accrued to the registration prior to issuance of the certificate of extension of protection.

[Added Nov. 2, 2003, 116 Stat. 1758, 1920]

Sec. 13403 (15 U.S.C. § 1141 note). Effective date

This subtitle and the amendments made by this subtitle shall take effect on the later of—

(1) the date on which the Madrid Protocol (as defined in section 60 of the Trademark Act of 1946) enters into force with respect to the United States; or

(2) the date occurring 1 year after the date of enactment of this Act.

[Added Nov. 2, 2003, 116 Stat. 1758, 1920]

Criminal Penalties

18 U.S.C

§2318. Trafficking in counterfeit labels, illicit labels, or counterfeit documentation or packaging

(a)(1) Whoever, in any of the circumstances described in subsection (c), knowingly traffics in—

(A) a counterfeit label or illicit label affixed to, enclosing, or accompanying, or designed to be affixed to, enclose, or accompany—

(i) a phonorecord;

(ii) a copy of a computer program;

(iii) a copy of a motion picture or other audiovisual work;

(iv) a copy of a literary work;

(v) a copy of a pictorial, graphic, or sculptural work;

(vi) a work of visual art; or

(vii) documentation or packaging; or

(B) counterfeit documentation or packaging,

shall be fined under this title or imprisoned for not more than 5 years, or both.

(b) As used in this section—

(1) the term ''counterfeit label'' means an identifying label or container that appears to be genuine, but is not;

(2) the term ''traffic'' has the same meaning as in section 2320(e) of this title;

(3) the terms ''copy'', ''phonorecord'', ''motion picture'', ''computer program'', and ''audiovisual work'', ''literary work'', ''pictorial, graphic, or sculptural work'', ''sound recording'', ''work of visual art'', and ''copyright owner'' have, respectively, the meanings given those terms in section 101 (relating to definitions) of title 17;

(4) the term ''illicit label'' means a genuine certificate, licensing document, registration card, or similar labeling component—

(A) that is used by the copyright owner to verify that a phonorecord, a copy of a computer program, a copy of a motion picture or other audiovisual work, a copy of a literary work, a copy of a pictorial, graphic, or sculptural work, a work of visual art, or documentation or packaging is not counterfeit or infringing of any copyright; and

(B) that is, without the authorization of the copyright owner—

(i) distributed or intended for distribution not in connection with the copy, phonorecord, or work of visual art to which such labeling component was intended to be affixed by the respective copyright owner; or

(ii) in connection with a genuine certificate or licensing document, knowingly falsified in order to designate a higher number of licensed users or copies than authorized by the copyright owner, unless that certificate or document is used by the copyright owner solely for the purpose of monitoring or tracking the copyright owner's distribution channel and not for the purpose of verifying that a copy or phonorecord is noninfringing;

(5) the term ''documentation or packaging'' means documentation or packaging, in physical form, for a phonorecord, copy of a computer program, copy of a motion picture or other audiovisual work, copy of a literary work, copy of a pictorial, graphic, or sculptural work, or work of visual art; and

(6) the term ''counterfeit documentation or packaging'' means documentation or packaging that appears to be genuine, but is not.

(c) The circumstances referred to in subsection (a) of this section are—

(1) the offense is committed within the special maritime and territorial jurisdiction of the United States; or within the special aircraft jurisdiction of the United States (as defined in section 46501 of title 49);

(2) the mail or a facility of interstate or foreign commerce is used or intended to be used in the commission of the offense;

(3) the counterfeit label or illicit label is affixed to, encloses, or accompanies, or is designed to be affixed to, enclose, or accompany—

(A) a phonorecord of a copyrighted sound recording or copyrighted musical work;

(B) a copy of a copyrighted computer program;

(C) a copy of a copyrighted motion picture or other audiovisual work;

(D) a copy of a literary work;

(E) a copy of a pictorial, graphic, or sculptural work;

(F) a work of visual art; or

(G) copyrighted documentation or packaging; or

(4) the counterfeited documentation or packaging is copyrighted.

(d) **Forfeiture and destruction of property; restitution.**—Forfeiture, destruction, and restitution relating to this section shall be subject to section 2323, to the extent provided in that section, in addition to any other similar remedies provided by law.

(e) Civil remedies.—

(1) In general.—Any copyright owner who is injured, or is threatened with injury, by a violation of subsection (a) may bring a civil action in an appropriate United States district court.

(2) Discretion of court.—In any action brought under paragraph (1), the court—

(A) may grant 1 or more temporary or permanent injunctions on such terms as the court determines to be reasonable to prevent or restrain a violation of subsection (a);

(B) at any time while the action is pending, may order the impounding, on such terms as the court determines to be reasonable, of any article that is in the custody or control of the alleged violator and that the court has reasonable cause to believe was involved in a violation of subsection (a); and

(C) may award to the injured party—

(i) reasonable attorney fees and costs; and

(ii) (I) actual damages and any additional profits of the violator, as provided in paragraph (3); or

(II) statutory damages, as provided in paragraph (4).

(3) Actual damages and profits.—

(A) In general.—The injured party is entitled to recover—

(i) the actual damages suffered by the injured party as a result of a violation of subsection (a), as provided in subparagraph (B) of this paragraph; and

(ii) any profits of the violator that are attributable to a violation of subsection (a) and are not taken into account in computing the actual damages.

(B) Calculation of damages.—The court shall calculate actual damages by multiplying—

(i) the value of the phonorecords, copies, or works of visual art which are, or are intended to be, affixed with, enclosed in, or accompanied by any counterfeit labels, illicit labels, or counterfeit documentation or packaging, by

(ii) the number of phonorecords, copies, or works of visual art which are, or are intended to be, affixed with, enclosed in, or accompanied by any counterfeit labels, illicit labels, or counterfeit documentation or packaging.

(C) Definition.—For purposes of this paragraph, the "value" of a phonorecord, copy, or work of visual art is—

(i) in the case of a copyrighted sound recording or copyrighted musical work, the retail value of an authorized phonorecord of that sound recording or musical work;

(ii) in the case of a copyrighted computer program, the retail value of an authorized copy of that computer program;

(iii) in the case of a copyrighted motion picture or other audiovisual work, the retail value of an authorized copy of that motion picture or audiovisual work;

(iv) in the case of a copyrighted literary work, the retail value of an authorized copy of that literary work;

(v) in the case of a pictorial, graphic, or sculptural work, the retail value of an authorized copy of that work; and

(vi) in the case of a work of visual art, the retail value of that work.

(4) Statutory damages.—The injured party may elect, at any time before final judgment is rendered, to recover, instead of actual damages and profits, an award of statutory damages for each violation of subsection (a) in a sum of not less than $2,500 or more than $25,000, as the court considers appropriate.

(5) Subsequent violation.—The court may increase an award of damages under this subsection by 3 times the amount that would otherwise be awarded, as the court considers appropriate, if the court finds that a person has subsequently violated subsection (a) within 3 years after a final judgment was entered against that person for a violation of that subsection.

(6) Limitation on actions.—A civil action may not be commenced under section unless it is commenced within 3 years after the date on which the claimant discovers the violation of subsection (a).

§2319. Criminal infringement of a copyright

(a) Any person who violates section 506(a) (relating to criminal offenses) of title 17 shall be punished as provided in subsections (b), (c), and (d) and such penalties shall be in addition to any other provisions of title 17 or any other law.

(b) Any person who commits an offense under section 506(a)(1)(A) of title 17—

(1) shall be imprisoned not more than 5 years, or fined in the amount set forth in this title, or both, if the offense consists of the reproduction or distribution, including by electronic means, during any 180-day period, of at least 10 copies or phonorecords, of 1 or more copyrighted works, which have a total retail value of more than $2,500;

(2) shall be imprisoned not more than 10 years, or fined in the amount set forth in this title, or both, if the offense is a second or subsequent offense under subsection (a); and

(3) shall be imprisoned not more than 1 year, or fined in the amount set forth in this title, or both, in any other case.

(c) Any person who commits an offense under section 506(a)(1)(B) of title 17—

(1) shall be imprisoned not more than 3 years, or fined in the amount set forth in this title, or both, if the offense consists of the reproduction or distribution of 10 or more copies or phonorecords of 1 or more copyrighted works, which have a total retail value of $2,500 or more;

(2) shall be imprisoned not more than 6 years, or fined in the amount set forth in this title, or both, if the offense is a second or subsequent offense under subsection (a); and

(3) shall be imprisoned not more than 1 year, or fined in the amount set forth in this title, or both, if the offense consists of the reproduction or distribution of 1 or more copies or phonorecords of 1 or more copyrighted works, which have a total retail value of more than $1,000.

(d) Any person who commits an offense under section 506(a)(1)(C) of title 17—

(1) shall be imprisoned not more than 3 years, fined under this title, or both;

(2) shall be imprisoned not more than 5 years, fined under this title, or both, if the offense was committed for purposes of commercial advantage or private financial gain;

(3) shall be imprisoned not more than 6 years, fined under this title, or both, if the offense is a felony and is a second or subsequent offense under subsection (a); and

(4) shall be imprisoned not more than 10 years, fined under this title, or both, if the offense is a second or subsequent offense under paragraph (2).

(e)(1) During preparation of the presentence report pursuant to Rule 32(c) of the Federal Rules of Criminal Procedure, victims of the offense shall be permitted to submit, and the probation officer shall receive, a victim impact statement that identifies the victim of the offense and the extent and scope of the injury and loss suffered by the victim, including the estimated economic impact of the offense on that victim.

(2) Persons permitted to submit victim impact statements shall include—

(A) producers and sellers of legitimate works affected by conduct involved in the offense;

(B) holders of intellectual property rights in such works; and

(C) the legal representatives of such producers, sellers, and holders.

(f) As used in this section—

(1) the terms "phonorecord" and "copies" have, respectively, the meanings set forth in section 101 (relating to definitions) of title 17;

(2) the terms "reproduction" and "distribution" refer to the exclusive rights of a copyright owner under clauses (1) and (3) respectively of section 106 (relating to exclusive rights in copyrighted works), as limited by sections 107 through 122 of title 17;

(3) the term "financial gain" has the meaning given the term in section 101 of title 17; and

(4) the term "work being prepared for commercial distribution" has the meaning given the term in section 506(a) of title 17.

§2319A. Unauthorized fixation of and trafficking in sound recordings and music videos of live musical performances

(a) Offense.—Whoever, without the consent of the performer or performers involved, knowingly and for purposes of commercial advantage or private financial gain—

(1) fixes the sounds or sounds and images of a live musical performance in a copy or phonorecord, or reproduces copies or phonorecords of such a performance from an unauthorized fixation;

(2) transmits or otherwise communicates to the public the sounds or sounds and images of a live musical performance; or

(3) distributes or offers to distribute, sells or offers to sell, rents or offers to rent, or traffics in any copy or phonorecord fixed as described in paragraph (1), regardless of whether the fixations occurred in the United States;

shall be imprisoned for not more than 5 years or fined in the amount set forth in this title, or both, or if the offense is a second or subsequent offense, shall be imprisoned for not more than 10 years or fined in the amount set forth in this title, or both.

(b) **Forfeiture and destruction of property; restitution.**—Forfeiture, destruction, and restitution relating to this section shall be subject to section 2323, to the extent provided in that section, in addition to any other similar remedies provided by law.

(c) Seizure and forfeiture.—If copies or phonorecords of sounds or sounds and images of a live musical performance are fixed outside of the United States without the consent of the performer or performers involved, such copies or phonorecords are subject to seizure and forfeiture in the United States in the same manner as property imported in violation of the customs laws. The Secretary of Homeland Security shall issue regulations by which any performer may, upon payment of a specified fee, be entitled to notification by United States Customs and Border Protection of the importation of copies or phonorecords that appear to consist of unauthorized fixations of the sounds or sounds and images of a live musical performance.

(d) Victim impact statement.—(1) During preparation of the presentence report pursuant to Rule 32(c) of the Federal Rules of Criminal Procedure, victims of the offense shall be permitted to submit, and the probation officer shall receive, a victim impact statement that identifies the victim of the offense and the extent and scope of the injury and loss suffered by the victim, including the estimated economic impact of the offense on that victim.

(2) Persons permitted to submit victim impact statements shall include—

(A) producers and sellers of legitimate works affected by conduct involved in the offense;

(B) holders of intellectual property rights in such works; and

(C) the legal representatives of such producers, sellers, and holders.

(e) Definitions.—As used in this section—

(1) the terms "copy", "fixed", "musical work", "phonorecord", "reproduce", "sound recordings", and "transmit" mean those terms within the meaning of title 17; and

(2) the term "traffic" has the same meaning as in section 2320(e) of this title.

(f) Applicability.—This section shall apply to any Act or Acts that occur on or after the date of the enactment of the Uruguay Round Agreements Act.

§2320. Trafficking in counterfeit goods or services

(a) Offenses.—Whoever intentionally—

(1) traffics in goods or services and knowingly uses a counterfeit mark on or in connection with such goods or services,

(2) traffics in labels, patches, stickers, wrappers, badges, emblems, medallions, charms, boxes, containers, cans, cases, hangtags, documentation, or packaging of any type or nature, knowing that a counterfeit mark has been applied thereto, the use of which is likely to cause confusion, to cause mistake, or to deceive,

(3) traffics in goods or services knowing that such good or service is a counterfeit military good or service the use, malfunction, or failure of which is likely to cause serious bodily injury or death, the disclosure of classified information, impairment of combat operations, or other significant harm to a combat operation, a member of the Armed Forces, or to national security, or

(4) traffics in a counterfeit drug,

or attempts or conspires to violate any of paragraphs (1) through (4) shall be punished as provided in subsection (b).

(b) Penalties.—

(1) In general.—Whoever commits an offense under subsection (a)—

(A) if an individual, shall be fined not more than $2,000,000 or imprisoned not more than 10 years, or both, and, if a person other than an individual, shall be fined not more than $5,000,000; and

(B) for a second or subsequent offense under subsection (a), if an individual, shall be fined not more than $5,000,000 or imprisoned not more than 20 years, or both, and if other than an individual, shall be fined not more than $15,000,000.

(2) Serious bodily injury or death.—

(A) Serious bodily injury.—Whoever knowingly or recklessly causes or attempts to cause serious bodily injury from conduct in violation of subsection (a), if an individual, shall be fined not more than $ 5,000,000 or imprisoned for not more than 20 years, or both, and if other than an individual, shall be fined not more than $15,000,000.

(B) Death.—Whoever knowingly or recklessly causes or attempts to cause death from conduct in violation of subsection (a), if an individual, shall be fined not more than $5,000,000 or imprisoned for any term of years or for life, or both, and if other than an individual, shall be fined not more than $15,000,000.

(3) Counterfeit military goods or services and counterfeit drugs.—Whoever commits an offense under subsection (a) involving a counterfeit military good or service or counterfeit drug—

(A) if an individual, shall be fined not more than $5,000,000, imprisoned not more than 20 years, or both, and if other than an individual, be fined not more than $15,000,000; and

(B) for a second or subsequent offense, if an individual, shall be fined not more than $15,000,000, imprisoned not more than 30 years, or both, and if other than an individual, shall be fined not more than $30,000,000.

(c) Forfeiture and destruction of property; restitution.—Forfeiture, destruction, and restitution relating to this section shall be subject to section 2323, to the extent provided in that section, in addition to any other similar remedies provided by law.

(d) Defenses.—All defenses, affirmative defenses, and limitations on remedies that would be applicable in an action under the Lanham Act shall be applicable in a prosecution under this section. In a prosecution under this section, the defendant shall have the burden of proof, by a preponderance of the evidence, of any such affirmative defense.

(e) Presentence report.—**(1)** During preparation of the presentence report pursuant to Rule 32(c) of the Federal Rules of Criminal Procedure, victims of the offense shall be permitted to submit, and the probation officer shall receive, a victim impact statement that identifies the victim of the offense and the extent and scope of the injury and loss suffered by the victim, including the estimated economic impact of the offense on that victim.

(2) Persons permitted to submit victim impact statements shall include—

(A) producers and sellers of legitimate goods or services affected by conduct involved in the offense;

(B) holders of intellectual property rights in such goods or services; and

(C) the legal representatives of such producers, sellers, and holders.

(f) Definitions.—For the purposes of this section—

(1) the term "counterfeit mark" means—

(A) a spurious mark—

(i) that is used in connection with trafficking in any goods, services, labels, patches, stickers, wrappers, badges, emblems, medallions, charms, boxes, containers, cans, cases, hangtags, documentation, or packaging of any type or nature;

(ii) that is identical with, or substantially indistinguishable from, a mark registered on the principal register in the United States Patent and Trademark Office and in use, whether or not the defendant knew such mark was so registered;

(iii) that is applied to or used in connection with the goods or services for which the mark is registered with the United States Patent and Trademark Office, or is applied to or consists of a label, patch, sticker, wrapper, badge, emblem, medallion, charm, box, container, can, case, hangtag, documentation, or packaging of any type or nature that is designed, marketed, or otherwise intended to be used on or in connection with the goods or services for which the mark is registered in the United States Patent and Trademark Office; and

(iv) the use of which is likely to cause confusion, to cause mistake, or to deceive; or

(B) a spurious designation that is identical with, or substantially indistinguishable from, a designation as to which the remedies of the Lanham Act are made available by reason of section 220506 of title 36;

but such term does not include any mark or designation used in connection with goods or services, or a mark or designation applied to labels, patches, stickers, wrappers, badges, emblems, medallions, charms, boxes, containers, cans, cases, hangtags, documentation, or packaging of any type or nature used in connection with such goods or services, of which the manufacturer or producer was, at the time of the manufacture or production in question, authorized to use the mark or designation for the type of goods or services so manufactured or produced, by the holder of the right to use such mark or designation;

(2) the term "financial gain" includes the receipt, or expected receipt, of anything of value;

(3) the term "Lanham Act" means the Act entitled "An Act to provide for the registration and protection of trademarks used in commerce, to carry out the provisions of certain international conventions, and for other purposes", approved July 5, 1946 (15 U.S.C. 1051 et seq.);

(4) the term "counterfeit military good or service" means a good or service that uses a counterfeit mark on or in connection with such good or service and that—

(A) is falsely identified or labeled as meeting military specifications, or

(B) is intended for use in a military or national security application;

(5) the term "traffic" means to transport, transfer, or otherwise dispose of, to another, for purposes of commercial advantage or private financial gain, or to make, import, export, obtain control of, or possess, with intent to so transport, transfer, or otherwise dispose of; and

(6) the term "counterfeit drug" means a drug, as defined by section 201 of the Federal Food, Drug, and Cosmetic Act, that uses a counterfeit mark on or in connection with the drug.

(g) Limitation on cause of action.—Nothing in this section shall entitle the United States to bring a criminal cause of action under this section for the repackaging of genuine goods or services not intended to deceive or confuse.

(h) Report to Congress.—**(1)** Beginning with the first year after the date of enactment of this subsection, the Attorney General shall include in the report of the Attorney General to Congress on the business of the Department of Justice prepared pursuant to section 522 of title 28, an accounting, on a district by district basis, of the following with respect to all actions taken by the Department of Justice that involve trafficking in counterfeit labels for phonorecords, copies of computer programs or computer program documentation or packaging, copies of motion pictures or other audiovisual works (as defined in section 2318 of this title), criminal infringement of copyrights (as defined in section 2319 of this title), unauthorized fixation of and trafficking in sound recordings and music videos of live musical performances (as defined in section 2319A of this title), or trafficking in goods or services bearing counterfeit marks (as defined in section 2320 of this title):

(A) The number of open investigations.

(B) The number of cases referred by the United States Customs Service.

(C) The number of cases referred by other agencies or sources.

(D) The number and outcome, including settlements, sentences, recoveries, and penalties, of all prosecutions brought under sections 2318, 2319, 2319A, and 2320 of title 18.

(2)(A) The report under paragraph (1), with respect to criminal infringement of copyright, shall include the following:

(i) The number of infringement cases in these categories: audiovisual (videos and films); audio (sound recordings); literary works (books and musical compositions); computer programs; video games; and, others.

(ii) The number of online infringement cases.

(iii) The number and dollar amounts of fines assessed in specific categories of dollar amounts. These categories shall be: no fines ordered; fines under $500; fines from $500 to $1,000; fines from $1,000 to $5,000; fines from $5,000 to $10,000; and fines over $10,000.

(iv) The total amount of restitution ordered in all copyright infringement cases.

(B) In this paragraph, the term "online infringement cases" as used in paragraph (2) means those cases where the infringer—

(i) advertised or publicized the infringing work on the Internet; or

(ii) made the infringing work available on the Internet for download, reproduction, performance, or distribution by other persons.

(C) The information required under subparagraph (A) shall be submitted in the report required in fiscal year 2005 and thereafter.

(i) Transshipment and exportation.—No goods or services, the trafficking in of which is prohibited by this section, shall be transshipped through or exported from the United States. Any such transshipment or exportation shall be deemed a violation of section 42 of an Act to provide for the registration of trademarks used in commerce, to carry out the provisions of certain international conventions, and for other purposes, approved July 5, 1946 (commonly referred to as the "Trademark Act of 1946" or the "Lanham Act").

Selected State Codes

California Business and Professions Code

§16600. Invalidity of Contracts

Except as provided in this chapter, every contract by which anyone is restrained from engaging in a lawful profession, trade, or business of any kind is to that extent void.

California Labor Code

§2870. Employment Agreements; Assignment of Rights

(a) Any provision in an employment agreement which provides that an employee shall assign, or offer to assign, any of his or her rights in an invention to his or her employer shall not apply to an invention that the employee developed entirely on his or her own time without using the employer's equipment, supplies, facilities, or trade secret information except for those inventions that either:

(1) Relate at the time of conception or reduction to practice of the invention to the employer's business, or actual or demonstrably anticipated research or development of the employer; or

(2) Result from any work performed by the employee for the employer.

(b) To the extent a provision in an employment agreement purports to require an employee to assign an invention otherwise excluded from being required to be assigned under subdivision (a), the provision is against the public policy of this state and is unenforceable.

California Civil Code

§980 Ownership; works not fixed in tangible medium; sound recordings; inventions or designs

(a) (1) The author of any original work of authorship that is not fixed in any tangible medium of expression has an exclusive ownership in the representation or expression thereof as against all persons except one who originally and independently creates the same or similar work. A work shall be considered not fixed when it is not embodied in a tangible medium of expression or when its embodiment in a tangible medium of expression is not sufficiently permanent or stable to permit it to be perceived, reproduced, or otherwise communicated for a period of more than transitory duration, either directly or with the aid of a machine or device.

(2) The author of an original work of authorship consisting of a sound recording initially fixed prior to February 15, 1972, has an exclusive ownership therein until February 15, 2047, as against all persons except one who independently makes or duplicates another sound recording that does not directly or indirectly recapture the actual sounds fixed in such prior sound recording, but consists entirely of an independent fixation of other sounds, even though such sounds imitate or simulate the sounds contained in the prior sound recording.

(b) The inventor or proprietor of any invention or design, with or without delineation, or other graphical representation, has an exclusive ownership therein, and in the representation or expression thereof, which continues so long as the invention or design and the representations or expressions thereof made by him remain in his possession.

§986 Work of fine art; sale; payment of percentage to artist or deposit for Arts Council; failure to pay; action for damages; exemptions

(a) Whenever a work of fine art is sold and the seller resides in California or the sale takes place in California, the seller or the seller's agent shall pay to the artist of such work of fine art or to such artist's agent 5 percent of the amount of such sale. The right of the artist to receive an amount equal to 5 percent of the amount of such sale may be waived only by a contract in writing providing for an amount in excess of 5 percent of the amount of such sale. An artist may assign the right to collect the royalty payment provided by this section to another individual or entity. However, the assignment shall not have the effect of creating a waiver prohibited by this subdivision.

(1) When a work of fine art is sold at an auction or by a gallery, dealer, broker, museum, or other person acting as the agent for the seller the agent shall withhold 5 percent of the amount of the sale, locate the artist, and pay the artist.

(2) If the seller or agent is unable to locate and pay the artist within 90 days, an amount equal to 5 percent of the amount of the sale shall be transferred to the Arts Council.

(3) If a seller or the seller's agent fails to pay an artist the amount equal to 5 percent of the sale of a work of fine art by the artist or fails to transfer such amount to the Arts Council, the artist may bring an action for damages within three years after the date of sale or one year after the discovery of the sale, whichever is longer. The prevailing party in any action brought under this paragraph shall be entitled to reasonable attorney fees, in an amount as determined by the court.

(4) Moneys received by the council pursuant to this section shall be deposited in an account in the Special Deposit Fund in the State Treasury.

(5) The Arts Council shall attempt to locate any artist for whom money is received pursuant to this section. If the council is unable to locate the artist and the artist does not file a written claim for the money received by the council within seven years of the date of sale of the work of fine art, the right of the artist terminates and such money shall be transferred to the council for use in acquiring fine art pursuant to the Art in Public Buildings program set forth in Chapter 2.1 (commencing with Section 15813) of Part 10b of Division 3 of Title 2, of the Government Code.

(6) Any amounts of money held by any seller or agent for the payment of artists pursuant to this section shall be exempt from enforcement of a money judgment by the creditors of the seller or agent.

(7) Upon the death of an artist, the rights and duties created under this section shall inure to his or her heirs, legatees, or personal representative, until the 20th anniversary of the death of the artist. The provisions of this paragraph shall be applicable only with respect to an artist who dies after January 1, 1983.

(b) Subdivision (a) shall not apply to any of the following:

(1) To the initial sale of a work of fine art where legal title to such work at the time of such initial sale is vested in the artist thereof.

(2) To the resale of a work of fine art for a gross sales price of less than one thousand dollars ($1,000).

(3) Except as provided in paragraph (7) of subdivision (a), to a resale after the death of such artist.

(4) To the resale of the work of fine art for a gross sales price less than the purchase price paid by the seller.

(5) To a transfer of a work of fine art which is exchanged for one or more works of fine art or for a combination of cash, other property, and one or more works of fine art where the fair market value of the property exchanged is less than one thousand dollars ($1,000).

(6) To the resale of a work of fine art by an art dealer to a purchaser within 10 years of the initial sale of the work of fine art by the artist to an art dealer, provided all intervening resales are between art dealers.

(7) To a sale of a work of stained glass artistry where the work has been permanently attached to real property and is sold as part of the sale of the real property to which it is attached.

(c) For purposes of this section, the following terms have the following meanings:

(1) "Artist" means the person who creates a work of fine art and who, at the time of resale, is a citizen of the United States, or a resident of the state who has resided in the state for a minimum of two years.

(2) "Fine art" means an original painting, sculpture, or drawing, or an original work of art in glass.

(3) "Art dealer" means a person who is actively and principally engaged in or conducting the business of selling works of fine art for which business such person validly holds a sales tax permit.

(d) This section shall become operative on January 1, 1977, and shall apply to works of fine art created before and after its operative date.

(e) If any provision of this section or the application thereof to any person or circumstance is held invalid for any reason, such invalidity shall not affect any other provisions or applications of this section which can be effected, without the invalid provision or application, and to this end the provisions of this section are severable.

(f) The amendments to this section enacted during the 1981–82 Regular Session of the Legislature shall apply to transfers of works of fine art, when created before or after January 1, 1983, that occur on or after that date.

§987 Preservation of works of art

(a) The Legislature hereby finds and declares that the physical alteration or destruction of fine art, which is an expression of the artist's personality, is detrimental to the artist's reputation, and artists therefore have an interest in protecting their works of fine art against any alteration or destruction; and that there is also a public interest in preserving the integrity of cultural and artistic creations.

(b) As used in this section:

(1) "Artist" means the individual or individuals who create a work of fine art.

(2) "Fine art" means an original painting, sculpture, or drawing, or an original work of art in glass, of recognized quality, but shall not include work prepared under contract for commercial use by its purchaser.

(3) "Person" means an individual, partnership, corporation, limited liability company, association, or other group, however organized.

(4) "Frame" means to prepare, or cause to be prepared, a work of fine art for display in a manner customarily considered to be appropriate for a work of fine art in the particular medium.

(5) "Restore" means to return, or cause to be returned, a deteriorated or damaged work of fine art as nearly as is feasible to its original state or condition, in accordance with prevailing standards.

(6) "Conserve" means to preserve, or cause to be preserved, a work of fine art by retarding or preventing deterioration or damage through appropriate treatment in accordance with prevailing standards in order to maintain the structural integrity to the fullest extent possible in an unchanging state.

(7) "Commercial use" means fine art created under a work-for-hire arrangement for use in advertising, magazines, newspapers, or other print and electronic media.

(c) (1) No person, except an artist who owns and possesses a work of fine art which the artist has created, shall intentionally commit, or authorize the intentional commission of, any physical defacement, mutilation, alteration, or destruction of a work of fine art.

(2) In addition to the prohibitions contained in paragraph (1), no person who frames, conserves, or restores a work of fine art shall commit, or authorize the commission of, any physical defacement, mutilation, alteration, or destruction of a work of fine art by any act constituting gross negligence. For purposes of this section, the term "gross negligence" shall mean the exercise of so slight a degree of care as to justify the belief that there was an indifference to the particular work of fine art.

(d) The artist shall retain at all times the right to claim authorship, or, for a just and valid reason, to disclaim authorship of his or her work of fine art.

(e) To effectuate the rights created by this section, the artist may commence an action to recover or obtain any of the following:

(1) Injunctive relief.

(2) Actual damages.

(3) Punitive damages. In the event that punitive damages are awarded, the court shall, in its discretion, select an organization or organizations engaged in charitable or educational activities involving the fine arts in California to receive any punitive damages.

(4) Reasonable attorneys' and expert witness fees.

(5) Any other relief which the court deems proper.

(f) In determining whether a work of fine art is of recognized quality, the trier of fact shall rely on the opinions of artists, art dealers, collectors of fine art, curators of art museums, and other persons involved with the creation or marketing of fine art.

(g) The rights and duties created under this section:

(1) Shall, with respect to the artist, or if any artist is deceased, his or her heir, beneficiary, devisee, or personal representative, exist until the 50th anniversary of the death of the artist.

(2) Shall exist in addition to any other rights and duties which may now or in the future be applicable.

(3) Except as provided in paragraph (1) of subdivision (h), may not be waived except by an instrument in writing expressly so providing which is signed by the artist.

(h) (1) If a work of fine art cannot be removed from a building without substantial physical defacement, mutilation, alteration, or destruction of the work, the rights and duties created under this section, unless expressly reserved by an instrument in writing signed by the owner of the building, containing a legal description of the property and properly recorded, shall be deemed waived. The instrument, if properly recorded, shall be binding on subsequent owners of the building.

(2) If the owner of a building wishes to remove a work of fine art which is a part of the building but which can be removed from the building without substantial harm to the fine art, and in the course of or after removal, the owner intends to cause or allow the fine art to suffer physical defacement, mutilation, alteration, or destruction, the rights and duties created under this section shall apply unless the owner has diligently attempted without success to notify the artist, or, if the artist is deceased, his or her heir, beneficiary, devisee, or personal representative, in writing of his or her intended action affecting the work of fine art, or unless he or she did provide notice and that person failed within 90 days either to remove the work or to pay for its removal. If the work is removed at the expense of the artist, his or her heir, beneficiary, devisee, or personal representative, title to the fine art shall pass to that person.

(3) If a work of fine art can be removed from a building scheduled for demolition without substantial physical defacement, mutilation, alteration, or destruction of the work, and the owner of the building has notified the owner of the work of fine art of the scheduled demolition or the owner of the building is the owner of the work of fine art, and the owner of the work of fine art elects not to remove the work of fine art, the rights and duties created under this section shall apply, unless the owner of the building has diligently attempted without success to notify the artist, or, if the artist is deceased, his or her heir, beneficiary, devisee, or personal representative, in writing of the intended action affecting the work of fine art, or unless he or she did provide notice and that person failed within 90 days either to remove the work or to pay for its removal. If the work is removed at the expense of the artist, his or her heir, beneficiary, devisee, or personal representative, title to the fine art shall pass to that person.

(4) Nothing in this subdivision shall affect the rights of authorship created in subdivision (d) of this section.

(i) No action may be maintained to enforce any liability under this section unless brought within three years of the act complained of or one year after discovery of the act, whichever is longer.

(j) This section shall become operative on January 1, 1980, and shall apply to claims based on proscribed acts occurring on or after that date to works of fine art whenever created.

(k) If any provision of this section or the application thereof to any person or circumstance is held invalid for any reason, the invalidity shall not affect any other provisions or applications of this section which can be effected without the invalid provision or application, and to this end the provisions of this section are severable.

§3344

(a) Any person who knowingly uses another's name, voice, signature, photograph, or likeness, in any manner, on or in products, merchandise, or goods, or for purposes of advertising or selling, or soliciting purchases of, products, merchandise, goods, or services, without such person's prior consent, or, in the case of a minor, the prior consent of his parent or legal guardian, shall be liable for any damages sustained by the person or persons injured as a result thereof. In addition, in any action brought under this section, the person who violated the section shall be liable to the injured party or parties in an amount equal to the greater of seven hundred fifty dollars ($750) or the actual damages suffered by him or her as a result of the unauthorized use, and any profits from the unauthorized use that are attributable to the use and are not taken into account in computing the actual damages. In establishing such profits, the injured party or parties are required to present proof only of the gross revenue attributable to such use, and the person who violated this section is required to prove his or her deductible expenses. Punitive damages may also be awarded to the injured party or parties. The prevailing party in any action under this section shall also be entitled to attorney's fees and costs.

(b) As used in this section, "photograph" means any photograph or photographic reproduction, still or moving, or any videotape or live television transmission, of any person, such that the person is readily identifiable.

(1) A person shall be deemed to be readily identifiable from a photograph when one who views the photograph with the naked eye can reasonably determine that the person depicted in the photograph is the same person who is complaining of its unauthorized use.

(2) If the photograph includes more than one person so identifiable, then the person or persons complaining of the use shall be represented as individuals rather than solely as members of a definable group represented in the photograph. A definable group includes, but is not limited to, the following examples: a crowd at any sporting event, a crowd in any street or public building, the audience at any theatrical or stage production, a glee club, or a baseball team.

(3) A person or persons shall be considered to be represented as members of a definable group if they are represented in the photograph solely as a result of being present at the time the photograph was taken and have not been singled out as individuals in any manner.

(c) Where a photograph or likeness of an employee of the person using the photograph or likeness appearing in the advertisement or other publication prepared by or in behalf of the user is only incidental, and not essential, to the purpose of the publication in which it appears, there shall arise a rebuttable presumption affecting the burden of producing evidence that the failure to obtain the consent of the employee was not a knowing use of the employee's photograph or likeness.

(d) For purposes of this section, a use of a name, voice, signature, photograph, or likeness in connection with any news, public affairs, or sports broadcast or account, or any political campaign, shall not constitute a use for which consent is required under subdivision (a).

(e) The use of a name, voice, signature, photograph, or likeness in a commercial medium shall not constitute a use for which consent is required under subdivision (a) solely because the material containing such use is commercially sponsored or contains paid advertising. Rather it shall be a question of fact whether or not the use of the person's name, voice, signature, photograph, or likeness was so directly connected with the commercial sponsorship or with the paid advertising as to constitute a use for which consent is required under subdivision (a).

(f) Nothing in this section shall apply to the owners or employees of any medium used for advertising, including, but not limited to, newspapers, magazines, radio and television networks and stations, cable television systems, billboards, and transit ads, by whom any advertisement or solicitation in violation of this section is published or disseminated, unless it is established that such owners or employees had knowledge of the unauthorized use of the person's name, voice, signature, photograph, or likeness as prohibited by this section.

(g) The remedies provided for in this section are cumulative and shall be in addition to any others provided for by law.

§3344.1

(a) (1) Any person who uses a deceased personality's name, voice, signature, photograph, or likeness, in any manner, on or in products, merchandise, or goods, or for purposes of advertising or selling, or soliciting purchases of, products, merchandise, goods, or services, without prior consent from the person or persons specified in subdivision (c), shall be liable for any damages sustained by the person or persons injured as a result thereof. In addition, in any action brought under this section, the person who violated the section shall be liable to the injured party or parties in an amount equal to the greater of seven hundred fifty dollars ($750) or the actual damages suffered by the injured party or parties, as a result of the unauthorized use, and any profits from the unauthorized use that are attributable to the use and are not taken into account in computing the actual damages. In establishing these profits, the injured party or parties shall be required to present proof only of the gross revenue attributable to the use and the person who violated the section is required to prove his or her deductible expenses. Punitive damages may also be awarded to the injured party or parties. The prevailing party or parties in any action under this section shall also be entitled to attorneys' fees and costs.

(2) For purposes of this subdivision, a play, book, magazine, newspaper, musical composition, audiovisual work, radio or television program, single and original work of art, work of political or newsworthy value, or an advertisement or commercial announcement for any of these works, shall not be considered a product, article of merchandise, good, or service if it is fictional or nonfictional entertainment, or a dramatic, literary, or musical work.

(3) If a work that is protected under paragraph (2) includes within it a use in connection with a product, article of merchandise, good, or service, this use shall not be exempt under this subdivision, notwithstanding the unprotected use's inclusion in a work otherwise exempt under this subdivision, if the claimant proves that this use is so directly connected with a product, article of merchandise, good, or service as to constitute an act of advertising, selling, or soliciting purchases of that product, article of merchandise, good, or service by the deceased personality without prior consent from the person or persons specified in subdivision (c).

(b) The rights recognized under this section are property rights, freely transferable, in whole or in part, by contract or by means of trust or testamentary documents, executed before or after January 1, 1985. The rights recognized under this section shall be deemed to have existed at the time of death of any deceased personality who died prior to January 1, 1985, and, except as provided in subdivision (*o*), shall vest in the persons entitled to these property rights under the testamentary instrument of the deceased personality effective as of the date of his or her death. In the absence of an express transfer in a testamentary instrument of the deceased personality's rights in his or her name, voice, signature, photograph, or likeness, a provision in the testamentary instrument that provides for the disposition of the residue of the deceased personality's assets shall be effective to transfer the rights recognized under this section in accordance with the terms of that provision. The rights established by this section shall also be freely transferable or descendible by contract, trust, or any other testamentary instrument by any subsequent owner of the deceased personality's rights as recognized by this section. Nothing in this section shall be construed to render invalid or unenforceable any contract entered into by a deceased personality during his or her lifetime by which the deceased personality assigned the rights, in whole or in part, to use his or her name, voice, signature, photograph, or likeness, regardless of whether the contract was entered into before or after January 1, 1985.

(c) The consent required by this section shall be exercisable by the person or persons to whom the right of consent, or portion thereof, has been transferred in accordance with subdivision (b), or if no transfer has occurred, then by the person or persons to whom the right of consent, or portion thereof, has passed in accordance with subdivision (d).

(d) Subject to subdivisions (b) and (c), after the death of any person, the rights under this section shall belong to the following person or persons and may be exercised, on behalf of and for the benefit of all of those persons, by those persons who, in the aggregate, are entitled to more than a one-half interest in the rights:

(1) The entire interest in those rights belong to the surviving spouse of the deceased personality unless there are any surviving children or grandchildren of the deceased personality, in which case one-half of the entire interest in those rights belong to the surviving spouse.

(2) The entire interest in those rights belong to the surviving children of the deceased personality and to the surviving children of any dead child of the deceased personality unless the deceased personality has a surviving spouse, in which case the ownership of a one-half interest in rights is divided among the surviving children and grandchildren.

(3) If there is no surviving spouse, and no surviving children or grandchildren, then the entire interest in those rights belong to the surviving parent or parents of the deceased personality.

(4) The rights of the deceased personality's children and grandchildren are in all cases divided among them and exercisable in the manner provided in Section 240 of the Probate Code according to the number of the deceased personality's children represented; the share of the children of a dead child of a deceased personality can be exercised only by the action of a majority of them.

(e) If any deceased personality does not transfer his or her rights under this section by contract, or by means of a trust or testamentary document, and there are no surviving persons as described in subdivision (d), then the rights set forth in subdivision (a) shall terminate.

(f) (1) A successor in interest to the rights of a deceased personality under this section or a licensee thereof shall not recover damages for a use prohibited by this section that occurs before the successor in interest or licensee registers a claim of the rights under paragraph (2).

(2) Any person claiming to be a successor-in-interest to the rights of a deceased personality under this section or a licensee thereof may register that claim with the Secretary of State on a form prescribed by the Secretary of State and upon payment of a fee as set forth in subdivision (d) of Section 12195 of the Government Code. The form shall be verified and shall include the name and date of death of the deceased personality, the name and address of the claimant, the basis of the claim, and the rights claimed.

(3) Upon receipt and after filing of any document under this section, the Secretary of State shall post the document along with the entire registry of persons claiming to be a successor in interest to the rights of a deceased personality or a registered licensee under this section upon the Secretary of State's Internet Web site. The Secretary of State may microfilm or reproduce by other techniques any of the filings or documents and destroy the original filing or document. The microfilm or other reproduction of any document under this section shall be admissible in any court of law. The microfilm or other reproduction of any document may be destroyed by the Secretary of State 70 years after the death of the personality named therein.

(4) Claims registered under this subdivision shall be public records.

(g) No action shall be brought under this section by reason of any use of a deceased personality's name, voice, signature, photograph, or likeness occurring after the expiration of 70 years from the death of the deceased personality.

(h) As used in this section, "deceased personality" means any natural person whose name, voice, signature, photograph, or likeness has commercial value at the time of his or her death, whether or not during the lifetime of that natural person the person used his or her name, voice, signature, photograph, or likeness on or in products, merchandise, or goods, or for purposes of advertising or selling, or solicitation of purchase of products, merchandise, goods, or services. A "deceased personality" shall include, without limitation, any such natural person who has died within 70 years prior to January 1, 1985.

(i) As used in this section, "photograph" means any photograph or photographic reproduction, still or moving, or any videotape or live television transmission, of any person, such that the deceased personality is readily identifiable. A deceased personality shall be deemed to be readily identifiable from a photograph when one who views the photograph with the naked eye can reasonably determine who the person depicted in the photograph is.

(j) For purposes of this section, a use of a name, voice, signature, photograph, or likeness in connection with any news, public affairs, or sports broadcast or account, or any political campaign, shall not constitute a use for which consent is required under subdivision (a).

(k) The use of a name, voice, signature, photograph, or likeness in a commercial medium shall not constitute a use for which consent is required under subdivision (a) solely because the material containing the use is commercially sponsored or contains paid advertising. Rather it shall be a question of fact whether or not the use of the deceased personality's name, voice, signature, photograph, or likeness was so directly connected with the commercial sponsorship or with the paid advertising as to constitute a use for which consent is required under subdivision (a).

(1) Nothing in this section shall apply to the owners or employees of any medium used for advertising, including, but not limited to, newspapers, magazines, radio and television networks and stations, cable television systems, billboards, and transit ads, by whom any advertisement or solicitation in violation of this section is published or disseminated, unless it is established that the owners or employees had knowledge of the unauthorized use of the deceased personality's name, voice, signature, photograph, or likeness as prohibited by this section.

(m) The remedies provided for in this section are cumulative and shall be in addition to any others provided for by law.

(n) This section shall apply to the adjudication of liability and the imposition of any damages or other remedies in cases in which the liability, damages, and other remedies arise from acts occurring directly in this state. For purposes of this section, acts giving rise to liability shall be limited to the use, on or in products, merchandise, goods, or services, or the advertising or selling, or soliciting purchases of, products, merchandise, goods, or services prohibited by this section.

(*o*) Notwithstanding any provision of this section to the contrary, if an action was taken prior to May 1, 2007, to exercise rights recognized under this section relating to a deceased personality who died prior to January 1, 1985, by a person described in subdivision (d), other than a person who was disinherited by the deceased personality in a testamentary instrument, and the exercise of those rights was not challenged successfully in a court action by a person described in subdivision (b), that exercise shall not be affected by subdivision (b). In that case, the rights that would otherwise vest in one or more persons described in subdivision (b) shall vest solely in the person or persons described in subdivision (d), other than a person disinherited by the deceased personality in a testamentary instrument, for all future purposes.

(p) The rights recognized by this section are expressly made retroactive, including to those deceased personalities who died before January 1, 1985.

New York General Business Law

Article 24. *Trademarks, Service-Marks and Business Reputation*

§349. Deceptive Acts and Practices Unlawful

(a) Deceptive acts or practices in the conduct of any business, trade, or commerce or in the furnishing of any service in this state are hereby declared unlawful.

(b) Whenever the attorney general shall believe from evidence satisfactory to him that any person, firm, corporation or association, or agent or employee thereof has engaged in or is about to engage in any of the acts or practices stated to be unlawful he may bring an action in the name and on behalf of the people of the state of New York to enjoin such unlawful acts or practices and to obtain restitution of any moneys or property obtained directly or indirectly by any such unlawful acts or practices. In such action preliminary relief may be granted under article sixty-three of the civil practice law and rules.

(c) Before any violation of this section is sought to be enjoined, the attorney general shall be required to give the person against whom such proceeding is contemplated notice by certified mail and an opportunity to show in writing within five business days after receipt of notice why proceedings should not be instituted against him, unless the attorney general shall find, in any case in which he seeks preliminary relief, that to give such notice and opportunity is not in the public interest.

(d) In any such action it shall be a complete defense that the act or practice is, or if in interstate commerce would be, subject to and complies with the rules and regulations of, and the statutes administered by, the federal trade commission or any official department, division, commission, or agency of the United States as such rules, regulations, or statutes are interpreted by the federal trade commission or such department, division, commission or agency or the federal courts.

(e) Nothing in this section shall apply to any television or radio broadcasting station or to any publisher or printer of a newspaper, magazine, or other form of printed advertising, who broadcasts, publishes, or prints the advertisement.

(f) In connection with any proposed proceeding under this section, the attorney general is authorized to take proof and make a determination of the relevant facts, and to issue subpoenas in accordance with the civil practice law and rules.

(g) This section shall apply to all deceptive acts or practices declared to be unlawful, whether or not subject to any other law of this state, and shall not supersede, amend, or repeal any other law of this state under which the attorney general is authorized to take any action or conduct any inquiry.

(h) In addition to the right of action granted to the attorney general pursuant to this section, any person who has been injured by reason of any violation of this section may bring an action in his own name to enjoin such unlawful act or practice, an action to recover his actual damages or fifty dollars, whichever is greater, or both such actions. The court may, in its discretion, increase the award of damages to an amount not to exceed three times the actual damages up to one thousand dollars, if the court finds the defendant willfully or knowingly violated this section. The court may award reasonable attorney's fees to a prevailing plaintiff.

§350. False Advertising Unlawful

False advertising in the conduct of any business, trade, or commerce or in the furnishing of any service in this state is hereby declared unlawful.

§350-a. False Advertising

(1) The term "false advertising" means advertising, including labeling, of a commodity, or of the kind, character, terms, or conditions of any employment opportunity if such advertising is misleading in a material respect. In determining whether any advertising is misleading, there shall be taken into account (among other things) not only representations made by statement, word, design, device, sound, or any combination thereof, but also the extent to which the advertising fails to reveal facts material in the light of such representations with respect to the commodity or employment to which the advertising relates under the conditions prescribed in said advertisement, or under such conditions as are customary or usual.

For purposes of this article, with respect to the advertising of an employment opportunity, it shall be deemed "misleading in a material respect" to either fail to reveal whether the employment available or being offered requires or is conditioned upon the purchasing or leasing of supplies, material, equipment, or other property or whether such employment is on a commission rather than a fixed salary basis and, if so, whether the salaries advertised are only obtainable if sufficient commissions are earned.

(2) An employer shall not be liable under this section as a result of a failure to disclose all material facts relating to terms and conditions of employment if the aggrieved person has not suffered actual pecuniary damage as a result of the misleading advertising of an employment opportunity or if the employer has, prior to the aggrieved person suffering any pecuniary damage, disclosed in writing to that person a full and accurate description of the kind, character, terms, and conditions of the employment opportunity.

(3) It shall constitute false advertising to display or announce, in print or broadcast advertising, the price of an item after deduction of a rebate unless the actual selling price is displayed or announced, and clear and conspicuous notice is given in the advertisement that a mail-in rebate is required to achieve the lower net price.

§350-c. Notice of Proposed Action

Before the attorney-general commences an action pursuant to section three hundred fifty-d of this article he shall be required to give the person against whom such action is contemplated appropriate notice by certified mail and an opportunity to show, either orally or in writing, why such action should not be commenced. In such showing, said person may present, among other things, evidence that the advertisement is subject to and complies with the rules and regulations of, and the statutes administered by, the Federal Trade Commission or any official department, division, commission, or agency of the state of New York.

§350-d. Civil Penalty

Any person, firm, corporation or association or agent or employee thereof who engages in any of the acts or practices stated in this article to be unlawful shall be liable to a civil penalty of not more than five hundred dollars for each violation, which shall accrue to the state of New York and may be recovered in a civil action brought by the attorney general. In any such action it shall be a complete defense that the advertisement is subject to and complies with the rules and regulations of, and the statutes administered by, the Federal Trade Commission or any official department, division, commission, or agency of the state of New York.

§350-e. Construction

(1) This article neither enlarges nor diminishes the rights of parties in private litigation except as provided in this section.

(2) This article does not repeal the provisions of subdivision twelve of section sixty-three of the executive law.

(3) Any person who has been injured by reason of any violation of section three hundred fifty or three hundred fifty-a of this article may bring an action in his own name to enjoin such unlawful act or practice, an action to recover his actual damages or fifty dollars, whichever is greater, or both such actions. The court may, in its discretion, increase the award of damages to an amount not to exceed three times the actual damages up to ten thousand dollars, if the court finds the defendant willfully or knowingly violated this section. The court may award reasonable attorney's fees to a prevailing plaintiff.

§350-f. Exceptions

Nothing in this article shall apply to any television or sound radio broadcasting station or to any publisher or printer of a newspaper, magazine, or other form of printed advertising, who broadcasts, publishes, or prints such advertisement.

§368-d. Injury to Business Reputation; Dilution [Eff. until Jan. 1,1997.]

Likelihood of injury to business reputation or of dilution of the distinctive quality of a mark or trade name shall be a ground for injunctive relief in cases of infringement of a mark registered or not registered or in cases of unfair competition, notwithstanding the absence of competition between the parties or the absence of confusion as to the source of goods or services.

New York Civil Rights Law

§50. Right of Privacy

A person, firm, or corporation that uses for advertising purposes, or for the purposes of trade, the name, portrait, or picture of any living person without having first obtained the written consent of such person, or if a minor of his or her parent or guardian, is guilty of a misdemeanor.

Texas Business and Commerce Code

§15.50. Criteria for Enforceability of Covenants Not to Compete

(a) Notwithstanding Section 15.05 of this code, and subject to any applicable provision of Subsection (b), a covenant not to compete is enforceable if it is ancillary to or part of an otherwise enforceable agreement at the time the agreement is made to the extent that it contains limitations as to time, geographical area, and scope of activity to be restrained that are reasonable and do not impose a greater restraint than is necessary to protect the goodwill or other business interest of the promisee.

(b) A covenant not to compete relating to the practice of medicine is enforceable against a person licensed as a physician by the Texas Medical Board if such covenant complies with the following requirements:

(1) the covenant must:

(A) not deny the physician access to a list of his patients whom he had seen or treated within one year of termination of the contract or employment;

(B) provide access to medical records of the physician's patients upon authorization of the patient and any copies of medical records for a reasonable fee as established by the Texas Medical Board under Section 159.008, Occupations Code; and

(C) provide that any access to a list of patients or to patients' medical records after termination of the contract or employment shall not require such list or records to be provided in a format different than that by which such records are maintained except by mutual consent of the parties to the contract;

(2) the covenant must provide for a buy out of the covenant by the physician at a reasonable price or, at the option of either party, as determined by a mutually agreed upon arbitrator or, in the case of an inability to agree, an arbitrator of the court whose decision shall be binding on the parties; and

(3) the covenant must provide that the physician will not be prohibited from providing continuing care and treatment to a specific patient or patients during the course of an acute illness even after the contract or employment has been terminated.

§15.51. Procedures and Remedies in Actions to Enforce Covenants not to Compete

(a) Except as provided in Subsection (c) of this section, a court may award the promisee under a covenant not to compete damages, injunctive relief, or both damages and injunctive relief for a breach by the promisor of the covenant.

(b) If the primary purpose of the agreement to which the covenant is ancillary is to obligate the promisor to render personal services, for a term or at will, the promisee has the burden of establishing that the covenant meets the criteria specified by Section 15.50 of this code. If the agreement has a different primary purpose, the promisor has the burden of establishing that the covenant does not meet those criteria. For the purposes of this subsection, the "burden of establishing" a fact means the burden of persuading the triers of fact that the existence of the fact is more probable than its nonexistence.

(c) If the covenant is found to be ancillary to or part of an otherwise enforceable agreement but contains limitations as to time, geographical area, or scope of activity to be restrained that are not reasonable and impose a greater restraint than is necessary to protect the goodwill or other business interest of the promisee, the court shall reform the covenant to the extent necessary to cause the limitations contained in the covenant as to time, geographical area, and scope of activity to be restrained to be reasonable and to impose a restraint that is not greater than necessary to protect the goodwill or other business interest of the promisee and enforce the covenant as reformed, except that the court may not award the promisee damages for a breach of the covenant before its reformation and the relief granted to the promisee shall be limited to injunctive relief. If the primary purpose of the agreement to which the covenant is ancillary is to obligate the promisor to render personal services, the promisor establishes that the promisee knew at the time of the execution of the agreement that the covenant did not contain limitations as to time, geographical area, and scope of activity to be restrained that were reasonable and the limitations imposed a greater restraint than necessary to protect the goodwill or other business interest of the promisee, and the promisee sought to enforce the covenant to a greater extent than was necessary to protect the goodwill or other business interest of the promisee, the court may award the promisor the costs, including reasonable attorney's fees, actually and reasonably incurred by the promisor in defending the action to enforce the covenant.

§15.52. Preemption of Other Law

The criteria for enforceability of a covenant not to compete provided by Section 15.50 of this code and the procedures and remedies in an action to enforce a covenant not to compete provided by Section 15.51 of this code are exclusive and preempt any other criteria for enforceability of a covenant not to compete or procedures and remedies in an action to enforce a covenant not to compete under common law or otherwise.

Paris Convention for the Protection of Industrial Property of March 20, 1883, as Revised at Brussels on December 14, 1900, at Washington on June 2, 1911, at the Hague on November 6, 1925, at London on June 2, 1934, at Lisbon on October 31, 1958, and at Stockholm on July 14, 1967

U.N.T.S. No. 11851, vol. 828, pp. 305-388

Article 1

(1) The countries to which this Convention applies constitute a Union for the protection of industrial property.

(2) The protection of industrial property has as its object patents, utility models, industrial designs, trademarks, service marks, trade names, indications of source or appellations of origin, and the repression of unfair competition.

(3) Industrial property shall be understood in the broadest sense and shall apply not only to industry and commerce proper, but likewise to agricultural and extractive industries and to all manufactured or natural products, for example, wines, grain, tobacco leaf, fruit, cattle, minerals, mineral waters, beer, flowers, and flour.

(4) Patents shall include the various kinds of industrial patents recognized by the laws of the countries of the Union, such as patents of importation, patents of improvement, patents and certificates of addition, etc.

Article 2

(1) Nationals of any country of the Union shall, as regards the protection of industrial property, enjoy in all the other countries of the Union the advantages that their respective laws now grant, or may hereafter grant, to nationals; all without prejudice to the rights specially provided for by this Convention. Consequently, they shall have the same protection as the latter, and the same legal remedy against any infringement of their rights, provided that the conditions and formalities imposed upon nationals are complied with.

(2) However, no requirement as to domicile or establishment in the country where protection is claimed may be imposed upon nationals of countries of the Union for the enjoyment of any industrial property rights.

(3) The provisions of the laws of each of the countries of the Union relating to judicial and administrative procedure and to jurisdiction, and to the designation of an address for service or the appointment of an agent, which may be required by the laws on industrial property are expressly reserved.

Article 3

Nationals of countries outside the Union who are domiciled or who have real and effective industrial or commercial establishments in the territory of one of the countries of the Union shall be treated in the same manner as nationals of the countries of the Union.

Article 4

A

(1) Any person who has duly filed an application for a patent, or for the registration of a utility model, or of an industrial design, or of a trademark, in one of the countries of the Union, or his successor in title, shall enjoy, for the purpose of filing in the other countries, a right of priority during the periods hereinafter fixed.

(2) Any filing that is equivalent to a regular national filing under the domestic legislation of any country of the Union or under bilateral or multilateral treaties concluded between countries of the Union shall be recognized as giving rise to the right of priority.

(3) By a regular national filing is meant any filing that is adequate to establish the date on which the application was filed in the country concerned, whatever may be the subsequent fate of the application.

B

Consequently, any subsequent filing in any of the other countries of the Union before the expiration of the periods referred to above shall not be invalidated by reason of any acts accomplished in the interval, in particular, another filing, the publication or exploitation of the invention, the putting on sale of copies of the design, or the use of the mark, and such acts cannot give rise to any third-party right or any right of personal possession. Rights acquired by third parties before the date of the first application that serves as the basis for the right of priority are reserved in accordance with the domestic legislation of each country of the Union.

C

(1) The periods of priority referred to above shall be twelve months for patents and utility models, and six months for industrial designs and trademarks.

(2) These periods shall start from the date of filing of the first application; the day of filing shall not be included in the period.

(3) If the last day of the period is an official holiday, or a day when the Office is not open for the filing of applications in the country where protection is claimed, the period shall be extended until the first following working day.

(4) A subsequent application concerning the same subject as a previous first application within the meaning of paragraph (2), above, filed in the same country of the Union, shall be considered as the first application, of which the filing date shall be the starting point of the period of priority, if, at the time of filing the subsequent application, the said previous application has been withdrawn, abandoned, or refused, without having been laid open to public inspection and without leaving any rights outstanding, and if it has not yet served as a basis for claiming a right of priority. The previous application may not thereafter serve as a basis for claiming a right of priority.

D

(1) Any person desiring to take advantage of the priority of a previous filing shall be required to make a declaration indicating the date of such filing and the country in which it was made. Each country shall determine the latest date on which such declaration must be made.

(2) These particulars shall be mentioned in the publications issued by the competent authority, and in particular in the patents and the specifications relating thereto.

(3) The countries of the Union may require any person making a declaration of priority to produce a copy of the application (description, drawings, etc.) previously filed. The copy, certified as correct by the authority which received such application, shall not require any authentication, and may in any case be filed, without fee, at any time within three months of the filing of the subsequent application. They may require it to be accompanied by a certificate from the same authority showing the date of filing, and by a translation.

(4) No other formalities may be required for the declaration of priority at the time of filing the application. Each country of the Union shall determine the consequences of failure to comply with the formalities prescribed by this Article, but such consequences shall in no case go beyond the loss of the right of priority.

(5) Subsequently, further proof may be required.

Any person who avails himself of the priority of a previous application shall be required to specify the number of that application; this number shall be published as provided for by paragraph (2), above.

E

(1) Where an industrial design is filed in a country by virtue of a right of priority based on the filing of a utility model, the period of priority shall be the same as that fixed for industrial designs.

(2) Furthermore, it is permissible to file a utility model in a country by virtue of a right of priority based on the filing of a patent application, and vice versa.

F

No country of the Union may refuse a priority or a patent application on the ground that the applicant claims multiple priorities, even if they originate in different countries, or on the ground that an application claiming one or more priorities contains one or more elements that were not included in the application or applications whose priority is claimed, provided that, in both cases, there is unity of invention within the meaning of the law of the country.

With respect to the elements not included in the application or applications whose priority is claimed, the filing of the subsequent application shall give rise to a right of priority under ordinary conditions.

G

(1) If the examination reveals that an application for a patent contains more than one invention, the applicant may divide the application into a certain number of divisional applications and preserve as the date of each the date of the initial application and the benefit of the right of priority, if any.

(2) The applicant may also, on his own initiative, divide a patent application and preserve as the date of each divisional application the date of the initial application and the benefit of the right of priority, if any. Each country of the Union shall have the right to determine the conditions under which such division shall be authorized.

H

Priority may not be refused on the ground that certain elements of the invention for which priority is claimed do not appear among the claims formulated in the application in the country of origin, provided that the application documents as a whole specifically disclose such elements.

I

(1) Applications for inventors' certificates filed in a country in which applicants have the right to apply at their own option either for a patent or for an inventor's certificate shall give rise to the right of priority provided for by this Article, under the same conditions and with the same effects as applications for patents.

(2) In a country in which applicants have the right to apply at their own option either for a patent or for an inventor's certificate, an applicant for an inventor's certificate shall, in accordance with the provisions of this Article relating to patent applications, enjoy a right of priority based on an application for a patent, a utility model, or an inventor's certificate.

Article 4*bis*

(1) Patents applied for in the various countries of the Union by nationals of countries of the Union shall be independent of patents obtained for the same invention in other countries, whether members of the Union or not.

(2) The foregoing provision is to be understood in an unrestricted sense, in particular, in the sense that patents applied for during the period of priority are independent, both as regards the grounds for nullity and forfeiture, and as regards their normal duration.

(3) The provision shall apply to all patents existing at the time when it comes into effect.

(4) Similarly, it shall apply, in the case of the accession of new countries, to patents in existence on either side at the time of accession.

(5) Patents obtained with the benefit of priority shall, in the various countries of the Union, have a duration equal to that which they would have, had they been applied for or granted without the benefit of priority.

Article 4*ter*

The inventor shall have the right to be mentioned as such in the patent.

Article 4*quater*

The grant of a patent shall not be refused and a patent shall not be invalidated on the ground that the sale of the patented product or of a product obtained by means of a patented process is subject to restrictions or limitations resulting from the domestic law.

Article 5

A

(1) Importation by the patentee into the country where the patent has been granted of articles manufactured in any of the countries of the Union shall not entail forfeiture of the patent.

(2) Each country of the Union shall have the right to take legislative measures providing for the grant of compulsory licenses to prevent the abuses which might result from the exercise of the exclusive rights conferred by the patent, for example, failure to work.

(3) Forfeiture of the patent shall not be provided for except in cases where the grant of compulsory licenses would not have been sufficient to prevent the said abuses. No proceedings for the forfeiture or revocation of a patent may be instituted before the expiration of two years from the grant of the first compulsory license.

(4) A compulsory license may not be applied for on the ground of failure to work or insufficient working before the expiration of a period of four years from the date of filing of the patent application or three years from the date of the grant of the patent, whichever period expires last; it shall be refused if the patentee justifies his inaction by legitimate reasons. Such a compulsory license shall be non-exclusive and shall not be transferable, even in the form of the grant of a sub-license, except with that part of the enterprise or goodwill which exploits such license.

(5) The foregoing provisions shall be applicable, mutatis mutandis, to utility models.

B

The protection of industrial designs shall not, under any circumstance, be subject to any forfeiture, either by reason of failure to work or by reason of the importation of articles corresponding to those which are protected.

C

(1) If, in any country, use of the registered mark is compulsory, the registration may be cancelled only after a reasonable period, and then only if the person concerned does not justify his inaction.

(2) Use of a trademark by the proprietor in a form differing in elements which do not alter the distinctive character of the mark in the form in which it was registered in one of the countries of the Union shall not entail invalidation of the registration and shall not diminish the protection granted to the mark.

(3) Concurrent use of the same mark on identical or similar goods by industrial or commercial establishments considered as co-proprietors of the mark according to the provisions of the domestic law of the country where protection is claimed shall not prevent registration or diminish in any way the protection granted to the said mark in any country of the Union, provided that such use does not result in misleading the public and is not contrary to the public interest.

D

No indication or mention of the patent, of the utility model, of the registration of the trademark, or of the deposit of the industrial design, shall be required upon the goods as a condition of recognition of the right to protection.

Article 5*bis*

(1) A period of grace of not less than six months shall be allowed for the payment of the fees prescribed for the maintenance of industrial property rights, subject, if the domestic legislation so provides, to the payment of a surcharge.

(2) The countries of the Union shall have the right to provide for the restoration of patents which have lapsed by reason of non-payment of fees.

Article 5*ter*

In any country of the Union the following shall not be considered as infringements of the rights of a patentee:

(1) the use on board vessels of other countries of the Union of devices forming the subject of his patent in the body of the vessel, in the machinery, tackle, gear, and other accessories, when such vessels temporarily or accidentally enter the waters of the said country, provided that such devices are used there exclusively for the needs of the vessel;

(2) the use of devices forming the subject of the patent in the construction or operation of aircraft or land vehicles of other countries of the Union, or of accessories of such aircraft or land vehicles, when those aircraft or land vehicles temporarily or accidentally enter the said country.

Article 5*quater*

When a product is imported into a country of the Union where there exists a patent protecting a process of manufacture of the said product, the patentee shall have all the rights, with regard to the imported product, that are accorded to him by the legislation of the country of importation, on the basis of the process patent, with respect to products manufactured in that country.

Article 5*quinquies*

Industrial designs shall be protected in all the countries of the Union.

Article 6

(1) The conditions for the filing and registration of trademarks shall be determined in each country of the Union by its domestic legislation.

(2) However, an application for the registration of a mark filed by a national of a country of the Union in any country of the Union may not be refused, nor may a registration be invalidated, on the ground that filing, registration, or renewal has not been effected in the country of origin.

(3) A mark duly registered in a country of the Union shall be regarded as independent of marks registered in the other countries of the Union, including the country of origin.

Article 6*bis*

(1) The countries of the Union undertake, ex officio if their legislation so permits, or at the request of an interested party, to refuse or to cancel the registration, and to prohibit the use, of a trademark which constitutes a reproduction, an imitation, or a translation, liable to create confusion, of a mark considered by the competent authority of the country of registration or use to be well known in that country as being already the mark of a person entitled to the benefits of this Convention and used for identical or similar goods. These provisions shall also apply when the essential part of the mark constitutes a reproduction of any such well-known mark or an imitation liable to create confusion therewith.

(2) A period of at least five years from the date of registration shall be allowed for requesting the cancellation of such a mark. The countries of the Union may provide for a period within which the prohibition of use must be requested.

(3) No time limit shall be fixed for requesting the cancellation or the prohibition of the use of marks registered or used in bad faith.

Article 6*ter*

(1)(a) The countries of the Union agree to refuse or to invalidate the registration, and to prohibit by appropriate measures the use, without authorization by the competent authorities, either as trademarks or as elements of trademarks, of armorial bearings, flags, and other State emblems, of the countries of the Union, official signs and hallmarks indicating control and warranty adopted by them, and any imitation from a heraldic point of view.

(b) The provisions of subparagraph (a), above, shall apply equally to armorial bearings, flags, other emblems, abbreviations, and names of international intergovernmental organizations of which one or more countries of the Union are members, with the exception of armorial bearings, flags, other emblems, abbreviations, and names that are already the subject of international agreements in force, intended to ensure their protection.

(c) No country of the Union shall be required to apply the provisions of subparagraph (b), above, to the prejudice of the owners of rights acquired in good faith before the entry into force, in that country, of this Convention. The countries of the Union shall not be required to apply the said provisions when the use or registration referred to in subparagraph (a), above, is not of such a nature as to suggest to the public that a connection exists between the organization concerned and the armorial bearings, flags, emblems, abbreviations, and names, or if such use or registration is probably not of such a nature as to mislead the public as to the existence of a connection between the user and the organization.

(2) Prohibition of the use of official signs and hallmarks indicating control and warranty shall apply solely in cases where the marks in which they are incorporated are intended to be used on goods of the same or a similar kind.

(3)(a) For the application of these provisions, the countries of the Union agree to communicate reciprocally, through the intermediary of the International Bureau, the list of State emblems, and official signs and hallmarks indicating control and warranty, which they desire, or may hereafter desire, to place wholly or within certain limits under the protection of this Article, and all subsequent modifications of such list. Each country of the Union shall in due course make available to the public the lists so communicated.

Nevertheless such communication is not obligatory in respect of flags of States.

(b) The provisions of subparagraph (b) of paragraph (1) of this Article shall apply only to such armorial bearings, flags, other emblems, abbreviations, and names of international intergovernmental organizations as the latter have communicated to the countries of the Union through the intermediary of the International Bureau.

(4) Any country of the Union may, within a period of twelve months from the receipt of the notification, transmit its objections, if any, through the intermediary of the International Bureau, to the country or international intergovernmental organization concerned.

(5) In the case of State flags, the measures prescribed by paragraph (1), above, shall apply solely to marks registered after November 6, 1925.

(6) In the case of State emblems other than flags, and of official signs and hallmarks of the countries of the Union, and in the case of armorial bearings, flags, other emblems, abbreviations, and names, of international intergovernmental organizations, these provisions shall apply only to marks registered more than two months after receipt of the communication provided for in paragraph (3) above.

(7) In cases of bad faith, the countries shall have the right to cancel even those marks incorporating State emblems, signs, and hallmarks, which were registered before November 6, 1925.

(8) Nationals of any country who are authorized to make use of the State emblems, signs, and hallmarks of their country may use them even if they are similar to those of another country.

(9) The countries of the Union undertake to prohibit the unauthorized use in trade of the State armorial bearings of the other countries of the Union, when the use is of such a nature as to be misleading as to the origin of the goods.

(10) The above provisions shall not prevent the countries from exercising the right given in paragraph (3) of Article 6quinquies, Section B, to refuse or to invalidate the registration of marks incorporating, without authorization, armorial bearings, flags, other State emblems, or official signs and hallmarks adopted by a country of the Union, as well as the distinctive signs of international intergovernmental organizations referred to in paragraph (1) above.

Article 6*quater*

(1) When, in accordance with the law of a country of the Union, the assignment of mark is valid only if it takes place at the same time as the transfer of the business or goodwill to which the mark belongs, it shall suffice for the recognition of such validity that the portion of the business or goodwill located in that country be transferred to the assignee, together with the exclusive right to manufacture in the said country, or to sell therein, the goods bearing the mark assigned.

(2) The foregoing provision does not impose upon the countries of the Union any obligation to regard as valid the assignment of any mark the use of which by the assignee would, in fact, be of such a nature as to mislead the public, particularly as regards the origin, nature, or essential qualities of the goods to which the mark is applied.

Article 6*quinquies*

A

(1) Every trademark duly registered in the country of origin shall be accepted for filing and protected as is in the other countries of the Union, subject to the reservations indicated in this Article. Such countries may, before proceeding to final registration, require the production of a certificate of registration in the country of origin, issued by the competent authority. No authentication shall be required for this certificate.

(2) The country of origin shall be considered the country of the Union where the applicant has a real and effective industrial or commercial establishment, or, if he has no such establishment within the Union, the country of the Union where he has his domicile, or, if he has no domicile within the Union but is a national of a country of the Union, the country of which he is a national.

B

Trademarks covered by this Article may be neither denied registration nor invalidated except in the following cases:

(1) when they are of such a nature as to infringe rights acquired by third parties in the country where protection is claimed;

(2) when they are devoid of any distinctive character, or consist exclusively of signs or indications which may serve, in trade, to designate the kind, quality, quantity, intended purpose, value, place of origin, of the goods, or the time of production, or have become customary in the current language or in the bona fide and established practices of the trade of the country where protection is claimed;

(3) when they are contrary to morality or public order and, in particular, of such a nature as to deceive the public. It is understood that a mark may not be considered contrary to public order for the sole reason that it does not conform to a provision of the legislation on marks, except if such provision itself relates to public order.

This provision is subject, however, to the application of Article 10*bis*.

C

(1) In determining whether a mark is eligible for protection, all the factual circumstances must be taken into consideration, particularly the length of time the mark has been in use.

(2) No trademark shall be refused in the other countries of the Union for the sole reason that it differs from the mark protected in the country of origin only in respect of elements that do not alter its distinctive character and do not affect its identity in the form in which it has been registered in the said country of origin.

D

No person may benefit from the provisions of this Article if the mark for which he claims protection is not registered in the country of origin.

E

However, in no case shall the renewal of the registration of the mark in the country of origin involve an obligation to renew the registration in the other countries of the Union in which the mark has been registered.

F

The benefit of priority shall remain unaffected for applications for the registration of marks filed within the period fixed by Article 4, even if registration in the country of origin is effected after the expiration of such period.

Article 6*sexies*

The countries of the Union undertake to protect service marks. They shall not be required to provide for the registration of such marks.

Article 6*septies*

(1) If the agent or representative of the person who is the proprietor of a mark in one of the countries of the Union applies, without such proprietor's authorization, for the registration of the mark in his own name, in one or more countries of the Union, the proprietor shall be entitled to oppose the registration applied for or demand its cancellation or, if the law of the country so allows, the assignment in his favor of the said registration, unless such agent or representative justifies his action.

(2) The proprietor of the mark shall, subject to the provisions of paragraph (1) above, be entitled to oppose the use of his mark by his agent or representative if he has not authorized such use.

(3) Domestic legislation may provide an equitable time limit within which the proprietor of a mark must exercise the rights provided for in this Article.

Article 7

The nature of the goods to which a trademark is to be applied shall in no case form an obstacle to the registration of the mark.

Article 7*bis*

(1) The countries of the Union undertake to accept for filing and to protect collective marks belonging to associations the existence of which is not contrary to the law of the country of origin, even if such associations do not possess an industrial or commercial establishment.

(2) Each country shall be the judge of the particular conditions under which a collective mark shall be protected and may refuse protection if the mark is contrary to the public interest.

(3) Nevertheless, the protection of these marks shall not be refused to any association the existence of which is not contrary to the law of the country of origin, on the ground that such association is not established in the country where protection is sought or is not constituted according to the law of the latter country.

Article 8

A trade name shall be protected in all the countries of the Union without the obligation of filing or registration, whether or not it forms part of a trademark.

Article 9

(1) All goods unlawfully bearing a trademark or trade name shall be seized on importation into those countries of the Union where such mark or trade name is entitled to legal protection.

(2) Seizure shall likewise be effected in the country where the unlawful affixation occurred or in the country into which the goods were imported.

(3) Seizure shall take place at the request of the public prosecutor, or any other competent authority, or any interested party, whether a natural person or a legal entity, in conformity with the domestic legislation of each country.

(4) The authorities shall not be bound to effect seizure of goods in transit.

(5) If the legislation of a country does not permit seizure on importation, seizure shall be replaced by prohibition of importation or by seizure inside the country.

(6) If the legislation of a country permits neither seizure on importation nor prohibition of importation nor seizure inside the country, then, until such time as the legislation is modified accordingly, these measures shall be replaced by the actions and remedies available in such cases to nationals under the law of such country.

Article 10

(1) The provisions of the preceding Article shall apply in cases of direct or indirect use of a false indication of the source of the goods or the identity of the producer, manufacturer, or merchant.

(2) Any producer, manufacturer, or merchant, whether a natural person or a legal entity, engaged in the production or manufacture of or trade in such goods and established either in the locality falsely indicated as the source, or in the region where such locality is situated, or in the country falsely indicated, or in the country where the false indication of source is used, shall in any case be deemed an interested party.

Article 10*bis*

(1) The countries of the Union are bound to assure to nationals of such countries effective protection against unfair competition.

(2) Any act of competition contrary to honest practices in industrial or commercial matters constitutes an act of unfair competition.

(3) The following in particular shall be prohibited:

(a) all acts of such a nature as to create confusion by any means whatever with the establishment, the goods, or the industrial or commercial activities, of a competitor;

(b) false allegations in the course of trade of such a nature as to discredit the establishment, the goods, or the industrial or commercial activities of a competitor;

(c) indications or allegations the use of which in the course of trade is liable to mislead the public as to the nature, the manufacturing process, the characteristics, the suitability for their purpose, or the quantity of the goods.

Article 10*ter*

(1) The countries of the Union undertake to assure to nationals of the other countries of the Union appropriate legal remedies effectively to repress all the acts referred to in Articles 9,10, and 10*bis*.

(2) They undertake, further, to provide measures to permit federations and associations representing interested industrialists, producers, or merchants, provided that the existence of such federations and associations is not contrary to the laws of their countries, to take action in the courts or before the administrative authorities, with a view to the repression of the acts referred to in Articles 9,10, and 10*bis*, insofar as the law of the country in which protection is claimed allows such action by federations and associations of that country.

Article 11

(1) The countries of the Union shall, in conformity with their domestic legislation, grant temporary protection to patentable inventions, utility models, industrial designs, and trademarks, in respect of goods exhibited at official or officially recognized international exhibitions held in the territory of any of them.

(2) Such temporary protection shall not extend the periods provided by Article 4. If, later, the right of priority is invoked, the authorities of any country may provide that the period shall start from the date of introduction of the goods into the exhibition.

(3) Each country may require, as proof of the identity of the article exhibited and of the date of its introduction, such documentary evidence as it considers necessary.

Article 12

(1) Each country of the Union undertakes to establish a special industrial property service and a central office for the communication to the public of patents, utility models, industrial designs, and trademarks.

(2) This service shall publish an official periodical journal. It shall publish regularly:

(a) the names of the proprietors of patents granted, with a brief designation of the inventions patented;

(b) the reproductions of registered trademarks.

Article 13

(1)(a) The Union shall have an Assembly consisting of those countries of the Union which are bound by Articles 13 to 17.

(b) The Government of each country shall be represented by one delegate, who may be assisted by alternate delegates, advisors, and experts.

(c) The expenses of each delegation shall be borne by the Government which has appointed it.

(2)(a) The Assembly shall:

(i) deal with all matters concerning the maintenance and development of the Union and the implementation of this Convention;

(ii) give directions concerning the preparation for conferences of revision to the International Bureau of Intellectual Property (hereinafter designated as "the International Bureau") referred to in the Convention establishing the World Intellectual Property Organization (hereinafter designated as "the Organization"), due account being taken of any comments made by those countries of the Union which are not bound by Articles 13 to 17;

(iii) review and approve the reports and activities of the Director General of the Organization concerning the Union, and give him all necessary instructions concerning matters within the competence of the Union;

(iv) elect the members of the Executive Committee of the Assembly;

(v) review and approve the reports and activities of its Executive Committee, and give instructions to such Committee;

(vi) determine the program and adopt the triennial budget of the Union, and approve its final accounts;

(vii) adopt the financial regulations of the Union;

(viii) establish such committees of experts and working groups as it deems appropriate to achieve the objectives of the Union;

(ix) determine which countries not members of the Union and which intergovernmental and international non-governmental organizations shall be admitted to its meetings as observers;

(x) adopt amendments to Articles 13 to 17;

(xi) take any other appropriate action designed to further the objectives of the Union;

(xii) perform such other functions as are appropriate under this Convention;

(xiii) subject to its acceptance, exercise such rights as are given to it in the Convention establishing the Organization.

(b) With respect to matters which are of interest also to other Unions administered by the Organization, the Assembly shall make its decisions after having heard the advice of the Coordination Committee of the Organization.

(3)(a) Subject to the provisions of subparagraph (b), a delegate may represent one country only.

(b) Countries of the Union grouped under the terms of a special agreement in a common office possessing for each of them the character of a special national service of industrial property as referred to in Article 12 may be jointly represented during discussions by one of their number.

(4)(a) Each country member of the Assembly shall have one vote.

(b) One-half of the countries members of the Assembly shall constitute a quorum.

(c) Notwithstanding the provisions of subparagraph (b), if, in any session, the number of countries represented is less than one half but equal to or more than one third of the countries members of the Assembly, the Assembly may make decisions but, with the exception of decisions concerning its own procedure, all such decisions shall take effect only if the conditions set forth hereinafter are fulfilled. The International Bureau shall communicate the said decisions to the countries members of the Assembly which were not represented and shall invite them to express in writing their vote or abstention within a period of three months from the date of the communication. If, at the expiration of this period, the number of countries having thus expressed their vote or abstention attains the number of countries which was lacking for attaining the quorum in the session itself, such decisions shall take effect provided that at the same time the required majority still obtains.

(d) Subject to the provisions of Article 17 (2), the decisions of the Assembly shall require two thirds of the votes cast.

(e) Abstentions shall not be considered as votes.

(5)(a) Subject to the provisions of subparagraph (b), a delegate may vote in the name of one country only.

(b) The countries of the Union referred to in paragraph (3) (b) shall, as a general rule, endeavor to send their own delegations to the sessions of the Assembly. If, however, for exceptional reasons, any such country cannot send its own delegation, it may give to the delegation of another such country the power to vote in its name, provided that each delegation may vote by proxy for one country only. Such power to vote shall be granted in a document signed by the Head of State or the competent Minister.

(6) Countries of the Union not members of the Assembly shall be admitted to the meetings of the latter as observers.

(7)(a) The Assembly shall meet once in every third calendar year in ordinary session upon convocation by the Director General and, in the absence of exceptional circumstances, during the same period and at the same place as the General Assembly of the Organization.

(b) The Assembly shall meet in extraordinary session upon convocation by the Director General, at the request of the Executive Committee or at the request of one fourth of the countries members of the Assembly.

(8) The Assembly shall adopt its own rules of procedure.

Article 14

(1) The Assembly shall have an Executive Committee.

(2)(a) The Executive Committee shall consist of countries elected by the Assembly from among countries members of the Assembly. Furthermore, the country on whose territory the Organization has its headquarters shall, subject to the provisions of Article 16(7) (b), have an ex officio seat on the Committee.

(b) The Government of each country member of the Executive Committee shall be represented by one delegate, who may be assisted by alternate delegates, advisors, and experts.

(c) The expenses of each delegation shall be borne by the Government which has appointed it.

(3) The number of countries members of the Executive Committee shall correspond to one fourth of the number of countries members of the Assembly. In establishing the number of seats to be filled, remainders after division by four shall be disregarded.

(4) In electing the members of the Executive Committee, the Assembly shall have due regard to an equitable geographical distribution and to the need for countries party to the Special Agreements established in relation with the Union to be among the countries constituting the Executive Committee.

(5)(a) Each member of the Executive Committee shall serve from the close of the session of the Assembly which elected it to the close of the next ordinary session of the Assembly.

(b) Members of the Executive Committee may be re-elected, but only up to a maximum of two thirds of such members.

(c) The Assembly shall establish the details of the rules governing the election and possible re-election of the members of the Executive Committee.

(6)(a) The Executive Committee shall:

(i) prepare the draft agenda of the Assembly;

(ii) submit proposals to the Assembly in respect of the draft program and triennial budget of the Union prepared by the Director General;

(iii) approve, within the limits of the program and the triennial budget, the specific yearly budgets and programs prepared by the Director General;

(iv) submit, with appropriate comments, to the Assembly the periodical reports of the Director General and the yearly audit reports on the accounts;

(v) take all necessary measures to ensure the execution of the program of the Union by the Director General, in accordance with the decisions of the Assembly and having regard to circumstances arising between two ordinary sessions of the Assembly;

(vi) perform such other functions as are allocated to it under this Convention.

(b) With respect to matters which are of interest also to other Unions administered by the Organization, the Executive Committee shall make its decisions after having heard the advice of the Coordination Committee of the Organization.

(7)(a) The Executive Committee shall meet once a year in ordinary session upon convocation by the Director General, preferably during the same period and at the same place as the Coordination Committee of the Organization.

(b) The Executive Committee shall meet in extraordinary session upon convocation by the Director General, either on his own initiative, or at the request of its Chairman or one fourth of its members.

(8)(a) Each country member of the Executive Committee shall have one vote.

(b) One-half of the members of the Executive Committee shall constitute a quorum.

(c) Decisions shall be made by a simple majority of the votes cast.

(d) Abstentions shall not be considered as votes.

(e) A delegate may represent, and vote in the name of, one country only.

(9) Countries of the Union not members of the Executive Committee shall be admitted to its meetings as observers.

(10) The Executive Committee shall adopt its own rules of procedure.

Article 15

(1)(a) Administrative tasks concerning the Union shall be performed by the International Bureau, which is a continuation of the Bureau of the Union united with the Bureau of the Union established by the International Convention for the Protection of Literary and Artistic Works.

(b) In particular, the International Bureau shall provide the secretariat of the various organs of the Union.

(c) The Director General of the Organization shall be the chief executive of the Union and shall represent the Union.

(2) The International Bureau shall assemble and publish information concerning the protection of industrial property. Each country of the Union shall promptly communicate to the International Bureau all new laws and official texts concerning the protection of industrial property. Furthermore, it shall furnish the International Bureau with all the publications of its industrial property service of direct concern to the protection of industrial property which the International Bureau may find useful in its work.

(3) The International Bureau shall publish a monthly periodical.

(4) The International Bureau shall, on request, furnish any country of the Union with information on matters concerning the protection of industrial property.

(5) The International Bureau shall conduct studies, and shall provide services, designed to facilitate the protection of industrial property.

(6) The Director General and any staff member designated by him shall participate, without the right to vote, in all meetings of the Assembly, the Executive Committee, and any other committee of experts or working group. The Director General, or a staff member designated by him, shall be ex officio secretary of these bodies.

(7)(a) The International Bureau shall, in accordance with the directions of the Assembly and in cooperation with the Executive Committee, make the preparations for the conferences of revision of the provisions of the Convention other than Articles 13 to 17.

(b) The International Bureau may consult with intergovernmental and international non-governmental organizations concerning preparations for conferences of revision.

(c) The Director General and persons designated by him shall take part, without the right to vote, in the discussions at these conferences.

(8) The International Bureau shall carry out any other tasks assigned to it.

Article 16

(1)(a) The Union shall have a budget.

(b) The budget of the Union shall include the income and expenses proper to the Union, its contribution to the budget of expenses common to the Unions, and, where applicable, the sum made available to the budget of the Conference of the Organization.

(c) Expenses not attributable exclusively to the Union but also to one or more other Unions administered by the Organization shall be considered as expenses common to the Unions. The share of the Union in such common expenses shall be in proportion to the interest the Union has in them.

(2) The budget of the Union shall be established with due regard to the requirements of coordination with the budgets of the other Unions administered by the Organization.

(3) The budget of the Union shall be financed from the following sources:

(i) contributions of the countries of the Union;

(ii) fees and charges due for services rendered by the International Bureau in relation to the Union;

(iii) sale of, or royalties on, the publications of the International Bureau concerning the Union;

(iv) gifts, bequests, and subventions;

(v) rents, interests, and other miscellaneous income.

(4)(a) For the purpose of establishing its contribution towards the budget, each country of the Union shall belong to a class, and shall pay its annual contributions on the basis of a number of units fixed as follows:

Class I 25
Class II 20
Class III 15
Class IV 10
Class V 5
Class VI 3
Class VII 1

(b) Unless it has already done so, each country shall indicate, concurrently with depositing its instrument of ratification or accession, the class to which it wishes to belong. Any country may change class. If it chooses a lower class, the country must announce such change to the Assembly at one of its ordinary sessions. Any such change shall take effect at the beginning of the calendar year following the said session.

(c) The annual contribution of each country shall be an amount in the same proportion to the total sum to be contributed to the budget of the Union by all countries as the number of its units is to the total of the units of all contributing countries.

(d) Contributions shall become due on the first of January of each year.

(e) A country which is in arrears in the payment of its contributions may not exercise its right to vote in any of the organs of the Union of which it is a member if the amount of its arrears equals or exceeds the amount of the contributions due from it for the preceding two full years. However, any organ of the Union may allow such a country to continue to exercise its right to vote in that organ if, and as long as, it is satisfied that the delay in payment is due to exceptional and unavoidable circumstances.

(f) If the budget is not adopted before the beginning of a new financial period, it shall be at the same level as the budget of the previous year, as provided in the financial regulations.

(5) The amount of the fees and charges due for services rendered by the International Bureau in relation to the Union shall be established, and shall be reported to the Assembly and the Executive Committee, by the Director General.

(6)(a) The Union shall have a working capital fund which shall be constituted by a single payment made by each country of the Union. If the fund becomes insufficient, the Assembly shall decide to increase it.

(b) The amount of the initial payment of each country to the said fund or of its participation in the increase thereof shall be a proportion of the contribution of that country for the year in which the fund is established or the decision to increase it is made.

(c) The proportion and the terms of payment shall be fixed by the Assembly on the proposal of the Director General and after it has heard the advice of the Coordination Committee of the Organization.

(7)(a) In the headquarters agreement concluded with the country on the territory of which the Organization has its headquarters, it shall be provided that, whenever the working capital fund is insufficient, such country shall grant advances. The amount of these advances and the conditions on which they are granted shall be the subject of separate agreements, in each case, between such country and the Organization. As long as it remains under the obligation to grant advances, such country shall have an ex officio seat on the Executive Committee.

(b) The country referred to in subparagraph (a) and the Organization shall each have the right to denounce the obligation to grant advances, by written notification. Denunciation shall take effect three years after the end of the year in which it has been notified.

(8) The auditing of the accounts shall be effected by one or more of the countries of the Union or by external auditors, as provided in the financial regulations. They shall be designated, with their agreement, by the Assembly.

Article 17

(1) Proposals for the amendment of Articles 13, 14, 15, 16, and the present Article, may be initiated by any country member of the Assembly, by the Executive

Committee, or by the Director General. Such proposals shall be communicated by the Director General to the member countries of the Assembly at least six months in advance of their consideration by the Assembly.

(2) Amendments to the Articles referred to in paragraph (1) shall be adopted by the Assembly. Adoption shall require three fourths of the votes cast, provided that any amendment to Article 13, and to the present paragraph, shall require four fifths of the votes cast.

(3) Any amendment to the Articles referred to in paragraph (1) shall enter into force one month after written notifications of acceptance, effected in accordance with their respective constitutional processes, have been received by the Director General from three fourths of the countries members of the Assembly at the time it adopted the amendment. Any amendment to the said Articles thus accepted shall bind all the countries which are members of the Assembly at the time the amendment enters into force, or which become members thereof at a subsequent date, provided that any amendment increasing the financial obligations of countries of the Union shall bind only those countries which have notified their acceptance of such amendment.

Article 18

(1) This Convention shall be submitted to revision with a view to the introduction of amendments designed to improve the system of the Union.

(2) For that purpose, conferences shall be held successively in one of the countries of the Union among the delegates of the said countries.

(3) Amendments to Articles 13 to 17 are governed by the provisions of Article 17.

Article 19

It is understood that the countries of the Union reserve the right to make separately between themselves special agreements for the protection of industrial property, insofar as these agreements do not contravene the provisions of this Convention.

Article 20

(1)(a) Any country of the Union which has signed this Act may ratify it, and, if it has not signed it, may accede to it. Instruments of ratification and accession shall be deposited with the Director General.

(b) Any country of the Union may declare in its instrument of ratification or accession that its ratification or accession shall not apply:

(i) to Articles 1 to 12, or

(ii) to Articles 13 to 17.

(c) Any country of the Union which, in accordance with subparagraph (b), has excluded from the effects of its ratification or accession one of the two groups of Articles referred to in that subparagraph may at any later time declare that it extends the effects of its ratification or accession to that group of Articles. Such declaration shall be deposited with the Director General.

(2)(a) Articles 1 to 12 shall enter into force, with respect to the first ten countries of the Union which have deposited instruments of ratification or accession without making the declaration permitted under paragraph (1)(b)(i), three months after the deposit of the tenth such instrument of ratification or accession.

(b) Articles 13 to 17 shall enter into force, with respect to the first ten countries of the Union which have deposited instruments of ratification or accession without making the declaration permitted under paragraph (1)(b)(ii), three months after the deposit of the tenth such instrument of ratification or accession.

(c) Subject to the initial entry into force, pursuant to the provisions of subparagraphs (a) and (b), of each of the two groups of Articles referred to in paragraph (1) (b) (i) and (ii), and subject to the provisions of paragraph (1)(b), Articles 1 to 17 shall, with respect to any country of the Union, other than those referred to in subparagraphs (a) and (b), which deposits an instrument of ratification or accession or any country of the Union which deposits a declaration pursuant to paragraph (1) (c), enter into force three months after the date of notification by the Director General of such deposit, unless a subsequent date has been indicated in the instrument or declaration deposited. In the latter case, this Act shall enter into force with respect to that country on the date thus indicated.

(3) With respect to any country of the Union which deposits an instrument of ratification or accession, Articles 18 to 30 shall enter into force on the earlier of the dates on which any of the groups of Articles referred to in paragraph (1) (b) enters into force with respect to that country pursuant to paragraph (2)(a), (b), or (c).

Article 21

(1) Any country outside the Union may accede to this Act and thereby become a member of the Union. Instruments of accession shall be deposited with the Director General.

(2)(a) With respect to any country outside the Union which deposits its instrument of accession one month or more before the date of entry into force of any provisions of the present Act, this Act shall enter into force, unless a subsequent date has been indicated in the instrument of accession, on the date upon which provisions first enter into force pursuant to Article 20(2) (a) or (b); provided that:

(i) if Articles 1 to 12 do not enter into force on that date, such country shall, during the interim period before the entry into force of such provisions, and in substitution therefor, be bound by Articles 1 to 12 of the Lisbon Act,

(ii) if Articles 13 to 17 do not enter into force on that date, such country shall, during the interim period before the entry into force of such provisions, and in substitution therefor, be bound by Articles 13 and 14(3), (4), and (5), of the Lisbon Act.

If a country indicates a subsequent date in its instrument of accession, this Act shall enter into force with respect to that country on the date thus indicated.

(b) With respect to any country outside the Union which deposits its instrument of accession on a date which is subsequent to, or precedes by less than one month, the entry into force of one group of Articles of the present Act, this Act shall, subject to the proviso of subparagraph (a), enter into force three months after the date on which its accession has been notified by the Director General, unless a subsequent date has been indicated in the instrument of accession. In the latter case, this Act shall enter into force with respect to that country on the date thus indicated.

(3) With respect to any country outside the Union which deposits its instrument of accession after the date of entry into force of the present Act in its entirety, or less than one month before such date, this Act shall enter into force three months after the date on which its accession has been notified by the Director General, unless a subsequent date has been indicated in the instrument of accession. In the latter case, this Act shall enter into force with respect to that country on the date thus indicated.

Article 22

Subject to the possibilities of exceptions provided for in Articles 20 (1) (b) and 28(2), ratification or accession shall automatically entail acceptance of all the clauses and admission to all the advantages of this Act.

Article 23

After the entry into force of this Act in its entirety, a country may not accede to earlier Acts of this Convention.

Article 24

(1) Any country may declare in its instrument of ratification or accession, or may inform the Director General by written notification any time thereafter, that this Convention shall be applicable to all or part of those territories, designated in the declaration or notification, for the external relations of which it is responsible.

(2) Any country which has made such a declaration or given such a notification may, at any time, notify the Director General that this Convention shall cease to be applicable to all or part of such territories.

(3)(a) Any declaration made under paragraph (1) shall take effect on the same date as the ratification or accession in the instrument of which it was included, and any notification given under such paragraph shall take effect three months after its notification by the Director General.

(b) Any notification given under paragraph (2) shall take effect twelve months after its receipt by the Director General.

Article 25

(1) Any country party to this Convention undertakes to adopt, in accordance with its constitution, the measures necessary to ensure the application of this Convention.

(2) It is understood that, at the time a country deposits its instrument of ratification or accession, it will be in a position under its domestic law to give effect to the provisions of this Convention.

Article 26

(1) This Convention shall remain in force without limitation as to time.

(2) Any country may denounce this Act by notification addressed to the Director General. Such denunciation shall constitute also denunciation of all earlier Acts and shall affect only the country making it, the Convention remaining in full force and effect as regards the other countries of the Union.

(3) Denunciation shall take effect one year after the day on which the Director General has received the notification.

(4) The right of denunciation provided by this Article shall not be exercised by any country before the expiration of five years from the date upon which it becomes a member of the Union.

Article 27

(1) The present Act shall, as regards the relations between the countries to which it applies, and to the extent that it applies, replace the Convention of Paris of March 20, 1883, and the subsequent Acts of revision.

(2)(a) As regards the countries to which the present Act does not apply, or does not apply in its entirety, but to which the Lisbon Act of October 31, 1958, applies, the latter shall remain in force in its entirety or to the extent that the present Act does not replace it by virtue of paragraph (1).

(b) Similarly, as regards the countries to which neither the present Act, nor portions thereof, nor the Lisbon Act applies, the London Act of June 2, 1934, shall remain in force in its entirety or to the extent that the present Act does not replace it by virtue of paragraph (1).

(c) Similarly, as regards the countries to which neither the present Act, nor portions thereof, nor the Lisbon Act, nor the London Act applies, the Hague Act of November 6, 1925, shall remain in force in its entirety or to the extent that the present Act does not replace it by virtue of paragraph (1).

(3) Countries outside the Union which become party to this Act shall apply it with respect to any country of the Union not party to this Act or which, although party to this Act, has made a declaration pursuant to Article 20 (1)(b) (i). Such countries recognize that the said country of the Union may apply, in its relations with them, the provisions of the most recent Act to which it is party.

Article 28

(1) Any dispute between two or more countries of the Union concerning the interpretation or application of this Convention, not settled by negotiation, may, by any one of the countries concerned, be brought before the International Court of Justice by application in conformity with the Statute of the Court, unless the countries concerned agree on some other method of settlement. The country bringing the dispute before the Court shall inform the International Bureau; the International Bureau shall bring the matter to the attention of the other countries of the Union.

(2) Each country may, at the time it signs this Act or deposits its instrument of ratification or accession, declare that it does not consider itself bound by the provisions of paragraph (1). With regard to any dispute between such country and any other country of the Union, the provisions of paragraph (1) shall not apply.

(3) Any country having made a declaration in accordance with the provisions of paragraph (2) may, at any time, withdraw its declaration by notification addressed to the Director General.

Article 29

(1)(a) This Act shall be signed in a single copy in the French language and shall be deposited with the Government of Sweden.

(b) Official texts shall be established by the Director General, after consultation with the interested Governments, in the English, German, Italian, Portuguese, Russian, and Spanish languages, and such other languages as the Assembly may designate.

(c) In case of differences of opinion on the interpretation of the various texts, the French text shall prevail.

(2) This Act shall remain open for signature at Stockholm until January 13, 1968.

(3) The Director General shall transmit two copies, certified by the Government of Sweden, of the signed text of this Act to the Governments of all countries of the Union and, on request, to the Government of any other country.

(4) The Director General shall register this Act with the Secretariat of the United Nations.

(5) The Director General shall notify the Governments of all countries of the Union of signatures, deposits of instruments of ratification or accession and any declarations included in such instruments or made pursuant to Article 20 (1) (c), entry into force of any provisions of this Act, notifications of denunciation, and notifications pursuant to Article 24.

Article 30

(1) Until the first Director General assumes office, references in this Act to the International Bureau of the Organization or to the Director General shall be deemed to be references to the Bureau of the Union or its Director, respectively.

(2) Countries of the Union not bound by Articles 13 to 17 may, until five years after the entry into force of the Convention establishing the Organization, exercise, if they so desire, the rights provided under Articles 13 to 17 of this Act as if they were bound by those Articles. Any country desiring to exercise such rights shall give written notification to that effect to the Director General; such notification shall be effective from the date of its receipt. Such countries shall be deemed to be members of the Assembly until the expiration of the said period.

(3) As long as all the countries of the Union have not become Members of the Organization, the International Bureau of the Organization shall also function as the Bureau of the Union, and the Director General as the Director of the said Bureau.

(4) Once all the countries of the Union have become Members of the Organization, the rights, obligations, and property of the Bureau of the Union shall devolve on the International Bureau of the Organization.

IN WITNESS WHEREOF, the undersigned, being duly authorized thereto, have signed this Act. DONE at Stockholm, on July 14, 1967.

Berne Convention for the Protection of Literary and Artistic Works

(Paris Text 1971)

Article 1

The countries to which this Convention applies constitute a Union for the protection of the rights of authors in their literary and artistic works.

Article 2

(1) The expression "literary and artistic works" shall include every production in the literary, scientific and artistic domain, whatever may be the mode or form of its expression, such as books, pamphlets and other writings; lectures, addresses, sermons and other works of the same nature; dramatic or dramatico-musical works; choreographic works and entertainments in dumb show; musical compositions with or without words; cinematographic works to which are assimilated works expressed by a process analogous to cinematography; works of drawing, painting, architecture, sculpture, engraving and lithography; photographic works to which are assimilated works expressed by a process analogous to photography; works of applied art; illustrations, maps, plans, sketches and three-dimensional works relative to geography, topography, architecture or science.

(2) It shall, however, be a matter for legislation in the countries of the Union to prescribe that works in general or any specified categories of works shall not be protected unless they have been fixed in some material form.

(3) Translations, adaptations, arrangements of music, and other alterations of a literary or artistic work shall be protected as original works without prejudice to the copyright in the original work.

(4) It shall be a matter for legislation in the countries of the Union to determine the protection to be granted to official texts of a legislative, administrative, and legal nature, and to official translations of such texts.

(5) Collections of literary or artistic works such as encyclopedias and anthologies which, by reason of the selection and arrangement of their contents, constitute intellectual creations shall be protected as such, without prejudice to the copyright in each of the works forming part of such collections.

(6) The works mentioned in this article shall enjoy protection in all countries of the Union. This protection shall operate for the benefit of the author and his successors in title.

(7) Subject to the provisions of Article 7(4) of this Convention, it shall be a matter for legislation in the countries of the Union to determine the extent of the application of their laws to works of applied art and industrial designs and models, as well as the conditions under which such works, designs and models shall be protected. Works protected in the country of origin solely as designs and models shall be entitled in another country of the Union only to such special protection as is granted in that country to designs and models; however, if no such special protection is granted in that country, such works shall be protected as artistic works.

(8) The protection of this Convention shall not apply to news of the day or to miscellaneous facts having the character of mere items of press information.

Article 2*bis*

(1) It shall be a matter for legislation in the countries of the Union to exclude, wholly or in part, from the protection provided by the preceding Article political speeches and speeches delivered in the course of legal proceedings.

(2) It shall also be a matter for legislation in the countries of the Union to determine the conditions under which lectures, addresses, and other works of the same nature which are delivered in public may be reproduced by the press, broadcast, communicated to the public by wire, and made the subject of public communication as envisaged in Article 11bis (1) of this Convention, when such use is justified by the informatory purpose.

(3) Nevertheless, the author shall enjoy the exclusive right of making a collection of his works mentioned in the preceding paragraphs.

Article 3

(1) The protection of this Convention shall apply to:

(a) authors who are nationals of one of the countries of the Union, for their works, whether published or not;

(b) authors who are not nationals of one of the countries of the Union, for their works first published in one of those countries, or simultaneously in a country outside the Union and in a country of the Union.

(2) Authors who are not nationals of one of the countries of the Union but who have their habitual residence in one of them shall, for the purposes of this Convention, be assimilated to nationals of that country.

(3) The expression "published works" means works published with the consent of their authors, whatever may be the means of manufacture of the copies, provided that the availability of such copies has been such as to satisfy the reasonable requirements of the public, having regard to the nature of the work. The performance of a dramatic, dramatico-musical, cinematographic, or musical work, the public recitation of a literary work, the communication by wire or the broadcasting of literary or artistic works, the exhibition of a work of art, and the construction of a work of architecture shall not constitute publication.

(4) A work shall be considered as having been published simultaneously in several countries if it has been published in two or more countries within thirty days of its first publication.

Article 4

The protection of this Convention shall apply, even if the conditions of Article 3 are not fulfilled, to:

(a) authors of cinematographic works the maker of which has his headquarters or habitual residence in one of the countries of the Union;

(b) authors of works of architecture, erected in a country of the Union or of other artistic works incorporated in a building or other structure located in a country of the Union.

Article 5

(1) Authors shall enjoy, in respect of works for which they are protected under this Convention, in countries of the Union other than the country of origin, the rights which their respective laws do now or may hereafter grant to their nationals, as well as the rights specially granted by this Convention.

(2) The enjoyment and the exercise of these rights shall not be subject to any formality; such enjoyment and such exercise shall be independent of the existence of protection in the country of origin of the work. Consequently, apart from the provisions of this Convention, the extent of protection, as well as the means of redress afforded to the author to protect his rights, shall be governed exclusively by the laws of the country where protection is claimed.

(3) Protection in the country of origin is governed by domestic law. However, when the author is not a national of the country of origin of the work for which he is protected under this Convention, he shall enjoy in that country the same rights as national authors.

(4) The country of origin shall be considered to be

(a) in the case of works first published in a country of the Union, that country; in the case of works published simultaneously in several countries of the Union which grant different terms of protection, the country whose legislation grants the shortest term of protection;

(b) in the case of works published simultaneously in a country outside the Union and in a country of the Union, the latter country;

(c) in the case of unpublished works or of works first published in a country outside the Union, without simultaneous publication in a country of the Union, the country of the Union of which the author is a national, provided that:

(i) when these are cinematographic works the maker of which has his headquarters or his habitual residence in a country of the Union, the country of origin shall be that country, and

(ii) when these are works of architecture erected in a country of the Union or other artistic works incorporated in a building or other structure located in a country of the Union, the country of origin shall be that country.

Article 6

(1) Where any country outside the Union fails to protect in an adequate manner the works of authors who are nationals of one of the countries of the Union, the latter country may restrict the protection given to the works of authors who are, at the date of the first publication thereof, nationals of the other country and are not habitually resident in one of the countries of the Union, If the country of first publication avails itself of this right, the other countries of the Union shall not be required to grant to works thus subjected to special treatment a wider protection than that granted to them in the country of first publication.

(2) No restrictions introduced by virtue of the preceding paragraph shall affect the rights which an author may have acquired in respect of a work published in a country of the Union before such restrictions were put into force.

(3) The countries of the Union which restrict the grant of copyright in accordance with this Article shall give notice thereof to the Director General of the World Intellectual Property Organization (hereinafter designated as "the Director General") by a written declaration specifying the countries in regard to which protection is restricted, and the restrictions to which rights of authors who are nationals of those countries are subjected. The Director General shall immediately communicate this declaration to all the countries of the Union.

Article 6*bis*

(1) Independently of the author's economic rights, and even after the transfer of the said rights, the author shall have the right to claim authorship of the work and to object to any distortion, mutilation or other modification of, or other

derogatory action in relation to the said work, which would be prejudicial to his honor or reputation.

(2) The rights granted to the author in accordance with the preceding paragraph shall, after his death, be maintained, at least until the expiry of the economic rights, and shall be exercisable by the persons or institutions authorized by the legislation of the country where protection is claimed. However, those countries whose legislation, at the moment of their ratification of or accession to this Act, does not provide for the protection after the death of the author of all the rights set out in the preceding paragraph may provide that some of these rights may, after his death, cease to be maintained.

(3) The means of redress for safeguarding the rights granted by this Article shall be governed by the legislation of the country where protection is claimed.

Article 7

(1) The term of protection granted by this Convention shall be the life of the author and fifty years after his death.

(2) However, in the case of cinematographic works, the countries of the Union may provide that the term of protection shall expire fifty years after the work has been made available to the public with the consent of the author, or, failing such an event within fifty years from the making of such a work, fifty years after the making.

(3) In the case of anonymous or pseudonymous works, the term of protection granted by this Convention shall expire fifty years after the work has been lawfully made available to the public. However, when the pseudonym adopted by the author leaves no doubt as to his identity, the term of protection shall be that provided in paragraph (1). If the author of an anonymous or pseudonymous work discloses his identity during the above-mentioned period, the term of protection applicable shall be that provided in paragraph (1). The countries of the Union shall not be required to protect anonymous or pseudonymous works in respect of which it is reasonable to presume that their author has been dead for fifty years.

(4) It shall be a matter for legislation in the countries of the Union to determine the term of protection of photographic works and that of works of applied art insofar as they are protected as artistic works; however, this term shall last at least until the end of a period of twenty-five years from the making of such a work.

(5) The term of protection subsequent to the death of the author and the terms provided by paragraphs (2), (3), and (4), shall run from the date of death or of the event referred to in those paragraphs, but such terms shall always be deemed to begin on the 1st of January of the year following the death or such event.

(6) The countries of the Union may grant a term of protection in excess of those provided by the preceding paragraphs.

(7) Those countries of the Union bound by the Rome Act of this Convention, which grant, in their national legislation in force at the time of signature of the

present Act, shorter terms of protection than those provided for in the preceding paragraphs, shall have the right to maintain such terms when ratifying or acceding to the present Act.

(8) In any case, the term shall be governed by the legislation of the country where protection is claimed; however, unless the legislation of that country otherwise provides, the term shall not exceed the term fixed in the country of origin of the work.

Article 7*bis*

The provisions of the preceding Article shall also apply in the case of a work of joint authorship, provided that the terms measured from the death of the author shall be calculated from the death of the last surviving author.

Article 8

Authors of literary and artistic works protected by this Convention shall enjoy the exclusive right of making and of authorizing the translation of their works throughout the term of protection of their rights in the original works.

Article 9

(1) Authors of literary and artistic works protected by this Convention shall have the exclusive right of authorizing the reproduction of these works, in any manner or form.

(2) It shall be a matter for legislation in the countries of the Union to permit the reproduction of such works in certain special cases, provided that such reproduction does not conflict with a normal exploitation of the work and does not unreasonably prejudice the legitimate interests of the author.

(3) Any sound or visual recording shall be considered as a reproduction for the purposes of this Convention.

Article 10

(1) It shall be permissible to make quotations from a work which has already been lawfully made available to the public, provided that their making is compatible with fair practice, and their extent does not exceed that justified by the purpose, including quotations from newspaper articles and periodicals in the form of press summaries.

(2) It shall be a matter for legislation in the countries of the Union, and for special agreements existing or to be concluded between them, to permit the utilization, to the extent justified by the purpose, of literary or artistic works by way of illustration in publications, broadcasts, or sound or visual recordings for teaching, provided such utilization is compatible with fair practice.

(3) Where use is made of works in accordance with the preceding paragraphs of this Article, mention shall be made of the source, and of the name of the author, if it appears thereon.

Article 10*bis*

(1) It shall be a matter for legislation in the countries of the Union to permit the reproduction by the press, the broadcasting or the communication to the public by wire, of articles published in newspapers or periodicals on current economic, political, or religious topics, and of broadcast works of the same character, in cases in which the reproduction, broadcasting or such communication thereof is not expressly reserved. Nevertheless, the source must always be clearly indicated; the legal consequences of a breach of this obligation shall be determined by the legislation of the country where protection is claimed.

(2) It shall also be a matter for legislation in the countries of the Union to determine the conditions under which, for the purpose of reporting current events by means of photography, cinematography, broadcasting, or communication to the public by wire, literary or artistic works seen or heard in the course of the event may, to the extent justified by the informatory purpose, be reproduced and made available to the public.

Article 11

(1) Authors of dramatic, dramatico-musical, and musical works shall enjoy the exclusive right of authorizing:

(i) the public performance of their works, including such public performance by any means or process;

(ii) any communication to the public of the performance of their works.

(2) Authors of dramatic or dramatico-musical works shall enjoy, during the full term of their rights in the original works, the same rights with respect to translations thereof.

Article 11*bis*

(1) Authors of literary and artistic works shall enjoy the exclusive right of authorizing:

(i) the broadcasting of their works or the communication thereof to the public by any other means of wireless diffusion of signs, sounds, or images;

(ii) any communication to the public by wire or by rebroadcasting of the broadcast of the work, when this communication is made by an organization other than the original one;

(iii) the public communication by loudspeaker or any other analogous instrument transmitting, by signs, sounds, or images, the broadcast of the work.

(2) It shall be a matter for legislation in the countries of the Union to determine the conditions under which the rights mentioned in the preceding paragraph may be exercised, but these conditions shall apply only in the countries where they have been prescribed. They shall not in any circumstances be prejudicial to the moral rights of the author, nor to his right to obtain equitable remuneration which, in the absence of agreement, shall be fixed by competent authority.

(3) In the absence of any contrary stipulation, permission granted in accordance with paragraph (1) of this Article shall not imply permission to record, by means of instruments recording sounds or images, the work broadcast. It shall, however, be a matter for legislation in the countries of the Union to determine the regulations for ephemeral recordings made by a broadcasting organization by means of its own facilities and used for its own broadcasts. The preservation of these recordings in official archives may, on the ground of their exceptional documentary character, be authorized by such legislation.

Article 11*ter*

(1) Authors of literary works shall enjoy the exclusive right of authorizing:

(i) the public recitation of their works, including such public recitation by any means or process;

(ii) any communication to the public of the recitation of their works.

(2) Authors of literary works shall enjoy, during the full term of their rights in the original works, the same rights with respect to translations thereof.

Article 12

Authors of literary or artistic works shall enjoy the exclusive right of authorizing adaptations, arrangements, and other alterations of their works.

Article 13

(1) Each country of the Union may impose for itself reservations and conditions on the exclusive right granted to the author of a musical work and to the author of any words, the recording of which together with the musical work has already been authorized by the latter, to authorize the sound recording of that musical work, together with such words, if any; but all such reservations and conditions shall apply only in the countries which have imposed them and shall not, in any circumstances, be prejudicial to the rights of these authors to obtain equitable remuneration which, in the absence of agreement, shall be fixed by competent authority.

(2) Recordings of musical works made in a country of the Union in accordance with Article 13 (3) of the Convention signed at Rome on June 2, 1928, and at Brussels on June 26, 1948, may be reproduced in that country without the

permission of the author of the musical work until a date two years after that country becomes bound by this Act.

(3) Recordings made in accordance with paragraphs (1) and (2) of this Article and imported without permission from the parties concerned into a country where they are treated as infringing recordings shall be liable to seizure.

Article 14

(1) Authors of literary or artistic works shall have the exclusive right of authorizing:

(i) the cinematographic adaptation and reproduction of these works, and the distribution of the works thus adapted or reproduced;

(ii) the public performance and communication to the public by wire of the works thus adapted or reproduced.

(2) The adaptation into any other artistic form of a cinematographic production derived from literary or artistic works shall, without prejudice to the authorization of the author of the cinematographic production, remain subject to the authorization of the authors of the original works.

(3) The provisions of Article 13(1) shall not apply.

Article 14*bis*

(1) Without prejudice to the copyright in any work which may have been adapted or reproduced, a cinematographic work shall be protected as an original work. The owner of copyright in a cinematographic work shall enjoy the same rights as the author of an original work, including the rights referred to in the preceding Article.

(2) (a) Ownership of copyright in a cinematographic work shall be a matter for legislation in the country where protection is claimed.

(b) However, in the countries of the Union which, by legislation, include among the owners of copyright in a cinematographic work authors who have brought contributions to the making of the work, such authors, if they have undertaken to bring such contributions, may not, in the absence of any contrary or special stipulation, object to the reproduction, distribution, public performance, communication to the public by wire, broadcasting or any other communication to the public, or to the subtitling or dubbing of texts, of the work.

(c) The question whether or not the form of the undertaking referred to above should, for the application of the preceding subparagraph (b), be in a written agreement or a written act of the same effect shall be a matter for the legislation of the country where the maker of the cinematographic work has his headquarters or habitual residence. However, it shall be a matter for the legislation of the country of the Union where protection is claimed to provide that the said undertaking shall be in a written agreement or a written act of the same effect. The countries whose legislation so provides shall notify the Director General by means of a written declaration, which will be immediately communicated by him to all the other countries of the Union.

(d) By "contrary or special stipulation" is meant any restrictive condition which is relevant to the aforesaid undertaking.

(3) Unless the national legislation provides to the contrary, the provisions of paragraph (2) (b) above shall not be applicable to authors of scenarios, dialogues, and musical works created for the making of the cinematographic work, nor to the principal director thereof. However, those countries of the Union whose legislation does not contain rules providing for the application of the said paragraph (2) (b) to such director shall notify the Director General by means of a written declaration, which will be immediately communicated by him to all the other countries of the Union.

Article 14*ter*

(1) The author, or after his death the persons or institutions authorized by national legislation, shall, with respect to original works of art and original manuscripts of writers and composers, enjoy the inalienable right to an interest in any sale of the work subsequent to the first transfer by the author of the work.

(2) The protection provided by the preceding paragraph may be claimed in a country of the Union only if legislation in the country to which the author belongs so permits, and to the extent permitted by the country where this protection is claimed.

(3) The procedure for collection and the amounts shall be matters for determination by national legislation.

Article 15

(1) In order that the author of a literary or artistic work protected by this Convention shall, in the absence of proof to the contrary, be regarded as such, and consequently be entitled to institute infringement proceedings in the countries of the Union, it shall be sufficient for his name to appear on the work in the usual manner. This paragraph shall be applicable even if this name is a pseudonym, where the pseudonym adopted by the author leaves no doubt as to his identity.

(2) The person or body corporate whose name appears on a cinematographic work in the usual manner shall, in the absence of proof to the contrary, be presumed to be the maker of said work.

(3) In the case of anonymous and pseudonymous works, other than those referred to in paragraph (1) above, the publisher whose name appears on the work shall, in the absence of proof to the contrary, be deemed to represent the author, and in this capacity be shall be entitled to protect and enforce the author's rights. The provisions of this paragraph shall cease to apply when the author reveals his identity and establishes his claim to authorship of the work.

(4)(a) In the case of unpublished works where the identity of the author is unknown, but where there is every ground to presume that he is a national of a country of the Union, it shall be a matter for legislation in that country to designate the competent authority who shall represent the author and shall be entitled to protect and enforce his rights in the countries of the Union.

(b) Countries of the Union which make such designation under the terms of this provision shall notify the Director General by means of a written declaration giving full information concerning the authority thus designated. The Director General shall at once communicate this declaration to all other countries of the Union.

Article 16

(1) Infringing copies of a work shall be liable to seizure in any country of the Union where the work enjoys legal protection.

(2) The provisions of the preceding paragraph shall also apply to reproductions coming from a country where the work is not protected, or has ceased to be protected.

(3) The seizure shall take place in accordance with the legislation of each country.

Article 17

The provisions of this Convention cannot in any way affect the right of the Government of each country of the Union to permit, to control, or to prohibit by legislation or regulation, the circulation, presentation, or exhibition of any work or production in regard to which the competent authority may find it necessary to exercise that right.

Article 18

(1) This Convention shall apply to all works which, at the moment of its coming into force, have not yet fallen into the public domain in the country of origin through the expiry of the term of protection.

(2) If, however, through the expiry of the term of protection which was previously granted, a work has fallen into the public domain of the country where protection is claimed, that work shall not be protected anew.

(3) The application of this principle shall be subject to any provisions contained in special conventions to that effect existing or to be concluded between countries of the Union. In the absence of such provisions, the respective countries shall determine, each insofar as it is concerned, the conditions of application of this principle.

(4) The preceding provisions shall also apply in the case of new accessions to the Union and to cases in which protection is extended by the application of Article 7 or by the abandonment of reservations.

Article 19

The provisions of this Convention shall not preclude the making of a claim to the benefit of any greater protection which may be granted by legislation in a country of the Union.

Article 20

The Governments of the countries of the Union reserve the right to enter into special agreements among themselves, insofar as such agreements grant to authors more extensive rights than those granted by the Convention, or contain other provisions not contrary to this Convention. The provisions of existing agreements which satisfy these conditions shall remain applicable.

Article 21

(1) Special provisions regarding developing countries are included in the Appendix.

(2) Subject to the provisions of Article 28(1)(b), the Appendix forms an integral part of this Act.

Article 22

(1)(a) The Union shall have an Assembly consisting of those countries of the Union which are bound by Articles 22 to 26.

(b) The Government of each country shall be represented by one delegate, who may be assisted by alternate delegates, advisors, and experts.

(c) The expenses of each delegation shall be borne by the Government which has appointed it.

(2)(a) The Assembly shall:

(i) deal with all matters concerning the maintenance and development of the Union and the implementation of this Convention;

(ii) give directions concerning the preparation for conferences of revision to the International Bureau of Intellectual Property (hereinafter designated as "the International Bureau") referred to in the Convention establishing the World Intellectual Property Organization (hereinafter designated as "the Organization"), due account being taken of any comments made by those countries of the Union which are not bound by Articles 22 to 26;

(iii) review and approve the reports and activities of the Director General of the Organization concerning the Union, and give him all necessary instructions concerning matters within the competence of the Union;

(iv) elect the members of the Executive Committee of the Assembly;

(v) review and approve the reports and activities of its Executive Committee, and give instructions to such Committee;

(vi) determine the program and adopt the triennial budget of the Union, and approve its final accounts;

(vii) adopt the financial regulations of the Union;

(viii) establish such committees of experts and working groups as may be necessary for the work of the Union;

(ix) determine which countries not members of the Union and which intergovernmental and international non-governmental organizations shall be admitted to its meetings as observers;

(x) adopt amendments to Articles 22 to 26;

(xi) take any other appropriate action designed to further the objectives of the Union;

(xii) exercise such other functions as are appropriate under this Convention;

(xiii) subject to its acceptance, exercise such rights as are given to it in the Convention establishing the Organization.

(b) With respect to matters which are of interest also to other Unions administered by the Organization, the Assembly shall make its decisions after having heard the advice of the Coordination Committee of the Organization.

(3)(a) Each country member of the Assembly shall have one vote.

(b) One-half of the countries members of the Assembly shall constitute a quorum.

(c) Notwithstanding the provisions of subparagraph (b), if, in any session, the number of countries represented is less than one-half but equal to or more than one-third of the countries members of the Assembly, the Assembly may make decisions but, with the exception of decisions concerning its own procedure, all such decisions shall take effect only if the following conditions are fulfilled. The International Bureau shall communicate said decisions to the countries members of the Assembly which were not represented and shall invite them to express in writing their vote or abstention within a period of three months from the date of the communication. If, at the expiration of this period, the number of countries having thus expressed their vote or abstention attains the number of countries which was lacking for attaining the quorum in the session itself, such decisions shall take effect provided that at the same time the required majority still obtains.

(d) Subject to the provisions of Article 26 (2), the decisions of the Assembly shall require two-thirds of the votes cast.

(e) Abstentions shall not be considered as votes.

(f) A delegate may represent, and vote in the name of, one country only.

(g) Countries of the Union not members of the Assembly shall be admitted to its meetings as observers.

(4)(a) The Assembly shall meet once in every third calendar year in ordinary session upon convocation by the Director General and, in the absence of exceptional circumstances, during the same period and at the same place as the General Assembly of the Organization.

(b) The Assembly shall meet in extraordinary session upon convocation by the Director General, at the request of the Executive Committee or at the request of one-fourth of the countries members of the Assembly.

(5) The Assembly shall adopt its own rules of procedure.

Article 23

(1) The Assembly shall have an Executive Committee.

(2)(a) The Executive Committee shall consist of countries elected by the Assembly from among countries members of the Assembly. Furthermore, the country on whose territory the Organization has its headquarters shall, subject to the provisions of Article 25 (7) (b), have an ex officio seat on the Committee.

(b) The Government of each country member of the Executive Committee shall be represented by one delegate, who may be assisted by alternate delegates, advisors, and experts.

(c) The expenses of each delegation shall be borne by the Government which has appointed it.

(3) The number of countries members of the Executive Committee shall correspond to one-fourth of the number of countries members of the Assembly. In establishing the number of seats to be filled, remainders after division by four shall be disregarded.

(4) In electing the members of the Executive Committee, the Assembly shall have due regard to an equitable geographical distribution and to the need for countries party to the Special Agreements which might be established in relation with the Union to be among the countries constituting the Executive Committee.

(5)(a) Each member of the Executive Committee shall serve from the close of the session of the Assembly which elected it to the close of the next ordinary session of the Assembly.

(b) Members of the Executive Committee may be re-elected, but not more than two-thirds of them.

(c) The Assembly shall establish the details of the rules governing the election and possible re-election of the members of the Executive Committee.

(6)(a) The Executive Committee shall:

(i) prepare the draft agenda of the Assembly;

(ii) submit proposals to the Assembly respecting the draft program and triennial budget of the Union, prepared by the Director General;

(iii) approve, within the limits of the program and the triennial budget, the specific yearly budgets and programs prepared by the Director General;

(iv) submit, with appropriate comments, to the Assembly the periodical reports of the Director General and the yearly audit reports on the accounts;

(v) in accordance with the decisions of the Assembly and having regard to circumstances arising between two ordinary sessions of the Assembly, take all necessary measures to ensure the execution of the program of the Union by the Director General;

(vi) perform such other functions as are allocated to it under this Convention.

(b) With respect to matters which are of interest also to other Unions administered by the Organization, the Executive Committee shall make its decisions after having heard the advice of the Coordination Committee of the Organization.

(7)(a) The Executive Committee shall meet once a year in ordinary session upon convocation by the Director General, preferably during the same period and at the same place as the Coordination Committee of the Organization.

(b) The Executive Committee shall meet in extraordinary session upon convocation by the Director General, either on his own initiative, or at the request of its Chairman or one fourth of its members.

(8)(a) Each country member of the Executive Committee shall have one vote.

(b) One half of the members of the Executive Committee shall constitute a quorum.

(c) Decisions shall be made by a simple majority of the votes cast.

(d) Abstentions shall not be considered as votes.

(e) A delegate may represent, and vote in the name of, one country only.

(9) Countries of the Union not members of the Executive Committee shall be admitted to its meetings as observers.

(10) The Executive Committee shall adopt its own rules of procedure.

Article 24

(1)(a) The administrative tasks with respect to the Union shall be performed by the International Bureau, which is a continuation of the Bureau of the Union united with the Bureau of the Union established by the International Convention for the Protection of Industrial Property.

(b) In particular, the International Bureau shall provide the secretariat of the various organs of the Union.

(c) The Director General of the Organization shall be the chief executive of the Union and shall represent the Union.

(2) The International Bureau shall assemble and publish information concerning the protection of copyright. Each country of the Union shall promptly communicate to the International Bureau all new laws and official texts concerning the protection of copyright.

(3) The International Bureau shall publish a monthly periodical.

(4) The International Bureau shall, on request, furnish information to any country of the Union on matters concerning the protection of copyright.

(5) The International Bureau shall conduct studies, and shall provide services, designed to facilitate the protection of copyright.

(6) The Director General and any staff member designated by him shall participate, without the right to vote, in all meetings of the Assembly, the Executive Committee, and any other committee of experts or working group. The Director

General, or a staff member designated by him, shall be ex officio secretary of these bodies.

(7)(a) The International Bureau shall, in accordance with the directions of the Assembly and in cooperation with the Executive Committee, make the preparations for the conferences of revision of the provisions of the Convention other than Articles 22 to 26.

(b) The International Bureau may consult with intergovernmental and international non-governmental organizations concerning preparations for conferences of revision.

(c) The Director General and persons designated by him shall take part, without the right to vote, in the discussions at these conferences.

(8) The International Bureau shall carry out any other tasks assigned to it.

Article 25

(1)(a) The Union shall have a budget.

(b) The budget of the Union shall include the income and expenses proper to the Union, its contribution to the budget of expenses common to the Unions, and, where applicable, the sum made available to the budget of the Conference of the Organization.

(c) Expenses not attributable exclusively to the Union but also to one or more other Unions administered by the Organization shall be considered as expenses common to the Unions. The share of the Union in such common expenses shall be in proportion to the interest the Union has in them.

(2) The budget of the Union shall be established with due regard to the requirements of coordination with the budgets of the other Unions administered by the Organization.

(3) The budget of the Union shall be financed from the following sources: (i) contributions of the countries of the Union; (ii) fees and charges due for services performed by the International Bureau in relation to the Union; (iii) sale of, or royalties on, the publications of the International Bureau concerning the Union; (iv) gifts, bequests, and subventions; (v) rents, interests, and other miscellaneous income.

(4)(a) For the purpose of establishing its contribution towards the budget, each country of the Union shall belong to a class, and shall pay its annual contributions on the basis of a number of units fixed as follows:

Class I 25
Class II 20
Class III 15
Class IV 10
Class V 5
Class VI 3
Class VII 1

(b) Unless it has already done so, each country shall indicate, concurrently with depositing its instrument of ratification or accession, the class to which it wishes to belong. Any country may change class. If it chooses a lower class, the country must announce it to the Assembly at one of its ordinary sessions. Any such change shall take effect at the beginning of the calendar year following the session.

(c) The annual contribution of each country shall be an amount in the same proportion to the total sum to be contributed to the annual budget of the Union by all countries as the number of its units is to the total of the units of all contributing countries.

(d) Contributions shall become due on the first of January of each year.

(e) A country which is in arrears in the payment of its contributions shall have no vote in any of the organs of the Union of which it is a member if the amount of its arrears equals or exceeds the amount of the contributions due from it for the preceding two full years. However, any organ of the Union may allow such a country to continue to exercise its vote in that organ if, and as long as, it is satisfied that the delay in payment is due to exceptional and unavoidable circumstances.

(f) If the budget is not adopted before the beginning of a new financial period, it shall be at the same level as the budget of the previous year, in accordance with the financial regulations.

(5) The amount of the fees and charges due for services rendered by the International Bureau in relation to the Union shall be established, and shall be reported to the Assembly and the Executive Committee, by the Director General.

(6)(a) The Union shall have a working capital fund which shall be constituted by a single payment made by each country of the Union. If the fund becomes insufficient, an increase shall be decided by the Assembly.

(b) The amount of the initial payment of each country to the said fund or of its participation in the increase thereof shall be a proportion of the contribution of that country for the year in which the fund is established or the increase decided.

(c) The proportion and the terms of payment shall be fixed by the Assembly on the proposal of the Director General and after it has heard the advice of the Coordination Committee of the Organization.

(7)(a) In the headquarters agreement concluded with the country on the territory of which the Organization has its headquarters, it shall be provided that, whenever the working capital fund is insufficient, such country shall grant advances. The amount of these advances and the conditions on which they are granted shall be the subject of separate agreements, in each case, between such country and the Organization. As long as it remains under the obligation to grant advances, such country shall have an ex officio seat on the Executive Committee.

(b) The country referred to in subparagraph (a) and the Organization shall each have the right to denounce the obligation to grant advances, by written notification. Denunciation shall take effect three years after the end of the year in which it has been notified.

(8) The auditing of the accounts shall be effected by one or more of the countries of the Union or by external auditors, as provided in the financial regulations. They shall be designated, with their agreement, by the Assembly.

Article 26

(1) Proposals for the amendment of Articles 22, 23, 24, 25, and the present Article, may be initiated by any country member of the Assembly, by the Executive Committee, or by the Director General. Such proposals shall be communicated by the Director General to the member countries of the Assembly at least six months in advance of their consideration by the Assembly.

(2) Amendments to the Articles referred to in paragraph (1) shall be adopted by the Assembly. Adoption shall require three-fourths of the votes cast, provided that any amendment of Article 22, and of the present paragraph, shall require four fifths of the votes cast.

(3) Any amendment to the Articles referred to in paragraph (1) shall enter into force one month after written notifications of acceptance, effected in accordance with their respective constitutional processes, have been received by the Director General from three fourths of the countries members of the Assembly at the time it adopted the amendment. Any amendment to the said Articles thus accepted shall bind all the countries which are members of the Assembly at the time the amendment enters into force, or which become members thereof at a subsequent date, provided that any amendment increasing the financial obligations of countries of the Union shall bind only those countries which have notified their acceptance of such amendment.

Article 27

(1) This Convention shall be submitted to revision with a view to the introduction of amendments designed to improve the system of the Union.

(2) For this purpose, conferences shall be held successively in one of the countries of the Union among the delegates of the said countries.

(3) Subject to the provisions of Article 26 which apply to the amendment of Articles 22 to 26, any revision of this Act, including the Appendix, shall require the unanimity of the votes cast.

Article 28

(1)(a) Any country of the Union which has signed this Act may ratify it, and, if it has not signed it, may accede to it. Instruments of ratification or accession shall be deposited with the Director General.

(b) Any country of the Union may declare in its instrument of ratification or accession that its ratification or accession shall not apply to Articles 1 to 21 and the Appendix, provided that, if such country has previously made a declaration under Article VI(1) of the Appendix, then it may declare in the said instrument only that its ratification or accession shall not apply to Articles 1 to 20.

(c) Any country of the Union which, in accordance with sub-paragraph (b), has excluded provisions therein referred to from the effects of its ratification or accession may at any later time declare that it extends the effects of its ratification or accession to those provisions. Such declaration shall be deposited with the Director General.

(2)(a) Articles 1 to 21 and the Appendix shall enter into force three months after both of the following two conditions are fulfilled:

(i) at least five countries of the Union have ratified or acceded to this Act without making a declaration under paragraph (1)(b).

(ii) France, Spain, the United Kingdom of Great Britain and Northern Ireland, and the United States of America, have become bound by the Universal Copyright Convention as revised at Paris on July 24,1971.

(b) The entry into force referred to in sub-paragraph (a) shall apply to those countries of the Union which, at least three months before the said entry into force, have deposited instruments of ratification or accession not containing a declaration under paragraph (1)(b).

(c) With respect to any country of the Union not covered by sub-paragraph (b) and which ratifies or accedes to this Act without making a declaration under paragraph (1) (b), Articles 1 to 21 and the Appendix shall enter into force three months after the date on which the Director General has notified the deposit of the relevant instrument of ratification or accession, unless a subsequent date has been indicated in the instrument deposited. In the latter case, Articles 1 to 21 and the Appendix shall enter into force with respect to that country on the date thus indicated.

(d) The provisions of sub-paragraphs (a) to (c) do not affect the application of Article VI of the Appendix.

(3) With respect to any country of the Union which ratifies or accedes to this Act with or without a declaration made under paragraph (1)(b), Articles 22 to 38 shall enter into force three months after the date on which the Director General has notified the deposit of the relevant instrument of ratification or accession, unless a subsequent date has been indicated in the instrument deposited. In the latter case, Articles 22 to 38 shall enter into force with respect to that country on the date thus indicated.

Article 29

(1) Any country outside the Union may accede to this Act and thereby become party to this Convention and a member of the Union. Instruments of accession shall be deposited with the Director General.

(2)(a) Subject to sub-paragraph (b), this Convention shall enter into force with respect to any country outside the Union three months after the date on which the Director General has notified the deposit of its instrument of accession, unless a subsequent date has been indicated in the instrument deposited. In the latter case, this Convention shall enter into force with respect to that country on the date thus indicated.

(b) If the entry into force according to sub-paragraph (a) precedes the entry into force of Articles 1 to 21 and the Appendix according to Article 28(2)(a), the said country shall, in the meantime, be bound, instead of by Articles 1 to 21 and the Appendix, by Articles 1 to 20 of the Brussels Act of this Convention.

Article 29*bis*

Ratification of or accession to this Act by any country not bound by Articles 22 to 38 of the Stockholm Act of this Convention shall, for the sole purposes of Article 14(2) of the Convention establishing the Organization, amount to ratification of or accession to the said Stockholm Act with the limitation set forth in Article 28(1)(b)(i) thereof.

Article 30

(1) Subject to the exceptions permitted by paragraph (2) of this article, by Article 28(1)(b), by Article 33(2), and by the Appendix, ratification or accession shall automatically entail acceptance of all the provisions and admission to all the advantages of this Convention.

(2)(a) Any country of the Union ratifying or acceding to this Act may, subject to Article V(2) of the Appendix, retain the benefit of the reservations it has previously formulated on condition that it makes a declaration to that effect at the time of the deposit of its instrument of ratification or accession.

(b) Any country outside the Union may declare, in acceding to this Convention and subject to Article V(2) of the Appendix, that it intends to substitute, temporarily at least, for Article 8 of this Act concerning the right of translation, the provisions of Article 5 of the Union Convention of 1886, as completed at Paris in 1896, on the clear understanding that the said provisions are applicable only to translations into a language in general use in the said country. Subject to Article I(6)(b) of the Appendix, any country has the right to apply, in relation to the right of translation of works whose country of origin is a country availing itself of such a reservation, a protection which is equivalent to the protection granted by the latter country.

(c) Any country may withdraw such reservations at any time by notification addressed to the Director General.

Article 31

(1) Any country may declare in its instrument of ratification or accession, or may inform the Director General by written notification at any time thereafter, that this Convention shall be applicable to all or part of those territories, designated in the declaration or notification, for the external relations of which it is responsible.

(2) Any country which has made such a declaration or given such a notification may, at any time, notify the Director General that this Convention shall cease to be applicable to all or part of such territories.

(3)(a) Any declaration made under paragraph (1) shall take effect on the same date as the ratification or accession in which it was included, and any notification given under that paragraph shall take effect three months after its notification by the Director General.

(b) Any notification given under paragraph (2) shall take effect twelve months after its receipt by the Director General.

(4) This article shall in no way be understood as implying the recognition or tacit acceptance by a country of the Union of the factual situation concerning a territory to which this Convention is made applicable by another country of the Union by virtue of a declaration under paragraph (1).

Article 32

(1) This Act shall, as regards relations between the countries of the Union, and to the extent that it applies, replace the Berne Convention of September 9, 1886, and the subsequent Acts of revision. The Acts previously in force shall continue to be applicable, in their entirety or to the extent that this Act does not replace them by virtue of the preceding sentence, in relations with countries of the Union which do not ratify or accede to this Act.

(2) Countries outside the Union which become party to this Act shall, subject to paragraph (3), apply it with respect to any country of the Union not bound by this Act or which, although bound by this Act, has made a declaration pursuant to Article 28(1)(b). Such countries recognize that the said country of the Union, in its relations with them:

(i) may apply the provisions of the most recent Act by which it is bound; and

(ii) subject to Article I(6) of the Appendix, has the right to adapt the protection to the level provided for by this Act.

(3) Any country which has availed itself of any of the faculties provided for in the Appendix may apply the provisions of the Appendix relating to the faculty or faculties of which it has availed itself in its relations with any other country of the Union which is not bound by this Act, provided that the latter country has accepted the application of the said provisions.

Article 33

(1) Any dispute between two or more countries of the Union concerning the interpretation or application of this Convention, not settled by negotiation, may, by any one of the countries concerned, be brought before the International Court of Justice by application in conformity with the Statute of the Court, unless the countries concerned agree on some other method of settlement. The country bringing the dispute before the Court shall inform the International Bureau; the International Bureau shall bring the matter to the attention of the other countries of the Union.

(2) Each country may, at the time it signs this Act or deposits its instrument of ratification or accession, declare that it does not consider itself bound by the provisions of paragraph (1). With regard to any dispute between such country and any other country of the Union, the provisions of paragraph (1) shall not apply.

(3) Any country having made a declaration in accordance with the provisions of paragraph (2) may, at any time, withdraw its declaration by notification addressed to the Director General.

Article 34

(1) Subject to Article 29*bis*, no country may ratify or accede to earlier Acts of this Convention once Articles 1 to 21 and the Appendix have entered into force.

(2) Once Articles 1 to 21 and the Appendix have entered into force, no country may make a declaration under Article 5 of the Protocol Regarding Developing Countries attached to the Stockholm Act.

Article 35

(1) This Convention shall remain in force without limitation as to time.

(2) Any country may denounce this Act by notification addressed to the Director General. Such denunciation shall constitute also denunciation of all earlier Acts and shall affect only the country making it, the Convention remaining in full force and effect as regards the other countries of the Union.

(3) Denunciation shall take effect one year after the day on which the Director General has received the notification.

(4) The right of denunciation provided by this article shall not be exercised by any country before the expiration of five years from the date upon which it becomes a member of the Union.

Article 36

(1) Any country party to this Convention undertakes to adopt, in accordance with its constitution, the measures necessary to ensure the application of this Convention.

(2) It is understood that, at the time a country becomes bound by this Convention, it will be in a position under its domestic law to give effect to the provisions of this Convention.

Article 37

(1)(a) This Act shall be signed in a single copy in the French and English languages and, subject to paragraph (2), shall be deposited with the Director General.

(b) Official texts shall be established by the Director General, after consultation with the interested Governments, in the Arabic, German, Italian, Portuguese, and Spanish languages, and such other languages as the Assembly may designate.

(c) In case of differences of opinion on the interpretation of the various texts, the French text shall prevail.

(2) This Act shall remain open for signature until January 31,1972. Until that date, the copy referred to in paragraph (1)(a) shall be deposited with the Government of the French Republic.

(3) The Director General shall certify and transmit two copies of the signed text of this Act to the Governments of all countries of the Union and, on request, to the Government of any other country.

(4) The Director General shall register this Act with the Secretariat of the United Nations.

(5) The Director General shall notify the Governments of all countries of the Union of signatures, deposits of instruments of ratification or accession and any declarations included in such instruments or made pursuant to Articles 28(1)(c), 30(2)(a) and (b), and 33(2), entry into force of any provisions of this Act, notifications of denunciation, and notifications pursuant to Articles 30(2)(c), 31(1) and (2), 33(3), and 38(1), as well as the Appendix.

Article 38

(1) Countries of the Union which have not ratified or acceded to this Act and which are not bound by Articles 22 to 26 of the Stockholm Act of this Convention may, until April 26, 1975, exercise, if they so desire, the rights provided under the said articles as if they were bound by them. Any country desiring to exercise such rights shall give written notification to this effect to the Director General; this notification shall be effective on the date of its receipt. Such countries shall be deemed to be members of the Assembly until the said date.

(2) As long as all the countries of the Union have not become Members of the Organization, the International Bureau of the Organization shall also function as the Bureau of the Union, and the Director General as the Director of the said Bureau.

(3) Once all the countries of the Union have become Members of the Organization, the rights, obligations, and property of the Bureau of the Union shall devolve on the International Bureau of the Organization.

Appendix

Article I

(1) Any country regarded as a developing country in conformity with the established practice of the General Assembly of the United Nations which ratifies or accedes to this Act, of which this Appendix forms an integral part, and which, having regard to its economic situation and its social or cultural needs, does not consider itself immediately in a position to make provision for the protection of all the rights as provided for in this Act, may, by a notification deposited with the Director General at the time of depositing its instrument of ratification or

accession or, subject to Article V(1)(c), at any time thereafter, declare that it will avail itself of the faculty provided for in Article II, or of the faculty provided for in Article III, or of both of those faculties. It may, instead of availing itself of the faculty provided for in Article II, make a declaration according to Article V(1)(a).

(2)(a) Any declaration under paragraph (1) notified before the expiration of the period of ten years from the entry into force of Articles 1 to 21 and this Appendix according to Article 28(2) shall be effective until the expiration of the said period. Any such declaration may be renewed in whole or in part for periods of ten years each by a notification deposited with the Director General not more than 15 months and not less than three months before the expiration of the ten-year period then running.

(b) Any declaration under paragraph (1) notified after the expiration of the period of ten years from the entry into force of Articles 1 to 21 and this Appendix according to Article 28(2) shall be effective until the expiration of the ten-year period then running. Any such declaration may be renewed as provided for in the second sentence of sub-paragraph (a).

(3) Any country of the Union which has ceased to be regarded as a developing country as referred to in paragraph (1) shall no longer be entitled to renew its declaration as provided in paragraph (2), and, whether or not it formally withdraws its declaration, such country shall be precluded from availing itself of the faculties referred to in paragraph (1) from the expiration of the ten-year period then running or from the expiration of a period of three years after it has ceased to be regarded as a developing country, whichever period expires later.

(4) Where, at the time when the declaration made under paragraph (1) or (2) ceases to be effective, there are copies in stock which were made under a license granted by virtue of this Appendix, such copies may continue to be distributed until their stock is exhausted.

(5) Any country which is bound by the provisions of this Act and which has deposited a declaration or a notification in accordance with Article 31(1) with respect to the application of this Act to a particular territory, the situation of which can be regarded as analogous to that of the countries referred to in paragraph (1), may, in respect of such territory, make the declaration referred to in paragraph (1) and the notification of renewal referred to in paragraph (2). As long as such declaration or notification remains in effect, the provisions of this Appendix shall be applicable to the territory in respect of which it was made.

(6)(a) The fact that a country avails itself of any of the faculties referred to in paragraph (1) does not permit another country to give less protection to works of which the country of origin is the former country than it is obliged to grant under Articles 1 to 20.

(b) The right to apply reciprocal treatment provided for in Article 30(2)(b), second sentence, shall not, until the date on which the period applicable under Article I(3) expires, be exercised in respect of works the country of origin of which is a country which has made a declaration according to Article V(1)(a).

Article II

(1) Any country which has declared that it will avail itself of the faculty provided for in this Article shall be entitled, so far as works published in printed or analogous forms of reproduction are concerned, to substitute for the exclusive right of translation provided for in Article 8 a system of non-exclusive and non-transferable licenses, granted by the competent authority under the following conditions and subject to Article IV.

(2)(a) Subject to paragraph (3), if, after the expiration of a period of three years, or of any longer period determined by the national legislation of the said country, commencing on the date of the first publication of the work, a translation of such work has not been published in a language in general use in that country by the owner of the right of translation, or with his authorization, any national of such country may obtain a license to make a translation of the work in the said language and publish the translation in printed or analogous forms of reproduction.

(b) A license under the conditions provided for in this Article may also be granted if all the editions of the translation published in the language concerned are out of print.

(3)(a) In the case of translations into a language which is not in general use in one or more developed countries which are members of the Union, a period of one year shall be substituted for the period of three years referred to in paragraph (2)(a).

(b) Any country referred to in paragraph (1) may, with the unanimous agreement of the developed countries which are members of the Union and in which the same language is in general use, substitute, in the case of translations into that language, for the period of three years referred to in paragraph (2)(a), a shorter period as determined by such agreement but not less than one year. However, the provisions of the foregoing sentence shall not apply where the language in question is English, French, or Spanish. The Director General shall be notified of any such agreement by the Governments which have concluded it.

(4)(a) No license obtainable after three years shall be granted under this Article until a further period of six months has elapsed, and no license obtainable after one year shall be granted under this Article until a further period of nine months has elapsed

(i) from the date on which the applicant complies with the requirements mentioned in Article IV(1); or

(ii) where the identity or the address of the owner of the right of translation is unknown, from the date on which the applicant sends, as provided for in Article IV(2), copies of his application submitted to the authority competent to grant the license.

(b) If, during the said period of six or nine months, a translation in the language in respect of which the application was made is published by the owner of the right of translation or with his authorization, no license under this Article shall be granted.

(5) Any license under this Article shall be granted only for the purpose of teaching, scholarship, or research.

(6) If a translation of a work is published by the owner of the right of translation or with his authorization at a price reasonably related to that normally charged in the country for comparable works, any license granted under this Article shall terminate if such translation is in the same language and with substantially the same content as the translation published under the license. Any copies already made before the license terminated may continue to be distributed until their stock is exhausted.

(7) For works which are composed mainly of illustrations, a license to make and publish a translation of the text and to reproduce and publish the illustrations may be granted only if the conditions of Article III are also fulfilled.

(8) No license shall be granted under this Article when the author has withdrawn from circulation all copies of his work.

(9)(a) A license to make a translation of a work which has been published in printed or analogous forms of reproduction may also be granted to any broadcasting organization having its headquarters in a country referred to in paragraph (1), upon an application made to the competent authority of that country by the said organization, provided that all of the following conditions are met:

(i) the translation is made from a copy made and acquired in accordance with the laws of said country;

(ii) the translation is only for use in broadcasts intended exclusively for teaching or for the dissemination of the results of specialized technical or scientific research to experts in a particular profession;

(iii) the translation is used exclusively for the purposes referred to in condition (ii) through broadcasts made lawfully and intended for recipients on the territory of said country, including broadcasts made through the medium of sound or visual recordings lawfully and exclusively made for the purpose of such broadcasts;

(iv) all uses made of the translation are without any commercial purpose.

(b) Sound or visual recordings of a translation which were made by a broadcasting organization under a license granted by virtue of this paragraph may, for the purposes and subject to the conditions referred to in subparagraph (a) and with the agreement of that organization, also be used by any other broadcasting organization having its headquarters in the country whose competent authority granted the license in question.

(c) Provided that all of the criteria and conditions set out in subparagraph (a) are met, a license may also be granted to a broadcasting organization to translate any text incorporated in an audio-visual fixation where such fixation was itself prepared and published for the sole purpose of being used in connection with systematic instructional activities.

(d) Subject to subparagraphs (a) to (c), the provisions of the preceding paragraphs shall apply to the grant and exercise of any license granted under this paragraph.

Article III

(1) Any country which has declared that it will avail itself of the faculty provided for in this Article shall be entitled to substitute for the exclusive right

of reproduction provided for in Article 9 a system of non-exclusive and non-transferable licenses, granted by the competent authority under the following conditions and subject to Article IV.

(2)(a) If, in relation to a work to which this article applies by virtue of paragraph (7), after the expiration of

(i) the relevant period specified in paragraph (3), commencing on the date of first publication of a particular edition of the work, or

(ii) any longer period determined by national legislation of the country referred to in paragraph (1), commencing on the same date,

copies of such edition have not been distributed in that country to the general public or in connection with systematic instructional activities, by the owner of the right of reproduction or with his authorization, at a price reasonably related to that normally charged in the country for comparable works, any national of such country may obtain a license to reproduce and publish such edition at that or a lower price for use in connection with systematic instructional activities.

(b) A license to reproduce and publish an edition which has been distributed as described in sub-paragraph (a) may also be granted under the conditions provided for in this article if, after the expiration of the applicable period, no authorized copies of that edition have been on sale for a period of six months in the country concerned to the general public or in connection with systematic instructional activities at a price reasonably related to that normally charged in the country for comparable works.

(3) The period referred to in paragraph (2)(a)(i) shall be five years, except that (i) for works of the natural and physical sciences, including mathematics, and of technology, the period shall be three years; (ii) for works of fiction, poetry, drama and music, and for art books, the period shall be seven years.

(4)(a) No license obtainable after three years shall be granted under this article until a period of six months has elapsed

(i) from the date on which the applicant complies with the requirements mentioned in Article IV(1), or

(ii) where the identity or the address of the owner of the right of reproduction is unknown, from the date on which the applicant sends, as provided for in Article IV(2), copies of his application submitted to the authority competent to grant the license.

(b) Where licenses are obtainable after other periods and Article IV(2) is applicable, no license shall be granted until a period of three months has elapsed from the date of the dispatch of the copies of the application.

(c) If, during the period of six or three months referred to in sub-paragraphs (a) and (b), a distribution as described in paragraph (2)(a) has taken place, no license shall be granted under this article.

(d) No license shall be granted if the author has withdrawn from circulation all copies of the edition for the reproduction and publication of which the license has been applied for.

(5) A license to reproduce and publish a translation of a work shall not be granted under this article in the following cases:

(i) where the translation was not published by the owner of the right of translation or with his authorization, or

(ii) where the translation is not in a language in general use in the country in which the license is applied for.

(6) If copies of an edition of a work are distributed in the country referred to in paragraph (1) to the general public or in connection with systematic instructional activities, by the owner of the right of reproduction or with his authorization, at a price reasonably related to that normally charged in the country for comparable works, any license granted under this article shall terminate if such edition is in the same language and with substantially the same content as the edition which was published under the said license. Any copies already made before the license terminates may continue to be distributed until their stock is exhausted.

(7)(a) Subject to sub-paragraph (b), the works to which this article applies shall be limited to works published in printed or analogous forms of reproduction.

(b) This article shall also apply to the reproduction in audio-visual form of lawfully made audio-visual fixations including any protected works incorporated therein and to the translation of any incorporated text into a language in general use in the country in which the license is applied for, always provided that the audio-visual fixations in question were prepared and published for the sole purpose of being used in connection with systematic instructional activities.

Article IV

(1) A license under Article II or Article III may be granted only if the applicant, in accordance with the procedure of the country concerned, establishes either that he has requested, and has been denied, authorization by the owner of the right to make and publish the translation or to reproduce and publish the edition, as the case may be, or that, after due diligence on his part, he was unable to find the owner of the right. At the same time as making the request, the applicant shall inform any national or international information center referred to in paragraph (2).

(2) If the owner of the right cannot be found, the applicant for a license shall send, by registered airmail, copies of his application, submitted to the authority competent to grant the license, to the publisher whose name appears on the work and to any national or international information center which may have been designated, in a notification to that effect deposited with the Director General, by the Government of the country in which the publisher is believed to have his principal place of business.

(3) The name of the author shall be indicated on all copies of the translation or reproduction published under a license granted under Article II or Article III. The title of the work shall appear on all such copies. In the case of a translation, the original title of the work shall appear in any case on all said copies.

(4)(a) No license granted under Article II or Article III shall extend to the export of copies, and any such license shall be valid only for publication of the translation or of the reproduction, as the case may be, in the territory of the country in which it has been applied for.

(b) For the purposes of sub-paragraph (a), the notion of export shall include the sending of copies from any territory to the country which, in respect of that territory, has made a declaration under Article I(5).

(c) Where a governmental or other public entity of a country which has granted a license to make a translation under Article II into a language other than English, French, or Spanish sends copies of a translation published under such license to another country, such sending of copies shall not, for the purposes of sub-paragraph (a), be considered to constitute export if all of the following conditions are met:

(i) the recipients are individuals who are nationals of the country whose competent authority has granted the license, or organizations grouping such individuals;

(ii) the copies are to be used only for the purpose of teaching, scholarship, or research;

(iii) the sending of the copies and their subsequent distribution to recipients is without any commercial purpose; and

(iv) the country to which the copies have been sent has agreed with the country whose competent authority has granted the license to allow the receipt, or distribution, or both, and the Director General has been notified of the agreement by the Government of the country in which the license has been granted.

(5) All copies published under a license granted by virtue of Article II or Article III shall bear a notice in the appropriate language stating that the copies are available for distribution only in the country or territory to which the said license applies.

(6)(a) Due provision shall be made at the national level to ensure

(i) that the license provides, in favor of the owner of the right of translation or of reproduction, as the case may be, for just compensation that is consistent with standards of royalties normally operating on licenses freely negotiated between persons in the two countries concerned; and

(ii) payment and transmittal of the compensation: should national currency regulations intervene, the competent authority shall make all efforts, by the use of international machinery, to ensure transmittal in internationally convertible currency or its equivalent.

(b) Due provision shall be made by national legislation to ensure a correct translation of the work, or an accurate reproduction of the particular edition, as the case may be.

Article V

(1)(a) Any country entitled to make a declaration that it will avail itself of the faculty provided for in Article II may, instead, at the time of ratifying or acceding to this Act:

(i) if it is a country to which Article 30(2)(a) applies, make a declaration under that provision as far as the right of translation is concerned;

(ii) if it is a country to which Article 30(2)(a) does not apply, and even if it is not a country outside the Union, make a declaration as provided for in Article 30(2)(b), first sentence.

(b) In the case of a country which ceases to be regarded as a developing country as referred to in Article I(1), a declaration made according to this paragraph shall be effective until the date on which the period applicable under Article I(3) expires.

(c) Any country which has made a declaration according to this paragraph may not subsequently avail itself of the faculty provided for in Article II even if it withdraws the said declaration.

(2) Subject to paragraph (3), any country which has availed itself of the faculty provided for in Article II may not subsequently make a declaration according to paragraph (1).

(3) Any country which has ceased to be regarded as a developing country as referred to in Article I(1) may, not later than two years prior to the expiration of the period applicable under Article I(3), make a declaration to the effect provided for in Article 30(2)(b), first sentence, notwithstanding the fact that it is not a country outside the Union. Such declaration shall take effect at the date on which the period applicable under Article I(3) expires.

Article VI

(1) Any country of the Union may declare, as from the date of this Act, and at any time before becoming bound by Articles 1 to 21 and this Appendix:

(i) if it is a country which, were it bound by Articles 1 to 21 and this Appendix, would be entitled to avail itself of the faculties referred to in Article I(1), that it will apply the provisions of Article II or of Article III or of both to works whose country of origin is a country which, pursuant to (ii) below, admits the application of those articles to such works, or which is bound by Articles 1 to 21 and this Appendix; such declaration may, instead of referring to Article II, refer to Article V;

(ii) that it admits the application of this Appendix to works of which it is the country of origin by countries which have made a declaration under (i) above or a notification under Article I.

(2) Any declaration made under paragraph (1) shall be in writing and shall be deposited with the Director General. The declaration shall become effective from the date of its deposit.

IN WITNESS WHEREOF, the undersigned, being duly authorized thereto, have signed this Act. DONE at Paris on July 24, 1971.

General Agreement on Tariffs and Trade — Multilateral Trade Negotiations (The Uruguay Round): Agreement on Trade-Related Aspects of Intellectual Property Rights, Including Trade in Counterfeit Goods

December 15, 1993

Agreement on Trade-Related Aspects of Intellectual Property Rights, Including Trade in Counterfeit Goods

Members,

Desiring to reduce distortions and impediments to international trade, and taking into account the need to promote effective and adequate protection of intellectual property rights, and to ensure that measures and procedures to enforce intellectual property rights do not themselves become barriers to legitimate trade;

Recognizing, to this end, the need for new rules and disciplines concerning:

(a) the applicability of the basic principles of GATT 1994 and of relevant international intellectual property agreements or conventions;

(b) the provision of adequate standards and principles concerning the availability, scope, and use of trade-related intellectual property rights;

(c) the provision of effective and appropriate means for the enforcement of trade-related intellectual property rights, taking into account differences in national legal systems;

(d) the provision of effective and expeditious procedures for the multilateral prevention and settlement of disputes between governments; and

(e) transitional arrangements aiming at the fullest participation in the results of the negotiations;

Recognizing the need for a multilateral framework of principles, rules, and disciplines dealing with international trade in counterfeit goods;

Recognizing that intellectual property rights are private rights;

Recognizing the underlying public policy objectives of national systems for the protection of intellectual property, including developmental and technological objectives;

Recognizing also the special needs of the least-developed country Members in respect of maximum flexibility in the domestic implementation of laws and regulations in order to enable them to create a sound and viable technological base;

Emphasizing the importance of reducing tensions by reaching strengthened commitments to resolve disputes on trade-related intellectual property issues through multilateral procedures;

Desiring to establish a mutually supportive relationship between the WTO and the World Intellectual Property Organization (WIPO) as well as other relevant international organisations;

Hereby agree as follows:

Part I. General Provisions and Basic Principles

Article 1. Nature and Scope of Obligations

1. Members shall give effect to the provisions of this Agreement. Members may, but shall not be obliged to, implement in their domestic law more extensive protection than is required by this Agreement, provided that such protection does not contravene the provisions of this Agreement. Members shall be free to determine the appropriate method of implementing the provisions of this Agreement within their own legal system and practice.

2. For the purposes of this Agreement, the term "intellectual property" refers to all categories of intellectual property that are the subject of Sections 1 to 7 of Part II.

3. Members shall accord the treatment provided for in this Agreement to the nationals of other Members.[4] In respect of the relevant intellectual property right, the nationals of other Members shall be understood as those natural or legal persons that would meet the criteria for eligibility for protection provided for in the Paris Convention (1967), the Berne Convention (1971), the Rome Convention, and the Treaty on Intellectual Property in Respect of Integrated Circuits, were all Members of the WTO members of those conventions.[5] Any Member availing itself of the possibilities provided in paragraph 3 of Article 5 or paragraph 2 of Article 6 of the Rome Convention shall make a notification as foreseen in those provisions to the Council for Trade-Related Aspects of Intellectual Property Rights.

Article 2. Intellectual Property Conventions

1. In respect of Parts II, III, and IV of this Agreement, Members shall comply with Articles 1-12 and 19 of the Paris Convention (1967).

2. Nothing in Parts I to IV of this Agreement shall derogate from existing obligations that Members may have to each other under the Paris Convention, the Berne Convention, the Rome Convention, and the Treaty on Intellectual Property in Respect of Integrated Circuits.

---

4. When "nationals" are referred to in this Agreement, they shall be deemed, in the case of a separate customs territory Member of the WTO, to mean persons, natural or legal, who are domiciled or who have a real and effective industrial or commercial establishment in that customs territory.

5. In this Agreement, "Paris Convention" refers to the Paris Convention for the Protection of Industrial Property; "Paris Convention (1967)" refers to the Stockholm Act of this Convention of 14 July 1967. "Berne Convention" refers to the Berne Convention for the Protection of Literary and Artistic Works; "Berne Convention (1971)" refers to the Paris Act of this Convention of 24 July 1971. "Rome Convention" refers to the International Convention for the Protection of Performers, Producers of Phonograms and Broadcasting Organisations, adopted at Rome on 26 October 1961. "Treaty on Intellectual Property in Respect of Integrated Circuits" (IPIC Treaty) refers to the Treaty on Intellectual Property in Respect of Integrated Circuits, adopted at Washington on 26 May 1989.

Article 3. National Treatment

1. Each Member shall accord to the nationals of other Members treatment no less favourable than that it accords to its own nationals with regard to the protection[6] of intellectual property, subject to the exceptions already provided in, respectively, the Paris Convention (1967), the Berne Convention (1971), the Rome Convention, and the Treaty on Intellectual Property in Respect of Integrated Circuits. In respect of performers, producers of phonograms, and broadcasting organizations, this obligation only applies in respect of the rights provided under this Agreement. Any Member availing itself of the possibilities provided in Article 6 of the Berne Convention and paragraph 1(b) of Article 16 of the Rome Convention shall make a notification as foreseen in those provisions to the Council for Trade-Related Aspects of Intellectual Property Rights.

2. Members may avail themselves of the exceptions permitted under paragraph 1 above in relation to judicial and administrative procedures, including the designation of an address for service or the appointment of an agent within the jurisdiction of a Member, only where such exceptions are necessary to secure compliance with laws and regulations which are not inconsistent with the provisions of this Agreement and where such practices are not applied in a manner which would constitute a disguised restriction on trade.

Article 4. Most-Favoured-Nation Treatment

With regard to the protection of intellectual property, any advantage, favour, privilege, or immunity granted by a Member to the nationals of any other country shall be accorded immediately and unconditionally to the nationals of all other Members. Exempted from this obligation are any advantage, favour, privilege, or immunity accorded by a Member:

(a) deriving from international agreements on judicial assistance and law enforcement of a general nature and not particularly confined to the protection of intellectual property;

(b) granted in accordance with the provisions of the Berne Convention (1971) or the Rome Convention authorizing that the treatment accorded be a function not of national treatment but of the treatment accorded in another country;

(c) in respect of the rights of performers, producers of phonograms, and broadcasting organizations not provided under this Agreement;

(d) deriving from international agreements related to the protection of intellectual property which entered into force prior to the entry into force of the agreement establishing the WTO, provided that such agreements are notified to the Council for Trade-Related Aspects of Intellectual Property Rights and do not

6. For the purposes of Articles 3 and 4 of this Agreement, protection shall include matters affecting the availability, acquisition, scope, maintenance, and enforcement of intellectual property rights as well as those matters affecting the use of intellectual property rights specifically addressed in this Agreement.

constitute an arbitrary or unjustifiable discrimination against nationals of other Members.

Article 5. Multilateral Agreements on Acquisition or Maintenance of Protection

The obligations under Articles 3 and 4 above do not apply to procedures provided in multilateral agreements concluded under the auspices of the World Intellectual Property Organization relating to the acquisition or maintenance of intellectual property rights.

Article 6. Exhaustion

For the purposes of dispute settlement under this Agreement, subject to the provisions of Articles 3 and 4 above, nothing in this Agreement shall be used to address the issue of the exhaustion of intellectual property rights.

Article 7. Objectives

The protection and enforcement of intellectual property rights should contribute to the promotion of technological innovation and to the transfer and dissemination of technology, to the mutual advantage of producers and users of technological knowledge and in a manner conducive to social and economic welfare, and to a balance of rights and obligations.

Article 8. Principles

1. Members may, in formulating or amending their national laws and regulations, adopt measures necessary to protect public health and nutrition, and to promote the public interest in sectors of vital importance to their socio-economic and technological development, provided that such measures are consistent with the provisions of this Agreement.

2. Appropriate measures, provided that they are consistent with the provisions of this Agreement, may be needed to prevent the abuse of intellectual property rights by right holders or the resort to practices which unreasonably restrain trade or adversely affect the international transfer of technology.

Part II. Standards Concerning the Availability, Scope, and Use of Intellectual Property Rights

Section 1. Copyright and Related Rights

Article 9. Relation to Berne Convention

1. Members shall comply with Articles 1-21 and the Appendix of the Berne Convention (1971). However, Members shall not have rights or obligations under

this Agreement in respect of the rights conferred under Article 6*bis* of that Convention or of the rights derived therefrom.

2. Copyright protection shall extend to expressions and not to ideas, procedures, methods of operation, or mathematical concepts as such.

Article 10. Computer Programs and Compilations of Data

1. Computer programs, whether in source or object code, shall be protected as literary works under the Berne Convention (1971).

2. Compilations of data or other material, whether in machine readable or other form, which by reason of the selection or arrangement of their contents constitute intellectual creations, shall be protected as such. Such protection, which shall not extend to the data or material itself, shall be without prejudice to any copyright subsisting in the data or material itself.

Article 11. Rental Rights

In respect of at least computer programs and cinematographic works, a Member shall provide authors and their successors in title the right to authorize or to prohibit the commercial rental to the public of originals or copies of their copyright works. A Member shall be excepted from this obligation in respect of cinematographic works unless such rental has led to widespread copying of such works which is materially impairing the exclusive right of reproduction conferred in that Member on authors and their successors in title. In respect of computer programs, this obligation does not apply to rentals where the program itself is not the essential object of the rental.

Article 12. Term of Protection

Whenever the term of protection of a work, other than a photographic work or a work of applied art, is calculated on a basis other than the life of a natural person, such term shall be no less than fifty years from the end of the calendar year of authorized publication, or, failing such authorized publication within fifty years from the making of the work or fifty years from the end of the calendar year of making.

Article 13. Limitations and Exceptions

Members shall confine limitations or exceptions to exclusive rights to certain special cases which do not conflict with a normal exploitation of the work and do not unreasonably prejudice the legitimate interests of the right holder.

Article 14. Protection of Performers, Producers of Phonograms (Sound Recordings), and Broadcasting Organizations

1. In respect of a fixation of their performance on a phonogram, performers shall have the possibility of preventing the following acts when undertaken without their authorization: the fixation of their unfixed performance and the reproduction

of such fixation. Performers shall also have the possibility of preventing the following acts when undertaken without their authorization: the broadcasting by wireless means and the communication to the public of their live performance.

2. Producers of phonograms shall enjoy the right to authorize or prohibit the direct or indirect reproduction of their phonograms.

3. Broadcasting organizations shall have the right to prohibit the following acts when undertaken without their authorization: the fixation, the reproduction of fixations, and the rebroadcasting by wireless means of broadcasts, as well as the communication to the public of television broadcasts of the same. Where Members do not grant such rights to broadcasting organizations, they shall provide owners of copyright in the subject matter of broadcasts with the possibility of preventing the above acts, subject to the provisions of the Berne Convention (1971).

4. The provisions of Article 11 in respect of computer programs shall apply mutatis mutandis to producers of phonograms and any other right holders in phonograms as determined in domestic law. If, on the date of the Ministerial Meeting concluding the Uruguay Round of Multilateral Trade Negotiations, a Member has in force a system of equitable remuneration of right holders in respect of the rental of phonograms, it may maintain such system provided that the commercial rental of phonograms is not giving rise to the material impairment of the exclusive rights of reproduction of right holders.

5. The term of the protection available under this Agreement to performers and producers of phonograms shall last at least until the end of a period of fifty years computed from the end of the calendar year in which the fixation was made or the performance took place. The term of protection granted pursuant to paragraph 3 above shall last for at least twenty years from the end of the calendar year in which the broadcast took place.

6. Any Member may, in relation to the rights conferred under paragraphs 1-3 above, provide for conditions, limitations, exceptions, and reservations to the extent permitted by the Rome Convention. However, the provisions of Article 18 of the Berne Convention (1971) shall also apply, mutatis mutandis, to the rights of performers and producers of phonograms in phonograms.

Section 2. Trademarks

Article 15. Protectable Subject Matter

1. Any sign, or any combination of signs, capable of distinguishing the goods or services of one undertaking from those of other undertakings, shall be capable of constituting a trademark. Such signs, in particular words including personal names, letters, numerals, figurative elements, and combinations of colours as well as any combination of such signs, shall be eligible for registration as trademarks. Where signs are not inherently capable of distinguishing the relevant goods or services, Members may make registrability depend on distinctiveness acquired through use. Members may require, as a condition of registration, that signs be visually perceptible.

2. Paragraph 1 above shall not be understood to prevent a Member from denying registration of a trademark on other grounds, provided that they do not derogate from the provisions of the Paris Convention (1967).

3. Members may make registrability depend on use. However, actual use of a trademark shall not be a condition for filing an application for registration. An application shall not be refused solely on the ground that intended use has not taken place before the expiry of a period of three years from the date of application.

4. The nature of the goods or services to which a trademark is to be applied shall in no case form an obstacle to registration of the trademark.

5. Members shall publish each trademark either before it is registered or promptly after it is registered and shall afford a reasonable opportunity for petitions to cancel the registration. In addition, Members may afford an opportunity for the registration of a trademark to be opposed.

Article 16. Rights Conferred

1. The owner of a registered trademark shall have the exclusive right to prevent all third parties not having his consent from using in the course of trade identical or similar signs for goods or services which are identical or similar to those in respect of which the trademark is registered where such use would result in a likelihood of confusion. In case of the use of an identical sign for identical goods or services, a likelihood of confusion shall be presumed. The rights described above shall not prejudice any existing prior rights, nor shall they affect the possibility of Members making rights available on the basis of use.

2. Article 6*bis* of the Paris Convention (1967) shall apply, mutatis mutandis, to services. In determining whether a trademark is well known, account shall be taken of the knowledge of the trademark in the relevant sector of the public, including knowledge in that Member obtained as a result of the promotion of the trademark.

3. Article 6*bis* of the Paris Convention (1967) shall apply, mutatis mutandis, to goods or services which are not similar to those in respect of which a trademark is registered, provided that use of that trademark in relation to those goods or services would indicate a connection between those goods or services and the owner of the registered trademark and provided that the interests of the owner of the registered trademark are likely to be damaged by such use.

Article 17. Exceptions

Members may provide limited exceptions to the rights conferred by a trademark, such as fair use of descriptive terms, provided that such exceptions take account of the legitimate interests of the owner of the trademark and of third parties.

Article 18. Term of Protection

Initial registration, and each renewal of registration, of a trademark shall be for a term of no less than seven years. The registration of a trademark shall be renewable indefinitely.

Article 19. Requirement of Use

1. If use is required to maintain a registration, the registration may be cancelled only after an uninterrupted period of at least three years of non-use, unless valid reasons based on the existence of obstacles to such use are shown by the trademark owner. Circumstances arising independently of the will of the owner of the trademark which constitute an obstacle to the use of the trademark, such as import restrictions on or other government requirements for goods or services protected by the trademark, shall be recognized as valid reasons for non-use.

2. When subject to the control of its owner, use of a trademark by another person shall be recognized as use of the trademark for the purpose of maintaining the registration.

Article 20. Other Requirements

The use of a trademark in the course of trade shall not be unjustifiably encumbered by special requirements, such as use with another trademark, use in a special form or use in a manner detrimental to its capability to distinguish the goods or services of one undertaking from those of other undertakings. This will not preclude a requirement prescribing the use of the trademark identifying the undertaking producing the goods or services along with, but without linking it to, the trademark distinguishing the specific goods or services in question of that undertaking.

Article 21. Licensing and Assignment

Members may determine conditions on the licensing and assignment of trademarks, it being understood that the compulsory licensing of trademarks shall not be permitted and that the owner of a registered trademark shall have the right to assign his trademark with or without the transfer of the business to which the trademark belongs.

Section 3. Geographical Indications

Article 22. Protection of Geographical Indications

1. Geographical indications are, for the purposes of this Agreement, indications which identify a good as originating in the territory of a Member, or a region or locality in that territory, where a given quality, reputation or other characteristic of the good is essentially attributable to its geographical origin.

2. In respect of geographical indications, Members shall provide the legal means for interested parties to prevent:

(a) the use of any means in the designation or presentation of a good that indicates or suggests that the good in question originates in a geographical area other than the true place of origin in a manner which misleads the public as to the geographical origin of the good;

(b) any use which constitutes an act of unfair competition within the meaning of Article 10*bis* of the Paris Convention (1967).

3. A Member shall, ex officio if its legislation so permits or at the request of an interested party, refuse or invalidate the registration of a trademark which contains or consists of a geographical indication with respect to goods not originating in the territory indicated, if use of the indication in the trademark for such goods in that Member is of such a nature as to mislead the public as to the true place of origin.

4. The provisions of the preceding paragraphs of this Article shall apply to a geographical indication which, although literally true as to the territory, region, or locality in which the goods originate, falsely represents to the public that the goods originate in another territory.

Article 23. Additional Protection far Geographical Indications for Wines and Spirits

1. Each Member shall provide the legal means for interested parties to prevent use of a geographical indication identifying wines for wines not originating in the place indicated by the geographical indication in question or identifying spirits for spirits not originating in the place indicated by the geographical indication in question, even where the true origin of the goods is indicated or the geographical indication is used in translation or accompanied by expressions such as "kind", "type", "style", "imitation" or the like.[7]

2. The registration of a trademark for wines which contains or consists of a geographical indication identifying wines or for spirits which contains or consists of a geographical indication identifying spirits shall be refused or invalidated, ex officio if domestic legislation so permits or at the request of an interested party, with respect to such wines or spirits not having this origin.

3. In the case of homonymous geographical indications for wines, protection shall be accorded to each indication, subject to the provisions of paragraph 4 of Article 22 above. Each Member shall determine the practical conditions under which the homonymous indications in question will be differentiated from each other, taking into account the need to ensure equitable treatment of the producers concerned and that consumers are not misled.

4. In order to facilitate the protection of geographical indications for wines, negotiations shall be undertaken in the Council for Trade-Related Aspects of Intellectual Property Rights concerning the establishment of a multilateral system of notification and registration of geographical indications for wines eligible for protection in those Members participating in the system.

Article 24. International Negotiations; Exceptions

1. Members agree to enter into negotiations aimed at increasing the protection of individual geographical indications under Article 23. The provisions of paragraphs

---

7. Notwithstanding the first sentence of Article 42, Members may, with respect to these obligations, instead provide for enforcement by administrative action.

4-8 below shall not be used by a Member to refuse to conduct negotiations or to conclude bilateral or multilateral agreements. In the context of such negotiations, Members shall be willing to consider the continued applicability of these provisions to individual geographical indications whose use was the subject of such negotiations.

2. The Council for Trade-Related Aspects of Intellectual Property Rights shall keep under review the application of the provisions of this Section; the first such review shall take place within two years of the entry into force of the Agreement Establishing the WTO. Any matter affecting the compliance with the obligations under these provisions may be drawn to the attention of the Council, which, at the request of a Member, shall consult with any Member or Members in respect of such matter in respect of which it has not been possible to find a satisfactory solution through bilateral or plurilateral consultations between the Members concerned. The Council shall take such action as may be agreed to facilitate the operation and further the objectives of this Section.

3. In implementing this Section, a Member shall not diminish the protection of geographical indications that existed in that Member immediately prior to the date of entry into force of the Agreement Establishing the WTO.

4. Nothing in this Section shall require a Member to prevent continued and similar use of a particular geographical indication of another Member identifying wines or spirits in connection with goods or services by any of its nationals or domiciliaries who have used that geographical indication in a continuous manner with regard to the same or related goods or services in the territory of that Member either (a) for at least ten years preceding the date of the Ministerial Meeting concluding the Uruguay Round of Multilateral Trade Negotiations or (b) in good faith preceding that date.

5. Where a trademark has been applied for or registered in good faith, or where rights to a trademark have been acquired through use in good faith either:

(a) before the date of application of these provisions in that Member as defined in Part VI below; or

(b) before the geographical indication is protected in its country of origin;

measures adopted to implement this Section shall not prejudice eligibility for or the validity of the registration of a trademark, or the right to use a trademark, on the basis that such a trademark is identical with, or similar to, a geographical indication.

6. Nothing in this Section shall require a Member to apply its provisions in respect of a geographical indication of any other Member with respect to goods or services for which the relevant indication is identical with the term customary in common language as the common name for such goods or services in the territory of that Member. Nothing in this Section shall require a Member to apply its provisions in respect of a geographical indication of any other Member with respect to products of the vine for which the relevant indication is identical with the customary name of a grape variety existing in the territory of that Member as of the date of entry into force of the Agreement Establishing the WTO.

7. A Member may provide that any request made under this Section in connection with the use or registration of a trademark must be presented within five years after the adverse use of the protected indication has become generally known in that Member or after the date of registration of the trademark in that Member provided

that the trademark has been published by that date, if such date is earlier than the date on which the adverse use became generally known in that Member, provided that the geographical indication is not used or registered in bad faith.

8. The provisions of this Section shall in no way prejudice the right of any person to use, in the course of trade, his name or the name of his predecessor in business, except where such name is used in such a manner as to mislead the public.

9. There shall be no obligation under this Agreement to protect geographical indications which are not or cease to be protected in their country of origin, or which have fallen into disuse in that country.

Section 4. Industrial Designs

Article 25. Requirements for Protection

1. Members shall provide for the protection of independently created industrial designs that are new or original. Members may provide that designs are not new or original if they do not significantly differ from known designs or combinations of known design features. Members may provide that such protection shall not extend to designs dictated essentially by technical or functional considerations.

2. Each Member shall ensure that requirements for securing protection for textile designs, in particular in regard to any cost, examination or publication, do not unreasonably impair the opportunity to seek and obtain such protection. Members shall be free to meet this obligation through industrial design law or through copyright law.

Article 26. Protection

1. The owner of a protected industrial design shall have the right to prevent third parties not having his consent from making, selling, or importing articles bearing or embodying a design which is a copy, or substantially a copy, of the protected design, when such acts are undertaken for commercial purposes.

2. Members may provide limited exceptions to the protection of industrial designs, provided that such exceptions do not unreasonably conflict with the normal exploitation of protected industrial designs and do not unreasonably prejudice the legitimate interests of the owner of the protected design, taking account of the legitimate interests of third parties.

3. The duration of protection available shall amount to at least ten years.

Section 5. Patents

Article 27. Patentable Subject Matter

1. Subject to the provisions of paragraphs 2 and 3 below, patents shall be available for any inventions, whether products or processes, in all fields of technology, provided

that they are new, involve an inventive step, and are capable of industrial application.[8] Subject to paragraph 4 of Article 65, paragraph 8 of Article 70, and paragraph 3 of this Article, patents shall be available and patent rights enjoyable without discrimination as to the place of invention, the field of technology, and whether products are imported or locally produced.

2. Members may exclude from patentability inventions, the prevention within their territory of the commercial exploitation of which is necessary to protect ordre public or morality, including to protect human, animal or plant life or health or to avoid serious prejudice to the environment, provided that such exclusion is not made merely because the exploitation is prohibited by domestic law.

3. Members may also exclude from patentability:

(a) diagnostic, therapeutic, and surgical methods for the treatment of humans or animals;

(b) plants and animals other than microorganisms, and essentially biological processes for the production of plants or animals other than nonbiological and microbiological processes. However, Members shall provide for the protection of plant varieties either by patents or by an effective sui generis system or by any combination thereof. The provisions of this sub-paragraph shall be reviewed four years after the entry into force of the Agreement Establishing the WTO.

Article 28. Rights Conferred

1. A patent shall confer on its owner the following exclusive rights:

(a) where the subject matter of a patent is a product, to prevent third parties not having his consent from the acts of: making, using, offering for sale, selling, or importing[9] for these purposes that product;

(b) where the subject matter of a patent is a process, to prevent third parties not having his consent from the act of using the process and from the acts of: using, offering for sale, selling, or importing for these purposes at least the product obtained directly by that process.

2. Patent owners shall also have the right to assign, or transfer by succession, the patent and to conclude licensing contracts.

Article 29. Conditions on Patent Applicants

1. Members shall require that an applicant for a patent shall disclose the invention in a manner sufficiently clear and complete for the invention to be carried out by a person skilled in the art and may require the applicant to indicate the best mode for carrying out the invention known to the inventor at the filing date or, where priority is claimed, at the priority date of the application.

---

8. For the purposes of this Article, the terms "inventive step" and "capable of industrial application" may be deemed by a Member to be synonymous with the terms "non-obvious" and "useful" respectively.

9. This right, like all other rights conferred under this Agreement in respect of the use, sale, importation, or other distribution of goods, is subject to the provisions of Article 6 above.

2. Members may require an applicant for a patent to provide information concerning his corresponding foreign applications and grants.

Article 30. Exceptions to Rights Conferred

Members may provide limited exceptions to the exclusive rights conferred by a patent, provided that such exceptions do not unreasonably conflict with a normal exploitation of the patent and do not unreasonably prejudice the legitimate interests of the patent owner, taking account of the legitimate interests of third parties.

Article 31. Other Use Without Authorization of the Right Holder

Where the law of a Member allows for other use[10] of the subject matter of a patent without the authorization of the right holder, including use by the government or third parties authorized by the government, the following provisions shall be respected;

(a) authorization of such use shall be considered on its individual merits;

(b) such use may only be permitted if, prior to such use, the proposed user has made efforts to obtain authorization from the right holder on reasonable commercial terms and conditions and that such efforts have not been successful within a reasonable period of time. This requirement may be waived by a Member in the case of a national emergency or other circumstances of extreme urgency or in cases of public non-commercial use. In situations of national emergency or other circumstances of extreme urgency, the right holder shall, nevertheless, be notified as soon as reasonably practicable. In the case of public non-commercial use, where the government or contractor, without making a patent search, knows or has demonstrable grounds to know that a valid patent is or will be used by or for the government, the right holder shall be informed promptly;

(c) the scope and duration of such use shall be limited to the purpose for which it was authorized, and in the case of semi-conductor technology shall only be for public non-commercial use or to remedy a practice determined after judicial or administrative process to be anti-competitive.

(d) such use shall be non-exclusive;

(e) such use shall be non-assignable, except with that part of the enterprise or goodwill which enjoys such use;

(f) any such use shall be authorized predominantly for the supply of the domestic market of the Member authorizing such use;

(g) authorization for such use shall be liable, subject to adequate protection of the legitimate interests of the persons so authorized, to be terminated if and when the circumstances which led to it cease to exist and are unlikely to recur. The competent authority shall have the authority to review, upon motivated request, the continued existence of these circumstances;

---

10. "Other use" refers to use other than that allowed under Article 30.

(h) the right holder shall be paid adequate remuneration in the circumstances of each case, taking into account the economic value of the authorization;

(i) the legal validity of any decision relating to the authorization of such use shall be subject to judicial review or other independent review by a distinct higher authority in that Member;

(j) any decision relating to the remuneration provided in respect of such use shall be subject to judicial review or other independent review by a distinct higher authority in that Member;

(k) Members are not obliged to apply the conditions set forth in sub-paragraphs (b) and (f) above where such use is permitted to remedy a practice determined after judicial or administrative process to be anti-competitive. The need to correct anti-competitive practices may be taken into account in determining the amount of remuneration in such cases. Competent authorities shall have the authority to refuse termination of authorization if and when the conditions which led to such authorization are likely to recur;

(1) where such use is authorized to permit the exploitation of a patent ("the second patent") which cannot be exploited without infringing another patent ("the first patent"), the following additional conditions shall apply:

(i) the invention claimed in the second patent shall involve an important technical advance of considerable economic significance in relation to the invention claimed in the first patent;

(ii) the owner of the first patent shall be entitled to a cross-license on reasonable terms to use the invention claimed in the second patent; and

(iii) the use authorized in respect of the first patent shall be non-assignable except with the assignment of the second patent.

Article 32. Revocation/Forfeiture

An opportunity for judicial review of any decision to revoke or forfeit a patent shall be available.

Article 33. Term of Protection

The term of protection available shall not end before the expiration of a period of twenty years counted from the filing date.[11]

Article 34. Process Patents: Burden of Proof

1. For the purposes of civil proceedings in respect of the infringement of the rights of the owner referred to in paragraph 1(b) of Article 28 above, if the subject matter of a patent is a process for obtaining a product, the judicial authorities shall have the authority to order the defendant to prove that the process to obtain an

11. It is understood that those Members which do not have a system of original grant may provide that the term of protection shall be computed from the filing date in the system of original grant.

identical product is different from the patented process. Therefore, Members shall provide, in at least one of the following circumstances, that any identical product when produced without the consent of the patent owner shall, in the absence of proof to the contrary, be deemed to have been obtained by the patented process:

(a) if the product obtained by the patented process is new;

(b) if there is a substantial likelihood that the identical product was made by the process and the owner of the patent has been unable through reasonable efforts to determine the process actually used.

2. Any Member shall be free to provide that the burden of proof indicated in paragraph 1 shall be on the alleged infringer only if the condition referred to in sub-paragraph (a) is fulfilled or only if the condition referred to in sub-paragraph (b) is fulfilled.

3. In the adduction of proof to the contrary, the legitimate interests of the defendant in protecting his manufacturing and business secrets shall be taken into account.

Section 6. Layout-Designs (Topographies) of Integrated Circuits

Article 35. Relation to IPIC Treaty

Members agree to provide protection to the layout-designs (topographies) of integrated circuits (hereinafter referred to as "layout-designs") in accordance with Articles 2-7 (other than paragraph 3 of Article 6), Article 12, and paragraph 3 of Article 16 of the Treaty on Intellectual Property in Respect of Integrated Circuits and, in addition, to comply with the following provisions.

Article 36. Scope of the Protection

Subject to the provisions of paragraph 1 of Article 37 below, Members shall consider unlawful the following acts if performed without the authorization of the right holder:[12] importing, selling, or otherwise distributing for commercial purposes a protected layout-design, an integrated circuit in which a protected layout-design is incorporated, or an article incorporating such an integrated circuit only insofar as it continues to contain an unlawfully reproduced layout-design.

Article 37. Acts Not Retiring the Authorization of the Right Holder

1. Notwithstanding Article 36 above, no Member shall consider unlawful the performance of any of the acts referred to in that Article in respect of an integrated

12. The term "right holder" in this Section shall be understood as having the same meaning as the term "holder of the right" in the IPIC Treaty.

circuit incorporating an unlawfully reproduced layout-design or any article incorporating such an integrated circuit where the person performing or ordering such acts did not know and had no reasonable ground to know, when acquiring the integrated circuit or article incorporating such an integrated circuit, that it incorporated an unlawfully reproduced layout-design. Members shall provide that, after the time that such person has received sufficient notice that the layout-design was unlawfully reproduced, he may perform any of the acts with respect to the stock on hand or ordered before such time, but shall be liable to pay to the right holder a sum equivalent to a reasonable royalty such as would be payable under a freely negotiated licence in respect of such a layout-design.

2. The conditions set out in sub-paragraphs (a)-(k) of Article 31 above shall apply mutatis mutandis in the event of any non-voluntary licensing of a layout-design or of its use by or for the government without the authorization of the right holder.

Article 38. Term of Protection

1. In Members requiring registration as a condition of protection, the term of protection of layout-designs shall not end before the expiration of a period of ten years counted from the date of filing an application for registration or from the first commercial exploitation wherever in the world it occurs.

2. In Members not requiring registration as a condition for protection, layout-designs shall be protected for a term of no less than ten years from the date of the first commercial exploitation, wherever in the world it occurs.

3. Notwithstanding paragraphs 1 and 2 above, a Member may provide that protection shall lapse fifteen years after the creation of the layout-design.

Section 7. Protection of Undisclosed Information

Article 39

1. In the course of ensuring effective protection against unfair competition as provided in Article 10*bis* of the Paris Convention (1967), Members shall protect undisclosed information in accordance with paragraph 2 below and data submitted to governments or governmental agencies in accordance with paragraph 3 below.

2. Natural and legal persons shall have the possibility of preventing information lawfully within their control from being disclosed to, acquired by, or used by others without their consent in a manner contrary to honest commercial practices[13] so long as such information:

13. For the purpose of this provision, "a manner contrary to honest commercial practices" shall mean at least practices such as breach of contract, breach of confidence and inducement to breach, and includes the acquisition of undisclosed information by third parties who knew, or were grossly negligent in failing to know, that such practices were involved in the acquisition.

(a) is secret in the sense that it is not, as a body or in the precise configuration and assembly of its components, generally known among or readily accessible to persons within the circles that normally deal with the kind of information in question;

(b) has commercial value because it is secret; and

(c) has been subject to reasonable steps under the circumstances, by the person lawfully in control of the information, to keep it secret.

3. Members, when requiring, as a condition of approving the marketing of pharmaceutical or of agricultural chemical products which utilize new chemical entities, the submission of undisclosed test or other data, the origination of which involves a considerable effort, shall protect such data against unfair commercial use. In addition, Members shall protect such data against disclosure, except where necessary to protect the public, or unless steps are taken to ensure that the data are protected against unfair commercial use.

Section 8. Control of Anti-Competitive Practices in Contractual Licences

Article 40

1. Members agree that some licensing practices or conditions pertaining to intellectual property rights which restrain competition may have adverse effects on trade and may impede the transfer and dissemination of technology.

2. Nothing in this Agreement shall prevent Members from specifying in their national legislation licensing practices or conditions that may in particular cases constitute an abuse of intellectual property rights having an adverse effect on competition in the relevant market. As provided above, a Member may adopt, consistently with the other provisions of this Agreement, appropriate measures to prevent or control such practices, which may include for example exclusive grantback conditions, conditions preventing challenges to validity and coercive package licensing, in the light of the relevant laws and regulations of that Member.

3. Each Member shall enter, upon request, into consultations with any other Member which has cause to believe that an intellectual property right owner that is a national or domiciliary of the Member to which the request for consultations has been addressed is undertaking practices in violation of the requesting Member's laws and regulations on the subject matter of this Section, and which wishes to secure compliance with such legislation, without prejudice to any action under the law and to the full freedom of an ultimate decision of either Member. The Member addressed shall accord full and sympathetic consideration to, and shall afford adequate opportunity for, consultations with the requesting Member, and shall co-operate through supply of publicly available non-confidential information of relevance to the matter in question and of other information available to the Member, subject to domestic law and to the conclusion of mutually satisfactory agreements concerning the safeguarding of its confidentiality by the requesting Member.

4. A Member whose nationals or domiciliaries are subject to proceedings in another Member concerning alleged violation of that other Member's laws and regulations on the subject matter of this Section shall, upon request, be granted an opportunity for consultations by the other Member under the same conditions as those foreseen in paragraph 3 above.

Part III. Enforcement of Intellectual Property Rights

Section 1. General Obligations

Article 41

1. Members shall ensure that enforcement procedures as specified in this Part are available under their national laws so as to permit effective action against any act of infringement of intellectual property rights covered by this Agreement, including expeditious remedies to prevent infringements and remedies which constitute a deterrent to further infringements. These procedures shall be applied in such a manner as to avoid the creation of barriers to legitimate trade and to provide for safeguards against their abuse.

2. Procedures concerning the enforcement of intellectual property rights shall be fair and equitable. They shall not be unnecessarily complicated or costly, or entail unreasonable time limits or unwarranted delays.

3. Decisions on the merits of a case shall preferably be in writing and reasoned. They shall be made available at least to the parties to the proceeding without undue delay. Decisions on the merits of a case shall be based only on evidence in respect of which parties were offered the opportunity to be heard.

4. Parties to a proceeding shall have an opportunity for review by a judicial authority of final administrative decisions and, subject to jurisdictional provisions in national laws concerning the importance of a case, of at least the legal aspects of initial judicial decisions on the merits of a case. However, there shall be no obligation to provide an opportunity for review of acquittals in criminal cases.

5. It is understood that this Part does not create any obligation to put in place a judicial system for the enforcement of intellectual property rights distinct from that for the enforcement of laws in general, nor does it affect the capacity of Members to enforce their laws in general. Nothing in this Part creates any obligation with respect to the distribution of resources as between enforcement of intellectual property rights and the enforcement of laws in general.

Section 2. Civil and Administrative Procedures and Remedies

Article 42. Fair and Equitable Procedures

Members shall make available to right holders[14] civil judicial procedures concerning the enforcement of any intellectual property right covered by this

14. For the purpose of this Part, the term "right holder" includes federations and associations having legal standing to assert such rights.

Agreement. Defendants shall have the right to written notice which is timely and contains sufficient detail, including the basis of the claims. Parties shall be allowed to be represented by independent legal counsel, and procedures shall not impose overly burdensome requirements concerning mandatory personal appearances. All parties to such procedures shall be duly entitled to substantiate their claims and to present all relevant evidence. The procedure shall provide a means to identify and protect confidential information, unless this would be contrary to existing constitutional requirements.

Article 43. Evidence of Proof

1. The judicial authorities shall have the authority, where a party has presented reasonably available evidence sufficient to support its claims and has specified evidence relevant to substantiation of its claims which lies in the control of the opposing party, to order that this evidence be produced by the opposing party, subject in appropriate cases to conditions which ensure the protection of confidential information.

2. In cases in which a party to a proceeding voluntarily and without good reason refuses access to, or otherwise does not provide necessary information within a reasonable period, or significantly impedes a procedure relating to an enforcement action, a Member may accord judicial authorities the authority to make preliminary and final determinations, affirmative or negative, on the basis of the information presented to them, including the complaint or the allegation presented by the party adversely affected by the denial of access to information, subject to providing the parties an opportunity to be heard on the allegations or evidence.

Article 44. Injunctions

1. The judicial authorities shall have the authority to order a party to desist from an infringement, inter alia to prevent the entry into the channels of commerce in their jurisdiction of imported goods that involve the infringement of an intellectual property right, immediately after customs clearance of such goods. Members are not obliged to accord such authority in respect of protected subject matter acquired or ordered by a person prior to knowing or having reasonable grounds to know that dealing in such subject matter would entail the infringement of an intellectual property right.

2. Notwithstanding the other provisions of this Part and provided that the provisions of Part II specifically addressing use by governments, or by third parties authorized by a government, without the authorization of the right holder are complied with, Members may limit the remedies available against such use to payment of remuneration in accordance with sub-paragraph (h) of Article 31 above. In other cases, the remedies under this Part shall apply or, where these remedies are inconsistent with national law, declaratory judgments and adequate compensation shall be available.

Article 45. Damages

1. The judicial authorities shall have the authority to order the infringer to pay the right holder damages adequate to compensate for the injury the right holder has suffered because of an infringement of his intellectual property right by an infringer who knew or had reasonable grounds to know that he was engaged in infringing activity.

2. The judicial authorities shall also have the authority to order the infringer to pay the right holder expenses, which may include appropriate attorney's fees. In appropriate cases, Members may authorize the judicial authorities to order recovery of profits and/or payment of pre-established damages even where the infringer did not know or had no reasonable grounds to know that he was engaged in infringing activity.

Article 46. Other Remedies

In order to create an effective deterrent to infringement, the judicial authorities shall have the authority to order that goods that they have found to be infringing be, without compensation of any sort, disposed of outside the channels of commerce in such a manner as to avoid any harm caused to the right holder, or, unless this would be contrary to existing constitutional requirements, destroyed. The judicial authorities shall also have the authority to order that materials and implements the predominant use of which has been in the creation of the infringing goods be, without compensation of any sort, disposed of outside the channels of commerce in such a manner as to minimize the risks of further infringements. In considering such requests, the need for proportionality between the seriousness of the infringement and the remedies ordered, as well as the interests of third parties, shall be taken into account. In regard to counterfeit trademark goods, the simple removal of the trademark unlawfully affixed shall not be sufficient, other than in exceptional cases, to permit release of the goods into the channels of commerce.

Article 47. Right of Information

Members may provide that the judicial authorities shall have the authority, unless this would be out of proportion to the seriousness of the infringement, to order the infringer to inform the right holder of the identity of third persons involved in the production and distribution of the infringing goods or services and of their channels of distribution.

Article 48. Indemnification of the Defendant

1. The judicial authorities shall have the authority to order a party at whose request measures were taken and who has abused enforcement procedures to provide to a party wrongfully enjoined or restrained adequate compensation for the injury

suffered because of such abuse. The judicial authorities shall also have the authority to order the applicant to pay the defendant's expenses, which may include appropriate attorney's fees.

2. In respect of the administration of any law pertaining to the protection or enforcement of intellectual property rights, Members shall only exempt both public authorities and officials from liability to appropriate remedial measures where actions are taken or intended in good faith in the course of the administration of such laws.

Article 49. Administrative Procedures

To the extent that any civil remedy can be ordered as a result of administrative procedures on the merits of a case, such procedures shall conform to principles equivalent in substance to those set forth in this Section.

Section 3. Provisional Measures

Article 50

1. The judicial authorities shall have the authority to order prompt and effective provisional measures:

(a) to prevent an infringement of any intellectual property right from occurring, and in particular to prevent the entry into the channels of commerce in their jurisdiction of goods, including imported goods immediately after customs clearance;

(b) to preserve relevant evidence in regard to the alleged infringement.

2. The judicial authorities shall have the authority to adopt provisional measures inaudita altera parte where appropriate, in particular where any delay is likely to cause irreparable harm to the right holder or where there is a demonstrable risk of evidence being destroyed.

3. The judicial authorities shall have the authority to require the applicant to provide any reasonably available evidence in order to satisfy themselves with a sufficient degree of certainty that the applicant is the right holder and that his right is being infringed or that such infringement is imminent, and to order the applicant to provide a security or equivalent assurance sufficient to protect the defendant and to prevent abuse.

4. Where provisional measures have been adopted inaudita altera parte, the parties affected shall be given notice, without delay after the execution of the measures at the latest. A review, including a right to be heard, shall take place upon request of the defendant with a view to deciding, within a reasonable period after the notification of the measures, whether these measures shall be modified, revoked or confirmed.

5. The applicant may be required to supply other information necessary for the identification of the goods concerned by the authority that will execute the provisional measures.

6. Without prejudice to paragraph 4 above, provisional measures taken on the basis of paragraphs 1 and 2 above shall, upon request by the defendant, be revoked or otherwise cease to have effect, if proceedings leading to a decision on the merits of the case are not initiated within a reasonable period, to be determined by the judicial authority ordering the measures where national law so permits or, in the absence of such a determination, not to exceed twenty working days or thirty-one calendar days, whichever is the longer.

7. Where the provisional measures are revoked or where they lapse due to any act or omission by the applicant, or where it is subsequently found that there has been no infringement or threat of infringement of an intellectual property right, the judicial authorities shall have the authority to order the applicant, upon request of the defendant, to provide the defendant appropriate compensation for any injury caused by these measures.

8. To the extent that any provisional measure can be ordered as a result of administrative procedures, such procedures shall conform to principles equivalent in substance to those set forth in this Section.

Section 4. Special Requirements Related to Border Measures[15]

Article 51. Suspension of Release by Customs Authorities

Members shall, in conformity with the provisions set out below, adopt procedures[16] to enable a right holder, who has valid grounds for suspecting that the importation of counterfeit trademark or pirated copyright goods[17] may take place, to lodge an application in writing with competent authorities, administrative or judicial, for the suspension by the customs authorities of the release into free circulation of such goods. Members may enable such an application to be made in respect of goods which involve other infringements of intellectual property rights,

15. Where a Member has dismantled substantially all controls over movement of goods across its border with another Member with which it forms part of a customs union, it shall not be required to apply the provisions of this Section at that border.

16. It is understood that there shall be no obligation to apply such procedures to imports of goods put on the market in another country by or with the consent of the right holder, or to goods in transit.

17. For the purposes of this Agreement: counterfeit trademark goods shall mean any goods, including packaging, bearing without authorization a trademark which is identical to the trademark validly registered in respect of such goods, or which cannot be distinguished in its essential aspects from such a trademark, and which thereby infringes the rights of the owner of the trademark in question under the law of the country of importation; pirated copyright goods shall mean any goods which are copies made without the consent of the right holder or person duly authorized by him in the country of production and which are made directly or indirectly from an article where the making of that copy would have constituted an infringement of a copyright or a related right under the law of the country of importation.

provided that the requirements of this Section are met. Members may also provide for corresponding procedures concerning the suspension by the customs authorities of the release of infringing goods destined for exportation from their territories.

Article 52. Application

Any right holder initiating the procedures under Article 51 above shall be required to provide adequate evidence to satisfy the competent authorities that, under the laws of the country of importation, there is prima facie an infringement of his intellectual property right and to supply a sufficiently detailed description of the goods to make them readily recognizable by the customs authorities. The competent authorities shall inform the applicant within a reasonable period whether they have accepted the application and, where determined by the competent authorities, the period for which the customs authorities will take action.

Article 53. Security or Equivalent Assurance

1. The competent authorities shall have the authority to require an applicant to provide a security or equivalent assurance sufficient to protect the defendant and the competent authorities and to prevent abuse. Such security or equivalent assurance shall not unreasonably deter recourse to these procedures.

2. Where pursuant to an application under this Section the release of goods involving industrial designs, patents, layout-designs, or undisclosed information into free circulation has been suspended by customs authorities on the basis of a decision other than by a judicial or other independent authority, and the period provided for in Article 55 has expired without the granting of provisional relief by the duly empowered authority, and provided that all other conditions for importation have been complied with, the owner, importer, or consignee of such goods shall be entitled to their release on the posting of a security in an amount sufficient to protect the right holder for any infringement. Payment of such security shall not prejudice any other remedy available to the right holder, it being understood that the security shall be released if the right holder fails to pursue his right of action within a reasonable period of time.

Article 54. Notice of Suspension

The importer and the applicant shall be promptly notified of the suspension of the release of goods according to Article 51 above.

Article 55. Duration of Suspension

If, within a period not exceeding ten working days after the applicant has been served notice of the suspension, the customs authorities have not been informed that proceedings leading to a decision on the merits of the case have been initiated by a party other than the defendant, or that the duly empowered authority has taken provisional measures prolonging the suspension of the release of the goods, the goods shall be released, provided that all other conditions for importation or exportation have been complied with; in appropriate cases, this time-limit may be extended by another ten working days. If proceedings leading to a decision on the merits of the case have been initiated, a review, including a right to be heard, shall take place upon request of the defendant with a view to deciding, within a reasonable period, whether these measures shall be modified, revoked, or confirmed. Notwithstanding the above, where the suspension of the release of goods is carried out or continued in accordance with a provisional judicial measure, the provisions of Article 50, paragraph 6 above shall apply.

Article 56. Indemnification of the Importer and of the Owner of the Goods

Relevant authorities shall have the authority to order the applicant to pay the importer, the consignee, and the owner of the goods appropriate compensation for any injury caused to them through the wrongful detention of goods or through the detention of goods released pursuant to Article 55 above.

Article 57. Right of Inspection and Information

Without prejudice to the protection of confidential information, Members shall provide the competent authorities the authority to give the right holder sufficient opportunity to have any product detained by the customs authorities inspected in order to substantiate his claims. The competent authorities shall also have authority to give the importer an equivalent opportunity to have any such product inspected. Where a positive determination has been made on the merits of a case, Members may provide the competent authorities the authority to inform the right holder of the names and addresses of the consignor, the importer, and the consignee and of the quantity of the goods in question.

Article 58. Ex Officio Action

Where Members require competent authorities to act upon their own initiative and to suspend the release of goods in respect of which they have acquired prima facie evidence that an intellectual property right is being infringed:

(a) the competent authorities may at any time seek from the right holder any information that may assist them to exercise these powers;

(b) the importer and the right holder shall be promptly notified of the suspension. Where the importer has lodged an appeal against the suspension with the competent authorities, the suspension shall be subject to the conditions, mutatis mutandis, set out at Article 55 above;

(c) Members shall only exempt both public authorities and officials from liability to appropriate remedial measures where actions are taken or intended in good faith.

Article 59. Remedies

Without prejudice to other rights of action open to the right holder and subject to the right of the defendant to seek review by a judicial authority, competent authorities shall have the authority to order the destruction or disposal of infringing goods in accordance with the principles set out in Article 46 above. In regard to counterfeit trademark goods, the authorities shall not allow the reexportation of the infringing goods in an unaltered state or subject them to a different customs procedure, other than in exceptional circumstances.

Article 60. De Minimis Imports

Members may exclude from the application of the above provisions small quantities of goods of a non-commercial nature contained in travellers' personal luggage or sent in small consignments.

Section 5. Criminal Procedures

Article 61

Members shall provide for criminal procedures and penalties to be applied at least in cases of wilful trademark counterfeiting or copyright piracy on a commercial scale. Remedies available shall include imprisonment and/or monetary fines sufficient to provide a deterrent, consistently with the level of penalties applied for crimes of a corresponding gravity. In appropriate cases, remedies available shall also include the seizure, forfeiture and destruction of the infringing goods and of any materials and implements the predominant use of which has been in the commission of the offence. Members may provide for criminal procedures and penalties to be applied in other cases of infringement of intellectual property rights, in particular where they are committed wilfully and on a commercial scale.

Part IV. Acquisition and Maintenance of Intellectual Property Rights and Related Inter-Partes Procedures

Article 62

1. Members may require, as a condition of the acquisition or maintenance of the intellectual property rights provided for under Sections 2-6 of Part II of this

Agreement, compliance with reasonable procedures and formalities. Such procedures and formalities shall be consistent with the provisions of this Agreement.

2. Where the acquisition of an intellectual property right is subject to the right being granted or registered, Members shall ensure that the procedures for grant or registration, subject to compliance with the substantive conditions for acquisition of the right, permit the granting or registration of the right within a reasonable period of time so as to avoid unwarranted curtailment of the period of protection.

3. Article 4 of the Paris Convention (1967) shall apply mutatis mutandis to service marks.

4. Procedures concerning the acquisition or maintenance of intellectual property rights and, where the national law provides for such procedures, administrative revocation and inter partes procedures such as opposition, revocation and cancellation, shall be governed by the general principles set out in paragraphs 2 and 3 of Article 41.

5. Final administrative decisions in any of the procedures referred to under paragraph 4 above shall be subject to review by a judicial or quasi-judicial authority. However, there shall be no obligation to provide an opportunity for such review of decisions in cases of unsuccessful opposition or administrative revocation, provided that the grounds for such procedures can be the subject of invalidation procedures.

Part V. Dispute Prevention and Settlement

Article 63. Transparency

1. Laws and regulations, and final judicial decisions and administrative rulings of general application, made effective by any Member pertaining to the subject matter of this Agreement (the availability, scope, acquisition, enforcement and prevention of the abuse of intellectual property rights) shall be published, or where such publication is not practicable made publicly available, in a national language, in such a manner as to enable governments and right holders to become acquainted with them. Agreements concerning the subject matter of this Agreement which are in force between the government or a governmental agency of any Member and the government or a governmental agency of any other Member shall also be published.

2. Members shall notify the laws and regulations referred to in paragraph 1 above to the Council for Trade-Related Aspects of Intellectual Property Rights in order to assist that Council in its review of the operation of this Agreement. The Council shall attempt to minimize the burden on Members in carrying out this obligation and may decide to waive the obligation to notify such laws and regulations directly to the Council if consultations with the World Intellectual Property Organization on the establishment of a common register containing these laws and regulations are successful. The Council shall also consider in this connection any action required regarding notifications pursuant to the obligations under this Agreement stemming from the provisions of Article 6*ter* of the Paris Convention (1967).

3. Each Member shall be prepared to supply, in response to a written request from another Member, information of the sort referred to in paragraph 1 above.

A Member, having reason to believe that a specific judicial decision or administrative ruling or bilateral agreement in the area of intellectual property rights affects its rights under this Agreement, may also request in writing to be given access to or be informed in sufficient detail of such specific judicial decisions or administrative rulings or bilateral agreements.

4. Nothing in paragraphs 1 to 3 above shall require Members to disclose confidential information which would impede law enforcement or otherwise be contrary to the public interest or would prejudice the legitimate commercial interests of particular enterprises, public or private.

Article 64. Dispute Settlement

1. The provisions of Articles XXII and XXIII of the General Agreement on Tariffs and Trade 1994 as elaborated and applied by the Understanding on Rules and Procedures Governing the Settlement of Disputes shall apply to consultations and the settlement of disputes under this Agreement except as otherwise specifically provided herein.

2. Sub-paragraphs XXIII:1(b) and XXIII:1(c) of the General Agreement on Tariffs and Trade 1994 shall not apply to the settlement of disputes under this Agreement for a period of five years from the entry into force of the Agreement establishing the Multilateral Trade Organization.

3. During the time period referred to in paragraph 2, the TRIPS Council shall examine the scope and modalities for Article XXIII:1 (b) and Article XXIII:1 (c)-type complaints made pursuant to this Agreement, and submit its recommendations to the Ministerial Conference for approval. Any decision of the Ministerial Conference to approve such recommendations or to extend the period in paragraph 2 shall be made only by consensus, and approved recommendations shall be effective for all Members without further formal acceptance process.

Part VI. Transitional Arrangements

Article 65. Transitional Arrangements

1. Subject to the provisions of paragraphs 2, 3 and 4 below, no Member shall be obliged to apply the provisions of this Agreement before the expiry of a general period of one year following the date of entry into force of the Agreement Establishing the WTO.

2. Any developing country Member is entitled to delay for a further period of four years the date of application, as defined in paragraph 1 above, of the provisions of this Agreement other than Articles 3, 4 and 5 of Part I.

3. Any other Member which is in the process of transformation from a centrally-planned into a market, free-enterprise economy and which is undertaking structural reform of its intellectual property system and facing special problems in the preparation and implementation of intellectual property laws, may also benefit from a period of delay as foreseen in paragraph 2 above.

4. To the extent that a developing country Member is obliged by this Agreement to extend product patent protection to areas of technology not so protectable in its territory on the general date of application of this Agreement for that Member, as defined in paragraph 2 above, it may delay the application of the provisions on product patents of Section 5 of Part II of this Agreement to such areas of technology for an additional period of five years.

5. Any Member availing itself of a transitional period under paragraphs 1, 2, 3 or 4 above shall ensure that any changes in its domestic laws, regulations and practice made during that period do not result in a lesser degree of consistency with the provisions of this Agreement.

Article 66. Least-Developed Country Members

1. In view of their special needs and requirements, their economic, financial, and administrative constraints, and their need for flexibility to create a viable technological base, least-developed country Members shall not be required to apply the provisions of this Agreement, other than Articles 3, 4, and 5, for a period of 10 years from the date of application as defined under paragraph 1 of Article 65 above. The Council shall, upon duly motivated request by a least-developed country Member, accord extensions of this period.

2. Developed country Members shall provide incentives to enterprises and institutions in their territories for the purpose of promoting and encouraging technology transfer to least-developed country Members in order to enable them to create a sound and viable technological base.

Article 67. Technical Cooperation

In order to facilitate the implementation of this Agreement, developed country Members shall provide, on request and on mutually agreed terms and conditions, technical and financial cooperation in favour of developing and least-developed country Members. Such cooperation shall include assistance in the preparation of domestic legislation on the protection and enforcement of intellectual property rights as well as on the prevention of their abuse, and shall include support regarding the establishment or reinforcement of domestic offices and agencies relevant to these matters, including the training of personnel.

Part VII. Institutional Arrangements; Final Provisions

Article 68. Council for Trade-Related Aspects of Intellectual Property Rights

The Council for Trade-Related Aspects of Intellectual Property Rights shall monitor the operation of this Agreement and, in particular, Members' compliance with their obligations hereunder, and shall afford Members the opportunity of consulting on matters relating to the trade-related aspects of intellectual property rights. It shall carry out such other responsibilities as assigned to it by the Members, and it shall, in particular, provide any assistance requested by them in the context of dispute settlement procedures. In carrying out its functions, the Council may consult with and seek information from any source it deems appropriate. In consultation with the World Intellectual Property Organization, the Council shall seek to establish, within one year of its first meeting, appropriate arrangements for cooperation with bodies of that Organization.

Article 69. International Cooperation

Members agree to cooperate with each other with a view to eliminating international trade in goods infringing intellectual property rights. For this purpose, they shall establish and notify contact points in their national administrations and be ready to exchange information on trade in infringing goods. They shall, in particular, promote the exchange of information and cooperation between customs authorities with regard to trade in counterfeit trademark goods and pirated copyright goods.

Article 70. Protection of Existing Subject Matter

1. This Agreement does not give rise to obligations in respect of acts which occurred before the date of application of the Agreement for the Member in question.

2. Except as otherwise provided for in this Agreement, this Agreement gives rise to obligations in respect of all subject matter existing at the date of application of this Agreement for the Member in question, and which is protected in that Member on the said date, or which meets or comes subsequently to meet the criteria for protection under the terms of this Agreement. In respect of this paragraph and paragraphs 3 and 4 below, copyright obligations with respect to existing works shall be solely determined under Article 18 of the Berne Convention (1971), and obligations with respect to the rights of producers of phonograms and performers in existing phonograms shall be determined solely under Article 18 of the Berne Convention (1971) as made applicable under paragraph 6 of Article 14 of this Agreement.

3. There shall be no obligation to restore protection to subject matter which on the date of application of this Agreement for the Member in question has fallen into the public domain.

4. In respect of any acts in respect of specific objects embodying protected subject matter which become infringing under the terms of legislation in conformity with this Agreement, and which were commenced, or in respect of which a significant investment was made, before the date of acceptance of the Agreement Establishing the WTO by that Member, any Member may provide for a limitation of the remedies available to the right holder as to the continued performance of such acts after the date of application of the Agreement for that Member. In such cases the Member shall, however, at least provide for the payment of equitable remuneration.

5. A Member is not obliged to apply the provisions of Article 11 and of paragraph 4 of Article 14 with respect to originals or copies purchased prior to the date of application of this Agreement for that Member.

6. Members shall not be required to apply Article 31, or the requirement in paragraph 1 of Article 27 that patent rights shall be enjoyable without discrimination as to the field of technology, to use without the authorization of the right holder where authorization for such use was granted by the government before the date this Agreement became known.

7. In the case of intellectual property rights for which protection is conditional upon registration, applications for protection which are pending on the date of application of this Agreement for the Member in question shall be permitted to be amended to claim any enhanced protection provided under the provisions of this Agreement. Such amendments shall not include new matter.

8. Where a Member does not make available as of the date of entry into force of the Agreement Establishing the WTO patent protection for pharmaceutical and agricultural chemical products commensurate with its obligations under Article 27, that Member shall:

(i) notwithstanding the provisions of Part VI above, provide as from the date of entry into force of the Agreement Establishing the WTO a means by which applications for patents for such inventions can be filed;

(ii) apply to these applications, as of the date of application of this Agreement, the criteria for patentability as laid down in this Agreement as if those criteria were being applied on the date of filing in that Member or, where priority is available and claimed, the priority date of the application;

(iii) provide patent protection in accordance with this Agreement as from the grant of the patent and for the remainder of the patent term, counted from the filing date in accordance with Article 33 of this Agreement, for those of these applications that meet the criteria for protection referred to in subparagraph (ii) above.

9. Where a product is the subject of a patent application in a Member in accordance with paragraph 8(i) above, exclusive marketing rights shall be granted, notwithstanding the provisions of Part VI above, for a period of five years after obtaining market approval in that Member or until a product patent is granted or rejected in that Member, whichever period is shorter, provided that, subsequent to the entry into force of the Agreement Establishing the WTO, a patent application has been filed and a patent granted for that product in another Member and marketing approval obtained in such other Member.

Article 71. Review and Amendment

1. The Council for Trade-Related Aspects of Intellectual Property Rights shall review the implementation of this Agreement after the expiration of the transitional period referred to in paragraph 2 of Article 65 above. The Council shall, having regard to the experience gained in its implementation, review it two years after that date, and at identical intervals thereafter. The Council may also undertake reviews

in the light of any relevant new developments which might warrant modification or amendment of this Agreement.

2. Amendments merely serving the purpose of adjusting to higher levels of protection of intellectual property rights achieved, and in force, in other multilateral agreements and accepted under those agreements by all Members of the WTO may be referred to the Ministerial Conference for action in accordance with Article X, paragraph 6, of the Agreement Establishing the WTO on the basis of a consensus proposal from the Council for Trade-Related Aspects of Intellectual Property Rights.

Article 72. Reservations

Reservations may not be entered in respect of any of the provisions of this Agreement without the consent of the other Members.

Article 73. Security Exceptions

Nothing in this Agreement shall be construed:

(a) to require any Member to furnish any information the disclosure of which it considers contrary to its essential security interests; or

(b) to prevent any Member from taking any action which it considers necessary for the protection of its essential security interests;

(i) relating to fissionable materials or the materials from which they are derived;

(ii) relating to the traffic in arms, ammunition and implements of war and to such traffic in other goods and materials as is carried on directly or indirectly for the purpose of supplying a military establishment;

(iii) taken in time of war or other emergency in international relations; or

(c) to prevent any Member from taking any action in pursuance of its obligations under the United Nations Charter for the maintenance of international peace and security.

Sherman Antitrust Act

Title 15, U.S.C.

§1. Trusts, etc., in Restraint of Trade Illegal; Penalty

Every contract, combination in the form of trust or otherwise, or conspiracy, in restraint of trade or commerce among the several States, or with foreign nations, is declared to be illegal. Every person who shall make any contract or engage in any combination or conspiracy hereby declared to be illegal shall be deemed guilty of a felony, and, on conviction thereof, shall be punished by fine not exceeding $100,000,000 if a corporation, or, if any other person, $1,000,000, or by imprisonment not exceeding 10 years, or by both said punishments, in the discretion of the court.

§2. Monopolizing Trade a Felony; Penalty

Every person who shall monopolize, or attempt to monopolize, or combine or conspire with any other person or persons, to monopolize any part of the trade or commerce among the several States, or with foreign nations, shall be deemed guilty of a felony, and, on conviction thereof, shall be punished by fine not exceeding $100,000,000 if a corporation, or, if any other person, $1,000,000, or by imprisonment not exceeding 10 years, or by both said punishments, in the discretion of the court.

§3. Trusts in Territories or District of Columbia Illegal; Combination a Felony

(a) Every contract, combination in form of trust or otherwise, or conspiracy, in restraint of trade or commerce in any Territory of the United States or of the

District of Columbia, or in restraint of trade or commerce between any such Territory and another, or between any such Territory or Territories and any State or States or the District of Columbia, or with foreign nations, or between the District of Columbia and any State or States or foreign nations, is declared illegal. Every person who shall make any such contract or engage in any such combination or conspiracy, shall be deemed guilty of a felony, and, on conviction thereof, shall be punished by fine not exceeding $100,000,000 if a corporation, or, if any other person, $1,000,000, or by imprisonment not exceeding 10 years, or both said punishments, in the discretion of the court.

(b) Every person who shall monopolize, or attempt to monopolize, or combine or conspire with any other person or persons, to monopolize any part of the trade or commerce in any Territory of the United States or of the District of Columbia, or between any such Territory and another, or between any such Territory or Territories and any State or States or the District of Columbia, or with foreign nations, or between the District of Columbia, and any State or States or foreign nations, shall be deemed guilty of a felony, and, on conviction thereof, shall be punished by fine not exceeding $10,000,000 if a corporation, or, if any other person, $350,000, or by imprisonment not exceeding three years, or by both said punishments, in the discretion of the court.

§4. Jurisdiction of Courts; Duty of United States Attorneys; Procedure

The several district courts of the United States are invested with jurisdiction to prevent and restrain violations of sections 1 to 7 of this title; and it shall be the duty of the several United States attorneys, in their respective districts, under the direction of the Attorney General, to institute proceedings in equity to prevent and restrain such violations. Such proceedings may be by way of petition setting forth the case and praying that such violation shall be enjoined or otherwise prohibited. When the parties complained of shall have been duly notified of such petition the court shall proceed, as soon as may be, to the hearing and determination of the case; and pending such petition and before final decree, the court may at any time make such temporary restraining order or prohibition as shall be deemed just in the premises.

§5. Bringing in Additional Parties

Whenever it shall appear to the court before which any proceeding under section 4 of this title may be pending, that the ends of justice require that other parties should be brought before the court, the court may cause them to be summoned, whether they reside in the district in which the court is held or not; and subpoenas to that end may be served in any district by the marshal thereof.

§6. Forfeiture of Property in Transit

Any property owned under any contract or by any combination, or pursuant to any conspiracy (and being the subject thereof) mentioned in section 1 of this title, and being in the course of transportation from one State to another, or to a foreign country, shall be forfeited to the United States, and may be seized and condemned by like proceedings as those provided by law for the forfeiture, seizure, and condemnation of property imported into the United States contrary to law.

§6a. Conduct Involving Trade or Commerce with Foreign Nations

Sections 1 to 7 of this title shall not apply to conduct involving trade or commerce (other than import trade or import commerce) with foreign nations unless—

(1) such conduct has a direct, substantial, and reasonably foreseeable effect—

(A) on trade or commerce which is not trade or commerce with foreign nations, or on import trade or import commerce with foreign nations; or

(B) on export trade or export commerce with foreign nations, of a person engaged in such trade or commerce in the United States; and

(2) such effect gives rise to a claim under the provisions of sections 1 to 7 of this title, other than this section.

If sections 1 to 7 of this title apply to such conduct only because of the operation of paragraph (1)(B), then sections 1 to 7 of this title shall apply to such conduct only for injury to export business in the United States.

§7. "Person" or "Persons" Defined

The word "person", or "persons", wherever used in sections 1 to 7 of this title shall be deemed to include corporations and associations existing under or authorized by the laws of either the United States, the laws of any of the Territories, the laws of any State, or the laws of any foreign country.

National Cooperative Research Act

Title 15, U.S.C.

§4301. Definitions

(a) For purposes of this chapter:

(1) The term "antitrust laws" has the meaning given it in subsection (a) of section 12 of this title, except that such term includes section 45 of this title to the extent that section 45 of this title applies to unfair methods of competition.

(2) The term "Attorney General" means the Attorney General of the United States.

(3) The term "Commission" means the Federal Trade Commission.

(4) The term "person" has the meaning given it in subsection (a) of section 12 of this title.

(5) The term "State" has the meaning given it in section 15g(2) of this title.

(6) The term "joint venture" means any group of activities, including attempting to make, making, or performing a contract, by two or more persons for the purpose of—

(A) theoretical analysis, experimentation, or systematic study of phenomena or observable facts,

(B) the development or testing of basic engineering techniques,

(C) the extension of investigative findings or theory of a scientific or technical nature into practical application for experimental and demonstration purposes, including the experimental production and testing of models, prototypes, equipment, materials, and processes,

(D) the production of a product, process, or service,

(E) the testing in connection with the production of a product, process, or service by such venture,

(F) the collection, exchange, and analysis of research or production information, or

(G) any combination of the purposes specified in subparagraphs (A), (B), (C), (D), (E), and (F),

and may include the establishment and operation of facilities for the conducting of such venture, the conducting of such venture on a protected and proprietary basis, and the prosecuting of applications for patents and the granting of licenses for the results of such venture, but does not include any activity specified in subsection (b) of this section.

(7) The term "standards development activity" means any action taken by a standards development organization for the purpose of developing, promulgating, revising, amending, reissuing, interpreting, or otherwise maintaining a voluntary consensus standard, or using such standard in conformity assessment activities, including actions relating to the intellectual property policies of the standards development organization.

(8) The term "standards development organization" means a domestic or international organization that plans, develops, establishes, or coordinates voluntary consensus standards using procedures that incorporate the attributes of openness, balance of interests, due process, an appeals process, and consensus in a manner consistent with the Office of Management and Budget Circular Number A-119, as revised February 10, 1998. The term "standards development organization" shall not, for purposes of this Act, include the parties participating in the standards development organization.

(9) The term "technical standard" has the meaning given such term in section 12(d)(4) of the National Technology Transfer and Advancement Act of 1995.

(10) The term "voluntary consensus standard" has the meaning given such term in Office of Management and Budget Circular Number A-119, as revised February 10, 1998.

(b) The term "joint venture" excludes the following activities involving two or more persons:

(1) exchanging information among competitors relating to costs, sales, profitability, prices, marketing, or distribution of any product, process, or service if such information is not reasonably required to carry out the purpose of such venture,

(2) entering into any agreement or engaging in any other conduct restricting, requiring, or otherwise involving the marketing, distribution, or provision by any person who is a party to such venture of any product, process, or service, other than—

(A) the distribution among the parties to such venture, in accordance with such venture, of a product, process, or service produced by such venture,

(B) the marketing of proprietary information, such as patents and trade secrets, developed through such venture formed under a written agreement entered into before June 10,1993, or

(C) the licensing, conveying, or transferring of intellectual property, such as patents and trade secrets, developed through such venture formed under written agreement entered into on or after June 10, 1993,

(3) entering into any agreement or engaging in any other conduct—

(A) to restrict or require the sale, licensing, or sharing of inventions, developments, products, processes, or services not developed through, or produced by, such venture, or

(B) to restrict or require participation by any person who is a party to such venture in other research and development activities,

that is not reasonably required to prevent misappropriation of proprietary information contributed by any person who is a party to such venture or of the results of such venture,

(4) entering into any agreement or engaging in any other conduct allocating a market with a competitor,

(5) exchanging information among competitors relating to production (other than production by such venture) of a product, process, or service if such information is not reasonably required to carry out the purpose of such venture,

(6) entering into any agreement or engaging in any other conduct restricting, requiring, or otherwise involving the production (other than the production by such venture) of a product, process, or service,

(7) using existing facilities for the production of a product, process, or service by such venture unless such use involves the production of a new product or technology, and

(8) except as provided in paragraphs (2), (3), and (6), entering into any agreement or engaging in any other conduct to restrict or require participation by any person who is a party to such venture, in any unilateral or joint activity that is not reasonably required to carry out the purpose of such venture.

(c) The term "standards development activity" excludes the following activities:

(1) Exchanging information among competitors relating to cost, sales, profitability, prices, marketing, or distribution of any product, process, or service that is not reasonably required for the purpose of developing or promulgating a voluntary consensus standard, or using such standard in conformity assessment activities.

(2) Entering into any agreement or engaging in any other conduct that would allocate a market with a competitor.

(3) Entering into any agreement or conspiracy that would set or restrain prices of any good or service.

§4302. Rule of Reason Standard

In any action under the antitrust laws, or under any State law similar to the antitrust laws, the conduct of—

(1) any person in making or performing a contract to carry out a joint venture, or

(2) a standards development organization while engaged in a standards development activity, shall not be deemed illegal per se; such conduct shall be judged on the basis of its reasonableness, taking into account all relevant factors affecting competition, including, but not limited to, effects on competition in properly defined, relevant research, development, product, process, and service markets. For the purpose of determining a properly defined, relevant market,

worldwide capacity shall be considered to the extent that it may be appropriate in the circumstances.

§4303. Limitation on Recovery

a. *Amount Recoverable*

Notwithstanding section 15 of this title and in lieu of the relief specified in such section, any person who is entitled to recovery on a claim under such section shall recover the actual damages sustained by such person, interest calculated at the rate specified in section 1961 of title 28 on such actual damages as specified in subsection (d) of this section, and the cost of suit attributable to such claim, including a reasonable attorney's fee pursuant to section 4304 of this title if such claim—

(1) results from conduct that is within the scope of a notification that has been filed under section 4305 (a) of this title for a joint venture, or for a standards development activity engaged in by a standards development organization against which such claim is made, and

(2) is filed after such notification becomes effective pursuant to section 4305(c) of this title.

b. *Recovery by States*

Notwithstanding section 15c of this title, and in lieu of the relief specified in such section, any State that is entitled to monetary relief on a claim under such section shall recover the total damage sustained as described in subsection (a) (1) of such section, interest calculated at the rate specified in section 1961 of title 28 on such total damage as specified in subsection (d) of this section, and the cost of suit attributable to such claim, including a reasonable attorney's fee pursuant to section 15c of this title if such claim—

(1) results from conduct that is within the scope of a notification that has been filed under section 4305 (a) of this title for a joint venture, or for a standards development activity engaged in by a standards development organization against which such claim is made and

(2) is filed after such notification becomes effective pursuant to section 4305(c) of this title.

c. *Conduct Similar Under State Law*

Notwithstanding any provision of any State law providing damages for conduct similar to that forbidden by the antitrust laws, any person who is entitled to recovery on a claim under such provision shall not recover in excess of the actual damages sustained by such person, interest calculated at the rate specified in section 1961 of title 28 on such actual damages as specified in subsection (d) of this section, and the cost of suit attributable to such claim, including a reasonable attorney's fee pursuant to section 4304 of this title if such claim—

(1) results from conduct that is within the scope of a notification that has been filed under section 4305 (a) of this title for a joint venture, or for a standards

development activity engaged in by a standards development organization against which such claim is made, and

(2) is filed after notification has become effective pursuant to section 4305 (c) of this title.

d. Interest

Interest shall be awarded on the damages involved for the period beginning on the earliest date for which injury can be established and ending on the date of judgment, unless the court finds that the award of all or part of such interest is unjust in the circumstances.

(e) Subsections (a), (b), and (c) shall not be construed to modify the liability under the antitrust laws of any person (other than a standards development organization) who—

(1) directly (or through an employee or agent) participates in a standards development activity with respect to which a violation of any of the antitrust laws is found,

(2) is not a fulltime employee of the standards development organization that engaged in such activity, and

(3) is, or is an employee or agent of a person who is, engaged in a line of commerce that is likely to benefit directly from the operation of the standards development activity with respect to which such violation is found.

f. Applicability

This section shall be applicable only if the challenged conduct of a person defending against a claim is not in violation of any decree or order, entered or issued after October 11, 1984, in any case or proceeding under the antitrust laws or any State law similar to the antitrust laws challenging such conduct as part of a joint venture, or of a standards development activity engaged in by a standards development organization.

§4304. Award of Costs, Including Attorney's Fees, to Substantially Prevailing Party; Offset

(a) Notwithstanding sections 15 and 26 of this title, in any claim under the antitrust laws, or any State law similar to the antitrust laws, based on the conducting of a joint venture, or of a standards development activity engaged in by a standards development organization, the court shall, at the conclusion of the action—

(1) award to a substantially prevailing claimant the cost of suit attributable to such claim, including a reasonable attorney's fee, or

(2) award to a substantially prevailing party defending against any such claim the cost of suit attributable to such claim, including a reasonable attorney's fee, if the claim, or the claimant's conduct during the litigation of the claim, was frivolous, unreasonable, without foundation, or in bad faith.

(b) The award made under subsection (a) of this section may be offset in whole or in part by an award in favor of any other party for any part of the cost of suit, including a reasonable attorney's fee, attributable to conduct during the litigation by any prevailing party that the court finds to be frivolous, unreasonable, without foundation, or in bad faith.

(c) Subsections (a) and (b) shall not apply with respect to any person who—

(1) directly participates in a standards development activity with respect to which a violation of any of the antitrust laws is found,

(2) is not a fulltime employee of a standards development organization that engaged in such activity, and

(3) is, or is an employee or agent of a person who is, engaged in a line of commerce that is likely to benefit directly from the operation of the standards development activity with respect to which such violation is found.

§4305. Disclosure of Joint Venture

a.(1) Written Notifications; Filing

Any party to a joint venture, acting on such venture's behalf, may, not later than 90 days after entering into a written agreement to form such venture or not later than 90 days after October 11, 1984, whichever is later, file simultaneously with the Attorney General and the Commission a written notification disclosing—

(A) the identities of the parties to such venture,

(B) the nature and objectives of such venture, and

(C) if a purpose of such venture is the production of a product, process, or service, as referred to in section 4301 (a) (6) (D) of this title, the identity and nationality of any person who is a party to such venture, or who controls any party to such venture whether separately or with one or more other persons acting as a group for the purpose of controlling such party.

Any party to such venture, acting on such venture's behalf, may file additional disclosure notifications pursuant to this section as are appropriate to extend the protections of section 4303 of this title. In order to maintain the protections of section 4303 of this title, such venture shall, not later than 90 days after a change in its membership, file simultaneously with the Attorney General and the Commission a written notification disclosing such change.

(2) A standards development organization may, not later than 90 days after commencing a standards development activity engaged in for the purpose of developing or promulgating a voluntary consensus standards or not later than 90 days after June 22, 2004, whichever is later, file simultaneously with *665 the Attorney General and the Commission, a written notification disclosing—

(A) the name and principal place of business of the standards development organization, and

(B) documents showing the nature and scope of such activity.

Any standards development organization may file additional disclosure notifications pursuant to this section as are appropriate to extend the protections of section 4303 to standards development activities that are not covered by the initial filing or that have changed significantly since the initial filing.

b. Publication; Federal Register; Notice

Except as provided in subsection (e) of this section, not later than 30 days after receiving a notification filed under subsection (a) of this section, the Attorney General or the Commission shall publish in the Federal Register a notice with respect to such venture that identifies the parties to such venture and that describes in general terms the area of planned activity of such venture, or a notice with respect to such standards development activity that identifies the standards development organization engaged in such activity and that describes such activity in general terms. Prior to its publication, the contents of such notice shall be made available to the parties to such venture or available to such organization, as the case may be.

c. Effect of Notice

If with respect to a notification filed under subsection (a) of this section, notice is published in the Federal Register, then such notification shall operate to convey the protections of section 4303 of this title as of the earlier of—(1) the date of publication of notice under subsection (b) of this section, or (2) if such notice is not so published within the time required by subsection (b) of this section, after the expiration of the 30-day period beginning on the date the Attorney General or the Commission receives the applicable information described in subsection (a) of this section.

d. Exemption; Disclosure; Information

Except with respect to the information published pursuant to subsection (b) of this section—

(1) all information and documentary material submitted as part of a notification filed pursuant to this section, and

(2) all other information obtained by the Attorney General or the Commission in the course of any investigation, administrative proceeding, or case, with respect to a potential violation of the antitrust laws by the joint venture, or the standards development activity, with respect to which such notification was filed,

shall be exempt from disclosure under section 552 of title 5, and shall not be made publicly available by any agency of the United States to which such section applies except in a judicial or administrative proceeding in which such information and material is subject to any protective order.

e. Withdrawal of Notification

Any person or standards development organization that files a notification pursuant to this section may withdraw such notification before notice of the joint venture involved is published under subsection (b) of this section. Any notification so withdrawn shall not be subject to subsection (b) of this section and shall not confer the protections of section 4303 of this title on any person or any standards development organization with respect to whom such notification was filed.

f. Judicial Renew; Inapplicable with Respect to Notifications

Any action taken or not taken by the Attorney General or the Commission with respect to notifications filed pursuant to this section shall not be subject to judicial review.

g. Admissibility into Evidence; Disclosure of Conduct; Publication of Notice; Supporting or Answering Claims Under Antitrust Laws

(1) Except as provided in paragraph (2), for the sole purpose of establishing that a person or standards development organization is entitled to the protections of section 4303 of this title, the fact of disclosure of conduct under subsection (a) of this section and the fact of publication of a notice under subsection (b) of this section shall be admissible into evidence in any judicial or administrative proceeding.

(2) No action by the Attorney General or the Commission taken pursuant to this section shall be admissible into evidence in any such proceeding for the purpose of supporting or answering any claim under the antitrust laws or under any State law similar to the antitrust laws.

§4306. Application of Section 4303 Protections to Production of Products, Processes, and Services

Notwithstanding sections 4303 and 4305 of this title, the protections of section 4303 of this title shall not apply with respect to a joint venture's production of a product, process, or service, as referred to in section 4301 (a) (6) (D) of this title, unless—

(1) the principal facilities for such production are located in the United States or its territories, and

(2) each person who controls any party to such venture (including such party itself) is a United States person, or a foreign person from a country whose law accords antitrust treatment no less favorable to United States persons than to such country's domestic persons with respect to participation in joint ventures for production.

Appendix A
Biotechnology: An Introduction to the Technology

A full understanding of the legal questions surrounding major new technologies — primarily computers and biotechnology — requires some knowledge of and appreciation for the technologies themselves. To assist the reader in this regard, this section offers a brief introduction to the basics of biotechnology.[1]

A. *Recombinant Protein Production*

By now most of us are familiar with certain basic facts about the genetic makeup of human beings: that our genes are "encoded" on chemical strands in the nuclei of our cells that wrap around each other in a double helix. A brief review of genetics will clear the way for a discussion of how scientists have learned to manipulate genes to produce the new products of the biotechnology industry.

The double-stranded molecule in the nuclei of all living things is deoxyribonucleic acid (DNA). DNA contains the instructions for making proteins — chemical products used by a cell. Thus DNA codes for every protein produced in the body, from those that make up cellular components, muscle fibers, and bones, to hormones, blood components, and enzymes.

1. This section owes a great deal to the contributions of Diedre Conley, J.D. 2000, Boalt Hall School of Law, Ph.D., Genetics, Stanford Univ.; Evelyn Findeis, J.D. 1993, Boston Univ., M.S. Biochemistry, M.I.T.; and Edwin Flores, J.D. 1997, Univ. of Texas, Ph.D. Immunology, Washington Univ.

DNA resembles a spiral staircase. The sides of the molecular staircase are made of sugar-phosphates, while the "steps" are made of chemical compounds known as nucleotide bases or simply bases. (This structure was the basic discovery of Nobel Prize winners James Watson and Francis Crick.) Each "step" has two halves—two separate bases joined by hydrogen bonds in the middle. The four bases that form the steps are adenine (A), thymine (T), guanine (G), and cytosine (C). Because of the asymmetry of the bases, only two pairings are possible: adenine with thymine (A-T) and guanine with cytosine (G-C). This is a key feature of the makeup of human genes; we will see below how scientists have used the "complementary" nature of DNA to manipulate genetic structure in useful ways. Figure A-1 shows the basic structure of DNA.

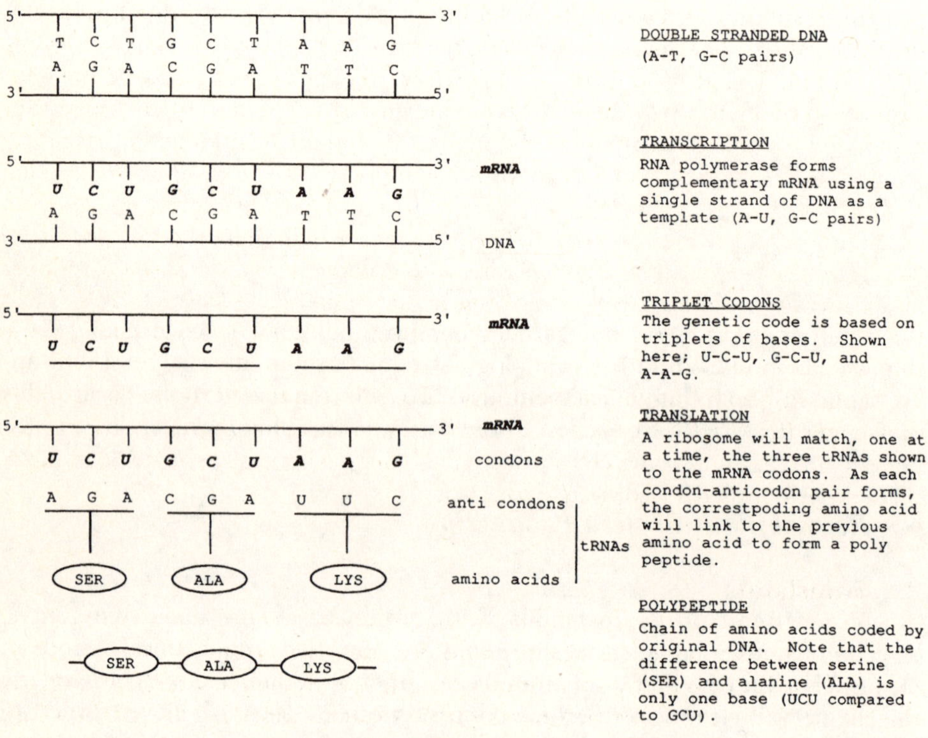

Figure A-1

In humans, DNA is arranged in 46 large units called chromosomes, with smaller units called genes on each chromosome. There are perhaps 30,000 genes in the entire collection of human DNA, which is collectively referred to as the genome.

A gene is a stretch of DNA that, in general, codes for a protein. To encode a protein the gene is divided into groups of three bases. Each set of three base pairs is called a codon. Each codon triplet codes for one of the chemical units that make up a protein; these units are called amino acids. A short chain of amino acids is called a peptide, and a long chain, a polypeptide or a protein. Thus genes contain codons, which code for amino acids, which form proteins.

The genetic instructions encoded by the DNA are carried out in a two-step process. In the first, transcription, the instructions are re-coded into a form more usable by the cell. In the second, translation, they are used as blueprints in the creation of proteins.

1. Transcription

When a cell calls for the production of a certain protein, a cellular enzyme complex called RNA polymerase forces the portion of the DNA that codes for that protein to split apart. One of the resulting single strands is the site of transcription. After the strand opens, the RNA polymerase "reads" the base sequence of the DNA strand. The RNA polymerase joins free ribonucleotides to create an RNA strand complementary to the DNA template strand. The transcribed strand thus assembled contains ribose rather than deoxyribose along its spine; hence its name, RNA. RNA also differs from DNA in that thymine is replaced by uracil (U), so that each time adenine, thymine's complementary base, is encountered on the DNA strand, the RNA strand contains a uracil base rather than thymine. The process of "reading" the DNA code and assembling a complementary RNA strand continues along the DNA molecule until reaching a stop signal. The strands of the DNA then rejoin as soon as the wave of RNA synthesis has passed, thus forming again the original double-stranded structure.

Following some modifications to its ends, the resulting RNA strand is called messenger RNA (mRNA), because it is now able to leave the nucleus and carry the genetic information to the cell's ribosomes. This is where protein synthesis takes place, via the process known as translation.

2. Translation

Translation commences when the mRNA sequence created by the transcription process is engaged by a ribosome. Meanwhile, free amino acids throughout the cell are joined to short fragments of RNA called transfer RNA (tRNA). Each tRNA binds a particular amino acid on one end and contains a triplet anticodon complementary to an mRNA codon on the other. Once bound, the tRNA/amino acid complex moves toward the ribosome. Here, the mRNA codon being processed matches up with a complementary tRNA anticodon. The amino acid attached to the tRNA is added to the amino acid chain that is being formed. Once the amino acid is joined, the empty tRNA drops off the chain and the empty slot is available for the next tRNA-amino acid in the chain. In this way the amino acids are strung together in the order dictated by the mRNA template to form the resultant protein. See Figures A-2 and A-3.

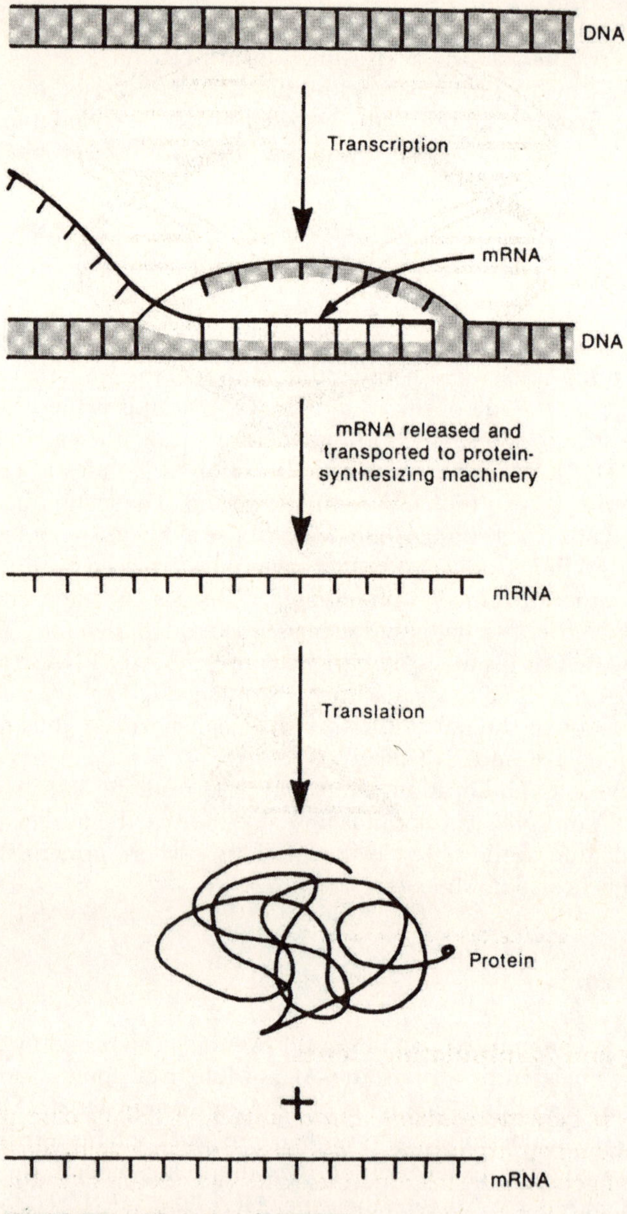

SOURCE: Office of Technology Assessment.

Figure A-2

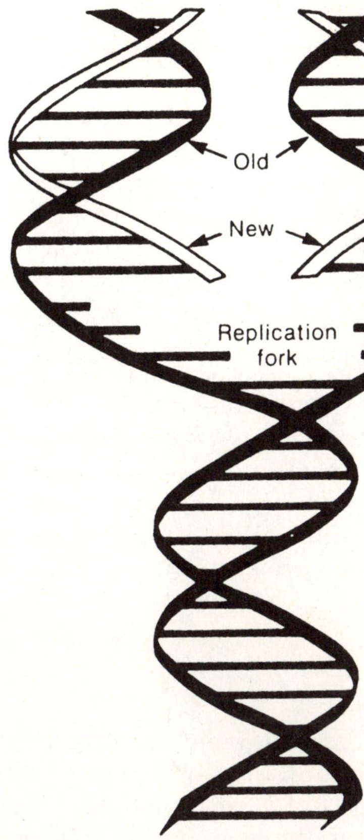

SOURCE: Office of Technology Assessment.

Figure A-3

3. Isolating and Manipulating Genes

The human genome contains approximately 3 billion base pairs with an average gene being approximately 1,200 base pairs in length. Since the gene is the basic unit of genetic information, researchers are usually interested in working with a single gene. But to do so, they must first locate the gene of interest and produce multiple copies of the gene for analysis and experimentation. Given the vast number of base pairs to choose from (1,200 out of 3 billion), powerful techniques are required to selectively isolate and multiply the genes of interest.

The first great advance in molecular genetics began with the discovery of natural enzymes that can identify precise locations at which to cut the DNA strand. By first characterizing these restriction enzymes and then using them to cut DNA into smaller pieces, scientists can routinely work with gene-sized fragments of

DNA. The key to this technique is that each restriction enzyme cuts the DNA strand only when it recognizes a specific nucleotide sequence. This is called the recognition site of the restriction enzyme.

In general, it is difficult to do much research with restriction enzymes working on a single DNA molecule. Because of the difficulties of working with such a small amount of DNA, scientists have developed techniques for isolating and multiplying DNA fragments cut by the restriction enzymes. We will consider one technique for making such copies — classic bacterial gene cloning — in the following paragraphs. In a later section, we will see that a significant new technique for multiplying specific DNA regions has been developed (and patented): the polymerase chain reaction, or PCR.

The second great advance in molecular genetics was the discovery and isolation of reverse transcriptase by David Baltimore and Howard Temin (for which they shared a Nobel Prize). Reverse transcriptase reads an RNA template and "reverse-transcribes" the message into a complementary DNA molecule (cDNA). A cDNA is basically the retrieved message from a mature mRNA that is ready for translation into a protein. Once the cDNA has been made, it is cloned into a cloning vector and can be easily manipulated using restriction enzymes and, as discussed below, PCR.

4. Gene Cloning

Gene cloning takes advantage of the rapid growth of bacteria, their small size, minimal nutritional requirements and viruses known as bacteriophages that infect bacteria. By inserting random DNA fragments into these bacterial viruses, scientists can quickly isolate and amplify the gene obtained from donor cells. The entire collection of viruses is called a gene library.[2] Researchers can test the library for viruses that contain the foreign DNA to see which clone or clones include a copy of the gene under study. When such a clone is identified, it is allowed to multiply to produce enough copies of the gene for it to be isolated, characterized, and manipulated. These DNA workhorses are known as cloning vectors.

Fortunately for science and medicine, the complementary coding scheme of DNA is conserved across species. Also fortunate is the fact that both bacteria and viruses are quite receptive to DNA from other organisms, such as humans, plants, or even other bacteria.

There are two primary techniques for inserting "foreign" DNA (i.e., DNA from another species, such as a human) into a bacterial cell: transfection and transformation. Transfection involves the use of tiny viruses that infect bacteria, called phages, engineered to "smuggle" specific genes into the DNA of bacterial cells. Transformation is similar, except that the DNA is cloned into small circular strands of "independent" (i.e., non-genomic) DNA called plasmids. Transformation occurs when a bacterial cell is induced to engulf a plasmid that contains

2. The term "gene library" originated in Cambridge, Massachusetts. In Switzerland a gene library is known as a gene bank, and in California it is referred to as a gene pool, perhaps suggesting certain cultural differences between the major areas of biotechnology research.

a foreign DNA sequence within its own sequence. Whether it is formed by transfection or transformation, the bacteria that contains the foreign genetic material is called a chimera, a reference to mythical beasts that combined traits from two different animals (e.g., a lion and a horse). Plasmids or viruses that contain a foreign DNA and are able to produce the protein product of the foreign gene are called expression vectors. See Figure A-4.

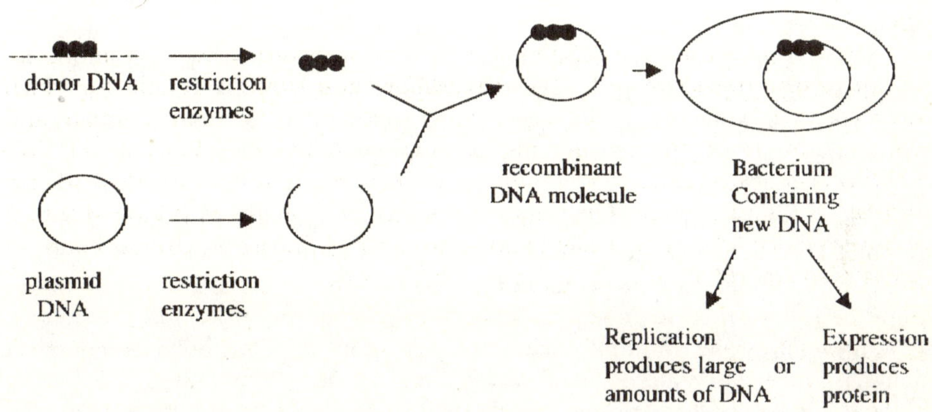

Figure A-4

As mentioned, transformation and transfection are "bulk" processes. Only a tiny number of the bacteria (or phage) from among thousands and even millions screened contain the specific genetic fragment that correponds to the desired gene. The trick is to find these and separate them from the others.

In practice, the cloning of large genomic DNA fragments entails creation of many DNA fragments by cutting the DNA with restriction enzymes or via mechanical shearing. Almost all of the fragments are useless, since restriction enzymes can cut the DNA in many possible locations and create random fragments of varying size. However, fragments of relatively uniform size can be selected by molecular weight. All the resulting fragments of uniform size are cloned into cloning vectors and introduced into bacterial cells, but only the few containing the desired genes will be produced. Hence, most of the bacteria or phage containing the foreign DNA will be discarded. Only those containing the desired gene will be preserved. On the other hand, cDNAs are relatively easy to insert into cloning vectors because they are small and relatively uniform in size.

In order to select the bacteria transformed with a cloning vector that contains the newly introduced gene or gene fragment, the cloning vector plasmids contain genes that code for proteins that confer resistance to an antibiotic such as penicillin.

These plasmids are then introduced into bacteria that are normally killed by penicillin. When the resulting bacterial colonies are exposed to penicillin, only those that have incorporated the desired cloning vector plasmid will survive.

Once the bacteria containing the desired plasmid are identified, the key ingredient—the copies of the gene desired for study—must be extracted. This involves many steps, but the basic idea is simple. The plasmid DNA must be separated from the bacteria's other, non-plasmid DNA. This is achieved through techniques that take advantage of the different properties of the two types of DNA.

When molecular biologists want to determine whether a known gene is present in this purified plasmid DNA, there are several alternative techniques available. One is to construct a probe that is complementary to the base pair sequence of the known gene. This probe will attach itself to the DNA that is inserted in the cloning vector (the plasmid or phage). By labelling the probe, for instance with radioactive material, scientists can determine whether the probe has located the DNA insert (or "hybridized"), which indicates that the DNA cloning vector contains the desired gene.

When the desired gene sequence is unknown, but the end-product (protein) sought is known, one approach to cloning involves adding a step to the preceding process. Beginning with the protein, scientists work backward—by means of what might be called "reverse genetic engineering"—from the protein to the genetic sequence. The order of amino acids in the protein is determined by chemical analysis. Since every amino acid corresponds to one of a limited number of codons, it is possible to predict the base pair sequence of the gene from the sequence of amino acids. However, the prediction is not exact, because several different codons can code for one amino acid. Next, the researchers synthesize a stretch of "artificial" DNA based on these predictions. Finally, the bacterial cells containing the cloning vectors are screened for the presence of the desired sequence or a closely related one.

The discussion so far has centered on techniques for producing multiple copies of a desired gene; that is, gene cloning. How does one go about producing bulk quantities of the protein product of that gene?

5. Industrial and Pharmaceutical Applications

Molecular biology has provided a new and exciting way to produce polypeptides and proteins. The technology joins precise knowledge of which DNA sequences code for which proteins with the ability to transfer foreign DNA into "workhorse" expression vectors such as certain bacteria. The term "expression" is used to describe production of the protein encoded by the gene or genes that has been inserted into an expression vector. Although the expression of foreign or heterologous genes is a function of a variety of complex factors, industrial and academic researchers exploit these factors to maximize the expression of cloned sequences. It is now possible to produce a wide variety of foreign proteins in certain well-understood "host cells" (or vectors)—most notably the E. coli bacterium, certain types of yeast, and certain mammalian cells. These proteins include polypeptides with human therapeutic applications, such as insulin, interferon, growth hormone, monoclonal antibodies and antibody fragments, and enzymes involved in the dynamics of blood coagulation (e.g., Factor VIII:C).

This large-scale exercise in protein expression is at the heart of the biotechnology industry. Human insulin manufactured through the use of genetically engineered E. coli bacteria has been safely used to treat diabetes. The recombinant anti-clotting agent tissue plasminogen activator (tPA) produced by expression in mammalian host cells is marketed for early treatment of heart attack. A monoclonal antibody against an estrogen receptor in breast tissue, anti-HER-2, is also expressed in mammalian host cells and marketed for treatment of breast cancer. Other proteins such as human growth hormone and interferon have found uses for treating a variety of human diseases. Research has been under way around the world to manufacture vaccines and many other therapeutic and diagnostic agents using recombinant DNA technology.

Techniques for "harvesting" relatively large quantities of protein from chimeric host cells often involve optimizing the placement of the foreign gene behind host genetic sequences that induce the transcription of an RNA message. These sequences are known as "promoters." A common technique employs specially constructed expression vectors made from "superproducer" plasmids that contain powerful promoters. It is possible using well-developed techniques of this sort to induce a transformed bacterial cell to produce so much of the desired protein that it comprises 40% of the total cellular protein. Human insulin, used daily by diabetics, is now made on a vast scale using these techniques.

Once the bacterial cell has been induced to express sufficient quantities of the protein, the protein must be extracted and isolated to be of any use. There are many ways this can be done. One method is to collect the bacteria from the liquid culture medium in which they grew and express the desired protein, then kill the bacteria cells and physically separate the desired protein from extraneous bacterial debris (e.g., through biochemical separation). A second method is to exploit cellular factors to induce secretion of the desired protein from the host cell into the liquid medium. Because the medium contains few cellular proteins, purification of the desired protein is simplified. A related technique is to keep the cells alive in a special medium and continuously "milk" the protein product. This can be done by attaching the cells to a porous surface and growing them in a "bioreactor." Of course, to use this method the desired protein must not only be expressed inside the cell, but the cell must be made to excrete the protein. Thus, there are a number of technical problems associated with such a technique. It does have the distinct advantage, however, of operating continuously and keeping the bacterial cells alive — an advantage that makes this approach superior in some respects.

Because these process technologies can be effectively protected by trade secrets, it is thought that many advances in this area have been made by biotechnology firms but not widely publicized. Because at least some biotechnology products require regulatory approval prior to marketing, however, the details of these advances may become known as more products reach the commercialization stage and undergo federal review.

B. Gene Therapy

In most organisms, growth takes place by cell division. When a cell divides, the double-stranded DNA in the nucleus is separated by a "DNA polymerase" along the bonds that hold the base pairs together. The two "single" strands of DNA are then "replicated" by the DNA polymerase using ambient bases in the nucleus to form two pairs of double-stranded DNA, each new strand being complementary to the parent strand. Because of the complementary nature of DNA, each pair will be identical. Each then goes to one of the new daughter cells.

Unlike bacteria, the reproduction of "higher" organisms requires the formation of gametes, reproductive cells with only half the usual complement of chromosomes. During sexual reproduction, these combine with the gametes of another similar species to form a new organism with genetic information from both parents.

When the parent DNA is free of functionally significant defects, reproduction simply creates a new healthy organism. But where one or both parents have genes that somehow cause disease, the offspring will inherit the disease genes and in the case of a homozygous organism (one in which both copies of the gene are defective) the organism may display the disease trait. A new horizon in biotechnology applications considers ways to treat these genetically inherited diseases.

Genetic diseases are usually due to the presence of an abnormal gene product, over- or under-expression of a gene product, or the absence of a necessary gene product. The presence of a single dominant abnormal gene is the basis of some diseases (e.g., osteogenesis imperfecta, neurofibromatosis); however, most human genetic diseases are caused by the unfortunate inheritance of two recessive genes. Initial proposals for gene therapy concentrated on the treatment of diseases such as adenosine deaminase (ADA) deficiency and hypoxanthine guanine phosphoribosyltransferase (HGPT) deficiency, which are caused by recessive genes that result in the production of an abnormal gene product. Such diseases could be corrected by inserting a single copy of the normal gene anywhere in the target genome. Other diseases are caused by defects in (or the absence of) proteins that regulate gene expression. Correction of these requires the precise placement of a copy of the normal gene, which is as yet very difficult. In the future, it may be possible to correct diseases caused by the presence of dominant genes by carefully and selectively removing the gene in question, perhaps prior to fertilization, in the early stages of development, or even in mature individuals. At the experimental level the current approaches have not proven fruitful; however, work continues to increase the efficiency of gene transfer.

The first step toward any form of gene therapy is the identification of the gene responsible for the disease. The list of diseases whose genes have been identified remains limited, although many are currently under investigation. Under first generation gene therapy approaches, a normal gene is inserted into diseased cells in the laboratory (i.e., outside the body, or ex vivo). The "treated" cells are then put back into the patient. For example, in one such therapy bone marrow cells are removed from a patient and transfected with the corrected gene. The successfully transfected cells are then reintroduced into the host organism. However, as is typical in first-generation gene therapy, this approach brings complications. The patient's own bone marrow must be destroyed before the transfected bone marrow is infused, so that the transfected bone marrow can grow. Another approach for delivering a gene is based on in vivo gene therapy, in which the viral vector containing the therapeutic gene is administered directly to the patient. In vivo gene therapy has been tested for the treatment of cystic fibrosis, cardiovascular diseases, infectious diseases such as AIDS, and cancer.

At present, first generation gene therapy involves physical (microinjection), chemical (charged lipid carriers of DNA), and viral methods to introduce the normal gene in chosen target cells. The viral "vectors" containing the inserted therapeutic DNA are attenuated or modified versions of viruses that are incapable of replicating in the patient, but retain the ability to efficiently deliver DNA to the cell. These "vectors" also contain promoters that turn the normal gene on and off. Much research has gone into characterizing precise promoters. For example, with the use of sophisticated promoters, the gene for hemoglobin (an oxygen-carrying component of blood) can be turned on in cells where it is needed, but turned off in other cells where it would be useless or possibly harmful. In the future, vectors may be developed that infect specific types of cells, such as liver cells or neurons. This would enhance the "targeting" of specific cells for treatment.

A second generation of gene therapy, still largely experimental, would take a different approach to introducing genes into the nucleus. Instead of using a cellular virus to transfect the cells, this approach would introduce gene sequences by "disguising" them as a molecule recognized by the cell and passing them through the cell membrane. Such an approach would combine a specific gene sequence with a receptor ligand that binds to a particular receptor on the outside of the cell. Once the genetic material was introduced, it may be possible to have it exist separately from the chromosomes as a free-standing piece of DNA. This would minimize the chances of the inserted sequence interfering with other sequences in the host cell's genome.

The main questions associated with gene therapy are how well the transfected gene will function and whether the presence of the normal gene will be enough to cure the patient. Because the vectors used to execute gene therapy are often derived from viral sources, they must be chosen and engineered carefully. In addition, a host of religious/ethical/policy concerns surround the technology. These concerns explain why, so far, gene therapy has been reserved for very serious diseases, affecting a limited number of individuals and used largely on otherwise terminal patients, who are often willing to consent to the use of experimental therapies.

The techniques described so far primarily involve infusing diseased bodies with normal cells. Another proposal is to introduce normal cells that actually incorporate into the germline or reproductive cells of the body—which means the "normal" gene would be passed on to the next generation, and indeed become a permanent part of the genome. A standard technique for making transgenic animals involves microinjecting a genetic construct containing a desired gene so that the resulting offspring express it. These animals are capable of passing the incorporated gene to their offspring. Researchers have successfully generated small transgenic animals such as mice. Efforts continue to produce transgenic livestock capable of expressing large amounts of a desired gene product—in milk, for example. For some, this implicates a number of very serious ethical issues. There is likely to be a good deal of discussion on these issues as the technology develops.

C. Polymerase Chain Reaction

The polymerase chain reaction (PCR) is an in vitro method for amplification of specific lengths of nucleic acid sequences, such as DNA, or in the case of RNA a cDNA created using reverse transcriptase. The PCR technique requires a researcher to first characterize two relatively short oligonucleotide sequences called primers that flank the targeted region. These primers are single-strand DNA oligonucleotides that are complementary to opposite strands (A and B) of the DNA and that function as delimiting markers which direct the synthesis (the extension of nucleic acid sequences) by a DNA polymerase. Primer A is complementary to a short section of Strand B; Primer B is complementary to a section of Strand A. The region of interest that will be amplified is between Primer A and Primer B.

The first step is denaturation of DNA, which causes the tightly coiled DNA to unwind and the two strands to separate. Denaturation is carried out by heating the DNA to approximately 95° Celsius in an aqueous solution that includes Taq DNA polymerase, a heat-stable enzyme that is able to withstand repeated cycles of heating and cooling. At 95° the two complementary strands of DNA, Strand A and Strand B, separate into single strands. The second step is cooling to approximately 50-60°, at which point the primers bind selectively and rapidly to the template DNA. Primer A binds to Strand B because Primer A is complementary to a specific sequence of Strand B, and since Strand A is complementary to Strand B, Primer B binds specifically to Strand A.

The third step is the addition of deoxyribonucleotide triphosphates (dNTPs) by Taq DNA polymerase, which occurs at 72°. Taq DNA polymerase lengthens Primer A by adding complementary dNTP's to one end of the bound primer (technically, the 3′ end, named for its position on the "backbone" of the DNA molecule), according and complementary to the template provided by Strand B. dNTP addition extends Primer A through the region of interest. After this cycle, Strand B is bound to a complementary Strand A′ (comprising the primer and the added dNTPs), that is identical to the portion of Strand A of interest. Similarly, at the end of one cycle, Strand A is bound to a complementary sequence Strand B′, identical to Strand B, but shorter because Primer B initiated synthesis at a particular point on Strand B.

The original DNA strands A-B have become two copies, A-B′ and A′-B. The cycle usually takes one or two minutes. During the second cycle, denaturation separates the two pairs into four separate strands. However, when two new primers bind the DNA strands, Primer A will bind to both strands B and B′. Therefore, extension of the Primer A bound to one of these template strands (B or B′) will result in assembly of another strand complementary to the target sequence. The extension of two B primers to strands A and A′ will result in the creation of two complementary target sequences. The extension of full-length Strands A-A′ and B-B′ will result in eight products similar to those from the first cycle.

Developed by Randall Saiki, Henry Erlich, and Kary Mullis at Cetus Corporation and patented in 1987, this iterative technique has become automated. Twenty cycles yields up to a millionfold amplification (220). The error rate for a single base pair substitution (adding the wrong nucleotide once) was 1 in 9,000 in one instance. Unless the error occurs in an early cycle, the presence of some incorrect sequences will not affect most applications. The technique requires very little material (Strand A-B), and has been used to amplify DNA from a single cell. This allows multiplication of sequences, which makes possible detection of certain bacterial and viral pathogens, such as HIV, hepatitis B, and chlamydia. Similarly, certain kinds of cancer with unique DNA sequences may be detected.

PCR has significant implications for researchers involved in mapping the human genome. For example, Lap-Chee Tsui and Francis Collins used PCR during their sophisticated search for the gene whose mutations cause cystic fibrosis, the most common genetic disorder among Caucasians. Their identification may allow prenatal diagnosis and development of treatments for the disease. PCR was also applied in a prenatal diagnosis of sickle-cell anemia, phenylketonuria mutations, Huntington's disease, diabetes, certain kinds of leukemia, B-thalassemia mutations, and other autosomal recessive and X-linked disorders. In addition, PCR may be used to amplify DNA found in a forensic sample, or archival material such as formalin-fixed paraffin-embedded tissue blocks, and old Guthrie cards (newborn blood spots stored on filters) for retrospective analysis. These diagnostic techniques involve starting with PCR, and then continuing with DNA sequencing, allele-specific oligonucleotides (ASO), or DNA probes to detect nucleotide substitutions, large deletions, chromosomal translocations, or the presence of a particular gene or polymorphism.

D. Monoclonal Antibodies

Antibodies are naturally occurring immune system proteins. An antibody binds to a specific site (the epitope or antigenic site) on an exposed surface of foreign material (e.g., a particular part of an invasive bacterium), marking that material for destruction or removal. Antibodies are produced by a subset of lymphocytes called B cells. These cells recognize and respond to an epitope on an antigen by reproducing or cloning themselves and then producing antibodies specific to that epitope. Because the body is exposed to many different antigens, the blood of a vertebrate will contain antibodies to many different antigenic substances.

Scientists have long used the ability of antibodies to recognize antigens as a tool to identify or label particular cells or molecules and to separate them from a mixture. Traditionally, antibodies used for these applications have been derived from the blood serum of an antibody-producing animal immunized or exposed to the antigen.

Recent advances have made it possible to isolate and cultivate a single clone of lymphocytes to obtain a virtually unlimited supply of antibodies specific to one particular antigenic site. These antibodies are known as monoclonal antibodies because they are produced by a laboratory culture of cells (a cell line), grown from a single antibody-secreting cell. Monoclonal antibodies are made from production vehicles known as hybridomas. These are produced by fusing a specific cancer cell, the myeloma cell, with spleen cells from an animal, generally a mouse, that has been immunized with the antigen. A subset of the fused hybrid spleen and myeloma cells produce antibodies to the antigen initially injected into the mouse. The hybridomas are then put into individual test tubes so that there is only one hybridoma per tube. Each hybridoma then reproduces itself and produces one type of monoclonal antibody. (The basic technique was developed in 1975 by G. Kohler and C. Milstein, who received the Nobel Prize in medicine in 1984 for their work.) The net effect is a virtually unlimited supply of identical antibodies, directed to only one epitope on an antigen. This is a major advance over the traditional isolation of antibodies from serum, because the traditional technique yields many different antibodies that bind to many different epitopes on many different antigens.

Many diagnostic tests use monoclonal antibodies to detect cells, infectious organisms, or minute levels of blood and fluid components (such as hormones and proteins). For example, home testing kits are now available that use monoclonal antibodies to identify the heightened hormone levels in blood or urine that signal the beginning of pregnancy. Many other laboratory applications have been developed that use monoclonal antibodies to screen for the presence of antigens signalling diseases such as hepatitis and malaria. In addition, researchers have used monoclonals to more precisely characterize and identify cells and other antigens of interest.

The binding specificity of monoclonal antibodies makes them useful as biopharmaceuticals. They can be developed to bind and inactivate a circulating compound or cell surface receptor that causes human disease. Monoclonal antibodies have shown promise for the treatment of Crohn's inflammatory disease, colorectal cancer, and breast cancer.

E. *The Human Genome Project*

Until recently, scientists have worked primarily with small portions of human genetic material, usually one or a few genes of interest. However, as this work has progressed, a consensus among scientists began to emerge concerning a major new project: collecting a list of all the genetic information in the human genome. The international, collaborative effort that has grown out of this consensus is now known as the Human Genome Project.

As currently envisioned, the Project has two stages. In the first stage, the genome was mapped at a relatively high level. This means that researchers systematically determined the linear order of genes on each chromosome using applied genomics to identify and localize genes in a process known as transcript mapping. Messenger RNA transcripts from expressed genes are converted to complementary DNAs (cDNAs), which are then sequenced and mapped to sites on particular chromosomes. An example of a genetic map is shown in Figure A-5.

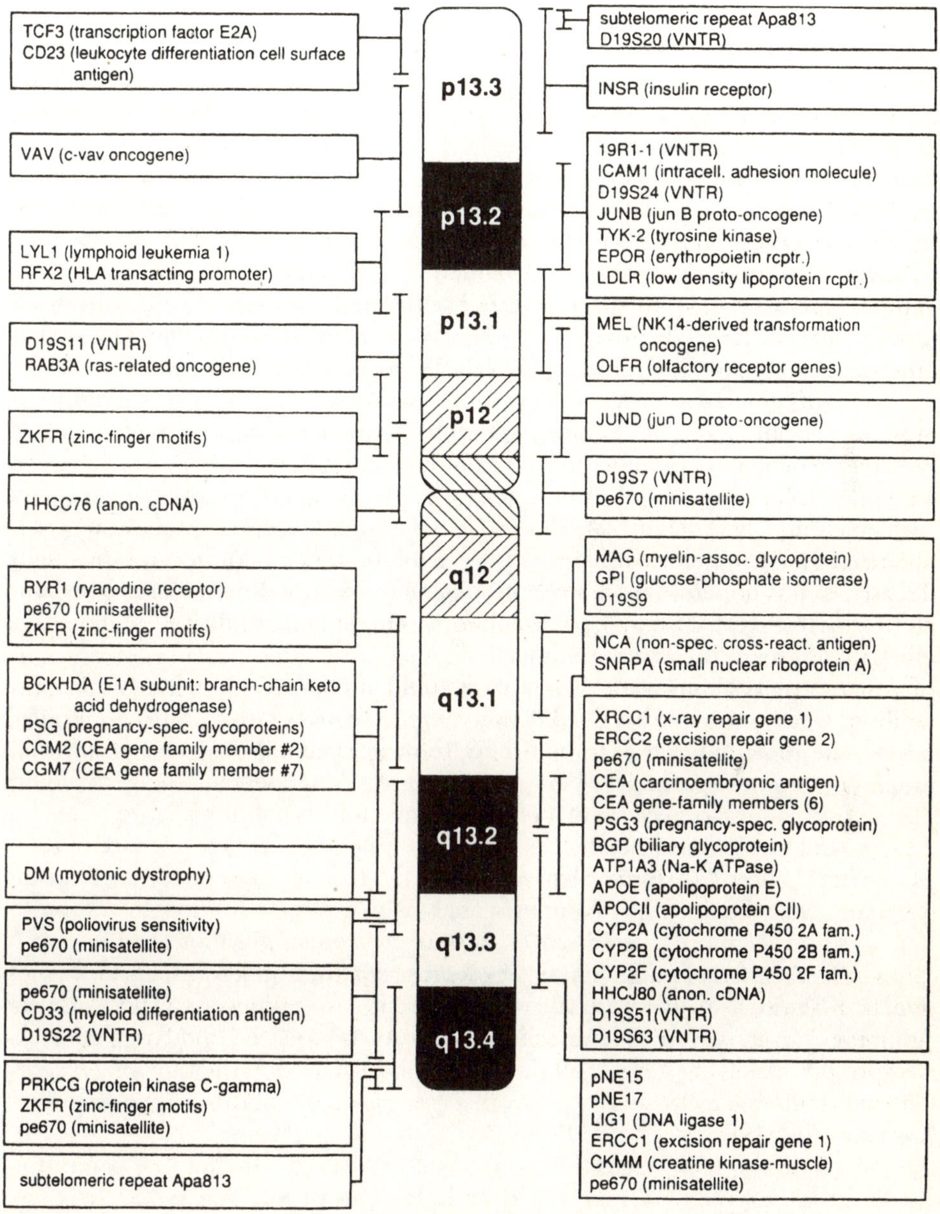
TCF3 (transcription factor E2A)
CD23 (leukocyte differentiation cell surface antigen)
VAV (c-vav oncogene)
LYL1 (lymphoid leukemia 1)
RFX2 (HLA transacting promoter)
D19S11 (VNTR)
RAB3A (ras-related oncogene)
ZKFR (zinc-finger motifs)
HHCC76 (anon. cDNA)
RYR1 (ryanodine receptor)
pe670 (minisatellite)
ZKFR (zinc-finger motifs)
BCKHDA (E1A subunit: branch-chain keto acid dehydrogenase)
PSG (pregnancy-spec. glycoproteins)
CGM2 (CEA gene family member #2)
CGM7 (CEA gene family member #7)
DM (myotonic dystrophy)
PVS (poliovirus sensitivity)
pe670 (minisatellite)
pe670 (minisatellite)
CD33 (myeloid differentiation antigen)
D19S22 (VNTR)
PRKCG (protein kinase C-gamma)
ZKFR (zinc-finger motifs)
pe670 (minisatellite)
subtelomeric repeat Apa813
p13.3
p13.2
p13.1
p12
q12
q13.1
q13.2
q13.3
q13.4
subtelomeric repeat Apa813
D19S20 (VNTR)
INSR (insulin receptor)
19R1-1 (VNTR)
ICAM1 (intracell. adhesion molecule)
D19S24 (VNTR)
JUNB (jun B proto-oncogene)
TYK-2 (tyrosine kinase)
EPOR (erythropoietin rcptr.)
LDLR (low density lipoprotein rcptr.)
MEL (NK14-derived transformation oncogene)
OLFR (olfactory receptor genes)
JUND (jun D proto-oncogene)
D19S7 (VNTR)
pe670 (minisatellite)
MAG (myelin-assoc. glycoprotein)
GPI (glucose-phosphate isomerase)
D19S9
NCA (non-spec. cross-react. antigen)
SNRPA (small nuclear riboprotein A)
XRCC1 (x-ray repair gene 1)
ERCC2 (excision repair gene 2)
pe670 (minisatellite)
CEA (carcinoembryonic antigen)
CEA gene-family members (6)
PSG3 (pregnancy-spec. glycoprotein)
BGP (biliary glycoprotein)
ATP1A3 (Na-K ATPase)
APOE (apolipoprotein E)
APOCII (apolipoprotein CII)
CYP2A (cytochrome P450 2A fam.)
CYP2B (cytochrome P450 2B fam.)
CYP2F (cytochrome P450 2F fam.)
HHCJ80 (anon. cDNA)
D19S51(VNTR)
D19S63 (VNTR)
pNE15
pNE17
LIG1 (DNA ligase 1)
ERCC1 (excision repair gene 1)
CKMM (creatine kinase-muscle)
pe670 (minisatellite)

Figure A-5

A complete map of the roughly 30,000 genes in the human body containing the exact nucleotide sequence of the human genome was produced in late 2000. The entire sequence of the genome provides a huge database of the approximately 3 billion base pairs of nucleotides in human DNA. Such a database will make it much easier to study and detect genetic abnormalities associated with diseases. This concept produced results, even before the sequencing of the entire human genome is complete. As a benchmark of "normal" gene sequences is cataloged, particular genetic mutations have been correlated with increased susceptibility to diseases from polycystic kidney disease and Alzheimer's to breast cancer, colon cancer, and diabetes. By late 1997, approximately 100 disease-related genetic loci had been mapped. An overview of the chromosomal location of genes and their variants associated with disease is provided by the Human Gene Mutation Database (Krawczak M., Cooper D.N. (1997) Trends Genet. 13:121-22). This information is of practical diagnostic importance to human molecular geneticists, physicians, and genetic counselors. By late 1998, approximately 16,000 mutations in over 800 genes had been discovered, indicating the sequence variability of disease-associated genetic lesions. Armed with this knowledge, researchers may be able to diagnose diseases more easily and ultimately develop therapies for them. Prenatal diagnosis of disease susceptibility, for instance, will likely become available on a far broader scale than now, which will alert parents and doctors to the need for increased surveillance of the child for early diagnosis and treatment. Moreover, it is hoped that once common genetic abnormalities are identified and understood, therapies may be developed to offset or neutralize the deleterious physical effects of the errant genes.

The final 3 billion base pair sequence would, if printed out on paper, take up a million pages and require a third of a lifetime to read. Genes constitute 2% of the whole; the remaining 98% contains hundreds of thousands of pages of information of unknown significance. Very rapid progress in sequencing the genome has been possible in part because of a new technique for assembling a large collection of gene fragments for all sequences that actually code for genes. As more of these fragments, called Expressed Sequence Tags (ESTs), are generated, overlapping sequences may be linked by computer analysis to generate full-length coding sequences for even more rapid discovery of genes. Patent applications on these ESTs pose difficult issues of patentability. See Rebecca S. Eisenberg & Robert P. Merges, Opinion Letter as to the Patentability of Certain Inventions Associated with the Identification of Partial cDNA Sequences, 23 AIPLA Q.J. 1 (1995).

Appendix B
Introduction to Computer Technology

Computer software and Internet commerce are among the fastest growing and most promising industries in the United States. A government report notes that more than half of United States nonfarm industries either produce information technology (IT) directly or invest in and use information technology products and services. *U.S. Commerce Department, The Emerging Digital Economy II* (1999). The information technology sector of the United States economy represented 8 percent of gross domestic product (GDP) in 1999, accounting for more than $700 billion. Computer software accounted for $200 billion of this total. The IT sector of the United States economy has steadily increased its share of the GDP in the 1990s and remains robust. These patterns can be seen throughout the global economy. *A World Gone Soft: A Survey of the Software Industry, The Economist* (May 25, 1996).

While firms such as Intel, Microsoft, Compaq, IBM, Cisco, AOL, and Amazon.com attract much of the attention in the IT marketplace, the IT industries touch almost all aspects of the modern economy. For example, traditional manufacturing firms, such as General Motors, make significant use of computers, computer software, and computer networks in their businesses. Automobile manufacturers use CAD ("computer-aided design") software to design new vehicles, CAM ("computer-aided manufacturing") software to assemble these designs, and digital networks to purchase component parts and to distribute vehicles to customers. Few businesses, government agencies, schools, or other organizations operate today without extensive use of computer technology and digital networks. An increasingly wide array of companies—whether they sell information, cars, or anything else—use digital networks, principally the Internet, to market products and transact business. While it is easy to scoff at estimates of the potential growth of global electronic commerce because they seem like (and probably are) rank guesswork, electronic commerce has surpassed what once seemed like exaggerated estimates. The *Emerging Digital Economy II* report notes that in 1997 "private analysts forecast that the value of Internet retailing could reach $7 billion by 2000—a level surpassed by nearly 50 percent in 1998." While the popular press has mainly concentrated on the growth of Internet business-to-consumer companies, such as Amazon.com, many industry experts believe that business-to-business ecommerce will be larger and have more far-reaching implications for the United States economy. Internet technology make possible great efficiencies in the ways businesses are structured, distribute product and service information, and conduct transactions. The extent of these possibilities are just beginning to emerge.

This appendix describes the early history of computing in section A. It explains how computers work in very basic terms in section B and introduces the principal models of software engineering in section C.

A brief note on methodology in this appendix is in order. Our goal in providing information about the computer industry is to offer students essential background on how computer software works and how markets for computer software function. We do not intend this introduction to provide a complete understanding of the field. We reference a number of excellent sources providing detailed background in our summaries below. The reader who is interested in more detail than we can possibly provide here should seek out these sources.

A. The Early History of Computers

This section introduces the reader to the early history of computer hardware and software. The purpose of this section is to describe the enormous changes that occurred in the early days of the computer industry in order to provide context for the discussions that will follow. This section does not describe events up to the present day. Rather, more recent developments (including the growth of the Internet) are discussed in the sections that follow, and in chapter VII of the main text.

Following the invention of the abacus approximately 5,000 years ago, the field of computing machines did not develop significantly until the 18th century. Leonardo da Vinci (1425-1519) sketched some designs for mechanical adding machines. Blaise Pascal (1623-1662) invented and built the "Pascaline," a sophisticated mechanical device for counting. Although not commercially successful because of its cost and delicate construction, the counting-wheel design served as the basis for most mechanical calculators until the 1960s. At the turn of the 19th century, Joseph-Marie Jaquard (1752-1834) introduced a new loom technology that used punched cards to control the movement of needles, thread, and fabric to create distinctive patterns through a binary mechanical automation technology. In the mid-19th century, Charles Babbage envisioned mechanical devices (the Difference Engine and the Analytical Engine) to perform arithmetic operations. His designs, involving thousands of gears, proved impractical. One of his students, Lady Ada August Lovelace, proposed the use of punched cards to automate the operation of such devices.

Toward the end of the 19th century, a United States Census Bureau agent named Herman Hollerith developed a punched-card tabulating machine to automate the census. Drawing upon the use of "punched photography" by railroads (to encript passengers' hair and eye color on tickets), Hollerith proposed the encoding of census data for each person on a separate card which could be tabulated mechanically. After developing this technology for the Census Bureau, he formed the Tabulating Machine Company in 1896 to serve the growing demand for office machinery, such as typewriters, record-keeping systems, and adding machines. The company grew through the expansion of its business and merger with other office supply companies and in 1924, Thomas J. Watson, the company's general manager, changed the company's name to International Business Machines Corporation (IBM). By the late 1920s, IBM was the fourth largest office machine supplier in the world, behind Remington-Rand, National Cash Register (NCR), and Burroughs Adding Machine Company. IBM made numerous improvements to tabulating technology during the 1920s and 1930s, eventually developing a machine that could compare cards, a significant innovation that enabled machines to perform simple logic (if-then) operations.

1. Computer Hardware

The critical breakthrough defining modern computers was the harnessing of electrical impulses to process information. In 1939, Professor John Vincent Atanasoff, with the help of his graduate student Clifford Berry, developed the first electronic calculating machine. This computer could solve relatively complicated physics computations. They built a more sophisticated version, the ABC (Atanasoff Berry Computer), in 1942. Shortly thereafter, driven in part by wartime demand for computing technology, Professor Howard Aiken, funded in substantial part by IBM, developed a massive electromechanical computer (MARK I). This machine contained three-fourths of a million parts, hundreds of miles of wire, and was 51 feet long, 8 feet high and 2 feet deep. It could perform three additions a second and one multiplication every six seconds. Although it used an electric motor and a serial collection of electromechanical calculators, the MARK I was in many respects similar to the design of Babbage's analytical engine.

At about this same time, Dr. John Mauchly persuaded the United States Army to fund the development of a new computing device to compute trajectory tables to improve the targeting of ordnance. Mauchly envisioned using vacuum tubes rather than mechanical relays to store binary information.

In collaboration with J. Presper Eckert, Jr., a young electrical engineer, Mauchly completed the Electronic Numerical Integrator and Computer (ENIAC) in 1946. This computer occupied 15,000 square feet, weighed 30 tons, and contained 18,000 vacuum tubes. It operated in decimal (rather than binary) and therefore needed 10 vacuum tubes to represent a single digit. The ENIAC could perform over 80 addition or 8 multiplication operations per second.

The flexibility provided by programmability greatly enhanced the utility of computers. In the early 1950s, Mauchly and Eckert developed the first commercially viable electronic computer, the Universal Automatic Computer (UNIVAC I) for Remington-Rand Corporation. Limitations on electronic technology, however, constrained the computing power of the first generation of computers. Vacuum tubes, which were bulky, failed frequently, consumed large amounts of energy, and generated substantial heat. This first generation of computers was programmed in binary code (zeros and ones), which could be understood by only a few specialists. IBM introduced its first commercial computer, the IBM 650, in 1954. IBM made incremental improvements to this technology and emerged as the market leader.

Because computers use binary electronic switches to store and process information, the great challenge for the computer industry was to reduce the size of these switches. The second generation of computers replaced vacuum tubes with transistors, which were smaller, required less power, and ran without generating significant heat. This and other innovations in data storage technology made computers smaller, faster, and more reliable. The first scientific computer using transistors was the IBM 7090. A second important innovation of this era was the development of high-level computer languages, which enabled computer specialists to write programs using coded instructions that resemble human language. The IBM 705, introduced in 1959, used the FORTRAN language processor. This model became the standard machine for large-scale data processing companies. Notwithstanding these innovations, computers of this generation remained complex and expensive because circuits had to be wired by hand.

The development of integrated circuits enabled computer manufacturers to incorporate many transistors within the layers of semiconductor material. The greater computing power and efficiency of computers brought the cost of data processing services within the reach of an increasing number of businesses. Many businesses contracted with companies specializing in data processing services. A few acquired their own computers. IBM's 360 series of mainframe computers emerged during this period as the market leader. These machines used a single machine language. As businesses upgraded their equipment within the 360 series, they could continue to use the same computer programs. This increased the benefit of owning a computer (rather than outsourcing data processing) and expanded the mainframe market. This larger market generated greater demand for computer programmers and spawned new companies to provide computer-related services. An independent software industry began to emerge. The 1960s and 1970s also witnessed the implementation of time-sharing and telecommunication technologies, which enabled multiple users to access a computer from remote terminals. In addition, computers developed during this period could handle multiple tasks simultaneously (parallel processing and multiprogramming).

In 1965, the Digital Equipment Corporation (DEC) introduced the first minicomputer, the PDP-8 (Programmed Data Processor). This machine was substantially smaller and about one-fourth the price of mainframe computers. Minicomputers substantially widened the market for computers and computer programmers. Domestic consumers purchased 260 minicomputers and 5,350 mainframes in 1965. Minicomputer unit sales surpassed mainframe unit sales by 1974. By the 1970s, computers incorporated "semiconductor chips" no larger than a human fingernail and containing more than 100,000 transistors. As chip technology advanced, the size of computers decreased while their computing power increased. Semiconductor chips today can hold many millions of transistors. For the past two decades, the memory capacity of a semiconductor chip has doubled approximately every 18 months.

In the early 1970s, Intel Corporation developed the microprocessor, a chip that contains the entire control unit of a computer. Very large-scale integration (VLSI) technology led to the development of the microcomputer. Originally oriented toward computer hobbyists, microcomputers came to dominate the computer industry by the mid-1980s. With its Apple II computer system, which included a keyboard, monitor, floppy disk drive, and operating system, Apple Computer vastly expanded the market for computers. Microcomputer unit sales surpassed minicomputer unit sales in 1976, their second year of production. By 1986, sales of microcomputers (costing less than $1,000) reached approximately 4 million units and produced revenues of almost $12 billion, giving microcomputers the largest share of computer industry revenues.

The rapid growth of the microcomputer sector of the industry spurred the emergence of independent software vendors (ISVs) who developed mass marketed programs for this growing market of versatile machines. Microcomputer owners were anxious to experiment with different programs. The cost of developing software for these machines was relatively low, product cycles were short, and there was constant pressure to upgrade products.

IBM entered the microcomputer market in 1981 with its PC (Personal Computer) product. The IBM PC utilized an Intel microprocessor (16-bit 8088 chip) and an operating system (PC-DOS (Disk Operating System)) licensed from Microsoft, then a fledgling company. Microsoft's MS-DOS is a single-tasking, single-user operating system with a command-line interface. Like other operating systems, MS-DOS oversees operations such as disk input and output, video support, keyboard control, and many internal functions related to program execution and file maintenance.

IBM's strong trademark in the business computer industry as well as its vast distribution network for computers enabled IBM to rapidly attract customers for its PC product. Many independent software vendors (ISVs) and hardware manufacturers developed and marketed software and peripheral products to run on the IBM PC. IBM actively encouraged independent software vendors and the makers of peripheral equipment (e.g., printers, monitors) to develop products for the PC. While promoting an "open architecture" with regard to these sectors of the industry, IBM included a specialized chip (BIOS)[1] for transferring data within the PC that hindered other OEMs from offering fully compatible computer systems. This enabled IBM to charge premium prices for its PC product.

---

1. The BIOS chip is the set of essential software routines that test hardware at startup, start the operating system, and support transfer of data among hardware devices. It is typically stored in read-only memory (ROM) so that it can be executed when the computer is turned on. The BIOS is usually invisible to computer users.

The rapid success of the IBM PC spurred independent software vendors (ISVs) to develop a wide range of programs to run on the IBM PC, including word processing, database, and spreadsheet software. For example, Lotus Corporation developed a version of the spreadsheet Visicalc (originally designed to run on the Apple II) to run on the IBM PC platform. Within a year of its introduction, Lotus 1-2-3 eclipsed Visicalc and became the spreadsheet market leader. Its success led to the label "killer app," to designate an application program of such widespread popularity that it spurs sales of a hardware/operating system platform. This reinforced the importance of owning an IBM PC, thereby adding further to the value of IBM's trademark in the microcomputer market. The powerful IBM trademark and the growing availability of software designed to run on the IBM/Microsoft platform catapulted IBM to a dominant position in the early microcomputer marketplace and greatly encouraged the dissemination of microcomputers. It also made Microsoft and Intel well-recognized trademarks in the microcomputer industry.

Microsoft and Intel retained rights to market their products to other OEMs (original equipment manufacturers) in the computer industry. Because of the availability of software designed to run on the IBM PC platform, other OEMs sought to develop computer systems that could run the growing supply of IBM-compatible software. Although Microsoft's MS-DOS operating system could be licensed in the marketplace, IBM refused to license its BIOS chip. As a result, other OEMs could not fully emulate the internal operations of the IBM PC readily and some software designed for the IBM PC did not operate satisfactorily on the computer systems of other OEMs. As a result, consumers strongly favored IBM PCs in the marketplace. Other computer companies had little choice but to offer IBM PC compatibility in order to compete effectively in the microcomputer marketplace. Computer manufacturers that developed their own platform did not fare well. With the exception of Apple Computer, which maintained a niche in the marketplace, no serious alternative to the IBM PC/MS DOS platform survived.

By 1984, Compaq developed a BIOS chip that successfully ran software developed for the IBM PC. Later that year, Phoenix Technologies Ltd. developed a fully IBM PC-compatible ROM BIOS which it licensed to a broad range of OEMs. Other OEMs entered the market for IBM PC-compatible computer systems. As consumers became increasingly confident that software application programs designed for the IBM PC would run on the computer systems of other OEMs, these PC "clone" computers eroded IBM's dominance of the marketplace by offering lower prices, wider selection, and additional features.

By 1986, numerous OEMs competed in the IBM-compatible/MS-DOS marketplace and IBM's hold on the market had significantly loosened. The broad range of software available for the IBM-compatible/MS-DOS platform enabled MS-DOS to emerge as the de facto operating system standard in the industry by the late 1980s. At about that time, Microsoft began developing the Windows operating system platform incorporating a graphical user interface. The Windows platform was backward-compatible with MS-DOS (i.e., applications designed to operate in the MS-DOS environment could run on the Windows platform as well). Most MS-DOS users as well as new computer users migrated to the Windows platform, which has been the dominant platform since the mid-1990s.

2. Computer Software

During the 1940s and 1950s, hardware and software innovation were integrated. The development of computer software was a highly specialized field of scientific research done by academic, government, and government-funded commercial research laboratories. Those who worked with computers had significant scientific and technical expertise. The computer languages and techniques for developing programs were just being created and tested. Computers had relatively narrow use in scientific, military, and space applications. Each computer was unique and programming was specialized for each machine.

IBM became and remained the dominant force in the commercial computer industry from the 1950s until the early 1980s. During the 1950s and 1960s, IBM and other mainframe manufacturers (e.g., Burroughs, Raytheon, RCA, Honeywell, General Electric, Remington Rand) bundled operating system and application software with hardware for the same price, commonly through a leasing arrangement. During the early stages of the industry, this bundling arrangement made economic sense because there were relatively few computer applications and the hardware manufacturers were able to support these uses of their systems. As the industry developed, manufacturers encouraged their customers to share software among themselves through software sharing institutions. IBM formed and supported a user group named SHARE, which served as a clearinghouse for programming information and software for computer users. SHARE distributed software programs, including libraries of subroutines, algorithms published in technical journals, computer code published in textbooks, and in some instances, programs written to solve problems in specific areas. Those companies contributing to the software sharing "bank" were entitled to borrow the works of others.

As the industry developed and computers became increasingly powerful, versatile, and affordable, the sharing model began to break down. Those companies making substantial investments in software development were less willing to share these innovations with others. In addition, computer technology was diffusing from governmental and scientific uses to commercial applications.

Specialty software supply houses, such as Applied Data Research, Inc. (incorporated in 1959), emerged to provide customized and general purpose software in direct competition with the hardware manufacturers on a fee basis. This early software industry offered specialized services on a contract basis. They competed with the bundled (and hence, unpriced) software programs provided by mainframe manufacturers through mainframe sales and leases.

The advent of less expensive minicomputers as well as the growing versatility and computing power of mainframes spurred the independent software industry. By 1965, there were approximately 40 to 50 independent software suppliers. F. Fisher, J. McKie, & R. Mancke, IBM, and the U.S. Data Processing Industry: *An Economic History* 322 (1983). Applied Data Research introduced Autoflow, a flow chart program, which was the first internationally marketed computer program. International Computer Programs, Inc. (ICP) published catalogs of software programs. The independent software industry grew quickly. There were almost

3,000 vendors by 1968. In 1969, contract programming produced revenues of $600 million; software products generated another $20-25 million (Id. at 323). Nonetheless, this accounted for less than 10% of the amount spent for programming. The remainder was spent on programmers working in-house.

Founded in 1959, Computer Sciences Corporation (CSC) became a successful software company during the mainframe era. CSC began its business by designing, developing, and implementing software systems for computer manufacturers. Over the course of the 1960s, its computer programming business branched out to serve large companies outside of the computer industry and federal, state, and local governmental agencies. During this same period, CSC increasingly shifted its software toward the development of software products. It developed a range of products generally directed to business uses, including tax, accounting, and personnel management software.

IBM's increasing dominance of the computer industry led to antitrust scrutiny by the federal government. In addition, the costs of software development within IBM increased dramatically and there was increasing pressure within the company to price software separately. Following the lodging of the government's antitrust complaint in 1969, IBM voluntarily unbundled its hardware from application programs effective in January 1970. This event greatly expanded the business opportunities for independent software vendors. By 1975, there were over 1,000 software firms in the United States offering more than 3,000 products.

The proliferation of minicomputers in the early 1970s fostered the growth of independent software vendors and the shift away from custom programming and support services toward pre-packaged software products. The introduction of the microcomputer in the mid- to late 1970s dramatically changed the software industry. With relatively small investments, computer programmers could develop software for the growing numbers of microcomputer users. Beginning in the late 1970s, independent software vendors (ISVs) began selling through retail and other channels pre-packaged (i.e., non-customized) software products for use on microcomputers. Wordstar, Visicalc, and other independently developed software products dominated the early microcomputer software marketplace.

As noted in the discussion of computer hardware, Lotus Corporation developed a version of the spreadsheet Visicalc to run on the IBM PC platform. The powerful IBM trademark and the growing availability of software designed to run on the IBM/MS-DOS platform catapulted IBM to a dominant position in the early microcomputer marketplace and greatly encouraged the dissemination of microcomputers. These factors stimulated rapid growth in the software industry.

By the late 1980s, Microsoft had emerged as a dominant force in the computer industry. Its MS-DOS operating system was installed on the majority of microcomputers and its Windows graphical user interface platform was gaining acceptance in the higher end of the microcomputer marketplace. By 1991, Microsoft's operating systems were installed on almost 90% of microcomputers in the world. Building upon this success, Microsoft began bundling its office software products into an office suite of products (Microsoft Word word processing software, Microsoft Excel spreadsheeet software, Microsoft Access database software, and Microsoft Powerpoint presentation software). This marketing strategy has enabled it to become the leading seller in each of these product categories.

B. An Introduction to Computer Technology

Virtually unknown 50 years ago, computers literally surround most Americans in their daily lives today. In their most easily recognized form, mainframe and mini-computers can be found in most businesses, government offices, and schools. Microcomputers can be found on most business and home desktops for use in word processing, information storage, entertainment games, and electronic shopping. Less commonly recognized, computers can also be found in many home appliances, hand-held organizers, telecommunication devices, automobile dashboards, and elevators, among other places.

1. Computer Hardware

Computers use a binary base. By setting electrical switches to "on" (electrical current is flowing) or "off" (current is not flowing), early computers could create a single "bit" of information. That piece of information is read as either a 1 ("on") or a 0 ("off"). By translating information into a series of such 1s and 0s, computers could perform mathematical operations.

The first computing machines did not utilize computer "programs" in a form that we would recognize today. These machines were in essence a series of hardwired circuits constructed to perform one particular computational task. That is, the mathematical function performed by the computer was determined by the physical arrangement and structure of the circuits. The computers had to be rewired in order to perform a different function. These machines were comprised solely of what we call today "hardware"—the physical circuits that make up the machine.

During the late 1940s, scientists developed the first machines that could store and use encoded instructions or programs. This set of innovations dramatically increased the flexibility and usefulness of computers. Users could perform a variety of computational tasks without having to rewire the basic hardware of the computer. Instead, they could simply direct the computer to perform one of the functions that it had stored in its memory. The actual computer in these programmable or "universal" machines is the central processing unit (CPU). The CPU has two principal components: an arithmetic logic unit (ALU) which performs a basic set of "primitive functions" such as addition and multiplication, and a control unit which directs the flow of electric signals within the computer. In essence, a computer processes data by performing controlled sequences of primitive functions. As computers grew more powerful and the tasks they performed grew more complex, computer scientists relied on increasingly complex sets of instructions that are executed automatically by the computer. These sets of instructions are known as computer programs, and they will be the focus of most of this book.

The basic hardware of a modern microcomputer system includes a CPU, internal memory storage, disk drives or other devices for physically transferring data and programs into and out of the internal memory, and telephone or network interconnections for linking the computer with other computers. The internal memory of the computer typically features three types of information storage: random access memory (RAM), read only memory (ROM), and data storage memory ("disk space"). Data can be input into RAM, erased, or altered. RAM chips serve two information storage functions: they act as temporary storage devices for programs and data currently "running" on the computer, and they also serve as permanent memory for data or programs. ROM chips have information permanently embedded in the architecture of the chip, and that information can only be read (not altered) by the computer. ROM chips are used primarily to direct certain basic operating functions of the computer.

Computer engineers design the programming capability of a computer to suit the user's needs. By building more of the desired functions directly into the wiring of the computer, they can achieve more efficient processing for certain applications. Such "pre-designed" computers are known as "special purpose computers," because they are designed to perform only certain specific tasks. Their greater speed comes at a cost of less flexibility — that is, less ability to run a wide range of programs. This technological trade-off harks back to the early days of computer technology when all programs were hard-wired into the computer.

Advances in computer technology have made greater efficiencies of processing possible without the need to hard-wire the computer. Most modern computers, particularly personal computers, are "general purpose computers" which feature a high degree of programming flexibility. When a user has only a few computing needs or desires high-speed processing, however, she may still prefer to rely heavily upon internal programming.

Besides the internal memory and processing chips, computers are composed of input and output devices (sometimes referred to as I/O devices) and peripherals. These devices control the transfer of information into and out of a computer. Early programmers "input" information into a computer by changing the physical structure of its circuits, or (in more sophisticated models) by using "punch cards," which allowed computer users to write data for the computer in the form of holes punched in special note cards, and then feed that data into the computer in the form of stacks of such cards. Most modern computers use the typewriter keyboard as their primary input device, allowing users to enter data into the computer by typing it. The typewriter keyboard has been supplemented with other input devices, including the computer "mouse," the telephone line, and microphones coupled with voice recognition software.

The output devices of computers have also changed. Computers originally communicated data to humans in the form of lights that turned on or off, representing the bits of data produced by some computer operation. Advances in computer outputs include the development of the printer, the introduction of television-like computer monitors and screen displays, and telephone and cable output which can "send" data to a remote location.

2. Computer Software

Computer programs are the instructions that allow general purpose computers to be many different types of machines. When a computer is running a video game, the computer is a videogame machine. When it executes program instructions to enable users to write letters or reports, the computer is a word processing machine. When it carries out a programmed search for data in a large repository of information, the computer is an information retrieval machine. Programs can also be written for special-purpose hardware (e.g., the semiconductor chip that monitors the functioning of your toaster) when this will best achieve the developer's objectives.

Computer programs that operate on a single machine fall into two basic categories: "operating systems" and "applications programs". Operating system programs manage the internal functions of the computer. They coordinate the reading and writing of data between the internal memory, CPU, and the external devices (e.g., disk drives, keyboard, and printer); perform basic housekeeping functions of the computer system; and facilitate use of application programs. In essence, the operating system prepares the computer to execute the application programs and serves as an intermediary between the application program and the hardware of the computer. An applications program may order the deletion of a file, but generally, the operating system will actually carry through the details of this function.

Every computer needs an operating system to direct its functions and to manage other software that is run on the computer. Computers do not need, and many do not have, more than one operating system. Because the operating system controls the interactions between the user, the software, and the computer itself, an applications program must be "compatible" with a particular operating system if it is to interoperate with that program and run on a computer using that operating system. This compatibility requirement means that the designers of operating systems have some degree of control over the applications programs that will work with that operating system. Although the specifications of applications programming interfaces ("APIs") are sometimes published freely, often they are licensed from an operating system program developer. Microsoft, for example, licenses its APIs for its operating system programs to applications developers. An alternative way to get access to APIs is through a laborious process of reverse-engineering the program (of which more in Chapter VII of the main text).

Operating systems may control the execution of instructions in the central processing unit of the computer, but they generally do not perform specific tasks of interest to end users. Applications programs enable users to accomplish specific tasks with computers. Bookkeeping, statistical and financial analysis, word processing, and video game programs are among the many types of application programs available today.

Application programs are often developed to run on particular operating systems. The task of adapting an application program designed to run on one operating system to another operating system is often technically complex, time consuming, and costly. In recent years, software developers have developed programs that run on more than one operating system. In addition, translator programs have been developed to allow users to move files from one application program to another. Nonetheless, the problem of "compatibility" between applications programs and operating systems remains a major concern in the computer software industry. There is some hope that the Java programming language (about which there is more in subsection C) will enable programmers to write a program once and have it run on many machines.

The line between operating systems and application programs, while sharp in particular instances, may blur when programs once distributed as applications programs are integrated into an operating systems program. Microsoft, for example, has integrated a number of "add-on" features (such as compression software and even rudimentary word processing) into its operating system over time. By selling a bundled product that includes both an operating system and certain applications, Microsoft arguably provides more value to consumers, but also eliminates a competitive market for such add-on products. Microsoft's decision to make its Internet Explorer web browsing software an integrated part of the Windows operating system contributed to the U.S. Justice Department's decision to charge Microsoft with antitrust violations in the late 1990s. (We discuss this suit in more detail in Chapter VIII of the main text.)

3. Computer Networks

Early computers were self-contained machines. Data could be input into them (usually laboriously, by punch card) and after the computer processed the data, output would be produced. But the data never left the physical environs of a single computer. To move data from one computer to another, one had to take the output of one computer and transport it physically to and then input it into the new computer. Even when input and output became somewhat more efficient, for example, by use of magnetic storage media such as "floppy disks," the necessity for physical transfer remained.

Computer engineers recognized early on that great benefits would flow from the networking of computers. Networking has been around in specialized computing environments since the late 1960s, but it was not until the 1990s that the Internet afforded easy widespread access to large computer networks. As of June 1999, more than 171 million people around the world had Internet access and 37 percent of the U.S. population had Internet access at home or at work.

Networking technology allows computers to communicate with each other. This communication capability enables dispersed computer users to exchange information, for example, by electronic mail ("e-mail"). It also reduces the time and cost of transferring information regardless of physical location of computers.

Networked computing also allows computers — and people — to work together to achieve certain tasks that would take much longer to do alone.

Networking computers requires some basic technologies. First, some form of hardware must connect the two computers, either physically or virtually. If the computers are physically close to one another, this can be accomplished by stringing a cable between the two computers. If the computers are physically separated by large distances, networking requires that they be connected either via telephone lines or by some form of wireless communication. Long-distance connections require some form of hardware to convert digital computer data into a form that can be transferred over the telephone lines or narrowcast over the airwaves. One common kind of hardware to enable telephone transmissions of digital data is known as a "modem." Modems take digital computer data, convert it to analog (sound) form for transfer over telephone lines, and then reconvert it to digital form. Networking software (and sometimes hardware components as well) is required to enable computers to communicate with each other. When one computer sends data, the other must be able to process it. Interoperability issues thus arise in the context of computer networking design. The development of standard protocols for exchanging information has fostered the growth of networking.

Local Area Networks. Many organizations find it useful to develop "local area networks" (LANs) to link a number of computers so that members of the organization can share information and information resources. Local area networks provide all of the communications advantages described above — they enable employees to communicate by e-mail, to send files to one another, and share computing resources. In addition, LANs have made possible a form of specialization or division of labor among computers. Because the computers in a LAN are linked together in real time, it is possible to store files in a single central location and allow any computer to access them at any time. Rather than requiring each computer to be self-sufficient, certain (generally more powerful) computers can be designated central "servers" where files or programs are located. Individual "client" computers in the network can call up the files or programs on an as-needed basis. Because the server machines are generally faster and more powerful than ordinary computers, LANs offer not only centralized access but also quicker processing time.

LANs also facilitate group projects. "Workgroups" can add to or change the same document simultaneously over the network. This is a particular advantage for companies that rely on large information databases which must be regularly updated (for example, sales companies, airlines, hotels). Documents or databases that previously had to be changed by one central programmer can now be altered instantaneously to reflect current information. Because the "workgroup" is organized on the computer network, and not by physical sharing of documents, LAN workgroups also contribute to a more flexible organizational structure.

The mass corporate movement towards LANs has important legal implications as well. Information that used to reside on a single computer is now accessible by dozens or hundreds of computers, many physically remote from the actual location of the data. LAN administrators must worry about limiting access to their networks, guaranteeing the security of their data, implementing version control on revised documents, and monitoring access to certain "controlled" data (particularly applications software licensed for limited uses from third parties).

Other kinds of private networks also exist. Anyone with a computer and a modem can connect to private "dial-up" networks. Such "dial-up" connections normally take place over the telephone lines between individual computers and one or more central "host" computers operated by the private network administrator. Some are operated by information providers who allow people to sign on to the network and download information of particular interest to them. Others may serve as communications forums for people interested in particular topics. Individuals dial into the host computer and communicate with each other through the host, either in real-time or by leaving e-mail messages.

Large-Scale Public Networks. In the mid-1960s, researchers working for the Department of Defense's Advanced Research Project Agency (ARPA) began working on the problem of connecting its computers scattered around the United States at various universities and research laboratories to enhance memory storage and time-sharing capabilities. An important design objective of this system was that it not be vulnerable to breaking down in the event of a nuclear attack or other widespread disruption of telecommunications systems. It was initially believed that this would require 136 separate communication lines (17 computers times 16 divided by 2). These lines would be expensive and the cost would grow geometrically as more computers were added to the network.

At about that time, researchers at Rand Corporation and the National Physical Laboratory in England independently developed the concept of "store-and-forward packet switching" which avoided the problem of having to connect independently each node of a network to each other node. Packet switching technology, using techniques similar to those developed in the telegraph industry, allows multiple messages to flow through a common "backbone" line by transmitting information in smaller packets that contain the address of the destination. Large messages are broken into streams of packets that are sent as individual items into the network. Switching nodes pass the messages along and the receiving computer reconstitutes the full message. Such a system minimizes the risk that the entire system could crash by allowing information to travel along a variety of paths.

In 1970, the ARPANET successfully implemented packet switching technology in a four-node network. The ARPANET grew to more than 20 sites by 1971 and over 200 sites by 1981. The ARPANET paved the way for the Internet, an international collection of interconnected computer networks based on an open technical standard known as Transmission Control Protocol/Internet Protocol (TCP/IP). Computers implementing this protocol may connect to the Internet.

By 1996, more than 9 million host computers were connected to the Internet and it is projected that this number will exceed 200 million by the year 2000. Computer users may gain access to the Internet through Internet Service Providers (ISPs) such as America Online (AOL), Microsoft Network (MSN), and Netcom, university and library connections, corporate and non-profit portals, and government nodes. Thus, the Internet is a largely decentralized system utilizing a common communication backbone.

Until recently, the Internet has been governed primarily by InterNIC (NSFnet Internet Network Information Center), a consortium involving the National Science Foundation, AT&T, General Atomics, and Network Solutions, Inc. (NSI). NSI has had principal authority to register Internet names and addresses, what have come to be known as domain names. Domain names are represented in a hierarchical format: server.organization.type, e.g., www.whitehouse.gov. In 1999, the assigning of domain names was opened up to a range of entities operating under the authority of the Internet Corporation for Assigned Names and Numbers (ICANN).

In 1990, Tim Berners-Lee, a researcher at CERN, a European particle physics laboratory, developed the World Wide Web (www or "Web"), a widespread means by which users of the Internet could share databases. The www makes accessible to Internet users hypertext documents residing on HTTP (Hypertext Transfer Protocol) servers throughout the world. These Web pages, identified by Uniform Resource Locators, such as www.whitehouse.gov, are written in HTML (Hypertext Markup Language). Codes embedded in Web pages can instantly access other documents on the World Wide Web. They may also activate embedded software programs and audiovisual images. The Web became a mass media phenomenon with the development of "Web browser" programs that permit personal computers to access a wide variety of files from Web sites. The Web has also become the transaction medium for most electronic commerce.

Users of the Internet can locate Web sites of interest in a number of ways. They can often find online merchants, educational and government entities, and companies by using trade and organization names, such as "Microsoft.com", "law.berkeley.edu", and "uspto.gov". They can also "surf" the Internet with the assistance of various search engines, such as Yahoo, Altavista, and Infoseek. These "search engines" look for keywords in domain names, actual text on Web pages, and metatags, HTML code created by a Web page developed to attract search engines looking for particular keywords. Search engines will then rank Web sites from all over the Web according to various algorithms designed to arrange indexed materials. Web users may create "bookmarks" on their computers in order to access their favorite Web sites quickly.

As computing shifted in the 1990s from a stand-alone activity to one conducted increasingly over networks, the way computers function changed as well. It no longer makes sense to talk about how "a computer" functions. Most computers used in business, and many used in the home, do not operate in self-contained fashion. Computer programs are distributed across local area networks (LANs) and over the Internet. And more and more programs are distributed in pieces on an as-needed basis over the Internet. We discuss this change and its implications for computer law in more detail in the sections that follow.

The Internet has also changed the way commerce is conducted. Virtually everyone was caught off guard by the speed with which the millions of people using the Internet began to use it to buy things. The development of such "ecommerce" poses important new challenges for the legal and technological framework of the Internet.

C. *How Software Is Made*

Software development has evolved from a significantly hardware-constrained activity to a highly flexible and sophisticated field of engineering. This section briefly summarizes the principal methods involved in modern software development. We note, however, that the proliferation of a wide range of computers — from mainframes to desktop units to a host of embedded systems — has spawned a great range of programming methods. We focus here upon the methods used in the development of relatively complex commercial operating systems and application programs, although many of the stages can be found in other programming contexts.

Software development is an inherently functional enterprise. Software provides the instructions that enable a computer to perform tasks that serve the users' needs. Software can control the relatively simple operation of a clock to the highly complex control of an airplane. Whereas the programs that operate a wristwatch may have 1,000 instructions, modern spreadsheet and word processing programs have well over a million lines of instructions or code. In order to make computers more effective and easy to use, software can become extraordinarily complex in design and implementation. To a large extent, software engineering has become an applied science of mastering complexity.

The engineering nature of the discipline is discussed in detail in Pamela Samuelson, Randall Davis, Mitchell D. Kapor, and J.H. Reichman, A Manifesto Concerning the Legal Protection of Computer Programs, 94 Colum. L. Rev. 2308, 2326-32 (1994). They argue:

> 1.4.3 Constructing Programs Is an Industrial Design Process
>
> Once one understands that programs are machines that happen to have been constructed in the medium of text, it becomes easier to understand that writing programs is an industrial design process akin to the design of physical machines. Each stage of the development process requires industrial design works: from identifying the constraints under which the program will operate, to listing the tasks to be performed (i.e., determining what behavior it should have), to deciding what component parts to utilize to bring about this behavior (which in the case of software includes algorithms and data structures), to integrating the component utilitarian elements in an efficient way.
>
> A substantial amount of skilled effort of program development goes into the design and implementation of behavior. Designing behavior involves a skilled effort to decompose the overall, complex task (e.g., word processing) into a set of simpler subtasks (e.g., deciding whether the 'delete word' command should be implemented as a sequence of delete character commands) also requires design skill. Knowing how and where to break up a complicated task and how to get the simpler components to work together is, in itself, an important form of the engineering design skill of programmers.
>
> The goal of a programmer designing software is to achieve functional results in an efficient way. While there may be elements of individual style present in program design, even those style elements concern issues of industrial design, e.g., the choice

of one or another programming technique or the clarity (or obscurity) of the functional purpose of a portion of the program. . . .

At the same time, others have characterized programming, particularly in its earlier days, as akin to an art form. Certainly it is true that while the ultimate goal of computer programming is the design of a functional work, programming has historically been less routinized than most engineering disciplines, and much computer programming has involved "reinventing the wheel."[2] We discuss some of the reasons for that, and modern ways of dealing with it, in the sections that follow.

Software development has traditionally been described as a multistage process often analogized to a waterfall.[3] *See generally* Stephen R. Schach, *Classical and Object-Oriented Software Engineering with UML and C++* (4th ed. 1999); Ben Schneiderman, *Designing the User Interface: Strategies for Effective Human-Computer Interaction* (3rd ed. 1998); *Grady Booch, Object-Oriented Analysis and Design with Applications* (2nd ed. 1994); Ian Somerville, *Software Engineering* (4th ed. 1992). All software development processes begin with a clear definition of the problem to be solved or task to be automated. This stage of the process identifies the goals of the users and the constraints of the hardware system. This stage is followed by the development of a system and software design. A user interface will also be designed to serve the needs of the target audience for the software product. After the software has been structured and an interface designed, programmers implement the design, test its performance and reliability, and fine tune the system. In addition, many software products require ongoing maintenance and updating to correct errors and enhance its capability.

Although the waterfall model implies a linear flow, modern software engineering typically has an iterative quality with many feedback loops along the development and use life cycle. That is, programming does not occur from the "top down," with ideas being turned into algorithms and thence into code. Rather, the development of lower-level program components can influence the design of higher-level components, and can even cause the programmer to rethink the goals of the program itself. One might think of this as "bottom up" programming. As the authors of the *Manifesto* note:

> Innovation in software development is typically incremental. Programmers commonly adopt software design elements—ideas about how to do particular things in software—by looking around for examples or remembering what worked in other programs. These elements are sometimes adopted wholesale, but often

---

2. The Manifesto observes that "[t]ypically, a programmer writes every line of code afresh, no matter how large the program is, or how common its tasks. To perceive the impact of this lack of standard building blocks, imagine trying to design an entire car in complete detail, down to the last fastener, without being able to assume the existence of any standard parts at all (not even nuts, bolts or screws)." As its authors observe, "[t]his is not a desirable state of affairs." Samuelson, Davis, Kapor, & Reichman, supra, 94 Colum. L. Rev. at 2322.

3. Other programming paradigms include the rapid prototyping model, exploratory programming, formal transformation, and system assembly from reusable components.

> they are adapted to a new context or set of tasks. In this way, programmers both contribute to and benefit from a cumulative innovation process. While innovation in program design occasionally rises to the level of invention, most often it does not. Rather, it is the product of the skilled use of know-how to solve industrial design tasks. . . .
>
> The technical community has recognized the cumulative and incremental nature of software development, and has welcomed efforts to direct the process of software development away from the custom-crafting that typified its early stages and toward a more methodical, engineering approach. The creation of a software engineering discipline reflects an awareness that program development requires skilled effort and applied know-how comparable to other engineering disciplines.
>
> The products of software engineering almost invariably contain admixtures of old and new elements. The innovation in such programs may lie in the manner in which the known elements have been combined in a new and efficient manner. Or it may come from combining some new elements with well-known elements in order to achieve the same result in a new way. When we speak of programs as "industrial compilations of applied know-how," it is in recognition of the frequency with which software engineering involves the reuse of known elements. Use of skilled efforts to construct programs brings about cumulative, incremental innovation characteristic of engineering disciplines. A well-designed program is thus akin to the work of a talented engineer whose skilled efforts in applying know-how, accumulated from years of experience and training, yields a successful design for a bridge or other useful product.

Pamela Samuelson, Randall Davis, Mitchell D. Kapor, and J.H. Reichman, *A Manifesto Concerning the Legal Protection of Computer Programs,* 94 Colum. L. Rev. 2308, 2326-32 (1994).

However programs are designed and written, a fundamental fact about computer programs is that they are functional. They are designed to carry out certain tasks and bring about certain behaviors. Consequently, programs are judged largely on how well they perform the tasks they have been programmed to accomplish. Because the computer instructions serve no function other than to accomplish the defined tasks, the overriding concern in designing a program is to meet the users' needs in the most efficient manner.

The concept of efficiency in this context is broad. Efficiency may mean one or more of the following: (1) code efficiency — maximizing the processing speed; (2) memory efficiency — using solution techniques and addressing methods to minimize the amount of memory needed to accomplish the desired tasks; (3) input/output efficiency — maximizing the quality and speed of information transmission between the computer and the user or external hardware devices (such as keyboards and printers); (4) stability — the program must be easy to maintain, upgrade, and adapt to new hardware platforms; and (5) usability — the ease of use by the intended audience. The software engineering field strives to develop methods for improving the efficiency and reliability of programs in a cost-effective manner.

The subsections that follow describe typical processes for designing software at different "levels" of abstraction, from high-level ideas down through the actual writing of program code.

1. Requirements Analysis and Program Specification

At the conceptual level, software development comprises several tasks. Of most importance, the design team must identify the goals of the design process. For example, in developing software to provide banking services through an automated teller machine (ATM), the design team will map out the data necessary, desired functionality, and hardware and security constraints. They might develop tables or flow charts to understand the flow of information needed to accomplish the various transactions. As another example, in developing software to run a law office, the designers must identify the various tasks that the computer program should handle — e.g., billing clients, filing documents, ordering supplies, preparing budgets, filing court documents. The design team would typically interview the prospective users of the software system about their needs and desires, study the way information flows in a law office, and develop a schematic representation of the tasks to be programmed. This abstract representation of tasks will map data inputs and outputs and assess the hardware requirements for possible systems.

We should emphasize, however, that flow charts are merely conceptual aids, and are by no means a necessary part of programming.

Writing an application program is a complex and iterative process. While some programs are small and relatively simple to write, most modern programs involve thousands and even millions of lines of computer code. They may be written not by a single person, but by teams of programmers working together over the course of months or years.

Partly to facilitate collaborative work within teams, programmers sometimes develop a "flowchart" to depict the logical structure of the program. This will not just show what the program is supposed to do, but also how it will carry out the desired tasks. A sample flowchart for a computer program is reproduced as Figure B-l.

2. Software Design

Processes for designing software systems have undergone significant change over the past four decades as computer hardware has become more powerful and versatile and the problems sought to be addressed have become more complex. Managing the complexity of large-scale software projects — for example, computer operating systems, avionics (programs to fly large commercial airplanes), robotics, spacecraft simulation, telecommunications, and air traffic control — entails significant technological creativity. Programmers have traditionally sought to divide complex problems into component parts and solve each component separately. This "procedure" oriented, "functional decomposition," or "top down" methodology served as the dominant paradigm for software design through at least the mid-1980s, and it continues to be widely used in solving some classes of problems. The programmer begins with a general description of the functions that a program is to perform. The programmer then outlines the program, specifying data structures and algorithms to be used. Such outlines are

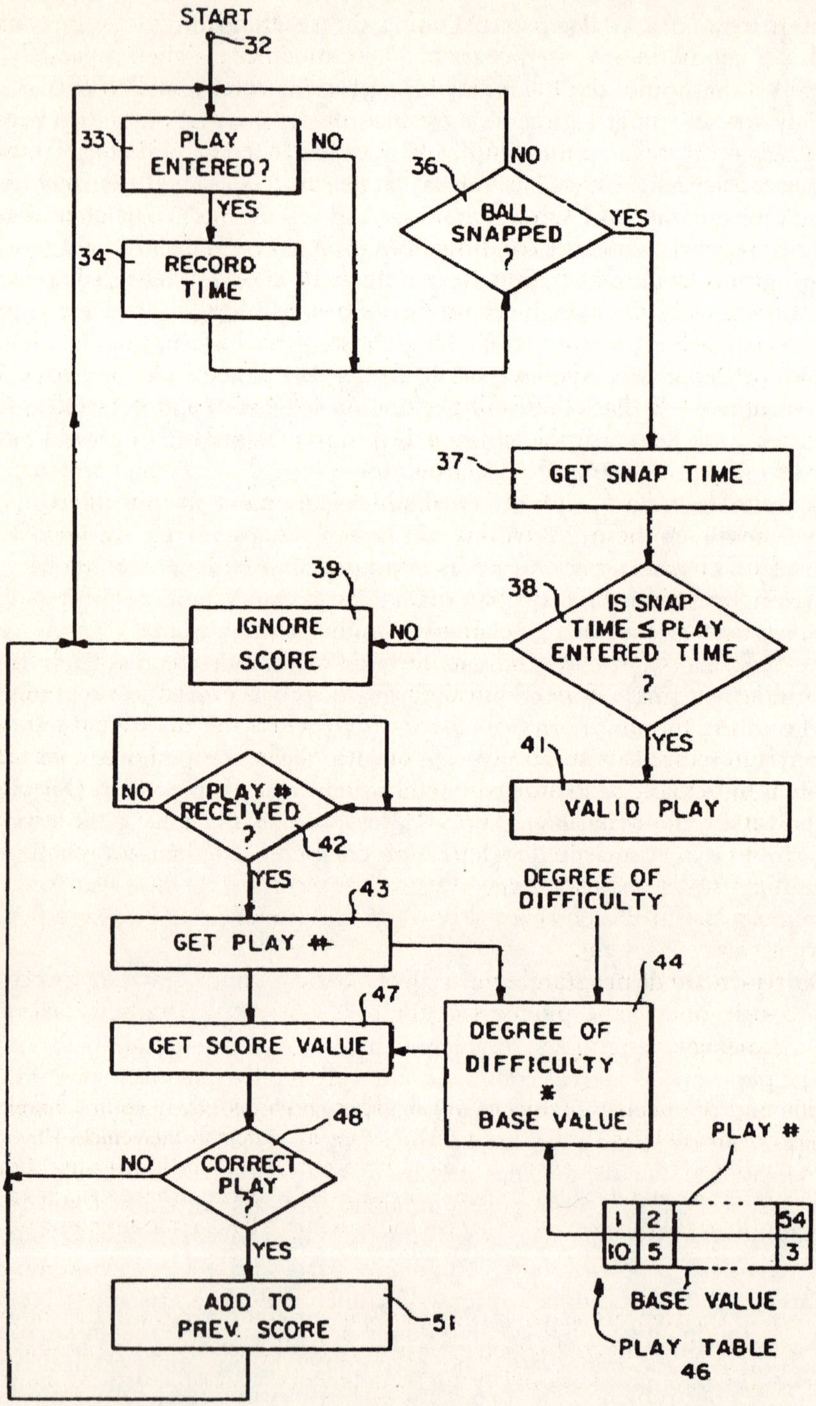
START
32
33
PLAY ENTERED?
NO
YES
34
RECORD TIME
36
BALL SNAPPED ?
NO
YES
37
GET SNAP TIME
38
IS SNAP TIME ≤ PLAY ENTERED TIME ?
39
IGNORE SCORE
NO
YES
41
VALID PLAY
PLAY # RECEIVED ?
NO
42
YES
43
GET PLAY #
DEGREE OF DIFFICULTY
44
47
GET SCORE VALUE
DEGREE OF DIFFICULTY * BASE VALUE
48
CORRECT PLAY ?
NO
YES
PLAY #
1 2 54
10 5 3
BASE VALUE
ADD TO PREV. SCORE
51
PLAY TABLE 46

Figure B-1

frequently expressed as flowcharts showing the relationship among the various modules or subroutines of the program. These modules are then separately further broken down until the full logic of the program can be spelled out.

While conceptually logical, the classical design methodology has become increasingly problematic as the complexity, scale, and need for updating of software projects has grown. The many parts of very large software systems often interact in a multitude of intricate and subtle ways. Classical top-down designs often require significant redundancies and are difficult to evolve. Programmers must typically build programs from scratch, often requiring re-invention of many components. This contrasts with other engineering disciplines, which typically reuse existing components. Such reuse can reduce design, testing, and other costs. In addition, top-down procedural techniques typically define data structures in one place. Various subroutines refer back to this single location. While this approach offers some efficiencies, it can lead to problems when the program is updated or revised. Whenever a data structure is revised, all subroutines or modules drawing upon that data structure must be suitably altered as well, which can cause a domino effect.

As a result of these limitations of classical programming methodologies, which tend to become more acute as systems grow larger, computer scientists developed the object-oriented paradigm. Programmers using this paradigm design software by structuring relationships among independent "objects," each of which represents a physical entity in the real world. Each object stores both data representing attributes of the physical objects as well as procedures representing actions that the physical object can perform. Whereas the traditional top-down program is structured around processing of data, object-oriented programs model a problem by actually simulating the interaction of physical entities. Objects are grouped within hierarchical structures. Higher level classes inherit the attributes of sub-classes. Object-oriented systems simplify complexity by encapsulating the internal data structures and procedures within objects. The data structures and procedures of particular objects can be altered without affecting other aspects of the larger software system.

Object-oriented programming reflects basic human learning processes. Grady Booch, one of the pioneers of the object-oriented paradigm, offers the following analogy:

> [W]ith just a few minutes of orientation, an experienced pilot can step into a multi-engine jet aircraft he or she has never flown before, and safely fly the vehicle. Having recognized the properties common to all such aircraft, such as the functioning of the rudder, ailerons, and throttle, the pilot primarily needs to learn what properties are unique to that particular aircraft. If the pilot already knows how to fly a given aircraft, it is far easier to know how to fly a similar one.[4]

Thus, a programmer starting with an object-oriented program that simulates a basic aircraft can simulate the behavior of a more sophisticated aircraft, such as a fighter jet, by adding additional features, behaviors, and processes of the more

4. Grady Booch, *Object-Oriented Design with Applications* 12 (1991).

advanced plane. The programmer would not need to begin the software design process from scratch.[5]

This example illustrates some of the ways in which object-oriented design systems economize on programming time and cost. Such designs are more readily adaptable to changes in data structures and new variables than traditional top-down approaches. The adaptability of object-oriented software programs substantially reduces the risk that a large investment in software design will be lost if new parameters or features need to be added to a program. Moreover, object-oriented designs tend to yield smaller systems through the reuse of common mechanisms, which reduces the complexity of the software design and the cost of writing code. See Mark A. Lemley and David W. O'Brien, *Encouraging Software Reuse,* 49 Stan. L. Rev. 255, 259-68 (1997). This reuse of "chunks" of computer code is not only valuable because it saves the cost of rewriting program code, but because it permits a particular software "object" to be refined and debugged, and then reused in a variety of different programs. This reuse of software objects across firms may enhance software compatibility, decrease programming costs, and even improve the safety and reliability of software (since mission-critical software will not be written from scratch for each new product, but can use code that has already been tested and proven to work).

As a result of these advantages, object-oriented methodologies have increasingly become the norm in designing highly complex computer programs.

3. User Interface Design

As computers have become more versatile and available to a broader and often less technically trained range of users, software development has increasingly focused upon tailoring the program to the particular goals and knowledge-base of the intended users. See generally Ben Schneiderman, *Designing the User Interface: Strategies for Effective Human-Computer Interaction* (3rd ed. 1998). The design of computer program user interfaces draws upon the field of computer-human interaction (CHI), a subfield of a more general field of "human factors" analysis that aims to understand how human beings process information so that products can better be designed to enhance usability. Human factors analysis brings together insights from the fields of education, graphic art, industrial design, industrial management, computer science, mechanical engineering, psychology, artificial intelligence, linguistics, information science, and sociology. Recognizing the many ways that the study of human factors can aid computer system design, the computer industry has taken a particularly strong interest in the effort to synthesize and expand this learning.

---

5. See Barkan, Software Litigation in the Year 2000: The Effect of Object-Oriented Design Methodologies on Traditional Software Jurisprudence, 7 High Tech. L.J. 315 (1992); Smith, Abstraction, Filtration, and Comparison in Computer Copyright Infringement: An Explanation and Update for the Object-Oriented Paradigm, 26 AIPLA Q.J. 1 (1998).

The field of computer-human interaction has profoundly changed the way in which application programs are both conceptualized and written. The field has identified five human factor goals that programmers should strive to achieve in designing application programs: (1) minimize learning time, (2) maximize speed of performance, (3) minimize rate of user errors, (4) maximize user satisfaction, and (5) maximize users' retention of knowledge over time.

Application programmers attempt to achieve these objectives in designing computer-human interfaces. As a result, many of their design choices are guided by principles of human factor analysis that have evolved over years of empirical research. Among the design issues that have been studied are the choice among command-based programs, menu-driven programs, and natural language programs; menu design; windowing versus scrolling; the choice of abbreviations and command names; the layout and scheme of graphical interfaces; the color and highlighting of video displays; the consistency of user interfaces; and the design of direct manipulation interfaces. To aid in the implementation of these ideas, many firms engaged in application program development have compiled computer-interface guidelines and criteria to guide their programmers. Apple Computer, which introduced the first commercially successful graphical user interface, developed the following guidelines for user-friendly software design:

1. *Metaphors*—the interface should embody plain, simple metaphors for accomplishing tasks with audio and visual effects to support the metaphor.
2. *Direct Manipulation*—the interface should allow users to directly manipulate items or actions on the screen, rather than indirectly through abstract commands.
3. *See and Point*—the interface should allow the user to effect action by a see-and-point mechanism that is intuitive, rather than by a remember-and-type mechanism that can be intimidating, too abstract, or unnatural.
4. *Consistency*—the symbols and mechanisms for accomplishing tasks should be consistent across all application programs through a consistent interface.
5. *WYSIWYG (What You See Is What You Get)*—the interface should present the information on the screen to the user in exactly the form that it will come out on the printer, removing the need to type in abstract formatting commands or to make mental calculations to envision how the screen translates to paper.[6]

These principles—and the basic concept of graphical user interfaces—are in widespread use in the software industry and can be seen in most application programming and Web-based user environments.

6. Written Submission of Apple Computer, Inc., U.S. Copyright Office Public Hearing on Registration and Deposit of Computer Screen Displays 5-6 (Sept. 4, 1987).

4. Generating Computer Code

After designing computer systems, including the user interface, a software engineer engages in a series of steps to translate the design into binary code (representing open and closed circuits) actually "readable" by computers. While it is possible to write programs directly into machine-level language, most programming is done in higher level languages that use abbreviations and short words to convey the action the program will need to carry out. Well-known programming languages include FORTRAN, COBOL, and Pascal. Special languages such as C++, Ada, Small-talk, Object Pascal, and Java have been developed more recently to implement object-oriented designs. These programming languages, which can be readily understood by skilled programmers, are often referred to as "source code."

Figure B-2 contains the source code version of a simple program written in PASCAL for determining and listing in five columns any number of prime numbers. PASCAL is a high-level programming language, and as you can see, it is relatively easy to comprehend. For example, the statement "Go to 40" instructs the computer to skip intervening steps and execute the instruction at line 40. But it is a specialized language all the same, and (because the programmer is writing for a computer "audience") its rules must be followed precisely. "Execute line 40 now" may mean the same thing to a human reader as "Go to 40," but it may be incomprehensible to a computer.

Traditionally, source code was written by computer programmers who had learned one of the specialized computer languages described above. Increasingly, however, the process of writing source code has itself become automated. Computer programmers can now "write" code with the help of a computer "wizard" that translates higher-level design concepts directly into code. Having a computer help write source code can be faster and more efficient than writing it by hand; it can also help prevent programmers from making mistakes that will interfere with the program. As more and more code is written with computer assistance, similarities in actual program code become less meaningful as evidence of copying. This is important to the treatment of software under copyright law.

In order to be executable by the central processing unit of a computer, source code must be transformed into a machine-executable form. This is generally accomplishing by using a special purpose computer program known as a "compiler" to process source code instructions into "object code," the binary code that is processed by the computer. In this machine language, a single "0" or "1" represents the absence or presence of an electrical charge. Each binary digit is referred to as a "bit." These electrical signals processed by a CPU will determine which functions of the CPU will be executed in what order. Computers do not process program instructions one bit at a time. Rather they process standard lengths of bits. A "word" is the "set of characters that occupies one storage location and is treated by the computer circuits as a unit and transported as such." Until the mid-1980s, most computers acted upon words of eight bits, known as a "byte" (e.g., "01101001"). More recently computers have been designed to act upon words of

```
          Program PrimeNum (Input, Output);
          {
* * * * * * * * * * * * * * * * * * * * * * * * * * * * * * * * * * * * * * * * * * * * * * * * * *
WRITTEN BY:       Vance F. Brown
DATE:             April 5, 1988
PROGRAM PURPOSE: This program allows user to enter the desired number of prime numbers
to be ascertained. The program then lists the prime numbers (starting with the prime number 2) in
numerical order in five columns. DEFINITION OF PRIME NUMBER: A number is prime if it
is not evenly divisible by any number other than itself and 1.
* * * * * * * * * * * * * * * * * * * * * * * * * * * * * * * * * * * * * * * * * * * * * * * * * *
 }
    Var {for variable function explanation, see below}
          num, count, primes, times, divisor: Integer; check: Boolean;
    Begin {of program}
          {***** initialize variables *****}
    num :=2;{the primenumber candidate}
    count :=0;{number of prime numbers found}
    times :=0;{number of columns before a carriage return}
    {***** ask user how many prime numbers to list *****}
    write ('How many prime numbers do you want to find?');
    read (primes);
    writeln;
    writeln;   {carriage returns}
    writeln;
 repeat {until the number of primes is found}
  check := false;  {initialize: assume first that number is not
                                     prime}
  divisor := 2;    {first number to be divided into candidate}
  divisor := succ(divisor);  {first number to be divided into candidate}
  {***** loop to determine if numbr is prime *****}
  while (divisor < num) and (not check) do begin

   check := num mod divisor = 0; {the remainder after dividing
                                                  divisor into num}
   divisor := succ(divsor);      {increment divisor}
  end; {of while loop}

  {***** if a prime number, then display it *****}
  If not check then begin
    write (num:10);
    count :=succ(count);  {increment number of primes
                                            found}
    times := succ(times);  {increment column number}

 {***** column adjustment, once five numbers displayed,
  carriage return*****}
     if times = 5 then begin
             writeln; {carriage return}
             times :=0; {start counter for column number over}
     end; {of column adjustment if statement)}
   end; {of display of if prime number statement}
   num := succ(num); {next candidate for prime number}
 until count = primes; {once the number of primes entered by user is
                              found, the program ends}
end.  {of program}
```

Figure B-2

11101001	01111001	00101100	10010000	10010000	11001101	10101011
01000011	01101111	01110000	01111001	01110010	01101001	01100111
01101000	01110100	00100000	00101000	01000011	00101001	00100000
00110001	00111001	00111000	00110101	00100000	01000010	01001111
01010010	01001100	01000001	01000100	01000100	00100000	01001001
01101110	01100011	00000010	00000100	00000000	10110001	01010111
00000000	00111100	00110011	00000000	00000000	00000000	00000000
00000000	00000000	00000000	00000000	00000000	00000000	00000000
00000000	00000000	00000000	00000000	00000000	00000000	00000000
00000000	00000000	00000000	00000000	00000000	00000000	00000000
00000000	00000000	00000000	00000000	00000000	00000000	00000000
00000000	00000000	00000000	00000000	00000000	00000000	00000000
00000000	00010100	01000100	01100101	01100110	01100001	01110101
01101100	01110100	00100000	01100100	01101001	01110111	01110000
01101100	01100001	01111001	00100000	01101101	01101111	01100100
01100101	01010000	00011001	00000001	11111111	11111111	00001111
00000111	00000111	01110000	00001111	00000111	00000111	01110000
00001110	00000111	00000111	01001111	00101110	10001010	00100111
00001010	11100100	11111001	01110100	00001110	01000011	00101110
10001010	00000111	01010000	11101000	11011000	00001000	01011000
11111110	11001100	01110101	11110011	11111000	11000011	01111010
00000000	11111111	01111011	00001000	00000000	00011111	11000111
00000110	00010010	00000000	001101110	00000000	00101110	11000110
00000110	10010100	00000001	00000000	10111110	00100000	00000000

Figure B-3

16, 32, or 64 bits. Figure B-3 contains 24 of 1,674 lines of the object code derived from the PASCAL program in Figure B-2.

Obviously, high-level programming languages are more compact than machine language, as well as being easier for humans to decipher. This results from the fact that the representation of complex information as either a 1 or a 0 requires an enormous number of 1s and 0s.

Because of the difficulty of using and deciphering object code, programmers developed an intermediate level language referred to as "assembly language" to facilitate the translation of higher level languages to object code. Assembly language uses abbreviated alphanumeric symbols such as "ADC," which means "add with carry." For older models of Apple computers, "ADC" translates to "01101001" in object code. An assembly program or assembler translates assembly language into object code. Figure B-4 contains 29 of 7,330 lines of assembly code produced

```
L05CF:      PUSH      AX                ;05CF   50
            PUSH      CX                ;05DO   51
            MOV       CL,4              ;05D1   B1 04
            SHR       AX,CL             ;05D3   D3 E8
            ADD       BX,AX             ;05D5   03 D8
            POP       CX                ;05D7   59
            POP       AX                ;05D8   58
            AND       AX,0F             ;05D9   25 0F 00
            RET_NEAR                    ;05DC   C3
                                        ;L05DD     L062A  CC
L05DD:      CMP       BX,DX             ;05DD   3B DA
            JNZ       L05E3             ;05DF   75 02
            CMP       AX,CX             ;05E1   3B C1
                                        L05E3      L05DF CJ
L05E3:      RET_NEAR                    ;05E3   C3
                                        ;L05E4  L0656 CC
L05E4:      ADD       AX,CX             ;05E4   03 C1
            ADD       BX,CX             ;05E6   03 DA
            JMP       SHORTL05CF    ;05E8   EB E5
                                        L05EA      L0625 CC
L05EA:      MOV       AX,ES:[DI+4]  ;05EA     26 8B 45 04
            MOV       BX,ES:[DI+6]  ;05EE     26 8B 5D 06
            PUSH      AX                ;05F2   50
            OR        AX,BX             ;05F3   OB C3
            POP       AX                ;05F5   58
            RET_NEAR                    ;05F6   C3
```

Figure B-4

by "disassembling" the object code. Like object code and unlike higher level languages, assembly language is specific to a particular computer system.

5. Software Validation and Maintenance

Although this discussion may suggest that programming is a linear process, in fact, it is typically an iterative process. Program design often occurs in a series of "feedback loops" in which each stage of the process influences all of the others. Thus, a software engineer may develop a design for a program and then write some code to implement it, but in the process of writing the code, the engineer may discover logical problems with the design which require rethinking the program's basic structure. Often, individual modules or components of a software program will be tested and refined before integrating the various components into a full working version.

Before a software program enters service or the marketplace more generally, the software design team will run the program through a suite of tests selected to ensure that the program operates properly and achieves the objectives set out in the requirement and specification stages of the process. After the various systems within the program are operating properly, the team conduct alpha (or acceptance) testing. Alpha testing involves running the software with actual data (rather than simulated data) from the intended user. For products that are to be marketed widely, software vendors typically engage a further level of testing commonly referred to as beta testing in which the program is provided on a limited basis to a range of target customers who agree to use the program and report problems. Beta testing may also identify features that users desire that were not initially included in the program. Based upon this input and further internal testing, the vendor typically makes some further refinements before releasing the product into the commercial marketplace.

All complex software programs have a range of errors (commonly referred to as "bugs"), and the validation stage can be one of the more challenging phases of the software life cycle. Because of the importance of reaching the market quickly, the design team is often under tremendous pressure to complete this phase.

Following the release of a product, software enters the last and what is often the longest phase of its life cycle: maintenance. Computer software users place great significance on a software vendor's ability to correct, improve, and adapt software products and services. Indeed, consumers are often wary of buying the first version of a new software product because it may be "buggy." The planning for maintenance begins early in the process with the documentation of the computer program. Most companies put out a series of "updates" or new releases of their programs to address maintenance problems and to enhance the program's functionality.

Some computer bugs may not manifest until the product has been in use for a long time. Software vendors' reputations depend significantly on their ability to support their products and users once they are in the marketplace.

COMMENTS AND QUESTIONS

1. Compare the process of writing a computer program to the following activities: writing a novel, writing a poem, writing a symphony, writing a cookbook, preparing a map, designing a building, engineering a bridge, inventing a new process for manufacturing steel. In what ways is computer programming similar to and different from these creative activities?

2. Is it possible to identify the "most significant" part of a computer program? Is it the idea for the program? The flowchart that structures the program? The source code that implements it? Which of these activities, if any, can be called the "heart" of programming?